Thermal Characterization
of Polymeric Materials

Contributors

Harvey E. Bair
P. K. Gallagher
M. Jaffe
Y. P. Khanna
J. J. Maurer
E. M. Pearce
R. Bruce Prime
D. Raucher
Shalaby W. Shalaby
W. W. Wendlandt
Bernhard Wunderlich

Thermal Characterization of Polymeric Materials

Edited by EDITH A. TURI

Corporate Research and Development
Allied Corporation
Corporate Headquarters
Morristown, New Jersey

 1981

ACADEMIC PRESS
A Subsidiary of Harcourt Brace Jovanovich, Publishers

New York   London
Paris   San Diego   San Francisco   São Paulo
Sydney   Tokyo   Toronto

6654-4439

CHEMISTRY

ACADEMIC PRESS, INC.
111 Fifth Avenue, New York, New York 10003

United Kingdom Edition published by
ACADEMIC PRESS, INC. (LONDON) LTD.
24/28 Oval Road, London NW1 7DX

Library of Congress Cataloging in Publication Data
Main entry under title:

Thermal characterization of polymeric materials.

Bibliography: p.
Includes indexes.
1. Polymers and polymerization--Thermal properties.
I. Turi, Edith A.
TA455.P58T5   620.1'9204296   81-17578
ISBN 0-12-703780-2   AACR2

PRINTED IN THE UNITED STATES OF AMERICA

82 83 84   9 8 7 6 5 4 3 2

Contents

Chapter 1. **Instrumentation**

W. W. Wendlandt and P. K. Gallagher

Chapter 2. **The Basis of Thermal Analysis**

Bernhard Wunderlich

Chapter 3. **Thermoplastic Polymers**

Shalaby W. Shalaby

299, 300 (handwritten margin note near V.)
337, 339 (handwritten margin note near VII.)

Chapter 4. **Block Copolymers and Polyblends**

Shalaby W. Shalaby and Harvey E. Bair

Chapter 5. **Thermosets**

R. Bruce Prime

460, 465-467, 470-473 (handwritten margin note)
489 491-493, 496 (handwritten margin note)
531 492 (handwritten margin note)

Chapter 6. **Elastomers**

J. J. Maurer

Contents

Chapter 7. **Fibers**

M. Jaffe

Chapter 8. **Thermal Analysis
in Polymer Flammability**

E. M. Pearce, Y. P. Khanna, and D. Raucher

Chapter 9. **Thermal Analysis of Additives
in Polymers**

Harvey E. Bair

List of Contributors

Numbers in parentheses indicate the pages on which the author's contributions begin.

HARVEY E. BAIR (365, 845), Bell Laboratories, Murray Hill, New Jersey 07974

P. K. GALLAGHER (1), Bell Laboratories, Murray Hill, New Jersey 07974

M. JAFFE (709), Celanese Research Company, Summit, New Jersey 07901

Y. P. KHANNA (793), Corporate Research and Development, Allied Corporation, Morristown, New Jersey 07960

J. J. MAURER (571), Corporate Research Laboratories, Exxon Research and Engineering Company, Linden, New Jersey 07036

E. M. PEARCE (793), Department of Chemistry, Polytechnic Institute of New York, Brooklyn, New York 11201

R. BRUCE PRIME (435), International Business Machines Corporation, San Jose, California 95193

D. RAUCHER* (793), Department of Chemistry, Polytechnic Institute of New York, Brooklyn, New York 11201

SHALABY W. SHALABY (235, 365), Research and Development Division, Ethicon, Inc., Somerville, New Jersey 08876

W. W. WENDLANDT (1), Department of Chemistry, University of Houston, Houston, Texas 77004

BERNHARD WUNDERLICH (91), Department of Chemistry, Rensselaer Polytechnic Institute, Troy, New York 12181

*Present address: Monsanto Company, St. Louis, Missouri.

ix

Preface

The concept of this book was born of a feeling of frustration I have shared with many of the scientists engaged in thermal analysis. This was caused by the lack of a comprehensive reference book covering the very important field of the thermal characterization of polymers and related materials.

Polymers, in the past decade, have gained more significance than any other class of materials. The exponential growth of polymer applications has led to the development of several new techniques for polymer characterization; no single technique has proved more useful than thermal analysis. This method delivers information often unobtainable by other means and has been used in many areas of basic and applied research, production, and quality control. New instrumentation, techniques, and applications have surfaced at an unusual pace; thousands of publications dealing with thermal analysis in polymer technology appear every year.

My association with national and international societies dedicated to promoting the theory and application of thermal analysis [such as the North American Thermal Analysis Society (NATAS) and the International Confederation of Thermal Analysis (ICTA)] presented me with an opportunity to discuss the idea of this book with prominent scientists in the field who later agreed to author one or more chapters.

I have designed this book with the objective of presenting a critical review of the literature and a concise evaluation of the application of thermal analysis in polymer science and engineering. To achieve this goal, I have asked a team of distinguished authors to undertake the difficult task of presenting in this volume the state of the art, with proper interpretation and critical assessment of selected experimental data. In addition, the authors have included many unpublished results of their relevant research.

The authors of this book offer expertise in every area of polymer characterization by thermal analysis. With considerable cooperative effort, this has been blended into a systematic review that includes the instrumentation and the basic principles of thermal analysis, the scientific background underlying the practical application of thermal analytical techniques, and a large and diversified selection of practical examples for problem solving. When applicable, nonthermal methods (such as microscopy and x-ray diffraction) are also mentioned, since they play a supportive role.

This book is to be considered a guideline and not an encyclopedia. Many of the methods and techniques described in this text may serve as models for the reader who wants to apply them to some other related problem. Special efforts were made to keep the references up-to-date; some of them were added in the last phases of proofreading. Here I wish to thank the publisher for the flexibility that allowed for these last-minute additions.

Who benefits from this book? Because of the ever-increasing use of polymeric materials, more chemists and chemical engineers work in polymer-related fields than in any other single area of industry. The book may serve as a reference and practical guide for polymer scientists, thermal analysts, engineers, and technologists. In the libraries of educational institutions, it will help both teachers and students.

The book provides an in-depth overview of thermal analysis, concentrating on the polymer field. The three major parts of the book deal with instrumentation, theory, and a wide variety of applications.

Instrumentation.   The first chapter is a broad treatment of virtually all aspects of thermal analytical methods and apparatus, with an emphasis on polymer applications. The present state of the art and the currently available commercial instruments are described. The subject is treated in such a way that professionals unfamiliar with the field can quickly gain a grasp of the techniques while experienced thermal analysts will find a current comprehensive treatment complete with suggestions and sources of value to them.

The Basis of Thermal Analysis.   The author of this unique chapter discusses the theory underlying the basic principles of thermal analysis. The theories and functions of state for thermometry, dilatometry, thermomechanical analysis, calorimetry, and thermogravimetry are explained. He describes single-component and multicomponent systems, and their phase transitions, as influenced by concentration, pressure, deformation, molecular weight, and copolymerization.

Thermoplastic Polymers.   In this section the reader is presented with an in-depth guide to the important chemical and physical parameters as they influence the glass transition, crystallization, and melting of thermoplastic materials. Factors affecting the thermal and thermo-oxidative degradations of thermoplastics are also considered. In addition, the chapter offers a general treatise on the applications of thermal analysis to some selected problems.

Block Copolymers and Polyblends. This chapter serves as a comprehensive review of the applications of the pertinent thermal characterization methods. The theoretical aspects of polymer–polymer compatibility, phase separation, and miscibility in mixed polymer systems are discussed in depth. Also stressed is the quantitative analysis based on specific heat measurements of block and graft copolymers and numerous polyblends.

Thermosets. In Chapter 5 practical applications of thermal analysis to thermosetting resins (such as epoxies, phenolics, and polyesters) are stressed. The theoretical and technical backgrounds for the widest range of applications are provided. The chapter also deals with curing, properties of uncured resins, cured thermosets, effects of catalysts and fillers, and other areas of practical importance.

Elastomers. This chapter provides a comprehensive review of the major thermal methods as applied to elastomers. Screening, development, and quality control analysis of practical rubber compounds and vulcanizates are presented. Both research and development applications of thermal analysis are covered.

Fibers. A practical approach for obtaining and interpreting thermal analysis data of fibers is introduced in this chapter. The origin of thermal analysis responses for fibers is explained from the point of view of problem solving and polymer physics. The relevance of fiber processing history to the observed fiber thermal analysis responses is stressed, and many examples are shown of how thermal history is manifested in the experimental results.

Thermal Analysis in Polymer Flammability. The authors demonstrate how thermal analysis techniques can be utilized in the polymer flammability area. Methods of obtaining information on the relative flammability properties of polymers, for screening fire retardant additives, and for studying the mechanism of flame inhibition are discussed. Application of thermal analysis to the flammability of some important polymeric materials is also covered.

Thermal Analysis of Additives in Polymers. The last chapter is useful not only in helping the plastics manufacturer develop effective additive systems for polymers but also in aiding the user to evaluate whether a polymeric material has the expected processing properties and performance characteristics. The additives that are treated in this chapter include antioxidants, stabilizers, lubricants, plasticizers, impact modifiers, and fire retardants.

The intention of this volume was to meet an urgent need and to bridge a gap between theory and practice. We hope that the users of this work will find the information and guidance required for achieving their goals in research, production, and application.

The cooperation of the authors in the editorial work was a rewarding and enjoyable experience. I am most grateful for their invaluable contributions.

The authors of the individual chapters and I have profited greatly from and appreciate the generous advice and comments of the reviewers.

I wish to express my gratitude to Professor Bernhard Wunderlich for his helpful suggestions in the various phases of the preparation of this book. My thanks are also due to a number of publishers and societies for the courtesy of permitting us to reproduce copyrighted material. My sincere appreciation is expressed to Academic Press for their professionalism in the production of this volume, and to the members of their editorial staff who prepared the Author Index and Subject Index.

Finally, I wish to thank my husband, Paul, a fellow chemist, for support, guidance, encouragement, and patience during the years of preparation of this book.

Morristown, New Jersey EDITH A. TURI

CHAPTER 1

Instrumentation

W. W. WENDLANDT
Department of Chemistry
University of Houston
Houston, Texas

P. K. GALLAGHER
Bell Laboratories
Murray Hill, New Jersey

1

I. Introduction

A. PURPOSE AND SCOPE

At the foundation of experimental science are the apparatus and techniques by which the necessary observations are made. This crucial interface between nature and its laws is in constant flux. Fortunately, the foundation spreads and becomes more entwined with time, which only serves to strengthen the pyramid of conclusions constructed upon this base. Few areas of the foundation have grown more rapidly in recent years than thermal analysis. The purpose of this chapter is to describe that portion of science and engineering.

As strong as the foundation is, it nevertheless is like a living creature in that cells are constantly dying, becoming obsolete in this case. Higher in the scientific pyramid valuable data and ideas seldom die, but rather are built upon; however, good apparatus and techniques frequently become obsolete and of value only to museums and scientifc historians. Consequently, it is most difficult to write about such things in a permanently lasting fashion. Instead one describes the present state of the art with the knowledge that what is said will change with time. This is joyful rather than distressful, however, because it is a statement of progress. It is hoped that the reader will recognize and appreciate this fact. Within this limitation the authors have attempted to describe the current methods and tools.

B. BRIEF HISTORICAL REVIEW

1. Thermogravimetry (TG)

The basic components of thermogravimetry (TG) have existed for thousands of years. Mastabas or tombs in ancient Egypt (2500 B.C.) have wall carvings and paintings depicting both the balance and fire (Vieweg, 1972). It was centuries, however, before the two were coupled in a procedure for the study of gold refining during the fourteenth century (Szabadvary, 1966). The main driving force for early classical TG was to determine the stability range for various analytical precipitates. This aspect reached its zenith under Duval (1963), who studied over 1000 gravimetric precipitates and developed an automated analytical method based on this technique.

Honda (1915) laid the somewhat broader foundation for modern TG when he first used the term *thermobalance* and concluded his TG investigations of $MnSO_4 \cdot H_2O$, $CaCO_3$, and CrO_3 with the modest statement, "All of the results given are not all together original; the present investigation with the thermobalance has however revealed the exact positions of the change in structure and also the velocity of the change in respective temperatures.'' Much of the early work of the Japanese school has been summarized by Saito (1962).

Other even earlier thermobalances had been constructed by Nernst and Riesenfeld (1903), Brill (1905), Truchot (1907), and Urbain and Boulanger (1912). The first commercial instrument in 1945 was based on the work of Chevenard *et al.* (1944). Most early thermobalances, except the Chevenard, had been built by the individual investigators. The Derivatograph, developed by Erdey *et al.* (1956), introduced the simultaneous measurement of thermogravimetry and differential thermal analysis. Garn *et al.* (1962) successfully adapted the commercial Ainsworth recording balance for TG to 1600°C in various controlled atmospheres. Similarly, a Sartorius recording balance was modified for TG (Gallagher and Schrey, 1963). The advent of the modern automatic thermobalance began with the introduction by Cahn and Schultz (1963) of the electrobalance. This balance had a sensitivity of 0.1 μg and a precision of one part in 10^5 of the mass change. Other commercial thermobalances followed: DuPont: Sarasohn and Tabeling (1964); Mettler: Wiedemann (1964); and Perkin-Elmer: Gray (1975).

2. *Differential Thermal Analysis (DTA) and Differential Scanning Calorimetry (DSC)*

Accurate temperature measuring techniques, such as the thermocouple, resistance thermometer, and optical pyrometer, were all firmly established in Europe by the late 1800s. As a result, it was inevitable that they soon would be applied to chemical systems at elevated temperatures. Thus, LeChatelier (1887), who was interested in both clay mineralogy and pyrometry, introduced the use of heating-rate-change curves dT_s/dt to identify clays.

The differential temperature method, in which the temperature of the sample is compared to that of an inert reference material, was conceived by an English metallurgist, Roberts-Austen (1889). This technique eliminated the effect of heating rate and other outside disturbances that could change the temperature of the sample. The second thermocouple was placed in a neutral body, which was sufficiently removed from the sample so as not to be influenced by it. The difference in temperature $T_s - T_r$ was observed directly on one galvanometer, whereas the sample temperature was obtained on another. Saladin (1904) carried this method one step further and developed a photographic recorder that recorded $T_s - T_r$ versus

T directly. A versatile photographic recorder, which used a rotating drum, was developed by Kurnakov (1904). This instrument was used extensively by Russian workers in DTA for 50 or 60 years, and was influential in starting the Russian school of DTA.

Burgess (1908) discussed the various methods of recording cooling curves, including the use of differential curves and derivative curves. His article is an excellent review of the state of the art up to 1908. Clays and silicate minerals were the subject of other early DTA papers, such as those by Ashley (1911), Brown and Montgomery (1912), and Fenner (1912). It should be noted that these materials were the subject of numerous future DTA studies and constituted the major use of this technique for the next 40 years. Thus, because of its predominantly geological applications, DTA was developed by geologists, clay mineralogists, ceramicists, soil scientists, and others who work in this area (see Mackenzie, 1957).

The modern era of DTA instrumentation began with the introduction of the dynamic gas atmosphere DTA system by Stone (1951). This system permitted the flow of gas or vapor through the sample during heating or cooling cycles. Reactions could be studied under various partial pressures of different gases, which could be changed in composition during the thermal cycle if so desired.

The 1960s saw the introduction of elaborate DTA instruments, including the development of differential scanning calorimetry (DSC) by Watson *et al.* (1964a,b). Other notable systems were developed by DuPont, Deltatherm, and Mettler, to name only a few. Most of these commercial systems are described in detail in a monograph by Wendlandt (1974a).

Perhaps the greatest impetus to the development of new instruments and thermal techniques was the interest shown in the techniques by polymer chemists. DTA or DSC is an ideal tool for use in polymer characterization—such parameters as melting-point glass transition temperatures, polymerization, thermal degradation, and oxidation reactions can be studied rapidly and conveniently. Determinations that once took hours or even a day to complete could be obtained in a matter of minutes. Not only were elevated temperatures required, but low-temperature studies were also necessary for the determination of glass transition temperatures and other data. Since small sample sizes were required, new furnace and sample holder designs were developed, as well as better dc amplifiers and linear temperature programmers. The most recent trend is toward computer interaction. Wendlandt (1973) has reviewed some of these early applications.

Many of the classic papers in TG, DTA, and DSC are collected in a monograph edited by Wendlandt and Collins (1976). A recent review that provides a historical perspective and insight into the development of thermal analysis is by Szabadvary and Buzagh-Gere (1979).

II. Thermal Analysis Techniques

The most frequently used thermal analysis techniques are given in Table I. For each technique, the parameter measured, the instrument employed, and the transducer or sensor of the measured parameter are given. It is difficult to classify differential scanning calorimetry (DSC) because of the extensive use of two types of instruments, both of which are given this name. One type is *power compensation DSC,* in which power or heat input is recorded; the other is the *heat-flux DSC* type in which the temperature difference $T_s - T_r$ is recorded as a function of temperature. Obviously, the second type will include differential thermal analysis (DTA) as well.

Besides the techniques listed in Table I, there are numerous less frequently used thermal analysis techniques. These include thermosonimetry, thermoacoustimetry, thermoluminescence thermomagnetometry, thermomanometry, dynamic reflectance spectrometry, and emanation thermal analysis. Frequently, new names are employed to describe existing techniques. An example is differential thermal gas analysis (DTGA), proposed by Mizutani and Kato (1975), which is actually a sophisticated evolved gas detection (EGD) apparatus. Numerous other examples could be cited for these "new" techniques.

Seldom is one thermoanalytical technique by itself sufficient to provide a complete description or solution. Generally several techniques are used along with other physiochemical methods. When working with magnetic phases or intermediates, the simple addition of a small external magnetic field can be exceedingly helpful in evaluating the nature and mechanisms of the processes being investigated (Gallagher and Warne, 1981). There are a number of commercially available instruments that will perform several techniques during the same experiment. When the techniques (e.g., TG, DTA, EGA) are performed on the same sample, they are called "simultaneous." When they are performed using separate samples subjected to the same thermal program and atmosphere, they are referred to as "combined." Naturally this saves time and equipment, but more important, it provides a firmer basis for associating the observed events with each other. Generally this outweighs any disadvantages of the added complexity in the design of equipment or possible sacrifice of sensitivity. Relevant aspects of these multipurpose instruments will be discussed under the appropriate sections.

TABLE I

MOST FREQUENTLY USED THERMAL ANALYSIS TECHNIQUES

Technique	Parameter measured	Instrument employed	Transducer or sensor
Thermogravimetry (TG)	Mass	Thermobalance	Recording balance
Derivative thermogravimetry (DTG)	dm/dt	Thermobalance	Recording balance
Differential thermal analysis (DTA)	T_s-T_r	DTA apparatus	Thermocouple thermistor
Differential scanning calorimetry (DSC)	dH/dt	DSC calorimeter	Pt resistance[a] thermometer
Thermomechanical analysis (TMA)[b] (dilatometry)	Deformation, volume or length	Dilatometer	LVDT
DYNAMIC thermo-mechanometry[c]	Modulus/damping	Many different instruments	Various
Evolved gas detection or analysis (EGD, EGA)	Thermal con-ductivity[d]	EGD or EGA apparatus	TC cell or other[d]
Thermooptometry[e]	Light emission or transmittance	Many different instruments[e]	Photodetector
Thermoelectrometry	Current or resistance	Electrical conductivity apparatus	Different types
Thermosonimetry	Sound, sound velocity	Many different instruments	Piezoelectric crystal

[a] Power compensation DSC only. Heat flux DSC employs thermocouples as detectors.

[b] TMA determines the deformation of a substance under nonoscillatory load. Dilatometry is concerned with the change in dimensions of the substance. Both use LVDT (linear voltage differential transformer) as the transducer.

[c] Frequently called dynamic mechanical analysis (DMA), the method applies to an oscillatory load and measures the modulus and/or damping of the substance as a function of temperature. Torsion braid analysis (TBA) is a particular case in which the sample is supported on a braid.

[d] For EGD measurements; EGA technique may employ various analytical techniques such as GC, MS, etc.

[e] Includes thermophotometry, thermospectrometry, thermoreflectometry, thermoluminescence.

There are numerous applications of the two major techniques, TG and DTA, some of which are listed in Tables II and III, respectively. Some of the phenomena that give rise to observable effects in DTA are given in Table IV along with the nature of the transition. Second-order phase transitions frequently give rise to changes in the slope rather than distinct peaks.

TABLE II

SOME APPLICATIONS OF THERMOGRAVIMETRY[a]

1. Thermal decomposition of inorganic, organic, and polymeric substances
2. Corrosion of metals in various atmospheres at elevated temperatures
3. Solid-state reactions
4. Roasting and calcining of minerals
5. Distillation and evaporation of liquids
6. Pyrolysis of coal, petroleum, and wood
7. Determination of moisture, volatiles, and ash contents
8. Rates of evaporation and sublimation
9. Dehydration and hygroscopicity studies
10. Phase diagrams and nonstoichiometry
11. Thermal oxidative degradation of polymeric substances
12. Decomposition of explosive materials
13. Development of gravimetric analytical procedures
14. Reaction kinetics studies
15. Discovery of new chemical compounds
16. Vapor pressure determination and heats of vaporization
17. Adsorption, desorption curves
18. Magnetic properties, e.g., Curie temperature, magnetic susceptibility

[a] From Wendlandt (1976b).

TABLE III

SOME APPLICATIONS OF DTA AND DSC[a]

Catalysts	Decomposition reactions
Polymeric materials	Phase diagrams
Lubricating greases	Reaction kinetics
Fats and oils	Solid-state reactions
Coordination compounds	Dehydration reactions
Carbohydrates	Radiation damage
Amino acids and proteins	Heats of adsorption
Metal salt hydrates	Heats of reaction
Metal and nonmetal oxides	Heats of polymerization
Coal and lignite	Heats of sublimation
Wood and related substances	Desolvation reactions
Natural products	Solid-gas reactions
Clays and minerals	Curie point determinations
Metals and alloys	Purity determinations
Soil	Thermal stability
Biological materials	Oxidation stability
Amorphous materials	Glass transition determinations
Cements	Comparison
Semiconductors	Hazards evaluation
Pharmaceuticals	Thermal conductivity
Explosives	Specific heats

[a] From Wendlandt (1974b).

TABLE IV

PHYSICOCHEMICAL ORIGINS OF PEAKS IN DTA AND DSC[a]

	Enthalpic change upon heating	
	Endothermal	Exothermal
Physical		
Crystalline transition	×	×
Fusion	×	
Crystallization		×
Vaporization	×	
Sublimation	×	
Adsorption		×
Desorption	×	
Absorption	×	
Curie point transition	×	
Glass transition	Change of baseline, no peaks	
Liquid crystal transition	×	
Heat capacity transition	Change of baseline, no peaks	
Chemical		
Chemisorption		×
Desolvation	×	
Dehydration	×	
Decomposition	×	×
Oxidative degradation		×
Oxidation in gaseous atmosphere		×
Reduction in gaseous atmosphere		×
Redox reactions	×	×
Solid-state reaction	×	×
Combustion		×
Polymerization		×
Precuring (resins)		×
Catalytic reactions		×

[a] From Wendlandt (1974b).

III. Thermogravimetry—the Thermobalance

The thermobalance is an instrument that permits the continuous weighing of a sample as a function of temperature. The sample may be heated or cooled at some selected rate or it may be isothermally maintained at a fixed temperature. Perhaps the most common mode of operation is heating the sample at furnace heating rates from 5 to 10°C min^{-1}. Almost all modern thermobalances are automatically recording instruments, although manual recording is still used occasionally for long-term isothermal measurements (e.g., with helix-type thermobalances).

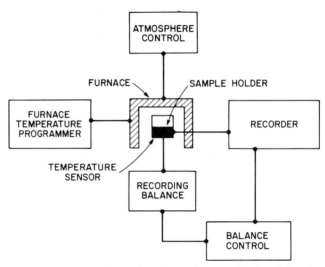

Fig. 1 Schematic diagram of a modern thermobalance. Reprinted from W. W. Wendlandt, ''Handbook of Commercial Scientific Instruments,'' Vol. 2, 1974, p. 144, by courtesy of Marcel Dekker, Inc.

The modern thermobalance is illustrated schematically in Fig. 1 (Wendlandt, 1974a). It generally consists of the following component parts: (a) recording balance; (b) furnace; (c) furnace temperature programmer or controller; and (d) recorder, either of the strip-chart or $X-Y$ function type. Modern digital data acquisition systems may substitute for, or be run in conjunction with, the recorder. The specific details of each component depend on the particular application that is required of the instrument. For example, furnaces can be obtained that operate up to 2400°C or more, and employ air, inert gases, hydrogen, nitrogen, vacuum, and even corrosive atmospheres. Likewise, for the recording balance, sensitivities from as low as 0.02 mg full-scale deflection to 100 g or more are available.

An attempt will be made here to discuss each component, as given above, of a modern thermobalance. This is a rather difficult task because of the wide variety of thermobalances, both commercial and noncommercial, that are available to the thermal analysis. Each thermobalance has available numerous sample containers, at least two different furnaces, and a choice of several different recording systems. The numerous sample containers permit the study of various types of samples in different furnace environments. Multiple furnaces, of course, permit the upper temperature limit to be extended since a high-temperature furnace may not operate efficiently at ambient and low temperatures. A choice of recording systems is necessary because of the number of channels of information required and/or the type of mass-change curve desired, such as mass versus time or mass versus temperature, and so on. Bradley and Wendlandt (1971) have described an apparatus capable of automatic sequential analysis of up to eight samples. There are no commercially available instruments that offer automated options such as are prevalent in other analytical methods, e.g., x-ray diffraction and fluorescence.

A. RECORDING BALANCES

The most important component of the thermobalance is the recording balance. Requirements for a suitable recording balance are essentially those for a good analytical balance, i.e., accuracy, precision, sensitivity, capacity, resistance to corrosion, rugged construction, and insensitivity to ambient temperature changes. In addition (Gordon and Campbell, 1960), the balance should have an adjustable range of mass change and a high degree of electronic and mechanical stability, be able to respond rapidly to changes in mass, be relatively unaffected by vibration, and be of sufficiently simple construction to minimize the initial cost and need for maintenance. From a practical viewpoint, the balance should be simple to operate and versatile so that it can be used for varied applications.

Recording balances can be divided basically into three general classifications based on their mode of operation: (a) deflection-type instruments, (b) null-type instruments, and (c) those based on changes in a resonance frequency.

Instruments of type c have tremendous sensitivity and can be highly miniaturized but are very specialized in their applications. The quartz crystal microbalance is the most common type. Mass deposits on, or lost from, the surface of the highly polished crystal (e.g., condensation of evaporation) are detected by the resulting shift in the oscillating frequently (thickness sheer). Such devices have sensitivities of the order of nanograms per square centimeter with accuracies of the order of 1% (Glassford, 1976). Complete balance packages without associated furnaces and temperature controllers can be obtained from Berkley Controls, Inc., or Celesco, Inc., to name a couple of suppliers. Glassford (1976) described both isothermal and dynamic measurements of water vaporization rates; they are also useful for a wide range of outgassing, contamination, and other surface studies.

The automatic null-type balance (type b) incorporates a sensing element that detects a deviation of the balance beam from its null position: horizontal for beam balances and vertical for electromagnetic-suspension types. A restoring force, of either electrical or mechanical mass loading, is then applied to the beam through the appropriate electronic or mechanical linkages, restoring it to the null position. This change in restoring force, which is proportional to the change in mass, is recorded directly or through an electromechanical transducer of some type.

Various deflection-type balances (type a) are shown in Fig. 2. These instruments involve the conversion of the balance-beam deflections about

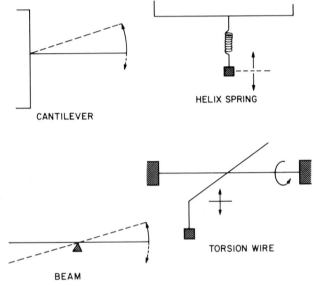

Fig. 2 Deflection-type balance principles.

the fulcrum into mass-change curves by (a) photographic recording, (b) recording electrical signals generated by an appropriate displacement measurement transducer, and (c) using an electromechanical device. The following are types of deflection balances:

(a) the helical spring, in which changes of mass are detected by contraction or elongation of the spring;

(b) the cantilevered beam, constructed so that one end is fixed and the other end, from which the sample is suspended, is free to undergo deflection;

(c) the suspension of a sample by an appropriately mounted strain gauge that stretches and contracts in proportion to mass changes;

(d) the attachment of a beam to a taut wire that serves as the fulcrum and is rigidly fixed at one or both ends so that deflections are proportional to changes in mass and torsional characteristics of the wire.

The last three types can be readily arranged to suspend the sample from above as well as below. Such an arrangement can be used to place the furnace above the balance where thermal convection to the balance will be less.

The null-type balance principle is now used in almost all commercially available thermobalances. Because of the maximum capacities of the balance, balance mechanisms differ widely, although most of them employ some type of a lamp-aperture–photodetector–servoloop arrangement. The restoring force is applied as a current passing through a coil which interacts with a magnetic field. This change in electrical current (or voltage) is proportional to the change in mass of the sample and can be recorded. Also, the analog signal output of the balance can be converted to a derivative of the mass change or digitized by an analog-to-digital converter for data processing by digital computers. If real-time computer processing is applied, a wide variety of on-stream process controls with feedback loops to control the experimental parameters and data collection are possible.

B. Sample Containers

Numerous sample containers are available for containment of the sampel in a thermobalance. The type employed usually depends on the nature, amount, and reactivity of the sample and the maximum temperature desired. Creep of liquid samples can be particularly bothersome. In many cases the choice of sample container will have an important influence on the mass-change curve and may even catalyze the decomposition of the sample. In some thermobalances, the sample container is fixed; hence, it must be used under all conditions of sampling and instrument parameters.

Representative sample containers are illustrated in Fig. 3. In some cases the sample may be secured or hung directly on the suspension wire. For crucible-type containers, volumes may range from 0.1 to 5 ml, thus permitting a wide choice of sample sizes. Common materials of construction include alumina, platinum, platinum,–10% rhodium, aluminum, quartz, glass, nickel, tungsten, and graphite. In (a), the sample container is a hemisphere, 5 to 9 mm in diameter, constructed of platinum or quartz. Sample sizes range from 1 to 20 mg, depending on the sample's density. Sample container (b), which is constructed of platinum foil, is triangular, with the sample being inserted through either open end. The remaining sample containers illustrated are used in the Mettler system. In (c), the open-crucible type, various sizes are available with capacities ranging from 0.1 to 5 ml. The flat-plate-type container (d) is used for high vacuum studies or when a large exposed surface area is desired. For larger samples the polyplate container in (e) may be employed. It exposes an extremely large surface area of the sample. Similar holders have been described by Erdey *et al.* (1966) and Paulik *et al.* (1966) for use in the

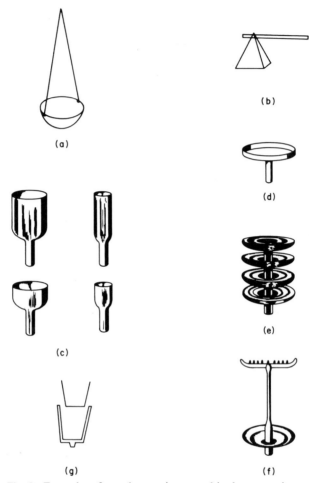

Fig. 3   Examples of sample containers used in thermogravimetry.

Derivatograph; one holder consisted of 10–20 plates spaced at 2-mm intervals, permitting the use of samples from 0.2 to 1 g. Thin strips or wire-
type samples may be conveniently studied using the sample container
shown in (f). Protective disposable liners are frequently employed, as illustrated in (g), if the sample reacts with the crucible materials. The liners
may be constructed of alumina or various metals.

For vapor pressure measurements the sample containers in Fig. 4 may
be used. In (a), a schematic cross section of the sample container using
the Knudsen method, as described by Wiedemann (1972), is shown. The
main body of the cell A is constructed of aluminum and is suspended on

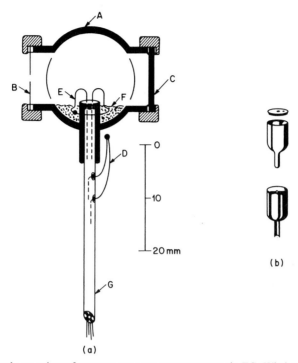

Fig. 4 Sample containers for vapor pressure measurements in TG (Wiedemann, 1972).

the thermobalance by a ceramic tube G. A copper–constantan thermocouple E brought into the cell through a vacuum-tight connection, detects the sample temperature F or the temperature of the vapor phase. A second thermocouple D serves to control the furnace programmer. The orifice B is made from 0.01-mm-thick Nichrome foil and contains a small hole. A screw fitting on the cell permits rapid removal or exchange of orifices of different diameters (1–3 mm). The other end plate C is used to load the sample into the cell. Both end plates are sealed to the body by Teflon O-rings. Sample volume is less than or equal to 1.5 ml and the cell can be used in the temperature range from − 100 to 200°C. A simpler version of a vapor pressure cell, which contains a removable lid, is shown in (b).

For studies involving dissociation reactions in "self-generated" atmospheres (Garn and Kessler, 1960), the sample holders in Fig. 5 may be used. In the piston and cylinder arrangement, shown in (a), the clearance between the two provides a long diffusion path and effectively prevents contamination from the atmosphere. The Forkel (1960) crucible, as shown in (b), has a lid containing a ball valve to ensure separation of the sample

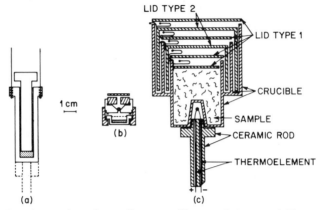

Fig. 5 Sample containers for "self-generated" atmosphere control (Garn and Kessler, 1960; Forkel, 1960; Newkirk, 1971; Paulik and Paulik, 1972).

atmosphere and the atmosphere in the furnace. Various other glass and quartz sample holders have been discussed by Newkirk (1971). Perhaps the most elaborate sample holder is the one shown in (c), which was described by Paulik and Paulik (1972). The sample crucible is covered with six close-fitting lids or covers in which the gaseous decomposition products are forced to escape through a long and narrow labyrinth. The advantages and disadvantages of self-generated atmospheres have been extensively discussed by Newkirk (1971).

C. Furnaces and Furnace Temperature Programmers

A wide variety of furnaces are available for thermobalances, each designed for a specific temperature range from $-150°C$ to over $2800°C$. Most of the furnaces have resistance heater elements, but some employ infrared heating for extremely rapid heating and cooling rates. Maximum temperature limits for the various resistance heater elements are shown in Table V. These limits are only approximate since they depend on the design of the furnace, the insulation, and the type of atmosphere employed. No one furnace design can be used over the entire temperature range from -150 to $2800°C$. Generally, there may be four different furnaces used to cover this range: (a) -150 to $500°C$, (b) 25 to 1000 or $1200°C$, (c) 25 to $1600°C$, and (d) 400 to $2800°C$. As expected, because of the complexity of the furnace design, the higher temperature furnaces are the most expensive. The most common type of furnace of the four listed is (b), although several commercial thermobalance manufacturers give the user a choice of all four temperature ranges. Because of thermal shock, oxidation, and

TABLE V

APPROXIMATE MAXIMUM TEMPERATURE LIMITS
FOR FURNACE RESISTANCE ELEMENTS

Element	Approximate temperature, °C	Required atmosphere[a]
Nichrome	1000	O_x
Chromel A	1100	O_x
Tantalum	1330	N_{ox}
Kanthal	1350	O_x
Platinum	1400	N_{ox}, O_x
Globar	1500	O_x
Platinum–10% rhodium	1500	N_{ox}, O_x
Platinum–20% rhodium	1500	N_{ox}, O_x
Kanthal Super	1600	O_x
Rhodium	1800	N_{ox}, O_x
Molybdenum	2200	N_{ox}, H_2
Tungsten	2800	N_{ox}, H_2

[a] O_x = oxidizing; N_{ox} = nonoxidizing (inert, vacuum).

evaporation, there is a finite life for the resistance elements and refractories that necessitates their ultimate replacement. This problem is most important for the higher temperature furnaces, where crystallization, oxidation, and/or evaporation of heater elements become important. Such things as current limiting, judicious programming rates, and conservative rating of the maximum temperature and the time spent there can greatly enhance furnace lifetimes. The design of the furnace is important if magnetic samples are to be used. If the furnace is wound correctly, there will be a minimum of interference from the magnetic field generated by the heater current (Gallagher and Gyorgy, 1979).

The furnace may be positioned above, below, or parallel to the balance, as shown in Fig. 6. Perhaps the two most common configurations are (a) and (b), although (c) is used in the DuPont thermobalance as well as others. Each configuration has its advantages and disadvantages; the furnace above the balance appears to be the preferred configuration for high temperature ranges, whereas for lower temperatures, the furnace below the balance is more convenient. The latter configuration is also less expensive to construct than the former because of the types of recording balances available.

The novel microfurnace of the Perkin-Elmer Model TGS-2 thermobalance is constructed from an alumina cylinder 12.7 mm in diameter by 19 mm in length mounted inside the sample space on a ceramic tube. It is wound noninductively with platinum wire, which functions as both the heater and temperature sensor. During one half of a cycle the resistance

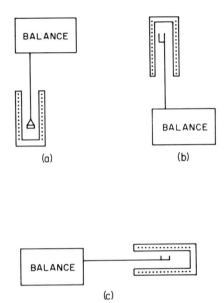

Fig. 6 Furnace position in thermobalances: (a) below balance; (b) above balance; (c) parallel to balance.

(related to its temperature) is determined and compared with the desired program temperature. In the other half of the cycle, electrical power is supplied to the wire such as to null the difference between the actual and desired temperature. A thermocouple in close proximity to the sample pan is also provided as an alternative measure of the sample temperature. This compact furnace is coaxially aligned with the sample in a glass hang-down tube that allows for easy viewing. The outer glass tube is forced air cooled. The low thermal mass of the microfurnace permits high cooling rates; the furnace is said to cool from 1000 to 100°C in about 10 min.

Even faster cooling rates are claimed for the infrared heating system employed in the ULVAC-RIKO TGD-3000-RH thermobalance. When the power is switched off, the heater cools to room temperature in 2–3 min. A cross section of the infrared image furnace is shown in Fig. 7. Radiation from infrared heaters, 150 mm in length, is concentrated by elliptical reflectors onto the hot zone 10 mm in diameter by 100 mm in length. Temperature is highest at the area approximately 5 mm in diameter in the horizontal direction and rapidly falls off along the perimeter of a circle 25 mm in diameter. Maximum temperature of the furnace is about 1400°C, with programmable heating rates of up to 999°C sec^{-1}.

The type of temperature controller varies from the simple variable-voltage transformer coupled to a synchronous motor to the more sophisti-

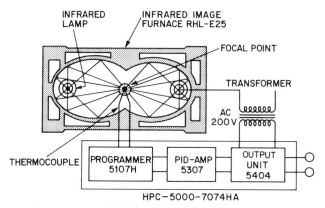

Fig. 7 Cross-section of ULVAC-RIKO TGD-3000-RH infrared furnace.

cated feedback, proportional-type controller. Controllers of the on–off type cannot be used, because the fluctuating power outputs give rise to severe thermal gradients in the furnace and sample holder system. The solid-state, feedback-type, proportional controller used in the DuPont thermal analysis instruments is shown in Fig. 8. In a proportional-type controller, when the error signal generated by the difference between the command voltage and the output of the control signal varies by an amount more than the "dead band" of the control amplifier, the error signal is

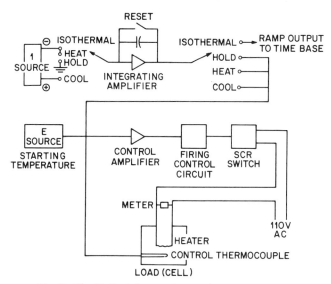

Fig. 8 The DuPont furnace temperature programmer.

amplified, and this power is applied to the heater. The power applied to the heater is proportional to the error signal at the input to the control amplifier. The heating-rate accuracy of this programmer is said to be ± 5% or 0.1°C min⁻¹, whichever is greater. Accuracy is governed mainly by the output of the control thermocouple. In the controller itself, the limiting factor is the adjustment and drift of the power supply that controls the current integrated by the integrating amplifiers. Reproducibility is 0.1°C min⁻¹, while the heating-rate linearity is ± 1% or 0.01°C min⁻¹. The former is dependent on the drift of the power supply and amplifier bias, whereas the latter depends on the output linearity of the control thermocouple.

Another very reliable thermocouple-feedback-type controller is the TECO Model TP-2000 Thermocouple Temperature Programmer, illustrated in Fig. 9. Power is applied to the furnace (load) by a phase-controlled bidirectional triac connected in series with the load. This results in a true proportioning of power to the load and more stable control than in conventional devices employing on–off or long-interval-time proportioning control.

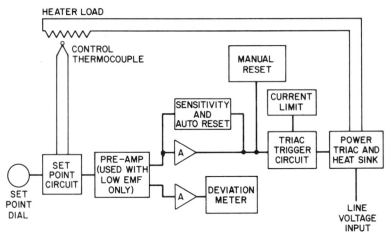

Fig. 9 TECO Model TP-2000 programmer.

The rate of temperature increase or decrease is controlled by a furnace temperature programmer. This device should be capable of linear temperature programming, i.e., the temperature of the furnace should be directly proportional to time over a number of different temperature ranges, and hence must be compatible with several different thermocouple types. Heating of the furnace system should be linear with time and must be reproducible. Output from the programmer should be stable with respect to line voltage variations and ambient temperature.

The desired temperature time profile can be obtained in various ways. Linear rates are generally achieved by motor-driven set points or potentiometers that provide a null point or electrical bias within the temperature controller for comparison with the thermocouple signal from the furnace. Varying linear rates have been achieved by circuits that interrupt the motor drive for different percentages of the time. Microswitches can then provide continuous interruption (hold) periods, reverse the motor's direction (cool/cycle), etc.

Nonlinear rates have been traditionally achieved by cam following devices or more recently by optically following a boldly delineated curve or electrically following a scribed line on a conducting surface. An example of the latter type of programmer is the Datatrac line. Variations are now achieved by changing the rotation rate of the cam or drum.

The availability of small, relatively inexpensive, reliable microprocessors has had a wide impact on most types of chemical instrumentation, particularly in the area of furnace programming. Besides those built into specific thermal analysis instruments (to be discussed later), there are a number available as simply programmers, e.g., Iveron Instruments, or as programmer controller packages, e.g., Leeds and Northrup (Model 1300) or Tetrahedron Co. (Wizard Model), to name a few of them. These instruments provide a tremendous range of linear ramps, holds, cycles, alarms, relay controls, etc. The relay closures, for example, can be used to operate valves, which in turn control the atmosphere. Some thought, however, must be given to the step size in these instruments. For example, the Iveron programmer has its range divided into 1000 bits, which for a typical range of room temperature to 1000°C means that it moves in approximately 1°C steps. This may be too coarse to obtain the smooth heating and cooling ramps desired for thermal analysis.

Besides the use of dedicated microprocessors as furnace controllers, it is also possible, through time sharing, to use a portion of a larger computer to program the temperatures and atmosphere based on either time or some feedback system. These flourishes, however, are beyond the scope of this chapter.

Exact determination of the heating rate of furnaces used in thermobalances (and other thermal analysis equipment) is not a simple matter. The usual time–temperature curves are insensitive to small changes in heating rate, especially at low rates of heating (or cooling). They are useful, however, to detect gross changes or interruptions in the heating rate. From a plot of temperature versus time, the heating rate can be estimated to ±5% or better. To determine minute fluctuations in the furnace heating-rate curve, the derivative method proposed by Wendlandt (1975) may be employed. In this method the first derivative of the furnace temperature, as detected by a thermocouple located within the furnace, is recorded as a function of time or temperature. The temperature–time curves and the derivative of the temperature (dT_s/dt)–time curves are illustrated in Fig. 10. Three furnace heating rates are shown for each set of curves, namely,

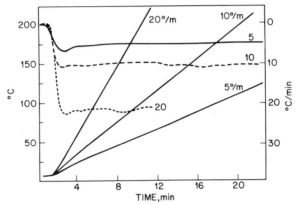

Fig. 10 Temperature-time and dT_s/dt curves for furnace at heating rates of 5, 10, and 20°C min^{-1} (Wendlandt, 1975).

5, 10, and 20°C min^{-1}. Although the temperature–time curves appear to be straight lines (above 2 min), the derivative curves contain numerous fluctuations after an initial equilibration period. If there were no heating-rate fluctuations, the derivative curves would appear as horizontal lines; thus the curve irregularities are indicative of small changes in the heating rate. These changes are not observable on the temperature–time curves. The smallest changes are found with the slower heating rate, 5°C min^{-1}, whereas the largest are at 20°C min^{-1}.

Most commonly used heating rates in thermogravimetry are linear, about 5–10°C min^{-1}. Faster heating rates are generally not employed unless very small samples are used or the investigation is a general survey of thermal properties.

Furnace programmers and controllers are, of course, available from a wide varity of manufacturers. Most commercially available thermal analysis equipment, however, comes with specifically matched and prepackaged controllers. An interesting alternative aimed at low-cost routine quality control applications is provided by Omnitherm Corporation. Their Model Q.C.25 controller is intended to be coupled with thermal analytical modules of other manufacturers to provide a low-cost alternative with limited capabilities. The Q.C.25 operates over the temperature range −90–1000°C with an accuracy of ± 1°C. There are four preset rates: 5, 10, 20, and 50°C min^{-1}, with one adjustable rate (0–50°C min^{-1}). A digital temperature readout is standard. Programmed cooling, isothermal step programming, and automatic gas switching are options. A recorder package, Model 715, is also available.

Traditionally, linear heating rates have been employed although there have been occasional suggestions for other temperature profiles for some kinetic studies. In the last decade considerable work has been done, particularly in Hungary, that uses an optical feedback mechanism to achieve "quasi-isothermal" (the heating rate controlled by the decomposition rate) conditions (Paulik and Paulik, 1972). Most modern commercial instruments have analog derivative computers suitable for following the rate of weight change, DTG. The output of these devices could be used in conjunction with a proportional controller to control the rate of heating to yield some preselected low rate of weight change. Paulik and Paulik (1972) also discuss the use of self-generated atmospheres to establish "quasi-isobaric" conditions (see Fig. 5c). The effects of these two techniques are clearly illustrated in Fig. 11. Curve 1 was done in both the

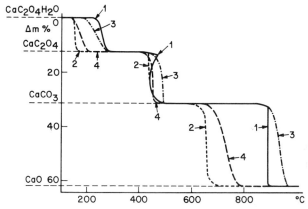

Fig. 11   Change in weight plotted as a function of temperature for the thermal decomposition of calcium oxalate monohydrate: (1) quasi-isothermal and quasi-isobaric; (2) quasi-isothermal; (3) quasi-isobaric and 10°C min⁻¹; (4) 10°C min⁻¹ (Paulik and Paulik, 1972).

quasi-isothermal and quasi-isobaric modes; curve 2 was done only in the quasi-isothermal mode; curve 3 was done only in the quasi-isobaric mode; and curve 4 done in the conventional way. The quasi-isobaric method strongly influences the temperature of reversible decompositions. This is reflected in the different temperatures for the loss of water and CO_2 between curves 1 and 2 or between curves 3 and 4. The irreversible loss of CO is not particularly affected by the atmosphere. The onset of the loss of water or CO_2 is not as abrupt as might be hoped in curves 1 and 3, because it is necessary for several percent of the reaction to occur in order to completely displace the original atmosphere from the sample holder and establish quasi-isobaric conditions. This enhanced resolution or separation evidenced in curve 1 can be of particular value in distinguishing between otherwise overlapping processes. As an example the accuracy in the analysis of dolomite is markedly improved by applying quasi-isothermal, quasi-isobaric methods. Figure 12 demonstrates how the exact amounts of Mg and Ca can be clearly established compared to the uncertainty present under normal TG conditions.

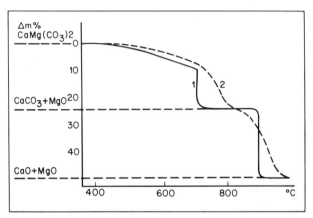

Fig. 12   TG curves of dolomite. (1) Recorded under quasi-isothermal and quasi-isobasic conditions. (2) Recorded under normal conditions at a heating rate of 10°C/min, in an open crucible.

Another example of a nonlinear heating rate is the parameter-jump and relaxation methods proposed by Flynn and Dickens (1976) and extended by Dickens *et al.* (1977). In this method, parameters such as temperature are changed in discrete steps, which has value in the kinetic interpretation of results. The field of nonisothermal kinetics, a major one, is not discussed here. Sestak *et al.* (1973) have extensively reviewed this field. Thermobalances and the related apparatus are also used for a host of isothermal studies that, strictly speaking, are not thermal analysis but practically speaking are generally included because of the great similarity in equipment and objectives.

D. RECORDERS

Many different types of recorders have been used with thermobalances, ranging from photographic light-beam galvanometer types to modern electronic potentiometric models. In general, three types of analog recorders are currently used in modern thermobalances—the time-base potentiometric strip chart and multipoint recorders, and the temperature-based $X-Y$ or $X-Y^1Y^2$ function plotters.

One advantage of time-base strip-chart recorders is that the furnace heating rate can be observed and checked for variations. The effects of a temperature perturbation on the mass-loss curves using both types of recording system are shown in Fig. 13. In (a), using a time-base recorder,

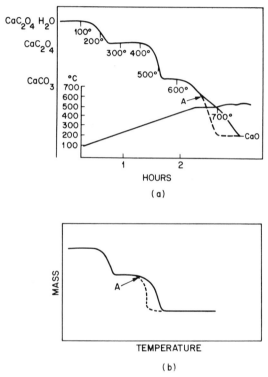

Fig. 13 Effect of furnace temperature perturbation on TG curves. (a) Time-base recorder, heating rate changed at A; (b) $X-Y$ plotter. Solid line: normal curve; dashed line: heating rate slowed down at A.

the temperature perturbation occurred at A, resulting in the recording of the solid-line curve. The normal curve, if no change in the heating rate took place, is shown by the dashed-line curve. Similarly, for the $X-Y$ recorder (b), the temperature perturbation occurred at A, resulting in the change to the dashed-line curve. The normal curve is indicated by the solid curve. However, in the time-base recorder, a curve of the system temperature was also recorded so that the change in heating rate could be immediately detected. With the $X-Y$ recorder, this change in heating rate would probably not be detectable unless the curve was repeated several times.

Most commercial thermobalances, with several exceptions, record the change in sample mass rather than the percent mass change; the latter is a more convenient form of presentation. From direct mass-change data, the sample mass must be known in order to calculate the percent mass change and/or other stoichiometry calculations. In percent mass-change data recording, the data are readily available for stoichiometry calculations of various types.

Modern digital data acquisition systems provide a wide range of flexibility and advantages for subsequent data analysis. This is particularly desirable if kinetic parameters (e.g., rate equations or activation energies) are to be derived. In addition, curve smoothing, differentiation, integration, and other mathematical manipulations can be accomplished by the computer with relative ease and greater accuracy. Data acquisition is a rapidly evolving field, and it would be impossible even to summarize it here. However, several examples of systems that have been successfully applied to thermal analysis are (1) Gallagher and Schrey (1970), in which data are collected and stored on punched paper tape; (2) Gallagher and Johnson (1972), in which the data are collected and stored on magnetic tape; and (3) Dickens *et al.* (1977), in which the data are fed directly to the computer and the computer feedback actually can control the experimental parameters. Use of a Hewlett Packard microcomputer for data acquisition during simultaneous TG-DTA is described by Yuen *et al.* (1980). Conventional recording devices are generally run in parallel with such systems in order to provide convenient visual monitoring of the course of the experiment.

E. Temperature Detection

As previously mentioned, in TG the mass change of the sample is continuously recorded as a function of temperature. The temperature, in this definition, may be that of the furnace chamber, the temperature near the sample (i.e., in close contact with the sample container), or the temperature of the sample. Typical arrangements for temperature detection are shown in Fig. 14. In (a) the thermocouple is near the sample container but not in contact with it. There is a correlation between the temperature of the container and that detected by the thermocouple, but the thermocouple will either lead or lag behind the sample temperature, depending on the thermochemistry of the reaction. Most thermobalances use this type of thermocouple arrangement even though it is a poor one. It is even worse when low-pressure atmospheres are employed, as in high-vacuum

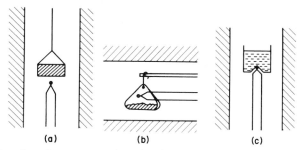

Fig. 14   Location of temperature-detection thermocouple in thermobalance furnace.

thermogravimetry; in this case, heat transfer is by radiation rather than by convection and conduction.

The thermocouple is close to the sample in (b), and is positioned inside, but not in contact with, the sample holder. This arrangement is better than (a) because the thermocouple will respond to changes of sample temperature; however, wild excursions will be noticed during sample combustion. The best method of sample temperature detection is to have the thermocouple in contact either with the sample or with the sample container, as shown in (c). In the latter, the temperature detected will be an integrated temperature. However, the main problem is that, with sensitive recording balances, the thermocouple leads can cause weighing errors or at least interference with the balance mechanism. MacKenzie and Luyet (1964) have minimized these problems by using very fine flexible thermocouple wires as the sample hangdown wire and taking the electrical pickoff very near the fulcrum of the balance.

One way to detect the actual bulk sample temperature and yet not interfere with the balance mechanism is to suspend an electronic device near the sample holder that will transmit the sample temperature to a fixed receiver located near the sample container. Manche and Carroll (1964) described a unijunction transistor relaxation oscillator that used a thermistor as the temperature detector. The frequency of oscillation, which is a function of sample temperature, was transmitted via a mutual inductance between two suspended coils to a receiver and counter. The device was limited, however, to a maximum temperature of about 150°C.

Sestak (1969) and Wendlandt (1977b) have proposed a simple method for recording the sample temperature in a thermobalance, using an auxiliary thermocouple, such as shown in Fig. 15. A second sample container (D), attached to a thermocouple and identical to the one suspended from the balance (C), is used to detect changes of the bulk sample temperature. This container is positioned as close as possible to the one attached to the recording balance. Obviously, two identical samples of approximately the

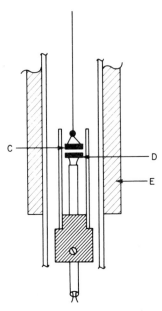

Fig. 15 Two sample containers used to record sample temperature in a thermobalance (Wendlandt, 1977b).

same mass must be employed, one for the balance container and the other for the fixed sample container. If the fixed container does not contain a sample, it detects the furnace rather than the sample temperature. The dual container system is convenient to use, although it does employ two separate samples. As Sestak (1969) points out, it is particularly valuable when working in vacuum where thermal transport is reduced. Most of the TG curves obtained by this method, at least for hydrated salts, have increased "resolution," which makes it easier to detect intermediate compounds that may be formed during the thermal decomposition process. It must be emphasized, however, that even the best of these measuring methods gives *only an estimate of the true temperature at the actual reaction interface*. Besides thermal transport factors, the enthalpy of the reaction is an important factor.

F. CALIBRATION OF FURNACE TEMPERATURE

The calibration of the temperature of the furnace and/or sample chamber has been discussed by Stewart (1969) and Norem *et al.* (1969, 1970). Stewart (1969) used a conventional thermobalance containing a thermocouple mounted external to the sample, whereas Norem *et al.* (1969, 1970) calibrated a furnace using a resistance element for temperature detection.

Stewart (1969) discussed three approaches to temperature calibration:

(1) the use of standard materials with reproducible mass-loss points that could be referred to the temperature;

(2) the use of materials having known reproducible (and reversible) temperature transitions and direct measurement of temperature;

(3) the use of materials with magnetic transitions that could be displayed on a mass-loss curve and be referred to the temperature.

The first approach is the most appealing, but the evolution of a volatile product is dependent not only on the temperature and rate of temperature change but also on the type and nature of the furnace atmosphere. The second method was used by Stewart (1969) in his temperature calibration scheme, but it was required that the thermocouple be in contact with the sample or the sample container during the calibration procedure. Compounds chosen for standards were those containing $solid_1 \rightleftarrows solid_2$ or solid $\rightleftarrows$ liquid type transitions, which were not atmosphere dependent. The standards used and their transition temperatures were potassium nitrate (129.5 and 333°C), potassium chromate (665°C), and tin (231.9°C). These standards require simultaneous TG–DTA because the transitions are not sensed by mass change but by DTA. Alternatively, the thermocouple may be placed in the sample and the weight signal simply disregarded during a calibration experiment under otherwise standard operating conditions.

Norem *et al.* (1969, 1970) used the third method, as previously discussed, to calibrate the temperature of their type of furnace and/or sample container. A ferromagnetic material was placed in the sample container and suspended within a magnetic field gradient. At the material's Curie temperature the magnetic effect diminishes to zero and the thermobalance indicates an apparent mass change. The magnetic field gradient should be kept small consistent with the balance sensitivity in order to maximize the sharpness of the effect. As the field gradient approaches saturation, the apparent loss of weight will start at 0 K for all materials and results will be more diffuse. For calibration over the temperature range from ambient temperature to 1000°C, it is obvious that a number of ferromagnetic materials must be used.

The criteria that were considered characteristic of an ideal standard were the following (Norem *et al.*, 1970):

(1) the transition must be sharp (i.e., its natural or true width should extend over a small temperature range).

(2) the energy required to effect the transition should be small under the dynamic scanning conditions of TG and when there is not a significant rate barrier, the "sharpness" of a transition is inversely proportional to

transition energy. Since magnetic transitions are second order, the ΔH values are negligible.

(3)   the transition temperature should be unaffected by the chemical nature of the atmosphere and independent of pressure. Precautions to prevent oxidation of the magnetic standards are necessary.

(4)   the transition should be reversible so that the sample can be run repeatedly to optimize or check the calibration.

(5)   the transition should be unaffected by the presence of other standards so that several samples can be run simultaneously to obtain a multipoint calibration in a single experiment.

(6)   the transition should be readily observable using standard samples in the milligram range—comparable to normal sample weights investigated with the apparatus.

A typical calibration curve using five standard ferromagnetic samples with a Perkin-Elmer TGS-1 system is shown in Fig. 16. The abscissa in Fig. 16a represents the output of the temperature transducer. Controls on the furance programmer allow the operator to make adjustments that will nearly linearize the transducer output with respect to the calibration points, i.e., $°C = aV + b$, where a and b are constants. Values in parentheses result from the least-squares analysis of the five calibration points and indicate a very good correlation with the reported values. This calibration is used in Fig. 16b to convert the abscissa to degrees Celsius (calibrated).

The Perkin-Elmer Co. has pioneered in the use of magnetic standards for calibration of the temperature axis in TG and has designed its instruments with this in mind. However, virtually all commercially available thermobalances can be calibrated by this technique. The magnet applying the field can be either a permanent magnet or an electromagnet located above or below the balance. Table VI is a list of potentially useful materials with suggested temperatures from either Perkin-Elmer or the ICTA committee on standardization (Garn *et al.*, 1980).

Transition temperatures are also dependent on the heating rate of the furnace, but this dependency is quite small in the 5–20°C min^{-1} range. The effect of heating rate on the observed initial transition temperature is given by

$$T_{\text{indicated}} = T_{\text{isothermal}} + RC_s T_n$$

where R is the effective thermal resistance between heat source and sample container, C_s is the effective heat capacity of the sample and its container, a most important parameter characterizing the performance of the instrument (or any thermal analysis treatment), and T_n is the heating rate.

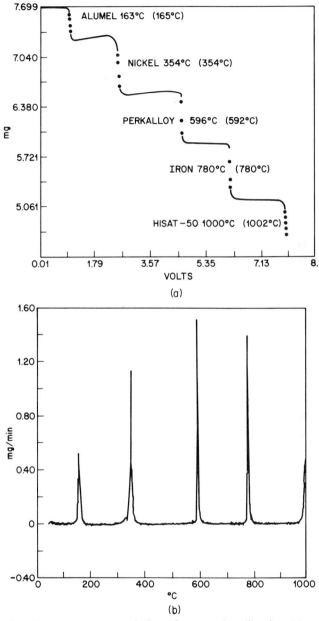

Fig. 16 Sample computer generated plots of a magnetic calibration: (a) weight versus voltage of temperature detector; (b) rate of weight loss versus calibrated temperature (°C min⁻¹ in N₂ (40 ml min⁻¹) (Gallagher and Schrey, 1970).

TABLE VI

Material	Recommended T_c (°C)	Source
Monel	65	Perkin-Elmer
Alumel	163	Perkin-Elmer
Permanorm 3	266.3 ± 6.6	ICTA–NBS
Nickel	354/354.4 ± 5.4	Perkin-Elmer/ICTA-NBS
Numetal	393/386.2 ± 7.4	Perkin-Elmer/ICTA-NBS
Nicroseal	438	Perkin-Elmer
Permanorm 5	458.8 ± 7.6	ICTA-NBS
Perkalloy	596	Perkin-Elmer
Trafoperm	753.8 ± 10.2	ICTA-NBS
Iron	780	Perkin-Elmer
Hisat-50	1000	Perkin-Elmer

[a] ICTA–NBS standards (GM-761) are supplied with tables describing observed differences based on the type of instrument and other variables.

[b] The temperatures listed are those associated with the extrapolated end point of the apparent weight change.

A calibration such as that in Fig. 16 will account quite well for this lag but requires a different calibration (values of a and b) for each heating rate or for those cases where the gaseous flow rate or heat capacity varies significantly.

The difference between the indicated temperature and the sample temperature is highly dependent on the geometrical relationship between the furnace, sample, and temperature sensor. It would be expected to be highest for those cases where the temperature gradient is steepest. With the trends toward miniaturization of furnaces and small sample sizes calibrations become essential. Figure 17 illustrates the temperature differentials that can occur for a small furnace (Gallagher et al., 1980). At zero rate there is still a significant effect under the experimental conditions indicated.

G. Some Potential Sources of Error

The preceding section has described problems associated with accurately defining the sample temperature. There are also potential difficulties in determining the sample weight under conditions normally imposed by TG experiments, e.g., changing temperature, flowing gases, and evolving gases. Besides the normal instrumentation errors (e.g., electronic stability of amplifiers and servomechanisms, chart paper slippage and expansion, weight calibration), there are also sources of error that are

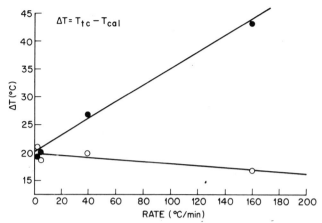

Fig. 17 Differences between the thermocouple temperature and the magnetic calibration value (596°C) at different heating and cooling rates for a Perkin-Elmer TGS-1 System. Nominal rates 160, 40, 5.0, 0.62 °C min⁻¹ in Ar (40 ml min⁻¹). ●: heating; ○: cooling. (Gallagher *et al.*, 1980.)

exaggerated by the changing temperature and atmospheres. Table VII lists some of these potential problems. Newkirk (1960) and Wendlandt (1974a) address these problems in greater detail. Norris *et al.* (1980) also discuss some of these problems with particular reference to isothermal kinetic studies.

TABLE VII

SOME POTENTIAL SOURCES OF ERROR IN THERMOGRAVIMETRY

1. Buoyancy effects
2. Aerodynamic forces
3. Condensation on the sample support system
4. Reaction of the sample or its products with the containment or support system
5. Spurious sample loss
6. Effects of electrostatic and magnetic fields
7. Temperature-induced changes in the balance

Buoyancy corrections can be very significant depending on the density of the atmosphere and the volume of the sample and suspension system. Calculations can be tedious, and a blank experiment under identical conditions is a more convenient method of correction. A number of balances have been built in which both the sample and the tare sections of the balance are heated simultaneously in an identical manner to minimize this effect.

Cahn and Schultz (1963) describe effects due to aerodynamic forces. These effects are induced by such factors as convection currents, turbulent flow, and thermomolecular flow. Obviously a great many factors enter into this problem. Most important are (1) shape and dimensions of the sample holder, suspension, hangdown tube, and baffles; (2) pressure and flow of the gaseous environment; (3) heating rate; and (4) temperature uniformity. Cox *et al.* (1973) describe several baffle designs that minimize the effects due to convection.

Condensation of a reactive atmosphere or gaseous products on cooler parts of the sample suspension system can be a problem. A flow away from the sample support system is advisable to counter this. Similarly, reaction with hot parts of the suspension system is also a potential problem. Oxidation is a typical example of the latter. Regardless of whether the oxide is volatile or adhering, an error will result. Occasionally, the sample container will have a catalytic effect on the decomposition that will lead to enhanced weight loss at misleading temperatures.

Samples can be inadvertently lost from the holder by (1) too great a flow of gas over the sample; (2) natural decripitation of a material, particularly minerals and single crystals; or (3) too rapid an evolution of product gases such as during a decomposition in vacuum.

One of the most distracting factors in TG is the influence of static charge on the sample. The frequent use of a carefully dried gas stream and air-conditioned, dehumidified laboratories only compounds the problem. The effect can be obvious, as when a sample clings to the side of the hangdown tube, or can go insidiously undetected. A grounded electrically conductive coating or mesh on the hangdown tube is very effective, but it must be able to stand the environment. The potential involvement of magnetic fields has been mentioned earlier in conjunction with magnetic temperature standards.

Manufacturers have gone to great lengths to minimize all of these problems. They have had varying degrees of success depending on the specific needs of the investigator. A blank run under the desired experimental conditions remains one of the better tools for evaluating an instrument. Convenience, cost, and versatility are also important but performance rates the "bottom line."

IV. Commercial Thermobalances

A. DuPont Model 951 Thermogravimetric Analyzer

The DuPont Model 951 Thermogravimetric Analyzer thermobalance is but one of a number of individual modules (see DTA–DSC, Section VI.A) that plug into the DuPont series of thermal analyzers (programmer and recorder consoles). The thermobalance consists of a furnace, sample holder, glass enclosure, and recording balance mechanism. The balance-to-furnace arrangement is as in Fig. 6c and the sample holder like that in Fig. 3b. The balance mechanism contains a null-balancing, taut-band meter movement with an optically actuated servoloop. Capacity of the balance is 1 g with mass ranges of 0.05 to 50 mg in $^{-1}$ in 10 steps with a sensitivity and precision of mass measurement of 0.2 and 0.4% of full scale, respectively, and an accuracy of mass measurement of ± 1% of full scale on the Model 900 or 990, 0.3% on the Model 1090, or 0.5% on the Model R90 analyzer. The atmosphere in the furnace chamber can be varied from 760 to 1 torr, with gas flow rates of up to 1000 cm³ min⁻¹ (depends on size and density of sample). A derivative circuit on the Model 990 thermal analyzer permits the recording of the first derivative of the mass change, with rates of mass change from 0.05 to 50 mg min⁻¹ in.⁻¹ in 10 steps. Derivatives are also available using the new 1090 (see Section VI.A) and R90 analyzers.

B. Perkin-Elmer Model TGS-2 Thermogravimetric System

The Perkin-Elmer Model TGS-2 Thermogravimetric System consists of the TGS analyzer unit (includes the thermobalance, balance electronic control unit, and furnace), furnace programmer unit, and a recorder. The configuration is like Fig. 6a. An accessory derivative unit permits recording the first derivative of the sample mass change or a separate DSC recording (with DSC-2 system). The Model AR-2 vacuum recording microbalance has a mass sensitivity of 0.1 μg with 16 full-scale recorder ranges from 0.01 to 1000 mg and electronic taring of 10, 100, or 1000 mg. The balance may be operated at pressures from 760 to 10⁻⁴ torr under oxidative, reductive, reactive, or inert atmospheres, at flow rates of up to 200 ml min⁻¹. A low-mass internal microfurnace with a platinum resistance wire heater is employed, which when coupled with the Model UU-1 temperature programmer controller permits heating and cooling rates from 0.312 to 320°C min⁻¹ in 11 steps, to a maximum temperature of 1000°C. The platinum resistance furnace element provides for both heating and temperature measurement. Heating equilibrium time is less than 1 min and the isothermal temperature control is to a few tenths of a degree.

The new Perkin-Elmer System 4 programmer controller illustrates the great usefulness and power of the microprocessor controller. This console is for use with their TMA and DTA units as well as the TGS-2. Simple and multiple ramp and hold periods can be programmed along with valve manipulations for purge gas control. The temperature axis can be automatically calibrated and linearized and the recorder scale adjusted. Temperature is displayed digitally to a tenth of a degree. Heating rates are adjustable over a wide range, 0.1–320°C min^{-1}. Besides the user-generated programs, which can be stored with protection, there are built-in, analyzer-specific, calibration, and diagnostic programs. Figure 18 shows a relatively simple but useful example of microprocessor control (Cassel, 1978). This multistep program provides automatic operation, faster turnaround time, and better accuracy for determining the various primary components of rubber.

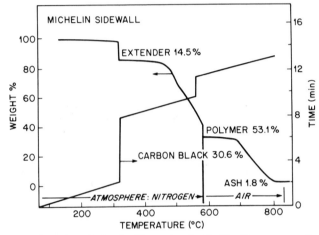

Fig. 18 Stepwise TG analysis of rubber using the Perkin-Elmer system 4 microprocessor controller (Cassel, 1978).

C. METTLER THERMOANALYZER SYSTEMS

1. TA-1 Thermoanalyzer (Simultaneous TG–DTG–DTA)

The Mettler TA-1 Thermoanalyzer is one of the most elaborate and versatile thermobalances commercially available at the present time. The Model TA-1 may be used for simultaneous TG–DTG–DTA under various conditions of furnace atmospheres and pressures to a maximum furnace temperature of 2400°C.

The balance and furnace assembly are in the arrangement shown in Fig. 6b and are mounted in a single cabinet; the control cabinet is separate. The balance cabinet also contains the vacuum system and all controls for the gas flow into the sampler chamber.

The control cabinet contains the multipoint recorder, temperature programmer, and measurement and control circuits for all the balance functions. The balance consists of an aluminum beam substitution-type balance using sapphire knife edges and planes. The mass measurement system consists of an electrical coil, attached to the end of the beam, which moves vertically around a permanent magnet attached to the balance housing. Changes in sample mass cause a beam deflection that moves a light shutter interrupting a light beam display on two photodiodes. The imbalance between the photodiode currents is amplified and fed back to the coil on the balance beam as a restoring force. This general system is used on the other Mettler balances as well. Total balance capacity is 16 g weighing above the beam (as in Fig. 6b) or 42 g below the beam. Vacuum-sealed control rods operate a weight set of 0–15.9 g, which is used as a tare for sample holder and support rod compensation. The electrical weight indication has a dual weighing range with three different sensitivities as standard (a fourth is optional). Two consecutive sensitivities, in the ratio of 1:10, are always recorded. One range is 0–1000 mg, recorded as 100 mg in.$^{-1}$, on the 10 in. chart. The second is the same 1000 mg but in steps of 100 mg full scale, or 10 mg in.$^{-1}$. This is done by recording the curve on the chart 10 times with the appropriate automatic shifts in the recorder zero from 0 to 900 mg in 100 mg steps. A more sensitive weight range of 0–100 mg is recorded in an identical manner as above. An additional high sensitivity weight range of 0–10 mg recorded, in the same manner, is available as an option. Sensitivity of the balance on the 0–100 mg range is 20 μg, with an overall accuracy of ± 50 μg.

For atmospheric or moderate vacuum applications, noncorrosive gases are passed through the balance housing and up into the sample container chamber. For high vacuum applications, two diffusion pumps are used to evacuate the system, each with a rated capacity of 60 liters sec^{-1}. One pump evacuates the balance compartment, the second the base of the sample tube. A cold trap is used with the second pump, giving the total system a capacity of about 200 liters sec^{-1}. The low pressure in the system is measured by a built-in ionization gauge located at the base of the sample tube. Ultimate pressure, achieved with a pump-down time of about 30 min, is of the order of 8×10^{-6} torr.

2. Mettler TA-2 Thermoanalyzer

The Mettler TA-2 Thermoanalyzer is a modified version of the TA-1 previously described. Although more compact in size, it offers the same advantages of simultaneous TGA–DTG–DTA. Analyses can be carried out in an accurately defined and controlled gaseous atmosphere, even in that of a corrosive nature, or under vacuum. The unit is designed as a tabletop laboratory instrument.

The weighing system is housed in a high vacuum-tight cell. By using an inert carrier gas, it is possible to work with corrosive gases without damage to the weighing system. The temperature programmer has 26 heating and cooling rates from 1 to 100°C min⁻¹. Or, the system may be operated isothermally at any temperature from 25 to 1000°C. Cycling between two preselected limits is also possible to repeat reversible processes.

3. Mettler TA-HE 20 Thermobalance

The Mettler TA-HE 20 Thermobalance is a general-purpose, modular instrument consisting of a recording balance and furnace, a temperature programmer, a balance control unit, and a recorder. The furnace, which is capable of a maximum temperature of 1000°C, is mounted on the top of the balance. A gas inlet and exhaust connections on the furnace permit the use of controlled gaseous atmospheres. The maximum sample mass is 40 g with recording mass ranges of 10, 100, or 1000 mg. Accessories include a DTA amplifier for TG–DTA, a digital display, BCD output, and others.

4. Mettler 2000C Thermoanalyzer (Simultaneous TG–DTG–DSC)

The newest model in the line has a modular tabletop design that emphasizes convenience and easy operation. Simple top loading of the sample is combined with pushbutton taring. The temperature range of the unit is from 20 to 1200°C with the standard furnace or to 1600°C with an optional furnace. Temperature is controlled by a microprocessor that allows heating rates from 0 to 30°C min⁻¹ in steps of 0.1°C. The sample size can be up to 6 g. The maximum sensitivity range displays 1 mg full scale (100 mV) with a precision of 10 μg. The applicable pressure range is similar to that of the TA-1 but the vacuum system operates in a pushbutton programmed mode for enhanced convenience. Corrosive atmospheres may be used with a special optional furnace.

Computer interfacing equipment is available. There is some software provided for specific use with a Hewlett Packard desk computer, Model HP9815A. The interface consists of a CT10 encoder, CT12 five-channel multiplexer, CT19 timer and pulse generator, CT25 A/D converter, and line filter.

5. Mettler TG50 Thermobalance

This is a modular component of the newest system in the line, the TA3000 system. The thermobalance has an electrical weighing range of 1 μg to 150 mg. The small overall size of the module is condusive to rapid heating and cooling rates. Forced cooling can bring the sample from 1000 to 100°C in only 8 min. Circulating gases may be used along with programmable valving. The temperature is controlled by the TC10 TA processor. A wide range of temperature programs may be used and stored. Data interfacing with a serial plotter/printer is present. Kinetic analysis routines are available.

D. NETZSCH THERMOGRAVIMETRIC SYSTEMS

1. STA 429 Simultaneous Thermal Analyzer

The Netzsch Model STA 429 Simultaneous Thermal Analyzer is a versatile simultaneous TG–DTA system. The system consists of a recording balance and vacuum pumps as part of the furnace cabinet and a cabinet containing the recorder, temperature programmer, pressure-measuring equipment, and other electronic units. The recording balance has a capacity of 10 g with selectable mass ranges of 5–500 mg full scale. The furnace is above the balance. Furnaces are available with temperature ranges of −150–420, 25–1000 (vapor atmosphere), 25–1350, and 25–2400°C. The temperature programmer permits the selection of furnace heating rates of 0.1–100°C min^{-1} in 10 ranges, as well as isothermal and cyclic operation. The furnace system can be flushed with inert or reactive gases at pressures from 760 to 1×10^{-6} torr. A six-channel multipoint recorder is used to record the parameters desired, i.e., mass, differential temperature, time derivatives of the preceding quantities, temperature, pressure, and/or other analytical data. Accessories for the system include an EGA unit based on thermal conductivity, an emanation thermal analysis (ETA) system, and a quadrapole mass spectrometer.

Another thermobalance, the Model STA 409, a simplified tabletop unit, is also available from Netzsch.

2. Former Sartorius Systems

The Netzsch Co. has recently assumed responsibility for the Sartorius thermobalance line. This includes instruments that are specially designed for gravimetric determination of surface area and pore size distribution and others for more conventional simultaneous TG–DTA or simply TG. A characteristic of these systems is that both the sample and the tare suspensions below the balance beam are heated similarly in order to minimize buoyancy corrections. The various systems are based on three recording balances. The Model 4102 is the most sensitive with a capacity of 2.5 g and a sensitivity of 0.1 μg. The Model 4104 has a capacity of 25 g and a sensitivity of 1.0 μg. Figure 19 shows a portion of Model 4201. There is complete isolation of the balance and sample compartments achieved by means of a clever magnetic suspension. Obviously, this offers greater opportunity for studies in corrosive atmospheres, higher

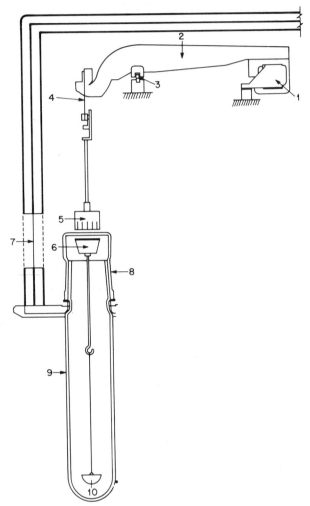

Fig. 19 Magnetic Suspension Balance Sartorius Model 4201. (1) Magnet and coil. (2) Beam. (3) Beam support. (4) Suspension ribbon. (5) Beam magnet. (6) Suspension magnet. (7) Observation window. (8) Upper glass body. (9) Lower glass body (removable). (10) Pan.

pressures, or better vacuum. The balance has a capacity of 30 g with a sensitivity of 10 µg.

E. Cahn Instruments

The Cahn Models RG and RH recording balances have been the bulwark of most investigator-built instruments for the last 20 years. Many of the commercial TG instruments during this era were constructed using the model RG as a nucleus, e.g., Perkin-Elmer TGS-1, Fisher Scientific TGA systems, Theta Instruments, Harrop Co., Technical Equipment Co., Tempress Co., and Columbia Scientific (earlier versions marketed by the R. L. Stone and Tracor companies). The Model RG has a maximum capacity of 1 g with 0.1 μg sensitivity (2.5 g maximum with 0.25 μg sensitivity on a secondary beam position) and the RH has a maximum capacity of 100 g with a sensitivity of 1 μg. Because of the excellent reliability and durability of these balances, they will continue in use for many years.

The manufacturer, however, has supplanted these models with models that have even greater stability and enhanced flexibility and convenience. Model 1000 replaces the RH. Its sensitivity is 0.5 μg with a maximum load including sample support of 100 g. Precision is given as 1.5 μg. There is an electronic tare or weight suppression of up to 10 g, which also corresponds to the maximum observable weight change. There are six weighing ranges in decades from 10 g to 10^{-4} g full scale. The standard housing allows vacuums of 10^{-6} torr and bakeout temperatures of 125°C. The Model 2000 replaces the RG. It has a maximum capacity of 1.5 g with a sensitivity of 0.05 μg (3.5 g and 0.5 μg on the less-sensitive beam position). It has analogous electronic features and similar vacuum capabilities with lower noise and better temperature stability than the RG. Cahn Instruments also markets complete TG instruments, including furnaces for room temperature to 1100°C operation and a microprocessor controller capable of uniform heating or cooling rates from 0 to 25°C min^{-1}. System 113 uses the Model 2000 balance, and System 114 uses the Model 1000 balance. Special hangdown systems minimize the noise levels. A wide variety of optional equipment is available to (a) minimize the temperature-induced drift in the mass signal, (b) provide a derivative signal, (c) determine gravimetric surface adsorption isotherms, (d) measure particle size distribution by sedimentation, (e) measure magnetic susceptibilities, (f) measure surface tension, and (g) determine permeation rates.

F. Theta Instruments

Theta has two styles of thermobalances. The first type, top-loading balances, has two models with capacities of 10 g and 200 g, respectively, and a sensitivity of 0.1 mg. They can be equipped with a 1000°C furnace for operation at atmospheric pressure or a 1200°C furnace capable of vacuum operation.

The second style of thermobalance uses the Cahn Model 1000 and 2000 balances having the specifications described earlier. The Theta thermobalance can be either a straightforward TG instrument (Gravitronic II) or a combined TG–DTA apparatus (Gravitronic III). In the former case it has either a 1400°C Kanthal furnace or a 1600° platinum–rhodium furnace. The combined TG–DTA model has only the 1600° furnace. Figure 20 shows the combination of the DTA instrument coming from below with the TG apparatus from above but both housed within the same furnace. A variety of forms of data presentation and acquisition and other options are available.

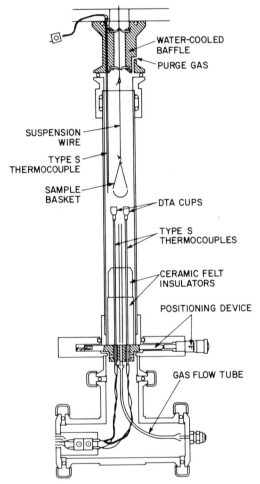

Fig. 20 Theta Gravitronic Combined TG/DTA.

G. Harrop Laboratories

TG packages based on the Cahn Model 1000 and 2000 balances are available. Alumina hangdown tubes are used. A 1200°C Kanthal furnace or a 1600°C Pt–Rh furnace is supplied. The control package provides heating rates of 0 to 75°C min⁻¹. Provisions are made for separate flow paths in case of reactive atmospheres.

H. Columbia Scientific Industries, Inc.

A variety of TG options are available from CSI. They are based on the Cahn Models 1000 and 2000 balances in conjunction with the CSI 202 series of recorder controllers. Heating rates range from 0.5 to 50°C min⁻¹ in addition to isothermal operation. Two water-cooled furnaces are available for ambient to 1000 or 1450°C operation.

I. SETARAM (Société d'Étude d'Automatisation de Régulation et d'Appareils de Mesures)

The SETARAM Model G70 Thermoanalyzer, their introductory instrument, consists of a knife-edge, null-balance, electronic-recording balance; a Kanthal element furnace (1000°C maximum temperature); a temperature programmer; and a two- or three-pen strip-chart recorder. The balance has a useful load of 10 g with an absolute sensitivity of 0.1 mg. Mass ranges, with the appropriate recorder, are 100 mg, 400 mg, or 2 g, full scale. The furnace heating rate, using the above programmer, can be varied from 50 to 500°C hr⁻¹.

There are two other classes of thermobalance (simultaneous or combined TG–DTA) available. The first class, simultaneous TG–DTA, uses the B70 balance, which is an improved version of that used in the G70. The capacity is 10 g with a sensitivity of 10 μg. Five furnace options cover the range of operation from − 196 to 2400°C with program rates of from 0 to 99°C min⁻¹. The second class, combined TG–DTA, uses the MTB series of balances where a torsion braid replaces the knife edge. Sensitivities range from 0.4 to 10 μg and capacities vary from 10 to 100 g depending on the model. A dual furnace arrangement is used so that the tare side of the instrument is also heated to minimize buoyancy variations. Three furnaces cover the range from − 196 to 1700°C with heating rates from 0 to 99°C min⁻¹. Both classes can be used in vacuum as well as controlled flow through atmospheres.

J. Shimadzu Thermobalances

Many different models of thermobalances and/or TG–DTA instruments are available from Shimadzu. The Model TGA-20 B Micro Thermobalance features a mass sensitivity of less than 1 μg and can determine the mass changes of a sample in an atmosphere of air, inert and reactive gases, and vacuum.

K. Rigaku TG, TG–DTA, and TG–DSC Systems

Of the numerous models of thermobalances available from this manufacturer, two of the most interesting are the simultaneous TG–DTA and
TG–DSC models of the Thermoflex series. The simultaneous TG–DTA
system is a desktop thermobalance and DTA unit that converts from TG
to TG–DTA by interchanging the sample holder. Two furnaces are available, with upper temperature limits of 1000 and 1500°C. Furnace temperature programming rates can be varied from 0.625 to 20°C min^{-1} in six
steps. Two TG ranges are employed: the micro-type range from 1 to
100 mg and the macro-type range from 2 to 500 mg. A two- or three-pen
recorder is used, the latter being required for simultaneous TG–DTA
measurements. The TG–DSC model has an upper limit of 800°C and a
sensitivity of 0.5 mcal sec^{-1} full scale.

Two new systems have been recently added. A different motion
(model DM) TG–DTA unit has the reference motival and holder as part of
the tare side of the balance so that corrections for buoyancy, convection,
and thermolecular flow are minimized. A microprocessor furnace controller is used in conjunction with the 1000 or 1500°C furnaces. Maximum
sensitivities are 0.5 mg and 5 μV full scale. Vacuums as low as 10^{-3} torr
can be employed.

The second system is a TG unit that incorporates a rapidly heating infrared furnace capable of 1200–1300°C. The furnace can heat from room
temperature to 900°C in as little as 30 sec in conjunction with the microprocessor control unit.

L. ULVAC–RIKO

As with the Rigaku thermobalances, there are a large number of different instruments available from ULVAC–RIKO (Sinku Riko Co.). One of
the more interesting thermobalances is the Model TGD-3000-RH, which
also employs an infrared image furnace for very high heating rates of
500°C min^{-1} or more, up to a maximum temperature of 1200 or 1400°C
(see Section III.C). Rapid cooling of the furnace is also possible: cooling
from 1000°C to ambient room temperature is said to take place in only a
few minutes. Simultaneous TG–DTA measurements can also be performed with the instrument.

M. Linseis Thermobalances

Numerous instruments are available from Linseis, the most elaborate
being the Model L81 simultaneous TG–DTA instrument. This instrument
consists of a microbalance (capacity up to 15 g sensitivity of 5 mg per

250 mm recording chart width) that can be used in either a horizontal or vertical arrangement, a Kanthal resistance element (1200°C) or a Crusilite furnace (1550°C), a furnace temperature programmer, a high vacuum system, and the appropriate DTA amplifier and sample container. A quadrupole mass spectrometer may be connected to the instrument to provide data on the composition of the evolved gases.

N. STANTON-REDCROFT TG-750 THERMOBALANCE

The Model TG-750 thermobalance features an electronic microbalance having a 500 mg capacity which gives a range of switch-selected sensitivities from 1 to 250 mg per full-scale recorder deflection. Because of the water-cooled furnace design, there is "no noticeable buoyancy effect" over the entire temperature range even when it is operating at maximum sensitivity. The small volume of the furnace makes it suitable for connection to a gas analysis system including mass spectrometry. Maximum temperature of the furnace is 1000°C.

The newer model STA-780 series provides simultaneous TG–DTA. The electronic balance has a capacity of 5 g, but, as is frequently the case, the sample volume is the limiting factor. The readability is 1 μg. When used in the TG mode only, a lighter sample suspension allows greater sample size. The provisions for control of the atmosphere are somewhat unique. A small thimblelike cup surrounds the sample from below and the gas enters at the bottom of this cup and exits out the top through the small space between the thermal baffles on the hangdown and the cup (see Fig. 21).

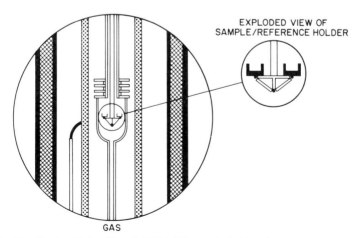

Fig. 21   Stanton Redcroft model STA-780 sample holder–furnace arrangement.

O. Hungarian Optical Works (MOM, Budapest)

The Derivatograph has long been the major thermal analytical instrument of Eastern Europe. Simultaneous TG–DTA can be accomplished under a wide variety of conditions. Thermal expansion can also be measured with a change of sample holder. An air-dampened analytical balance has a useful capacity of 10 g with a maximum sensitivity of 0.1 mg mm^{-1} of chart deflection and a minimum sensitivity of 10 mg mm^{-1} of chart deflection. A wide range of sample holders is available, e.g., Fig. 2d, e, and f and Fig. 5c. The recording device uses a photographic drum and projection galvanometers. Maximum furnace temperature is 1100°C. Heating may be linear (0–20°C min^{-1} in seven steps) or quasi-isothermal (see Section III.C).

P. Spectrum Products, Inc.

The Spectrum Model TGA-1 is a thermobalance designed for high-pressure, primarily isothermal, operation (Gardner *et al.*, 1974). Two versions are available with either 3- or 6-g capacity. The samples are loaded in an antichamber that is subsequently pressurized to the reaction pressure. A ball valve is then opened and the sample lowered into the high-temperature (up to 1000°C) pressurized (up to 1500 100 atm) reactor. The lowering rate determines the heating rate but weight data can not be obtained during that period. The weight transducer provides an output of 4 V g^{-1}. Operation can be performed in H_2, CO, CO_2, H_2O or < 1% H_2S atmospheres.

V. Differential Thermal Analysis (DTA) and Differential Scanning Calorimetry (DSC)

A typical DTA apparatus is illustrated schematically in Fig. 22. The apparatus generally consists of (a) a furnace or heating device, (b) a sample holder, (c) a low-level dc amplifier, (d) a differential temperature detector, (e) a furnace temperature programmer, (f) a recorder, and (g) control equipment for maintaining a suitable atmosphere in the furnace and sample holder. Many modifications of this basic design have been made, but all instruments measure the differential temperature of the sample as a function of temperature or time (assuming that the temperature rise is linear with respect to time).

There have been numerous DTA systems described in the literature, many of which employ novel designs of the sample container, furnace or heating device, differential temperature $T_s - T_r$ detection devices, atmo-

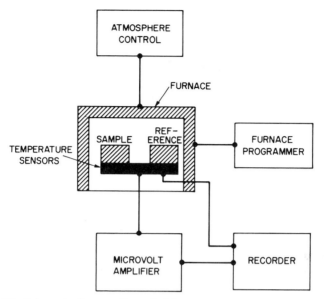

Fig. 22 Schematic diagram of a typical DTA apparatus (Wendlandt, 1974b).

sphere control, and so on. It is now possible to determine routinely the DTA curve of a sample in the temperature range from -190 to $2400°C$ at extremes of pressure from 10^{-6} torr to hundreds of atmospheres. With the wide variety of instrumentation available, it is difficult to describe a "typical" DTA apparatus in the short space allocated here. Instrumentation, especially the commercially available types, has been extensively discussed by Wendlandt (1972, 1974a,b).

A. SAMPLE CONTAINERS

One of the most important components of a DTA apparatus is the type of sample and reference material container employed. There are a wide variety of sample containers available in commercial instruments or that have been described. The type of sample container used depends, of course, on the nature and quantity of the sample and also on the maximum temperature to be investigated. Sample containers have been constructed from alumina, zirconia, borosilicate glass, Vycor glass, fused quartz, beryllia, boron nitride, graphite, stainless steel, nickel, aluminum, platinum or platinum alloys, silver, copper, tungsten, the sample itself, and numerous other materials. Some typical sample containers used in DTA are shown in Fig. 23. In (a), the sample (0.1–100 mg) is pressed into a closed-end tube and the tube is placed over the ceramic insulator tube containing the thermojunction (Garn, 1962, 1965). This type of sample container will permit the sample to dissociate in a self-generated atmosphere. It cannot be used with samples that fuse on heating, however. In determining the heats of explosion for a number of explosive materials, the isochoric sample container in (b) was used by Bohon (1963). It consisted of a stainless-steel body and cap that was sealed with a screw cap and a copper gasket. The internal volume was about 0.085 ml and contained about 25 mg of sample. By means of a loading chamber, the sample

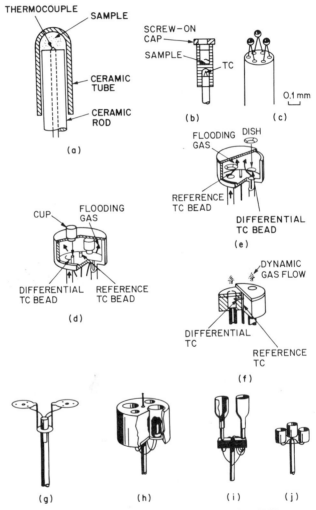

Fig. 23 DTA sample containers (Wendlandt, 1974b).

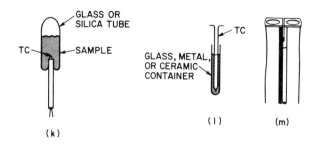

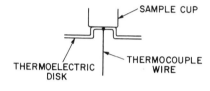

Fig. 23 Cont'd.

container could be charged with a gas at pressures up to 100 atm. Mazieres (1964) developed the microscample container in (c) for use with samples 1–200 μg in mass. The sample was contained in a chamber drilled into the thermojunction itself. A similar cup-type container was also described that could be used for samples 0.1–10 mg in mass. Some difficulty would certainly be experienced in handling microgram quantities of sample.

Sample containers used in the Columbia Scientific (earlier Stone or Tracor) instruments are shown in (d)–(f). Small cups are used in (d) to contain samples of 10–20 mg; the cups are constructed from aluminum, stainless steel, nickel, or platinum and/or palladium alloys. For smaller samples, 0.1–20 mg, the highly sensitive ring thermocouple holder, as shown in (e), is used. Sample dishes can be made from aluminum, stainless steel, or platinum by the investigator using a simple press and die. True dynamic gas atmosphere control is featured in the sample holder in (f). The gas flow is through the sample and reference materials; it cannot be used with samples that fuse, however.

Sample containers illustrated in (g)–(j) are used in the Mettler thermoanalyzer systems. In (g), the sample is in a small cup or crucible which is placed on the small circular disk containing the thermojunction. A block-type sample container is shown in (h) in which an alumina block is employed. The sample is contained in a crucible which may be constructed from platinum or other metals. For macro amounts of sample, the container in (i) may be used. It is constructed of alumina or of different metals. If only small amounts of sample are to be studied, the microcrucible sample container in (j) may be used.

For sealed-tube DTA studies, the sample is enclosed in a sealed glass or quartz tube, as shown in (k). The glass capillary tube sample container, illustrated in (l), consists of a 1.5–2.5 mm diameter disposable glass capillary tube. Sample temperature is detected by a thermocouple that is inserted in the sample from the open end of the tube.

A sample container for use at very high temperatures (2200°C) is illustrated in (m). It is constructed of tungsten–tungsten–26% rhenium and is for use in the Mettler instruments.

A very convenient sample container is shown in (n). The sample is contained in a small cup (aluminum, platinum, gold, etc.) which is placed on a thermoelectric disk that has a thermocouple welded to it. By means of an identical reference container (not shown), a differential temperature is detected between the sample and reference materials. A sample container used by Stanton Redcroft was shown earlier in Fig. 21.

The sample holder can have a profound effect upon the observation. Dollimore and Mason (1981) have noted catalytic effects, and Garn and Menis (1980) discuss some differences observed during certification.

B. DIFFERENTIAL TEMPERATURE DETECTION SYSTEMS

The choice of a temperature detection device depends on the maximum temperature desired, the chemical reactivity of the sample, and the combined sensitivity of the dc amplifier and recording equipment. The most common means of differential temperature detection is with thermocouples, although thermopiles, thermistors, and resistance elements have been employed. For high-temperature studies, an optical pyrometer may also be practical.

Thermocouples normally used in DTA instruments are shown in Table VIII. Temperature limits listed are for relatively accurate measurements with 20-gauge wire in air. The copper–constantan thermocouple is commonly used in the −159 to 250°C temperature range. Noble metals and tungsten–rhenium alloys are used at temperatures up to 2400°C and beyond. For extremely high temperature, up to 3000°C, tantalum carbide versus graphite has been suggested (Brewer and Zaritsanos, 1957).

Kollie et al. (1975) have discussed the precision and accuracy of annealed type K (Chromel versus Alumel) thermocouples. The manufacturer specifies that the thermal emf versus temperature for the thermocouple is within ±0.375% or ±0.750% of that given by ASTM E-230.

TABLE VIII

THERMOCOUPLES COMMONLY USED IN DTA AND DSC

ISA type	Positive metal	Negative metal	Thermoelectric power (μV °C^{-1})^a	Approximate maximum temperature (°C)
T	Copper	Constantan	40	250
Y, J	Iron	Constantan	51	450
E	Chromel	Constantan	59	1000
K	Chromel	Alumel	41	1350
–	Platinel I	Platinel II	31	1370
S	Platinum	Platinum– 10% rhodium	5.5	1600
–	Tungsten	Tungsten– 26% rhenium	3.3	2400

^{a} At 25°C.

However, during use at temperatures above 200°C, the Chromel thermo-element undergoes a solid-state transformation that causes deviations of up to 1.3% or more.

To increase the emf from differential thermocouples without the use of increased amplification, thermopiles have been suggested (Cox and McGlynn, 1957; Lodding and Sturm, 1957; Joncich and Bailey, 1960). The advantage of such a system is the greater output signal with a lower noise level, due to lack of electronic amplification. A five-thermocouple thermopile used in the Mettler TA 2000 system has a ΔT sensitivity of 115 μV °C^{-1}.

Unusual thermocouple configurations that have been proposed are the thin-film type (King *et al.*, 1968; King, 1968), and the disk (Yamamoto *et al.*, 1969; Baxter, 1969) type. Thin-film thermocouples eliminate the problems of attempting to match the thermojunctions formed from wire elements. The former can be made light in weight and can be matched exactly by the evaporation of thin films of dissimilar metals that overlap to form the thermojunction. The thin-film thermocouple described by King *et al.* (1968; King, 1968) consisted of gold and nickel deposited on a quartz plate. It could be employed in the temperature range from -125 to 500°C and had a thermoelectric power of 10 μV °C^{-1} at 25°C to about 25 μV °C^{-1} at 200°C. A similar thin-layer thermocouple was described by Audiere *et al.* (1974), in which the gold–nickel thermojunctions were deposited on a 2.54-cm diameter sapphire disk. They found a thermoelectric power of about 23 μV °C^{-1} in the temperature range of 20–300°C.

The disk-type thermocouple and sample container are shown in Fig. 24. The disk is made of constantan and serves as the major path of heat

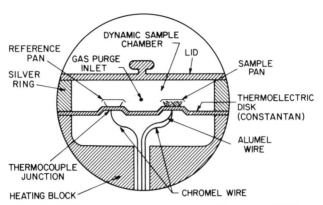

Fig. 24 Disk-type thermocouple described by Baxter (1969).

transfer to and from the sample and also as one-half of the ΔT-measuring thermocouple. A Chromel wire is connected to each raised indentation, thus forming a Chromel–constantan differential thermocouple system. This system is usable in the temperature range from -150 to 600°C. Yamamoto *et al.* (1969) used a dumbbell-shaped piece of Chromel that consisted of two circular disks connected by a narrow strip. Alumel wire, welded to the center of each disk, served as the other thermocouple junction.

Sheffield *et al.* (1979) have discussed and demonstrated enhanced sensitivity in DTA by obtaining the differential signal from mathematical differentiation of the single thermocouple (thermal analysis curve). Their data acquisition system records the thermocouple voltage to ± 0.1 μV at a rate of 4 sec^{-1}. A computer regression analysis subtracts the background, and the resulting peak area is also integrated. In this manner the reference material is actually the sample. Other advantages involve the use of only one couple, which should reduce noise and drift.

Since thermocouples for DTA use are difficult to calibrate to a precision of greater than ± 0.01°C, Mashiko *et al.* (1971) developed a high-precision DTA apparatus containing Degussa platinum resistance thermometers. These detectors could be calibrated with a precision of better than ± 0.004°C against a standard platinum resistance thermometer. The apparatus was used to determine the purity of organic compounds to a stated accuracy of ± 0.004 mole % in the temperature range 200–200°C.

Fairly high-resistance thermistors, 100,000 Ω at ambient temperature, connected in a bridge circuit have been used to detect the differential temperature (Paulik and Leonard, 1959; Weaver and Keim, 1960). This method does not normally require the use of a dc amplifier. Because their resistance decreases rapidly with increase in temperature, thermistors are generally useful only up to about 300°C (Weaver and Keim, 1960).

Calibration of the temperature axis in DTA–DSC has been the subject of considerable effort. If simultaneous TG–DTA instruments are used or if the DTA–DSC instruments are sensitive enough to detect the small thermal effects associated with second-order phase transitions, then the Curie temperature magnetic standards described earlier (Table VI) can be used. A number of carefully chosen first-order phase transitions have been studied in a cooperative effort by numerous laboratories (McAdie *et al.*, 1972; Garn and Menis, 1980). This effort led to a series of NBS–ICTA Standard Reference Materials (757, 758, 759, 760). These results have also been accepted by ASTM (E474-73). Figure 25 shows how the extrapolated onset and peak temperatures are defined and Table IX lists the recommended values. These values can also be used to help calibrate simultaneous TG–DTA instruments.

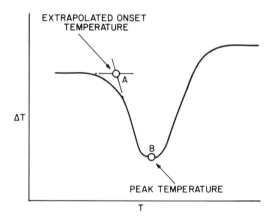

Fig. 25 Points for calibration of DTA instruments.

C. FURNACES AND FURNACE TEMPERATURE PROGRAMMERS

Furnaces required for DTA systems are similar to those described for thermobalances (Section III.C). The choice of furnace heating element depends on the temperature range desired, which is usually from − 190 to 1600°C or higher. The furnace may be mounted vertically or horizontally; it may be heated by a resistance element, by high-frequency rf oscillation (Brewer and Zaritsanos, 1957), or by a coil of tubing through which a heated or cooled liquid or gas is circulated (Clampitt, 1963).

A wide variety of DTA furnace configurations have been described, perhaps more than for thermobalance furnaces. Vassallo and Harden (1962) described a furnace block assembly that is heated by means of a low-wattage heater cartridge. It had provision for rapid cooling or for use

TABLE IX

RECOMMENDED CALIBRATION TEMPERATURES FOR DTA STANDARDS
(NBS CERTIFICATES 757, 758, 759, AND 760)

	DTA mean values		
Material	Equilibrium value (°C)	Extrapolated onset (°C)	Peak (°C)
1,2-Dichloroethane	−35.6	−35.8	−31.5
Cyclohexane	−86.9	−86.1	−31.5
	6.7	4.8	7.0
Phenylether	26.9	25.4	28.7
o-Terphenyl	56.2	55.0	57.9
KNO_3	127.7	128	135
In (metal)	157	154	159
Sn (metal)	231.9	230	237
$KClO_4$	299.5	299	309
Ag_2SO_4	~430	424	433
SiO_2	573	571	574
K_2SO_4	583	582	588
K_2CrO_4	665	665	673
$BaCO_3$	810	808	819
$SrCO_3$	925	928	938

below room temperature by passing a coolant through the coils surrounding the furnace. Sample and reference materials were placed in glass capillary tubes.

More sophisticated furnaces, for use in the temperature range from − 150 to 2400°C, are used in the Mettler thermoanalyzer. The furnace for use from − 150 to 400°C has a Kanthal resistance wire heater element. The 25 to 1000°C temperature range furnace is constructed from quartz and uses a Kanthal heater element in the form of a helix. The high-temperature furnace, 25 to 1600°C, has a super-Kanthal heater element and, like the 1000°C furnace, can be used at low pressure. The superhigh temperature furnace, for use in the temperature range from 400° to 2400°C, uses tungsten heater elements.

The great versatility of modular furnaces for DTA is illustrated by the four DuPont furnaces. The standard cell, containing a general-purpose block-type furnace, covers the temperature range from − 150 to 500°C. Samples are placed in disposable 2- and 4-mm diameter glass tubes into which the thermocouples are inserted. The intermediate cell, 25 to 850°C, uses quartz tubes rather than glass to contain the samples. Two high-temperature cells are available, one with a maximum temperature of 1200°C, the other 1600°C. Both employ platinum or ceramic sample containers.

The DuPont Series 99 Thermal Analyzer is a microprocessor-based programmer for use with the full range of DuPont modules. Two versions are available: an extended-range Series 99 (XR) for the full breadth of research applications, and a limited-range Series 99 (LR) for repetitive analyses and quality control applications.

The Series 99 is used in conjunction with a 28×43 cm $X-Y-Y'$ two-pen recorder. It is designed for ease of use and features microprocessor control with a nine-method memory and automatic accessory switching for gas and cooling. The temperature programs can be linked for annealing profiles and can control at heating and cooling rates from 0.1 to 100°C min^{-1} (depending on the module and programmer used).

The full range of DuPont modules can be interfaced and controlled by the Series 99 (DSC, Pressure DSC, DTA, high-temperature DTA, TG, TMA and DMA).

Requirements for a good DTA furnace include symmetry in heating and the ability of the heater elements to heat uniformly. The furnace temperature distribution must be uniform in the area of the sample container for good results. Wiedemann (1964) has reported the temperature distribution curves of the Mettler furnace. A temperature distribution study has also been given by Yamamoto *et al.* (1969) for their furnace.

For operation at low temperature, the furnace may be surrounded by a Dewar flask and precooled with liquid nitrogen. Another method is to use a gas as a heat-exchange medium. Most temperature programmers do not function efficiently unless a thermal reservoir at least 30°C below the program temperature is available. The ultimate in low-temperature control should perhaps be thermoelectric cooling; however, such a system is not yet available.

Temperature programmers for DTA furnaces are similar to those described for thermobalance furnaces (Section III.C) so they will not be discussed further here. Also, the fluctuations in furnace heating rates can be determined using the methods described in thermogravimetry (see Section III.C).

Again there is some interest in isothermal operation of DTA or DSC equipment. Howard (1973) has used DTA equipment operated in an isothermal model to evaluate antioxidants. In this technique the time (induction time) to an exothermic indication at different temperatures or even atmospheres indicates the stability of the particular composition. Similar approaches have been taken for the evaluation of hazardous materials. However, an even better method has been devised that uses an adiabatic rate calorimeter (ARC) (Townsend and Tou, 1980). This technique surrounds the sample cell with an adiabatic jacket eliminating the loss of heat that occurs in an isothermal DTA or DSC experiment. This adiabatic condition more closely simulates the potential thermal runaway reaction for the storage of hazardous materials.

D. Low-Level dc Voltage Amplifiers

The output voltage from a differential thermocouple is of the order of $0.1-100\ \mu V$, depending on the type of thermocouples used (see Table VIII) and the temperature difference between them. Hence, unless a very sensitive recording system is used ($< 100\ \mu V$ full scale), the T_s—T_r signal must be amplified. The amplifier must have low noise, low drift, and high stability to be used for DTA instrumentation. Instability of the amplifier will result in an unstable baseline (Theall, 1969), and drift by either input voltage or ambient temperature changes will cause output fluctuations. Pickup of 60 Hz ac by the input wiring can cause output noise as well as an unstable base line.

Many times, in an effort to reduce amplifier noise, capacitors are added across the output of the amplifier, and occasionally at the input (Theall, 1969). These capacitors frequently dampen the response of the amplifier, which causes a shift in the curve peaks and also a loss of peak resolution. A proper value of capacitor must be used, if noise is a problem, to form a compromise between noise reduction and loss of peak resolution. The Spectrum Scientific Co. manufactures a line of active electronic filters that provide a wide range of adjustable time constants and some amplification. These can be used for TG or other thermal analytical signals as well as for DTA.

E. Recorders

The types of recorders and data acquisition systems used in DTA are similar to those employed in TG (Section III.D). Modern instruments employ $X-Y$, $X-Y^1Y^2$, or two or more single-channel strip-chart recorders. The DTA curve is recorded as a function of sample (T_s), reference T_r, furnace (T_f) temperature, or of time. If T_r is used, the actual sample temperature can be determined by adding the ΔT value. The use of T_r will give a smoother curve without the possibility of retrograde temperature changes observed with large values of ΔT. For quantitative DTA studies, a time basis is preferred, whereas for precise determination of transition temperatures, a temperature basis is used.

F. DSC Instruments

Two types of DSC instrument have been widely used: the heat-flux DSC (e.g., DuPont 910 DSC and Mettler DSC 20 or 30) and the power-compensational DSC (Perkin-Elmer and Setarum 101). The basic principle of operation of these two types of DSC instrument is best described in terms of these two major commercial systems.

1. Heat-Flux DSC

Figure 24 shows a schematic diagram of the DuPont DSC cell. The cell uses a constantan disk as its primary means of heat transfer to the sample and reference positions and as one element of the temperature measuring thermoelectric junctions. The sample of interest and a reference are placed in pans that sit on raised platforms on the constantan disk. Heat is transferred from the disk and up into the sample and reference via the sample pans. The differential heat flow to the sample and reference is monitored by chromel–constantan area thermocouples formed by the function of the constantan disk and a chromel wafer that covers the underside of each platform. Chromel and alumel wires are connected to the underside of the chromel wafers, and the resultant chromel–alumel thermocouple is used to monitor the sample temperature directly. Constant calorimetric sensitivity is maintained throughout the usuble range of the cell via electronic linearization of the cell calibration coefficient E. Temperature ranges from -180 to $725°C$ are possible in inert atmospheres ($600°C$ maximum in oxidizing atmospheres).

Heat flow is measured to a sensitivity of 0.01 mW cm^{-1} in the calibrated mode, and enthalpy values are calculated with a calorimetric precision of $\pm 0.35\%$ using the 1090 Thermal Analyzer with DSC data analysis software. Isothermal stability to $\pm 0.1°C$ is feasible with this system at temperatures above ambient, and heat rates are controlled from 0.1 to $100°C$ min^{-1} in $0.1°C$ increments.

2. Power Compensated DSC

The Perkin-Elmer DSC system is of this type. Models DSC-1 and DSC-1B, introduced in 1963 (O'Neill, 1964; Watson *et al.*, 1964), have been superseded by the Model DSC-2, which has a temperature range extended to $725°C$, improved baseline reproducibility and linearity, and higher calorimetric sensitivity. A comparison of the sample holders used in the DSC-1, DSC-1B, and DSC-2 instruments is shown in Fig. 26. In the DSC-1 cell, the sample and reference holder consists of a stainless-steel cup and support, a platinum-wire sensor, an etched Nichrome heater, and other thermal parts. These components are mechanically crimped together in a very tight sandwich. This sample holder operates well over the temperature range -125 to $500°C$. In the DSC-2 sample holder, the materials of construction used are a platinum–iridium alloy for the body and structural members of the holder, a platinum wire for both the heater and sensor, and α-alumina for electrical insulation. All structural parts of the holder are spot welded together.

A schematic diagram of the calorimeter is shown in Fig. 27. The appa-

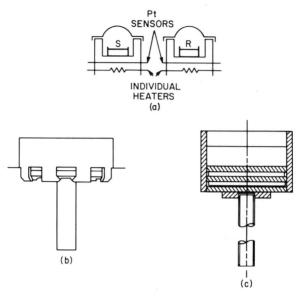

Fig. 26 Sample containers used in Perkin-Elmer DSC instruments. (a) DSC sample holder (schematic). (b) DSC-1B holder construction. (c) DSC-2 holder.

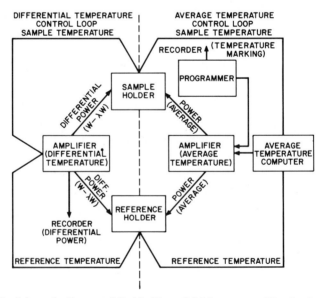

Fig. 27 Schematic diagram of Perkin-Elmer DSC instrument (Wendlandt, 1974b).

ratus, unlike DTA, maintains a sample temperature equal to a reference substance (or furnace block) by supplying heat to the sample or reference material. The amount of heat required to maintain these isothermal conditions is then recorded as a function of time (or temperature). In addition to recording the enthalpy curve, if the sample evolved a volatile material during the heating process, the gas evolved by the sample can be measured by passing the exit gas stream through an appropriate detector.

The instrument contains two *control loops,* one for the average-temperature control and the other for the differential-temperature control. In the former, a programmer provides an electrical output signal that is proportional to the desired temperature of the sample and reference holders. The programmer signal that reaches the average-temperature amplifier is compared with signals received from platinum resistance thermometers permanently embedded in the sample and reference holders and the necessary power supplied equally to both heaters.

In the differential temperature loop, signals representing the sample and reference temperatures, as measured by the platinum resistance thermometers, are fed to the differential-temperature amplifier via a comparator circuit, which determines whether the reference or the sample temperature is greater. The differential-temperature-amplifier output then supplies power to the reference or sample heater as necessary to correct any temperature difference between them. A signal proportional to this differential power is also transmitted to the pen of the recorder, giving a curve of differential power versus time or temperature. The area under a peak is then directly proportional to the thermal energy absorbed or liberated in the transition.

Besides measuring enthalpy the DSC unit is useful to measure specific heat, particularly with the software package available from the manufacturers. Accuracy of at least 1% can be achieved by comparison with known materials (Mraw and O'Rourke, 1981). At lower temperatures He is a suitable ambient; however, for higher temperature work Ar is preferable because of its lower thermal conductivity (Gallagher *et al.,* 1979).

Various sample holders have been described for the Perkin-Elmer DSC instrument. A sealed metal cell with a removable screw-on cap has been described by Freeberg and Alleman (1966). Metals used were brass, stainless steel, and aluminum. Wendlandt (1970b) described a capillary-tube sample holder that used 2-mm-diameter capillary tubes. The tubes were contained in aluminum holders that were set in the sample and reference cells of the calorimeter. Sample holders for measuring the vapor pressure of a liquid (Farritor and Tao, 1970) as well as for heats of mixing (Mita *et al.,* 1971) have been described. Enclosure of the sample-holder chamber in a vacuum chamber has been described by Mori *et al.* (1972).

Flynn (1974a,b) reviewed the underlying theory of DSC and discussed some of the uses. The use of DSC to measure heat capacity has been recently described by Wiedemeir *et al.* (1981) along with a digital data-acquisition system to facilitate the determination.

VI. Commercial DTA and DSC Instruments

A. DuPont DSC and DTA Cells

1. DuPont DSC Cell Model 910

The DuPont cell, which is of the heat flow type, has been previously described earlier, and is a modular cell that must be used with the DuPont cell base, which in turn is connected to one of the DuPont thermal analyzers. The Model 990 Thermal Analyzer contains the programmer and recorder. The programmer can control the temperature in various DTA and DSC cells over the range $-190-1600°C$, while the recorder is a one or two pen, $X-Y$ or $X-Y^1Y^2$ type.

The Model 1090 control console offers much greater flexibility in furnace programming and particularly in data processing and presentation. Data are collected at maximum sensitivity and stored on a floppy disk, and can be monitored simultaneously on an $X-Y$ recorder. Alternatively, data from a previous run may be manipulated and plotted simultaneously with data collection of the run in progress. Recorder scales can be designated or automatically scaled based on the data. For example, the temperature axis can be an expanded 4.6°C segment or the full range. Multiple plots of expanded segments or thermal analytical data from different devices can be placed on the same graph span. The data can be permanently stored on the floppy disk or transferred to permanent files on other devices.

A continuously expanding library of software programs is provided for the 1090 on dual-sided floppy disks. Examples of the existing DSC software are transition temperatures (onset and inflection), heats of transition, percent crystallinity, degree of cure, dynamic purity (Van't Hoff), partial areas, kinetics, generic equation calculations, and automatic calorimetric calibration.

In addition, the 1090 is equipped with an RS-232-C Serial communications interface for two-way interconnection with an external terminal computer system. This permits either data transmission from the 1090 for subsequent processing using self-generated programs or external control of the 1090 via the remote computer.

The multiple microprocessor architecture of the 1090 permits the user to perform all four basic tasks simultaneously:

(1) setup of experimental conditions;
(2) real-time data storage from an experiment;
(3) data playback on the *XY* printer plotter;
(4) data analysis of a previously recorded experiment.

The DSC cell, which is illustrated in Fig. 24, can be used up to 600°C in oxidizing and 725°C in inert or reducing atmospheres. Samples are placed in aluminum, gold, copper, platinum, or graphite sample containers of 0.05 ml capacity. The calorimetric sensitivity is given at 3.7 μcal sec^{-1} in.$^{-1}$ (6 μW cm^{-1}) and the precision is said to be $\pm 1\%$, based on metal fusion standards. Chiu and Fair (1979) describe a simple modification to the instrument to allow the determination of thermal conductivity with a precision of $\pm 3\%$. Slight modifications to the cell also permit simultaneous EGA for catalytic studies (Gallagher *et al.,* 1976).

Also available is a pressure DSC cell that incorporates the DSC cell described here in a gas-tight enclosure. The cell may be used over the preceding temperature range at pressures from 0.01 torr to 70 atm. This high-pressure system has been described by Levy *et al.* (1970) and found useful for studies of hydrogenation catalysts (Kosak, 1976).

2. DuPont DTA Cells

Two DTA cells cover the temperature range from about 60 to 1600°C. Each cell plugs into the DuPont cell base with data readout and temperature programming and control from the DuPont Model 990, 1090, or R90 Thermal Analyzer. The two furnaces were described earlier (Section V.C) and the sample holder consists of aluminum, platinum, or alumina cups resting directly on the thermocouple junctions.

B. PERKIN-ELMER DSC AND DTA

1. Perkin-Elmer DSC-2 Calorimeter

The Perkin-Elmer DSC-2 differential scanning calorimeter has been discussed in Section V.F and in Figs. 26 and 27. Samples are placed in low thermal mass aluminum or platinum containers of 0.05 ml capacity, which may be hermetically sealed to withstand an internal pressure of 3 atm. Temperature detection is with platinum resistance thermometers giving a temperature accuracy of $\pm 1.0°C$ and a precision of $\pm 0.1°C$. Differential power ranges are from 0.1 to 20.0 mcal sec^{-1} full scale in eight ranges, with from ± 0.002 to ± 0.004 mcal sec^{-1} noise. The atmosphere in the furnace compartment may consist of nitrogen, air, oxygen, or argon, static or dynamic, to 200 ml min^{-1} at pressures from 380 torr to 3 atm. The temperature range of the instrument is -175 to 725°C. Helium is required for low-temperature operation. An optional dry box is available that allows loading the sample in a moisture-free environment in order to prevent condensation during subambient operation. It is also valuable for handling moisture sensitivity materials or studying moisture-related process (Bair and Johnson, 1977). An optional improved cooling system is also available.

The Perkin-Elmer TADS-101 thermal analysis data station is designed to couple with the DSC-2c. The experimental parameters (e.g., heating rate and temperature limits) can be entered via the keyboard or recalled from memory. A CRT unit displays the thermogram during the course of the experiment. Expanded plotting, comparison with previously stored data, or data processing can be achieved using stored or operator-generated programs. Automatic scaling is available and the entire screen along with the relevant experimental parameters can be reproduced by the plotter/printer or stored on floppy disks. Various complete programs are supplied with the TADS-101 but the instrument is fully programmable using operator-oriented BASIC language. This enables users to modify the manufacturers programs or readily devise their own.

The manufacturer's readily available software determines such things as peak search, peak separation, peak areas and temperatures, derivative, glass transition temperature, polymer blend analysis, percent crystallinity, percent solids, and purity analysis. Perkin-Elmer provides a Laboratory Information Management System (LIMS), which is a minicomputer at yet a higher level. This LIMS is designed to control a variety of data stations, such as the TADS.

2. Perkin-Elmer DTA-1700

The new Perkin-Elmer high-temperature DTA (or heat flow DSC) instrument operates over the temperature range from 20 to 1500°C extendable to 1700°C with the appropriate options. Sensitivity is 0.5°C or 5 mcal sec^{-1} full scale and the System-4 microprocessor makes corrections to linearize both the T and ΔT scale and facilitates integration for energy measurement. There is a continuous temperature display with automatic temperature calibration. Multistep programming is possible with heating and cooling rates from 0.5 to 100°C min^{-1}. Forced-air cooling enables rapid turnaround times. Vacuum or purge gases can be used and there are gas switching capabilities. The sample containers of alumina or platinum come in two sizes, 60 and 100 mm^3.

C. METTLER DSC AND DTA SYSTEMS

1. Mettler TA 2000 DSC System

The Mettler TA 2000 DSC (heat flow) system, which previously was usable over the temperature range from -20 to 500°C, has been extended to -170 to 1200°C by use of the optional accessories. Differential temperature is detected by a unique five-thermojunction thermopile, vapor-deposited on a ceramic disk. Temperature accuracy is ± 0.5°C from -100 to 550°C and ± 1.0°C from -170 to -100°C, both at a precision of ± 0.1°C. Calorimetric sensitivity is given as ~ 60 μV mcal^{-1} sec^{-1} at a precision and accuracy of ± 0.5 and $\pm 2\%$, respectively. Heating and cooling rates are adjustable from 0.1 to 29.9°C min^{-1} in steps of 0.1°C min^{-1}. Samples are placed in aluminum or Nimonic 80A crucibles each with a capacity of 40 μl and 0.5 ml, respectively. The latter sample crucibles coupled with gold seals have a rated bursting strength of 100 atm.

The TZ 2000 and TA 2000Z are two computer interface systems designed for use with the TA 2000 DSC. The current software provides calibration, peak integration, several methods of purity analysis, specific heat calculations, and applied kinetic analysis.

2. Mettler 2000C Thermoanalyzer (Simultaneous TG-DSC)

The general description of the instrument and its TG capabilities were discussed earlier (see Section IV.C.4). The temperature range is from room temperature to 1200°C standard or to 1600°C with optional equipment. Heating rates are in 0.1°C steps from 0 to 29.9°C min^{-1}. Temperature accuracy is ± 2.0°C with a precision of ± 0.2°C. Calorimetric sensitivity is about 2.5 μV mW^{-1} with an accuracy of $\pm 5\%$ and a precision of $\pm 2\%$. Vacuum or flowing gas atmosphere can be employed. The computer equipment described for the TG is useful for the DSC (heat flow) as well, including the Model 2000D.

3. Mettler DSC 20 and DSC 30

These modules are part of the latest Mettler thermal analysis system, the TA3000. The DSC 20 covers the temperature range from -20 to 600°C. Controlled cooling is facilitated by a fan. The DSC-30 uses liquid-nitrogen cooling and covers the range from -170 to 600°C. Both units are controlled by the TC10 TA processor. Heating rates from 0 to 100°C min^{-1} are used and cooling rates up to 20°C min^{-1} (to 0°C) and 10°C min^{-1} (below 0°C) are possible with the appropriate module. The microprocessor allows for a variety of temperature programs and data presentations. Software is available for integration, purity analysis, oxidative stability, degree of crystallinity, glass transition temperature, reaction kinetics, heat capacity, liquid fraction, and enthalpy–temperature function.

D. STANTON-REDCROFT DTA SYSTEMS

The five complete DTA systems available from Stanton-Redcroft are as follows: DTA 671 B, -150–500°C; DTA 672, 25–500°C, DTA 673, 25–

1000°C; DTA 674, 25–1500°C; and DTA 675, 25–1650°C. Each system consists of four modules: (a) DTA cell, (b) temperature programmer, (c) dc microvolt amplifier, and (d) potentiometric recorder. Various sample container configurations are available, depending on the maximum temperature range desired. The temperature programmer has rates from 1 to 20°C min^{-1} in eight steps and can be used in a programmed cool or isothermal mode. Besides these specific DTA systems there is also the simultaneous TG-DTA instrument described earlier (see Section IV.N). This instrument has a capability of 25 to either 1000° or 1500°C depending on the choice of furnace. DTA sensitivity is 0.005 to 0.25°C mm^{-1}. The sample holder–furnace arrangement is shown in Fig. 21.

E. NETZSCH DSC and DTA SYSTEMS

There are nine DTA systems available from this manufacturer plus those for simultaneous TG–DTA previously described in Section IV.D.1. Temperature ranges covered by these models are from − 180 to 1600°C and include many different types of furnace assemblies, sample containers, and accessory equipment. A high-pressure DTA cell is also available for use up to 500 atm.

The general-purpose DTA 404 S incorporates a six-channel multipoint recorder and can be used from − 160 to 1600°C, depending on the choice of furnace and sample container selected. Ten different heating rates can be selected, from 0.1 to 100°C min^{-1}, as well as four different modes (heating, cooling, isothermal, and cycle). Optional accessories include two EGA systems, a vacuum system, and others.

The Model DSC M444 (heat flow) has a temperature range from − 180 to 500°C. Aluminum crucibles are used with a maximum capacity of 400 mm^3. An electronic microheater is used for calibration. The sensitivity is 20 μV mW^{-1}. Heating rates vary from 0 to 19.9°C min^{-1} in steps of 0.1°C. A variety of programs are offered and there is a digital temperature display. Linearity in the temperature control system is ± 0.1% and the reproducibility is ± 0.2°C.

F. SETARAM DTA AND DSC SYSTEMS

There are numerous DTA–DSC systems currently available from SETARAM, not including the simultaneous or combined TG–DTA systems (see Section IV.I). Sensitivity of the dual systems is 50 μcal. The DTA systems include the following: (a) DTA D80, 25–1000°C, sensitivity 150 μcal; (b) DTA Model 2000k, 25–1750°C, sensitivity 70 μcal; (c) Model

1500k, 25–1250°C, sensitivity 50 μcal; (d) Micro DTA M-5, $-170-$ 1500°C, sensitivity 10 μcal; (e) DSC, $-123-827°$C, sensitivity 15 μW, and (f) DSC 101, -123 to 527°C, sensitivity 15 μW. The SETARAM DSC uses a horizontally mounted programmable reference block contained in an enclosure whose exterior is at ambient temperature. Two refractory tubes transverse the middle of the block, the midsection of each containing the sample holder and detector. This midsection is covered by a heat flow calorimetric detector formed from thermocouples thermally connected to the reference block, the same principle as used in the Tian-Calvert calorimeter. Useful capacity of the tubes is 380 mm^3 each, with a maximum sample size of 7–10 mm. Furnace programming rates are from 30°C min^{-1} to a minimum of 1°C hr^{-1}. Normal calorimetric detection sensitivity is 15 μW isothermal or less than 30 μW in the scanning mode (0.05 μW mm^{-3} isothermal or less than 1 μW mm^{-3} scanning).

G. Harrop DTA System

The Harrop High Temperature DTA system consists of a Model TA-700 control console and a DT module. The control console includes the temperature programmer and controller and an $X-Y$ recorder. The DT module includes the furnace, furnace stand, sample container, and thermocouple assembly. Furnaces are available for use to maximum temperatures of 1200 and 1600°C, at furnace programmer rates of 0.1–40°C min^{-1}. Furnace atmosphere control permits the use of dynamic or static noncorrosive gases at pressures from 1 × 10^{-3} torr to 3 atm. The maximum sensitivity is 5 μV in.$^{-1}$.

The Harrop Low Temperature DTA module (Model DT-707) covers the range from -150 to 700°C. The TA-700 control console is also used with this module giving the same sensitivity and programming characteristics as the high-temperature unit. A low-cost series 600 DTA unit is also available, which covers a temperature range of $-100-1000°$C.

H. TECO Thermit 10-C DTA System

The TECO Thermit 10-C is a low-cost DTA system consisting of a DTA module and furnace, a control console, and a 10-in. strip-chart recorder. Two furnaces are available, covering the range from 25 to 500°C or 25 to 1000°C. Furnace programming rates are from 1 to 20°C min^{-1} in seven steps with two modes of control. Various sample containers are available so that sample sizes from a few micrograms to 250 mg may be studied.

I. Rigauk Thermoflex DTA and DSC System

Several DTA and DSC systems are available from this manufacturer, including models employing simultaneous TG–DTA and TG–DSC capability. The desktop DTA system can be used to cover the temperature range from $-185°$ to 1500°C, depending on the furnace selected. The desktop low-temperature DSC system only covers the range from -175 to 100°C.

There are three models of the desktop DTA series: a low-temperature model (-185 to 100°C); a macro-DTA system; and a micro-DTA system. Each of the macro- and micro-DTA systems can be equipped with standard (25–1000°C) or high-temperature (25–1500°C) furnaces. A six-step heating rate programmer covering the range from 0.625 to 100°C min^{-1} is employed in all the systems.

J. Columbia Scientific Industries DTA and DSC Systems

The Columbia Scientific Industries DTA–DSC (heat flow) systems, which are based on the original design of Stone, consist of a recorder-controller console with choice of strip-chart or $X–Y$ recorder, a furnace platform, and a furnace and sample container module. Three furnaces are available, covering the temperature range from 25 to 1600°C. There are six different sample containers, several of which have been illustrated in Fig. 23. The furnace programmer covers the range of rates from 0.5 to 50°C min^{-1}, infinitely adjustable plus 10 switch selected. Optional accessories include an EGA, a high-vacuum DTA, a high-pressure (to 3000 psig), and a subambient sample container module ($-160–300°C$, 10^{-2} torr to 18 atm.)

K. Theta DTA Systems

Most of the Theta DTA instruments share a common amplifier and programmer so that the general specifications include ΔT amplification in steps from 1 to 100 with offset and heating rates from 0.5 to 60°C min^{-1} in 12 steps. The combined TG–DTA Model Gravitronic (25–1600°C) has been described earlier (Section IV.F, Fig. 19). The Labtronic II is basically the DTA portion without the TG capability. Besides a 25–1600°C furnace, there is also a 25–1400°C option.

The Labtronic IIIR is a combined DTA–thermodilatometer (25–1600°C). The Labtronic IIIE is only the DTA portion of the equipment. Similarly, the Labtronic IVS is a combined DTA–thermodilatometer (25–1000°) and the IVE is the DTA portion alone. The Labtronic 2000 system contains a wide range of modules, one of which is a DTA. Three temperature ranges are available from ambient to 600, 1000, or 1600°C. Atmospheres can range down to 10^{-6} torr. A multiple sample head available as an option will handle six measurements simultaneously subject to the same temperature and atmosphere program. The furnace programmer for this model has 10 steps from 0.1 to 10°C min^{-1}.

L. LINSEIS CO. DTA AND DSC SYSTEMS

The simultaneous TG–DTA Model L81 has three temperature ranges: −170–450, 20–1250, and 20–1550°C. The ΔT signal can be amplified by factors of 100 or 200. There are also several DTA units that cover temperature ranges of −150–400, 20–1000, 20–1550°C, and a high-temperature unit from 20–2300°C. Two DSC units (heat flow) are available (−50–400°C) with differing sensitivities.

M. HUNGARIAN OPTICAL WORKS (MOM, BUDAPEST)

The Derivatograph simultaneous TG–DTA has been described earlier. Amplification of the ΔT signal ranges in steps from 1 to 200. The furnace covers from ambient to 1100° at six rates from 0.5 to 20°C min^{-1}.

N. TETRAHEDRON ASSOCIATES, INC.

The Unirelax system is a modular one designed to measure a wide range of properties and processes associated with polymers and plastics. The DTA module is also used for dielectric testing so that combined DTA–EC can be accomplished (see Section VII.D). The sensitivity of the DTA is 10 mV °C^{-1} over the temperature range −200–500°C.

The DSC module (heat flow) utilizes copper–constantan thermopiles. Aluminum or glass holders are used. The instrument operates within the temperature range −200–500°C in a variety of atmospheres including vacuum. The sensitivity is 0.1 mcal over the operating range of temperature. Thermal conductivity measurements can also be made using this module.

O. ULVAC (SINKU-RIKO) DTA and DSC SYSTEMS

All the ULVAC DTA systems are used in conjunction with other thermal analysis techniques such as the Model HPTGD-3000-M high-pressure TG–DTA system and the Model TDG-3000/MSQ-500 combined TG–DTA–MS system. There is also available a SH-3000 series adiabatic scanning calorimeter, which is based on the continuous measurement of changes in the specific heat of the sample. The two models of the instrument cover the temperature ranges 25–800°C and −175–227°C.

VII. Thermomechanical Analysis (TMA)

A. THERMOMECHANICAL ANALYZERS

Thermomechanical analysis (TMA) is a technique in which the deformation of a substance is measured under nonoscillatory load as a function of temperature as the substance is subjected to a controlled temperature program. Thermodilatometry, on the other hand, is a technique in which the dimensions of a substance are measured as a function of temperature as the substance is subjected to a controlled temperature program (Lombardi, 1980). Both of these techniques may be obtained on the same apparatus; only the sample probes and/or probe loadings are different.

A typical thermomechanical analyzer, such as is incorporated in the Perkin-Elmer TMS-2, is shown in Fig. 28. In the penetration and expansion modes, the sample is placed on the platform of a quartz sample tube. The appropriate quartz probe is connected to the armature of a linear variable differential transformer (LVDT) and any change in the position of the armature results in an output voltage from the transformer which is then recorded. The probe assembly includes a weight tray, which permits a choice of loadings on the sample surface. All components on the assembly are supported by a plastic float rigidly fixed to the shaft and totally immersed in a high-density fluid. This method of support has the advantage that true loading on the sample is essentially independent of the probe position over the range where the float remains totally immersed. This is particularly advantageous for work on fibers. The sensitivity of the apparatus provides an amplification in displacement of 4×10^{-5} on a 10-mV recorder. Two furnaces are employed to cover the ranges $-150–325°C$ and $25–725°C$.

A first-derivative computer accessory can be added to the Perkin-Elmer TMS-2 for the simultaneous recording of TMA and DTMA curves. The system will measure expansion and extension as well as penetration and compression modes. The system temperature range is $-170–325°C$ (high-temperature furnace to 725°C). The System 4 programmer and data presentation (see Section IV.B) are also useful for TMA work.

Sample probes that are available for use in TMA are illustrated in Fig. 29. Probe (a) is used for measurements of thermal expansion, whereas (b) is used to determine sample penetration, softening points, etc. For viscous softening or heat distortion under load, the probe in (c) is used. Special-purpose probes for textile fibers and films are shown in (d) and (e). In (e), the sample is placed between slots in the clips, which are then crimped closed. Very thin samples can be conveniently handled and studied under uniform tension. Differential measurements (i.e., change relative to a known standard) is also possible on some dilatometers.

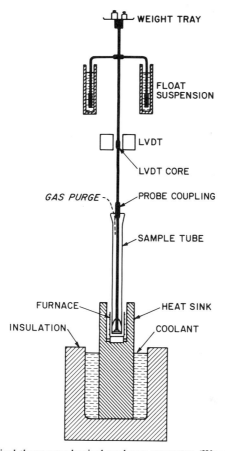

WEIGHT TRAY

FLOAT
SUSPENSION

LVDT

LVDT CORE

GAS PURGE -

PROBE COUPLING

SAMPLE TUBE

FURNACE

HEAT SINK

INSULATION

COOLANT

Fig. 28 Typical thermomechanical analyzer apparatus (Wendlandt, 1974b).

The DuPont Model 943 Thermomechanical Analyzer module is capable of handling samples in the form of plugs, films, pelletized powders, or fibers in the temperature range −180–800°C (an optional accessory permits measurements from 25° to 1200°C). The sensitivity is 50 μin. in^{-1} of recorder chart and this model also has a calibrated simultaneous derivative readout. Interchangeable sample probes permit the determinations of penetration, expansion, tension, and dilatometry of the samples. A spring is used for support in place of the float arrangement shown in Fig. 28.

The 943 TMA has available a wide selection of optimal accessories, such as:

(1) *Parallel Plate Rheometry*—for low shear viscosity analysis of viscous materials (10–10 Pa s). This is particularly useful in following

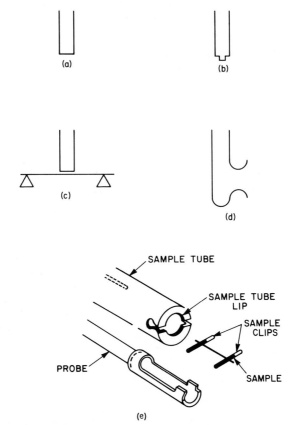

Fig. 29 TMA sample probes: (a) expansion mode; (b) compression mode; (c) flexure mode; (d) extension mode; (e) extension mode (Wendlandt, 1974b).

cure advancement of thermosets to optimize processing times and temperatures.

(2) *Fiber Tension Spectrometry*—which is used to measure shrinkage tension of fibers as a function of temperature held under constant elongation. In this case, the primary transducer is a load cell of low spring constant in series with the LVDT sensor of the TMA. A very useful accessory for investigating thermal history of fibers (heat set, texturing, and draw ratio effects).

(3) *Stress Relaxation Spectrometry*—used for measurement of relaxation modulus versus time on visoelastic materials. A similar transducer to the fiber tension spectrometer is used, but with a high force load cell and using invar steel probes for up to 1-kg load.

The flexibility in data presentation, furnace programming, available software, etc., achieved through the Model 1090 (see Section IV.A) is also available in TMA and DMA as well.

Thermomechanical analysis or thermodilatometric analysis systems are available from many other manufacturers. A partial list of the instruments available is given in Table X.

TABLE X

TMA or TDA Systems from Other Manufacturers

Manufacturer	Model number	Temperature range (°C)	Mode[a]
Stanton-Redcroft	TMA 691	−180–500	TDA, TMA
Rigaku	TMA	−60–1500	TMA
Netzsch	402	−160–1700	TDA
ULVAC (Sinku-Riko)	DL-1500	−100–1500	TDA
Harrop	700	20–1600	TDA
Theta	Numerous	−150–3000	TDA, simultaneous TDA–DTA
Orton	Numerous	20–1500	TDA
Linseis	Numerous	−270–2200	TDA, TMA
Hungarian Optical Co.		20–1100	TDA, simultaneous TDA–TG

[a] TMA = thermomechanical analysis; TDA = thermodilametric analysis.

B. DuPont Model 981 Dynamic Mechanical Analysis System

Closely related to thermomechanical analysis is the technique of dynamic mechanical analysis (DMA). This technique, which is called dynamic thermomechanometry by Lombardi and the ICTA (1980), is a technique in which the dynamic modulus and/or damping of a substance is measured under oscillatory load as a function of temperature as the substance is subjected to a controlled temperature program.

In the DuPont Model 981 DMA module, illustrated schematically in Fig. 30, the sample is oscillated at its resonant frequency and an amount of energy, equal to that lost by the sample, is added on each cycle to keep the sample in oscillation at constant amplitude. The logarithm of this extra energy (called damping) and the frequency are displayed digitally as well as recorded as a function of sample temperature using the Model 990 console. With the Model 1090 the energy display is either linear or logarithmic.

The sample is clamped between two arms, one of which is a passive support, while the other is driven by the electromechanical transducer.

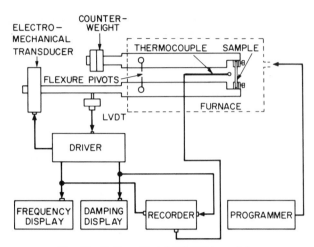

Fig. 30   DuPont Model 981 DMA module.

Rotation of the driven arm a few tenths of a millimeter puts the sample in flexure stress so that when the displacing force is released, the deflection energy stored in the sample causes it to go into resonant oscillation. The frequency and amplitude of this oscillation are detected by a linear voltage differential transformer (LVDT) positioned at the opposite end of the active arm. This LVDT signal is fed to the driver circuit, which then feeds back enough energy to the electromechanical transducer to keep the sample in oscillation at constant amplitude. The temperature range is $-150-500°C$ (with the optional cooling accessory) with temperature programming rates depending on the choice of console. Software is available, directly on the Model 1090 or through a Tektronix Model 31 programmable calculator for the Model 990, to convert the dampening value to the mechanical loss, tan δ. The instrument is useful for a wide range of materials (modulus range from 100 GPa to 10 MPa).

C. POLYMER LABORATORIES, LTD., DYNAMIC MECHANICAL THERMAL ANALYZER

A sinusoidal current drives an electromechanical vibrator and compares the amplitude and phase of the resulting sample motion with that of the input. Various clamping arrangements are available to accommodate different samples. The frequency range is selected by push button from eight values in the range from 0.033 to 90 Hz. Strains can be set to a maximum level of 0.25 mm displacement. The log modulus and tan δ are displayed versus temperature on an Y_1-Y_2-X recorder. An IEEE 488 computer interface is optional. The controller programmer uses a platinum resistance temperature sensor and operates over the range -150–$300°C$. Linear programming rates (heat/cool/cycle/isothermal) vary from 0.1 to $20°C$ min^{-1}.

D. Tetrahedron Universal Relaxation Spectrometer

The Tetrahedron Universal Relaxation Spectrometer is an extremely versatile system that can determine the dynamic mechanical and thermomechanical properties of a polymer as well as its dynamic dielectric, dynamic depolarization, viscometric properties (Yalof and Brisbin, 1973; Yalof and Hedvig, 1976). The cited references should be consulted for a description of the apparatus.

E. Torsional Braid Analysis

The technique of torsional braid analysis (TBA) was introduced by Lewis and Gillham (1962) and Gillham (1966) for the investigation of the mechanical properties of polymeric substances. It permits thermomechanical "fingerprints" of polymer transitions in the temperature range -190 to $500°C$ in controlled atmospheres. The sample is prepared by impregnating a glass braid or thread substrate with a solution of the material to be tested, followed by evaporation of the solvent. During the heating of the sample impregnated braid, it is subjected to free torsional oscillations. From these oscillations the relative rigidity parameter I/P^{-2}, where P is the period of oscillation, is used as a measure of the shear modulus. The mechanical damping index n^{-1} is used as a measure of the logarithmic decrement, where n is the number of oscillations between two arbitrary but fixed boundary conditions in a series of waves. Changes in the relative rigidity and damping and damping index are interpreted as far as possible in terms of changes in the polymer. Major and secondary transitions, such as melting or glass transitions, are readily revealed, as are the effects of many other chemical and degradative reactions. The technique has been the subject of a review by Gillham (1974).

The apparatus used in TBA determines the frequency (less than 1 Hz) and decay of a freely oscillating pendulum, which provides information on the modulus and mechanical damping of the polymer under examination (Gillham, 1979). An electrical analog of the decaying pendulum oscillation is obtained by attenuating light with a circular transmission disk, which features a linear relationship between light transmission and displacement angle. Manzione and Stillwagon (1980) have developed an automated thermoset characterization laboratory controlled interactively through a

HP9845 minicomputer. Among the numerous devices controlled in this manner is a TBA apparatus.

Potential temperature standards for use in TBA have been proposed by Takahashi and Ozawa (1975).

VIII. Evolved Gas Detection and Evolved Gas Analysis

Evolved gas detection (EGD) is a technique in which the evolution of gas from a substance is detected as a function of temperature as the substance is subjected to a controlled temperature program. The technique of evolved gas analysis (EGA) is similar except that the nature and/or amount of volatile products are determined (Lombardi, 1980). The instrumentation for EGD and EGA has been discussed in detail by Lodding (1967), Wendlandt (1974a,b), Wendlandt and Collins (1978), and Langer (1980). In many cases, EGD–EGA is determined simultaneously with other thermal analysis techniques such as DTA and TG (Wendlandt, 1974a,b). It is a simple matter to add a thermal conductivity detector to a DTA or TG apparatus and hence determine the EGD curve of the sample. Such as apparatus is shown in Fig. 31, where Emmerich and Bayreuther

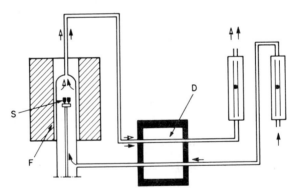

Fig. 31 Typical DTA-EGD system (Netzsch).

(1975) connected a thermal conductivity detector to a Netzsch DTA system. The evolved gases from the sample (*S*) are swept from the furnace (*F*) by means of a carrier gas and detected by the thermal conductivity detector (*D*). A flow meter permits precise adjustment of the carrier gas flow rate. The carrier gas is either inert or active, in reference to its reactivity with the sample (e.g., hydrogen with metal oxides). The output voltage of the detector is proportional to the concentration of the reaction products (evolved gases or carrier gas absorbed).

A gas regulating system and thermal conductivity detector bridge (EGA 403/1) is available from Netzsch. Columbia Scientific Industries also has available a simultaneous DTA–EGD sample holder and accessory equipment.

Langer (1980) has discussed in considerable detail the various types and applications of EGA, involving gas chromatography or mass spectroscopy. There are pyrolysis probes available for coupling with gas chromatography to provide the temperature capabilities. An example of these would be the Pyroprobe Models 100 or 120 manufactured by Chemical Data Systems, Inc.; the main element is platinum (ribbon or coil) which serves as sample holder, temperature sensor and heater simultaneously. The platinum is one leg of a Wheatstone bridge circuit and the balancing leg is the temperature control point. The rapid response allows for heating rates of 0.1 to 20°C sec^{-1} up to 1200°C maximum. Without the linear control in operation the full temperature can be achieved in various intervals of 20 msec to 20 sec, leading to very fast, albeit nonlinear heating rates.

The use of mass spectrometers for EGA has been particularly rewarding because of their high sensitivity, which gives them an ability to directly identify the vapor species. This approach is also useful for kinetic studies that involve the evolution of gaseous products (Price *et al.*, 1980). There are numerous examples of quadrupole mass spectrometers having been coupled with conventional TG or DTA apparatus. They are useful only for the detection of the permanent gases or highly volatile species, however, because the path between the furnace and the mass spectrometer allows for condensation. If the apparatus is designed solely for EGA it is relatively simple to place the furnace in the vacuum system directly under the ionizer of the mass spectrometer (Gallagher, 1978). In this manner gaseous species that would normally condense prior to detection can be measured; for example, Fig. 32 shows the evolution of arsenic arising from the reaction between GaAs and its gold electrical contacts (Kinsbron *et al.*, 1979). The Netzsch simultaneous TG–DTA–EGA apparatus provides a line-of-sight path so that readily condensable species can be measured in this simultaneous fashion. Figure 33 shows the differentially pumped double orifice interface between the sample and the mass spectrometer. The TG–DTA experiment can be performed at atmospheric pressure in a flowing gas while the mass spectrometer is operating in a vacuum of less than 10^{-4} torr. An intermediate chamber is operated around 1 torr.

A time-of-flight mass spectrometer has been used by Friedman (1970) to obtain information on the ablation of polymers. This method is capable of much faster response. The most elegant application of this type is described by Lum (1977). A schematic of the apparatus is shown in Fig. 34.

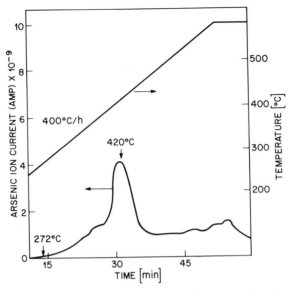

Fig. 32 The mass spectrometer output for a repeated scan through a short-range around As. As starts to appear at ~250°C and reaches a maximum rate of evolution at 450°C. Uncoated GaAlAs starts to decompose at ~260°C. 1000 Å Au on GaAlAs. Contact area 0.144 cm² (Kinsbron *et al.*, 1979).

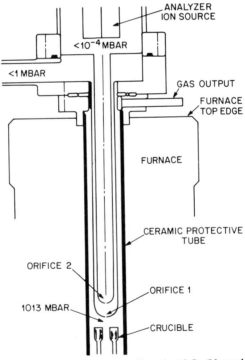

Fig. 33 Gas inlet system with two orifices in Al₂O₃ (Netzsch).

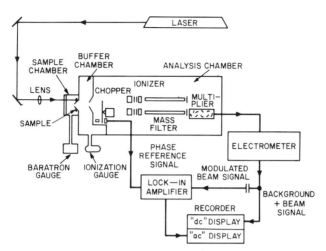

Fig. 34 Schematic of laser-probe–molecular-beam apparatus (Lum, 1977).

Laser heating is used in conjunction with differential pumping to create a molecular beam of the decomposition products. The resulting beam is chopped to allow for complete elimination of the background signal and also to permit a phase analysis of the modulated beam. Coupling this instrument with modern data acquisition and processing equipment provides three-dimensional plots that describe the course of the decomposition or other process in great detail. An example is given in Fig. 35 showing the temperature profiles for a number of volatile products from the thermal decomposition of a plastic encapsulent used for semiconductor devices (Lum and Feinstein, 1980).

Water is a particularly interesting decomposition product, and a number of specific techniques have been developed to determine the moisture content of materials. The DuPont Model 902 Moisture Evolution Analyzer utilizes an electrochemical cell to measure coulometrically the amount of water absorbed from the gas stream after it passes over the heated sample. A sample (5–6 cm³ maximum) is heated in a stream of dry gas, usually nitrogen. Two temperature regimes are possible, ambient to 300 or 1000°C. The instrument has high sensitivity, 0.1 μg of H_2O. A meter constantly reads the integrated total moisture content as the temperature is programmed, and there are outlets on the rear of the instrument that will give the unintegrated value so that rate plots can be obtained.

There is a wide variety of moisture sensors that could be conveniently coupled with other thermal analysis instruments. An example is the sen-

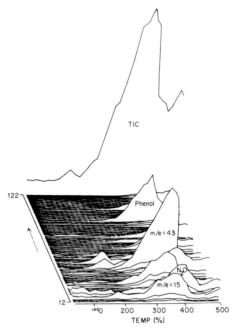

Fig. 35   Isometric plot of mass spectral ion profiles from novolac epoxy (Lum and Feinstein, 1980).

sor line made by Panametrics, Inc. The sensor is a thin aluminum–alumina–aluminum sandwich whose electrical impedance depends on the moisture adsorbed in the pores of the alumina. The response is rapid (67% change in 7 sec) and covers the range from 1 ppb to 0.2% of water (dew–frost point of −110 to 20°C). Gallagher and Gyorgy (1980) have shown that the kinetic parameters evaluated based on such an analysis can give values in excellent agreement with such data based on weight loss experiments. Figure 36 shows a comparison of the integrated water loss with the TG curve for $Ba(OH)_2 \cdot H_2O$. Such devices are useful not only for the cases where water is a directly evolved species, e.g., for a hydrate, hydroxide, or adsorbed moisture, but also for those instances where it is the product of a heterogeneous reaction such as the hydrogen reduction of an oxide.

In EGA there are precautions and caveats that must be considered. Most of these center on what happens between the evolution and the detection steps. Condensation, time lag, reaction between gases or with the walls and solid products, sensor selectivity, differences in diffusion rates, etc., are possible problems.

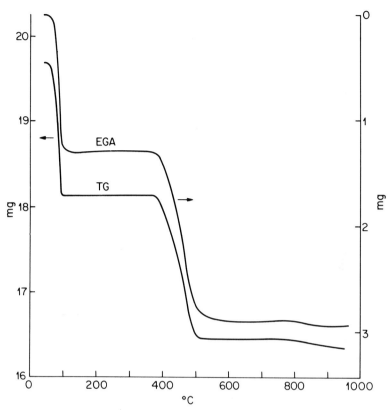

Fig. 36 Comparison of TG and EGA curves for $Ba(OH)_2 \cdot H_2O$ heated at 1°C min^{-1} in N_2 (Gallagher and Gyorgy, 1980).

IX. Miscellaneous Thermal Analysis Techniques

A large number of other thermal analytical techniques, most of them summarized in Table I, are used to investigate a wide variety of chemical and technological problems. Unfortunately, space limitations prevent their instrumentation from being discussed in detail here. Many of these techniques are discussed in detail elsewhere (Wendlandt, 1974a,b).

A. THERMOSONIMETRY

Thermosonimetry measures the noise signals produced by the displacement of a component in the crystal structure. Such movements may be caused by mechanical fracture, eruption of inclusions, phase transformations, plastic deformations, etc. (Lonvik, 1975; Clark, 1978; Clark and Garlick, 1979; Betteridge *et al.*, 1981).

The apparatus used in thermosonimetry is shown schematically in Fig. 37. A quartz or other piezoelectric transducer is used to convert vibra-

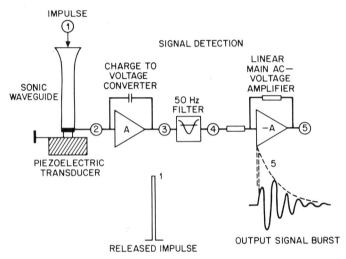

Fig. 37   Thermosonimetry apparatus.

tions in the sample to the audio range. These signals are then converted to a voltage, filtered by a 50 Hz filter, and then passed through an ac amplifier before display on a recorder. The amplitude of these signals is then plotted as a function of temperature to give the thermosonimetric curve. The technique has been combined with DTA by Clark and Garlick (1979).

B. THERMOACOUSTIMETRY

This technique is based on the change in velocity and amplitude of a constant frequency, sonic pulse through a medium as a function of temperature. From such measurements the viscoelastic properties and orientation of the polymer medium can be deduced. Figure 38 illustrates the apparatus used by Chatterjee (1974) to perform combined thermoacoustic and DTA experiments on polymeric fibers. A DuPont Model 900 DTA apparatus was modified. *A* represents the heating block and *H* the heater. The control thermocouple is placed in a dummy hole. The fiber is suspended under light tension by the pully, weight, piezoelectric transducer, support arrangements (*P, W, Z, S*). The DTA uses short lengths of fiber in the well associated with thermocouple T_2 and glass beads in that contain-

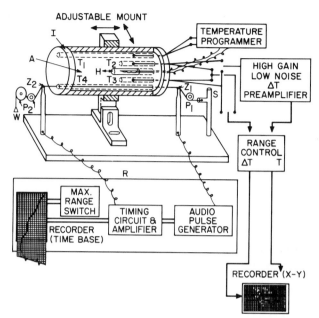

Fig. 38 A schematic representation of dynamic thermoacoustical system. Reprinted from P. K. Chatterjee (1974), *J. Macromol. Sci* **A8**, 191–209, by courtesy of Marcel Dekker, Inc.

ing T_3. The delay time of a 7 kHz pulse between generation at Z_1 and detection at Z_2 is measured while the temperature is raised at 20°C min⁻¹ in an air atmosphere.

Besides yielding viscoelastic constants of the material it is also useful as an analytical tool to distinguish between fibers that would be difficult to distinguish by normal TG or DTA techniques. An example from Chatterjee (1974) is shown in Fig. 39. The differences between rayon and cotton are prominent in the thermoacoustical curves. Chatterjee (1980) has de-

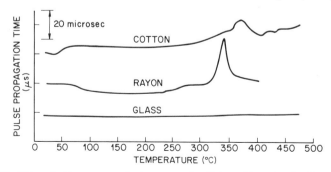

Fig. 39 Dynamic thermoacoustical curves of cellulose fibers in air. Reprinted from P. K. Chatterjee (1974), *J. Macromol. Sci.* **A8**, 191–209, by courtesy of Marcel Dekker, Inc.

scribed the technqiue in detail and interested readers are directed to that review.

C. THERMOELECTROMETRY

Thermoelectrometry, also called electrical conductivity (EC) and amperometric thermal analysis (ATA), is a technique in which the electrical characteristics of a substance are measured as a function of temperature as the substance is subjected to a controlled temperature program (Lombardi, 1980). This method can be used to supplement other thermal methods such as DTA or calorimetry and may be more useful than the preceding methods in certain cases. In general, the instrumentation is more elementary than DTA but the applicability is not as great as the latter. Combined DTA–electrical conductivity systems have been described by Berg and Burmistrova (1960), David (1970), Chiu (1967), Halmos and Wendlandt (1973), and others. Still other EC systems have been described by Williams and Wendlandt (1973) and by Wendlandt (1970a, 1977a).

The EC–DTA apparatus used by Halmos and Wendlandt (1973) is shown in Fig. 40. Current (dc) flowing through the sample is detected by a microammeter and recorded as a function of temperature on an X–Y recorder. The DTA part of the system is conventional in design.

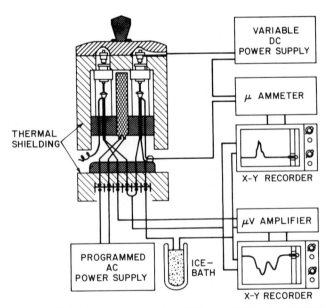

Fig. 40   EC–DTA apparatus used by Halmos and Wendlandt (1973).

Besides measurements of dc conductivity described in the preceding paragraph, there is also interest in measuring ac properties such as dielectric constant and loss as a function of temperature and frequency. Such information helps elucidate conduction mechanisms, molecular orientation, polarization, relaxation, etc. (Rajeshwar *et al.*, 1979; Yalof, 1975; Yalof and Zika, 1975).

D. Emanation Thermal Analysis (ETA)

The technique of emanation thermal analysis (ETA) is based on the introduction of inert radioactive gases into a solid and the measurement of their subsequent liberation from the substance as it is heated (Balek, 1969, 1970). Release of the radioactive gas makes possible the monitoring of various types of changes taking place during the thermal cycle. These include chemical reactions such as dehydration, thermal dissociation and synthesis, polymorphic transformations, melting, conversion of metastable amorphous structures into crystalline compounds, sintering, and changes in the concentrations of lattice defects. ETA possesses several advantages over conventional TG and DTA. Under dynamic conditions it permits the study of structural changes of compounds even when these changes are not related to significant weight changes or enthalpic effects. In other cases, when finely crystalline or amorphous phases are formed, ETA is more sensitive than x-ray methods.

In the ETA apparatus (Balek, 1969), which also permits the recording of the DTA and dilatometric curves concurrently, a labeled sample (generally 100 mg), a DTA reference material (Al_2O_3), and a sample for dilatometric measurements are placed in the heated chamber. Temperature measurement is by thermocouples embedded directly into the samples. A typical heating rate is 8–10°C min^{-1}, since this is an optimum rate for DTA as well as ETA measurements. The radioactive gas released from the solid sample is carried by a carrier-gas stream, which flows at a constant rate into the cells for radioactivity measurements. The apparatus simultaneously registers the α-activity of radon and the β-activity of xenon introduced by ion bombardment. An ETA curve is recorded together with the DTA and dilatometric curves using a multipoint recorder. Emmerich and Balek (1973) have also described an apparatus that records the DTA, TG, and DTG curves as well as the ETA curve of the sample and is available as part of the Netzsch apparatus (see Section IV.D.1).

E. Thermophotometry

The technique of thermoluminescence involves the measurement of the light emitted from a sample as a function of temperature as it is heated to elevated temperatures at a slow, constant rate, below incandescence. The curve so obtained, called a glow curve, consists of peaks or maxima caused by the emission of light at various elevated temperatures. The intensity and temperatures at which the peaks appear may be used to characterize the sample or may be used for other applications. These applications include the study of geological activity, dating of ancient pottery, lunar materials, evaluation of catalysts, and solid-state radiation dosimetry.

Various instruments have been described for the measurement of the thermoluminescence of solid samples. Instrumental arrangements have been described by Urbach (1930), Randall and Wilkins (1945), Boyd (1949), Daniels *et al.* (1953), Saunders (1953), Parks (1953), and Lewis (1956). Basically, the apparatus consists of the following components: (a) a heated sample block; (b) a sample-temperature programmer or power supply; (c) a photomultiplier tube and power supply; and (d) a recorder, either of the two-pen strip chart or the $X-Y$ function type. The glow curves are normally obtained at a heating rate of $1°C$ min^{-1}. As the sample temperature is increased, the sample emits light that is detected by the photomultiplier tube whose output current varies linearly with the intensity of the emitted light. This output current is converted into a voltage signal and recorded on one channel of the strip-chart recorder. The other channel of the recorder is connected to a thermocouple embedded in the sample chamber and hence records the sample temperature. The temperature range of the apparatus is from ambient to 500°C. Glow curves can be obtained on 50 mg of finely powdered sample.

Thermophotometry has been used to study polymers by Wynne and Wendlandt (1976) and coordination compounds by Wendlandt (1980a,b).

F. THERMOANALYTICAL MICROSCOPY (TAM)

Hot-stage microscopy is a common and useful technique. It is particularly valuable in a complementary role to confirm or refute conclusions drawn from other thermoanalytical techniques. It is especially useful in verifying reaction mechanisms inferred from kinetic data. The technique has progressed to the point that simultaneous DTA–TAM methods have been developed by several authors (Kunihisa, 1979; Van Tets and Wiedemann, 1969; Perron *et al.*, 1980).

G. THERMAL CONDUCTIVITY AND DIFFUSIVITY

The thermal transport properties of materials are important for a wide range of applications from encapsulation or heat sinks for electronic devices, to establishing processing conditions for the recovery of oil from shale or the curing of resins and rubbers. Thermal diffusivity α is generally measured and related to thermal conductivity K by the equation for heat flow, $K = \alpha\rho C$, where ρ is density and C is the heat capacity.

A wide variety of instrumentation is available from manufacturers such as Dynatec Co. or Netzsch Co., to name two. Lamoureux (1979) has recently constructed automated equipment run by microprocessors that greatly speeds up the measurement with a claimed potential accuracy of $\pm 1\%$.

Laser flash techniques have also been developed (Parker *et al.*, 1961; Taylor, 1973, Wang *et al.*, 1979). Wang *et al.* (1980) have constructed a computer-controlled version of such an apparatus. Commercial apparatus to measure thermal conductivity by the laser flash method is available from Theta Instrument Co.

Several authors, most recently Chiu and Fair (1979), have successfully adapted DSC units to measure thermal conductivity. Generally, an isothermal mode of operation is preferred and a precision of $\pm 3\%$ is achieved.

X. Conclusions

The tremendously wide interdisciplinary application of thermal analysis bodes well for its future. It also makes forecasting impractical. It is scientific need coupled with an investigator's perception that a thermoanalytical technique can be devised or modified to contribute to the solution that ultimately leads to progress in thermal analysis. Journals, meetings, and societies for thermal analysis exist so that the necessary cross-fertilization between scientific disciplines can occur. To some extent manufacturers also serve this purpose in their attempts to expand the market.

Although it is these unforeseeable breakthroughs that result in major progress, there is nevertheless continuing investigation in predictable areas that will lead to less dramatic but steady improvement. Some of these areas seem on contradictory courses, but the emphasis depends on the particular objective desired. For example, some instrumentation will evolve toward the use of larger samples to assure a representative sample for notoriously inhomogeneous materials such as coal or minerals. Larger samples might also be desired for synthetic purposes or to more closely simulate industrial processes. However, the frontiers will also be expanded in the other direction in an effort to conserve sample, improve resolution, achieve faster heating and cooling rates, or minimize the physical limitations of thermal and mass transport.

Another example of mutually exclusive goals is that of simplicity versus complexity. The potential value of thermoanalytical techniques for quality control leads to the construction of simpler, cheaper, more portable, and faster equipment, perhaps with greater automation, aimed at a specific application and capable of operation by less-skilled workers. In contrast, the demands of research are such that more complex and versatile instruments are constantly required. These will perform simultaneous measurements, with greater accuracy and over wider ranges of conditions and environments. Speed and cost will be less important than ultimate performance and skilled scientists will ply the controls.

One very obvious direction of instrumentation will be toward increased microprocessor or computer interaction. The beauty of this interaction is that it can be used to satisfy the needs of both simplicity or complexity. It can take the place of the sophisticated scientist (or serve as a slave) at the control of routine measurements or it can assist in achieving sophisticated or even otherwise impossible measurements and mathematical analyses. The software associated with these applications is probably the portion of thermal analysis that is most vulnerable to obsolescence and it is hoped that the manufacturers will make every effort to provide proper support and constant upgrading with a minimum of hardware changes.

Finally, it is to be expected that thermoanalytical measurements will be required in more hostile environments demanding greater resistance to corrosion or the ability ot function in higher vacuums and temperatures.

ACKNOWLEDGMENTS

The authors are very grateful to Dr. P. D. Garn and Dr. E. A. Turi for helpful discussions, suggestions, and criticisms of the manuscript.

References

Ashley, H. E. (1911). *J. Ind. Eng. Chem.* **3**, 91–110.
Audiere, J. P., Mazieres, C., Carballes, J. C., and de Cremoux, B. (1974). *J. Phys.* **7**, 355–365.
Bair, H. E., and Johnson, G. E. (1977). *Anal. Calorim.* **4**, 219–227.
Balek, V. (1969). *J. Mater. Sci.* **4**, 919–924.
Balek, V. (1970). *Anal. Chem.* **42**, 16A–18A.
Baxter, R. A. (1969). *In* "Thermal Analysis" (R. F. Schwenker, and P. D. Garn, eds.) Vol. **1**, pp. 65–84. Academic Press, New York.
Berg, L. G., and Burmistrova, N. P. (1960). *Russ. J. Inorg. Chem. (Engl. Transl.)* **5**, 326–329.
Betteridge, D., Joslin, M. T., and Lilley, T. (1981). *Anal. Chem.* **53**, 1064–1073.
Bohon, R. L. (1963). *Anal. Chem.* **35**, 1845–1848.
Boyd, C. A. (1949). *J. Chem. Phys.* **17**, 1221–1228.
Bradley, W. S., and Wendlandt, W. W. (1971). *Anal. Chem.* **43**, 223–226.
Brewer, L., and Zaritsanos, P. (1957). *J. Phys. Chem. Solids* **2**, 284–288.

Brill, O. (1905). *Z. Anorg. Chem.* **45**, 275–291.

Brown, G. H., and Montgomery, E. T. (1912). *Trans. Am. Ceram. Soc.* **14**, 709–725.

Burgess, G. K. (1908). *Bull. Bur. Stand.* **5**, 199–225.

Cahn, L., and Schultz, H. (1963). *Anal. Chem.* **35**, 1729–1732.

Cassel, B. (1978). *Ind. Res. Dev.* **20**, 135–139.

Chatterjee, P. K. (1974). *J. Macromol. Sci.* **A8**, 191–209.

Chatterjee, P. K. (1980). *In* "Treatise on Analytical Chemistry" (I. M. Kolthoff, P. J. Elving, and E. B. Sandell, eds.), Part I, Chapter 16. Wiley, New York.

Chevenard, P., Wache, X., and De La Tullaye, R. (1944). *Bull. Soc. Chim. Fr.* **11**, 41–48.

Chiu, J. (1967). *Anal. Chem.* **39**, 861–867.

Chiu, J., and Fair, P. G. (1979). *Thermochim. Acta* **34**, 267–273.

Clampitt, B. H. (1963). *Anal. Chem.* **35**, 577–582.

Clark, G. M. (1978). *Thermochim. Acta* **27**, 19–25.

Clark, G. M., and Garlick, R. (1979). *Thermochim. Acta* **34**, 365–375.

Cox, D. B., and McGlynn, J. F. (1957). *Anal. Chem.* **29**, 960–961.

Cox, M. G. C., McEnaney, B., and Scott, V. D. (1973). *Progr. Vac. Microbalance Tech.* **2**, 27–33.

Daniels, F., Boyd, C. A., and Saunders, D. F. (1953). *Science* **117**, 343–345.

David, D. J. (1970). *Thermochim. Acta* **1**, 277–287.

Dickens, B., Pummer, W. J., and Flynn, J. H. (1977). *In* "Analytical Pyrolosis" (C. E. R. Jones, ed.), pp. 383–391. Am. Elsevier, New York.

Dollimore, D., and Mason, J. (1981). *Thermochim. Acta* **43**, 183–187.

Duval, C. (1963). "Inorganic Thermogravimetric Analysis," 2nd ed., pp. 4–8. Elsevier, Amsterdam.

Emmerich, W. D., and Balek, V. (1973). *High Temp.—High Pressures* **5**, 67–72.

Emmerich, W. D., and Bayreuther, K. (1975). *In* "Thermal Analysis" (I. Buzas, 4th, ed.), Vol. 3, pp. 1017–1025. Heyden, London.

Erdey, L., Paulik, F., Paulik, J. (1956). *Acta Chim. Acad. Sci. Hung.* **10**, 61–80.

Erdey, L., Paulik, F., and Paulik, J. (1966). *Mikrochim. Acta* pp. 699–710.

Farritor, R. E., and Tao, L. C. (1970). *Thermochim. Acta* **1**, 297–302.

Fenner, C. N. (1912). *Am. J. Sci.* **36**, 331–342.

Flynn, J. H. (1974a). *Thermochim. Acta* **8**, 69–81.

Flynn, J. H. (1974b). *Anal. Calorim.* **3**, 17–44.

Flynn, J. H., and Dickens, B. (1976). *Thermochim. Acta* **15**, 1–16.

Forkel, W. (1960). *Naturwissenschaften* **47**, 10–25.

Freeberg, F. E., and Alleman, T. G. (1966). *Anal. Chem.* **38**, 1806–1807.

Friedman, H. L. (1970). *Thermochim. Acta* **1**, 199–221.

Gallagher, P. K. (1978). *Thermochim. Acta* **26**, 175–183.

Gallagher, P. K., and Gyorgy, E. M. (1979). *Thermochim. Acta* **31**, 380–382.

Gallagher, P. K., and Gyorgy, E. M. (1980). *In* "Thermal Analysis" H. G. Wiedemann, ed.,), Vol 1, pp. 113–118, Birkhaeuser, Basel.

Gallagher, P. K., and Johnson, D. W. (1972). *Thermochim. Acta* **4**, 283–289.

Gallagher, P. K., and Schrey, F. (1963). *J. Am. Ceram. Soc.* **46**, 567–573.

Gallagher, P. K., and Schrey, F. (1970). *Thermochim. Acta* **1**, 465–476.

Gallagher, P. K., and Warne, S. St. J. (1981). *Thermochim. Acta* **43**, 297–303.

Gallagher, P. K., Johnson, D. W., and Vogel, E. M. (1976). *In* "Catalysis in Organic Synthesis—1976" (P. N. Rylander and H. Greenfield, eds.), pp. 113–136. Academic Press, New York.

Gallagher, P. K., Gyorgy, E. M., and Bair, H. E. (1979). *J. Chem. Phys.* **71**, 830–835.

Gallagher, P. K., Coleman, E., Jin, S., and Sherwood, R. C. (1980). *Thermochim. Acta* **37**, 291–300.

Gardner, N., Samuels, E. and Wilks, K. (1974). *Adv. Chem. Ser.* **131,** 217–236.
Garn, P. D. (1962). *Proc. Int. Symp. Microchem. Tech., 1961* pp. 1105–1109.
Garn, P. D. (1965). *Anal. Chem.* **37,** 77–80.
Garn, P. D., and Kessler, J. E. (1960). *Anal. Chem.* **32,** 1563–1565.
Garn, P. D., and Menis, O. (1980). *Thermochim. Acta* **42,** 125–134.
Garn, P. D., Geith, C. R., and DeBala, S. (1962). *Rev. Sci. Instrum.* **33,** 293–297.
Garn, P. D. Menis, O., and Wiedemann, H. G. (1980). *In* "Thermal Analysis" (H. G. Wiedemann, ed.), Vol. 1, pp. 201–206. Birkhaeuser, Basel.
Gillham, J. K. (1966). *Appl. Polymer. Symp.* **2,** 45–60.
Gillham, J. K. (1974). *AIChE J.* **20,** 1066–1079.
Gillham, J. K. (1979). *Polym. Eng. Sci.* **19,** 676–682.
Glassford, A. P. M. (1976). *AIAA Thermophy. Conf., 11th, 1976* Paper 76-438.
Gordon, S., and Campbell, C. (1960). *Anal. Chem.* **32,** 271R–280R.
Gray, A. P. (1975). *Pittsburgh Conf. Appl. Spectrosc. Anal. Chem., 1975.*
Halmos, Z., and Wendlandt, W. W. (1973). *Thermochim. Acta* **7,** 95–102.
Honda, K. (1915). *Sci. Rep. Tohoku Univ.* **4,** 97–105.
Howard, J. B. (1973). *Polym. Eng. Sci.* **13,** 429–434.
Joncich, M. J., and Bailey, D. R. (1960). *Anal. Chem.* **32,** 1578–1582.
King, W. H. (1968). *Anal. Calorim.* **1,** 261–265.
King, W. H., Camilli, C. T., and Findeis, A. P. (1968). *Anal. Chem.* **40,** 1330–1334.
Kinsbron, E., Gallagher, P. K., and English, A. T. (1979). *Solid-State Electron.* **22,** 517–524.
Kollie, T. G., Horton, J. L., Carr, K. R., Herskovitz, M. G., and Mossman, C. A. (1975). *Rev. Sci. Instrum.* **46,** 1447–1455.
Kosak, J. R. (1976). *In* "Catalysis in Organic Synthesis—1976" (P. N. Rylander and H. Greenfield, eds.), pp. 137–151. Academic Press, New York.
Kunihisa, K. S. (1979). *Thermochim. Acta* **31,** 1–11.
Kurnakov, N. S. (1904). *Z. Anorg. Chem.* **42,** 184–201.
Lamoureux, R. T. (1979). *Thermochim. Acta* **34,** 127–132.
Langer, H. G. (1980). *In* "Treatise on Analytical Chemistry" (I. M. Kolthoff, P. J. Elving, and E. B. Sandell, eds.), Part I, Chapter 15. Wiley, New York.
LeChatelier, H. (1887). *Bull. Soc. Fr. Mineral. Cristallogr.* **10,** 204–211.
Levy, P. F., Nievweboer, G., and Semanski, L. C. (1970). *Thermochim. Acta* **1,** 429–435.
Lewis, A. F., and Gillham, J. K. (1962). *J. Appl. Polym. Sci.* **6,** 422–429.
Lewis, D. R. (1956). *J. Phys. Chem.* **60,** 698–705.
Lodding, W., ed. (1967). "Gas Effluent Analysis." Dekker, New York.
Lodding, W., and Sturm, E. (1957). *Am. Mineral.* **42,** 780–780.
Lombardi, G. (1980). "For Better Thermal Analysis," pp. 18–19. ICTA, Rome.
Lonvik, K. (1975). *In* "Thermal Analysis" (I. Buzas, ed.), Vol. 3, pp. 1089–1105. Heyden, London.
Lum, R. M. (1977). *Thermochim. Acta* **18,** 73–94.
Lum, R. M., and Feinstein, L. G. (1980). *Proc.—Electron. Components Conf.* **30,** 113–120.
McAdie, H. G., Garn, P. D., and Menis, O. (1972). *NBS Spec. Publ. (U.S.)* **260.**
MacKenzie, A. P., and Luyet, B. J. (1964). *Biodynamica* **9,** 194–206.
MacKenzie, R. C. (1957). "The Differential Thermal Investigation of Clays." Minerological Society, London.
Manche, E. P., and Carroll, B. (1964). *Rev. Sci. Instrum.* **35,** 1486–1488.
Mashiko, Y., Shinoda, T., and Enokido, H. (1971). *Bull. Chem. Soc. Jpn.* **44,** 888–892.
Mazieres, C. (1964). *Anal. Chem.* **36,** 602–610.
Mita, I., Imai, I., and Kambe, H. (1971). *Thermochim. Acta* **2,** 337–340.

Mizutani, N., and Kato, M. (1975). *Anal. Chem.* **47**, 1389–1394.

Mori, G. P., Power, T. A., and Glover, C. A. (1972). *Thermochim. Acta* **3**, 259–265.

Mraw, S. C., and O'Rourke, D. F. (1981). *J. Chem. Thermodynam.* **13**, 199–200.

Nernst, W., and Riesenfeld, E. H. (1903). *Ber. Dtsch. Chem. Ges.* **36**, 2086–2091.

Newkirk, A. E. (1960). *Anal. Chem.* **32**, 1558–1563.

Newkirk, A. E. (1971). *Thermochim. Acta* **2**, 1–23.

Norem, S. D., O'Neill, M. J., and Gray, A. P. (1969). *Proc. Toronto Symp. Therm. Anal., 3rd, 1969*, pp. 221–232.

Norem, S. D., O'Neill, M. J., and Gray, A. P. (1970). *Thermochim. Acta* **1**, 29–34.

Norris, A. C., Pope, M. I., and Selwood, M. (1980). *Thermochim. Acta* **35**, 11–22.

O'Neill, M. J. (1964). *Anal. Chem.* **36**, 1238–1248.

Parker, W. J., Jenkins, R. J., Butler, C. P., and Abbot, G. L. (1961). *J. Appl. Phys.* **32**, 1679–1683.

Parks, J. M. (1953). *Am. Assoc. Pet. Geol. Bull.* **37**, 125–127.

Paulik, F., and Paulik, J. (1972). *Thermochim. Acta* **4**, 189–199.

Paulik, F., Paulik, J., and Erdey, L. (1966). *Talanta* **13**, 1405–1421.

Paulik, J. M. and Leonard, G. W. (1959). *Anal. Chem.* **31**, 1037–1045.

Price, D., Dollimore, D., Fatemi, N. S., and Whitehead, R. (1980). *Thermochim. Acta* **42**, 323–332.

Perron, W., Bayer, G., and Wiedemann, H. G. (1980) *In* "Thermal Analysis" (H. G. Wiedemann, ed.), Vol. 1, pp. 279–284. Birkhaeuser, Basel.

Rajeshwar, K., Nottenburg, R., Freeman, M., and Dubow, J. (1979). *Thermochim. Acta* **33**, 157–168.

Randall, J. T., and Wilkins, M. H. F. (1945). *Proc. R. Soc. London A Ser.* **184**, 347–355.

Roberts-Austen, W. C. (1889). *Metallographist* **2**, 186–195.

Saito, H. (1962). "Thermobalance Analysis." Technical Books Publ. Co., Tokyo.

Saladin, V. S. (1904). *Iron Steel Metall. Metallogr.* **7**, 237–245.

Sarasohn, I. M., and Tabeling, R. W. (1964). *Pittsburgh Conf. Appl. Spectrosc. Anal. Chem., 1964.*

Saunders, D. F. (1953). *Am. Assoc. Pet. Geol. Bull.* **37**, 114–121.

Sestak, J. (1969). *In* "Thermal Analysis" (R. F. Schwenker and P. D. Garn, eds.), Vol. 2, pp. 1085–1093. Academic Press, New York.

Sestak, J., Satava, V., and Wendlandt, W. W. (1973). *Thermochim. Acta* **7**, 333–556.

Sheffield, G. S., Hare, T. M., and McGaughey, G. S. (1979). *Thermochim. Acta* **32**, 45–52.

Stewart, L. N. (1969). *Proc. Toronto Symp. Therm. Anal. 3rd, 1969* p. 205.

Stillwagon, L. E., and Manzione, L. T. (1980). *Proc. NATS Conf., 10th, 1980*, pp. 89–94.

Stone, R. L. (1951). *Bull—Ohio State Univ., Eng. Exp. Stn.* **146**, 1–77.

Szabadvary, F. (1966). "History of Analytical Chemistry," pp. 16–17. Pergamon, Oxford.

Szabadvary, F., Buzágh-Gere, É. (1979). *J. Therm. Anal.* **15**, 389–398.

Takahashi, Y., and Ozawa, T. (1975). *J. Therm. Anal.* **8**, 125–135.

Taylor, R. E., and Clark, L. M. (1974). *High Temp. High Pres.* **6**, 65–72.

Theall, G. G. (1969). *In* "Thermal Analysis" (R. F. Schwenker and P. D. Garn, eds.), Vol. 1, pp. 97–111. Academic Press, New York.

Townsend, D. I., and Tou, J. (1980). *Thermochim. Acta* **37**, 1–30.

Truchot, P. (1907). *Rev. Quim. Pura Apl.* **10**, 2–12.

Urbach, F. (1930). *Sitzungsber. Akad. Wiss. Wien, Math.-Naturwiss. Kl., Abt. IIa* **139**, 363–372.

Urbain, G., and Boulanger, C. (1912). *C. R. Hebd. Seances Acad. Sci.* **154**, 347–349.

VanTets, A., and Wiedemann, H. G. (1969). *In* "Thermal Analysis" (R. F. Schwenker and P. D. Garn, eds.), Vol. 1, pp. 121–135. Academic Press, New York.

Vassallo, D. A., and Harden, J. C. (1962). *Anal. Chem.* **34**, 132–133.

Vieweg, R. (1972). *Prog. Vac. Microbalance Tech.* **1**, 1–24.

Wang, Y., Dubow, J., Rajeshwar, K., and Nottenburgh, R. (1979). *Thermochim Acta* **28**, 23–35.

Wang, Y., Freeman, M., Nottenburg, R., Rajeshwar, K., and Dubow, J. (1980). *Thermochim. Acta* **37**, 287–290.

Watson, E. S., O'Neill, M. J., Justin, J., and Brenner, N. (1964). *Anal. Chem.* **36**, 1233–1238.

Weaver, E. E., and Keim, W. (1960). *Proc. Indiana Acad. Sci.* **70**, 123–128.

Wendlandt, W. W. (1970a). *Thermochim. Acta* **1**, 11–21.

Wendlandt, W. W. (1970b). *Anal. Chim. Acta* **49**, 187–196.

Wendlandt, W. W. (1972). *J. Chem. Educ.* **49**, A571–580, A623–630, A671–678.

Wendlandt, W. W. (1973). *Thermochim. Acta* **5**, 225–242.

Wendlandt, W. W. (1974a). "Handbook of Commercial Scientific Instruments," Vol. 2. Dekker, New York.

Wendlandt, W. W. (1974b). "Thermal Methods of Analysis," 2nd ed., Vol. 19. Wiley, New York.

Wendlandt, W. W. (1975). *Thermochim. Acta* **12**, 109–115.

Wendlandt, W. W. (1977a). *Thermochim. Acta* **21**, 291–300.

Wendlandt, W. W. (1977b). *Thermochim. Acta* **21**, 295–296.

Wendlandt, W. W. (1980a). *Thermochim. Acta* **35**, 247–253.

Wendlandt, W. W. (1980b). *Thermochim. Acta* **35**, 255–257.

Wendlandt, W. W., and Collins, L. W. (1976). "Benchmark Papers in Analytical Chemistry," Vol. 2. Dowden, Hutchinson & Ross, Inc., Stroudsburg, Pennsylvania.

Wendlandt, W. W., and Collins, L. W. (1981). "Evolved Gas Detection and Analysis." Elsevier, Amsterdam, to be published.

Wiedemann, H. G. (1964). *Chem.-Ing.-Tech.* **36**, 1105–1120.

Wiedemann, H. G. (1972). *Thermochim. Acta* **3**, 355–366.

Wiedemeir, H., Pultz, G., Gaur, U., and Wunderlich, B. (1981). *Thermochim. Acta* **43**, 297–303.

Williams, J. R., and Wendlandt, W. W. (1973). *Thermochim. Acta* **7**, 261–269.

Wynne, A. M., and Wendlandt, W. W. (1976). *Thermochim. Acta* **14**, 61–69.

Yalof, S. A. (1975). *Chemtech.* pp. 165–170.

Yalof, S. A., and Brisbin, D. (1973). *Am. Lab. (Fairfield, Conn.)* **5**, 65.

Yalof, S. A., and Hedvig, P. (1976). *Thermochim. Acta* **17**, 301–308.

Yalof, S. A., and Zika, K. (1975). *CHEMTECH* pp. 682–687.

Yamamoto, A., Yamada, K., Maruta, M., and Akiyama, J. (1969). *In* "Thermal Analysis" (R. F. Schwenker and P. D. Garn, eds.), Vol. 1, pp. 105–111. Academic Press, New York.

Yuen, H. K., Grote, W. A., and Young, R. C. (1980). *Thermochim. Acta* **42**, 305–314.

CHAPTER 2

The Basis of Thermal Analysis

BERNHARD WUNDERLICH

Department of Chemistry
Rensselaer Polytechnic Institute
Troy, New York

I. Introduction

Thermal characterization of polymeric materials covers many aspects of materials science. In this chapter an effort is made to collect the basic knowledge required to derive and interpret thermal analysis data.

Today it is impossible to discuss the results of a macroscopic measurement, such as thermal analysis, without understanding of the molecular structure of matter, the microscopic structure. The first part of the chapter is thus a brief summary of the microscopic structure of macromolecules. The second part contains a discussion of the theoretical basis needed for the understanding of the variables used in thermal analysis. The theories of equilibrium thermodynamics, irreversible thermodynamics, and kinetics are developed.

A collection of formulas with references might have been sufficient. Checking these references, it became clear, however, that none covers all topics and each one covers a much larger list of topics, so that even looking up the simplest equation can become a major undertaking. For this reason, and also to point out the simplicity and limited scope of the thermal analysis theory, the chapter was designed to be largely self-contained. References are given for further study, but are usually not needed for initial understanding.

In the early part of the chapter the conceptionally simple quantity total energy U, which leads to the Helmholtz free energy F is used. After calorimetry is discussed, this changes to enthalpy H, which is more convenient for the interpretation of thermal analysis. The enthalpy leads to the Gibbs free energy G, which is called the free enthalpy.

The final two parts of the chapter cover the description of single and multicomponent systems. Only few examples have been given for each topic; many more are found in the later chapters. Also, it must be noted that some thermal analysis techniques have a simple theoretical basis and are thus mentioned less often than others. This should not be taken as a value judgment. The reverse is often true; the less complicated the theoretical basis, the easier is the interpretation of the measurements and the more valuable are the results.

Throughout the chapter an effort is made to point out applications or misapplications. With help of the subject index, connections to the various topics throughout the book can be made.

II. The Microscopic Structure of Materials

For the description of matter one can use two approaches: the microscopic description, which considers matter on an atomic level; and the macroscopic description, which deals with phenomena directly observable with our senses. In this introductory section the microscopic description of matter is briefly given, to form the basis for the following macroscopic description.

A. A System of Classifying All Materials

It is convenient for the microscopic description of materials to distinguish three major classes of molecules:

Class 1   rigid macromolecules
Class 2   flexible macromolecules
Class 3   small molecules

The term *macromolecule* was coined by Staudinger (1950). A macromolecule must at least contain 1000 atoms in an identifiable structure.

Several conclusions can be made about each class of molecules. Rigid macromolecules may extend strong bonds in one, two, or three directions of space. On melting (or sublimation), they always lose their molecular integrity. Once molten, the strong bonds are not permanent, but interchange frequently, although at any moment only a small fraction of bonds is actually broken (a reasonable guess is 5–10% at the melting temperature). Since the bonds that set up the molecular structure are strong, the melting temperatures are relatively high. The number of different molecules in this class of matter is determined by variation in atoms and types of bonding. The detailed molecular structure is, however, of less importance than the crystal or glassy structure, which is based on the knowledge of a much smaller unit than the whole molecule. More detailed knowledge requires understanding of the grain size and shape, the boundary type and configuration, and the defect structure and concentration. Three-dimensional macromolecules are possible with all three types of strong bonding (ionic, covalent, and metallic). Rigid macromolecules are found among salts, silicate glasses, ceramics, molecular solids, and metals.

Small molecules behave quite differently from rigid macromolecules. Their crystal structure is set up by weak forces, so that on melting, the molecular integrity is often preserved. This leads to low melting temperatures. Many small molecules can at somewhat higher temperatures even be brought into the gaseous state without decomposition. Examples in this class are not only the noble gases and inorganic, small molecules (like H_2O, NH_3 and CO_2), but also all organic molecules of less than 1000 atoms. Taking the listing of Beilstein (1918) as a guide, one may guess at 10^6–10^7 known different small molecules.

Flexible macromolecules are intermediate between the molecules of the two other classes. Their structure is usually largely linear to obtain the needed flexibility. They can often be molten or dissolved without loss of molecular integrity and have for this reason an intermediate melting temperature. Since all degrees of molecular rigidity can be produced, linear macromolecules can span melting temperatures from below room temperature, as in *cis*-1,4-polybutadiene (T_m = 11.5°C) to temperatures above the thermal stability of the C—C bond, as in polyphenylene. Because of their size, flexible macromolecules cannot be brought into the gaseous state without loss of molecular integrity. The number of possible linear macromolecules that maintain their molecular identity during melting is practically unlimited. Any number of repeating units can be linked in any number of ways. This multiplicity lies at the root of the importance of macromolecular science. If one takes only 20 different repeating units and links them to linear macromolecules of 100 units, one can make 20^{100} (i.e., 10^{130}) distinguishable molecules. This is a number so large that all atoms in the known universe are not enough to make one molecule of each. This immense multiplicity forces one to predict properties in order to advance knowledge. It is an impossible undertaking to make even a small fraction of these molecules to check their properties and decide their use to man.

B. THE FLEXIBLE, LINEAR MACROMOLECULES

The materials treated in this book are flexible linear macromolecules. They are frequently less accurately called polymers.* Flexible macromolecules represent the largest class of matter. Naturally occurring macromolecules, as well as a few accidentally discovered synthetic linear macromolecules, were used industrially for some time before the understanding of their molecular nature. In fact, since proteins and polysaccharides are most prominent in nature, they were used already by early man (in the form of wood, bark, leaves, fibers, skin, hair, etc.). Modern industrial use of macromolecules had its start around 1850 with larger-scale production of vulcanized rubber and cellulose nitrate and acetate. Today more than 50 million tons of scores of synthetic linear macromolecules are produced. This growth occurred exponentially, starting with the recognition of the molecular structure of linear macromolecules in the 1920s by Staudinger [see Staudinger (1961) for a summary of his work and Flory (1953) for a general historic summary].

* Molecules smaller than macromolecules may also be "polymers."

C. The Chemical Structure of Linear Macromolecules

A linear macromolecule has a continuous backbone of covalently bonded atoms that may have side groups attached to it. Taking into account valence restrictions and bond length and angle, one can construct the many possible molecules. Recognizable units that are the building blocks of the molecule are called the repeating units. For example, the propylene unit CH_2—$CHCH_3$— is easily recognized as the only repeating unit in polypropylene:

$$\cdots CH_2\text{—}CHCH_3\text{—}CH_2\text{—}CHCH_3\text{—}CH_2\text{—}CHCH_3\text{—}CH_2\text{—}CHCH_3\text{—} \cdots$$

The naming is in this case simply derived from the monomer out of which the macromolecule can be made by adding the syllable *poly* ("many")

$$x\ CH_2\text{=}CHCH_3 \xrightarrow{\hspace{1cm}} (CH_2\text{—}CHCH_3\text{—})_x \tag{1}$$
$$\text{propylene} \qquad \text{polymerization} \qquad \text{polypropylene}$$

In case the monomer name has two or more words, they are enclosed in parentheses, as, for example, for poly(ethylene terephthalate)

$$(CH_2\text{—}CH_2\text{—}O\text{—}CO\text{—}C_6H_4\text{—}CO\text{—}O\text{—})_x$$

For a structure-based (rather than synthesis-based) nomenclature and more detailed naming rules see Brandrup and Immergut (1975). Macromolecules that are made of only one type of repeating unit are called homopolymers. Their widespread use is a sign of our lack of knowledge of the properties of more complicated molecules.

As various repeating units are linked together, one produces copolymers. Their nomenclature is only simple as long as the number of different comonomers is small and the order in the molecule is a simple repetition. In fact, the homopolymer poly(ethylene terephthalate) is in the strict sense already an alternating copolymer of dimethylene and terephthaloyl units, or one can even break it into four smaller repeating units coupled in the sequence A—A—B—C—D—C—B—. A true alternating copolymer is made, for example, by polymerization of isobutene and tetrafluoroethylene

$$[CH_2\text{—}C(CH_3)_2\text{—}CF_2\text{—}CF_2\text{—}]_x$$

[*alt*-poly(isobutene-*co*-tetrafluoroethylene)]. Another simple order in polymerization is achieved in the block-copolymers, as in diblock poly(styrene-*co*-methylmethacrylate)

$$(CH_2\text{—}CHC_6H_5\text{—})_x(CH_2\text{—}CCH_3COOCH_3\text{—})_y$$

One can imagine that multiplicity of repeating units and order of linking make names of more complicated copolymers page-long, unwielding expressions. Trivial names, abbreviations, letter codes, or worse, trade names are then often used. A general description of many linear macromolecules, of their structure, properties, and synthesis can be found in the "Encyclopedia of Polymer Science and Technology," edited by Mark *et al.* (1964–1972).

D. The Type, Size, and Shape of Linear Macromolecules

The specification of the chemical structure of a flexible macromolecule is not sufficient for full characterization. One needs also a characterization of the type, size, and shape of the molecule. Only strictly linear macromolecules can be specified simply by their molecular weight. Figure 1

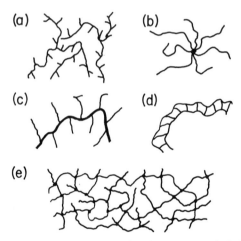

Fig. 1 Examples of flexible macromolecules that are not strictly linear. (a) Branched macromolecule (with short and long branches). (b) Star macromolecule. (c) Graft copolymer (with branches grafted on a backbone; the branches may be regularly spaced and of regular length forming a comb-type macromolecule or irregularly spaced as shown). (d) Ladder polymer (e) Cross-linked, network polymer.

shows a series of types of macromolecules that deviate from linearity. In these cases information about branch points, length, and network characteristics must be known. The nonlinear, flexible macromolecules have largely different rheological behavior from similar linear molecules. The sliding of molecules past each other on sample deformation is more or less hindered (A–C), or the flexibility is reduced (D). For the cross-linked sample (E) all chains are connected to a giant network. If the number of network points is not too large, the sample shows considerable elasticity due to the possible reversible deformation of the flexible, connecting segments. These materials are elastomers, with lightly cross-linked natural rubber [*cis*-1,4-poly(2-methyl butadiene)] the most well-known representative.

All samples of synthetic macromolecules contain more or less broad distributions of molecules caused by the randomness of initiation, termination, and propagation of the polymerization reactions. To characterize these distributions one specifies the number average molecular weight

$$M_n = \Sigma \, n_x M_x = M_0 \, \Sigma(N_x x)/\Sigma \, N_x \qquad (2)$$

weight average molecular weight

$$M_w = \Sigma \, w_x M_x = M_0 \, \Sigma(N_x x^2)/\Sigma(N_x x) \qquad (3)$$

and possibly even averages higher in the power of x. In Eqs. (2) and (3) n_x and w_x stand for the mole and weight fractions of molecules, x repeating units in length. M_x is the molecular weight of a molecule x units in length. The sums in Eqs. (2) and (3) go over all species. M_0 is the molecular weight of the repeating unit. The number average molecular weight of Eq. (2) indicates the abscissa of the center of gravity of a number of molecules versus molecular weight plot. Since the longer molecules contribute much more weight to the distribution, the weight average molecular weight is in many applications more important than the number average. The ratio of the weight average to the number average molecular weight, the polydispersity, is a measure of the broadness of the molecular-weight distribution. Its value is always higher than 1.

Left to itself, a dissolved or molten macromolecule will take a random coil macroconformation, as is shown in Fig. 2. The reason for the coiling is the possibility of rotation around the backbone bonds of a flexible molecule. The positions about each bond are randomized, just as the positions

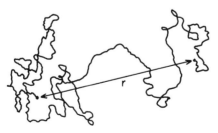

Fig. 2 Schematic drawing of a random coil macroconformation of a flexible linear macromolecule projected into two dimensions. The chain ends are marked by circles. The end-to-end distance is indicated by the length r.

of gas molecules in a given volume are randomized. Trying to stretch such a molecule causes a retractive force, just as a gas resists compression. For carbon, oxygen, and nitrogen atoms in the backbone chain this rotation is hindered. The three more favored conformations are called rotational isomers. To give the specific shape or macroconformation of a molecule, the conformation about each bond must be known. In addition, the molecule changes continuously its macroconformation. The information of value is thus the average shape, as given, for example, by the root-mean-square end-to-end distance (see also Fig. 2).

The calculation of the root-mean-square end-to-end distance is quite simple for a freely jointed chain, i.e., a molecule that has no bond angle or hindered rotation restrictions. The problem was solved as the random flight problem. If one marks each bond in sequence by vectors, l_1, l_2, l_3, etc., the end-to-end distance, as indicated in Fig. 2, is simply the vector sum over all l_1

$$\mathbf{r} = \sum_{i=1}^{x} \mathbf{l}_i \tag{4}$$

For the evaluation of the root mean square, it is sufficient to know the square of Eq. (4):

$$\mathbf{r} \cdot \mathbf{r} = r^2 = \sum_{i,j} \mathbf{l}_i \cdot \mathbf{l}_j \tag{5}$$

The mean of Eq. (5) is easy to guess at. Any two bonds can take any direction relative to each other that averages the product $\mathbf{l}_i \cdot \mathbf{l}_j = l^2 \cos \sphericalangle i,j$ to zero as long as $i \neq j$. For $i = j$, $\sphericalangle i,j$ is zero, i.e., $\cos \sphericalangle i,j$ is 1 and the contribution to Eq. (5) is l^2. For every i there is one nonzero term in the mean, leading to the important relation for the root-mean-square end-to-end distance of a freely jointed chain

$$\langle r^2 \rangle^{1/2} = x^{1/2} l \tag{6}$$

The root-mean-square end-to-end distance increases with the square root of the number of freely jointed segments of the molecule. In case the joining of the chain atoms is not completely free, the calculations are more involved. The most powerful method is in this case to sum up over the possible rotational isomers. The method is described by Flory (1969). As an example, various length calculations for polyethylene of 20,000 chain atoms are given in Table I. The refinement of the calculation by considering fixed bond angles and hindering of rotation around bonds leads to a minor expansion of the random coil. The further expansion, listed under excluded volume calculation, results from the fact that different parts of

TABLE I

ROOT-MEAN-SQUARE END-TO-END DISTANCES
OF (CH_2-) 20,000

Contour length[a] $(20,000 \times 0.154)$	3080.0 nm
Freely jointed chain $[141.4 \times 0.154$, Eq. (6)]	21.8 nm
Rotational isomer calculation	49.3 nm
Excluded volume calculation	62.8 nm
Experimental data (at the Θ temperature)	52.9 nm

[a] The carbon–carbon bond length is 0.154 nm (1 nm = 10 Å).

the molecule cannot occupy the same position. Experimental data are obtained from light scattering experiments in solution. Since in solution there are interactions between solvent and macromolecule, the end-to-end distance changes with the degree of interaction. If the solvent is preferentially in contact with the macromolecule (good solvent), the end-to-end distance increases, the random coil expands. If the solvent is repelled by the macromolecule (poor solvent), the random coil contracts. Since higher temperature usually increases the solvent power, the root-mean-square end-to-end distance can be varied with temperature. It is possible in this way to find a temperature Θ at which the excluded volume is just compensated by the contraction of the random coil due to solvent repulsion. The Θ temperature of a macromolecule in solution is similar to the Boyle temperature of a gas in which the attraction due to intermolecular forces is just compensated by the excluded volume of the gas molecules and seemingly an ideal gas results that follows the ideal gas law $PV = nRT$. Table I shows that the experimental value for the root-mean-square end-to-end distance at the Θ temperature is in good agreement with the rotational isomer calculation.

Cross-linked samples of flexible linear macromolecules may not be dissolved, but if the chain segments between network-points are long enough, the sample swells, expanding the network. The work of expansion must be balanced by the driving force to dissolution.

E. THE SOLID-STATE MACROCONFORMATIONS OF LINEAR MACROMOLECULES

The macroconformation of the linear macromolecule in the solid state may vary considerably from the random coil. Schematically, the possible macroconformations are shown in Fig. 3. Area A indicates the macroconformation in the glassy state. No change of macroconformation has oc-

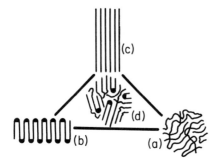

Fig. 3 Schematic drawing of the macro-conformations of a flexible linear macromolecule in the solid state. (a) Random coil in the glassy state. (b) Folded chain, best found in solution-grown crystals (see Fig. 6). (c) Extended chain crystals (equilibrium crystals, see Fig. 5). (d) Fringed micelle macroconformation.

curred on solidification of the liquid. The macromolecules are still in the random coil macroconformations. In the glassy solid state no change of the macroconformation occurs with time. The macroconformation is frozen in at the transition temperature. Macromolecules that have insufficient order along the chain or that are quickly cooled, so that they cannot crystallize, form glasses.

More regular macromolecules crystallize on cooling. The crystallization process is complicated because of the size and flexibility of the molecule. The equilibrium macroconformation in the crystal must have the lowest free energy. Each backbone bond should be in the most favorable rotational position. This leads to a fully extended macroconformation in form of a planar zig-zag chain or a regular helix. These extended chains are then packed so that a maximum of interaction is possible (close packing). A typical crystal structure is shown in Fig. 4 through a drawing of its unit cell. The equilibrium crystal derives from it by continuing the unit cell in all three directions of space to macroscopic dimensions. In the crystallographic c-direction this lengthens the molecule; in the other directions new molecules are added. Crystals of such large size and perfection are rare. Figure 5 shows an example of an extended chain crystal.

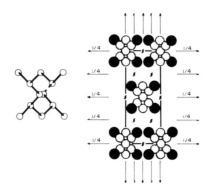

Fig. 4 Crystal structure of syndiotactic polypropylene $(CH_2—CHCH_3—)_x$ projected onto the ab plane (normal to the chain axis). A slightly oblique projection is shown to the left. The black circles indicate $CH_3—$ groups. The heavy outline marks the unit cell. Full arrows indicate twofold axes, half arrows screw axes, $\mathcal{S}$ vertical screw axes, each chain coinciding with an additional screw axis, $\frac{1}{4}$ indicates that the symmetry axis is found at $\frac{1}{4}$ and $\frac{3}{4}$ elevation in the unit cell. Space group $C222_1$, orthorhombic. From Wunderlich (1973).

Fig. 5 Extended-chain crystals of polyethylene. Electron micrograph of a replica of a fracture surface of a sample crystallized at elevated pressure. The molecular chains lie parallel to the striation. Scale bar 1 μm. Reprinted from Wunderlich and Melillo (1968, Fig. 11), by courtesy of Hüthig and Wept Verlag, Basel.

Crystallization from melt or solution is restricted by the chain-folding principle. A sufficiently mobile and flexible linear macromolecule grows initially always into 5–50 nm thick folded-chain crystals. The reason for this chain folding lies in the nature of the crystallization process. Ordering (extending) of the chains and packing occurs in close succession for thermodynamic reasons. For long molecules this is not possible in a finite time, so that a compromise is found in chain folding. Rather perfect chain-folded crystals are grown from dilute solution. A typical example is shown in Fig. 6. The chain-folded crystals are metastable because of their small size. As a result, annealing may occur at any time when sufficient mobility exists in the solid state.

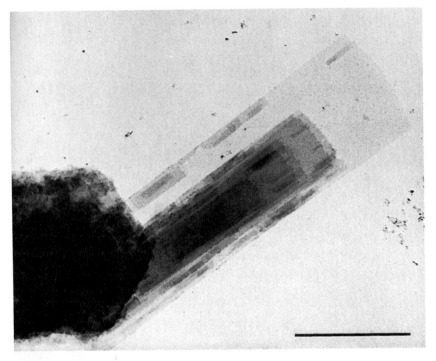

Fig. 6   Chain-folded crystal of poly-p-xylylene $(CH_2—CH_2—C_6H_4—)_x$ grown from solution (0.1% in α-chloronaphthalene). Scale bar 0.5 μm. Lamellae are about 10 nm thick with the chains oriented normal to the plane of the figure. From Wunderlich (1973).

The three areas A, B, and C of Fig. 3 represent limiting macroconformations. Typical semicrystalline solids have macroconformations that contain elements of all three limits. They are best described as fringed micellar. Because of the large molecular length, the molecules participate in several crystalline and amorphous areas. A detailed description of such a structure requires a knowledge of the weight fraction of crystallinity [see Eq. (125a)], the crystal structure, crystal morphology (shape) and size, the defects in the crystals, and the deviations in the amorphous areas from typical melt or glass structure because the molecules are tied to crystalline areas. A full description of crystal structure, morphology, and defects is given in the treatise by Wunderlich (1973).

F. The Crystals of Linear Macromolecules

Equilibrium crystals should be of similar dimensions in all directions (isometric crystals to minimize the surface free energy). Crystals of linear macromolecules are, in contrast, frequently extremely thin lamellae or fibrils, or crystalline regions that are small in all three dimensions. The small dimension may be only 5–10 nm. The lamellar crystals can often be linked to the chain-folding principle, which produces lamellar top and bottom surfaces that contain chain folds and ends that cannot grow further. The fibrillar crystals can be formed, for example, by extreme deformation. The small crystalline regions are observed on fast crystallization from the melt. A detailed description of the various nucleation, crystal growth, and annealing processes is given by Wunderlich (1976).

The three basic crystal habits—lamellar, fibrillar, and fringed micellar —may be connected to a higher level of organization. In Fig. 7 an example is given for the formation of bundles of larger fibrils out of small fibrils. Lamellar crystals may change on fast crystallization to dendrites, which are skeletonized crystals, often grown by repeated twinning. An example of a dendrite is shown in Fig. 8. Since the basic unit of the dendrite is very

Fig. 7   Bundles of fibrils as revealed on etching of natural cellulose. Etchant HCl, 2.5 *N* at 105°C for 15 min. Scale bar 1 μm. Electron micrograph from Battista (1965). Reprinted by permission of *American Scientist,* journal of Sigma Xi, The Scientific Research Society.

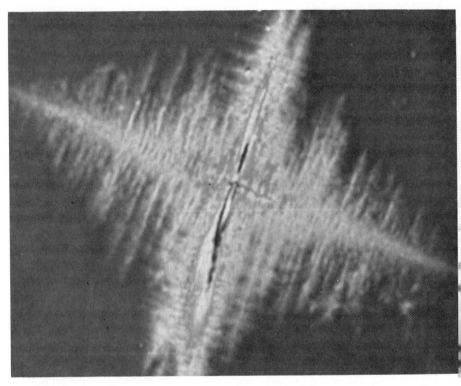

Fig. 8 Dendrite of polyethylene. Interference micrograph of a dendrite grown from 0.02% solution in toluene by cooling to room temperature. Length of the dendrite 0.1 mm. Micrograph from Wunderlich and Sullivan (1962). See also Wunderlich (1963). Reprinted by Courtesy of J. Wiley and Sons, Inc.

thin and flexible, one finds that the various layers of their structure may fill space by bending. Through multiple branching during crystal growth, spherulites, as shown in Fig. 9, result that are so fine in structure that the lamellae or skeletonized lamellae are not visible at the given magnification. Reducing the lamellae or fibrils further in size, similar superstructures, but with increasingly fringed micellar internal character, arise.

G. The Techniques of Microscopic Structure
   Determination

The concern of this book is the thermal analysis, which is a macroscopic technique. Full understanding of a macroscopic observation can, however, be achieved only when the microscopic origin of the effect is understood. For example, the force needed to expand a rubber band was

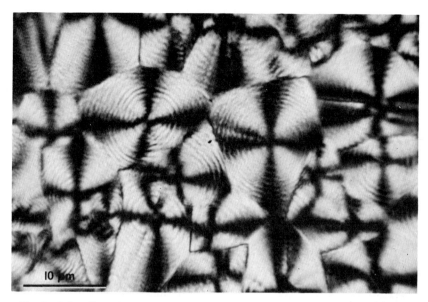

Fig. 9 Spherulites of polyethylene. Optical micrograph between crossed polarizer and analyzer. Polymer, melt-cooled to room temperature. The lamellae marking up the spherulitic superstructure are too small to be seen. From Wunderlich (1963). Reprinted by Courtesy of J. Wiley and Sons, Inc.

found in 1805 to cause substantial heating of the rubber band (Gough, 1805); steel, however, cools somewhat on expansion. Since the process is reversible, this means the entropy or disorder decreases. It took, however, until about 1940 (see, for example, Treolar, 1949) to understand the molecular origin of rubber elasticity, namely, the extension of the end-to-end distance of the rubber molecules (see Fig. 2). It is thus necessary to find out the microscopic structure, at least on some example materials. In this section a brief list of experiments and references is given.

As soon as sufficient order is present in a solid, x-ray diffraction can give information about crystal structure, size, and defects (see, for example, Alexander, 1969). The molten and glassy states are considerably more difficult to analyze. A new analysis technique involves coherent, elastic neutron scattering from the molecules immersed in a matrix of fully deuterated molecules (see Ballard and Schelten, 1980; Benoit, 1974). The technique is similar to light scattering in solution, which gives microscopic information on end-to-end distances and molecular weights (Huglin, 1972). Solid samples can also be analyzed by light scattering relative to density fluctuations and orientation (Stein, 1973).

Electron microscopy is naturally a direct tool to see structures down to the smallest crystal or phase-separated structures. It may be necessary, however, to enhance the contrast by selective staining with electron-heavy stains or by differential etching with suitable oxidants to see the detailed structures (see, for example, Valdré, 1971; Hall, 1966).

Infrared and Raman analysis has often been found useful not only to look for chemical groups of characteristic vibrational frequencies, but also to give information on crystal normal mode frequencies and force constants, packing and folding in crystals, and determination of lamellar thickness (Zbinden, 1964). Similarly, nuclear magnetic resonance has, in addition to chemical information, given information on mobility and phase structure in macromolecular materials (Slonim and Lyubinov, 1970; Bovey, 1974).

The optical microscope should naturally not be overlooked as a tool that may give valuable structure information. The crystalline superstructures, dendrites and spherulites, are easily detected by polarizing microscopy, but single lamellae and fibrils also can be made visible by interference microscopy. Furthermore, any orientation in the sample can be observed by birefringence studies and is quantitatively coupled to the microscopic behavior (see, for example, Shelley, 1975).

For successful interpretation of thermal analysis, one or more of the preceding methods that give direct information on the microscopic structure are of great value.

III. The Macroscopic Description of Materials

The theories needed for the macroscopic description of materials are discussed briefly in the first parts of this section. This is followed with a more detailed summary of the variables that are measured in each of the thermal analysis techniques. This section serves thus as reference for the subsequent characterization of single- and multiple-component systems.

The macroscopic description of materials can be given without reference to the microscopic structure, and many pride themselves on this fact. For deeper understanding, however, a problem is only solved if both the macroscopic description and the microscopic cause are understood. Sections II and III thus contain the basic information for successful thermal analysis of macromolecules.

A. Delineation of Systems and Subsystems

To delineate precisely the portion of matter to be discussed, one erects real or imaginary boundaries and calls the content within the boundaries the system. For example, in calorimetry all the material within the calorimeter proper is our system. Careful attention is then given to energy and matter gained or lost by the system. Three types of systems are distinguished: *open systems, closed systems,* and *isolated systems.* The open system exchanges both matter and energy. The experimental setup of the thermogravimetry represents an example of an open system. It is clear from the preceding that a shortcoming of thermogravimetry is the lack of direct information on the second variable, energy. Coupled calorimetry and thermogravimetry is thus a goal of more complete thermal analysis of an open system. If the lost matter is unknown, techniques such as gas chromatography of mass spectrometry are employed together with thermogravimetry.

The closed system is closed only to matter exchange. Heat and/or other forms of energy are permitted to be gained or lost. The closed system is represented by calorimetry.

Dilatometry, which is used to measure volume, deals also with closed systems. Changes in heat content have to be obtained from separate calorimetry for full description.

Thermomechanical analysis, which measures length and force, deals also with a closed system. Again, coupling with calorimetry should be a goal for complete thermal analysis.

The isolated system may have no matter or energy exchange and its thermal properties can be analyzed only with difficulty.

The rest of the universe, which lies outside of the system investigated, is called the surroundings. Since the main concern is with the system, all quantities added to the system are positive, all quantities lost from the system are negative.

In case the system is not homogeneous, i.e., it can be recognized to contain areas or phases that need a different description, it may be necessary to break the system into subsystems. The subsystem is chosen such that it is in itself homogeneous. Naturally, one has to make sure that the subdivision is not so fine that the molecular structure by itself causes recognizable inhomogeneity. For macromolecules this may cause occasional difficulties, since, for example, in the fringed micellar structure the molecule may go through several crystalline and amorphous areas, each of which represents a different subsystem. Careful bookkeeping on a repeating unit scale is needed in such a case, and sometimes a macroscopic description may become impossible. Assuming calorimetry as an example, an unknown of 100 small crystals would be a system of at least 100 subsystems. Only if all crystals are identical in macroscopic behavior can one expect a clear answer from a single simple experiment.

B. Equilibrium Thermodynamics

The state of a system, delineated as discussed in Section A, can be described in terms of functions of state such as total energy U, temperature T, volume V, pressure P, number of moles N, and mass M. These functions of state can be divided into two classes, extensive and intensive. Extensive functions of state are proportional to the amount of matter (such as U, V, N, and M), and intensive functions of state do not change with changing amount of matter (T, P). In case the system is in equilibrium, the interrelations between the functions of state are given by equilibrium thermodynamics. A detailed description of equilibrium thermodynamics is given in any textbook on physical chemistry (see, for example, Moore, 1972). The basis of equilibrium thermodynamics is given in three experimental laws.*

The first law of thermodynamics describes the fact that energy is conserved in all experiments. In other words, the total energy U of a system in equilibrium is a function of state, and its value is fixed for a given set of variables of state. For a change of U one can write

$$dU = \delta Q + \delta w \tag{7}$$

$$dU = (\partial U/\partial T)_{V,N}\, dT + (\partial U/\partial V)_{T,N}\, dV + (\partial U/\partial N)_{T,V}\, dN + \cdots \tag{8}$$

where δQ and δw are changes in the two forms of energy, heat and work. In case additional variables of state are necessary, Eq. (8) can easily be expanded. The first term of the right-hand side can be evaluated by calorimetry if volume and number of moles are kept constant. Both $\partial U/\partial T$ and dT are available through calorimetry. In case volume and number of moles are to be changed at the same time, dilatometry and thermogravimetry would have to be performed simultaneously with calorimetry to obtain information on dV and dN. If the coefficients $\partial U/\partial V$ and $\partial U/\partial N$ are known, dU could then be calculated. This example illustrates that it is a rather complicated task to change several variables at the same time. In general, experimental conditions are chosen that all but one variable are constant.

The second law of thermodynamics puts restrictions on the proportioning of work and heat during changes of state. For reversible changes δQ and δw become also functions of state, which permits the introduction

---

* In not so precise terms one can describe the three laws by saying (see Moore, 1972, p. vi): (1) The first law says you cannot win; the best you can do is break even. (2) The second law says you can break even only at absolute zero. (3) The third law says you can never reach absolute zero.

of two new functions of state, the entropy, defined at constant temperature by

$$dQ/T \equiv dS \tag{9}$$

and the Helmholtz free energy, defined through Eqs. (7) and (9) as

$$F \equiv U - TS \tag{10}$$

Since both new functions are made up of functions of state, one can write

$$dS = \left(\frac{\partial S}{\partial T}\right) dT + \left(\frac{\partial S}{\partial V}\right) dV + \left(\frac{\partial S}{\partial N}\right) dN \cdots \tag{11}$$

$$dF = \left(\frac{\partial F}{\partial T}\right) dT + \left(\frac{\partial F}{\partial V}\right) dV + \left(\frac{\partial F}{\partial N}\right) dN + \cdots \tag{12}$$

For an isolated system, experience has shown that dS is always zero for an equilibrium processes. For nonequilibrium, spontaneous processes one finds that dS must be positive. Macroscopic processes with a negative entropy change in isolated systems have never been observed. Experimentally, S cannot be determined directly, but calorimetry is able to give a measure of Q, which under proper circumstances can be related to S.

Most well known are the formulations of the second law by Thomson: It is impossible to devise an engine that, working in a cycle, will produce no effect other than the extraction of heat from a reservoir and the performance of an equal amount of work; and by Clausius: It is impossible to devise an engine that, working in a cycle, will produce no effect other than the transfer of heat from a colder to a hotter body.

In both formulations one assumes the engine to be a closed system. Performing one cycle of the process would bring the closed system (engine) back to its original state so that its functions of state suffer no change. The first law provides, then, that the total change in energy must be zero, or the change in work must be equal in magnitude and opposite in sign to the change in heat [Eq. (7)]. This means that at constant temperature work could only be produced from the reservoir if the entropy decreases [Eq. (10)], something that has not been observed (content of the Thomson formulation). Similarly, for the Clausius formulation the first law would require, since no work is involved, that the changes in heat of the hot and cold body must be equal in absolute magnitude and opposite in sign. The magnitude of the entropy change would, however, be smaller in magnitude at the higher temperature, as indicated by Eq. (9). Going from a lower temperature T_1 to a higher temperature T_2 would then again require a negative change in entropy

$$\Delta S = |Q|/T_2 - |Q|/T_1 < 0 \tag{13}$$

which has never been observed.

More convenient than the entropy statement

$$dS \geq 0 \tag{14}$$

which holds for an isolated system (or open or closed systems including surroundings), is the statement

$$dF \leq 0 \tag{15}$$

which holds also for a closed system. For both Eqs. (14) and (15), the equal sign applies for equilibrium changes of state, and the unequal sign, for permitted irreversible changes (spontaneous changes). For the isolated system the equivalence of Eqs. (14) and (15) is obvious from Eqs. (7) and (10) (since $dU = 0$). For a closed system, flux of energy in or out of the system would change the total energy exactly by the amount of the flux, so that dF for the system is unaffected according to Eq. (10).

The third law of thermodynamics is easy to understand if one couples the macroscopic definition of entropy [Eq. (9)] with the statistical interpretation through the Boltzmann equation

$$S = k \ln W \tag{16}$$

where W is the number of ways a certain system can be arranged (thermodynamic probability) and k is the Boltzmann constant. One can state the third law of thermodynamics (which was first proposed by Nernst) as follows: "If the entropy of each element in some crystalline state be taken as zero at the absolute zero of temperature, every substance has a finite positive entropy; but at the absolute zero of temperature the entropy may become zero, and does so become in case of perfect crystalline substances" (Lewis and Randall, 1923). It is obvious from Eq. (16) that in a perfect crystal (at absolute zero) there is only one way to arrange the atoms and thus S is zero. With this statement one has fixed an absolute value of the entropy and it would remain only for us to prove that Eq. (16) is indeed a representation of the entropy, a fact that can be looked up in any text of statistical thermodynamics (see, for example, Dole, 1954).

The three laws of thermodynamics form the basis for all thermal analysis and will be made use of frequently.

C. IRREVERSIBLE THERMODYNAMICS

Most of the systems to be studied by thermal analysis, particularly those of macromolecules in the solid state, are not in equilibrium. It is thus of importance to learn how to handle the thermodynamics of irreversible processes (see, for example, Prigogine, 1967, or de Groot and Mazur, 1962).

Some of the functions of state are explicitly defined only for the state of equilibrium. This applies particularly to functions of state like entropy, free energy, temperature, and pressure. On the other hand, functions of state like mass, volume, and total energy, are indifferent to the state of equilibrium. To use equilibrium functions of state for the description of nonequilibrium systems, one postulates that these systems can be subdivided into a sufficient number of subsystems, each of which is in local equilibrium at a given time. The extensive functions of state are then just the sum of the functions of state of the subsystems. Naturally, such description holds only as long as changes do not occur so quickly that measurements cannot establish the various thermodynamic functions in the subsystems.

An overall system, which varies throughout in temperature and pressure, for example, is thus subdivided into a sufficient number of subsystems so that each is of constant temperature and pressure. As the system approaches equilibrium, the fate of each subsystem is followed by observing the changes as a function of time. The changes in the extensive functions of state, such as entropy, are divided into flux terms d_eS/dt and production terms d_iS/dt. The flux accounts for all changes due to flow across the subsystem boundary, whereas the production term accounts for changes within the subsystem. For the overall changes in entropy we can thus write

$$\frac{dS}{dt} = \frac{d_eS}{dt} + \frac{d_iS}{dt} \tag{17}$$

The entropy flux can be written, since the subsystem is assumed to be at constant temperature and pressure, as follows:

$$\frac{d_eS}{dt} = \frac{1}{T}\frac{dQ}{dt} + \sum S_i^* \frac{d_eN_i}{dt} \tag{18}$$

where S_i^* is the molar change in entropy due to flux of matter from the surroundings (open system). Both quantities dQ/dt and the various d_eN_i/dt can be measured by thermal analysis (calorimetry and thermogravimetry). The entropy production term, in contrast, cannot be measured directly.

Since all flux is evaluated separately, the remaining production terms can be thought of as belonging to an isolated system. The thermodynamic conclusions about isolated systems of the last section apply thus fully to the production terms. Because of Eq. (14), we can write

$$d_iS \geq 0 \tag{19}$$

Thus production can only be positive or zero. Similarly, for energy production, the first law of thermodynamics (conservation of energy) requires for a one-component system that $d_iU = 0$. Finally, the conservation of mass requires for any subsystem that $\Sigma\, d_iM = 0$.

The driving force for nonequilibrium processes can be derived from the overall change of the free energy of the system. In general, one distinguishes between transport phenomena (which involve nonequilibria of the intensive variables like temperature, pressure, and composition and are balanced by the transport of heat, momentum, and mass) and relaxation phenomena (where successive nonequilibrium states are produced by the change of internal degrees of freedom of the system). Taking a relaxation process as an example, one can write (if T, V, and N are constant) at any time t

$$\frac{dF}{dt} = \left(\frac{\partial F}{\partial \zeta}\right)\frac{d\zeta}{dt} \tag{20}$$

where ζ is the internal variable—for example, a degree of order of the system. In case of several internal variables, Eq. (20) has to be expanded into a sum. The coefficient $\partial F/\partial \zeta$ is the driving force of the process that vanishes when equilibrium has been reached. At distances not too far from equilibrium, one may assume the following linear connection between the driving force and the change in internal variable and expand $\partial F/\partial \zeta$ into a Taylor series, dropping higher terms of the expansion

$$\frac{d\zeta}{dt} = -L\left(\frac{\partial F}{\partial \zeta}\right) = -L\left(\frac{\partial^2 F}{\partial \zeta^2}\right)_e [\zeta(t) - \zeta(e)] \tag{21}$$

In Eq. (21) L is a constant, the so-called phenomenological coefficient, $\zeta(t)$ is the value of ζ at time t, and $\zeta(e)$ is the equilibrium value. The solution of Eq. (21) is

$$\zeta(t) = [\zeta(0) - \zeta(e)]e^{-t/\tau} + \zeta(e) \tag{22}$$

where the time constant τ is called the relaxation time and is given by

$$1/\tau = L(\partial^2 F/\partial \zeta^2)_e \tag{23}$$

Equations (17) and (21) give the tools necessary to apply thermodynamics to nonequilibrium and time-dependent processes, which are often encountered in thermal analysis.

D. KINETICS

The time dependence of macroscopic changes or their kinetics could be described through irreversible thermodynamics, as is shown, for example, by Eqs. (22) and (23). However, one often makes use of a microscopic model to derive the macroscopic time dependence. One must distinguish various kinetic models. Most common are those involved in chemical reactions (see, for example, Frost and Pearson, 1953).

Reaction kinetics is based on the model that the reaction partners (A and B) have to come together in order to react to the products (C and D). The kinetic equation of two reactants A and B that form products C and D is written

$$A + B \rightarrow C + D \tag{24}$$

with the kinetic rate equation based on this mechanism

$$-d[A]/dt = -d[B]/dt = d[C]/dt = d[D]/dt = k[A][B] \tag{25}$$

where the bracketed terms represent the concentrations in suitable units and k is the specific rate constant.

From this basic and simple equation one can generalize and write

$$-d[A]/dt = k[A]^n \tag{26}$$

where n is called the order of the reaction. In Eq. (24) the rate would be first order in [A], but also first order in [B], so that the overall order of the rate as given in Eq. (25) is second. The simplified Eq. (26) can only be used if either B is held constant, by, for example, using a large excess of B or by making $[A] = [B]$. In the first case $k[B]$ of Eq. (25) is a constant and takes the place of k in Eq. (26) with $n = 1$. In the second case, $n = 2$. Equation (26) can be integrated for various values of n:

for $n = 1$

$$[A] = [A]_0 e^{-kt} \tag{27}$$

for $n = 2$

$$1/[A] = (1/[A]_0) - kt \tag{28}$$

for $n = 3$

$$1/[A]^2 = (1/[A]_0)^2 - 2kt \tag{29}$$

The concentration of A at the beginning of the reaction that is conveniently chosen to occur at time zero is given by $[A]_0$. Complications arise if several reactions are coupled or if chain reactions occur. Detailed mechanisms have been worked out for many possible kinetic paths (Frost and Pearson, 1953). Thermal analysis can follow a reaction either by mass determination, if one or more of the products or reactants is volatile, or by following the heat of reaction, which is directly proportional to the number of molecules which have reacted.

For the description of the kinetics it has been assumed that the reactions are carried out isothermally. Changing the temperature will change the rate constant k. More or less empirically, and in parallel with the temperature dependence of the equilibrium constant for chemical reactions, Arrhenius found that

$$d \ln k/dT = E_a/(RT^2)$$ (30)

or

$$k = Ae^{-E_a/(RT)}$$ (31)

where E_a is the Arrhenius activation energy and A the so-called preexponential factor. In other words, a plot of the logarithm of the rate constant as a function of $1/T$ should give a straight line with E_a/R as the slope and A as the intercept.

Knowing A and E_a in the range of interest permits not only the calculation of the reaction kinetics at any temperature, but also the calculation of nonisothermal kinetics, as would occur, for example, on reaction during a constant heating rate experiment of calorimetry or thermogravimetry. Equation (31) is for this purpose inserted into the appropriate rate equation, Eq. (26), giving

$$\frac{-d[A]}{dt} = A \exp\left[-\frac{E_a}{RT}\right] [A]^n$$ (32)

Freeman and Carroll (1958) analyze Eq. (32), for example, by taking the logarithm of both sides and differentiating with respect to ln[A] to obtain

$$\frac{d \ln(-d[A]/dt)}{d \ln[A]} = -\frac{E_a}{R} \frac{d(1/T)}{d \ln[A]} + n$$ (33)

A plot of the ratio of the change in logarithm of the rate to the logarithm of the concentration versus the ratio of the change in $(1/T)$ to the logarithm of concentration yields n as an intercept and E_a/R as slope. The concentration could be measured directly by thermogravimetry through weight loss or by calorimetry through the proportionality with the heat of reaction. Other analyses of Eq. (32) have been derived by Coats and Redfern (1964) and Archar *et al.* (1966); see also Sharp and Wentworth (1969). All analysis methods rely necessarily on the basic premise that the reaction actually follows over the whole temperature range the kinetics described by Eq. (26). This is rarely true. Additional instrumentation problems such as sample environment and diffusion of reactants and products make the application of Eq. (33) without isothermal check at several temperatures questionable. For some detailed criticism see, for example, Garn (1972), Johnson and Gallagher (1972), and Ball and Casson (1978).

A better understanding of the rate constant that is similar in nature to the reciprocal of the relaxation time introduced in Eq. (22) is possible through the introduction of more detailed microscopic models. The collision theory and the absolute reaction rate theory are the best known of such models.

The collision theory is applicable to bimolecular collisions. The rate constant is derived by calculating the rate of collisions of molecules of sufficient energy to overcome the hindrance to rearrange the electronic structure to that of the products. The total collision number in gases expressed as number of collisions of a given number of molecules A with those of B per second is given by

$$Z_{AB} = d^2_{AB}\pi \bar{c}_{AB}[A][B] \tag{34}$$

where d_{AB} is the collision diameter of the two colliding molecules, approximated by the sum of the two radii, and $\bar{c}_{AB}$ is the average speed of molecules A relative to molecules B, which is given by $[8RT/\pi\mu]^{1/2}$ [with μ representing the reduced mass $M_A M_B/(M_A + M_B)$]. The concentrations of A and B must in this case be given in number of molecules per unit volume. Z_{AB} is the number of all collisions, regardless of whether they lead to a successful reaction or not. To count only successful collisions, one must multiply Eq. (34) with a factor $e^{-E_a/RT}$ to obtain the number of sufficiently energetic collisions to lead to reaction. This factor can easily be derived from the velocity distribution function. Furthermore, one would have to exclude collisions that are not properly oriented. Glancing collisions or collisions of larger molecules that do not hit the proper reactive portion of the molecules must be excluded. This is done by a steric factor p. Comparing now Eq. (34) with Eq. (25) leads to an expression for k:

$$k = (d^2_{AB}\pi) \left(\frac{8RT}{\pi\mu}\right)^{1/2} e^{-E/(RT)}p \tag{35}$$

A comparison with Eq. (31) shows that considerable detail for the preexponential factor is provided, but the steric factor p, which takes values of the magnitude 0.01 on comparison with experiments, is more an empirical fitting factor than derivable from the actual steric collision configuration. For other than bimolecular reaction the detailed kinetic steps must be evaluated. Applying Eq. (35) to reactions in the liquid state leads to even more qualitative results. For surface reactions, appropriate modifications must be made in the collision parameters.

The absolute reaction rate theory makes use of a model of statistical thermodynamics. The actual potential energy contour map of a collision is approximated by the plot shown in Fig. 10. The process is described by the kinetic equation

$$A + B \rightarrow [AB^\ddagger] \rightarrow C + D \tag{36}$$

In comparison with Eq. (24) the intermediate $[AB^\ddagger]$, the activated com-

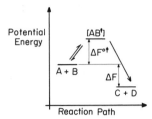

Fig. 10 Potential energy diagram for a chemical reaction as shown in Eq. (36). The free energy of reaction ΔF is $F_C^\circ + F_D^\circ - F_A^\circ - F_B^\circ$. For the free energy of activation $\Delta F^{\circ\ddagger}$ see Eq. (39).

plex, is written as a separate, identifiable entity. The key assumption of the absolute reaction rate theory is that equilibrium exists between reactants and activated complex. This assumption permits the introduction of an equilibrium constant $K^\ddagger$, which can be calculated, at least approximately, from statistical thermodynamics (see, for example, Dole, 1954; Frost and Pearson, 1953). The kinetic rate equation is, according to Eq. (25),

$$-d[A]/dt = k[A][B] = [AB^\ddagger] \times \nu^\ddagger \tag{37}$$

where $\nu^\ddagger$ is the frequency of passage over the barrier of Fig. 10. The rate constant can be expressed as

$$k = \nu^\ddagger K^\ddagger = \left(\frac{kT}{h}\right) \frac{z^\ddagger}{z_A z_B} \exp\left(-E_0/RT\right) \kappa \tag{38}$$

where the far right-hand side is obtained through standard statistical thermodynamic reasoning. The first factor $kT/h = \nu^\ddagger$ is a measure of the time needed to cross the potential energy barrier. The second factor $z^\ddagger/z_A z_B$ is the ratio of appropriately chosen partition functions. The exponent E_0 is a measure of the potential energy barrier of Fig. 10. The final quantity κ is the transmission coefficient, which represents a probability of occurrence of forward or reverse reaction. Frequently its value is between 0.5 and 1.0. Since $K^\ddagger$ is an equilibrium constant, one can express it in terms of the free energy $\Delta F^{\circ\ddagger}$, the Helmholtz free energy of activation

$$\Delta F_{AB}^{\circ\ddagger} = F^{\circ\ddagger} - F_A^\circ - F_B^\circ \tag{39a}$$

Furthermore, using the equation of the type of Eq. (10), one can express the free energy of activation in terms of an energy of activation and an entropy of activation

$$\Delta F^{\circ\ddagger} = \Delta U^{\circ\ddagger} - T\Delta S^{\circ\ddagger} \tag{39b}$$

and finally, the rate constant is given from the left half of Eq. (38) as

$$k = (kT/h)\exp(\Delta S^{\circ\ddagger}/R)\exp(-\Delta U^{\circ\ddagger}/RT) \tag{40}$$

To relate Eq. (40) to the Arrhenius equation, Eq. (31), and the experimentally available activation energy E_a, one can perform the differentiation suggested by Eq. (30) and equate the result with the right-hand side of Eq. (30).

$$d \ln k/dT = (RT + \Delta U^{\circ\ddagger})/RT^2 = E_a/RT^2 \qquad (41)$$

The preexponential factor A of Eq. (31) is, as can be seen by comparison of Eqs. (31) and (40), related mainly to the activation entropy (for limitations see also Sections V.A.3 and V.D).

A completely different kinetic problem is the description of a cooperative process such as crystallization. The growth of a crystal occurs in at least two stages, nucleation and growth (see Wunderlich, 1976). The nucleation kinetics is in its limits athermal or thermal. In the first case, a fixed number of nuclei start growing at the beginning of the experiment with no additional nuclei forming at a later time. In the second case, new nuclei are forming throughout the crystallization experiment. Athermal nucleation may be observed when heterogeneous particles initiate crystal growth. Thermal nucleation is always found on homogeneous nucleation, and often one assumes that the nucleation rate is a constant. Most frequently, nucleation is, however, intermediate between the two limits. For this and other reasons it is certainly important for any experiment to see how far the nucleation rate is constant. Because of their small heat effect, nucleation rates cannot be determined by thermal analysis. Optical microscopy is most suited for counting the number of appearing new crystals as a function of time. Even a simple check of the resultant crystal morphology is often sufficient to evaluate the nucleation kinetics. If all crystals are of the same size, nucelation was certainly athermal. If a continuous size distribution is present, nucleation was thermal. The spherulite growth shown in Fig. 9, for example, must be described as athermal. Detailed kinetic expressions for nucleation based on the transition state theory are available (Wunderlich, 1976). The activation barriers are caused by the surface free energy of the nucleus and the hindered segmental motion across the phase boundary. The former becomes very large close to the melting temperature. The latter increases rapidly in the vicinity of the glass transition temperature. Nucleation can thus occur only somewhere between the melting and the glass transition temperature.

The linear growth rate of a crystal (i.e., the rate of advance of a crystal surface normal to its surface) is constant as long as the crystal growth is not transport controlled (transport of crystallizing matter to the crystal surface or heat from the crystal surface). Depending on the type of growth, the temperature dependence is linear or quadratic in supercooling (distance in temperature from the equilibrium melting temperature), or, as always in linear macromolecules, it can have an exponential dependence. Again, experimental observation is done by optical microscopy, simply by observing the actual progress of an identified crystal surface. Different types of surfaces on the same crystal have normally different growth rates, giving rise to the characteristic crystal habits. Isometric crystals are found if linear crystal growth rates are similar for all growing faces. Fibrillar crystals need particularly fast growth in one direction, which for macromolecules is, under proper conditions, the molecular chain direction (see Fig. 5). Lamellar crystals result when growth in one direction is retarded, as in the folded-chain lamellae of macromolecules (see Fig. 6). Detailed expressions for linear crystal growth rates are available (see, for example, Wunderlich, 1976).

Thermal analysis is capable of following the overall crystallization rate through measurement of the volume change on crystallization (dilatometry) or the evolution of the heat of crystallization (calorimetry). Quite frequently the change in crystallinity can be expressed in terms of the so-called Avrami equation

$$v^c = 1 - \exp[-Kt^n] \qquad (42)$$

where v^c is the volume fraction crystallinity; K, a temperature-dependent constant that contains nucleation, linear crystal growth rate, and crystal geometry information. The exponent of time n is most easily extracted from experimental data by double logarithmic plotting. Typical curves of the change in crystallinity are shown in Fig. 11. A list of typical exponents

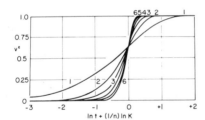

Fig. 11 Increase in volume crystallinity as given by the Avrami equation for different values of n. The different values of n are listed next to the curves. See Eq. (42), $v^c = (\rho - \rho_1)/(\rho_c - \rho_1)$ with ρ, ρ_a and ρ_1 representing the sample, completely liquid, and crystalline sample densities.

for different nucleation and crystal growth as well as crystal morphology is given in Table II. Unfortunately, the exponents are not unique. In addition, complications such as volume change on crystallization, incomplete crystallization, change in mechanisms during crystallization, and two-stage crystallizations are not considered. All of these do occur, however, to various degrees in crystallization of linear macromolecules, weakening the interpretation of the macroscopic analysis in terms of microscopic mechanisms without detailed microscopic analysis.

A typical derivation of the Avrami equation is given next for the case of athermal nucleation and constant linear crystal growth rate. For simplicity, spherical crystals are chosen. Each crystal has then at time t the volume

$$v_i = 4\pi c^3 t^3 / 3 \qquad (43)$$

TABLE II

EXPONENTS OF TIME IN THE AVRAMI EQUATION

Type of crystallization	Nucleation	n
Fibrillar	Athermal	≤ 1
Fibrillar, branching	Athermal	1, increasing with time
Spherical, diffusion controlled	Athermal	$\frac{3}{2}$
Fibrillar	Thermal	≤ 2
Circular lamellar	Athermal	≤ 2
Fibrillar, branching	Thermal	2, increasing with time
Truncated sphere	Athermal	2–3
Spherical, diffusion controlled	Thermal	$\frac{5}{2}$
Spherical	Athermal	3
Truncated sphere	Thermal	3–4
Spherical	Thermal	4
Solid sheaf	Athermal	≥ 5
Solid sheaf	Thermal	≥ 6

as long as crystals are not hindered on growth by impingement with other crystals. The linear growth rate of the crystal is given the symbol c in Eq. (43). If N is the number of nuclei per unit volume and V_0 is the total experimental volume at the beginning of the crystallization, the total crystallized volume in case of no impingement would be (free growth approximation)

$$\sum V_i = \frac{4\pi N V_0 c^3 t^3}{3} \tag{44}$$

The condition of constant mass during the crystallization leads to the following relationship

$$V_0 \rho_l = V v^c \rho_c + V(1 - v^c)\rho_l \tag{45}$$

out of which the volume of the partially crystallized sample V can be calculated

$$V = V_0 \rho_l / [v^c \rho_c + (1 - v^c)\rho_l] \tag{46}$$

where ρ_l and ρ_c are the densities of the liquid and crystalline samples, respectively. The chance X of any nucleus to reach through growth a randomly chosen point P in the sample in time t is

$$X = \sum \frac{V_i}{V} \tag{47}$$

i.e., after half crystallization ($\sum V_i = 0.5V$), X is $\frac{1}{2}$. The chance that the point P is not reached by any crystal is then P_0 or the fraction of noncrystallized material $(1 - v^c)$. The probability situation is, in turn, just the condition described by the Poisson distribution

$$P_m = \exp(-X)(X^m/m!) \tag{48}$$

where X is the expected value for an event to occur and m is the number of times the event should occur at point P with probability P_m up to time t. For our experiment $m = 0$ so that

$$P_0 = 1 - v^c = \exp(-X) \qquad (49)$$

or

$$v^c = 1 - \exp(-\Sigma \, V_i/V) \qquad (50)$$

For changing volume V with time, no explicit expression for v^c can be obtained, which is one of the major problems of the evaluation of Eq. (50). Assuming, however, no volume change on crystallization, V is during the whole crystallization equal to v_0 and an equation identical in form to Eq. (42) results

$$v^c = 1 - \exp(-4\pi N c^3 t^3/3) \qquad (51)$$

This derivation makes it clear that a kinetic analysis can be successful only with a proved microscopic model. In most cases the thermal analysis can give only a guide to which type of kinetic process may be occurring and exclude obviously wrong kinetic models. For full analysis, as far as it is possible at all, additional experiments with more direct microscopic information are needed. For macromolecules crystallized from the melt, nucleation rate, linear crystal growth rate, and crystal morphology are needed for the interpretation of Eq. (42). The degree of crystallization and its change during or after crystallization through reorganization or secondary crystallization can then be derived by thermal analysis (for more details, see Wunderlich, 1976).

E. Functions of State Needed for Thermal Analysis

The needed functions of state for thermal analysis are derived in this section and their units are given in the Système International d'unités (SI units; see, for example, Page and Vigoureux, 1971, or any modern text on physical chemistry for a detailed history of units; see also Klein, 1974). The order of techniques chosen is such that each subsequent technique makes use of part or all of the prior described functions of state.

1. Thermometry

Thermometry is the simplest form of thermal analysis. It consists of a plain measurement of temperature. The most common applications are the determination of a single temperature in the study of the phase transition, or the recording of temperature as a function of time in form of heating or cooling curves (usually to study phase diagrams).

Temperature is a basic thermodynamic function. Its meaning goes back to the change in sensations from pain to comfort to pain in going from cold to warm to hot (physiological temperature scale). Modern quantization of temperature determination was possible since the invention of the capillary glass thermometers in the seventeenth century (see also Knowles Middleton, 1966). The measurement in a liquid in glass thermometer rests with a quantitative volume determination. By observing the liquid in a constant diameter capillary communicating with a bulb, the volume change measurement is reduced to a length measurement l. Calibration is based simply on an arbitrary linear relationship.

$$\theta = a + bl \tag{52}$$

To establish temperature scales, one must agree on two fixed temperatures and on the thermometer liquid. Mercury was universally accepted as the most convenient thermometer liquid. The two fixed temperatures chosen, varied, however, widely. Most well known are the scales of Fahrenheit and Celsius.

Progress to an absolute temperature (kelvin) that is independent of the material chosen as thermometric liquid and unlimited in range of measurement* came after development of thermodynamics. The definition of temperature is coupled with the entropy definition [Eq. (9)] and based on the second law of thermodynamics (see Section II.B). One can state without specifying a temperature scale that a system going through a reversible cycle between two temperatures θ_2 and θ_1 exchanges the heats Q_2 and Q_1 at θ_2 and θ_1 according to the relationship

$$|Q_2|/|Q_1| = \phi(\theta_2)/\phi(\theta_1) \tag{53}$$

where the $\phi(\theta)$ are unspecified functions of temperature, free to be chosen. The simplest choice is

$$\phi(\theta) \equiv T \tag{54}$$

This choice can also be written

$$\partial U/\partial S \equiv T \quad \text{(at const } V \text{ and } N) \tag{55}$$

[because at constant volume and composition no work can be done and for a reversible process $dU = dQ_{rev}$, see Eq. (7)].

The units of the kelvin scale are fixed by specifying one temperature, the triple-point temperature of water (equilibrium between gaseous, liquid, and solid water). Using the unit spacing of the celsius scale, the triple-point temperature is 273.16 K. Since the triple point of water is in the

---

* The mercury in glass thermometer is limited by the freezing of Hg at $-38.87°C$ and its boiling temperature $356.58°C$ (at atmospheric pressure).

celsius scale at $+0.01°C$, the interrelation between celsius and kelvin scales is

$$\text{temperature (in K)} = \text{temperature (in °C)} + 273.15 \qquad (56)$$

Fortunately the subdivision of the base interval of the celsius scale, 100 degrees from the freezing point of water to the boiling point of water, is sufficiently linear to agree degree for degree with an absolute thermometer, such as the gas thermometer.* Only at higher temperatures are there deviations of several degrees that must be corrected by calibration of high-temperature mercury in glass thermometers relative to the absolute high temperature, i.e., above 100°C the mercury in glass thermometer has unequal spacing along the glass capillary.

The absolute thermometers such as the gas thermometer, are, however, not the most precise thermometers. For highest precision one makes use of platinum resistance thermometers (13.81–903.89 K) and platinum–platinum/rhodium thermocouples (up to 1337.58 K). For such high precision, which is practically never needed for thermal analysis, a series of fixed points have been internationally agreed on. Their temperatures have been chosen to be as close to the absolute temperature as possible and values are corrected toward the absolute temperature as soon as more accurate measurements become available (International Practical Temperature Scale IPTS (1968).

A continuing series of publication on temperature and its measurement by the American Institute of Physics (1941, 1955, 1962, 1972) and a general description by Sturtevant (1971b) provide further information.

The unit of the measurement of time is the second (s), which has its base in the earth's motion. The unit chosen is the ephemeris second, which represents a fixed fraction of the tropical year 1900. The SI unit is chosen to match the ephemeris second as closely as possible using a Cesium 133 clock, which has the precision of one part in 10^{11}. The astronomical standard is thus supplanted. For all thermal analysis, normal clock accuracy is sufficient.

2. Dilatometry and Thermomechanical Analysis

Dilatometry involves the measurement of volume at constant pressure. Many experimental details on pressure and volume measurement are given by Thomson and Douslin (1971). Thermomechanical analysis requires the measurement of length in a penetration, expansion, or extension mode. Constant and variable force experiments are possible. The

---

* A gas thermometer is based on the ideal gas law $PV = nRT$. It is identical to the thermodynamic temperature but limited to the region of existence of gases and containers, i.e., from a few kelvin to perhaps 1000 K.

variables time and temperature for these experiments were described in Section III.E.1. The new functions of state—pressure, volume, and length—are described in the following discussion. When force is treated as a variable in thermomechanical analysis, the mechanical properties need to be understood as described, for example, by Ferry (1970).

Pressure is defined as the force per unit area. The SI unit is thus the newton per square meter (N/m^2), also called the pascal (Pa). Relative to the older units atmosphere and bar, the pascal is relatively small and has found only slow acceptance. Some useful conversion factors are

To go from	to Pa (or N/m^2) multiply with
atm	101,325.0
bar	100,000 (exact)
mm Hg (torr)	133.3224
dyne/cm²	0.1 (exact)
lb/in.²	6894.76

The most useful unit is perhaps the megapascal (MPa), since it has a reasonable conversion factor to the standard atmosphere (1 atm = 0.101325 MPa). If no further information is given, measurements are usually carried out close to the standard atmosphere pressure (101,325.0 Pa).

The measurement of length has also undergone many changes in units. The SI unit is the meter, based originally on a length of $1/(4 \times 10^7)$ of the earth's circumference. A standard meter stick made of platinum was given this length. Naturally, the standard meter was, as soon as more accurate measurements of the earth's circumference became available, dissociated from its origin. (The present best value of the earth's circumference at the equator is 4.007464×10^7 m.) The SI unit of meter is not even based on the platinum standard meter anymore; it is based on the more precisely measurable 1,650,763.73 wavelengths of the krypton 86 radiation in vacuum on transition between the $2p_{10}$ to $5d_5$ level (orange–red). It still matches as closely as possible the old standard meter, so that no conversions are needed to go from older measurements to new ones.

The volume to be derived from the meter is the cubic meter. Using the cubic meter as the basic volume unit makes for relatively large subdivisions for dilatometry. The density, which used to be given in grams per cubic centimeter, is now, to give the same numerical value, to be given in megagrams per cubic meter (Mg/m^3).

Derived quantities of the volume with respect to temperature and pressure are the thermal expansivity α and the isothermal compressibility β defined by

$$\alpha \equiv (1/V)(\partial V/\partial T) \tag{57}$$

$$\beta \equiv -(1/V)(\partial V/\partial P) \tag{58}$$

The reciprocal of Eq. (58) is the isothermal elasticity coefficient or bulk modulus. Since both volume and pressure are functions of state, one can write (for a one-component closed system, $dN = 0$)

$$dV = (\partial V/\partial T)_P \, dT + (\partial V/\partial P)_T \, dP \tag{59}$$

$$dP = (\partial P/\partial T)_V \, dT + (\partial P/\partial V)_T \, dV \tag{60}$$

By rewriting Eq. (59)

$$dP = dV/(\partial V/\partial P)_T - [(\partial V/\partial T)_P/(\partial V/\partial P)_T] \, dT \tag{61}$$

and comparing coefficients with Eq. (60), it is obvious that there is a connection between Eqs. (59) and (60), which can be expressed as

$$(\partial P/\partial T)_V = -(\partial V/\partial T)_P/(\partial V/\partial P)_T = \alpha/\beta \tag{62}$$

Referred to the units of pressure, Eq. (62) is called the thermal pressure coefficient

$$\gamma \equiv (1/P)(\partial P/\partial T)_V \tag{63}$$

Analogous relationships exist for the length l of a sample with respect to temperature and tensile force f (for a one-component closed system at constant pressure $dN = 0$, $dP = 0$)

$$dl = (\partial l/\partial T)_f \, dT + (\partial l/\partial f)_T \, df \tag{64}$$

$$df = (\partial f/\partial T)_l \, dT + (\partial f/\partial l)_T \, dl \tag{65}$$

The linear thermal expansivity

$$\alpha_L \equiv (1/l)(\partial l/\partial T)_f \tag{66}$$

is analogous to Eq. (57), whereas the isothermal extensivity is usually written as the linear isothermal elasticity coefficient, or Young's modulus E:

$$E = l(\partial f/\partial l)/A \tag{67}$$

where A is the cross-sectional area so that $\partial f/A$ represents the change in stress and $\partial l/l$ the change in strain.

The linear and volume expansivity can be related. Most easily, this is done for a material that expands isotropically, i.e., the linear thermal expansivity is equal in all three directions of space. Assuming a cube of side length l and an expansivity that is temperature independent, the volume at temperature T can be calculated, using Eq. (66), to be

$$V(T) = l^3 = l_0^3[1 + \alpha_L(T - T_0)]^3 \tag{68}$$

or

$$V(T) = l_0^3[1 + 3\alpha_L(T - T_0) + 3\alpha_L^2(T - T_0)^2 + \alpha_L^3(T - T_0)^3] \tag{69}$$

Since the expansivity of liquids and solids is relatively small, terms higher than first power in α_L may be neglected, resulting in the simple equation

$$V(T) = l_0^3[1 + 3\alpha_L(T - T_0)] \tag{70}$$

which through Eq. (57) can also be expressed in terms of the volume expansivity

$$V(T) = V(T_0)[1 + \alpha(T - T_0)] \tag{71}$$

or within the chosen assumptions

$$\alpha = 3\alpha_L \tag{72}$$

Similarly, the bulk modulus $1/\beta$ [Eq. (58)] and Young's modulus E [Eq. (67)] are related through the equation (see, for example, Ferry, 1970)

$$E = 3(1/\beta)(1 - 2\sigma) \tag{73}$$

where σ is the Poisson ratio, which indicates the negative ratio of the linear contraction relative to the extension in a tensile extension experiment. The numerical value of σ varies from zero (in case there is no lateral contraction on extension) to 0.5 (in case there is no volume change on extension). For most crystals and glasses Poisson's ratio is between 0.2 and 0.35. For rubbery materials σ approaches 0.5, so that Young's modulus is much smaller than the bulk modulus, which is easily rationalized from the change in shape of macromolecules on extension (see Section II.D).

3. Calorimetry

The central function of state of calorimetry without chemical or phase changes in the heat capacity. It is the heat absorbed by a system (which is assumed to be closed and constant in composition) on raising the temperature 1 K

$$C = Q/\Delta T \tag{74}$$

For a more precise correlation with the thermodynamic functions of Section III.B we notice that a closed system with constant composition can only exchange heat and volume work $P\,dV$ so that both right-hand terms of Eq. (7) become functions of state. (The negative sign of $P\,dV$ results from a loss of work for the system on expansion.)

$$dU = dQ - P\,dV \tag{75}$$

Combining now Eqs. (75) and (8), one obtains

$$dQ = (\partial U/\partial T)_{V,N}\,dT + [(\partial U/\partial V)_{T,N} + P]\,dV \tag{76}$$

Measuring the heat capacity at constant volume leads to a simple correlation of the measured quantity dQ/dT and the change in internal energy

$$dQ/dT = (\partial U/\partial T)_{V,N} \equiv C_V$$ (77)

Unfortunately, this type of measurement is quite impossible for condensed matter because of a rather large thermal pressure coefficient [Eq. (63)]. Liquid mercury, for example, has a value of γ of 46.4 K^{-1}, meaning that for each degree of temperature rise the pressure must be increased by about 4.7 MPa (46 atm) to keep the volume constant.

Description of thermal analysis is greatly simplified by derivation of thermodynamic functions that can easily be represented at constant (atmospheric) pressure. The internal energy U is not such a function. Expressing U as a function of the independent variables P and T, instead of V and T (assuming again a closed system of constant composition), one gets

$$dU = (\partial U/\partial T)_{P,N}\, dT + (\partial U/\partial P)_{T,N}\, dP$$ (78)

which is analogous to Eq. (8). The combination of Eqs. (78) and (75) that is needed to get to an expression for dQ must now, however, take into account that dV is dependent on changes in pressure as expressed by Eq. (59). Combining all these leads to

$$dQ = \left[\left(\frac{\partial U}{\partial T}\right)_{P,N} + P\left(\frac{\partial V}{\partial T}\right)_{P,N}\right] dT + \left[\left(\frac{\partial U}{\partial P}\right)_{T,N} + P\left(\frac{\partial V}{\partial P}\right)_{T,N}\right] dP \quad (79)$$

and

$$\frac{dQ}{dT} = \left(\frac{\partial U}{\partial T}\right)_{P,N} + P\left(\frac{\partial V}{\partial T}\right)_{P,N}$$ (80)

Introducing a new function of state, the enthalpy

$$H \equiv U + PV$$ (81)

simplifies matters considerably. Since H is made up of other functions of state, it must itself also be a function of state. Enthalpy is the internal energy to which is added the work needed for the expansion at the experimental pressure from volume zero to the actual volume of the system. For the condensed state, the system volume is relatively small and, accordingly, the difference between H and U is small so that H is still an approximate measure of U. The simplification becomes clear on insertion of Eq. (81) into Eq. (80)

$$dQ/dT = (\partial H/\partial T)_{P,N} - (\partial PV/\partial T)_{P,N} + P(\partial V/\partial T)_{P,N}$$ (82)

or

$$dQ/dT = (\partial H/\partial T)_{P,N} \equiv C_P$$ (83)

The heat capacity at constant pressure C_P is the calorimetrically measured quantity. The connection between C_P and C_V is obvious on substitution of Eqs. (77) and (59) into Eq. (76)

$$dQ = \left\{ C_V + \left[\left(\frac{\partial U}{\partial V} \right)_{T,N} + P \right] \left(\frac{\partial V}{\partial T} \right)_{P,N} \right\} dT$$

$$+ \left[\left(\frac{\partial U}{\partial V} \right)_{T,N} + P \right] \left(\frac{\partial V}{\partial P} \right)_{T,N} dP \qquad (84)$$

$$C_P \equiv \frac{dQ}{dT} = C_V + \left[\left(\frac{\partial U}{\partial V} \right)_{T,N} + P \right] \left(\frac{\partial V}{\partial T} \right)_{P,N} \qquad (85)$$

The difference between the two heat capacities arises thus from two contributions: the volume dependence of the internal energy and the work performed by volume expansion. For gases the second outweighs by far the first ($\partial U / \partial V \approx 0$); for condensed systems the first contribution outweighs the second. For solids at low temperatures both portions may be neglected and $C_P \approx C_V$.

Based on the enthalpy definition of Eq. (81) all prior thermodynamic relationships can be rewritten. Instead of the Helmholtz free energy [Eq. (10)] we introduce the free enthalpy (or Gibbs free energy) G

$$G \equiv H - TS \qquad \text{[see Eq. (10)]} \qquad (86)$$

$$dG = \left(\frac{\partial G}{\partial T} \right)_{P,N} dT + \left(\frac{\partial G}{\partial P} \right)_{T,N} dP + \left(\frac{\partial G}{\partial N} \right)_{P,T} dN + \cdots$$

$$\text{[see Eq. (12)]} \qquad (87)$$

For a closed system the second law requires

$$dG \leq 0 \qquad \text{[see Eq. (15)]} \qquad (88)$$

and the irreversible processes described by Eqs. (20)–(23) can analogously be written in terms of G instead of F. Finally, the equilibrium between activated state and reactants during a chemical reaction can be expressed as

$$\Delta Gq^{\ddagger} = -RT \ln K^{\ddagger} \qquad \text{[see Eq. (38)]} \qquad (89)$$

In a development similar to Eqs. (39) to (40) one finds now that

$$k = (kT/h) \exp(\Delta S^{\circ \ddagger}/R) \exp(-\Delta H^{\circ \ddagger}/RT) \qquad \text{[see Eq. (40)]} \qquad (90)$$

With an experimental heat capacity at constant pressure as a function of temperature, one can calculate the enthalpy at any temperature using Eq. (83).

$$H(T) = H(T_0) + \int_{T_0}^{T} C_P \, dT \qquad (91)$$

Furthermore, since $(\partial G/\partial T)_{P,N}$ can be shown to be equal to $-S,$* Eq. (86) can be written

$$H = G - T(\partial G/\partial T)_{P,N} \tag{92}$$

and

$$C_P = (\partial H/\partial T)_{P,N} = -T(\partial^2 G/\partial T^2)_{P,N} \tag{93}$$

or

$$C_P = T(\partial S/\partial T)_{P,N} \tag{94}$$

Measurement of heat capacity as a function of temperature thus permits one also, through Eqs. (93) and (94), to evaluate entropy and free enthalpy. Commonly one calculates

$$S(T) = S(T_0) + \int_{T_0}^{T} \frac{C_P}{T} \, dT \tag{95}$$

and uses Eq. (86) for the free enthalpy calculation

$$G(T) = H(T) - TS(T) \tag{96}$$

A typical plot of H, TS, and G is shown in Fig. 12.

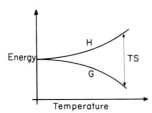

Fig. 12   Schematic drawing of the change of H, TS, and G as a function of temperature.

To describe $U(T)$ and $F(T)$ from the measured heat capacities at constant pressure, use must be made of Eq. (85) to calculate C_V first. Additional thermodynamic relationships permit the change of Eq. (85) to more accessible experimental quantities. Similar to the partial derivatives of the free enthalpy (see footnote, below) one can show that $(\partial F/\partial T)_{V,N} = -S$ and $(\partial F/\partial V)_{T,N} = -P$. From Eq. (10) one can thus derive that

$$\left(\frac{\partial U}{\partial V}\right)_{T,N} = \left(\frac{\partial F}{\partial V}\right)_{T,N} - T\left(\frac{\partial (\partial F/\partial T)_{V,N}}{\partial V}\right)_{T,N} \tag{97}$$

* From Eq. (86): $dG = dH - T \, dS - S \, dT$; from Eq. (81) $dH = dU + P \, dV + V \, dP$. For dU, one can make use of Eqs. (9) and (75) to derive that $dU = T \, dS - P \, dV$. Combining all these expressions leads to $dG = -S \, dT + V \, dP$, from which the partial derivatives of G with respect to T and P can be found by comparison of coefficients with Eq. (87).

and because $\partial^2 F/(\partial V\,\partial T) = \partial^2 F/(\partial T\,\partial V)$,

$$\left(\frac{\partial U}{\partial V}\right)_{T,N} = -P + T\left(\frac{\partial P}{\partial T}\right)_{V,N} \tag{98}$$

Substituting Eq. (98) into Eq. (85) yields

$$C_P = C_V + T\left(\frac{\partial P}{\partial T}\right)_{V,N}\left(\frac{\partial V}{\partial T}\right)_{P,N} \tag{99}$$

or with Eqs. (57), (58), and (62)

$$C_P = C_V + TV\alpha^2/\beta \tag{100}$$

Thus, if the thermal expansivity and the isothermal compressibility are known, exact calculations of C_V are possible. Furthermore, one finds that frequently for solids at not too high temperature, the heat capacity at constant pressure is approximately proportional to the thermal expansion coefficient and the ratio of the volume to the isothermal compressibility so that

$$C_P = C_V + AC_P^2T \tag{101}$$

where A is a constant, specific to the material analyzed.

The SI unit of heat is the joule (J), based on the mechanical energy definition (work done on a mass of 1 kg when it is moved 1 m in a force field of 1 N: m^2 kg s^{-2}). The historical unit the calorie was based on the specific heat capacity of one gram of water. Since the specific heat capacity varies with temperature,* different calories have been defined. Some useful conversion factors are given in the accompanying tabulation.

To go from	to joule multiple by
calorie (U.S. Bureau of Standards defined)	4.18400
calorie (International Steam Table Conference, 1923)	4.18674
calorie (mean)	4.19002
calorie (15°C)	4.18580
calorie (20°C)	4.18190

Calorimetry at constant pressure and temperature involves the measurement of latent heats or chemical reactions. The variables are the number of molecules of the species. N_i. The change in enthalpy can be written analogously to Eq. (8) as

$$dH = (\partial H/\partial T)_{P,N}\,dT + (\partial H/\partial P)_{T,N}\,dP + \Sigma(\partial H/\partial N_i)_{T,P,N\neq Ni}\,dN_i \tag{102}$$

The first two terms are zero at constant temperature and pressure and the

---

* The specific heat capacity at constant pressure of water is (Osborn *et al.*, 1939)

0°C	4.2177	J/(Kg)	15°C	4.1858	J/(Kg)
5°C	4.2022	J/(Kg)	20°C	4.1819	J/(Kg)
10°C	4.1922	J/(Kg)	25°C	4.1796	J/(Kg)

sum over the various species must be evaluated. Using for example the reaction

$$H_2(\text{gas}) + \tfrac{1}{2}O_2 \ (\text{gas}) \rightarrow H_2O \ (\text{gas}) \tag{103}$$

one would write

$$dH = \left(\frac{\partial H}{\partial N_{H_2}}\right)_{P,T} dN_{H_2} + \left(\frac{\partial H}{\partial N_{O_2}}\right)_{P,T} dN_{O_2} + \left(\frac{\partial H}{\partial N_{H_2O}}\right)_{P,T} dN_{H_2O} \tag{104}$$

On integration to produce 1 mole of H_2O, one obtains (assuming the temperature to be 298 K)

$$\Delta H(298) = -H^\circ_{H_2}(298) - 0.5 \, H^\circ_{O_2}(298) + H^\circ_{H_2O}(298) \tag{105}$$

Assuming all participating gases to be ideal, one can use Eq. (81) to obtain

$$\Delta H(298) = \Delta U(298) - 0.5RT \tag{106}$$

Since the reaction was carried out from the elements, the enthalpy for the production of 1 mole of water according to Eq. (103) is also called the enthalpy of formation. The experimental value is given in Eq. (107).

$$\Delta H^\circ_f(298) = -241.83 \quad \text{kJ/mole} \tag{107}$$

Heats of formation of many compounds are widely tabulated (see, for example, Stull *et al.*, 1969). Since H is a function of state, the enthalpy of a reaction can be obtained by summation of any sequence of reactions that lead to the same product, but that may be more convenient to measure. For example, the reaction of formation of ethane

$$2C \ (\text{graphite}) + 3H_2 \ (g) \rightarrow C_2H_6 \ (g) \tag{108}$$

is not measurable directly, but the enthalpies of combustion are easily determined. To evaluate the enthalpy of formation of ethane one would thus carry out the following reactions:

(1) $C_2H_6(g) + \tfrac{7}{2}O_2(g) \rightarrow 2CO_2(g) + 3H_2O(l)$ $\Delta H^\circ(298) = -1560.1$ kJ/mole
(2) $C \ (\text{graphite}) + O_2(g) \rightarrow CO_2(g)$, $\Delta H^\circ_f(298) = -393.5$ kJ/mole   (109)
(3) $H_2(g) + \tfrac{1}{2}O_2(g) \rightarrow H_2O(l)$, $\Delta H^\circ_f(298) = -285.8$ kJ/mole

The two bottom equations are also enthalpies of formations. Subtracting the first line from twice the second and three times the third line gives Eq. (108) and the enthalpy of formation

$$\Delta H^\circ_f(298)[C_2H_6] \ (\text{gaseous}) = -84.3 \quad \text{kJ/mole} \tag{110}$$

To evaluate ΔH at different temperatures, one makes use of the first part of Eq. (102). The difference between the heat of reaction at 298 K and T is equal to the integrated difference in heat capacities of products and reactants ΔC_P

$$\Delta H(T) = \Delta H(298) + \int_{298}^{T} \Delta C_P \, dT \qquad (111)$$

The required heat capacities are usually tabulated as linear, quadratic, or cubic empirical functions of T, so that ΔC_P can easily be formed and integrated.

Heats of transition can be treated similarly. The heat of evaporation of water can, for example, be written

$$H_2O \text{ (liquid)} \rightarrow H_2O \text{ (gas)} \qquad (112)$$

$$\Delta H_v(298) = 43.97 \text{ kJ/mole} \qquad (113)$$

The value of Eq. (113) is the difference between the heats of formation of water vapor [Eq. (107)] and liquid water [reaction 3 of Eq. (109)]. At different temperatures the heats of vaporization can be calculated using Eq. (111).

4. Thermogravimetry

The basic function of state for thermogravimetry, in addition to time and temperature, is the mass of the sample. The SI unit of mass is the kilogram (kg). The international prototype, a platinum cylinder, is kept at the International Bureau of Weights and Measures at Sèvres, near Paris. Copy number 20 is kept at the U.S. Bureau of Standards. Originally (in 1795) the gram was chosen to be the weight unit of 1 cm³ of pure water at the temperature of melting ice. In 1799 the change to 1000 cm³ of water at its density maximum (3.98°C) was made. After some additional changes the mass of the standard weight serves as the mass prototype.

As long as nuclear reactions are excluded, mass is conserved and does not change with temperature, pressure, or volume. In an isolated or closed system there is, by definition, also no change in mass on chemical reaction. Changes in mass can only occur in an open system by flux out of or into the system.

Changes in mass dM are generally measured directly, but for interpretation of data one usually converts into changes in number of moles of elements or compounds dN by division through the molecular mass MW

$$dM/MW = dN \qquad (114)$$

The molecular mass, when entered into any calculation, should be expressed in kilogram to agree with other SI units. The mole is defined as the number of atoms in exactly 12 g of ^{12}C isotope (Avogadro's number: 6.02217×10^{23}); i.e. the atomic weight of ^{12}C is to be used as 0.012 kg.

IV. Characterization of One-Component Systems

A one-component system is made up of only one type of molecule. In this section the thermal characterization of macromolecular, one-component systems is discussed, making use of the general descriptions derived in Sections II and III. It is of interest to note that linear macromolecules can exist only in the liquid or solid states. Evaporation can occur only on decomposition to small molecules or in case of dissolved low-molecular-weight components (such as residual monomer, plasticizer, antioxidants, or other additives). Special consideration must be given to the existence of molecules of widely different lengths (molecular-weight distributions, see Section II.D). Strictly speaking, each molecule of a different length is a different component, as is understood when discussing smaller molecules (such as ethane and propane, for example). The definition of a macromolecule was, however, chosen purposely such that differences due to molecular weight are relatively small. The melting temperature of a small 1000-atom macromolecule changes, for example, usually by less than 10 K when going to very large molecules. For many thermal properties it is thus often the practice to discuss a macromolecule as a one-component system and make no reference to the molecular weight or its distribution. To be precise, however, one should supply more detailed information on the chemical nature of the system under investigation. In Sections A–C, it is assumed that one deals with rather high molecular weight, monodisperse samples (usually above 10^5 MW). In Section D, molecular weight and molecular weight distribution effects are discussed.

A. SINGLE-PHASE SYSTEMS

Thermal analysis of single-phase systems can give information on H, S, and G through measurement of heat capacity by calorimetry (see Section III.E.3) or on V through volume measurement by dilatometry (see Section III.E.2). Thermomechanical analysis can give information on mechanical properties. No useful information on single-phase systems can be extracted from the thermogravimetry (see Section III.E.4). Its use lies in the study of multiphase systems as is described in Section V.D. For full characterization, the functions of interest must be determined not only at various temperatures, but also at different pressures, and in some cases, in addition, at different shapes (see Section III.E.2). In case the system is not in equilibrium, further internal parameters (such as the degree of imperfection) must also be determined (see Section III.C), and the time dependence (kinetics) of the approach to equilibrium needs to be studied (see Section III.D).

1. The Solid States

Two solid states can be distinguished, the crystalline and the glassy state. Microscopically, both states show no large-distance molecular motion. The main molecular motion in solids is vibrational motion. Only in some solid macromolecules are isolated side-group rotations possible (such as in polypropylene CH_2—$CHCH_3$—). The equilibrium position about which the vibrations or rotations of the atoms of the molecules are carried out, however, is fixed. In crystals, these equilibrium positions are regularly arranged as determined by the crystal structure (see Fig. 4). In glasses, the equilibrium positions have only a short-range order; their macroconformation is the random coil, similar to that of liquid somewhat above the glass transition temperature (see Fig. 3, area A). Glasses are distinguished from liquids only by their inability to change their molecular macroconformations.

The basic thermal measurement is heat capacity. At not too low temperatures (above 20–50 K) glassy and crystalline heat capacities are similar. Furthermore, an addition scheme for heat capacities of various chemical groups that, when linked together, give the macromolecular backbone has been derived (Wunderlich and Jones, 1969; Wunderlich and Gaur, 1980). Table III gives some typical examples. The heat capacity of solid polypropylene is simply obtained by adding columns 2 and 6; of polybutene, by adding columns 2, 6, and 7; of poly(vinyl chloride), by adding columns 2 and 4; etc. The data so obtained are good to about ±5% and can serve until scanning calorimetry of the material in question is available. Such data are also of use for the quantitative baseline estimation for heats of fusion and chemical reaction data.

The reason behind this additivity lies in the similarity of the vibrational frequency spectra of linear macromolecules (Wunderlich and Baur, 1970). It is usually possible to describe the heat capacity from 0 to perhaps 200 K with only two parameters, the characteristic temperatures Θ_1 and Θ_3, which enter into the Tarasov (1953, 1955, 1965) equation

$$C_V \approx C_P = 3R \left\{ D_1 \left(\frac{\theta_1}{T} \right) - \frac{\theta_3}{\theta_1} \left[D_1 \left(\frac{\theta_3}{T} \right) - D_3 \left(\frac{\theta_3}{T} \right) \right] \right\} \qquad (115)$$

The function $D_3(\Theta_3/T)$ is the three-dimensional Debye heat capacity function based on an isotropic continuous medium. For very low temperatures, the vibrations have wavelengths long enough that the detailed structure is not recognized and $D_3(\Theta_3/T)$ alone describes the heat capacity. For linear macromolecules the region of fit reaches usually from zero to less than 10 K, whereas for three-dimensional macromolecules, like diamond, the three-dimensional Debye function fits reasonably well over

TABLE III

HEAT CAPACITY CONTRIBUTIONS OF SOME TYPICAL
C-BACKBONE REPEATING UNITS[a]

T (K)	CH_2	CF_2	CHCl	CCl_2	$CHCH_3$	CH_2 (side chain)
50	4.6	10.5	10.5	12.6	9.6	5.0
60	5.9	12.1	12.1	15.9	11.3	6.3
70	7.1	14.2	13.8	18.8	13.0	7.5
80	7.9	15.9	15.1	22.2	15.1	8.4
90	8.8	18.0	15.9	24.7	16.7	9.6
100	9.6	19.7	17.2	27.2	19.2	10.5
110	10.5	21.3	18.0	28.9	19.7	11.3
120	10.9	23.0	18.8	30.1	20.9	12.1
130	11.7	24.7	19.7	31.4	22.6	13.0
140	12.6	25.9	20.5	32.6	24.3	13.8
150	13.0	27.6	21.3	34.3	25.1	14.6
160	13.8	29.3	21.8	35.6	26.8	15.5
170	14.6	30.5	22.6	37.2	28.0	16.3
180	15.5	32.2	23.4	38.1	29.3	17.2
190	15.9	33.5	24.3	39.7	30.5	18.0
200	16.7	34.7	24.7	41.0	32.2	18.8
210	17.2	36.0	25.9	42.7	33.5	20.1
220	17.6	37.7	26.8	44.4	34.3	20.9
230	18.0	38.9	27.6	46.0	36.0	21.8
240	18.4	40.2	28.5	47.7	37.7	23.0

[a] In J (K mole)$^{-1}$.

the whole temperature region [Θ_3 for diamond is 2050 K, a tabulation of $D_3(\Theta_3/T)$ has been made by Beattie, 1926–1927]. The function $D_1(\Theta_3/T)$ is the corresponding one-dimensional Debye heat capacity function based on an isotropic continuous medium. Its values are tabulated by Wunderlich (1962). Some typical data are (in K)

Poly(vinyl chloride)	$\Theta_3 = 175,$	$\Theta_1 = 350$
Polyethylene	$\Theta_3 = 145,$	$\Theta_1 = 540$
Polytetrafluoroethylene	$\Theta_3 = \ \ 46,$	$\Theta_1 = 270$

The additivity of heat capacities, discussed earlier, rests with the values of Θ_1. As a measure of the average intramolecular vibrational frequencies, Θ_1 is proportional to the square root of the molecular weight (Wunderlich and Baur, 1970). The values for Θ_3 depend more on the intermolecular vibrations and thus the intermolecular forces. Values of Θ_3 show thus no additivity of the constituent parts of the molecule. Since glasses usually have a less dense packing than crystals, their intermolecular forces are somewhat less and, as a result, Θ_3 values are lower, and their

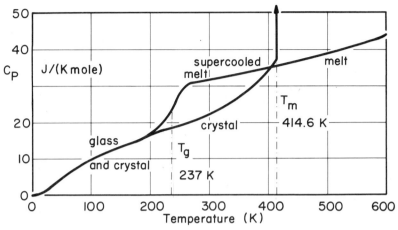

Fig. 13 Heat capacity of polyethylene. Data as collected by Wunderlich and Baur (1970) and Bares and Wunderlich (1973a). The glass transition is at about 250 K, the melting transition at 415 K.

heat capacities at lower temperatures are higher. Figure 13 illustrates the heat capacity of a typical solid. The low temperature region of higher heat capacity of glasses (< 50 K) is outside the accuracy of the drawing. The data on the liquid are described below. The transitions are further discussed in Section IV.B. A series of critical review papers covering all measured heat capacities of linear macromolecules was prepared by Gaur *et al.* (1981–1982).

The calorimetrically determined heat capacity forms the basis for the evaluation of enthalpy, entropy, and free enthalpy (see Section III.E.3). A typical set of equilibrium data on polyethylene is shown in Fig. 14. Below the melting temperature T_m only the crystal is stable; above T_m only the melt is stable. To combine all three functions in one graph, the enthalpy H and free enthalpy G at absolute zero temperature have been moved to zero (by subtraction of the heat of formation at 0 K, $H°$). Furthermore, the free enthalpy is plotted as its negative value to indicate the influence of the entropy up to the melting temperature T_m (see Fig. 12). On heating through the melting temperature, H and S change discontinuously, whereas G changes only its slope (see Section IV.B).

Data on the glassy state, which is not stable relative to the crystal, can be obtained only by measurement on quenched samples or by extrapolation of data on partially crystallized samples to the completely amorphous state. The glassy state is less ordered and has a finite entropy at absolute zero of temperature. This residual entropy is mainly due to conformational contributions. The conformations that can adjust in the liquid state,

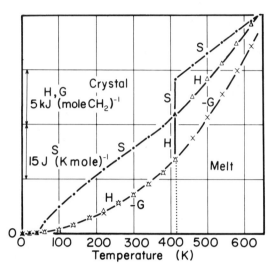

Fig. 14 Enthalpy, entropy, and free enthalpy of polyethylene as derived from heat capacities. The origin of the H and G ordinate has been set arbitrarily to zero at absolute zero in temperature.

freeze on going through the glass transition temperature. An estimate of the residual entropy of polyethylene at 0 K is 2.6 J (K mole)$^{-1}$. The enthalpy of the glass is a continuation of the values of the liquid without any heat of transition. The glassy enthalpy is thus higher than that of the crystal. As a metastable state, the free enthalpy of the glass is also higher than that of the crystal.

The volume of a solid is a measure of the packing of the molecules. Dilatometry is the basic thermal analysis technique for volume (or density) determination. The crystalline volume is more accurately determined by x-ray diffraction than by dilatometry. The analysis of macromolecular crystals has shown that for many classes of crystals the packing fractions are varying within narrow limits. The packing fraction k is defined as

$$k = V_{\text{unit cell}}/V_{\text{v.d. Waals}} \qquad (116)$$

where $V_{\text{unit cell}}$ is the unit cell volume per repeating unit and $V_{\text{v.d. Waals}}$ is the volume of the repeating unit as defined by its van der Waals radii.* For macromolecules in the planar zig-zag conformation, the crystalline packing fraction is frequently 0.7 or somewhat larger, whereas for the

---

* The van der Waals volume can be easily calculated from the covalent bond lengths and the van der Waals radii of the atoms. See, for example, the tabulations of Wunderlich (1973) and Bondi (1968).

more open, helical macroconformation, it can decrease to values as little as 0.6 (Wunderlich, 1973).

The glassy volume is usually measured by dilatometry. In most cases the specific volume (volume per unit mass) of glasses is larger than that of crystals (notable exception: poly-4-methyl-1-pentene). Since the freezing of a glass occurs at a temperature fixed by the cooling rate (and pressure), as is discussed in Section IV.B.4, there is a multiplicity of different glasses of a given chemical composition. Slowly cooled or annealed glasses have the lowest specific volumes. The mechanical properties of glasses vary even more with thermal history then the volume. It is of interest to note that all amorphous hydrocarbon macromolecules have about the same packing fraction at room temperature (0.6). Polyoxides, polyesters, and polyamides have a higher packing fraction, which is thought to be connected with a higher cohesive energy density [for cohesive energy densities, see van Krevelen and Hoftyzer (1972)].

The expansion coefficients of crystals and glasses [see Eqs. (57) and (66)] are linked to the anharmonicity of the vibrations of the solid, which for macromolecules is still poorly understood. The linear expansion coefficient is also accessible through thermomechanical analysis. More important for single-phase systems is, however, the modulus, tensile strength, and relaxation behavior as measured by thermomechanical analysis (see Section III.E.2).

Almost all macromolecular solids analyzed to date are more or less defective. Most of them are even best described as two-phase structures (see Section IV.B). Within the crystalline phase, defects decrease the density, whereas within the glassy phase, defects may either increase or decrease the density. The heat capacities change only little with defects and no general trends can be given. The measurements of volume or heat capacity are thus not very suited to detect defects. Structure sensitive properties, such as ultimate strength, diffusion, or chemical reactivity, are much more affected by defects. Thermomechanical analysis may be of use to characterize the defect state. Other thermal analysis techniques may still be used by studying the phase transitions (see Section IV.B), which show larger effects in the presence of defects. A detailed description of the microscopic structure of defects is given by Wunderlich (1973).

2. The Liquid States

The liquid states are somewhat more difficult to describe microscopically. The high degree of order in the crystal, which allowed easy microscopic description, is absent. The key property of the liquid is its large-distance mobility, which is macroscopically noted by the low viscosity and relative ease of diffusion. At the glass transition temperature the viscosity frequently approached 10^{13} P,* which makes a convenient approximate upper limit for the viscosity of liquids. For reference, the viscosity of water at room temperature is 10^{-2} P. The measurement of viscosity and comparison with 10^{13} P is, however, not an operation suitable for identifying the liquid state, since some liquids have higher viscosities and some solids have lower viscosities. A better definition is to call a liquid a mobile, condensed phase, limited at low temperatures by crystallization or vitrification. For linear macromolecules one is not concerned about the change into the gaseous state.

The microscopic structure of the liquid close to the glass transition temperature is similar to the structure of the glass (see Section IV.A.1). At the melting temperature, it is often possible to describe the short-range order in liquids as "quasi-crystalline" (Ubbelohde, 1965, 1978). With increasing temperature, the structure changes continuously and reaches for small molecules a "gaslike" structure in the vicinity of the critical temperature. For linear macromolecules it is not clear how much of a "quasi-crystalline" structure can be assumed in the liquid state because of the drastically different macroconformations in the two phases (see Fig. 2). It may be that the liquid macromolecules always have for conformational reasons an "anticrystalline" structure.

The basic calorimetric observation on heat capacities of macromolecular liquids is their almost linear temperature dependence over a rather large temperature range (see Fig. 13). In addition to the vibrational contribution to the heat capacities, there is a free volume contribution. Its magnitude can be estimated from the jump in C_P at the glass transition (see Section IV.B.4; Wunderlich and Baur, 1970; Bares and Wunderlich, 1973b). The free volume contribution to the heat capacity is not fully additive for different repeating units of a macromolecule. As a result the additivity of heat capacities is not as well observed as for the solid state. The changes of enthalpy, entropy, and free enthalpy can be seen from the polyethylene example of Fig. 14. At present such information relies on calorimetric measurement for each separate macromolecule.

The volume and thermal expansivity of liquids are practically always larger than of solids. Measurements are made by dilatometry (volume expansion) or by thermomechanical analysis (linear expansion). Special interest is found in the rubber elastic and viscoelastic behavior of liquid linear macromolecules. Tensile, shear, and bulk deformation or relaxation experiments in the proper time scale can be performed with the appropriate configuration of a thermomechanical analyzer. Experiments are best carried out at constant temperature and pressure. The equilibrium change

* The poise is not an SI unit; dimension g cm^{-1} s^{-1}, 1 P = 0.1 Pa s

of free enthalpy or extension is under such conditions simply the work of extension $f\,dl$. For a molar volume of $V_0(T)$ (which for ideally rubbery materials does not change on extension) the change in free enthalpy on extension is given by

$$\Delta G = V_0(T) \int_1^\alpha \frac{f}{A}\, d\alpha \tag{117}$$

where f/A is the equilibrium stress at temperature T and a given extension α [see also Eq. (67)]. The extension α is the ratio of the observed length l to the relaxed length l_0. The equilibrium stress for an ideal rubber network (Gaussian) has been expressed as [see Treolar (1958); for a more recent review of rubber elasticity, see Dunn (1974) and Mark (1981)].

$$f/A = (\rho RT/M_c)(\alpha - \alpha^{-2}) \tag{118}$$

where M_c is the molecular weight between network points. For the statistical segment, the random walk equation, Eq. (6), holds. To relate n to the number of repeating units, the detailed molecular chain statistics must be known. For *cis*-1,4-poly-2-methylbutadiene (natural rubber) one usually assumes that $1.5n = x$. For *trans*-1,4-poly-2-chlorobutadiene (chloroprene), $3n = x$. Combining Eqs. (118) and (117) gives the change of free enthalpy on extension

$$\Delta G = (RT/n)(\tfrac{1}{2}\alpha^2 + \alpha^{-1} - \tfrac{3}{2}) \tag{119}$$

where n is the number of statistical segments between network points.

Thermomechanical analysis under controlled stress and strain conditions is thus able to give information on network structure (through n) as well as free enthalpy (and entropy) information [for further discussion of rubber elasticity and general viscoelastic behavior of macromolecules see, for example, Ferry (1970)]. If the different conformations of the stretched conformations also have different energies, calorimetry must be performed to separate enthalpy from entropy effects [see, for example, the stretch calorimeter of Müller and Engelter (1957); for newer measurements see Godovsky (1981) and Göritz and Müller (1970, 1973)].

Besides the isotropic liquid state, discussed earlier, one finds also "liquid crystalline states" (mesomorphic or paracrystalline states) which only at higher temperatures undergo a transition to the isotropic liquid, and at lower temperatures crystallize in three-dimensional order or freeze into a "glassy mesophase." Liquid crystals tend to occur with small molecules which are rigid rods, such as p-azoxyanisole:

and linear macromolecules which are made of flexibly linked sections of such rigid rods, as for example copolymers of ethylene terephthalate and *p*-oxybenzoate:

$$\left[CH_2 - CH_2 - O - CO - \text{⟨⟩} - CO - O \right]_{x_1} \quad \left[\text{⟨⟩} - CO - O \right]_{x_2}$$

The degree of order in liquid crystals is insufficient to fix the molecular segments at equilibrium positions. Long-range mobility is still possible. It asserts itself in liquidlike low viscosity and fast diffusion. Small external shear leads to large-scale, macroscopic parallelizing of the molecules. The two basic liquid crystalline structures are illustrated in Fig. 15. In smetic

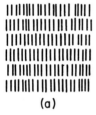
(a)

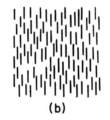
(b)

Fig. 15 Schematic drawing of the order in liquid crystals (mesophases). (a) Smectic structure (order by arrangement in planes, but no periodicity within the planes). (b) Nematic structure (order by parallel arrangement without any additional periodicity).

liquid crystals the molecules or molecular segments have two-dimensional order. They are parallel with their long axes and are arranged in successive planes. Within the plane around the molecular axes there is, however, disorder. In nematic liquid crystals the order is only one dimensional. The molecules or molecular segments have only a common parallel direction. Work on liquid crystals has, for example, been summarized by Johnson and Porter (1970) and Liebert (1978). An initial report on crystallization of polymeric liquid crystals was given by Warner and Jaffe (1980). The glass transition has been discussed by Menczel and Wunderlich (1980, 1981).

To have one or several liquid crystalline states, it is necessary that at the temperature of interest the free enthalpy of the partially ordered state be lower than that of the fully ordered crystal and that of the isotropic melt. According to Eq. (86) this is possible only when the decrease in entropy on partial ordering is compensated by the gain in interaction (decrease in H) caused by the improved contact of the elongated molecules at a temperature above the crystalline melting temperature. Experience tells us that for close to spherical molecules, or molecular segments, partial ordering is insufficient to overcome the entropy term at temperatures above the melting temperature. By adjusting the length and interaction, one can design molecules likely to show liquid crystalline behavior. The state of the highest degree of order must be stable at the lowest temperature, and the isotropic state, stable at the highest temperature, because of the fact that entropy is multiplied by temperature in Eq. (86). An initial calculation of the influence of a rodlike morphology on formation of a partially ordered, mesomorphic phase was given by Flory (1956).

The main thermal characterization of liquid crystals is done through the study of their transitions. The crystal–liquid crystal transition, transitions between different liquid crystal structures, and the liquid-crystal–liquid transition are usually first-order transitions with the largest heat of transition occurring in the crystal–liquid crystal transition (see also Section IV.B).

B. Two-Phase Systems

The presence of more than one phase is usually macroscopically recognizable by an inhomogeneity of the sample. The description of a two-phase system needs a subdivision into separate subsystems for each phase area (see Section III.A). Not only must each subsystem be studied as shown in Section IV.A, but as a new handle on the system, one can study the equilibrium between the phases. The study of one-component materials through their phase transitions is the main subject of this section.

According to Ehrenfest (1933) one can distinguish the phase transition types shown in Fig. 16. On traversing a first-order transition, the free

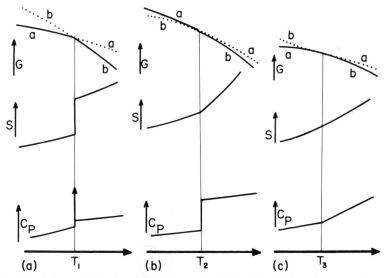

Fig. 16 Schematic drawing of the changes in G, S, and C_P during phase transitions according to Ehrenfest (1933). (a) First-order transition. (b) Second-order transition. (c) Third-order transition.

enthalpy G is a continuous function of the variables T and P, but it suffers a break at the transition temperature. As a consequence, at least one, or both, of the first derivatives of G

$$(\partial G/\partial T)_P = -S \quad \text{or} \quad (\partial G/\partial P)_T = V \quad (120)$$

show a discontinuous change on going through the transition. The main example of first-order transitions for macromolecular systems is the melting transition described in Section IV.B.1. The detailed shape of the C_P curve is usually that of a broad melting peak rather than the sharp infinity point indicated in Fig. 16a.

For a second-order transition, both G and the first derivatives are continuous functions of T and P, as shown in Fig. 16b, but at least one of the first derivatives has a break at the transition temperature. Discontinuities appear, accordingly, in the second derivation of G

$$(\partial^2 G/\partial T^2) = -C_P/T, \quad (\partial^2 G/\partial P\,\partial T) = V\alpha, \quad (\partial^2 G/\partial P^2) = -V\beta \quad (121)$$

i.e., in the heat capacity C_P, thermal expansivity α, and (or) in β, the isothermal compressibility. Note that on both sides of the transition, phase 1 is the phase with the lower free enthalpy, i.e., the more stable phase. A second-order transition is thus possible only if the continuations of phases a and b beyond the transition temperature are not physically realizable states. A typical example of an order–disorder transition is found in the thermal analysis of β-brass at 742 K. Below the transition, Cu and Zn occupy the corners and centers of a cubic unit cell, respectively. Above, the Cu and Zn atoms take on random positions in the same unit cell. For the thermal analysis of linear macromolecules the formalism of the second-order transition is, with some reservations, applied to the glass transition in Section IV.B.4. The detailed shape of C_P is often more complicated than that given in Fig. 16b.

The classification scheme can be continued to nth order transitions for which one finds breaks in the $(n - 1)$th derivative of G with respect to T and P and discontinuities in the nth derivative. An example of a third-order transition is shown in Fig. 16c. The higher order transitions are, however, increasingly more subtle and more difficult to establish.

Thermal analysis is ideally suited to detect the discontinuous changes at the transitions. Besides the study of the transitions under close to equilibrium conditions that are characteristic of a given material (see Sections IV.B.1 and 4), it is possible to study nonequilibrium samples. In these cases information on the thermal, mechanical, and even electrical history of the sample may be obtained, as will be shown in Sections IV.B.2–IV.B.4, and has been discussed by Wunderlich (1981).

1. Equilibrium Melting

The melting and crystallization equilibrium is a first-order transition (see Fig. 16). It can be interpreted as an intersection of the free enthalpies G_a and G_b as shown in Fig. 16a. The branches of higher free enthalpy lose their stability at T_m. At T_m, both phases coexist, so that ΔG, the change in free enthalpy on melting must be zero and one can, using Eq. (86), write for the equilibrium melting temperature

$$T_m = \Delta H_f / \Delta S_f \qquad (122)$$

where ΔH_f is the heat of fusion and ΔS_f, the entropy of fusion.

The study of equilibrium melting is one of the most useful techniques for thermal characterization and can be carried out by most thermal analysis techniques. Complications exist in the study of the melting of linear macromolecules, as illustrated by the early statement of Stuart (1955): "Crystallizing high polymers have no sharp melting point. . . . At equal thermal history, the disappearance of the last crystalline portions of spherulites and other regions of order, as observable with the polarizing microscope, is on sufficiently slow heating, however, reasonably sharp, and reproducible to 1°C." This statement holds true for many of the nonequilibrium crystals of linear macromolecules, and up to about 1960 hardly any linear macromolecular crystals approaching equilibrium, i.e., of extended chain macroconformation and macroscopic size (larger than about 1 μm in each dimension) had been grown and analyzed with respect to their melting behavior. Thus, it was thought earlier that flexible linear macromolecules always end up in a fringed micellar, nonequilibrium, broadly melting crystal (see Section IV.B.2). Equilibrium considerations serve for such nonequilibrium crystals only to set boundaries and define forces for the actual, irreversible processes. Local conformation equilibria, size restrictions, impurities, and sample heterogeneities were thus recognized as the major problems in melting of macromolecules.

Since the 1960s, it has become clear that close to equilibrium crystals of flexible macromolecules can be grown. This permits the discussion of equilibrium melting from an experimental and a theoretical point of view.

Early observations on melting of rigid macromolecules and small molecules (see Section II.A) revealed that this is a process that always occurs with an increase in disorder; i.e., the entropy increases. Quantitatively this increase is much less than the increase in entropy on evaporation. Early comparison of the entropies of melting of simple solids led to the realization that for these

$$\Delta S_f = \Delta H_f / T \approx 7\text{–}14 \quad \text{J (K mole)}^{-1} \qquad (123)$$

[Richards's rule (Richards 1897)]. This entropy increase compares with the molar entropy increase on evaporation of about 90 J $(K \, mole)^{-1}$ for simple liquids (Trouton's rule).

A second general observation on melting is the change in volume. For most substances dilatometry shows an increase in volume on melting. Typically, this volume increase is from 1 to 20% with the smaller volume changes (1–5%) occurring mostly for metals. There is, however, a small group of crystals of all types that shows a decrease in volume on melting (H_2O, -8%; Sb, -1%; Bi, -3%; and Ga, -3%). Trying to find corresponding state ratios of volume was not very successful, even for relatively similar molecules.

A third parameter of interest for thermal analysis is the heat capacity. As in the case of volume changes on melting, there are no general statements that can be made about heat capacity changes. Perhaps the broadest conclusion is that changes are relatively small. In more cases there is an increase in heat capacity than a decrease. In metals, typical changes range from -10 to $+10\%$. Ionic solids often show larger increases (up to $+50\%$). The increase in heat capacity may be caused by a decrease in vibrational frequencies due to weaker interactions and a greater rate of increase in the number of defects with temperature in the liquid. Particularly the faster increase in volume (hole formation, see Section IV.B.4) causes a substantial heat capacity contribution. Processes that may cause a decrease in heat capacity include changes of vibrations to rotations or even translations. The balance between the effects causing increases and decreases in heat capacity seems often to lead to little change.

The parameter that changes most for different materials is the heat of fusion. Assuming a largely constant entropy of fusion [Eq. (123)], higher melting temperatures can be caused only by higher heats of fusion. Indeed, melting temperatures vary from close to absolute zero to above 4000 K, roughly as the forces between the molecules, ions, or atoms increase (H_2, 14 K; and TaC, 4150 K; for example).

Based on these and similar observations, general phenomenological theories of fusion have been derived. Most prominent is the thermodynamic two-phase interpretation, developed to a large degree already by Tammann (1903) and used largely in the present discussion.

Various attempts have, however, also been made to characterize melting on the basis that a crystal becomes unstable above its melting temperature without the presence of a second, more stable phase. The first widely discussed and further developed attempt is the theory of Lindemann (1910), which suggests that fusion is caused by a vibrational instability of the crystal. As the temperature increases, the vibrational amplitudes increase, finally reaching a critical fraction of the lattice distance that renders the crystal unstable. Similarly, it was later suggested that a crystal may become mechanically unstable because the shear modulus decreases with increasing temperature. At the melting temperature a zero shear modulus is reached, which is a characteristic of the liquid. These are theories of melting that depend only on one phase; they predict instability regardless of the surrounding phase. The major difficulty of these theories is thus to explain solvent effects on the melting process. According to Ubbelohde (1965, 1978) it appears that the liquid states always become thermodynamically preferred before the vibrational or mechanical limiting temperatures are reached. The thermodynamic melting process seems, in addition, rapid enough so that even in extreme superheating experiments the vibrational or mechanical melting could not be accomplished (for superheating, see Section IV.B.2).

To find a model to explain Richards's rule [Eq. (123)], one may postulate the melt has short-range "quasi-crystalline" structure. The melting process is then broken down into three major steps. First, the crystal lattice is expanded to give the average separation of the molecules in the melt. Then, the positions of the motifs that make up the crystal are disordered with respect to the crystal lattice. Finally, additional defects are introduced into the quasi-crystalline melt. Relative to the degree of order, this model describes only positional disordering on melting. It should apply thus only to spherical motifs. To find an empirical value for the entropy of melting under such conditions, Table IV lists melting temperatures and entropies of fusion for a series of crystals with spherical or close to spherical motifs of various natures. All seem to obey Richards's rule within the rather wide limits of Eq. (123).

A more detailed suggestion for the origin of the entropy increase has been made by Hirschfelder *et al.* (1937). They showed that the entropy increase in going from motifs fixed in a small volume element of the crystal lattice to the total volume of the liquid is R, or 8.31 J $(K \text{ mole})^{-1}$ (communal entropy). Indeed, this value is close to the lowest values in Table IV, but it is difficult to see that all communal entropy should become available at the melting temperature. The noble gases, in turn, show that other contributions, such as those indicated earlier, must also be considered. A more specific model is the Lennard-Jones–Devonshire (1939a,b) model. A cubic, close-packed crystal is in this case assumed, which on melting, disorders by insertion of interstitial atoms. The three stages of melting are then calculated. For argon, for example, an entropy of fusion of 1.70 R, or 14.13 J $(K \text{ mole})^{-1}$, results, which fits the experimental value well.

The second group of materials in Table IV consists of metals. Compared to the noble gas crystals, they seem to have exceptionally low en-

TABLE IV

MELTING TEMPERATURES AND ENTROPIES OF FUSION
OF CRYSTALS OF SPHERICAL MOTIFS[a,b,c]

Motif	T_m	ΔS_f	Motif	T_m	ΔS_f
Ar	83.8	14.0	Ne	24.6	13.6
Kr	116.0	14.1	Xe	161.4	14.2
Ag	1234	9.2	Li	452	10.2
Al	932	11.5	Mo	2895	9.5
Au	1336	9.5	Na	371	7.1
Ba	998	7.7	Ni	1725	10.2
Ca	1124	8.3	Pb	600	8.5
Cd	594	10.3	Pt	2043	9.6
Co	1763	8.6	Sn	505	14.2
Cr	2163	7.1	Sr	1030	8.9
Cs	301	6.9	Ti	2073	10.1
Cu	1356	9.6	U	1406	11.0
Fe	1803	8.3	V	2190	8.0
Hg	234	10.0	W	3660	9.6
K	337	7.1	Zn	693	9.6
La	1193	8.4	Zr	2130	10.8
CF_4	84.5	8.4	H_2S	187.6	12.7
CO	68.2	12.2	N_2	63.2	11.4
H_2	13.9	8.4	O_2	54.4	8.1
HBr	186.2	12.9	P_4	317.2	7.9
HCl	158.8	12.5	PH_3	139.4	8.0
HI	222.2	12.9	SiH_4	88.5	8.4
CH_4	90.7	10.3	Pseudocumene	279.8	13.4
CH_3OH	371.0	8.6	cis-1,2-Dimethylcyclohexane	223.2	7.4
$C(CH_3)_4$	256.5	12.6	Cyclohexane	279.8	9.4
			Camphor	451.6	14.3

[a] Temperatures in K; entropies in J $(K\ mole)^{-1}$.
[b] A motif is the atom, ion, molecule, or part of a molecule that can be identified as the basic building block of the crystal.
[c] Data from Clusius and Weigand (1940) for noble gases; "Handbook of Chemistry and Physics" (1975) for metals and small organic molecules; see also Ubbelohde (1965), Tables 5.1 and 5.5.

tropies of fusion (and also volume changes on fusion). No simple explanation beyond positional disordering seems available. Effects such as changes of electronic band structure, abnormal changes in vibrational contributions, and possible clustering need to be considered for a description of the entropy change.

The third group of materials in Table IV refers to small inorganic molecules. They are close to spherical and start disordering their alignment within the crystal below the melting temperature, usually at a well-defined first- or second-order transition. On melting, they experience then again only positional disordering, which leads to entropies of fusion close to those of the noble gases.

The last group of materials in Table IV consists of more complex organic molecules. Again, they act at the melting temperature as practically spherical molecules and show entropy changes similar to the noble gases.*

Molecules with nonspherical shape have a definitely higher entropy of melting, as is shown by the data in Table V. To account for the increase in

TABLE V

MELTING TEMPERATURES AND ENTROPIES OF FUSION
OF CRYSTALS OF NONSPHERICAL MOTIFS[a,b]

Motif	T_m	ΔS_f	Motif	T_m	ΔS_f
HCF_3	118.0	34.4	C_2H_6	89.9	31.8
CO_2	216.5	38.7	SO_2	197.6	37.4
N_2O	182.3	35.9	$CHCl_3$	210.0	45.2
COS	134.3	35.2	$SiCl_4$	203.4	37.9
CS_2	161.1	27.2	C_6H_6	278.5	35.3
Br_2	267.0	40.4	SF_6	218.0	21.8
Cl_2	172.1	37.2	H_2O	273.2	22.0
C_2N_2	245.3	33.1	NH_3	195.4	28.9
HCN	259.8	32.3	NO	109.4	21.0
Diphenyl	438.6	44.7	Phenanthrene	369.4	50.5
Anthracene	489.6	58.9	Thiophene	233.8	21.2
2-Methylnaphthalene	238.8	50.1	Dioxane	284.2	45.2

[a] Temperatures in K; entropies in J (K mole)$^{-1}$.
[b] Data collected from Ubbelohde (1965) and the "Handbook of Chemistry and Physics" (1975).

entropy of melting, an additional mechanism is added to the positional disordering present in spherical or close to spherical motifs. This additional mechanism is called orientational disordering. The orientational disordering may add a larger contribution to the melting than the positional disordering, as can be seen by comparing the entropy of fusion of CF_4 (Table IV) with that of HCF_3 (Table V). The second group of motifs in Table V shows that, as in positional disordering, there seems little, if any, change in ΔS_f with size.

* Because of the large size of some of these substances, they give a particularly large freezing point lowering for dissolved impurities. Camphor, $C_{10}H_{16}O$, for example, lowers the freezing point by 40 K for 1 mole kg^{-1} impurity; the water freezing point constant is, in contrast, only 1.85 K [see Eq. (191)].

For nonrigid molecules a third mechanism needs to be considered: the mechanism of conformational disordering (for the description of conformations see Section II.D). The overall entropy of fusion is then

$$\Delta S_f = \Delta S_{pos} + \Delta S_{or} + \Delta S_{conf} \qquad (124)$$

Table VI presents a list of the melting temperatures and entropies of fusion of the first 12 paraffins. The jump on going from purely positional melting in methane to positional and orientational melting in ethane is clearly visible. The further increase in the entropy of fusion for higher paraffins is an indication of conformational changes on melting. Irregularities in the increase in ΔS_f are caused by changes in the crystal structures in going from one paraffin to the next. It is clear, however, that the conformational contribution to melting is, in contrast to the other two contributions, continuously growing with increasing numbers of bonds about which rotation can occur. The recalculated values per mole of carbon atoms show that a molecule with sufficient conformational disorder in the melt reaches a ΔS_f predicted by Richards's rule, but now per flexible backbone atom rather than per overall molecule (values in parentheses in Table VI).

TABLE VI

MELTING TEMPERATURES AND ENTROPIES OF FUSION
OF CRYSTALS OF LINEAR HYDROCARBONS[a,b,c]

Motif	T_m	ΔS_f	Motif	T_m	ΔS_f
CH_4	90.7	10.3 (10.3)	C_7H_{16}	182.6	77.6 (11.1)
C_2H_6	89.8	31.8 (15.9)	C_8H_{18}	216.4	95.4 (11.9)
C_3H_8	91.5	38.5 (12.9)	C_9H_{20}	219.6	70.5 (7.8)
C_4H_{10}	134.8	34.6 (8.6)	$C_{10}H_{22}$	243.4	118.2 (11.8)
C_5H_{12}	143.5	58.7 (11.7)	$C_{11}H_{24}$	247.6	90.1 (8.1)
C_6H_{14}	177.8	73.6 (12.3)	$C_{12}H_{26}$	263.6	138.8 (11.6)

[a] Temperatures in K; entropies in J (K mole)$^{-1}$.
[b] Data collected from the "Handbook of Chemistry and Physics" (1975).
[c] Values in parentheses are recalculated for 1 mole of carbon atoms.

No such general statements can be made about the volume or heat capacity changes on melting. Details about solid and liquid state must be known for the explanation of the specific observed changes. With a rough value of the entropy of fusion it is, however, possible to link through Eq. (122) heat of fusion with melting temperature. High-melting crystals can be obtained either by crystals with high heats of fusion, which involves introducing strong bonds between the motifs, or by crystals with low entropies of fusion, which involves reducing the conformational disordering by stiffening the melting molecules and changing the shape to spheres if possible.

For flexible linear macromolecules it is still difficult to collect equilibrium data. Inaccurate data or data collected on metastable crystals have in the past hindered a detailed discussion of the effect of chemical structure on melting. Frequently, it has been necessary to extrapolate the equilibrium data from low-molecular-weight solids that can easily be obtained as equilibrium crystals (see Section V.F). Other methods to obtain equilibrium parameters involve extrapolations as a function of crystal perfection (for detailed discussion, see Wunderlich, 1980). A dilatometric melting experiment on close-to-equilibrium crystals of polyethylene is shown in Fig. 17. About 80% of the whole sample melts within 1 K. The last-melting crystals are largest in number, giving a sharp melting end at 414.6 K. Overall, the melting of these linear macromolecular crystals is as sharp as normally found for small organic molecules.

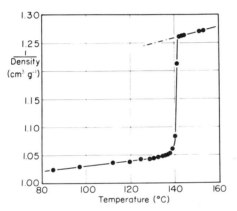

Fig. 17 Dilatometry of extended-chain crystals of polyethylene on melting. The polyethylene was produced by decomposition of diazomethane and had an estimated molecular weight of 1.4×10^7 with no molecules less than 10^5 molecular weight. Crystallization was done at about 480 MN m^{-2} at 500 K for 20 h. At room temperature the crystallinity was 0.98. The biggest crystals were 10 μm in the molecular chain direction. Data by Arakawa and Wunderlich (1967).

This example shows that close-to-equilibrium melting can be achieved, and a discussion of equilibrium melting behavior is thus meaningful.

Melting data on a series of flexible linear macromolecules for which equilibrium crystals were available or for which they could be extrapolated are listed in Table VII. Column 3 gives the melting temperature in K; column 4 gives the entropy of fusion per rigid backbone chain group (CR_1R_2—, O—, CR=CH—, COO—, C_6H_4—, and CONH—). The num-

TABLE VII

EQUILIBRIUM MELTING DATA OF FLEXIBLE LINEAR MACROMOLECULES

No.	Macromolecule	T_m° (K)	Δs_f^a (J (K mole)$^{-1}$)		Δh_f^a (kJ mole^{-1})		Packing^b fraction Δ (%)	kl	Cohesive energy density (kJ mole^{-1})
1	Polyethylene	414.6	9.91	(1)	4.11	(1)	14	0.60	4.18
2	Polytetrafluoroethylene	600	5.69	(1)	3.42	(1)	15	0.68	3.35
3	Selenium	490	12.72	(1)	6.23	(1)	11	0.76	—
4	Polypropylene	460.7	7.55	(2)	2.31	(3)	9	0.60	4.74
5	Poly-1-butene	411	8.50	(2)	1.75	(4)	9	0.60	4.60
6	Poly-1-pentene	403	7.80	(2)	1.26	(5)	9	0.59	4.52
7	Poly-4-methyl-1-pentene	523	9.50	(2)	1.66	(6)	−2	0.59	4.74
8	Poly-4-phenyl-1-butene	439	5.00	(2)	0.44	(10)	3	0.64	4.14
9	Polystyrene	516	9.70	(2)	1.25	(8)	6	0.63	4.13
10	1,4-Polybutadiene, *cis*	284.7	10.67	(3)	2.30	(4)	10	0.61	4.18
11	1,4-Polybutadiene, *trans*	415	2.90	(3)	0.90	(4)	13	0.60	4.18
12	1,4-Poly-2-methylbutadiene, *cis*	310	4.80	(3)	0.87	(5)	9	0.62	3.93
13	1,4-Poly-2-methylbutadiene, *trans*, α	352.7	12.13	(3)	2.57	(5)	14	0.61	3.93
14	1,4-Poly-2-methylbutadiene, *trans*, β	356	9.90	(3)	2.11	(5)	12	0.61	3.93
15	Polyoxymethylene	457	10.70	(2)	4.89	(2)	10	0.70	5.23
16	Poly(ethylene oxide)	342.1	8.43	(3)	2.88	(3)	10	0.65	4.88
17	Poly(tetramethylene oxide)	330	8.74	(5)	2.88	(5)	11	0.62	4.60
18	Polyglycolide	506	11.0	(2)	3.70	(3)	11	0.72	5.86
19	Poly-β-propiolactone	357	8.5	(3)	2.27	(4)	—	0.71	5.44
20	Poly-α,α'-dimethyl propiolactone	518	9.6	(3)	2.47	(6)	11	0.63	5.02
21	Poly-ε-caprolactone	337	8.0	(6)	2.31	(7)	9	0.64	4.90
22	Poly(ethylene adipate)	338	7.8	(8)	2.10	(10)	11	0.67	5.19
23	Poly(ethylene suberate)	348	7.7	(10)	2.22	(12)	11	0.66	5.02
24	Poly(ethylene sebacate)	356	7.5	(12)	2.28	(14)	12	0.64	4.90
25	Poly(ethylene terephthalate)	553	9.7	(5)	2.24	(12)	9	0.68	5.02
26	Poly(4,4'-isopropylidene carbonate)	553	12.6	(5)	1.86	(18)	10	0.65	4.58
27	Nylon 6, α	533	8.1	(6)	3.71	(7)	12	0.66	11.7
28	Nylon 8, γ	491	4.5	(8)	1.98	(9)	8	0.65	10.0
29	Nylon 6.6, α	553	10.2	(12)	4.85	(14)	12	0.66	11.7

^{a} The number of units in the repeating unit is given in parenthesis. Flexible chain units for Δs_f, and total interacting groups for Δh_f.

^{b} Calculated using typical van der Waals and covalent radii; see Eq. (116).

^{c} Calculated using Table 6.1, van Krevelen and Hoftyzer (1972), calculated in moles of total interacting groups as given in the Δh_f column.

ber of rigid chain groups per repeating unit is given in parentheses. Column 5 shows the heat of fusion per mole of interacting groups, which is the total number of interacting groups per repeating unit, given in parentheses. It involves counting all large atoms, disregarding only H, and counting CO— as one unit. Column 6 shows the percentage change in packing fraction and the packing fraction of the amorphous macromolecule at room temperature [see Eq. (116)]. The last column contains a calculation of the cohesive energy density per mole of amorphous, interacting groups (as in column 5) using the table of van Krevelen and Hoftyzer (1972).

The entropy of fusion of the majority of macromolecules is 9.5 J (K mole)$^{-1}$ of rigid backbone chain groups. This unique value supports the empirical subdivision of the molecule in "rigid" backbone groups. There are six exceptional macromolecules (2, 8, 11, 12, 14, 28) with frequently much smaller entropies of fusion. A check of the crystal structures or melting behavior of these macromolecules suggests that these crystals have high-temperature crystal forms that contain already increased disorder. For these six macromolecular crystals, details about the mobility in the crystal are needed for an understanding of their melting behavior. The existence of high-temperature crystals of increased mobility is based on a relatively smaller increase in enthalpy than entropy on going from the low-temperature crystal form to the high-temperature polymorph. In the extreme, this increased mobility without loss of parallel arrangement of the macromolecules leads to liquid-crystallike structures (see also Section IV.A.2).

Turning now to the macromolecules for which the chains are practically in a fixed, perfectly ordered conformation before melting, one expects a large portion of the entropy of fusion to be of purely conformational origin. Comparing the experimental data with calculated conformational entropies of fusion, one finds typically 65–85% for the contribution of conformational entropy to the total entropy of fusion.

Among the macromolecules with a high entropy of fusion further trends can be seen on inspection of Table VII. Macromolecules with phenylene groups within the backbone (25,26) seem to have a somewhat increased entropy over similar molecules that contain CH_2— groups. The same seems to hold true for polystyrene (9). When CH_2— and other functional groups such as O—, COO— alternate, and in selenium (15,18,4) one also has higher entropies of fusion. Much of this difference may be connected with the change in crystal and liquid structure relative to an increased nonpolar CH_2— environment (as in 16,17,19, and 21, for example). At present it is not possible to decide whether these trends are based on conformational or nonconformational effects.

The difference in volume between crystal and melt (at room temperature) also shows a correlation with structure. Columns 6 and 7 in Table VII contain the pertinent data. Among the all-carbon backbone macromolecules, those with close to all-trans conformations in the crystal show the largest percentage difference in packing fraction (1,2,11,13,14). The more helical, all-carbon backbone macromolecules show a distinctly lower percentage difference (4–9). For poly-4-methyl-1-pentene (7) this difference is even negative at room temperature, although at the temperatures above 50°C the change is positive. The polyoxides (15–17), polyesters (18–26), and polyamides (27–29) distinguish themselves by a high packing fraction in the liquid and in the melt, approaching the hydrocarbon level only for long CH_2— sequences. Their percentage change, however, does not differ much from the pure hydrocarbons. All pure hydrocarbon macromolecules (1,4–14) have quite similar packing fractions in the liquid state of 0.61 ± 0.02 regardless of the molecular structure. The liquid packing fraction increases almost linearly with the cohesive energy (1,4–25), but amide groups add an extraordinary amount to the cohesive energy density that is not matched by an increase in packing fraction (27–29). As the sequences of CH_2— increase in length in the polyethers, polyesters, and polyamides (17,24,27–29), the liquid packing fraction of polyethylene is approached, but at different rates for different types of macromolecules.

The heats of fusion in Table VII have been calculated per mole of interacting groups, and not per mole of rigid chain units. This is done to account for the major contribution to the heat of fusion that should be connected with the loss of interaction on expansion to the liquid. A quite uniform heat of fusion of about 2.4 kJ/mole results for many entries in Table VII (4,10,13,16,17,19–25), with others showing remarkable deviations. All macromolecules with exceptionally small entropies of fusion show also a 1–2 kJ mole^{-1} lower heat of fusion than expected, much in line with a high-temperature crystal form of higher enthalpy (2,8,11,12,14,-28). The polyamides (27–29) have a 1–2 kJ mole^{-1} higher heat of fusion than the average, which goes parallel with their much higher cohesive energy density. The macromolecules with larger side chain (5–9) have also a lower heat of fusion than the average when calculated per total of single interacting group. This loss in heat of fusion is also indicated to some degree in the small packing fraction difference. Finally, polyethylene, polytetrafluoroethylene, selenium, and the first polyoxide and polyester (1–3, 5,18) have a 1–2 kJ mole^{-1} higher heat of fusion.

Although the cohesive energy densities and changes in packing fraction can explain some of the changes in heat of fusion with chemical structure, they are not sufficient. One must add the intramolecular contributions to the heat of fusion that results from the change to some high-energy conformations on fusion. A particularly clear example is polyethylene, where in the crystal, only the low-energy trans conformation is found. In the melt, however, each carbon–carbon bond may also be in one of the two gauche conformations that are higher in energy by about 2 KJ mole^{-1}. At the melting temperature, the high-temperature limit of equal distribution among the three rotational isomers is not quite reached, but even on equal distribution among trans and gauche conformations, the intramolecular heat of fusion would be 1.0 kJ mole^{-1}, or about 25% of the total heat of fusion.

Combining now all the conclusions on entropy of fusion, packing fraction differences, cohesive energy density, and heat of fusion, one can make some general statements about the melting temperatures of flexible linear macromolecules:

The high melting temperature of polyethylene (1) relative to *cis*-polybutadiene (10) does not rest with a higher conformational entropy of fusion, but rather with the higher heat of fusion of polyethylene, a good part of which must result from intramolecular contributions and the larger volume change on fusion. Even the high-temperature form of *cis*-poly-2-methylbutadiene (12) does not reach the polyethylene melting temperature. The *trans*-polydienes, particularly in their high-temperature form (11) with an all-trans chain conformation in the crystal, come close to the polyethylene melting temperature, as one would expect.

The particularly high-melting vinyl polymers are polytetrafluoroethylene (2), poly-4-methyl-1-pentene (7), and polystyrene (8). The polytetrafluoroethylene melting temperature is elevated due to the high mobility in the crystal and, perhaps to a lesser degree, to a lower entropy in the melt. The other two macromolecules have, in contrast, a high melting temperature because of a high molar heat of fusion. Despite less dense packing in the crystal than other vinyl polymers, there is a large enough side group to more than compensate the packing defect. For these two macromolecules it also becomes clear that the present discussion is still too simplified, since side-group entropy gain on fusion is completely neglected, but should be included in a next step of refinement.

Selenium (3) is a macromolecule that needs more information for full discussion. Its melting temperature represents a balance of high heat of fusion and high enthalpy of fusion. The complicating factor is in this case the partial depolymerization to Se$_8$ rings on melting, a process that is endothermic and naturally goes also with an increase in entropy.

Polyoxymethylene (15) does not have, as was frequently suggested, a low entropy of fusion; instead its relatively high-equilibrium melting temperature must be connected, as in polyethylene, with an exceptionally high heat of fusion. Only a part of the increased heat of fusion can be accounted for by the higher cohesive energy contribution of the oxide group. Some of the increase must also come from intramolecular sources. The crystal consists practically of all gauche conformations. The trans conformations, which must also be present in the melt, are of higher energy. The high packing fraction may indicate an additional heat of fusion contribution from the dipole interaction that is still not too well understood. The higher analogs (16,17) drop to levels much more in line with ''normal'' macromolecules.

The series of polyesters (18–26) is normal in its melting behavior for the aliphatic members. The ester group has higher cohesive energy contribution than an oxide. Furthermore, dipole interactions have to be considered. Again, the first member (18) is exceptional, most likely because of the largely different crystal structure. The other high melting polyesters (20,25,26) are clearly resulting from the large differences between numbers of rigid backbone chain groups and total numbers of interacting groups.

The polyamides (27–29) owe their high melting temperatures to their high heats of fusion. The amide group has a cohesive energy density per backbone chain atom 7.3 times that of CH_2—. This contrasts the ester and oxide groups, which are only 1.6 and 1.5 times higher in cohesive energy density per chain atom than CH_2—.

This more detailed discussion of equilibrium melting has shown that analysis of the thermal behavior can lead to a rather detailed understanding of a material. Out of discussions like this has evolved the system of classifying materials described in Section II.A.

2. Nonequilibrium Phase Changes

The analysis of phase changes of a system not in equilibrium is more complex than the equilibrium melting just given. One can distinguish two situations. First, the system is already a nonequilibrium system, as one finds it normally in semicrystalline macromolecules (see Section II.F). The analysis of phase transitions must then involve a discussion of the change in metastability during any change of state of the system. This case will be discussed for the example of melting of a chain-folded lamellar crystal, making use of irreversible thermodynamics. Second, the system may be initially in equilibrium, but the phase transition does not occur at the equilibrium temperature and/or may not go to completion. The cases of interest here are the superheating of equilibrium crystals and the crystallization of solutions or melts. These cases are often treated using kinetic arguments (Section II.D). Both fields are still in the developmental stage. The major thermal analytical tools are the calorimeter

(DTA, DSC) and the dilatometer, although thermometry and thermomechanical analysis can also give information on nonequilibrium melting.

A typical macromolecular, semicrystalline solid such as polyethylene is best described as a two-phase metastable system. The two phases are crystalline and amorphous. Both phases are, however, neither independent of each other nor homogeneous. Figure 18 shows a free enthalpy distribution of crystalline subsystems of a melt-crystallized polyethylene.

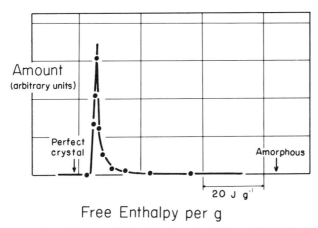

Fig. 18 Crystalline subsystem distribution for polyethylene at 300 K. Derived from heat of fusion measurements of a sample that showed only a minimum of reorganization. Drawn after Wunderlich (1964).

There are no subsystems close to equilibrium. A sharp peak centers around crystals of close to average fold length, but a broader tail stretches to higher free enthalpies of more defect crystalline subsystems. The distribution of defect subsystems does not, however, reach to the amorphous subsystem. The state of the amorphous subsystems is much less well known. That the amorphous phase is not uniform and a mere supercooled melt is obvious from the fact that there is no continued crystallization. Melts of polyethylene cannot be kept from crystallizing at room temperature. A distribution of amorphous subsystems can also be inferred from a broad glass transition interval. For simplicity (and out of lack of information) one usually still assumes the amorphous subsystem to be close to equilibrium as long as there is no orientation frozen in.

Based on Fig. 18, one simplifies the description of a semicrystalline, metastable macromolecular sample by assuming only two subsystem types: crystalline subsystems and amorphous subsystems. The increase of the free enthalpy of the average crystalline subsystem above the free enthalpy of the equilibrium crystal is mainly caused by the large surface area of the small crystals (see Section II.F). Properties that are little affected by surface effects are density, x-ray diffraction, heat of fusion, and the infrared absorption of the crystal. These properties are called structure insensitive and can be used to get an estimate of the overall fraction of crystalline subsystem or the fraction crystallinity. The mass fraction crystallinity w^c is given, for example, through the ratio of specific volumes v ($= 1/$density)

$$w^c = (v_a - v)/(v_a - v_c) \tag{125a}$$

The subscripts a and c refer to the specific volumes of completely amorphous and crystalline subsystems, respectively. X-ray diffraction can give information in crystallinity through the ratio of scattering intensities I of the crystalline subsystems (sharp characteristic diffraction) and the amorphous subsystems (broad halo). After correction of the raw scattering data for air scattering, differences in primary beam intensities, and differences in sample scattering masses, the crystallinity ratio of two samples can be expressed as

$$w_1^c/w_2^c = I_{c_1}/I_{c_2} \tag{125b}$$

$$(1 - w_1^c)/(1 - w_2^c) = I_{a_1}/I_{a_2} \tag{125c}$$

The total intensity ratios are usually approximated by the intensities at the maximum scattering angle. The heat of fusion of a semicrystalline sample Δh_f is simply related to the crystallinity through the equation

$$w^c = \Delta h_f/\Delta h_f^\circ \tag{125d}$$

where Δh_f° is the heat of fusion of the completely crystallized (equilibrium) sample as discussed in Section IV.B.1. In case infrared absorption bands characteristic of the amorphous or crystalline phase can be identified, measurement of the optical density (absorbancy) D, defined as $\log(I_0/I)$, can lead to a determination of the volume fraction crystallinity

$$v^c = D/(\epsilon_c \rho_c t) \tag{125e}$$

$$1 - v^c = D/(\epsilon_a \rho_a t) \tag{125f}$$

where ϵ is the absorbancy index (extinction coefficient) at the appropriate absorption frequency and t is the path length. The volume fraction crystallinity is given by the ratio of the density differences

$$v^c = (\rho - \rho_a)/(\rho_c - \rho_a) \tag{125g}$$

It is simply related to the weight fraction crystallinity through the equation

$$v^c = (\rho/\rho_c)w^c \tag{125h}$$

For most melt-crystallized samples all different methods of crystallinity determinations give reasonable agreement because of the relative structure insensitivity of the measured property (see Wunderlich, 1973). For samples that have been deformed, such as drawn fibers, this agreement between the different methods is poor. In such cases each method must be discussed separately. In drawn samples the density of the amorphous subsystems is, for example, increased, whereas the crystalline subsystems have a decreased density.

Structure-sensitive properties such as ultimate strength and gas permeability cannot be used for crystallinity estimates. Similarly, in Fig. 18 it is shown that the free enthalpy of the crystalline subsystems of a semi-crystalline sample is substantially higher than the free enthalpy of equilibrium crystals. Free enthalpy is thus also not a measure of crystallinity, but it can be coupled with crystallinity measurements through any of the other techniques to give a quantitative value to metastability. [See also Richardson and Savill (1980) for methods to establish free enthalpy from nonequilibrium measurements.]

The melting temperature determination is directly linked to free enthalpy, and its measurement in conjunction with the crystallinity serves as an improved description of metastable, semicrystalline macromolecules. As an example, the melting behavior of a lamella of a homopolymer is described. The crystallinity at low temperature is a measure of amount of crystalline and amorphous subsystems. The entropy changes per unit time are indicated in Eq. (17). Figure 19 shows schematically the melting

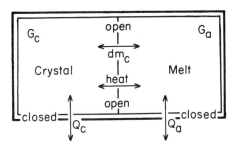

Fig. 19 Schematic drawing of a two-subsystem melting situation.

experiment. A crystalline, metastable subsystem of free enthalpy G_c forms an open boundary with an amorphous subsystem of free enthalpy G_a, which is assumed to correspond to a supercooled liquid. Both subsystems are closed from the surroundings and at constant temperature and pressure. One can assume they are together contained in a calorimeter so that the combined heat flux $Q_a + Q_c$ into the subsystems can be measured directly. The free enthalpy of an otherwise perfect lamellae can be expressed as

$$G_c = m_c g_c + (2m_c/\rho l)\gamma \qquad (126a)$$

where g_c is the bulk free enthalpy per unit mass of a large equilibrium crystal m_c the mass of the lamella, γ the surface free energy, ρ the density, and l the lamellar thickness. It can easily be seen that $2m_c/\rho l$ is the area of the two large lamellar surfaces. The side surfaces are assumed to be so much smaller as to be negligible. All metastability of the system rests thus with the nonequilibrium shape of the crystal as expressed by the lamellar thickness l. Any other irreversible effects such as side surfaces and interior defects would have to be added to Eq. (126a). One can see that for a full description of a metastable system, more than one experiment is usually necessary. The amorphous subsystem free enthalpy, since it is assumed to be a supercooled liquid, is simply

$$G_a = m_a g_a \qquad (126b)$$

The bulk free enthalpy of fusion is $\Delta g_f = g_a - g_c$, and one can write, if the entropy of fusion at the analysis temperature T is not much different from that at the equilibrium melting temperature of large crystals, T_m°,

$$\Delta g_f = \Delta h_f - T\,\Delta s_f = \Delta h_f\,\Delta T/T_m^\circ \qquad (127)$$

where ΔT is $T_m^\circ - T$. Permitting now melting ($dm_a = -dm_c$) and changes in thickness of the lamellar crystal (dl), one finds the following equation for the overall change of the entropy of the two-subsystem unit in the time interval dt [see Eqs. (17) and (86), remembering that $\Sigma\,d_iH$ and $\Sigma\,d_im$ must be zero in a closed system]:

$$dS = \left[\frac{dQ_c + dQ_a}{T}\right] + \left[\left(\frac{\Delta h_f\,\Delta T}{TT_m^\circ} - \frac{2\gamma}{T\rho l}\right)dm_c\right] + \left[\frac{2m_c\gamma}{T\rho l^2}\,dl\right] \qquad (128)$$

The first term in brackets is the entropy flux d_eS, which is directly measurable by calorimetry. The second term in brackets expresses the contribution of the free energy change on melting, and the third term is caused by reorganization, which is treated in Sect. IV.B.3. The second and third bracketed terms together represent the entropy production d_iS.

Equation (128) reverts, as expected, to the equilibrium expression when l becomes large and melting occurs at temperature T_m°. In this case the second and third bracketed terms are zero and the total entropy change dS is caused by flux d_eS, i.e. there is no entropy production. All heat dQ_c and dQ_a, which is measured by the calorimeter, is used for equilibrium melting. The entropy gain of the two subsystems is $(\Delta h_f/T_m^\circ)\,dm_c$. Equilibrium data can be derived from such experiments (heat of fusion, entropy of fusion, and melting temperature).

Retaining the assumption of thick crystals, so that all terms that contain l in the denominator are negligible, but undergoing the phase change at a temperature different from T_m°, leads to a nonzero contribution in the second bracketed term. The second law of thermodynamics prohibits negative values of $d_i S$ [see Eq. (19)]. Consequently, the only possible irreversible processes involve melting above T_m° ($dm_c < 0$, $\Delta T < 0$), or crystallization below T_m° ($dm_c > 0$, $\Delta T > 0$). The first process is called melting under condition of superheating; the second, crystallization at a given supercooling.

Figure 20 shows the melting of large, extended-chain polyoxymethylene crystals measured by DTA. Figure 21 illustrates isothermal dilato-

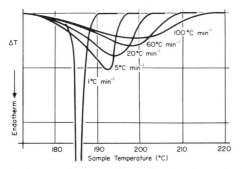

Fig. 20 Melting of polyoxymethylene with superheating. Increasing heating rate continues the melting far above the equilibrium melting temperature. Drawn after Jaffe and Wunderlich (1967). The curves were normalized to constant heats of fusion.

Fig. 21 Isothermal melting curves of high molecular weight extended chain crystals of polyethylene. The melting of 1.4×10^7 molecular weight polyethylene in the form of 0.98 crystallinity extended-chain crystals grown at 500 K and 480 MN m^{-2} for 20 h at various temperatures. Curve 1: 421.7 K; Curve 2: 419.2 K; Curve 3: 417.7 K; Curve 4: 416.7; Curve 5: 414.7 K; all temperatures ± 0.5 K. m_w indicates the mass fraction molten. From Hellmuth and Wunderlich (1965). Reprinted with permission of the American Institute of Physics.

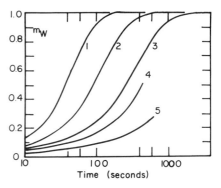

metry of extended-chain crystals of polyethylene above the melting temperature. Clear signs of superheating are visible. For quantitative analysis, a detailed kinetic mechanism must be assumed, involving the crystal geometry and the linear melting velocities, quantities usually not available. To date only linear melt velocities of large polyethylene and selenium crystals have been measured by optical microscopy (Czornyj and

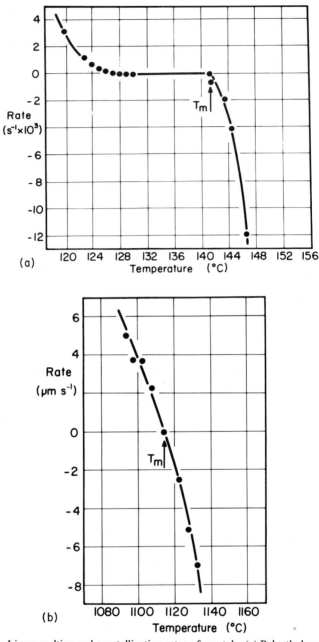

Fig. 22 Linear melting and crystallization rates of crystals. (a) Polyethylene, plotted is the inverse half-time for melting and crystallization of a high-molecular-weight sample. Melting data by Hellmuth and Wunderlich (1965). Crystallization data by Mandelkern (1959). (b) Germanium dioxide, rigid macromolecule, data by Vergano and Uhlmann (1970).

Wunderlich, 1977; Wunderlich and Shu, 1980). All other, more detailed information on superheating comes from crystals of rigid macromolecules or small molecules. In most cases the linear melt velocity approached T_m° with a linear or quadratic ΔT dependence. Figure 22 shows the linear crystal melting rates of polyethylene and germanium dioxide. It illustrates the similarity of the superheating behavior of the two compounds. Thermal analysis of samples that show superheating to obtain equilibrium data is thus only possible on very slow heating, as is possible, for example, in dilatometry. On the other hand, additional information on the structure of the crystals is available through the study of the melting kinetics as illustrated in Figs. 20 and 21. Crystals of different perfection give different degrees of superheating (see also Fig. 29).

Figure 22 illustrates also the crystallization rates. Here great differences exist between the two compounds. The germanium dioxide curve is continuous through T_m°, indicating that one can go with a given crystal reversibly through the equilibrium melting point. The polyethylene curve shows a sharp break. This break is connected with the long-chain nature of flexible linear macromolecules and their principle of chain folding (see Sections II.E and II.F). Once a macromolecule is completely removed from a crystal surface, it cannot recrystallize by a simple one-unit addition, as small motifs can. Linear macromolecules must undergo a molecular nucleation step before they can grow on a crystal surface (see Wunderlich, 1976). Reversibility exists only on a one-molecule level, i.e., a partially crystallized molecule can at T_m crystallize or melt on lowering or raising the temperature by a small amount. Crystallization close to T_m does not go, however, beyond the already partially crystallized molecules. This is an amount that is too small to be detected macroscopically. Molecular nucleation is a process that is proportional to $\exp(-1/\Delta T)$, similar in description to secondary or surface nucleation in crystals. At greater supercooling, the linear crystallization rate decreases, in contrast to the melting rate. This decrease, which finally stops crystallization, is caused by the slowing of the long-range molecular motion when approaching the glass transition temperature. Detailed crystallization studies can be carried out as discussed in Section III.D. Most useful are isothermal studies by dilatometry or calorimetry (DSC, DTA) supported by optical microscopy. Nonisothermal crystallization kinetics, as results from DSC or DTA on constant cooling rate, can in principle be described through evaluation of the temperature dependence of nucleation and growth expressions (see, e.g., Baro *et al.*, 1977). The resulting expressions contain, however, several uncertain parameters so that to date nonisothermal crystallization data have found usually only qualitative interpretations. Most common is the analysis of the effectiveness of nucleating agents by cooling large numbers of samples at a given rate from the molten state and noting the crystallization exotherm through DTA, DSC, or other thermal analysis techniques. The lower the exotherm, the less effective is the nucleating agent (see, e.g., Binsbergen, 1970). The crystallization can also be studied on heating after quenching to the glassy state (cold crystallization). In this case major nucleation has occurred during cooling and crystallization occurs faster on heating. Multiple crystallization exotherms on heating or on cooling may indicate nonhomogeneous samples (or additional solid–solid transitions, mesophase transitions, etc.)

Turning now to thin crystals, one can derive from Eq. (128) conclusions about reorganization without melting ($dm_c = 0$). The condition $d_iS > 0$ requires, within our assumption, that the thin lamella can only increase in thickness (see Section IV.B.3). The third term in brackets is positive only for positive values of dl. Melting and crystallization without change in lamellar thickness are described again by the second bracketed term, and one can distinguish, as before, melting with superheating and crystallization at a given supercooling, but entropy production is now modified by the surface free energy term. The thinner the crystal, the lower is the temperature where crystallization can take place. A good portion of the break in the crystal growth rate of Fig. 22a can be connected to the fact that the crystals growing at lower temperatures are thin, whereas the data for melting were taken on extended-chain crystals.

Of particular interest is the melting without superheating and without reorganization. Under such conditions the metastable crystal forms a melt of the same metastability and the entropy production is zero. This melting under zero entropy production condition can be used to measure the metastability by thermal analysis. In this case, either surface free energy or crystal thickness can be evaluated by setting the parenthesis of the second bracketed term equal to zero, or write

$$\Delta T = 2\gamma T_m^\circ / \Delta h_f \rho l \tag{129}$$

which is a form of the Thomson–Gibbs equation.

A typical example of the increasing shift of the melting transition to lower temperature on increasingly more metastable crystals is shown in Fig. 23. The major changes in the crystals is their decreasing size, as assumed in the derivation of Eq. (128); other defects should, however, not be neglected. This type of thermal analysis thus permits direct structural information for an identified system. The melting temperature of lamellar crystals measured under zero entropy production conditions is directly linked, through Eq. (129), to lamellar thickness.

A summary in form of a free enthalpy diagram of the discussed cases is shown in Fig. 24. The lower dotted circle indicates the equilibrium melt-

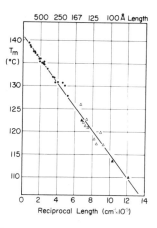

Fig. 23 The melting temperature of lamellar crystals of polyethylene as a function of reciprocal lamellar thickness. Equation for the straight line $T_m = 414.2(1 - 0.627/l) \pm 0.8$ K (l in nanometer); see Eqs. (128) and (129). Data from a literature survey of Wunderlich (1980).

ing that can give information on the chemical nature of an unknown sample and permits discussion of order and stability of crystal and melt (see Section IV.B.1). The upper dotted circle marks the area of zero entropy production melting of a metastable crystal. The melting temperature under such a condition can give quantitative information on metastability (level of the metastable crystal free enthalpy above that of the equilibrium crystal). The two crystallization circles indicate growth of equilibrium crystals (lower circle) and growth of metastable crystals (upper circle). Both can be followed by the standard techniques discussed in Section III.D. Both crystallization on cooling as well as crystallization on heating (cold crystallization) are possible. At least from isothermal experiments (as indicated), information on nucleation and linear crystal growth rates can be obtained. The two melting circles indicate areas of superheating that could be treated similar to the crystallization. The discussion cen-

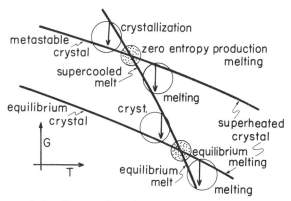

Fig. 24   Free enthalpy diagram of melting and crystallization. The heavy lines indicate the free enthalpies of the melt and the crystalline states. The circles illustrate the permitted phase transitions.

tered around the melting–crystallization transition; other first-order transitions, such as evaporation–condensation or solid–solid transitions, can
be treated analogously. Up to now these latter transitions have only been
analyzed using equilibrium assumptions.

3. Reorganization, Recrystallization, and Annealing

A metastable state is under a constant driving force to change to a
more stable state. Quantitatively, the driving force is expressed by Eq.
(21). Qualitatively the changes are indicated by the downward-pointing
arrows in Figs. 24 and 25.

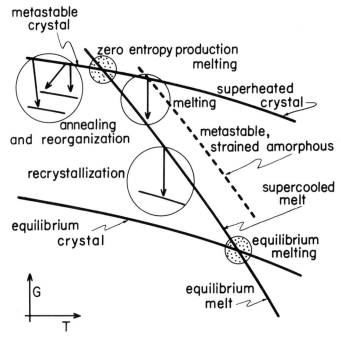

Fig. 25 Free enthalpy diagram of possible paths of annealing and reorganization of metastable crystals.

A crystal may be metastable because of internal defects, because it is
in a crystal form which is less stable, or just because of its small size as
assumed in the derivation of Eq. (128). An amorphous region is frequently
metastable beyond the supercooled liquid because of frozen-in strains. On
drawing, the free enthalpy of an amorphous system increases (see Section
IV.E). If crystallization occurs before a drawn sample can relax, the
higher free enthalpy of the amorphous portion of the molecules may
freeze in (dashed line in Fig. 25). Calorimetric estimates of the higher

enthalpy in amorphous regions of drawn samples were made by Peterlin and Meinel (1965), Fischer and Hinrichsen (1966), and Sumita *et al.* (1977).

The various processes of reorganization, recrystallization, and annealing are shown schematically within the circles of Fig. 25. The term *reorganization* is applied to the improvements of initially grown (metastable) crystals, either at the crystallization temperature or on cooling. Recrystallization is applied to the case of melting of a metastable crystal that is followed by renewed crystallization to a more perfect crystal. Annealing, finally, is the process of crystal improvement by heating to temperatures below the melting point. A special case of reorganization is finally observed in case the metastable crystals do not melt when reaching the free enthalpy of the supercooled melt because of the presence of metastable, strained amorphous regions. The strain in the amorphous regions is maintained by the crystals, which act as cross-links. On partial melting, the strain relaxes and melting accelerates. Initial melting occurs, however, at a higher temperature, and thermal analysis reveals a behavior analogous to the superheating illustrated in Fig. 24. Experimentally the two types of superheating can be distinguished by etching of the two materials. Even light etching will break tie molecules in metastable, fringed micellar aggregates that are needed to support the strain. On breaking of the tie molecules on etching, the melting temperature will drop rapidly to the zero entropy production melting temperature. Crystals in contact with a relaxed melt, in contrast, will be little affected by etching until major changes in molecular weight have been produced (Wunderlich, 1980).

Figure 26 shows several thermal analyses as examples of the various processes. Equation (128) with additional terms for the particular type of metastability can give the framework for a discussion of the various phenomena. The most efficient analysis is the melting at various heating rates. With present commercial equipment (see Chapter 1) it is easy to cover rates from about 0.1 K min^{-1} to 100 K min^{-1}. Higher rates up to several 10,000 K min^{-1} can be reached with a foil calorimeter (Hager, 1972), for example.

A typical sign for reorganization and annealing is the increase in melting temperature on leaving the sample at constant temperature for increasing lengths of time (Fig. 26a). If annealing occurs during the temperature scanning of a thermal analysis experiment, faster heating rates will show lower melting temperatures (Fig. 26b). Often the reorganization can be outrun by fast enough heating. In the example of Fig. 26b constant melting is observed beyond about 50 K min^{-1}. Methods of fixing crystal dimensions that eliminate reorganization are cross-linking of the amorphous areas by radiation, or immobilizing the amorphous areas by

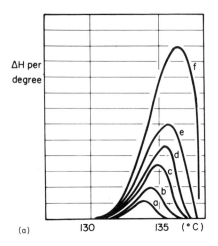

(a)   130   135   (°C)

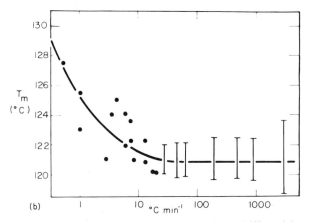

(b)   1   10 °C min⁻¹ 100   1000

Fig. 26 Thermal analysis of irreversibly melting crystals. (a) Differential scanning calorimeter curves, showing the change in melting peak position during the process of crystallization. Besides the increase in heat of fusion due to further crystallization, there is continued perfection, indicated by the increase in melting peak temperature. (Crystallization of
polyethylene from the melt at 128°C for a, 5 min; b, 10 min; c, 15 min; d, 20 min; e, 30 min; f,
69 min. Data from Wunderlich *et al.* (1967) measured at 5°C/min). Reprinted from Wunderlich *et al.* (1967), p. 485, by courtesy of Marcel Dekker, Inc. (b) Melting temperatures of
solution-grown polyethylene lamellae (grown from 0.05 wt % solution in toluene at 81°C,
lamellar thickness 13 nm). Data taken by thermometry on the hot stage of an interference
microscope. The melting temperature at low heating rate is lower because of reorganization.
From Hellmuth and Wunderlich (1965). Reprinted with permission of the American Institute
of Physics. (c) Differential thermal analysis of polyoxymethylene hedrites at 10°C min⁻¹.
The crystals were grown from 0.25 wt % solution in dimethylformamide on slow cooling.
The exotherm at about 165°C is clearly recognizable as recrystallization. From Jaffe and

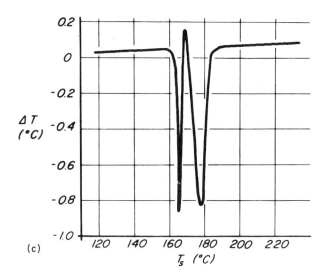

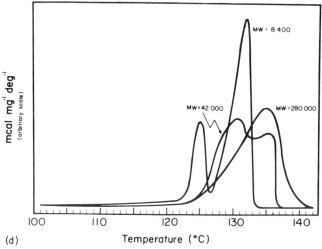

Wunderlich (1967). Reprinted with permission by courtesy of Dr. D. Steinkopf Verlag. (d) Differential scanning calorimetry curves of polyethylene single crystals of different molecular weight showing multiple melting peaks caused by reorganization and recrystallization. Samples similar to (b), grown at $T_c = 84.5°C$ from dilute solution, 5°C min^{-1} heating rate.) Only the lowest melting peak of the 8400 weight sample is indicative of the originally present crystals. All other melting peaks are shifted by reorganization and recrystallization. The higher molecular weights crystallize initially poorer and, as a result, perfect on heating. Reprinted with permission from F. Hamada, B. Wunderlich, T. Sumida, S. Hayashi, and A. Nakajima, *J. Phys. Chem.* **72**, 178. (1968). Copyright by the American Chemical Society.

chemical reaction. The major process of reorganization and annealing in most flexible macromolecules is the fold length increase. Changes in defect concentration are usually of less importance (Wunderlich, 1973, 1976, 1980).

Recrystallization during temperature scanning is an exothermic process, as shown in Fig. 26c. Often, however, initial melting, recrystallization, and crystal reorganization combine to yield no more than a broad melting peak with a surprisingly high melting end. Again, methods of cross-linking or etching are useful in this case to avoid a second crystallization. If melting and recrystallization occur only partially, multiple melting peaks may be observed, as shown in Fig. 26d. To distinguish these from samples of more than one crystal form at low temperature or from samples that contain a multimodal crystal size distribution, thermal analysis must be carried out at different heating rates as is shown in Fig. 27.

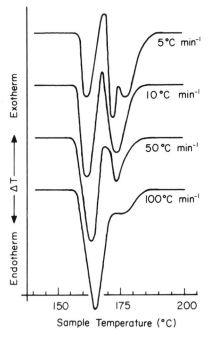

Fig. 27 Differential thermal analysis of polyoxymethylene hedrites. After initial melting, reorganization occurs, giving rise to an exotherm, followed by a second melting. At low heating rate (5°C min⁻¹) a third melting is visible. At high heating rate (100°C min⁻¹) the second melting is only a shoulder. Variation of heating rate permits the recognition of recrystallization and distinction from originally present bi- or trimodal crystal distributions. Data from Jaffe and Wunderlich (1967).

The superheating due to strained amorphous molecules is illustrated in Fig. 28 on the example of poly(ethylene terephthalate). Most of the samples analyzed to date have only been described qualitatively (see Wunderlich, 1980).

Whereas the study of the equilibrium changes of state were of use for the identification and characterization of materials, study of nonequilibrium processes permit the evaluation of the thermal history, for exam-

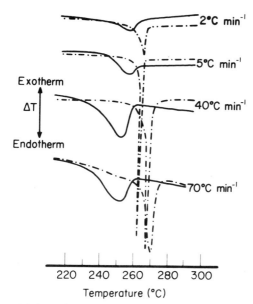

Fig. 28 Differential thermal analysis of drawn poly(ethylene terephthalate) at various heating rates. The drawn-out curves represent the fibers unrestrained (free to shrink). The melting peaks show signs of reorganization (see Fig. 26b). The broken lines represent the fibers held at fixed length until ultimate fusion. There are now signs of superheating. Of interest is also the drop in temperature at 5°C min⁻¹, indicating a loss of restraint after partial melting. From Miyagi and Wunderlich (1972). Reprinted with permission, courtesy J. Wiley and Sons, Inc.

ple, in terms of perfection of crystallization and orientation. Figure 29 shows the full spectrum of melting temperatures for good crystals of polyethylene. A range of 25°C in melting temperature and the typical changes with heating rate permit a fuller characterization of these samples. More extreme crystallization conditions can broaden this range of melting temperatures even further. Polyethylene crystallized during rapid stirring was shown, for example, by Pennings and Zwijinenburg (1979) to be able to exist in crystalline form for short periods of time up to 220°C due to strained amorphous tie molecules. Poorly melt-crystallized polyethylene shows, on the other hand, the first signs of melting around 0°C, as was revealed by an analysis of heat capacity measurements by Wunderlich (1963).

4. Glass Transition

The glass transition temperature is the main characteristic temperature of the amorphous solid and liquid states (see Section IV.A.). A liquid becomes a solid on cooling through the glass transition temperature. The

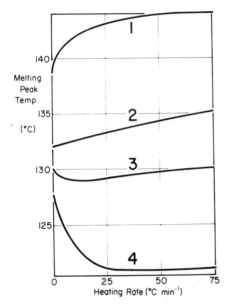

Fig. 29   Melting temperature of poly-ethylene as a function of heating rates. Data from Hellmuth and Wunderlich (1965). 1, Extended chain crystals grown at 227°C and elevated pressure. 2, Folded-chain crystals grown on cooling from the melt at 0.6°C min⁻¹. 3, Poorer folded-chain crystals grown on faster cooling from the melt at 450°C min⁻¹. 4, Single crystals grown from dilute solution (M_n = 8200, M_w = 80,000).

microscopic process involved is the freezing of large-scale molecular motion without change in structure. Since the heat capacity of the glass is always lower than that of the liquid at the same temperature and since there is no latent heat in stopping molecular motion, the glass transition takes the appearance of a thermodynamic second order transition [Eq. (121) and Fig. 16b]. The freezing of molecular motion is, however, time dependent, so that the glass transition must be called an irreversible process. The glass–liquid transition occurs at a recognizable "transition temperature" because of a rather large temperature dependence of the relaxation time for large-scale molecular motion.

The most precise determination of the glass transition temperature is done by cooling an equilibrium melt (which is, however, supercooled relative to crystals) at a specified rate and finding the temperature of half-freezing as it can be measured by heat capacity, expansion coefficient, or compressibility measurement [see Eq. (121)]. This temperature is close to the inflection point (see Fig. 31). On plots of the integral quantities volume, enthalpy, or entropy this temperature is close to the break in the curve (see top of Fig. 30). Experiments of faster or slower cooling rates will give higher or lower glass transition temperatures, respectively. Polystyrene, for example, shows a change in glass transition temperature from 365 K at a cooling rate of 1 K h⁻¹ to 380 K at a cooling rate of 1 K s⁻¹ (Wunderlich et al., 1964). Even for a cooling rate of 1 K yr⁻¹ the observed glass transition does not decrease below 351 K, and for a cooling rate of a

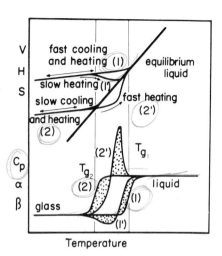

Fig. 30 Schematic drawing of changes in enthalpy, entropy, volume, and the derivative properties heat capacity, expansion coefficient, and compressibility on heating and cooling through the glass transition region. From Weitz and Wunderlich (1974). Reprinted with permission, courtesy J. Wiley and Wiley, Inc.

kelvin per century, 344 K is expected for the glass transition temperature. Because of this small variation of the transition temperature with cooling rate, the specification of time scale is often neglected. A large collection of glass transitions of linear macromolecules is given by Lee and Rutherford (1975).

Detailed and sometimes conflicting theories of the glass transition, mainly applied to linear macromolecules, are discussed, for example, by Kanig (1969), Petrie (1974), Someynsky and Simha (1971), Gibbs and DiMarzio (1958), Adam and Gibbs (1965), Wunderlich *et al.* (1964), Boyer (1976), and Shen and Eisenberg (1970). The theories are all simplified and do not describe the glass transition fully. A major shortcoming of all theories is the omission of the cooperative nature of the glass transition. Unfreezing one configuration significantly helps the neighboring molecular segments to move.

Schematically the changes in volume V, enthalpy H, and entropy S as well as the derivative heat capacity C_P and thermal expansivity α on cooling and heating are shown in Fig. 30. Coupling equal heating and cooling rates (paths 1 and 2) leads to close to "normal" glass transition behavior and permits measurement on heating as well as on cooling. The thermodynamic quantities change almost reversibly between the values characteristic for the glass and the melt at a temperature determined by the time scale of measurement. Coupling fast cooling with slow heating permits the system to drift toward the equilibrium melt before the temperature is reached, at which the system froze on cooling (path 1'). In C_P or α this behavior is characterized by a minimum before the increase to the value of the liquid. On coupling slow cooling with fast heating (path 2') equilibrium melt cannot be achieved at the low temperature where the system froze on cooling. As a result, one obtains a superheated glass that drifts quickly toward equilibrium as soon as the time scale of heating permits. In C_P or α this behavior is characterized by a maximum. Since the integrals over C_P and α over a cyclic path must give zero, because the integral functions H and V are functions of state, the corresponding dotted areas of Fig. 30 must be equal. This behavior of glasses is called the hysteresis behavior and can be used to obtain information on the thermal history. Such curves in heat capacity or expansion coefficient do not immediately reveal the position of the glass transition since their behavior is governed by both heating rate and thermal history. Richardson and Savill (1975) suggest ignoring in these cases the detailed shape of the C_P curve for the glass transition temperature determination; instead, one should extrapolate the integral curves of liquid and glassy enthalpy (or volume) to the point of intersection. The point of intersection is identical to the break in the enthalpy curve that one would have obtained on cooling of the sample at a rate that yields the same stability glass (see also Flynn, 1974).

Figure 31 illustrates a series of DTA traces on polystyrene cooled at various rates but heated at the same heating rate. The change from a glass transition coupled with an endothermic hystersis peak to a glass transition

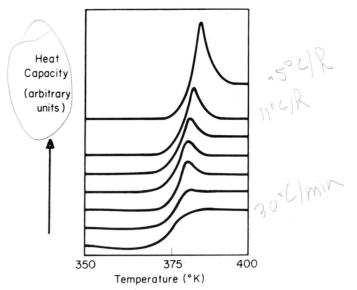

Fig. 31 Heat capacity of polystyrene in the glass transition region, cooled at different rates, measured by differential thermal analysis at a heating rate of 5°C min⁻¹. Cooling rates (from top to bottom): 0.5°C h⁻¹, 11.5°C h⁻¹, 0.5°C min⁻¹, 2.5°C min⁻¹, 5.0°C min⁻¹, 30°C min⁻¹. Data from Wunderlich *et al.* (1964).

proceeded by a shallow exotherm can be seen. The "normal" glass transition is observed as mentioned, only when cooling and heating rates are approximately equal. Time-dependent glass transition temperatures measured by thermal mechanical analysis are described, for example, by Schwartz (1978).

In case strain of any kind is frozen into the glass, it is released in the vicinity of the glass transition and can be measured by the heat, volume, or length effects. An example is shown in Fig. 32, where a polystyrene

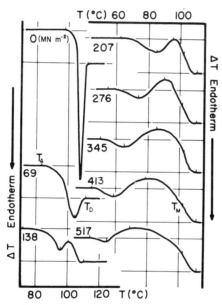

Fig. 32 Differential thermal analysis curves of polystyrene glasses produced by slow cooling (5°C h⁻¹) at various pressures (listed in megapascals next to the curves). Analysis after release of pressure at 5°C min⁻¹. Curves normalized to constant ΔC_p. The increasing exotherm indicates different amount of internal stress release. From Weitz and Wunderlich (1974). Reprinted with permission, courtesy J. Wiley and Sons, Inc.

sample cooled under various pressures is heated to the liquid state after release of pressure. It can be seen that several effects operate during the glass transition. Only the end of the glass transition region remains fixed, which agrees with the reaching of overall conformational equilibrium at that point. No strain is maintained at this temperature without external stress. The frozen strain does start relaxing, however, at increasingly lower temperatures for higher experimental pressures. Experiments of this type indicate that much more information can be deduced from a quantitative analysis of the glass transition region than is presently customary.

For strained fibers and films thermomechanical analysis is particularly suited to study the shrinkage in the glass transition region (see, for example, Eisenberg and Trepman, 1980).

A method of thermal analysis that relies on the disappearance of birefringence of strained samples that relax at the glass transition was developed by Kovacs and Hobbs (1972) (thermo-optical analysis).

The relaxation of frozen-in dielectric stresses can be analyzed by measuring the thermally stimulated discharge current (TSDC analysis) (van Turnhout, 1975; Carr, 1980).

Once cooled, glasses can, similar to crystals, change their perfection through annealing to a state of lower free enthalpy. Figure 33 shows such an annealing sequence on one of the pressure-cooled glasses of Fig. 32. Volume and enthalpy relaxation did not go parallel in these experiments, an indication that several ordering parameters are needed for a full description of the glass transition. For a more extensive discussion of annealing effects of glasses see, for example, Petrie (1974).

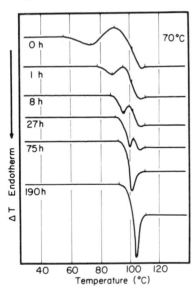

Fig. 33 Differential thermal analysis curves of polystyrene glass produced by slow cooling (5°C h⁻¹) at 417 MPa after different annealing times at 70°C. Annealing times given next to the curves. Compare with Fig. 32. From Weitz and Wunderlich (1974). Reprinted with permission, courtesy J. Wiley and Sons, Inc.

C. Three-Phase Systems

A system of three phases of one component in equilibrium should be possible only at one temperature, the triple point (Section V.B.1, phase rule). For equilibrium among the three phases α, β, and γ one can write

$$G^\alpha = G^\beta = G^\gamma \tag{130}$$

where G^α, G^β, and G^γ are the molar free enthalpies of the respective phases. The only variables in a one-component system in equilibrium are

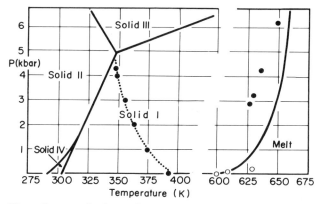

Fig. 34 Phase diagram of polytetrafluoroethylene after data from Hirakawa and Take-mura (1969). The dotted line indicates where the average untwisting of the helical conformation is complete.

temperature and pressure. Both are fixed, however, by the two equations in Eq. (130). A typical phase diagram is shown in Fig. 34 for polytetrafluoroethylene. The solid forms I, II, and III are in equilibrium at about 345 K and 5 kb (500 MPa) and the forms I, II, and IV, at about 315 K and 1.5 kb (150 MPa). Another triple point is expected betweeen the forms I, III, and melt. Triple points can be determined easily by thermal analysis (see also Section V.B.1 and Fig. 42).

As in two-phase systems, equilibrium is not always reached. Figure 35 shows the shift in the phase boundary between orthorhombic polyethylene crystals and melt on going to metastable folded-chain crystals. To a

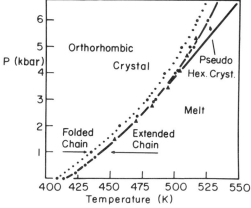

Fig. 35 Phase diagram of polyethylene. Data collected from the literature by Wunder-lich (1980). Typical broad-molecular-weight samples. The metastable folded-chain phase boundary lies above but largely parallel to the extended-chain, equilibrium line.

large degree the shift is parallel, as expressed by Eq. (129). The triple point must necessarily also change under such conditions. Quantitative data are still rare.

Nonequilibrium systems may contain more than the three phases permitted by the phase rule. As described in Section III.C, the system must then be broken down into subsystems that are treated separately (see also Fig. 18). The treatment of each subsystem, pair of subsystems, or triple of subsystems would follow the descriptions given previously.

D. PRESSURE DEPENDENCE OF PHASE TRANSITIONS

At an equilibrium first-order phase transition of a one-component system, ΔG is zero. The transition temperature is fixed by the enthalpy of transition ΔH and the entropy of transition ΔS [See Eq. (122)]. To find the pressure dependence of the transition, the free enthalpy changes in both phases (dG^{α} and dG^{β}) must be the same in order to maintain equilibrium. One can thus write for the example of fusion, making use of Eqs. (87) and (120),

$$dG_{\text{melt}} - dG_{\text{cryst}} = 0 = -\Delta S_{\text{f}}\, dT_{\text{m}} + \Delta V_{\text{f}}\, dP \qquad (131)$$

which is a form of the Clausius–Clapeyron equation

$$dT_{\text{m}}/dP = \Delta V_{\text{f}}/\Delta S_{\text{f}} = \Delta V_{\text{f}} T_{\text{m}}/\Delta H_{\text{f}} \qquad (132)$$

In Figs. 34 and 35 the pressure dependence of melting polytetrafluoroethylene and polyethylene are shown, respectively. Also shown are solid–solid transitions that can be treated analogously. In both cases the initially rapid increase of T_{m} with pressure slows down after reaching 1–2 kb of pressure (100–200 MPa). Furthermore, the metastable, folded-chain lamellar crystals of polyethylene seem to have a similar change in melting temperature with pressure.

Thermometry at different pressures thus allows the evaluation of the ratio $\Delta V_{\text{f}}/\Delta H_{\text{f}}$ and gives a check on the dilatometric ΔV_{f} and the calorimetric ΔH_{f}. For polyethylene, for example, the molar heat of fusion is 4.109 kJ and the molar volume change on fusion is 3.535×10^{-6} m³, which results at the equilibrium melting temperature, 414.6 K, in a pressure coefficient of the melting temperature of 33.8 K kbar⁻¹,* which compares favorably with the measured value of 35.2 K kbar⁻¹.

The change of melting temperature with pressure of several other macromolecules in the vicinity of atmospheric pressure is listed in Table VIII. At higher pressures, the slopes always decrease as in the polyethylene and polytetrafluoroethylene cases. In the case of poly-3,3-bis(chloro-

* To convert the non-SI pressure bar into pascal multiply by 10⁵.

TABLE VIII

CHANGE OF MELTING TEMPERATURE WITH PRESSURE[a]

Macromolecule	T_m° (K)	dT_m/dP (K kbar^{-1})
Polyethylene	414.6	35.2
Polytetrafluoroethylene	600	152
Polypropylene	460.7	36
Poly-1-butene	411	51
Poly-4-methyl-1-butene	523	56
Poly(vinylidene fluoride)	480	37.5
1,2-Poly-2-methylbutadiene	301	37
Poly(ethylene oxide)	342.1	15.7
Poly(tetramethylene oxide)	330	18.5
Poly-3,3-bis(chloromethyl)-oxacyclobutane	(455)	26
Poly(2-6-dimethylphenylene oxide)	545	25
Poly(ethylene adipate)	338	15
Poly(hexamethylene succinate)	(330)	16
Poly(4,4'-isopropylidenediphenylene carbonate)	(533)	42
Nylon 6	533	16

[a] Collected by Wunderlich (1980) from literature data.

methyl)oxacylobutane, there may even be a maximum in the melting temperature–pressure curve. Early experiments showed a broadening of the melting ranges of poorly crystallized samples with increasing pressure. This would, in retrospect, indicate either that poor metastable crystals are less affected by pressure than good crystals (since the Clausius–Clapeyron equation seems to be valid at the melt-end; as shown in Fig. 35) or that perfection of poor crystals is faster at atmospheric pressure. (Thermal analysis would in this case show final melting of better crystals than were initially present.) Experiments by Jenckel and Rinkens (1956) on poly(ethylene adipate) and Nylon 6 and by Parks and Richards (1949) on branched polyethylene show this broadening of the melting range particularly clearly. Newer experiments on better crystals have shown none of this broadening (see, for example, Yasuniwa *et al.*, 1973), or Takamizawa *et al.*, 1975).

E. DEFORMATION DEPENDENCE OF PHASE TRANSITIONS

Crystals of rigid macromolecules and small molecules, as defined in Section II.A, show no known equilibrium effects of mechanical deformation on melting transitions, since a melt of rigid macromolecules or small molecules cannot maintain a tensile stress under equilibrium conditions. Flexible linear macromolecules behave differently when they form networks through cross-linking between chains. In this case a state of stress can be maintained indefinitely in the melt, and a clear effect on the melting temperature of crystals in equilibrium with such melt is expected. Even if the molecules are not permanently linked through covalent cross-links, entanglements or crystals connecting portions of molecules may set up a network for a sufficiently long time. Furthermore, flexible linear macromolecules may be maintained in a state of deformation by extensional flow. The latter two effects are nonequilibrium effects.

The effect of equilibrium deformation on the melting temperature can be understood when comparing it to the effect of pressure discussed in Section IV.D. The pressure P is to be replaced by the tensile force f, and the volume change on fusion is to be replaced by a change in length Δl of the sample. The equilibrium change in melting temperature with tensile force at constant pressure is then expressed by an equation analogous to the Clausius–Clapeyron equation [Eq. (132)]

$$dT_m/df = -\Delta l/\Delta S \tag{133}$$

where ΔS is the entropy change on fusion at the given force f. The $P–T$-phase diagrams as shown, for example, in Figs. 34 and 35 find their equivalence in the $f–T$ phase diagram for cross-linked natural rubber, *cis*-1,4-poly-2-methylbutadiene, shown in Fig. 36. A relatively sharp equilibrium melting temperature is found on heating at constant tensile force f (following line 1–2), or a well-defined tensile melting force is found on decreasing the force at constant temperature (following line 3–4). Since the equilibrium crystal contains fully extended chain molecules and the molecular

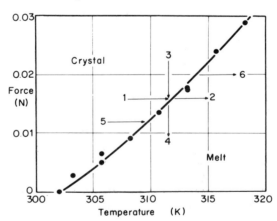

Fig. 36 Force–temperature phase diagram of *cis*-1,4-poly-2-methylbutadiene. Radiation cross-linked sample of amorphous relaxed dimensions of 7.63×10^{-3} cm² cross section and 2.08 cm length and extended, zero force crystalline length 8.30 cm at about 300 K. Cross-linked fraction of repeating units 0.0156. Crystallinity 0.24. Data from Oth and Flory (1958).

extension may be assumed to be proportional to the sample length, Δl is negative on fusion, the sample contracts. When held at constant length (following line 5–6), the force must thus increase on melting, raising the melting temperature of the remaining crystals. It is a general observation that sufficiently oriented crystallized fibers of flexible linear macromolecules shrink on fusion. The change of entropy on fusion ΔS should be positive, so that T_m should increase with increasing tensile force, as shown in Fig. 36.

Plotting length versus temperature of an oriented crystalline fiber, the equivalent of a volume–temperature phase diagram for melting at various pressures is obtained, as is shown schematically in Fig. 37. Heating a

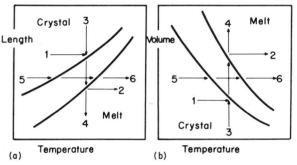

Fig. 37 (a) Schematic length–temperature diagram for a stretched sample. (b) Schematic volume–temperature diagram of a compressed sample.

stretched sample at constant length (following line 5–6), a melting temperature range in which the crystallinity varies is expected. Similarly, on relaxing the length at constant temperature, a length and crystallinity range for fusion is expected. The analogous paths on compression are indicated in Fig. 37b. Also shown are the paths 1–2 for heating at constant tensile force or pressure and path 3–4 at constant temperature for comparison with the diagram Fig. 36. Similar to the decrease in volume change on fusion at increasing pressures, the length difference decreases as larger tensile forces are applied.

Phenomenologically the deformation effect is thus well understood. Experimentally, crystallization on stretching was first observed and studied by Katz (1925a,b) and Hauser and Mark (1926a,b). Attempts at more detailed descriptions revealed, however, that the "simultaneity of crystallization and deformation is not conducive to the attainment of equilibrium, as would be required for a thermodynamic analysis for the fusion process. . ." (quote from Smith, 1976; see also Flory, 1947). In particular, at lower extension, the obtained crystals are not fully oriented; they are, as usual for flexible linear macromolecules, spherulitic in nature with chain folding inhibiting attainment of equilibrium.

A more detailed statistical treatment of the crystallization of an isolated, stretched macromolecule was first given by Alfrey and Mark (1942). Along the same line, Flory (1947) developed the semiquantitative equilibrium theory of crystallization and melting of a network of cross-linked molecules most commonly used. The theory is based on perfect crystal orientation in the direction of stress. The molecular chains of the crystals are connected at the appropriate surface with the network of Gaussian chains in an equilibrium strain field.* Conformational energy changes and effects of lateral crystal sizes are neglected, and it is assumed that a molecule makes only a single pass through a given crystal in a direction that will finally lead to ultimate equilibrium. For the melting temperature T_m° of the last trace of crystal of such a distribution (or the incipient crystallization temperature), one assumes that the crystal and amorphous free enthalpies are equal

$$G_c(T_m) = G_a(\alpha, T_m) \tag{134}$$

where α is the extension ratio of the sample. The crystal is assumed not to be affected by the extension. Expanding both sides of Eq. (134) with the free enthalpy of the amorphous network at $\alpha = 1$, $G_a(1, T_m)$, leads to

$$G_a(1, T_m) - G_c(T_m) = G_a(1, T_m) - G_a(\alpha, T_m) \tag{134a}$$

The left-hand side of Eq. (134a) represents just the change in free enthalpy on fusion of a crystal at T_m under no deformation as given by Eq. (126). The right-hand side of Eq. (134a) is just the change in free enthalpy on stretching the network from the extension ratio 1 to α. Combining both expressions yields

$$\frac{1}{T_m} = \frac{1}{T_m^\circ} - \left(\frac{V_0}{\Delta H_f}\right)\left(\frac{1}{T_m}\right)\int_1^\alpha \frac{f}{A}\,d\alpha \tag{135}$$

A measurement of stress as a function of extension ratio of the amorphous network on initial crystallization should thus be able to predict the equilibrium melting temperature.

Figure 38 shows a plot of the melting temperature as a function of extension ratio α for *trans*-1,4-poly-1-chlorobutadiene. Also shown in the figure are the expected melting temperatures from Eqs. (135) and (119). At low extension, large deviations are obvious, as one expects from the simplifying assumptions of the model.

---

* For a review of the structure and elasticity of polymer networks, see, for example, Dusek and Prins (1969) and Treolar (1958).

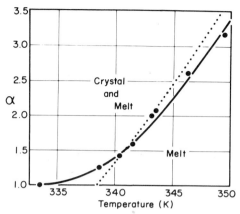

Fig. 38 Extrapolated melting temperatures as a function of extension ratio α of *trans*-1,-4-poly-2-chlorobutadiene. Dotted curve according to Eqs. (135) and (119), $n = 110$, $T_m^\circ = 333.45$ K, $\Delta H_f = 8.37$ kJ mole^{-1}. Data from Krigbaum *et al.* (1966).

One may conclude that the general features of equilibrium deformation effects are understood. Experimentally, it is possible to establish with reasonable precision the force–temperature phase diagram of Fig. 36. Of the length–temperature phase diagram shown schematically in Fig. 37a, only the liquidus curve is available experimentally (see Fig. 38). For the solidus curve, equilibrium would have to be established at high crystallinity, which is not possible.

Nonequilibrium effects can at present be understood only qualitatively. The extension of any amorphous subsystem increases its free enthalpy as schematically shown in Fig. 25. Higher melting temperatures can thus be observed. The level of the melting temperature increase is, however, critically dependent on the thermal history of the network and changes during the partial melting (see also Section V.B.2). The thermal analysis is carried out by stretch calorimetry (see Göritz and Müller, 1970) or thermomechanical analysis. In both instrumentations a measured force can be applied to the sample. Standard calorimetry, DSC, and DTA are applicable to analysis of metastable, strained samples either under conditions of free relaxation or constant length (see also Fig. 28).

F. Molecular-Weight Dependence of Phase Transitions

Molecules of different size can be treated as different components, so that we need at this point only to treat the changes one observes in going from one molecular weight to another. Mixtures of different molecular weight will be treated as multicomponent systems in Section V.C. Furthermore, in the definition of a macromolecule we required at least 1000 atoms in the molecule. This makes the changes in properties with molecular weight per unit mass small.

Most simply, one breaks down all extensive variables X into contributions from the repeating units X_a and separate excess contributions from the chain ends X_e. Any such quantity can then for a molecule of x repeating units be written

$$X = xX_a + 2X_e \qquad (136)$$

or per repeating unit

$$(X/x) = X_a + (2X_e/x) \qquad (137)$$

As x increases, the second term becomes increasingly negligible. Because the chain ends occupy a larger volume, the specific volume v_a of larger macromolecules is usually smaller. For liquid polyethylene and polypropylene, for example, the following empirical equation was derived by Wilski (1964):

$$v_a = (1.142 + 212/\overline{M}_\eta) + (9.25 \times 10^{-4} + 0.777/\overline{M}_\eta)t \qquad (138)$$

where v_a is expressed in cubic meters per megagram and t in degrees Celsius, and $\overline{M}_\eta$ is the viscosity average molecular weight, which is proportional to x. Because of the larger volume requirement of the chain ends one would also expect a somewhat reduced enthalpy for longer molecules. The greater mobility of chain ends in the liquid phase would yield a greater entropy for smaller molecules.

The entropy of liquid macromolecules shows, however, an additional length dependence in Eqs. (136) and (137). This length dependence arises from the number of ways ν the first segment of a molecule of x repeating units can be placed on a lattice of Nx positions, namely,

$$\nu_i = [Nx - (i - 1)x] \qquad (139)$$

where $(i - 1)x$ represents the number of lattice positions already occupied by $i - 1$ molecules. The thermodynamic probability of placing the first segments of all N molecules of x repeating units (i.e., filling the complete lattice) is the product over all ν_i, or

$$W = \prod_1^N \times \frac{N - (i - 1)}{N!} = x^N \qquad (140)$$

where the factor $N!$ arises because all N molecules are indistinguishable against exchange. Using Eq. (16) shows that placing the first segment of each molecule has the entropy contribution

$$s_1 = nk \ln x \qquad (141)$$

Per mole this contribution is $R \ln x$ and the amorphous entropy per repeating unit must be written with Eq. (137)

$$s = s_a + (2s_e + R \ln x)/x \qquad (142)$$

Equation (142) would approach a constant value more slowly than Eq. (137). Crystalline macromolecules show no added entropy term, since the positions of all segments of the molecules are fixed. Small molecule liquids or liquids of large molecules with interchanging bonds show also no similar term for the placement of the first segment, since all segments are interchangeable, not just the molecules.

Of particular interest is the change in melting temperature with molecular length which is measureable by thermometry and other thermal analysis techniques. The change in melting temperature can be derived from the molar free enthalpy of fusion ΔG_f, in which for the enthalpies of melt and crystal and the entropy of the crystals, simple expressions of the type of Eq. (137) are used, and for the liquid entropy Eq. (142) is inserted.

$$\Delta G_f = x \, \Delta g_f^\circ + \Delta g_e - RT \ln x \qquad (143)$$

where Δg_f° is the free enthalpy of fusion per repeating unit for a molecule of infinite molecular weight. At the melting temperature ΔG_f is zero and with Eq. (127) one finds that

$$T_m = T_m^\circ [1 - (RT \ln x - \Delta g_e)/(x \, \Delta h_f^\circ)] \qquad (144)$$

where T_m° is the melting temperature of a crystal of infinite molecular weight. For polyethylene the equation could be shown to be of the proper form to describe melting temperatures from $x = 12$ to infinity (Wunderlich and Czornyj, 1977). Restricting ourselves to linear macromolecules, i.e., to polyethylene of 1000 atoms or more ($x > 333$), one finds only a small change in temperature left as is shown in Fig. 39. For this region of molecular weight a simplified equation of the type

$$T_m = T_m^\circ - a/x \qquad (145)$$

is sufficient to represent the experimental data.*

Another macromolecule with a sufficiently large experimental body of information for discussion is poly(ethylene oxide). The melting data from six laboratories on 42 fractions of extended-chain crystals or close to ex-

---

* Equation (145) approaches for large x the equation $1/T_m = a + b/x$ suggested by Meyer and van der Wyk (1937) as an empirical rule for oligomer melting temperatures. Also shown in Fig. 39 is the equation $T_m = 414.3 \, (x - 1.5)/(x + 5.0)$, which was first derived by Broadhurst (1962) and fits the data also well.

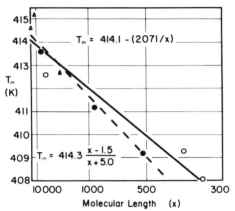

Fig. 39 Experimental equilibrium melting temperatures of polyethylenes of different molecular weights. Data collected by Wunderlich (1980).

tended-chain crystals were collected by Buckley and Kovacs (1975). Figure 40 shows the upper end of the experimental data. To compare the data with polyethylene, one should note that the repeating unit of poly(ethylene oxide) has three chain atoms rather than one. Furthermore, most data on poly(ethylene oxide) are still in the low-molecular-weight range, so that the fit of Eq. (145) is less satisfactory. Better results are obtained using the full Eq. (144), which is drawn in Fig. 40 as a dashed line. The

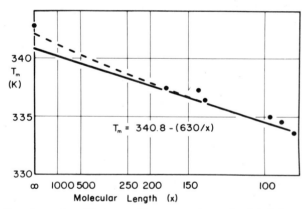

Fig. 40 Experimental equilibrium melting temperatures of poly(ethylene oxide)s of different molecular weights. Data collected by Wunderlich (1980). The dashed line is based on Eq. (145).

calculated curve shows, relative to the straight line of Eq. (145), a clear upward trend at high molecular weights, because of the logarithmic term in Eq. (144), and a downward trend starting at about 2000 MW, resulting from the entropy contribution to Δg_e.

The crystal structure of macromolecules is not expected to change with molecular weight. For equilibrium crystals only the crystal thickness in the molecular chain direction would increase linearly with chain length. For paraffins the number of chain atoms is still important to decide the type of stacking of the molecules; such effects are not expected for macromolecules. The specific volume change on fusion is thus mainly dependent on changes of the specific volume in the molten phase. In addition, it must be corrected for the temperature change of the melting temperature.

The glass transition temperature as a function of molecular weight is coupled with the free volume in the liquid state and follows a molecular weight dependence as given by (Fox and Loshaek, 1955)

$$T_g = T_g^\circ - a/x \tag{146a}$$

This equation is identical to Eq. (145) and easily derivable on a free volume argument based on higher free volume contribution by chain ends [see Eq. (136)]. Ueberreiter and Kanig (1951) suggested for wider ranges of molecular weights the equation [see footnote on p. 183]

$$1/T_g = 1/T_g^\circ + a/x \tag{146b}$$

V. Characterization of Multicomponent Systems

For the description of multicomponent systems the number of moles N_i or concentration of each of the components is added to temperature and pressure as variables of state [see, for example, Eq. (87)]. For each of the extensive functions of state such as V, U, S, G, H, C_P, and C_V, for example, we write for each component A, B, etc., the derivatives

$$(\partial X/\partial N_A)_{T,P,N_B} = X_A, \qquad (\partial X/\partial N_B)_{T,P,N_A} = X_B \tag{147}$$

and call the quantities X_A and X_B the partial molar quantities. The partial molar free enthalpies G_A and G_B are the chemical potentials and are written commonly μ_A and μ_B.

$$(\partial G/\partial N_A)_{T,P,N_B} = G_A \equiv \mu_A, \qquad (\partial G/\partial N_B)_{T,P,N_A} = G_B \equiv \mu_B \tag{148}$$

In general, the partial molar quantities are not additive, but one can derive the following relationship between the different components (shown for two components, but easily generalized for multicomponent systems; shown for volume, but easily also written for all other extensive functions of state):

$$dV = V_A \, dN_A + V_B \, dN_B \tag{149}$$

On integrating Eq. (149) without changing the composition, one obtains

$$V = N_A V_A + N_B V_B \tag{150}$$

The total differential of Eq. (150) must be

$$dV = V_A \, dN_A + N_A \, dV_A + V_B \, dN_B + N_B \, dV_B \tag{151}$$

Comparison of the coefficients of Eqs. (151) and (149) reveals the important Gibbs–Duhem equations

$$N_A \, dV_A + N_B \, dV_B = 0 \qquad \text{or} \qquad dV_A = -(N_B/N_A) \, dV_B \tag{152}$$

which permit the calculation of the change in partial molar quantities of one component from the other.

The thermodynamic description of the formation of a multicomponent system, the solution, can be written, for example, for the free enthalpy

$$N_A A + N_B B \rightarrow \text{solution} \tag{153}$$

with the minor component often called the solute and the major component the solvent.

$$\Delta G_{\text{solution}} = G_{\text{solution}} - N_A \mu_A^\circ - N_B \mu_B^\circ \tag{154}$$

where μ_A° and μ_B° are the chemical potentials of the pure solvent and solute. Making use of Eq. (150), one can write

$$\Delta G_{\text{solution}} = N_A \mu_A + N_B \mu_B - N_A \mu_A^\circ - N_B \mu_B^\circ \tag{155}$$

or

$$\Delta G_{\text{solution}} = N_A \, \Delta\mu_A + N_B \, \Delta\mu_B$$

The characterization of multicomponent systems through thermal analysis can be done by any of the thermal analytical techniques following the descriptions given in Section III.E. In this section single-phase systems will be treated first, followed with a discussion of multiple-phase systems, to be completed by a discussion of molecular-weight distributions that are common in multicomponent systems and the characterization of systems that undergo chemical reactions.

A. SINGLE-PHASE SYSTEMS

The basic question of a single-phase, multicomponent system is the compatibility of the components that keeps them in solution. For linear macromolecules this compatibility must be discussed in three stages: compatibility of small molecules with the macromolecule, compatibility within the macromolecules on a repeating unit scale, and compatibility among macromolecules. The thermodynamics is analogous to the single-phase system description given in the corresponding Section IV.A, with the addition of the evaluation of the partial molar quantities defined earlier. Figure 41 illustrates the determination of the partial molar volumes at

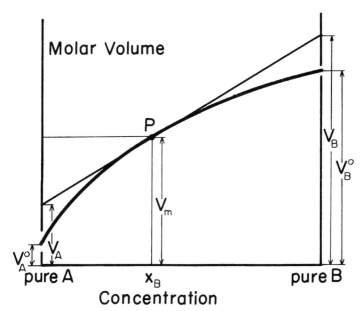

Fig. 41   Schematic drawing for the evaluation of partial molar quantities. Example: volume.

point P in a two-component single-phase system at constant temperature and pressure. Dilatometry can easily determine the mean molar volume to be

$$V_m = V/(N_A + N_B) \tag{156}$$

Similarly, calorimetry would have to be used for heat-related partial molar quantities. From Eq. (156) one can get an expression for the partial molar volume V_A by carrying out the differentiation suggested by Eq. (147)

$$V_A = V_m + (N_A + N_B)(\partial V_m/\partial N_A)_{T,P,N_B} \tag{157a}$$

Obviously, the change in mean molar volume with N_A must also be determined experimentally. Changing in Eq. (157) to concentration, the mole fraction of B [$x_B = N_B/(N_A + N_B)$] leads to

$$V_A = V_m - x_B \, dV_m/dx_B \tag{157b}$$

which is seen geometrically from Fig. 41 and derived by recognizing that

$$\left(\frac{\partial V_m}{\partial N_A}\right)_{N_B} = \frac{dV_m}{dx_B}\left(\frac{\partial x_B}{\partial N_A}\right)_{N_B} \quad \text{with} \quad \left(\frac{\partial x_B}{\partial N_A}\right)_{N_B} = -\frac{N_B}{(N_A + N_B)^2}$$

Measurement of the mean molar volume and the change of the mean molar volume with mole fraction is thus sufficient to calculate the partial molar volumes.

1. The Solid States

The multicomponent glassy state has again a structure similar to the liquid state, described in Section V.A.2. The characterization at a given concentration of the components follows the outline of Section IV.A. Measurements as a function of concentration can be used to evaluate the partial molar quantities as shown in Fig. 41. A distinction between single or multiphase nature is made by the study of the glass transition, given in Section V.B.2 (see also Section IV.B.4). A single-phase glass shows on heating only a single glass transition temperature at which large-scale motion becomes possible. The effect of composition on the glass transition temperature $T_g(w)$ of a two-component, one-phase system has been described empirically by Jenckel and Heusch (1958) as

$$T_g(w) = wT_g^1 + (1 - w)T_g^2 + w(1 - w)K \qquad (158a)$$

where w is the mass fraction of component 1 and K is an empirical parameter chosen for good fit over the whole concentration range. If K is constant, it can be calculated from the glass transition at concentration $w = 0.5$ or from the initial slope

$$T_g(0.5) = 0.5(T_g^1 + T_g^2) + 0.25K \qquad (158b)$$

$$dT_g(0)/dw = T_g^1 - T_g^2 + K \qquad (158c)$$

Another empirical equation makes use of the adjustable parameter L

$$T_g(w) = [wT_g^1 + L(1 - w)T_g^2]/[w - L(1 - w)] \qquad (159)$$

Both expressions have been linked to the conformational entropy theory of glass formation (Gordon *et al.*, 1977). For many systems Eqs. (158) and (159) have been expressed equally well with mole or volume fractions rather than weight fractions (using appropriately different constants). Couchman and Karasz (1978) and Couchman (1978, 1979, 1980) analyzed the correlation between the various expressions.

A multicomponent system that crystallizes with a common crystal structure is said to form mixed crystals. Mixed crystal formation is often possible for components that crystallize in their pure state with the same crystal shapes. For this reason this phenomenon is known as *isomorphism*. For spherical motifs (see also Section IV.B.1) the two main conditions for isomorphism are chemical compatibility, which usually means similar bonding between the different atoms, ions, or molecules in the crystal, and compatibility in size. The size restriction is much more severe than in the liquid state, so that the cases of isomorphism are fewer than the cases of solubility in the liquid. On going to irregular motifs, shape considerations have to be added to size compatibility for isomorphism. For linear macromolecules a further condition must be fulfilled, the chain conformations of the components must match. For energy reasons crystals can usually only be obtained with macromolecules close to their equilibrium conformations. In this conformation, close packing must be achievable with the second component (see Wunderlich, 1973).

Unlimited isomorphism of linear macromolecules with small molecules is not possible for geometrical reasons. Small molecules can, at best, be placed in special positions within the macromolecular crystal, as, for example, in poly(trimethylene oxide), which crystallizes with one water closely associated with each oxygen of the macromolecule in its crystal form I. The water does not replace the macromolecular motif, and this is thus not a true isomorphism. Similarly, globular large motifs, as in protein and virus crystals, exist often only if a variable amount of water fills the interstices between the motifs (Blundell and Johnson, 1976). In reverse, clathrate compounds of urea and thiourea, which are small molecules, with macromolecules, like poly(vinyl chloride) and *trans*-1,4-poly-2,3-dimethylbutadiene, have a joint crystal structure with fixed proportions of the two components (see Wunderlich, 1976).

Isomorphism between complete homopolymer chains is not very common because of the condition of matching chain conformations. The obvious cases of isomorphism are between molecules like deuterated and hydrogen substituted macromolecules. A true case of isomerism exists between poly(vinyl fluoride) $(CH_2—CHF—)_x$ and poly(vinylidene fluoride) $(CH_2—CF_2—)_x$. Both homopolymers have the same crystal structure and are chemically similar. Mixed crystals are found over the whole concentration range (Natta *et al.*, 1965).

Seventeen other systems of isomorphism between complete homopolymer chains of different configuration, conformation, and chemical structure are listed by Allegra and Bassi (1975).

Repeating unit isomorphism is somewhat more common in linear macromolecules. Its importance lies in the area of crystalline copolymers. In contrast to small molecule solvents, which can separate on crystallization, comonomer sequences in macromolecules cannot redistribute on crystallization. Crystallization is limited either to the homopolymer sequences that are long enough to overcome the nucleation barrier or to isomorphic substitution. Isomorphism on a repeating unit scale is accomplished by a marked increase in crystallinity over the case of no isomorphism. Natta *et al.* (1961) distinguished three types of repeating unit isomorphism:

Type 1. Homopolymers of both types of repeating units have crystals of similar structure. In this case a continuous change in lattice parameters from one homopolymer toward the other is expected for the copolymer as a function of crystallization. Examples are copolymers of styrene and *o*-fluorostyrene and copolyamides of comparable sequence length between the amide groups.

Type 2. Homopolymers of both types of repeating units have different crystal structures. In this case one repeating unit must be substituted into a crystal structure different from its own homopolymer. At some intermediate concentration the copolymer crystal structure changes from one crystal structure (with changed lattice parameters due to substitution) to the other (also with changed lattice parameters due to substitution). Frequently, both crystal structures may be found side by side in the intermediate concentration range. Examples are more frequent than for type I isomorphism, but the solubility is often limited. Many vinyl copolymers with different side groups fall into this category. Another example is the copolymer of tetrafluoroethylene and vinyl fluoride.

Type 3. Only one homopolymer of the pair is crystalline. The other repeating unit is not capable of crystallization of its own but shows isomorphism with the crystallizable repeating unit. In this case one expects a decrease in crystallinity on high concentration of the second component; however, the decrease in crystallinity is more gradual than in the case of random copolymers without isomorphism. A typical example is the copolymer of 4-methyl-1-pentene and 1-hexene. The homopolymer of the latter monomer does not crystallize, but isomorphism between the two is possible.

Additional examples can be found in the treatise by Wunderlich (1973) and the data collection of Allegra and Bassi (1975). As in the glassy solids, characterization of these bicomponent, and similarly multicomponent, systems, follows the outline of IV.A. Partial molar quantities may be evaluated using the scheme outlined in Fig. 41. Most characteristic are, however, the changes in the phase transitions discussed in Section V.B.

2. The Liquid States

The formation of a multicomponent homogenous liquid (i.e., a solution) requires the compatibility of the various components. As in the description of the multicomponent solid state, one can distinguish between solutions in which at least one component is a small molecule and solutions of macromolecules only.

Low-molecular-weight solvents for linear macromolecules may be found empirically. Extensive lists of solvents and nonsolvents have been published by Fuchs and Suhr (1975). They made the following generalization from their studies: Solubility increases often with temperature; an increase in molecular weight decreases the solubility; branching of the macromolecule increases the solubility when compared with a strictly linear macromolecule of the same molecular weight; combinations of solvents may become nonsolvents and vice versa; and copolymers with random, small repeating units resemble often the homopolymer of the dominating comonomer.

A more quantitative treatment is possible making use of the solubility parameter δ defined as (Hildebrand and Scott, 1936)

$$\delta = (\Delta E_{\text{vap}}/V)^{1/2} \tag{160}$$

where ΔE_{vap} is the energy of evaporation to a gas of zero pressure (i.e., to an infinite separation) and V is the molar volume. The enthalpy of mixing per unit volume is given by (if there are no specific forces such as strongly polar groups, hydrogen bonds, or largely different geometries in solution and pure components)

$$\Delta h = v_1 v_2 (\delta_1 - \delta_2)^2 \tag{161}$$

where v_1 and v_2 are the volume fractions of the two components. Since the overall thermodynamic stability relative to the pure states is given by [see also Eqs. (154) and (155)]

$$\Delta g = \Delta h - T \Delta s \tag{162}$$

which must be negative for a stable solution, the positive value of Eq. (161) must be overcome by the increase in entropy on mixing. For nearly equal values of δ of the components in a liquid, solubility is thus assured because of the large ΔS on dissolution of a macromolecule in a small molecule solvent. Several schemes of calculation of the solubility parameter δ based on heat of evaporation, surface tension, and molecular structure have been developed and extensive tables are available for small molecules and macromolecules (see Burrell, 1975; van Krevelen and Hoftyzer, 1972).

It is rare to find multicomponent systems where all components are macromolecular, since ΔS becomes much smaller for larger molecular sizes. The positive enthalpy of mixing [see Eq. (161)], which is more closely proportional to the number of repeating units than the number of total molecules, will usually overcompensate the entropy term in Eq. (162) and prohibit solution. A table of about 40 macromolecular systems of partial or total solubility is given by Bohn (1975). For a review see Krause (1972, 1978). Full miscibility has, for example, been observed for poly-α-methylstyrene and poly(2,6-dimethylphenylene oxide), and poly(isopropyl methacrylate) and poly(isopropyl acrylate). The influence of steric considerations, orientation of polar bonds, and saturation of hydrogen bonding, which in special cases may lead to a negative heat of mixing and enhance solubility, has not been analyzed in detail.

Of special interest is the solubilization by copolymerization of repeating units that as monomers would be incompatible. The introduction of the covalent bond prohibits separation into separate phases as long as the incompatible sequences are sufficiently short. As soon as microphase separation can occur, large changes in the properties of the material are observed, particularly if the two phases are different in character (such as rubbery and glassy).

The thermal analysis of liquid as well as glassy solutions goes parallel with the one-component systems for a given concentration. Useful techniques are calorimetry (and DTA and DSC), dilatometry, and thermomechanical analysis. Since no weight changes are observed in one-phase systems, thermogravimetry cannot be applied for such systems. Whenever series of measurements are made as a function of concentration, the partial molar quantities can be obtained as illustrated by Fig. 41. A fast characterization of the solutions is possible through their phase changes, which are characteristically affected by concentration as discussed in Section V.B.

3. Chemical Reactions

Chemical reactions in single-phase multicomponent systems can be recognized by their heat of reaction through calorimetry. In case of volume changes, it may also be possible to use dilatometry. As long as the system remains single phase, no thermogravimetry is applicable. Thermogravimetry becomes the major analysis technique as soon as weight losses are observed (see Section V.D). The thermal analysis of chemical reactions may have several major goals. Of initial interest may be the identification of an unknown system through its characteristic reactions. For a known system the evaluation of the thermodynamic functions as outlined in Section III.E.3 are of basic importance. A further analysis step is the evaluation of kinetic parameters, as outlined in Section III.D.

Analysis of the reaction written

$$aA + bB \rightleftharpoons cC + dD \tag{163}$$

would first consist of a determination of the enthalpy of reaction ΔH. With a knowledge of the heat of reaction it is possible with the help of tables or measurement of secondary reactions [see Eqs. (109) and (110)] to evaluate the enthalpy of formation of the product of interest. Heat capacities for all components enable us to calculate the change of the enthalpies at all temperatures [Eq. (111)]. With the help of equilibrium data (concentrations or electrochemical potentials) free enthalpies and entropies can be evaluated. The same is also possible if enthalpies and entropies of the appropriate species are known from zero kelvin to the temperature in question.

A study of the kinetics leads to macroscopic rate laws [Eqs. (27)–(29)] and activation energies [Eqs. (30) and (31)], which then can be used for discussion of the activated state (see Section III.D). Special applications of the kinetic data are stability and lifetime predictions. Irreversible thermodynamics as described in Section III.C may also be used for a less detailed discussion.

Despite development of differential scanning calorimetry to a precision of 1–5% in heats of reaction or transition and 0.1% in heat capacity, it is possible to reach one tenth of these values by standard isothermal or adiabatic calorimetry (see Sturtevant, 1971a). Also, development of kinetic data by following concentrations through spectroscopic or chromatographic analytical techniques may in some cases be more precise than differential scanning calorimetry (see Frost and Pearson, 1953). The justification in using differential scanning calorimetry lies in the speed of operation, small sample size, and low capital investment in instrumentation. The speed of operation permits the scanning of many samples, and, in particular, the analysis of metastable systems that often cannot be analyzed by classical calorimetry. Furthermore, surveys of larger bodies of data have shown that for many macromolecular samples reproducibility due to metastability is of the same order of magnitude as that found in differential scanning calorimetry (see Wunderlich and Baur, 1970).

Nonisothermal differential scanning calorimetry is experimentally easier than isothermal operation. Evaluation of the kinetic parameters from nonisothermal experiments is, however, more involved, as outlined in Section III.D [Eqs. (32)–(33)]. Heats of reaction measured by integrating heats beyond the heat capacity baseline may also suffer from excessive temperature ranges over which the heat capacity of the changing species is not constant. This problem of drawing of the baseline arises in the analysis of heats of transition that are broadened as in multiple-component systems or in metastable subsystem distributions. An understanding of the given reaction is necessary for the proper baseline choice. The simplest case arises when products and reactants have similar heat capacities and straight line continuations can be found to match initial and final heating periods. Changes in ordinate of the initial and final baseline may arise particularly if one of the species is volatile and decreases the mass within the calorimeter. The heat capacity change must then be apportioned proportional to the measured heat of reaction. A discussion of the baseline problem for sharp transitions has been given but Guttman and Flynn (1973).

Isothermal experimentation with a scanning calorimeter is easier to evaluate because of the absence of the baseline problem. The interpretation of kinetic data is furthermore simplified because isothermal kinetics can be used. For full characterization, reactions must be studied, however, at multiple temperatures, which is more time-consuming. The experimental difficulty lies in the initiation of the reaction which can usually only be done by quick heating to the reaction temperature. Such a procedure allows only poor fixing of the starting time for a reaction, a parameter of importance for the integration of rate expressions (see Section III.D). The inherent time for temperature equilibrium of the calorimeter limits also the quantitative kinetics studies to reactions that are not too fast. (Total reaction time should be more than about 1 min.) Slow reactions are also difficult to study directly because of the small heat evolution (or absorption) per unit time. (Total reaction time should not be more than about 1 h.) Faster reactions, such as those found for explosive decomposition, are best studied nonisothermally with information limited to temperature of fast reaction and total heat effect. Slower reactions can be studied by sampling the isothermal reaction at suitable time intervals to measure the remaining reactants or the formed products by nonisothermal analysis.

Recently it was found that the overall Arrhenius model (see Section III.D) has added problems in its application (Arnold *et al.*, 1980; Šestak, 1980). Because of its mathematical form, the parameters cannot be well estimated from the experimental curves and there does not even exist a definite one-to-one correspondence between experiment and calculation. In addition, there exists a mathematical correlation between A and E_a/T [Eqs. (30) and (31)].

B. Multiple-Phase Systems

A system that contains several components and more than one phase represents the most complex thermal analysis problem. As long as equilibrium persists between all phases, the description is simplified through the use of the phase rule (Section V.B.1). Phase diagrams can be established with a minimum of variables, as is shown in Section V.B.2. A special application is the purity analysis (Section V.B.3), which finds its application for small molecule characterization but may in the future also be used for detailed analysis of a broad-range melting peak of metastable macromolecular crystals. Such nonequilibrium systems will need the same analysis techniques, but additional characterization of the degree of stability is necessary as outlined in Section III.C. Special multicomponent, multiphase systems are the copolymers, discussed in Section V.B.4.

1. Phase Rule

The phase rule is based on an accounting of the equilibrium conditions between the various phases. The total number of variables of an equilibrium system are temperature, pressure, and the concentration of the various components c in each phase p:

$$\text{total variables} = c \times p + 2 \qquad (164)$$

The total number of thermodynamic equations that describe the phase equilibria between the various phases $\alpha, \beta, \gamma, \ldots, p$ and components* $1, 2, 3, \ldots, c$ can be calculated from the set

$$
\begin{aligned}
\mu_1^\alpha &= \mu_1^\beta = \mu_1^\gamma = \cdots = \mu_1^p \\
\mu_2^\alpha &= \mu_2^\beta = \mu_2^\gamma = \cdots = \mu_2^p \\
\mu_3^\alpha &= \mu_3^\beta = \mu_3^\gamma = \cdots = \mu_3^p \\
&\quad\cdot \qquad \cdot \qquad \cdot \qquad\qquad \cdot \\
&\quad\cdot \qquad \cdot \qquad \cdot \qquad\qquad \cdot \\
&\quad\cdot \qquad \cdot \qquad \cdot \qquad\qquad \cdot \\
\mu_c^\alpha &= \mu_c^\beta = \mu_c^\gamma = \cdots = \mu_c^p
\end{aligned}
\qquad (165)
$$

The total number of separate equations is $c \times (p - 1)$. In each phase p, one has, in addition, one materials balance equation that expresses that the sum of all mole fractions is 1. The number of independent variables f is now the difference between total number of variables and total number of equations

$$f = c \times p + 2 - c \times (p - 1) - p \qquad (166)$$

$$f = c - p + 2 \qquad (167)$$

The number of independent variables are commonly called the degrees of freedom and Eq. (167) is the well-known phase rule. For one component it makes, for example, at equilibrium a one-phase system bivariant, a two-

---

* The number of components is the minimum of chemically distinct constituents which can be varied independently. In the system

$$
\begin{aligned}
NaCl + KBr &\rightleftharpoons NaBr + KCl \\
KBr + H_2O &\rightleftharpoons KBr \cdot H_2O \\
NaCl + H_2O &\rightleftharpoons NaCl \cdot H_2O \\
NaBr + H_2O &\rightleftharpoons NaBr \cdot H_2O
\end{aligned}
$$

one can recognize eight chemically distinct constituents, for example. The number of components depends, however, on the makeup of the system. Using NaCl, KBr, and H_2O as starting materials leads obviously to three components (eight constituents: four chemical equations and one materials balance between NaCl, KBr, NaBr, and KCl). Addition of extra NaBr or KCl, or both, removes the materials balance restriction between the simple salts and the number of components increases to four, the maximum component number.

phase system univariant, a three-phase system invariant; higher numbers of phases cannot exist in equilibrium (see Section IV.C).

2. Phase Diagrams

A typical phase diagram is shown in the bottom half of Fig. 42. Besides the pure components 1 and 2, there are two compounds at positions B and E, so that at low temperature four constituents are recognized. The phase

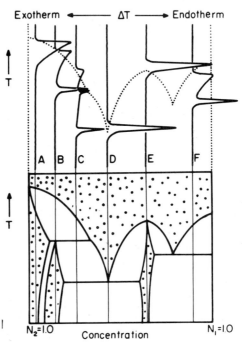

Fig. 42 Schematic phase diagram, bottom half, showing two eutectic points, compound formation, and one peritectic point. Selected differential thermal analysis curves are shown at the upper half of the drawing.

rule, Eq. (146) permits a maximum of three degrees of freedom in the presence of one phase. Keeping the pressure constant leaves concentration and temperature as independent variables, as indicated in the diagram. The one-phase areas are marked as the dotted areas in the phase diagram. They are the *melt* (at high temperature) and the *ranges of solid solution* (close to pure 2 and the two compounds). Component 1 does not show any solubility for compound E.

The two-phase areas are reduced in degrees of freedom by 1. For any given temperature and pressure the composition is thus fixed by the boundaries of the two-phase areas. The amount of each phase can be esti-

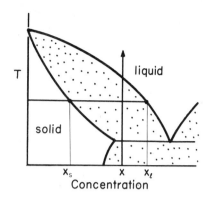

Fig. 43   Illustration of the lever rule.

mated by the lever rule, as indicated in Fig. 43. The overall concentration x in mole fraction at temperature T is broken into a solid solution of concentration x_s and a liquid solution of concentration x_l. The materials balance for one component can be written as

$$x(n_s + n_l) = x_s n_s + x_l n_l \tag{168}$$

where n_s and n_l are the total number of moles of solid and liquid, respectively. Division of Eq. (168) by n_l and rearrangement leads to the lever rule

$$n_s/n_l = (x_l - x)/(x - x_s) \tag{169}$$

The total number of moles of solid is proportional to the length of the distance between the liquid composition and the overall composition.

The three points in Fig. 42 where three phases are stable can only be changed by changing pressure. These triple points are the two eutectic points at concentration D and between E and F and the peritectic point at B. At the eutectic point D solid solutions of the two compounds are in equilibrium with the liquid solution. The other eutectic point occurs at the equilibrium between the compound E, pure component 1, and the liquid solution. The peritectic point at B indicates that the compound low in component 1 is not stable to its own melting temperature, but breaks up at the peritectic temperature into solid solution rich in component 2, and liquid solution.

Four phases in equilibrium are permitted according to Eq. (167), but only without any degrees of freedom. Temperature, pressure, and concentrations in all phases must be fixed. Figure 42 could show such quadruple points on expansion to the pressure variable.

The upper half of Fig. 42 indicates how thermal analysis could establish the phase diagram. The pure components, the stable compound E, and the eutectic compositions are recognized by their single, sharp-melting transitions. The solid solution A is recognized by a single, broad-melting transition. The off-eutectic two-phase solids C and F are characterized by a sharp, partial melting at the eutectic temperature to be followed by a broad melting of the excess solid. The peritectic point introduces an additional sharp transition at the peritectic reaction temperature, as can be seen at concentrations B and C.

A more quantitative analysis of the eutectic phase diagram, particularly considering the special nature of macromolecules, was carried out by Huggins (1942a,b,c) and Flory (1942). [For a detailed discussion see also Flory (1953).] The conditions for a eutectic phase diagram are complete miscibility in the melt and complete immiscibility in the solid state. Under these conditions it is possible to calculate the free enthalpy of mixing ΔG_M by evaluating separately an entropy of mixing and an excess free enthalpy of mixing that contains largely the contributions of the interaction between the components. To take into account large differences in size of the components, it is necessary to use the volume fractions v_1 and v_2 as the measure of concentration. Assuming component 2 to be the macromolecule, its partial molar volume V_2 is taken to be

$$V_2 = xV_1 \tag{170}$$

where V_1 is the partial molar volume of the second component of smaller size. This leads to the volume fractions

$$\begin{aligned} v_1 &= n_1V_1/(n_1V_1 + n_2V_2) = n_1/(n_1 + n_2x) \\ v_2 &= n_1V_2/(n_1V_1 + n_2V_2) = n_2x/(n_1 + n_2x) \end{aligned} \tag{171}$$

Assuming now, following the derivation by Hildebrand (1947), that in an ideal solution the fraction of free volume v_f is the same in both pure components and in the solution, there is an easy calculation of the entropy of mixing. It is simply for each component the entropy gain on expansion of its free volume to the free volume in solution.

$$V_{f1,2} = (n_1V_1 + n_2V_2)v_f \tag{172}$$

$$\Delta S = nR \ln (V_{\text{final}}/V_{\text{initial}}) \tag{173}$$

$$\Delta S_1 = n_1R \ln[(n_1V_1 + n_2V_2)v_f/n_1V_1v_f] \tag{174}$$

$$\Delta S_2 = n_2R \ln[(n_1V_1 + n_1V_2)v_f/n_2V_2v_f] \tag{175}$$

$$\Delta S_M = \Delta S_1 + \Delta S_2 = -n_1R \ln v_1 - n_2R \ln v_2 \tag{176}$$

The free enthalpy of mixing can then be written

$$\Delta G_M = \Delta H_M - T \Delta S_M = \Delta G_M^{\text{excess}} - T \Delta S_M^{\text{ideal}} \tag{177}$$

or

$$\Delta G_M = RT(n_1 \ln v_1 + n_2 \ln v_2 + n_1v_2\chi) \tag{178}$$

The excess free enthalpy has been rationalized as being based on the pair interaction on mixing

$$\Delta W_{1,2} = W_{1,2} - \tfrac{1}{2}(W_{1,1} + W_{2,2}) \tag{179}$$

which leads to the concentration dependence above by setting

$$\Delta G_M^{\text{excess}} = zxn_2v_1 \, \Delta W_{1,2} = n_1v_2\chi RT \tag{180}$$

where z is the coordination number for neighbors around the macromolecule, so that zxn_2 represents the maximum number of contacts possible with low-molecular-weight component 1 for all macromolecules; the fraction of the total contacts actually made is approximated by the volume fraction v_1. The constant χ collects all concentration independent terms in a convenient form to lead to the simple expression Eq. (178). Note that

$$xn_2v_1 \equiv n_1v_2 \tag{181}$$

With the help of Eq. (178) the solubility curve (liquidus) of the eutectic phase diagram can be evaluated. From the equilibrium condition Eq. (165) one can write

$$\mu_2^{\text{solution}} = \mu_2^{\text{crystal}} \tag{182}$$

which can be expanded to

$$\mu_2^{\text{solution}} - \mu_2^{\text{pure melt}} = \mu_2^{\text{crystal}} - \mu_2^{\text{pure melt}} \tag{183}$$

which, in turn, is from the definition of the chemical potential [Eq. (169)]:

$$(\partial G_M/\partial n_2) = \Delta G^{\text{crystallization}} = -\Delta H_{f_2} + T \, \Delta S_{f_2} \tag{184}$$

where ΔH_{f_2} and ΔS_{f_2} are the molar heat and entropy of fusion of component 2. Carrying out the suggested differentiation on the left-hand side and substituting on the right-hand side $-\Delta H_{f_2} \, \Delta T_{m_2}/T_{m_2}$ [see Eq. (127)] leads to the expression for the melting point lowering of component 2 due to component 1

$$\Delta T_{m_2} = -(RTT_{m_2}/\Delta H_{f_2})[\ln v_2 + (1 - v_2)(1 - x) + \chi x(1 - v_2)^2] \tag{185}$$

A plot of Eq. (185) is shown in Fig. 44 using $\chi = 0.4$. As long as x is small, there is only a small melting temperature lowering because of the large molar heat of fusion of macromolecules. Only as v_2 approaches zero is ΔT increased because of the logarithmic dependence on v_2.

For the melting temperature lowering of the smaller component 1 due to the macromolecular component 2, one can derive similarly

$$\Delta T_{m_1} = -(RTT_{m_1}/\Delta H_{f_1})\left[\ln v_1 + (1 - v_1)\left(1 - \frac{1}{x}\right) + \chi(1 - v_1)^2\right] \tag{186}$$

in which T_{m_1} and ΔH_{f_1} refer to the melting temperature and the molar heat

Concentration

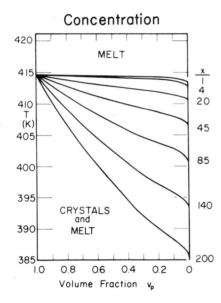

Fig. 44 Phase diagrams of a polyethylene of 24,500 MW with various second components of the given volume ratios x. Calculated using Eq. (185) with $\chi = 0.4$. Here $x = 1, 4, 20, 45, 85, 140$, and 200 and it corresponds to the molecular weights 24,500, 6125, 1225, 544, 288, 175, and 122, or an approximate number of carbon atoms of 1740, 437, 87, 39, 20, 12, and 9 in the second component.

of fusion of component 1, and x and χ have the same definitions as those given in Eqs. (170) and (180).

Various simplifications of Eqs. (185) and (186) are possible by assuming one or several of the following: (1) similar sizes [$x = 1$, Eqs. (185) and (186) revert to the common van't Hoff equation usually applied to thermal analysis of small molecules; see Sturtevant, 1971a)]; (2) no interaction ($\chi = 0$, ideal solution behavior); (3) large disparity in size ($x \gg 1$, typical macromolecular applications); (4) approaching dilute solution (v_2 or $v_1 \rightarrow 1.0$).

For not too small values of v_2 and a reasonably large macromolecule ($x \gg 1$), Eq. (185) becomes with the substitutions $(1 - v_2) = v_1$ and $TT_m \approx T^2$

$$\Delta T_{m_1} = (RT^2/\Delta H_{f_1})x[v_1 - \chi v_1^2] \qquad (187)$$

For low concentrations v_2, the logarithm of v_1 can be expanded

$$\ln v_1 = \ln(1 - v_2) = -v_2 - (v_2^2/2) - (v_2^3/3) - \cdots \qquad (188)$$

which when inserted into Eq. (186) leads to

$$\Delta T_{m_1} = (RT^2/\Delta H_{f_1})[(v_2/x) + (\tfrac{1}{2} - \chi)v_2^2 + (v_2^3/3) + \ldots] \qquad (189)$$

The first terms in both Eqs. (187) and (189) express the initial slope of the respective branches of the eutectic liquidus line of the phase diagram Fig. 41. Higher terms are needed for nonideal solutions and in the area of greater curvature. Applications of Eqs. (181) and (189) involve, besides the mathematical description of the phase diagram, the evaluation of molecular weight, heat of fusion, concentration, and the interaction parameter.

The molecular weight is derived with the help of Eq. (189) by measuring the freezing-point lowering of a low-molecular-weight solvent on adding decreasing concentrations c_2 of the macromolecule. The concentration is usually expressed in gram per kilogram of solvent. From Eq. (171) one can see that

$$\frac{v_2}{x} = \frac{n_2}{n_1 + n_2 x} = \frac{c_2/M_2}{(1000/M_1) + (c_2/M_2)x} \tag{190}$$

where M_1 and M_2 are the molecular weights of components 1 and 2. The term $(c_2/M_2)x$ is the number of moles of segments of the macromolecule of size V_1 per kilogram of solvent, in dilute solution a negligible amount relative to $(1000/M_1)$. The first term of the series Eq. (189) yields then

$$\lim_{c_2 \to 0} (\Delta T_{m_1}/c_2) = (RT^2/L_1)(1/M_2) \tag{191}$$

where L_1 is the heat of fusion of 1 kg of the solvent (1000 $\Delta H_{f_1}/M_1$). Some examples for the freezing-point constant are 1.853 for water, 5.085 for benzene, 20.5 for cyclohexane, and 40.0 for camphor [RT^2/L_1, in kelvin per molality].

The heat of fusion of macromolecules can be derived from Eq. (187). Although for crystallization of the low-molecular-weight components in the preceding molecular-weight determination, equilibrium is still easily approached, the use of Eq. (187) is more problematic. Equation (187) can only be used as long as the parameters for metastable macromolecular crystals are similar to those of the equilibrium crystals and no change in metastability occurs on melting (zero entropy production melting, see Section IV.B.2). In this case, however, the sample does not have to be completely crystalline to give the heat of fusion of the equilibrium crystal, a major advantage over direct calorimetry. Experimentally it is most advantageous to determine the upper melting region of well-grown crystals with decreasing amounts of low-molecular-weight diluent. The purpose is to have similar-fold-length crystals melt in each experiment. The data are then evaluated by plotting $\Delta T/T^2$ versus v_1, which gives for the zero concentration the slope $Rx/\Delta H_{f_2}$. Similarly, the interaction parameter can be evaluated by plotting $(\Delta T/T^2)/v_1$ versus v_1. Mandelkern (1964) showed that, indeed, a large number of macromolecular crystals–diluents systems yield such straight line relationships. For polyethylene, for example,

277, 278, 295, and 289 J g^{-1} were obtained as heats of fusion from ethyl benzoate, o-nitrotoluene, tetralin, and α-chloronaphthalene as diluents. The use of Eqs. (187) and (189) for concentration determination is particularly useful in purity analysis and will be described in Section V.B.3.

For specific phase diagrams one can now distinguish between systems where both components are large in molecular weight and the case where one component is of lower molecular weight.

The phase diagram of a macromolecule that forms a solution with another macromolecule in the melt but is immiscible in the crystalline state has been given by Nishi and Wang (1975). The melting-point lowering equation can be derived from Eq. (185) as

$$\frac{1}{T_m} = \frac{1}{T_m^\circ} = -\frac{RV_{2u}}{\Delta H_{2u}V_{1u}} \left[\frac{\ln v_2}{m_2} + \left(\frac{1}{m_2} - \frac{1}{m_1} \right) (1 - v_2) + \chi(1 - v_2)^2 \right]$$

(192)

where the subscript 2 refers to the crystallizing macromolecule and the subscript 1 refers to the macromolecular diluent; V_{2u} and V_{1u} are the repeating unit molar volumes; ΔH_{2u}, the repeating unit molar heat of fusion; m, the degrees of polymerization; and χ is the interaction parameter, as before. Since both degrees of polymerization are large, only the interaction parameter contributes significantly to the melting temperature lowering, so that

$$\frac{1}{T_m} - \frac{1}{T_m^\circ} = -(RV_{2u}/\Delta H_{2u}V_{1u}) \chi(1 - v_2)^2 \qquad (193)$$

which means that a sizable melting-point lowering is only expected with a negative value of χ. This is not a restriction, since the condition of miscibility in the liquid phase requires also a negative χ for high-molecular-weight components (see Section V.2).

An example of the melting-point lowering by a noncrystallizable macromolecular diluent is the effect of poly(methyl methacrylate) on the melting temperature of poly(vinylidene fluoride) (Nishi and Wang, 1975). The poly(vinylidene fluoride) had number and weight average molecular weights of 216,000 and 404,000 and the poly(methyl methacrylate) of 36,600 and 91,500, respectively. The mixtures were prepared originally by cocasting from dimethylformamide. The extrapolated equilibrium melting temperatures for the following weight ratios were

100.0 : 0	173.8°C
82.5 : 17.5	169.8°C
67.5 : 32.5	168.5°C
50 : 50	165.2°C

The total amount of sample crystallized decreased linearly to zero for decreasing amount of poly(vinylidene fluoride). The nonequilibrium melting data, which were taken over the whole melting range, followed also the relationship required by Eq. (193), indicating that the stable and metastable equilibrium phase diagrams are largely parallel ($\chi = -0.295$ at 160°C, using $V_{1u} = 84.9$ cm^3 mole^{-1}, $V_{2u} = 36.4$ cm^3 mole^{-1}, $\Delta H_{2u} = 6.7$ kJ mole^{-1}). Using faster cooling rates for the crystallizations, particularly at higher poly(methyl methacrylate) concentrations, crystallization was increasingly impeded. For example, at a 10 K min^{-1} cooling rate of a 60 : 40 sample, no crystallization was observable. To approximate equilibrium data, careful isothermal crystallization at high temperatures was necessary to overcome the increasing retardation of phase separation. A similar crystallizable–noncrystallizable pair of macromolecules soluble in the melt is poly-ϵ-caprolactone–poly(vinyl chloride); however, no equilibrium phase diagram has been established.

Isotactic polystyrene and poly(2,6-dimethylphenylene oxide) form a solution at high temperatures from which both components can crystallize; the molecules show no isomorphism. Hammel *et al.* (1975) have analyzed this system. The isotactic polystyrene could be crystallized isothermally at 170°C; the poly(2,6-dimethyl phenylene oxide) had to be induced to crystallization by exposure to 2-butanone vapor at 75°C. The overall crystallinity of the system, when crystallized to the fullest, remained at about 0.3 for the whole composition range, but the isotactic polystyrene crystallized only down to an equal weight fraction mixture. The poly(2,6-dimethylphenylene oxide) crystals were present from 20 wt % concentration on. Melting experiments were not performed.

On the example of isotactic polystyrene mixed with atactic polystyrene, the effect of a noncrystallizable stereoisomer on crystallization and melting was studied by Yeh and Lambert (1972). For equal molecular weight, almost no melting-point lowering was observed, as is expected from Eq. (192) and Fig. 44.

For the phase diagrams of macromolecules with low-molecular-weight diluents, one can imagine three basically different low-molecular-weight diluents: those that do not crystallize and those with higher melting and lower melting temperatures than the macromolecule. Orthmann and Ueberreiter (1957) analyzed all three cases on the examples polyethylene with amorphous hexachlorodiphenyl, crystalline anthracene, and crystalline naphthalene, respectively. Reasonable agreement with Eqs. (185) and (186) was observed. The eutectic compositions for the two crystalline diluents were above 0.9 weight fraction of the higher melting component.

Full phase diagrams have been established on systems with components of closer melting temperatures in the pure states. Smith and Pennings (1974, 1976) analyzed the binary-phase diagrams of polyethylene with 1,2,4,5-tetrachlorobenzene and hexamethylbenzene shown in Figs. 45 and 46. The low-molecular-weight component of the phase diagram be-

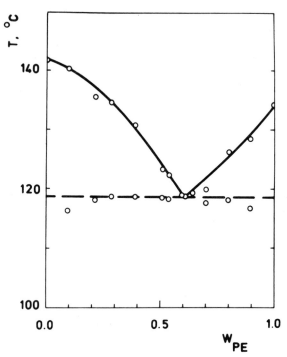

Fig. 45   Phase diagram of the system polyethylene–1,2,4,5-tetrachlorobenzene. From Smith and Pennings (1976). Polyethylene $\bar{M}_n = 8000$, $\bar{M}_w = 130{,}000$. Measurements by thermal analysis after cooling from the melt. Melting temperatures extrapolated to zero heating rate. The weight fraction of polyethylene is indicated by W_{PE}. Eutectic temperature 118°C, composition 0.6. See also Smith and Pennings (1974).

haves in both cases in close agreement with the equilibrium calculation [Eq. (186)]. The macromolecular component depends, however, largely on the crystallization condition. Figure 47 shows phase diagrams of the polyethylene-1,2,4,5-tetrachlorobenzene system crystallized at 81 and 105°C and the phase diagram determined from the crystallization temperatures. The high monomer concentration side shows little change. Even on cooling, one observes only about 0.5°C supercooling. At the eutectic composition, however, the system supercools by about 15.5°C relative to the data of Fig. 45. This is even more than observed for the pure polyethylene (13°C). It is also important to point out the change in slope of the polyethylene solidus line with crystallization conditions, which indicates de-

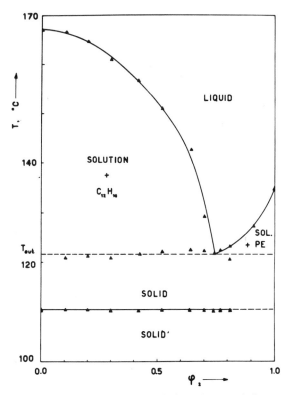

Fig. 46 Phase diagram of the system polyethylene–hexamethylbenzene. Diagram by Smith and Pennings (1974). Polyethylene $\bar{M}_n = 8000$, $\bar{M}_w = 130,000$. Measurements by thermal analysis after crystallization at 81°C. Melting temperatures extrapolated to zero heating rate (melting end temperatures). The volume fraction of polyethylene is given by ϕ_2. Eutectic temperature 122°C, composition 0.73 volume fraction.

viations from the earlier observed parallel behavior of equilibrium and metastable crystals. Changes are usually an indication that the macromolecular crystal perfection changes with concentration.

A phase diagram for the system polypropylene–pentaerythrityltetrabromide is shown in Fig. 48. It was obtained by crystallization rather than melting. The eutectic horizontal on the high diluent concentration side is displaced by 15°C to higher temperature, exceeding even the crystallization temperature of a pure polypropylene by 8°C. The diluent crystals seem to act as nucleating agents for the polypropylene. Changing the cooling rates from 0.5 to 32°C min⁻¹, it was observed by Smith and Pennings (1977) that the eutectic composition was not changed; the temperature at which polypropylene and diluent crystallized simultaneously, however, changed from 122°C to 102°C over the cooling rate range.

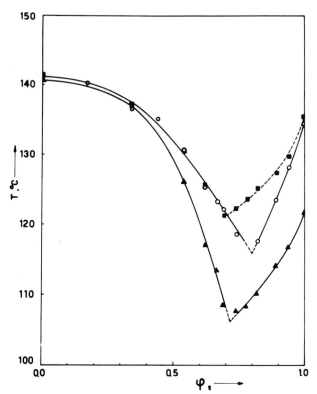

Fig. 47 Phase diagram of the system polyethylene–1,2,4,5,-tetrachlorobenzene. Materials as in Fig. 45. Open circles, diagram after crystallization at 81°C; filled squares, diagram after crystallization at 105°C; filled triangles, diagram as determined on cooling at 2°C min⁻¹ from the crystallization temperatures. The polymer concentration is given in volume fraction ϕ_2. Diagram by Smith and Pennings (1974).

The discussion of diluent effects as carried out previously rests with the assumption of uniform repeating unit distribution throughout the liquid. For dilute solutions this condition is not true. For concentrations below about 1 wt % of macromolecules of typical molecular weights, one can consider the individual molecules isolated by pure diluent. Further dilution will then be unable to change the repeating unit concentration within the domain of the molecule; it can only separate the molecules further from each other. As a result, in this concentration range only transport-controlled processes are affected by concentration (for example, crystallization). The dissolution temperature of polyethylene in tetrachloroethylene was, indeed, observed to be temperature independent below a concentration of 1 wt % (Koenig and Carrano, 1968; metastable crystals but dissolved at rates that did not permit recrystallization). As a

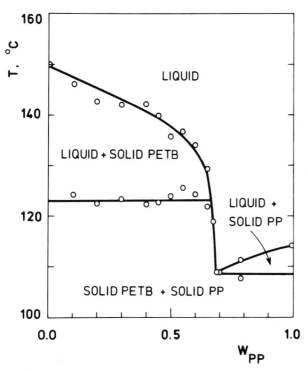

Fig. 48 Phase diagram of the system polypropylene–pentaerythrityltetrabromide. Diagram by Smith and Pennings (1977). Polypropylene $\bar{M}_n = 16,000$, $\bar{M}_w = 160,000$. Measurements by thermal analysis on cooling at 8°C from the melt. The weight fraction of polypropylene is indicated by W_{PP}. Eutectic temperature 108°C, composition 0.68. The initial crystals of the pentaerythrityltetrabromide seem to act as nuclei for polypropylene crystallization, causing the shift of 15°C in the solidus line at the eutectic concentration.

consequence, dissolution temperatures at high dilution can be treated similarly to melting temperatures.

Phase diagrams for the case of partial or complete solubility in both the crystalline and liquid phases are more difficult to calculate. Parallel to the derivation of Eqs. (182)–(187) one can find for large values of x

$$\Delta T_{m_2} = \frac{RT^2 x}{\Delta H_{f_2}} \left[\left(1 - \frac{v_1^s}{v_1^l} \right) v_1^l - \chi^l v_1^{l2} + \chi^s v_1^{s2} \right] \tag{194}$$

where the superscript s refers to the solid solution and the superscript l refers to the liquid solution (melt). Two interaction parameters need now to be known. The ratio v_1^s/v_1^l is called the distribution coefficient of the solute between solid and liquid. No experimental data have been analyzed to date.

A different type of metastable phase diagram arises in case the liquid phase solidifies to a glass, rather than crystallizes. The liquid-phase region is in this case extended below the equilibrium crystallization region and terminates in the glass transition region, where the liquid changes to the single-phase glass. The variation of the glass transition temperature with concentration is discussed in Section V.A.1. The usual thermal analysis techniques for glass transition measurement can be employed to establish the phase diagram (see Section IV.B.4).

Liquid–liquid phase separations are also possible. The case of solutions of macromolecules in low-molecular-weight solvents is described, for example, by Flory (1953) and Morawetz (1965). The major application of such phase separations is for molecular-weight fractionation. Thermal analysis techniques have not been applied to analyze such separations because of the relatively small heat effect involved.

Liquid–liquid phase separations of two or more polymeric components are frequent because of the limited solubility of macromolecules (see Section V.A.2). Each separate liquid phase will show its own phase transitions. In case of no crystallization multiple glass transitions indicate the presence of multiple phases. From the temperature and temperature range of the glass transition, information about partial solubility of the respective phases can be derived (see Gaur and Wunderlich, 1980).

3. Purity Analysis

One application of thermal analysis of small molecules has been the quantitative purity analysis. For flexible macromolecules this method has not been usable because of the lack of equilibrium on crystallization. A similar technique may, however, in the future be of value in the detailed analysis of melting ranges using the conditions of zero entropy production melting. The analysis is based on the discussion of the phase diagrams of the previous section. The calculations are conditional on the attainment of equilibrium, or zero entropy production melting in the case of nonequilibrium analysis (see also Fig. 18). For the equilibrium case the initial crystals must already correspond to the phase diagram, and melting must be carried out such that equilibrium is maintained at each increasing temperature. A measurement of the melting temperature lowering then permits, through Eqs. (185), (186), and (194), the evaluation of the concentration as long as all constants are known. Measurement or extrapolation to the last trace melting gives the overall concentration of the impurity. The equations contain the often unknown interaction parameter χ, but since it enters with a quadratic proportionality of the impurity volume fraction, its effect may be negligible for small impurity concentrations. The remaining constants that must be known are the heat of fusion of the pure compound, the volume ratio x, and, in case of partial solubility in the solid, the distribution coefficient. Since the volume ratio and the distribution coefficient are different for each impurity, one expects quantitative results only for known impurities.

For the case where the partial molar volume of the impurity and the major component are equal and where there is no solubility in the solid state (eutectic phase diagram, far right-hand-side of Fig. 42), Eqs. (185) or (186) can be simplified such that even unknown impurities can be determined quantitatively. This case is clearly the most desirable, but note the long list of conditions that must be fulfilled before such an analysis is possible.

Under the appropriate conditions, the first terms of Eqs. (187) and (189), which represent the slope of the liquidus line close to the pure compound concentration, can be written

$$T_m = T_0 - (RT_0^2/\Delta H_f)v_1/r \qquad (195)$$

where v_1 is the total volume fraction of the impurity (which is equal to the mole fraction because of the assumption $x = 1$) and r is the fraction of the sample molten at T_m. Observing only the melting temperature of the last trace crystal melting ($r = 1$) permits the calculation of v_1 with a knowledge of T_0, the melting temperature of the pure compound, and ΔH_f, its heat of fusion. A plot of the observed T_m as a function of the inverse of the various fractions of the sample molten ($1/r$) should give a straight line with the intercepts T_0 at $1/r = 0$ and T_m for the last trace melting at $1/r = 1$, and the slope $-(RT_0^2/\Delta H_f)v_1$. Any curvature can be taken as an indication of nonfulfillment of one or several the conditions discussed earlier.

Although the method seems extremely simple, there are experimental difficulties in the evaluation of the fraction of material molten. Using calorimetry, differential thermal analysis, or differential scanning calorimetry for the determination of r, the purity analysis is approximated by curve F of Fig. 42. The sharp low-temperature peak, which for relatively pure samples may be small, represents the eutectic melting. In case the eutectic melting peak is not observed, it is a good indication that the sample was not in equilibrium (or it does not follow a eutectic phase diagram) and application of Eq. (195) is questionable. At the eutectic melting peak, all impurity must melt, but in addition, an amount of the solvent melts that is given by the lever rule [Eq. (169)]. Since in the eutectic phase diagram the solid phases are pure, the fraction molten is approximately

$$n_l/n_s = v_1/v_1^{(\text{eutectic})} \qquad (196)$$

where v_1 is the total impurity volume (mole) fraction and $v_1^{(\text{eutectic})}$ is the fraction of component 1 at the eutectic point. For accurate determination of r from the heat of fusion measurement up to a temperature T_m, the contribution ΔH_{fi} of the impurity must be subtracted from the measured heat of fusion ΔH_{fm}.

$$r = (\Delta H_{fm} - \Delta H_{fi})/\Delta H_f \tag{197}$$

Usually not enough information is available for this correction, and only approximate data are obtained by plotting T_m as a function of the inverse of ΔH_{fm}.

A more precise method avoiding the explicit calculation of Eq. (197) involves the stepwise melting in fixed intervals ΔT as described by Gray and Fyans (1972). In this case the fraction molten is given by

$$r = (A + \Sigma \, \alpha_n)/\Delta H_f \tag{198}$$

where A represents the unmeasured heat of fusion up to an arbitrary starting temperature T_x of the stepwise melting sequence. The incremental heats of fusion of successive steps are $\alpha_1, \alpha_2, \ldots, \alpha_n$. By using such stepwise experiments it is not only possible to eliminate A from the calculation, but to avoid many experimental problems due to sluggishness of instrumental response as well as slow attainment of melting equilibrium. Equation (195) takes for step n of the melting experiment the form

$$T_n = T_0 - RT_0^2 v_1 (A + \Sigma \, \alpha_n)^{-1} \tag{199}$$

which can be rearranged to

$$A + \Sigma \, \alpha_n = RT_0^2 v_1 (T_0 - T_n)^{-1} \tag{200}$$

Between any two successive melting steps the unknown A can be eliminated when one remembers that $T_n - T_{n-1} = \Delta T$

$$\alpha_n = RT_0^2 v_1 \, \Delta T / [(T_0 - T_n)(T_0 - T_n + \Delta T)] \tag{201}$$

Furthermore, $RT_0^2 v_1$ can be eliminated by forming the ratio

$$\alpha_n/\alpha_{n-1} = 1 + 2\Delta T/(T_0 - T_n) \tag{202}$$

From Eq. (202) T_0 can be calculated and used for the evaluation of v_1, the impurity mole fraction by substitution of $T_0 - T_n$ in Eq. (201)

$$v_1 = \frac{2 \, \Delta T}{RT_0^2} \frac{\alpha_n \alpha_{n-1}(\alpha_n + \alpha_{n-1})}{(\alpha_n - \alpha_{n-1})^2} \tag{203}$$

Experimentally one needs thus only two successive melting steps of equal temperature interval to establish v_1. To reach highest accuracy, these two steps should be chosen close to but not at the final melting peak.

Extensive reviews of purity determinations are given by De Angelis and Papariello (1968), Plato and Glasgow (1969), Joy *et al.* (1971), Marti (1973), and Palermo and Chiu (1976). General discussions of method and accuracy have been given by Barrall and Diller (1970) and McCollough and Waddington (1957). Automated analysis systems have been described by Zynger (1975) and Moros and Stewart (1976).

4. Copolymers

A copolymer is described in Section II.C as a molecule consisting of different repeating units. Their sequence is, however, fixed during the polymerization, so that separation of the components into different phases is hampered. The nearest neighbors along the chain direction remain fixed as long as the molecule keeps its integrity. If the sequence length of identical repeating units is large (block copolymers), phase separation is possible and multiphase behavior as described in Section V.B.2 is approached. Random copolymerization does not contain sufficiently long sequences for phase separation. A single-phase liquid results, whose properties may be derived from the components similarly to multicomponent single-phase systems (see Section V.A.2). Measurement of the glass transition can give information on the number of phases and their structure as described in Sections V.A.1 and V.B.2. Of interest in this section is the case when one of the components of the random copolymer may crystallize. In this case, the behavior of the system is critically dependent on the degree of isomorphism possible between the copolymer repeating units (see Section V.A.1).

The basic two-phase equilibrium theory of crystallization and melting of copolymers was developed by Flory (1955). It contains the stringent conditions that all crystallizable repeating units, called A units, are freely available to add to crystals of A. No other units can be included in the crystal of A (no isomorphism). The noncrystallizable repeating units, called B units, can thus at best be located at the A crystal surface (outside the crystal proper). The A units are now characterized by their occurrence in sequences ζ in length. Every time a noncrystallizable B unit terminates an A sequence, the crystal must also be terminated. Ultimate equilibrium at low temperature is reached when all sequences of crystallizable A units are in crystals of lengths matching the sequence lengths ζ. Such morphology could be characterized as ideal fringed micellar (see Fig. 3 area *D* without chain folds). Added complications arise if equilibrium would require at a sufficiently low temperature that B units crystallize also. Actual copolymers do not reach equilibrium. They freeze before reaching the eutectic temperature, so that crystallization of more than one component has only been observed for block copolymers. The equilibrium calculations outlined below are only of use to provide an expression for the thermodynamic driving force toward crystallization.

The quantitative description of a copolymer melt is based on the probability P_ζ that a given A unit is located in a sequence at least ζ A units in length. Next we represent by w_j the probability that a randomly chosen repeating unit is an A unit and is also a member of a sequence of j A units terminated on both ends by B units. Finally $P_{\zeta,j}$ represents the probability that a specific A unit is a member of a sequence j in length and is followed in a given direction by at least $\zeta - 1$ additional A units. With these definitions the following relationships become obvious:

$$P_{\zeta,j} = w_j(j - \zeta + 1)/j, \qquad j \geq \zeta, \tag{204}$$

$$P_{\zeta,j} = 0, \qquad\qquad\qquad j < \zeta, \tag{205}$$

$$P_\zeta = \sum_{j=\zeta}^{\infty} P_{\zeta,j} \tag{206}$$

Writing out Eq. (206) for successive values of ζ reveals to us the validity of the following difference equation:

$$w_\zeta = \zeta(P_\zeta - 2P_{\zeta+1} + P_{\zeta+2}) \tag{207}$$

The quantities P_ζ and w_ζ must now be related to the structure of the two-phase system. In a copolymer before crystallization, the mole fraction of A units is n_A, the number of A units is N_A, the number of sequences of length ζ is N_ζ^0. Knowledge of these quantities permits the calculation of P_ζ^0 and w_ζ^0, the initial probabilities. Choosing a random copolymer, we find that the probability p of the occurrence of an A unit next to a chosen repeating unit of the polymer is equal to n_A, so that we can easily see that

$$P_\zeta^0 = p^\zeta \tag{208}$$

and

$$w_\zeta^0 = \zeta(1 - p)^2 p^\zeta \tag{209}$$

For nonrandom copolymer

$$P_\zeta^0 = n_A p^{\zeta-1} \tag{208a}$$

and

$$w_\zeta^0 = n_A \zeta N_\zeta^0 / N_A \tag{209a}$$

Equilibrium between a crystal consisting of sequences of A units ζ in length [which has ΔG_ζ^0 as its free enthalpy of fusion into a melt of sequences of length ζ only ($P_\zeta = 1$)] is reached under ideal solution conditions whenever melting can occur without change in free enthalpy or

$$\Delta G_\zeta = \Delta G_\zeta^0 + kT \ln P_\zeta^e = 0 \tag{210}$$

where P_ζ^e is the temperature-dependent equilibrium probability. (Note that all sequences of A units ζ or longer contribute to P_ζ.) For any ΔG_ζ^0 one can write analogously to the melting equations for lamellae Eqs. (125) to (128)

$$\Delta G_\zeta^0 = a_0 b_0 c_0 \zeta \, \Delta g_f - 2a_0 b_0 \gamma_e \tag{211}$$

where Δg_f is the free enthalpy of fusion per cubic centimeter and a_0, b_0, and c_0 are the appropriately chosen crystallographic dimensions of a repeating unit. The parameter γ_e represents a surface free energy of a fringed-micelle-type crystal.

A typical equilibrium distribution of w_ζ^e for different temperatures is shown in Fig. 49. The plotted data are calculated using Eqs. (207), (210),

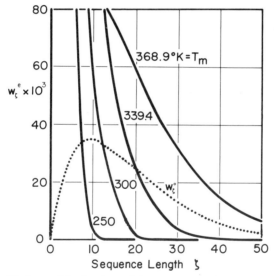

Fig. 49 Equilibrium weight fractions of A sequences in the melt. Dotted curve: initial distribution. The equilibrium temperatures are written next to the curves.

and (211). Whenever A and B units are of equal weight, w_ζ^e represents also the weight fraction of A sequences ζ in length remaining in the melt at equilibrium. Also shown in Fig. 49 is the initial distribution for a random copolymer melt computed from Eq. (209). Given both the initial and the equilibrium distributions of the different sequences ζ, the driving force toward crystallization can be calculated. From Eq. (210) one concludes that a necessary and sufficient condition for crystallization is that

$$P_\zeta^0 > P_\zeta^e \tag{212}$$

for one or more values of ζ (Flory, 1955). Comparing the initial values of P_ζ^0 (Eq. 208) with the equilibrium values P_ζ^e fixed by Eqs. (210) and (211), one can see that at any one temperature a critical length ζ^* exists below which no crystal growth can start and above which there is a driving force toward crystallization. At the length ζ^*, P_ζ^e and P_ζ^0 are equal, so that [from Eqs. (208) and (210)]

$$kT \ln P_{\zeta^*}^e = kT\zeta^* \ln p = - \Delta G_{\zeta^*}^0 \qquad (213)$$

The temperature for which ζ^* becomes very large is the equilibrium melting temperature T_m of the copolymer in question. When large crystals are contained in a crystallized copolymer, this temperature would be realized as the melt end. It can be calculated from Eqs. (213) and (211), letting ζ^* increase to large values

$$1/T_m - 1/T_m^0 = -(R/\Delta H_f) \ln p \qquad (214)$$

where T_m^0 is the equilibrium melting temperature of the pure homopolymer of A units only, and ΔH_f is the heat of fusion per mole of repeating units of A units.

Equation (214) would thus be the analogous expression to the two-phase–two-component melting-point-lowering expressions of Sections V.B.2 and 3. Its most important application would be in the evaluation of the copolymer concentration from the freezing-point lowering. Unfortunately, on crystallization, equilibrium is never achieved. An analysis of the crystallization path as predicted by irreversible thermodynamics showed that freezing into a metastable system is likely (Baur, 1965). Hoping for approximate use at small concentration of noncrystallizable units at the last trace of crystal melting is also unsuccessful, since large sequences of A units would, because of the chain-folding principle, fold into metastable lamellae as described in Section IV.B.2. Efforts of coping with these difficulties were made, for example, by Kilian (1965) and Glenz *et al.* (1977); see also Baur (1966). A general discussion of copolymer melting based on the equilibrium picture is also given by Mandelkern (1964). The case of crystallization has been treated by Sanchez and Eby (1973), with the simple assumption that only the heat of fusion is affected by copolymerization, and by Helfand and Lauritzen (1973), who based their description on the general fluctuation theory of chain folding by Lauritzen and Passaglia (1967).

C. MOLECULAR-WEIGHT DISTRIBUTIONS

Since most synthetic macromolecules are prepared in broad molecular-weight distributions (see Section II.C.), one must look on them as inherent multicomponent systems. The question of the effect of the distribution on the thermal characterization should be of great interest. As for other multicomponent systems, most important is the effect on the phase transitions.

Restricting our discussion to equilibrium melting, it turns out that only little experimental information is available. The reason for this lack of data is found in the difficulty of producing equilibrium crystals from a broad molecular-weight distribution. Two possible arrangements of mixtures of two lengths of chain molecules in crystals are shown in Fig. 50. In

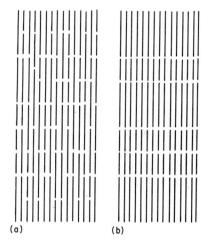

Fig. 50   Arrangement of macromolecules of different length in equilibrium crystals, (a) Solid solution. (b) Eutectic separation.

(a)   (b)

crystal A there is no separation of the molecules, making such crystals a solid solution of the two components. The joining of two molecules within the crystal would require, however, rather special geometrical conditions. Furthermore, the crystal growth requires a rather special mechanism to match the chain ends. The only currently known technique that is likely to lead to such crystals is the crystallization during polymerization. Until now, however, no detailed analyses of solid solution crystals and their subsequent melting have been reported. In case the ends are not placed in register, the observed disorder is sufficient to destroy the crystal. Arrangement B corresponds to a separation of each chain length into a different crystal. This represents the limit of eutectic separation of a solution on crystallization, as described in Section V.B.2. The two components are miscible in the melt, but not in the crystalline state. One would expect this arrangement to be the most frequent equilibrium limit, since, because of end groups on the crystal surface it has no geometrical restrictions. Realization of such crystallization for macromolecules is, however, difficult because of the chain-folding principle (Section II.E.), which limits crystal growth from the melt or from solution. Chain folding permits also the entry of various molecular lengths into the same crystal, since rejection of molecules is governed largely by the molecular nucleation process, rather than equilibrium crystallization. A qualitative effort was made by Lindenmeyer (1972) to account for the actually found chain-folded metastable crystals of broad molecular-weight distribution in terms of multicomponent phase diagrams.

To gain some experimental insight into equilibrium phase diagrams of linear macromolecules, one may review the behavior of binary oligomer mixtures. As long as the chain lengths of, for example, paraffins are largely different, true eutectic phase diagrams are observed. An example is shown in Fig. 51. Note that the eutectic temperature is close to the

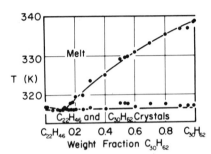

Fig. 51 Phase diagram of a paraffin mixture showing a eutectic separation. Data from Nechitailo *et al.* (1960). The polymorphic transition in $C_{30}H_{62}$ crystals at about 330 K has been omitted. Similarly there are two further solid-phase transitions at lower temperature.

melting temperature of the short-chain, pure component and that the eutectic mixture contains only a small amount of the long-chain component. Figure 52 illustrates an experimental phase diagram of paraffins with more closely similar chain lengths. Now the phase diagram is of the solid solution type. Most likely, even this phase diagram does not represent ultimate equilibrium, but the $C_{22}H_{44}$ crystal rejects the four-carbon atom excessive chain length of $C_{26}H_{24}$ into the interlamellar region, and at higher $C_{26}H_{54}$ concentration, $C_{22}H_{54}$ molecules are introduced with defects in the $C_{26}H_{54}$ lamellae. The driving force toward eutectic separation, which should represent the ultimate equilibrium, is too small to overcome defect

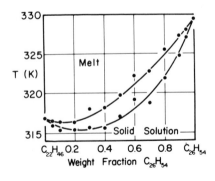

Fig. 52 Phase diagram of a paraffin mixture forming a solid solution. Data from Nechitailo *et al.* (1960). Solid-phase transitions have been omitted.

formation. The observed minimum in melting temperature is typical for many of the solid solutions described in the literature and may be as large as 4°C.

Similar experiments on binary mixtures of narrow molecular-weight fractions of low-molecular-weight poly(ethylene oxide)s (990–4000) were made by Gilg and Skoulios (1971). It could be clearly demonstrated that solid solution or eutectic separation depended on the difference in chain length for a given crystallization temperature. At a crystallization temperature of 25°C, for example, 1420/2000 and 1500/2000 molecular-weight mixtures resulted in solid solution crystals with linearly concentration-dependent x-ray long spacings between the fully extended chain limits of the pure components (1420, 8.9nm; 1500, 9.8 nm; 2000 13.1 nm). At the same temperature, mixtures like 990/1500 and 990/2000 showed two clearly separated thicknesses of lamellae for all concentrations corresponding to the pure components (990, 6.9 nm; 1500, 9.8 nm; 2000, 13.1 nm). The higher molecular weights would chain-fold once at the crystallization temperature and thus form metastable crystals, but would still depend on the fold length differences, form solid solutions, or eutectic separations. The 3200/2000 mixture thus formed solid solutions with linearly concentration-dependent x-ray long spacings varying between 10.8 nm, the once-folded 3200 molecular-weight spacing, and 13.1 nm, the fully extended 2000 molecular-weight spacing. This mixture is thus comparable in lamellar spacings to 1500/2000 mixture, which also formed solid solutions. On crystallization at higher temperature, where no chain folding would exist, one expects for this system also eutectic separation of the components. The opposite effect, conversion of a eutectic system into a solid solution, was demonstrated with the 990/1500 mixture by crystallization at −30°C instead of 25°C. The solid solutions demonstrated in these systems are all metastable and not of type A in Fig. 50, but rather accommodate the excess chain length as surface defects.

Based on the preceding observation, one may assume that an equilibrium phase diagram of binary macromolecular mixtures is typically that of a eutectic system. Figure 44 is such a phase diagram, calculated for polyethylene with various length components of equal or lower molecular weights. For all significantly shorter second components, the eutectic point is found at a concentration close to zero. On cooling of the mixture, under continuous equilibrium conditions, the high molecular weight should crystallize practically completely before the low molecular weight crystallizes in its thinner, fully extended chain lamellae. This is an experiment that cannot be performed because of the chain-folding principle on crystallization. A qualitative check could, however, be carried out by dissolving extended-chain crystals of polyethylene in polymer fractions (Sullivan, 1965). Pressure-crystallized, extended-chain crystals of broad molecular weight were mixed with molten polymer fractions in a ten-fold volume excess. By repeated trials on fresh samples at slightly different temperatures, the dissolution temperatures of the extended-chain crystals could be established to ± 0.5°C or better by optical microscopy (heating in steps of 2°C h⁻¹). For increasingly lower molecular weight diluents, the melting point decreased as expected from Fig. 44. More significantly, by using higher-molecular-weight diluent than in the average molecular weight of the crystals, the melting temperature could also be increased by approximately 0.5°C over that of the crystals by themselves, proving that there is a melting point depression in molecular-weight distribution because of lower-molecular-weight components melting earlier.

Turning now to multicomponent systems, Fig. 44 and Eq. (185) must be generalized by replacement of x by the appropriate average:

$$\bar{x} = \sum v_i x_i \Big/ \sum v_i \qquad (215)$$

where v_i and x_i refer to the volume fractions and volume ratios relative to the crystallizing species 2. Some experimental information is available from the study of melting of extended-chain crystals of polyethylene grown at elevated pressure by Prime and Wunderlich (1969). Figure 53

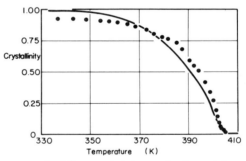

Fig. 53 Melting curve of a fully extended, low-molecular-weight, pressure-crystallized polyethylene. The drawn-out curve was calculated assuming complete eutectic separation of all molecular weight fractions [Eq. (185)]. Crystallized at 493 K and 4.8 kbar pressure for 21 h, followed by slow cooling (2–4 degrees per hour). $\bar{M}_n = 1640$, $\bar{M}_w = 2550$, measured by slow dilatometry. From Prime and Wunderlich (1969); see also Prime *et al.* (1969).

shows the good match possible between experimental data and the calculated melting, assuming eutectic separation for a relatively fully extended chain form. The sample had number and weight average molecular weights of 1540 and 25,550, respectively. Crystallizing the high-molecular-weight portion of a broad-molecular-weight distribution polyethylene (precipitated at 85°C, from p-xylene, to yield $\bar{M}_n = 14,800$, $\bar{M}_w = 78,300$) led to incomplete separation into extended-chain lamellae of different thickness above a molecular weight of about 12,000. The segregation on crystallization was governed by molecular nucleation, but on slow cooling after initial crystallization, full extension of the segregated lower molecular weights was achieved. Figure 54 illustrates the melting behavior. The

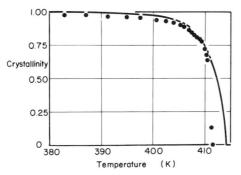

Fig. 54   Melting curve of a higher-molecular-weight polyethylene crystallized under elevated pressure. The drawn-out curve was calculated assuming complete separation of all molecular-weight fractions [Eq. (185)]. Crystallized at 498 K and 4.8 kbar pressure for 12 h followed by slow cooling (2–4 degrees per hour). $\bar{M}_n = 14,800$, $\bar{M}_w = 78,300$. Measured by slow dilatometry. From Prime and Wunderlich, (1969); see also Prime *et al.* (1969).

calculated curve for eutectic separation is followed down to about 411 K and 0.7 crystallinity, which agrees with the molecular-weight distribution and crystal distribution, assuming eutectic separation up to 12,000 molecular weight. Beyond this limit, melting occurs at a lower temperature and in a narrower temperature range than predicted by eutectic separation. It was concluded by Prime and Wunderlich (1969) that the higher molecular weights formed metastable mixed crystals.

The change of the melting curves of a broad molecular-weight distribution of polyethylene as a function of crystallization behavior is illustrated in Fig. 55. Curve 2 is similar to the curve shown in Fig. 54. The crystallization conditions were similar in the two cases. Curves 3–6 represent increasingly poorer crystallization conditions. One can now observe, in addition to the formation of metastable mixed crystals, the formation of lower crystallinity samples at low temperature, which is indicative of increasing amounts of chain folding and crystalline defects. Poor crystals would lower the melting temperature because of size and defect effects in addition to the lowering caused by mixed-crystal formation, as discussed in Sections III.C and IV.B.2.

D. CHEMICAL REACTIONS

A special case of characterization of multicomponent, multiphase systems arises when one of the components produced is gaseous. In this case thermogravimetry is best suited for analysis. The sample in the instru-

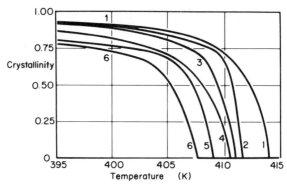

Fig. 55 Comparison of experimental melting curves of differently crystallized, identical molecular weight distributions of polyethylene with the equilibrium eutectic crystals of the same molecular weight distribution. Typical broad molecular distribution of linear polyethylene $\bar{M}_n = 8530$, $\bar{M}_w = 153,000$. Curve 1 calculated from Eq. (185). Successively lower curves refer to the sample crystallized at increasingly poorer crystallization conditions. All samples were kept at the crystallization temperature for long times and then cooled slowly ($2-4°/h$; sample 6, $3°/day$).

Curve	Crystal temp.	Crystal press.	Density, room temp.
	(kb)	(kb)	(g/cm^3)
2	489	4.9	0.994
3	484	4.8	0.993
4	474	3.5	0.991
5	413	0.0	0.982
6	413	0.5	0.980

Data from Prime and Wunderlich (1969).

ment holder of the thermogravimetry apparatus is an open system. The weight loss, which is coupled with a chemical reaction, is continuously monitored. The major applications are the identification of unknown systems through degradation or pyrolysis and comparison to known thermogravimetry curves, the study of degradation and pyrolysis products and the kinetics of the processes, and the study and prediction of lifetime and stability as a function of atmosphere, time, and temperature. Coupling thermogravimetry with calorimetry (see Section V.A.3) permits a full thermal analysis. Coupling thermogravimetry with standard chemical gas analysis techniques, such as the various forms of spectroscopy and chromatography, permits identification of the gases evolved. Jellinek (1978), Reich and Stivala (1971), and Madorsky (1964) described thermogravimetry in greater detail. The kinetic description in multiphase systems goes far beyond the difficulties of the homogeneous reactions discussed in Section III.D. Only in rare cases does homogeneous kinetics describe a thermogravimetric experiment (see the following discussion). The problem of the volatitlity of degradation products is discussed, for example, by Schneider (1980).

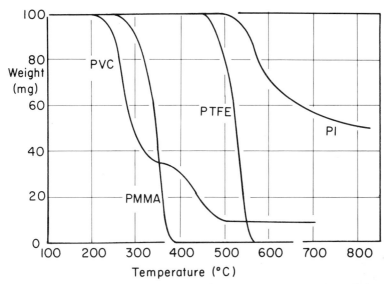

Fig. 56 Thermogravimetry curves of poly(vinyl chloride) (PVC), poly(methyl methacrylate) (PMMA), polytetrafluoroethylene (PTFE), and a polyimid (PI). Curves drawn after Chiu (1966). Experimental conditions, 10 mg sample heated at 5 K min^{-1} in N$_2$ atmosphere.

Several typical thermogravimetric analysis curves are shown in Fig. 56. The four macromolecules show a wide spread of stability and different types of decomposition. The polyimides exceed aluminum in their zero-strength temperature. Poly(methyl methacrylate) and polytetrafluoroethylene decompose largely by depolymerization to the volatile monomer, whereas poly(vinyl chloride) and the polyimid show irreversible losses of initially mainly HCl and a mixture of CO and CO$_2$, respectively. A detailed system of identification of polymeric samples by characteristic temperatures from thermogravimetry curves has been described by Pastor *et al.* (1978). Besides the possibility of identification and judgement on temperature stability of homopolymers, experiments of this type enable the characterization of copolymers and blends, the presence of fillers, additives, accelerators, etc. (Chiu, 1966).

The quantitative analysis of thermogravimetry involves the problems of nonisothermal kinetics, discussed in Section III.D. Of particular difficulty is the possibility of changing reaction mechanisms with temperature. Experimental difficulties arise from the need for a reproducible atmosphere and surface structure of the sample, which is complicated by the developing gas. Simple reaction kinetics are expected only if surface and diffusion effects are not rate determining over the whole temperature range of analysis. Some typical discussions of data treatment and experiment design are given by Broido (1969) and Flynn and Dickens (1976). A concise summary of the complex kinetics is given by Boyd (1970). Comparison with quantitative analysis of scanning calorimetry (Section V.A.3) shows that the baseline problem does not exist for thermogravimetry because of the continuous measurement of the total weight. Additional derivative recording, which can be directly compared to scanning calorimetry, often simplifies the interpretation.

Isothermal operation of thermogravimetry enhances the chances of development of kinetic parameters, as is also found in calorimetry (Section V.A.3). Thermogravimetry is furthermore not limited in the analysis of slow reaction, except by the patience of the operator. Determination of the isothermal kinetics at various temperatures is the best method for the development of nonisothermal expressions. With several kinetic expressions at widely spaced higher temperatures, where reactions are reasonably fast, one can make cautious extrapolations to lower temperatures to estimate aging and lifetime of polymeric materials. For a comparison of various thermogravimetric methods of lifetime prediction see Flynn and Dickens (1979).

A typical application of thermogravimetry in conjunction with scanning calorimetry and optical microscopy is illustrated in Figs. 57–60,

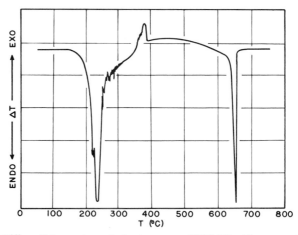

Fig. 57 Differential scanning calorimetry curve of LiH_2PO_4. 20 mg sample weight, 20°C min^{-1} heating rate, N_2 atmosphere, saturated with H_2O at room temperature. Predried samples. From Benkhoucha and Wunderlich (1978), courtesy J. A. Barth Verlag, Leipzig.

which deal with the polymerization and crystallization of lithium polyphosphate

$$xH_2PO_4^- \rightarrow H—(PO_3^{3-}—)_xOH + (x - 1)H_2O \qquad (216)$$

Figure 57 represents a differential scanning calorimetry curve. Two major endotherms and a broad intermediate exotherm characterize the curve.

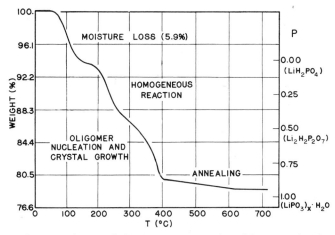

Fig. 58 Thermogravimetry of LiH_2PO_4, 20 mg sample weight, 20°C min^{-1} heating rate, N_2 atmosphere, saturated with H_2O at room temperature. Sample not predried. From Benkhoucha and Wunderlich (1978), courtesy J. A. Barth Verlag, Leipzig.

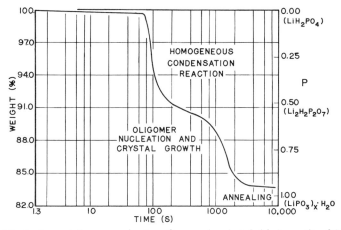

Fig. 59 Isothermal thermogravimetry of a powdered and dried sample of LiH_2PO_4. 13.38 mg, sample weight N_2 atmosphere, saturated with H_2O at room temperature, 300°C reaction temperature. P indicates the fractional progress of the reaction. From Benkhoucha and Wunderlich (1978), courtesy J. A. Barth Verlag, Leipzig.

Optical microscopy identifies the first exotherm as the monomer LiH_2PO_4 melting temperature. The broadness and large size, as well as the small irregularities on the endotherm, indicate that besides melting, decomposition occurs in this temperature range with a loss of water. Figure 58 confirms this interpretation. The major weight loss starts in the region of the endotherm, but it continues in a two-step process up to about 400°C. The initial moisture loss of 6% shown in the thermogravimetry was avoided in

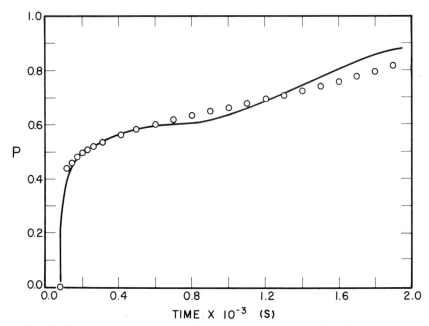

Fig. 60 Experimental and calculated fractional conversion of the polymerization and crystallization process of LiH_2PO_4. Solid line: experimental value of the fractional process of the reaction as shown in Fig. 59. Open circles are calculated as outlined in the text [combination of Eqs. (216) and (217) using the third-order reaction kinetics of autocatalyzed condensation and the Avrami expression for crystallization]. From Benkhoucha and Wunderlich (1978), courtesy J. A. Barth Verlag, Leipzig.

the scanning calorimetry by predrying. The last stage in thermogravimetry is a slow weight loss, labeled in Fig. 58 as annealing. The second endotherm at about 650°C shows up only in calorimetry and can be identified by microscopy as melting of the macromolecular crystals that were initiated in the intermediate temperature (close to the small sharp exotherm). A quantitative interpretation of the nonisothermal thermogravimetry is hopeless because of the number of overlapping stages. For detailed analysis crystal nucleation rates were determined by counting the growing macromolecular crystals under the microscope. Linear crystal growth rates were measured also by optical microscopy. Both these data were combined in an Avrami-type expression [see Eq. (42) and Fig. 11]. This was followed with an isothermal thermogravimetry study, as shown in Fig. 59. The processes are now clearly separated. A simple mechanism was then proposed after it was found that the initial reaction was third order [see Eq. (29)], which is characteristic for many condensation reactions.

$$
\begin{aligned}
\text{monomer} + \text{monomer} &\xrightarrow{k_a} \text{dimer} + \text{water} \\
\text{monomer} + \text{dimer} &\xrightarrow{k_b} \text{trimer} + \text{water} \\
\vdots \quad\quad \vdots \quad\quad &\;\; \vdots \quad\quad \vdots \quad\quad \vdots \\
\text{dimer} + \text{dimer} &\xrightarrow{k_b} \text{tetramer} + \text{water} \\
\text{dimer} + \text{trimer} &\xrightarrow{k_b} \text{pentamer} + \text{water} \\
\vdots \quad\quad \vdots \quad\quad &\;\; \vdots \quad\quad \vdots \quad\quad \vdots
\end{aligned}
\tag{217}
$$

This set of simultaneous reactions could describe the first, homogeneous stage of the reaction with two empirical rate constants k_a and k_b, computer fitted to combinations of rate expressions up to the triacontamer [which includes 635 rate expressions for reactions of the type given in Eq. (217)]. After oligomers of sufficient length and concentration for polymer crystal nucleation are produced (penta-, hexa-, and octadecamer at 300, 290, and 276°C, respectively), the Avrami equation took over describing the solid-phase development of the heterogeneous stage of the reaction through the equation

$$\text{crystal} + \text{dimer} \rightarrow \text{enlarged crystal} \tag{218}$$

Figure 60 illustrates the fit of calculated and measured conversion. Remaining discrepancies could probably be removed by introducing more than two different rate constants and by considering reverse reactions. Additional description of the annealing step of the once-grown crystals remains to be done.

This example illustrates the large amount of information that can be generated from a combination of thermal analysis techniques. Since measurements were made at several temperatures, curves can be generated for the whole range of temperatures for which the mechanism remains unchanged [see Eq. (31)].

VI. Conclusions

In this chapter the basic theory of thermal analysis of linear macromolecules has been displayed with a selected number of examples. Many more applications are shown in the following chapters. The major conclusion must be that the macroscopic thermal analysis is in many aspects well understood, and it takes only little effort for full description using thermodynamics, irreversible thermodynamics, and kinetics. The link between microscopic cause and macroscopic effect can in most cases be established. Once this is accomplished, thermal analysis not only can serve as an instrument of empirical characterization, but can yield detailed information on composition, structure, and history of the sample. It permits, furthermore, prediction of sample behavior under other conditions of time, temperature, and environment.

Of the five thermal analysis techniques (thermometry; dilatometry; the various forms of thermomechanical analysis; calorimetry, including differential thermal analysis and differential scanning calorimetry; and thermogravimetry), the last three have seen recently extensive development. A large factor in this development was the availability of commercial instruments of high quality. Their application to characterization problems is complementary and overlapping.

Thermomechanical analysis and differential scanning calorimetry and differential thermal analysis can be employed for identification of materials through "fingerprinting" using the various transitions (melting, crystallization, solid–solid, and glass transition). Differential scanning calorimetry, differential thermal analysis, and thermogravimetry, in turn, can be used also for identification of materials through characteristic chemical reactions.

Quantitative composition analysis is possible through direct determination of mass (thermogravimetry) or heat effects (integration of heats of transition in calorimetry) and through indirect determination of concentrations from transition temperature behavior such as in phase diagram and purity analysis (all thermal analysis techniques, sensitive to the given transition).

The detailed study of transitions and reactions relative to their thermodynamic equilibrium parameters (heats of reaction or transition, composition, molecular weight or volume changes), kinetic quantities (rate constants, order of reactions, mechanisms and activation energy and entropy), and deviations from equilibrium (metastability) is possible through any technique sensitive to the process in question. In these applications the special features of the various techniques become of importance. Thermomechanical analysis in its simple penetration, expansion, or extension mode, and even more so in its dynamic mode, is connected directly to the mechanical properties. As such, immediate connections to performance criteria can be made, such as expansion shrinkage, softening, strength. Some dynamic experiments and ultimate-strength tests are especially sensitive to small changes in the structure of a solid ("structure-sensitive" properties). Calorimetry is coupled to heat effects, and since no molecular changes can occur without readjustment of the heat content, calorimetry is perhaps most widely applicable. Its shortcoming may be that most heat effects can be considered "structure insensitive," i.e., the effects are largely only linearly dependent on the structure changes. Thermogravimetry, coupled to the mass of the system, relies on loss or gain of molecules due to evolution of or reaction with a gas. Its main advantage is in the absolute information it produces. It is perhaps the most easily interpreted thermal analysis technique. Complications arise from unreproducible surface and gas phase composition in the vincinity of the sample.

VII. Recommendations

After completing a study of this chapter one may ask the question, What is most urgently needed at present and what may be the development recommended for the future? Quite clearly, a check of the experimental results displayed in this chapter and, even more so, of those in the following chapters shows that there is frequently a gap between theory and application. This gap is often unnecessary, since the theoretical tools are available for a more detailed analysis. For thermal analysis to be more and more accepted as a primary analysis technique, this gap must be narrowed. Of favorable influence may be the increasing use of data handling by computer and the increasing precision of the instruments. The efforts of teaching applied thermal analysis as part of regular college curricula and in continued education must in this connection be increased.

Of need for further development is the interpretation of irreversible, often time-dependent effects in the analysis of macromolecules. Misinterpretation of melting temperatures has led in the past, for example, to the erroneous idea that macromolecules may not have a sharp equilibrium melting temperature. With increasing understanding of irreversible effects during melting, one expects not only improvement in characterization of linear macromolecules, but also a feedback to analysis of other materials. In this area of irreversible crystallization and melting, further theoretical developments toward a more quantitative description are hoped for. The importance of the more subtle transitions such as glass transitions and mesophase transitions are only just being understood, and further analysis to link experiment and theory should be undertaken.

In the area of instrumentation, one needs a close connection of theoretical understanding and performance design. In particular, as more intelligent computer coupling becomes available, it is of importance to introduce not only statistical data treatment, but also thermodynamic and kinetic theory intelligence and flexibility to the computer. For example, reporting arbitrary baselines should be avoided when it is possible with little more effort to report heat capacities. Multiple tests on identical samples, naturally, is an old goal that becomes increasingly more realizable. As one is able to distinguish better between irreversible and equilibrium processes, time becomes an increasingly important variable. Both faster and slower techniques should be developed. The slower techniques require simultaneous multiple sample analysis to be economical. The faster techniques must go to smaller sample size (micrograms). Heating rates in the 10^2-10^4 K s^{-1} ranges are recommended as a goal to instrument development. Two important areas of study would be opened by fast heating rate instruments: the study of increasingly metastable states and the simulation of actual production processes in the fiber, film, and molding industry.

A final field that has been neglected in the past to such a degree that I have not even included it in this chapter is the study of thermal conductivity. This is a field that may have much to offer to thermal characterization. More work in this area is highly recommended.* Only recently has modern laser flash equipment been introduced, permitting easy measurement.

In brief, the recommendations can be summarized: close the gap between theory and application, avoid misinterpretation, teach others about thermal analysis, improve the instruments by making them faster and slower and more theory intelligent, and perhaps remember thermal conductivity, since it is involved in every thermal analysis experiment and deserves more than just being eliminated as variable to have a constant temperature environment.

ACKNOWLEDGMENTS

I should like to thank the various reviewers for their comments and suggestions. Particularly appreciated is the help and advice from the editor, Dr. E. Turi, and from Dr. W. Wrasidlo and Dr. J. Flynn during the revisions and updating of the manuscript. I should also like to acknowledge the support the research in our laboratory has received from the Polymers Program of the National Science Foundation.

References

Adam, G., and Gibbs, J. H. (1965). *J. Chem. Phys.* **43,** 139.
Alexander, L. E. (1969). "X-Ray Diffraction Methods in Polymer Science." Wiley (Interscience), New York.
Alfrey, T., and Mark, H. (1942). *J. Phys. Chem.* **46,** 112.
Allegra, G., and Bassi, I. W. (1975). *In* "Polymer Handbook" (J. Brandrup and E. H. Immergut, eds.), 2nd ed.) p. III–205. Wiley (Interscience), New York.

* Some initial study may be started from the following references: Choy (1977), Knappe (1971), and Kline and Hansen (1970).

American Institute of Physics (1941). "Temperature: Its Measurement and Control in Science and Industry," Vol. 1. Van Nostrand Reinhold, New York.

American Institute of Physics (1955). "Temperature: Its Measurement and Control in Science and Industry," Vol. 2. Van Nostrand-Reinhold, Princeton, New Jersey.

American Institute of Physcis (1962). "Temperature: Its Measurement and Control in Science and Industry," Vol. 3. Van Nostrand Reinhold, New York.

American Institute of Physics (1972). "Temperature: Its Measurement and Control in Science and Industry," Vol. 4. Instrument Society of America, Pittsburgh, Pennsylvania.

Arakawa, T., and Wunderlich, B. (1967). *J. Polym. Sci., Part C* **16**, 653.

Archar, B. N. N., Brindley, G. W., and Sharp, J. H. (1966). *Proc. Int. Clay Conf., 2nd, 1967* Vol. 1, p. 67.

Arnold, M., Veress, G. E., Paulik, J., and Paulik, F. (1980). *In* "Thermal Analysis" (H. G. Wiedemann, ed.), Vol. 1, p. 69. Birkhaeuser, Basel.

Ball, M. C., and Casson, M. J. (1978). *Thermochim. Acta* **27**, 387.

Ballard, D. G. H., and Schelten, J. (1980). *J. Cryst. Growth* **48**, 169.

Bares, V., and Wunderlich, B. (1973a). *J. Polym. Sci., Polym. Phys. Ed.* **11**, 861.

Bares, V., and Wunderlich, B. (1973b). *J. Polym. Sci., Polym. Phys. Ed.* **11**, 1301.

Baro, M. D., Clavguera, N., Bordas, S., Clavaguera-Mora, M. T., and Casas-Vazaques, J. (1977). *J. Therm. Anal.* **11**, 271.

Barrall, E. M., and Diller, R. D. (1970). *Thermochim. Acta* **1**, 509.

Battista, O. A. (1965). *Am. Sci.* **53**, 151.

Baur, H. (1965). *Kolloid Z. Z. Polym.* **203**, 97.

Baur, H. (1966). *Kolloid Z. Z. Polym.* **212**, 92.

Beattie, J. A. (1926–1927). *J. Math. Phys. (Cambridge, Mass.)* **6**, 1.

Beilstein, F. K. (1918). "Beilstein's Handbuch der organischen Chemie," 4th ed., Suppls. I–IV. Literature up to 1959. Springer-Verlag, Berlin and New York.

Benkhoucha, R., and Wunderlich, B. (1978). *Z. Allg. Anorg. Chem.* **444**, 256, 267.

Benoit, H. (1974). *Angew Chem., Int. Ed. Engl.* **13**, 412.

Binsbergen, F. L. (1970). *Polymer* **11**, 253.

Blundell, T. R., and Johnson, L. N. (1976). "Protein Crystallography." Academic Press, New York.

Bohn, L. (1975). *In* "Polymer Handbook" (J. Brandrup and E. H. Immergut, eds.), 2nd ed., p. III–211. Wiley (Interscience), New York.

Bondi, A. (1968). "Physical Properties of Molecular Crystals, Liquids, and Glasses." Wiley, New York.

Bovey, F. A. (1974). *Macromol. Rev.* **9**, 1.

Boyd, R. H. (1970). *In* "Thermal Stability of Polymers" (R. T. Conley, ed.), p. 125. Dekker, New York.

Boyer, R. (1976). *J. Macromol. Sci., Phys.* **B12**, 253.

Brandrup, J., and Immergut, E. H., eds. (1975). "Polymer Handbook," 2nd ed. Wiley (Interscience), New York.

Broadhurst, M. G. (1962). *J. Chem. Phys.* **36**, 2578.

Broido, A. (1969). *J. Polym. Sci., Polym. Phys. Ed.* **7**, 1761.

Buckley, C. P., and Kovacs, A. J. (1975). *Prog. Colloid Polym. Sci.* **58**, 44.

Burrell, H. (1975). *In* "Polymer Handbook" (J. Brandrup and E. H. Immergut, eds.), 2nd ed., p. IV–337 Wiley (Interscience, New York.

Carr, S. H. (1980). *Proc. NATAS Conf., 10th, 1980* p. 1.

Chiu, J. (1966). *Appl. Polym. Symp.* **2**, 25.

Choy, C. L. (1977). *Polymer* **18**, 984.

Clusius, F., and Weigand, K. (1940). *Z. Phys. Chem., Abt. B* **46**, 1.

Coates, A. W., and Redfern, J. P. (1964) *Nature* (*London*) **201**, 68.

Couchman, P. R. (1978). *Macromolecules* **11**, 1156.

Couchman, P. R. (1979). *Phys. Lett.* **70A**, 155.

Couchman, P. R. (1980). *Proc. NATAS Conf., 10th, 1980* p. 301.

Couchman, P. R., and Karasz, F. E. (1978). *Macromolecules* **11**, 117.

Czornyj, G., and Wunderlich, B. (1977). *J. Polym. Sci., Polym. Phys. Ed.* **15**, 1905.

De Angelis, N. J., and Papriello, G. J. (1968). *J. Pharm. Sci.* **57**, 1868,

de Groot, S. R., and Mazur, P. (1962). "Non-Equilibrium Thermodynamics." North-Holland Publ., Amsterdam.

Dole, M. (1954). "Introduction to Statistical Thermodynamics." Prentice-Hall, Englewood Cliffs, New Jersey.

Dunn, A. S., ed. (1974). "Rubber and Rubber Elasticity," *J. Polym. Sci., Polym. Symp. No.* 48. Wiley, New York.

Dusek, K., and Prins, W. (1969). *Adv. Polym. Sci.* **6**, 1.

Ehrenfest, P. (1933). *Proc. Acad. Sci., Amsterdam* **36**, 153.

Eisenberg, A., and Trepman, E. (1980). *Proc. NATAS Conf., 10th, 1980* p. 39.

Ferry, J. D. (1970). "Viscoelastic Properties of Polymers," 2nd ed. Wiley, New York.

Fischer, E. W., and Hinrichsen, G. (1966). *Kolloid Z. Z. Polym.* **213**, 28.

Flory, P. J. (1942). *J. Chem. Phys.* **10**, 51.

Flory, P. J. (1947). *J. Chem. Phys.* **15**, 397.

Flory, P. J. (1953). Principles of Polymer Chemistry." Cornell Univ. Press, Ithaca, New York.

Flory, P. J. (1955). *Trans. Faraday Soc.* **51**, 848.

Flory, P. J. (1956). *Proc. R. Soc. London Ser.* A **234**, 73.

Flory, P. J. (1969). "Statistical Mechanics of Chain Molecules." Wiley (Interscience), New York.

Flynn, J. H. (1974). *Thermochim. Acta* **8**, 69.

Flynn, J. H., and Dickens, B. (1976). *Thermochim. Acta* **15**, 1.

Flynn, J. H., and Dickens, B. (1979). *ACS Symp. Ser.* **95**.

Fox, T. G., and Loshaek, S. (1955). *J. Polym. Sci.* **15**, 371.

Freeman, E. S., and Carrol, B. (1958). *J. Phys. Chem.* **62**, 394.

Frost, A., and Pearson, R. G. (1953). "Kinetics and Mechanism." Wiley, New York.

Fuchs, O., and Suhr, H.-H. (1975). *In* "Polymer Handbook" (J. Brandrup and E. H. Immergut, eds.), 2nd ed., p. IV–241.

Garn, P. D. (1972). *CRC Crit. Rev. Anal. Chem.* **3**, 65–111.

Gaur, U., Wunderlich, B., Shu, H.-C. Mehta, A., Lau, S.-F., and Wunderlich, B. B. (1981–1982). *J. Phys. Chem. Ref. Data* (to be published).

Gibbs, J. H., and DiMarzio, E. A. (1958). *J. Chem. Phys.* **28**, 373.

Gilg, B., and Skoulios, A. (1971). *Makromol. Chem.* **140**, 149.

Glenz, W., Kilian, H. G., Klattenhoff, D., and Stracke, F. (1977). *Polymer* **18**, 685.

Godorsky, Yu. K. (1981). *Polymer* **22**, 75.

Göritz, D., and Müller, F. H. (1970). *Kolloid Z. Z. Polym.* **251**, 1075.

Göritz, D., and Müller, F. H. (1973). *Kolloid Z. Z. Polym.* **251**, 879, 892.

Gordon, J. M., Rouse, G. B., Gibbs, J. H., and Risen, W. M., Jr., (1977). *J. Chem. Phys.* **66**, 4971.

Gough, J. (1805). *Proc. Lit. Philos. Soc., Manchester* [2] **1**, 288.

Gray, A. P., and Fyans, R. L. (1972). Some comments and calculations on a method of purity determination by stepwise melting. "Thermal Application Study," Vol. 3. Perkin-Elmer, Norwalk.

Guttman, C. M., and Flynn, J. H. (1973). *Anal. Chem.* **45**, 408.

Hager, N. E. (1972). *Rev. Sci. Instrum.* **43**, 116.

Hall, C. E. (1966). "Introduction to Electron Microscopy." McGraw-Hill, New York.

Hamada, F., Wunderlich, B., Sumida, T., Hayashi, S., and Nakajima, A. (1968). *J. Phys. Chem.* **72**, 178.

Hammel, R., MacKnight, W. J., and Karasz, F. E. (1975). *J. Appl. Phys.* **46**, 4199.

"Handbook of Chemistry and Physics" (1975). 55th ed. CRC Press, Cleveland, Ohio.

Hauser, E. A., and Mark, H. (1926a). *Kolloid-Beih.* **22**, 63.

Hauser, E. A., and Mark, H. (1926b). *Kolloid-Beih.* **23**, 64.

Helfand, E., and Lauritzen, J. I., Jr. (1973). *Macromolecules* **6**, 631.

Hellmuth, E., and Wunderlich, B. (1965). *J. Appl. Phys.* **36**, 3039.

Hildebrand, J. H. (1947). *J. Chem. Phys.* **15**, 225.

Hildebrand, J. H., and Scott, R. L. (1936). "The Solubility of Non-Electrolytes," Van Nostrand-Reinhold, Princeton, New Jersey (3rd ed., 1959).

Hirakawa, S., and Takemura, T. (1969). *Jpn. J. Appl. Phys.* **8**, 635.

Hirschfelder, J., Stevenson, D., and Eyring, H. (1937). *J. Chem. Phys.* **5**, 896.

Huggins, M. L. (1942a). *J. Phys. Chem.* **46**, 151.

Huggins, M. L. *Ann. N. Y. Acad. Sci.* **41**, 1.

Huggins, M. L. (1942c). *J. Am. Chem. Soc.* **64**, 1712.

Huglin, M. B. (1972). "Light Scattering from Polymer Solution." Academic Press, New York.

International Practical Temperature Scale IPTS (1968). *ASTM Spec. Tech. Publ.* **STP 565.**

Jaffe, M., and Wunderlich, B. (1967). *Kolloid Z. Z. Polym.* **216–217**, 203.

Jellinek, H. H. G. (1978). "Aspects of Degradation and Stabilization of Polymers." Elsevier, Amsterdam.

Jenckel, E., and Heusch, R. (1958). *Kolloid-Z.* **130**, 89.

Jenckel, E., and Rinkens, H. (1956). *Z. Elektrochem.* **60**, 970.

Johnson, D. W., and Gallagher, P. K. (1972). *J. Phys. Chem.* **76**, 1474.

Johnson, J. F., and Porter, R. S., eds. (1970). "Liquid Crystals and Ordered Fluids." Plenum, New York.

Joy, E. F., Bonn, J. D., and Barnard, A. J. (1971). *Thermochim. Acta* **2**, 57.

Kanig, G. (1969). *Kolloid Z.Z. Polym.* **233**, 54.

Katz, J. R. (1925a). *Kolloid-Z.* **36**, 300.

Katz, J. R. (1925b). *Kolloid-Z.* **37**, 19.

Kilian, H. G. (1965). *Kolloid Z. Z. Polym.* **202**, 97.

Klein, H. A. (1974). "The World of Measurement." Simon & Schuster, New York.

Kline, D. E., and Hansen, D. (1970). *In* "Thermal Characterization Techniques" (P. E. Slade and L. J. Jenkins, eds.), Vol. 2, p. 247. Dekker, New York.

Knappe, W. (1971). *Adv. Polym. Sci.* **7**, 477.

Knowles Middleton, W. E. (1966). "A History of the Thermometer". John Hopkins Press, Baltimore, Maryland.

Koenig, J. L., and Carrano, A. J. (1968). *Polymer* **9**, 359.

Kovacs, A. J., and Hobbs, S. Y. (1972). *J. Appl. Polym. Sci.* **16**, 301.

Krause, S. (1972). *J. Macromol. Sci., Rev. Macromol. Chem.* **C7**, 251.

Krause, S. (1978). *In* "Polymer Blends" (D. R. Paul and S. Newman, eds.). Academic Press, New York.

Krigbaum, W. R., Dawkins, J. V., Via, G. H., and Balta, Y. I. (1966). *J. Polym. Sci., Polym. Phys. Ed.* **4**, 475.

Lauritzen, J. I., Jr., and Passaglia, E. (1967). *J. Res. Natl. Bur. Stand., Sect. A* **71**, 261.

Lee, W. A., and Rutherford, R. A. (1975). *In* "Polymer Handbook" (J. Brandrup and E. H. Immergut, eds.), Wiley (Interscience), New York.

Lennard-Jones, J. E., and Devonshire, A. F. (1939). *Proc. R. Soc. London, Ser.* **A 169,** 317.

Lennard-Jones, J. E., and Devonshire, A. F. (1939b). *Proc. R. Soc. London, Ser.* **A 170,** 464.

Lewis, G. N., and Randall, M. (1923). "Thermodynamics," p. 448. McGraw-Hill, New York (2nd ed., revised by K. S. Pitzer and L. Brewer, 1961 p. 130).

Liebert, L., ed. (1978). "Liquid Crystals." Academic Press, New York.

Lindemann, F. A. (1910). *Phys. Z.* **11,** 609.

Lindenmeyer, P. H. (1972). *Polym. J.* **3,** 507.

McCullough, J. P. and Waddington, G. (1957). *Anal. Chim. Acta* **17,** 80.

Madorsky, S. L. (1964). "Thermal Degradation of Organic Polymers." Wiley (Interscience), New York.

Mandelkern, L. (1959). *SPE J.* **15,** 63.

Mandelkern, L. (1964). "Crystallization of Polymers." McGraw-Hill, New York.

Mark, J. E. (1981). *J. Chem. Ed.* (to be published).

Mark, H. F., Gaylord, N. G., and Bikales, N. M., eds. (1964–1972). "Encyclopedia of polymer Science and Technology," Vols. 1–16. Wiley (Interscience), New York.

Marti, E. E. (1973). *Thermochim. Acta* **5,** 173.

Menczel, J., and Wunderlich, B. (1980). *J. Polym. Sci. Polym. Phys. Ed.* **18,** 1433.

Menczel, J., and Wunderlich, B. (1981). *J. Polym. Sci. Polym. Phys. Ed.* (to be published).

Meyer, K. H., and van der Wyk, A. (1937). *Helv. Chim. Acta* **20,** 1313.

Miyagi, A., and Wunderlich, B. (1972). *J. Polym. Sci., Polym. Phys. Ed.* **10,** 1401.

Moore, W. J. (1972). "Physical Chemistry," 4th ed. Prentice-Hall, Englewood Cliffs, New Jersey.

Morawetz, H. (1965). "Macromolecules in Solution." Wiley (Interscience), New York.

Moros, S., and Stewart, D. (1976). *Thermochim. Acta* **14,** 13.

Müller, F., and Engelter, A. (1957). *Kolloid-Z.* **251,** 892.

Natta, G., Corradini, P., Sianesi, D., and Morero, D. (1961). *J. Polym. Sci.,* **51,** 527.

Natta, G., Allegra, G., Bassi, I. W., Sianesi, D., Caporiccio, G., and Torti, E. (1965). *J. Polym. Sci., Part A* **3,** 4263.

Nechitailo, N. A., Topchiev, A. V., Rozenberg, L. M., and Terenteva, E. M. (1960). *Zh. Fiz. Khim.* **34,** 2694.

Nishi, T., and Wang, T. T. (1975). *Macromolecules* **8,** 909.

Orthmann, H. J., Ueberreiter, K. (1957). *Z. Electrochem.* **61,** 106.

Osborne, N. S., Stimson, H. F., and Ginnings, D. C. (1939). *J. Res. U.S. Bur. Stand.* **23,** 238.

Oth, J. F. M., and Flory, P. J. (1958). *J. Am. Chem. Soc.* 0, 1297.

Page, C. H., and Vigoureux, P., eds., (1971). "The International System of Units (SI)," NBS Spec. Publ. (U.S.) No. 330. Natl. Bur. Stand., Washington, D. C.

Palmermo, E., and Chiu, J. (1976). *Thermochim. Acta* **14,** 1.

Parks, W., and Richards, R. B. (1949). *Trans. Faraday Soc.* **45,** 203.

Pastor, J., Pauli, A. M., and Arfi, C. (1978). *Analysis* **6,** 121.

Pennings, A. J., and Zwijinenburg, A. (1979). *J. Polym. Sci., Polym. Phys. Ed.* **17,** 1011.

Peterlin, A., and Meinel, G. (1965). *J. Polym. Sci., Part B* **2,** 751.

Petrie, S. E. B. (1974). *In* "Polymeric Materials: Relationships between Structure and Mechanical Behavior" (E. Baer and S. V. Radcliffe, eds.), p. 257. Am. Soc. Metals, Metals Park, Ohio.

Plato, C., and Glasgow, A. R. (1969). *Anal. Chem.* **41,** 330.

Prigogine, I. (1967). "Introduction to Thermodynamics of Irreversible Processes." Wiley, New York.

Prime, R. B., and Wunderlich, B. (1969). *J. Polym. Sci., Polym. Phys. Ed.* **7**, 2073.

Prime, R. B., Wunderlich, B., and Melillo, L. (1969). *J. Polym. Sci., Polym. Phys. Ed.*, **7**, 2091.

Reich, L., and Stivala, S. S. (1971). "Elements of Polymer Degradation." McGraw-Hill, New York.

Richards, J. W. (1897). *Chem. News J. Phys. Sci.* **75**, 278.

Richardson, M. J., and Savill, N. G. (1975). *Polymer* **16**, 753.

Richardson, M. J., and Savill, N. G. (1980). *In* "Thermal Analysis" (W. Hemminger, ed.), Vol. 2, p. 475. Birkhaeuser, Basel.

Sanchez, I. C., and Eby, R. K. (1973). *J. Res. Natl. Bur. Stand., Sect. A* **77**, 353.

Schneider, I. A. (1980). *In* "Thermal Analysis" (W. Hemminger, ed.), Vol. 2, p. 387. Birkhaeuser, Basel.

Schwartz, A. (1978). *J. Therm. Anal.* **13**, 489.

Sesták, J. (1980). *In* "Thermal Analysis" (H. G. Wiedemann, ed.), Vol. 1, p. 29. Birkhaeuser, Basel.

Sharp, J. H., and Wentworth, S. A. (1969). *Anal. Chem.* **41**, 2060.

Shelley, D. (1975). "Manual of Optical Mineralogy." Elsevier, Amsterdam.

Shen, M. C., and Eisenberg, A. (1970). *Rubber Chem. Technol.* **43**, 95.

Slonim, I. Ya., and Lyubimov, A. N. (1970). "The NMR of Polymers." Plenum, New York.

Smith, K. J., Jr. (1976). *Polym. Eng. Sci.* **16**, 168.

Smith, P., and Pennings, A. J. (1974). *Polymer* **15**, 413.

Smith, P., and Pennings, A. J. (1976).

J. Mater. Sci. **11**, 1450.

Smith, P., and Pennings, A. J. (1977). *J. Polym. Sci., Polym. Phys. Ed.* **15**, 523.

Someynsky, T., and Simha, R. (1971). *J. Appl. Phys.* **42**, 4545.

Staudinger, H. (1950). "Organische Kolloidchemie," 3rd ed. Vieweg Verlag, Braunschweig.

Staudinger, H. (1961). "Arbeitserinnerungen." Hüthig Verlag, Heidelberg.

Stein, R. S. (1973). *Appl. Polym. Symp.* **20**, 347.

Stuart, H. A. (1955). "Die Physik der Hochpolymeren," Vol. 3. Springer-Verlag, Berlin and New York.

Stull, D. R., Westrum, E. F., and Sinke, G. C. (1969). "The Chemical Thermodynamics of Organic Compounds." Wiley, New York.

Sturtevant, J. M. (1971a). *In* "Technique of Chemistry" (A. Weissberger and B. W. Rossiter, eds.), Vol. 1, Part V, p. 347. Wiley (Interscience), New York.

Sturtevant, J. M. (1971b). *In* "Technique of Chemistry" (A. Weissberger and B. W. Rossiter, eds.), Vol. 1, Part V, p. 1. Wiley (Interscience), New York.

Sullivan, P. (1965). "High Polymer Crystals," Ph. D. Thesis. Dept. of Chemistry, Rensselaer Polytechnic Inst., Troy, New York.

Sumita, M., Miyasaka, K., and Ishikawa, K. (1977). *J. Polym. Sci., Polym. Phys. Ed.* **15**, 837.

Takamizawa, K., Ohno, A., and Urabe, Y. (1975). *Polym. J.* **7**, 342.

Tammann, G. (1903). "Kristallisieren and Schmelzen." Barth, Leipzig (see also a later translation by R. F. Mehl "The State of Aggregation." Van Nostrand-Reinhold, Princeton, New Jersey, 1925).

Tarasov, V. V. (1953). *Zh. Fiz. Khim.* **24**, 1430.

Tarasov, V. V. (1955). *Dokl. Akad. Nauk SSSR* **100**, 307.

Tarasov, V. V. (1965). *Zh. Fiz. Khim.* **39**, 2077.

Thomson, G. W., and Douslin, D. R. (1971). *In* "Technique of Chemistry" (A. Weissberger and B. W. Rossiter, eds.), Vol. 1, Part V, p. 23. Wiley (Interscience), New York.

Treolar, L. R. G. (1949, 1958). "The Physics of Rubber Elasticity," Oxford Univ. Press (Clarendon), London and New York. 1st ed.

Treolar, L. R. G. (1958). "The Physics of Rubber Elasticity," 2nd ed. Oxford Univ. Press (Clarendon), London and New York.

Ubbelohde, A. R. (1965). "Melting and Crystal Structure." Oxford Univ. Press (Clarendon), London and New York.

Ubbelohde, A. R. (1978). "The Molten State of Matter." Wiley, New York.

Ueberreiter, K., and Kanig, G. (1951). *Z. Naturforsch., A* **6A**, 551.

Valdré, U., ed. (1971). "Electron Microscopy in Material Science." Academic Press, New York.

van Krevelen, D. W., and Hoftyzer, P. J. (1972). "Properties of Polymers, Correlation with Chemical Structure." Elsevier, Amsterdam.

van Turnhout, J. (1975). "Thermally Stimulated Discharge of Polymer Electrets." Esevier, Amsterdam.

Vergano, P. J., and Uhlmann, D. R. (1970). *Phys. Chem. Glasses* **11**, 30, 39.

Warner, S. B., and Jaffe, M. (1980). *J. Cryst. Growth* **48**, 184.

Weitz, A., and Wunderlich, B. (1974). *J. Polym. Sci., Polym. Phys. Ed.* **12**, 2473.

Wilski, H. (1964). *Kunststoffe* **54**, 10, 90.

Wunderlich, B. (1962). *J. Chem. Phys.* **37**, 1207.

Wunderlich, B. (1963). *J. Polym. Sci., Part C* **1**, 41.

Wunderlich, B. (1964). *Polymer* **5**, 125, 611.

Wunderlich, B. (1973). "Macromolecular Physics," Vol. 1. Academic Press, New York.

Wunderlich, B. (1976). "Macromolecular Physics," Vol 2. Academic Press, New York.

Wunderlich, B. (1980). "Macromolecular Physics," Vol. 3. Academic Press, New York.

Wunderlich, B. (1981). *In* "Thermal Analysis in Polymer Characterization" (E. Turi, ed.). Heyden, New York.

Wunderlich, B., and Baur, H. (1970). *Adv. Polym. Sci.* **7**, 306.

Wunderlich, B., and Czornyj, G. (1977). *Macromolcules* **10**, 906.

Wunderlich, B., and Gaur, U. (1980). *In* "Thermal Analysis" (W. Hemminger, ed.), p. 409. Birkhaeuser, Basel.

Wunderlich, B., and Jones, L. D. (1969). *J. Macromol. Sci., Phys.* **B3**, 67.

Wunderlich, B., and Melillo, L. (1968). *Makromol. Chem.* **118**, 250.

Wunderlich, B., and Shu, H.-C. (1980). *J. Cryst. Growth* **48**, 227.

Wunderlich, B., and Sullivan, P. (1962). *J Polym. Sci.* **61**, 195.

Wunderlich, B., Bodily, D. M., and Kaplan, M. H. (1964). *J. Appl. Phys.* **35**, 95.

Wunderlich, B., Melillo, L., Cormier, C. M., Davidson, T., and Snyder, G. (1967). *J. Macromol. Sci., Phys.* **B1**, 485.

Yasuniwa, M., Nakafuku, C., and Takemura, T. (1973). *Jpn. J. Appl. Phys.* **4**, 526.

Yeh, G. S. Y., and Lambert, S. L. (1972). *J. Polym. Sci., Polym. Phys. Ed.* **10**, 1183.

Zbinden, R. (1964). "Infrared Spectroscopy of High Polymers." Academic Press, New York.

Zynger, J. (1975). *Anal. Chem.* **47**, 1390.

CHAPTER 3

Thermoplastic Polymers

SHALABY W. SHALABY

Research and Development Division
Ethicon, Inc.
Somerville, New Jersey

235

I. Introduction

Polymers may be divided into thermoplastics and thermosets. The polymers of the first group deform plastically and flow on heating. The second group of polymers are infusable and insoluble because of cross-linking between the molecules, which often is done only after some manufacturing step. Thermosets can still be deformed after cross-linking, but their response is largely elastic, so that they tend to regain their shape after removal of the stress. To these two basic groups of polymers the hard segment elastomers have been added lately. These respond to stress in much the same way as cross-linked thermosets; but when a certain temperature is exceeded, these materials deform and flow as expected from thermoplastics.

This chapter will be concerned with the application of thermal analysis methods to the characterization of thermoplastic materials. Hard segment materials will be discussed more fully in Chapter 4. The author will not attempt an encyclopedic coverage of the enormous fields of thermoplastics. This would require a number of volumes to approximate completeness and should soon become outdated. Instead, a presentation of the various thermal properties of thermoplastics as they are presently understood will be followed by some selected applications to specific systems of importance. Where possible, thermal properties will be discussed in terms of the chemical and physical structures that give rise to them.

In the past two decades the number of completely new commercial polymers has not increased much. Proliferation and diversification have occurred through increasing control and modification of polymers originally synthesized in the second quarter of this century. The mastery of large-scale production and control of complex structures has largely been made possible by an increase in our basic understanding of the chemistry of polymerization and the physics of polymer molecules. Improved means of routine analysis of large molecules through various types of spectroscopy, chromatography, and calorimetry lies at the foundation of this new knowledge. Although these analytical methods are not new, their routine application is new. This routine application became possible by reduction of previously tedious and complex techniques to simple analyses through advance of instrumentation.

The modern chemist in a modest laboratory has at his disposal methods and instrumentation that a few years ago could be found only in the most specialized research facilities. This progress in instrumentation is expected to continue for many years to come. Unfortunately, possession of an instrument does not guarantee its intelligent application. Showing how to use advanced instrumentation in thermal analysis intelligently is the second theme of this chapter. The discussion will be limited to the following thermal techniques: thermogravimetry (TG), thermomechanical analysis (TMA), differential thermal analysis (DTA) and differential scanning calorimetry (DSC). Useful combinations of these techniques with the various chromatographic and spectroscopic methods will also be considered.

II. Material Classification and Relevant Chemical and Physical Parameters to the Thermal Behavior of Thermoplastics

A. MATERIAL CLASSIFICATION

Polymers can be classified according to the basic chemical composition of the monomer, e.g., as polyolefins, polyesters, polyamides or polyethers, or on the basis of their formation mode, e.g., chain-growth and step-growth polymers. Neither method is fully consistent, since some polymers can be made by chain as well as step reactions. Similarly, the same chemical compositions can occasionally be reached by different classes of monomers. A more comprehensive classification of polymers was described a few years ago (Shalaby and Pearce, 1974a). This classification took into account (a) the type of atoms constituting the main chain, i.e., homo- or heterochain; (b) the type of monomers used for the polymer formation and their functionalities, e.g., homofunctional (amino acid), heterofunctional (dicarboxylic acid), homocyclic (cyclopropane), or heterocyclic (caprolactam); and (c) mechanism of chain formation, e.g., acyclization (caprolactam polymerization), carbonyl addition (aldehyde polymerization), and heterofunctional condensation (formation of polyesters from diols and diacids). A fourth classification, based on the physical properties of the polymers, is commonly used by polymer engineers and materials scientists. According to the latter classification, polymers can be described as (a) thermoplastics, (b) thermosets, (c) elastomers or rubbers, (d) elastoplastics, and (e) thermostable polymers. Unfortunately none of these classifications is in the author's opinion useful for the thermal analyst. This leads to a proposal of what may be donoted as the "thermal analyst classification" (TAC). In the TAC certain features of the third and fourth classification were adopted and the role of chemical and physical inhomogeneities were taken into account. Since this chapter deals with thermoplastics, the application of TAC to thermosets will not be treated.

The three basic molecular and bulk properties on which the TAC is based are the (1) constitutive chemical and physical parameters of the polymer chain, (2) intermolecular interactions in the solid state, and (3) bulk heterogeneity. Further discussion of these properties is given in the next few paragraphs.

1. Constitutive Chemical and Physical Chain Properties

The thermal behavior of a chain molecule is dictated primarily by (a) the nature and sequence of atoms constituting the main chain (or backbone) in heterochain polymers as compared with homochain (usually made of carbon–carbon links) polymers; (b) the presence of side groups and their steric arrangements; (c) the presence of chemical chain defects such as branches, cross-links, and chain defects caused by chemical, oxidative, thermal, and/or thermooxidative reactions; (d) the nature of end groups, which may affect molecular end-to-end association and hence an apparent increase in molecular weight; (e) chain length, or molecular weight and molecular-weight distribution—above a certain molecular weight thermal properties are only little affected by chain length; and (f) thermodynamic instability associated with the presence of a monomer–polymer equilibrium that can lead to appreciable monomer generation (by chain unzipping) at moderate to high temperatures.

2. Intermolecular Cooperative Interaction in the Solid State

For symmetric linear chains, intermolecular forces such as van der Waals interactions, dipole interactions, and hydrogen bonding, and, in a few cases, ionic bonding are capable of maintaining cohesively aligned chains in a crystalline structure.

3. Bulk Heterogeneity

Bulk heterogeneity can be intrinsic or extrinsic, depending on whether one is dealing with a chemically uniform or nonuniform system, respectively. Intrinsic bulk heterogeneity, which is characteristic of a chemically uniform system, consists of (a) oriented and unoriented amorphous regions and/or (b) crystalline and noncrystalline phases. Systems with extrinsic bulk heterogeneity are those containing (a) extraneous chemical species, other than the principal macromolecular chains (plasticizers, stabilizers, solvents, and different forms of additives); (b) two or more types of chain molecules aggregated in micro- or macrophases as in block copolymers or polymer blends, respectively; and (c) random copolymers which are nonuniform, with wide distributions in chain composition.

The proposed thermal analyst classification (TAC) of polymeric thermoplastic materials is outlined below and some examples are given.

Classifications of Thermoplastics according to Bulk Properties

(A)   Bulk homogeneous thermoplastics.

(1)   Amorphous homochain homopolymer, e.g., atactic polypropylene.
(2)   Amorphous homochain alternating copolymers, e.g., poly(styrene-*co*-maleic anhydride).
(3)   Amorphous heterochain homopolymers, e.g., poly(*N*-ethyllaurolactam) and the phosphonitrile-type polymers.

(B)   Intrinsically bulk heterogeneous thermoplastics.

(1)   Partially oriented, amorphous homochain homopolymers and alternating copolymers.
(2)   Partially oriented, amorphous heterochain homopolymers and alternating copolymers.
(3)   Semicrystalline homochain homopolymers and alternating copolymers.
(4)   Semicrystalline heterochain homopolymers and alternating copolymers.

(C)   Extrinsically bulk heterogeneous thermoplastics.

(1)   Polymers containing plasticizers and/or other additives at $\geq 5\%$ concentration.
(2)   Polymers with two or more macrophases, e.g., polymer blends.
(3)   Polymers with two or more microphases, e.g., graft and block copolymers.

Classification of Thermoplastics according to Chain Thermoreactivity

(1)   Chain depolymerizable homochain polymers, e.g., poly(methyl methacrylate).
(2)   Chain depolymerizable heterochain polymers, e.g., Nylon 6.

(3) Randomly degradable homochain polymers, e.g., polypropylene.
(4) Randomly degradable heterochain polymers, e.g., poly(ethylene terephthalate).
(5) Homochain polymers with ring-forming side chains, e.g., polyacrylonitrile.
(6) Cross-linkable polymers.
(7) Homochain polymers with thermolabile side groups, e.g., poly(vinyl chloride).
(8) Heterochain polymers with reactive end groups, e.g., Nylon 6.

B. CHEMICAL AND PHYSICAL PARAMETERS RELEVANT TO
   THE THERMAL BEHAVIOR OF THERMOPLASTICS

1. Thermal Behavior of Homochain Homopolymers

Linear polyethylene (PE) is an ideal example of a symmetrical, highly crystallizable homochain polymer. Thus, one easily detects a melting transition in linear PE. Because of its high crystallization rate from the melt, it is rather difficult—i.e., until recently (Guar and Wunderlich, 1980)—to detect, directly, the glass transition by conventional techniques (Chang, 1972; Simon *et al.*, 1975; Stehling and Mandelkern, 1969). Substitution of one methyl group into the polyethylene repeat unit as in polypropylene (PP) creates a more complicated dependence of the thermal transitions on the chain physical parameters. Now depending on the steric order (or tacticity) of the methyl group about the PP backbone, the polymer is described as (a) isotactic when all the methyl-substituted carbons have the same configuration; (b) syndiotactic when two configurations of methyl-substituted carbons alternate regularly; and (c) atactic when a random distribution of two configurations of the methine group prevails. The atactic polymer is an amorphous "rubbery gum" that forms a glass at low temperatures (Kamide and Yamaguchi, 1972). On the other hand, the stereoregular, isotactic PP is crystalline and displays a melting transition well above polyethylene. An excellent discussion of melting temperatures of crystalline isotactic and syndiotactic polypropylene and their comparison with that of polyethylene is given in Chapter 2. Similarly, the effect of tacticity can be illustrated by comparing (a) the amorphous atactic and the crystalline isotactic polystyrene (T_m = 240°C) and (b) the amorphous atactic and crystalline syndiotactic (T_m = 160°C) poly(methyl methacrylate) (PMMA).

The effect of structure on the T_g of homochain polymers is less dramatic than that encountered with the T_m. Thus for simple polyolefins the T_g usually increases with the increase of the steric requirement of the chain substituents. For instance, PE, PP, and PS undergo glass transitions at about -90, -18, and 100°C, respectively. On the other hand, if the side groups are composed of polar moieties as in polyacrylonitrile (PAN, $T_g =$ 100°C), isotactic PMMA ($T_g = 45$°C) and polytetrafluoroethylene ($T_g =$ 127°C), the intermolecular cohesion will increase. The T_g now increases with polarity and the steric effectiveness of these groups.

Chemical structure and percent crystallinity are important parameters affecting the thermal and oxidative stability. To illustrate this, one may compare the properties of polyethylene (PE), isotactic polypropylene (i-PP), and atactic polypropylene (a-PP). The thermal and thermooxidative stability of PE is higher than that of i-PP and a-PP, because of the presence of the thermooxidatively sensitive methine group (CH) in the latter two polymers. Meanwhile, the crystalline i-PP is considerably more stable toward oxidative (and radiative) degradation than the amorphous, more permeable, a-PP.

2. Thermal Behavior of Heterochain Homopolymers

The thermal properties of useful heterochain homopolymers are determined by the polarity of the main chain and the presence of functional groups capable of hydrogen bonding. To illustrate the effect of these two parameters on the T_g and T_m, two series of typical heterochain polymers, namely, polyamides and polyesters, are considered. The effect of polarity will be pertinent to T_g and T_m of both types of polymers. On the other hand, the effect of hydrogen bonding is limited to polyamides. The T_g and T_m of representative examples of aliphatic polyesters and polyamides are summarized in Table I. These data indicate that in both types of polymers, T_g and T_m increase with the decrease in the number of methylene groups per ester or amide group in the repeat unit (poly-β-propiolactone versus poly-ϵ-caprolactone and poly-β-propiolactam versus poly-ϵ-caprolactam).

TABLE I

EFFECT OF CHAIN POLARITY AND HYDROGEN BONDING
ON T_g AND T_m IN POLYESTERS AND POLYAMIDES

Polymer	T_g (°K)	T_m (°K)
Nylon 6, 6	322	500
Nylon 6, 10	318	499
Nylon 12, 2	332	509
Nylon 10, 2	329	515
Nylon 8, 2	366	552
Nylon 6, 2	432	599
Poly(ethylene adipate)	210	325
Poly(ethylene terephthalate)	342	529

Methylation of the amide group of polylactams to eliminate hydrogen bonding causes a distinct decrease of both T_g and T_m, as in the case of polylaurolactam and poly-N-methyllaurolactam (Shalaby *et al.*, 1972, 1973b). A more extensive treatment of the effect of partial and complete N-alkylation of nylons was given by Boyer (1977). Table II lists some of the T_g and T_m values compiled by Boyer for various nylons and their methyl derivatives. Figure 1 is a plot of T_g versus T_m for nylons with 6–12

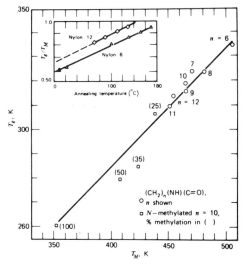

Fig. 1 T_g–T_m relationship for a series of aliphatic nylons and four N-methylated Nylon 11, using data of Champetier and Pied (1961). Inset shows T_g/T_m ratio for Nylons 6 and 12 as a function of annealing temperature, according to Northolt *et al.* All annealed specimens were x-ray amorphous. (From Boyer, 1977.)

methylene groups in their repeat units and also for N-methylated Nylon 11. The data in Table II and Fig. 1 show a reasonably good correlation between the availability of hydrogen-bonding sequences and the increase in T_g and T_m.

The inset of the figure shows the effect of annealing temperature on the $T_g : (T_m)$ ratio for Nylon 6 and 12 (Northolt *et al.*, 1975). Since T_gs as a function of annealing temperature lie on a common line for all the nylons, it seems clear that annealing is affecting a common feature, namely, perfection of the hydrogen-bonded structure that gives rise to T_g. Both Gordon (1971) and Northolt *et al.* (1975) noted the disappearance of T_g upon quenching molten nylons. Since any polymer free from very high crystallinity will show a T_g, it is evident that none of these investigators went to a sufficiently low temperature to find a T_g characteristic of a given composition in the absence of hydrogen bonds. Figure 2 is a plot of T_g as a function

TABLE II

Tʀᴀɴsɪᴛɪᴏɴ Dᴀᴛᴀ ғᴏʀ Sᴏᴍᴇ Nʏʟᴏɴs[a,b]

—(CH₂)ₙ—CO—NH—n=	T_g (K)	T_m (K)	N-methylation (%)	Hydrogen bonding[c] index
6	335	505	0	25
7	324	471	0	18
8	324	482	0	20
9	316	465	0	16.4
10	319	465	0	16.6
10	307	438	25	12.4
10	285	423	35	10.7
10	280	408	50	8.3
10	261	353	100	0
11	310	452	0	13.6
12	314	455	0	14.2

[a] From Boyer (1977).
[b] Thermal expansion data of Champetier and Pied (1961).
[c] 100 [2/(n + 2)] for even nylons, 85 [2/(n + 2)] for odd nylons.

of hydrogen-bonding index, according to Champetier and Pied (1961). Boyer (1977) has shown that the extrapolation to zero hydrogen-bonding content should give the T_g of polyethylene. This can be estimated to be about 285 K, which is much higher than the 237 K value proposed recently by Gaur and Wunderlich (1980). The correct extrapolation would

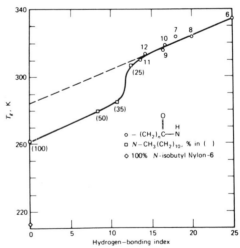

Fig. 2 T_g for the aliphatic nylons and N-methylated Nylon 11 as a function of hydrogen-bonding index according to Champetier and Pied (1961). The single point at 213 K for completely N-isobutylated Nylon 6 is by Lord (1974). (From Boyer, 1977.)

be on the basis of the (nonavailable) T_gs of nonhydrogen-bonding nylons; but not using the N-methylated derivatives that introduce a large side group.

3. Effect of Chemical Composition and Sequence Distribution in Copolymers

The effect of composition on T_g and T_m of heterochain polymers has been illustrated by several examples of copolyamides reported by this author over the last few years (Shalaby *et al.*, 1973a, b, 1974a, b, 1975, 1976a, b, c, 1978a, b). The effect of sequence distribution on the T_m of heterochain copolymers was demonstrated in high-temperature systems, which are beyond the scope of this chapter.

Johnson (1976) has reviewed the effect of sequence distribution on the T_g of several homochain copolymers. He demonstrated the possible prediction of T_g of copolymers, using an equation based on probability considerations.

III. Thermally Activated Processes and the Effect of Molecular and Morphological Parameters

On heating, many processes may occur in thermoplastic polymers. Those pertinent to thermal analysis include glass transition, crystallization, melting, and chain dissociation, cross-linking, oxidation, and side-group stripping or condensation.

A. Glass Transition

Glass transition has usually been associated with the onset of segmental mobility in the amorphous phase of an amorphous or semicrystalline polymer. Although one can qualitatively define the glass transition temperature (T_g) as the temperature at which segmental mobility becomes possible, the well-documented dependence of T_g on several structural, morphological, and experimental parameters has made quantitative studies of the glass transition rather difficult. Many of the factors that affect the value of T_g have been discussed in Chapter 2 and several earlier reviews (Billmeyer, 1971; Boyer, 1977; Bueche, 1962; Karasz and Mac-Knight, 1968; Mears, 1965; Shalaby *et al.*, 1978a, b; Wunderlich and Baur, 1970). The effect of chemical structure, molecular weight, and molecular-weight distribution, stereochemical features about the main chain, and intermolecular interactions as well as a few recent developments regarding glass transition constitute the framework of the present section. Other factors will be discussed later in the chapter.

1. Effect of Chemical and Physical Variables on T_g

a. Chemical Structure. The influence of chemical structure on the glass transition is not understood in all detail. The general rule, however, is that any structure that reduces chain mobility will increase T_g. This can be illustrated by the following examples. Homochain carbon backbone polymers without side groups such as polyethylene and *trans*-1,4-polybutadiene, have very low T_gs. Substitution of methyl groups, phenyl groups and both methyl and carbomethoxy groups leads to an increase in T_g due to the increase of the steric requirements about the main chain, as in polystyrene and poly(methyl methacrylate). The effect of chemical structure is illustrated by the data in Table III.

TABLE III

GLASS TRANSITION TEMPERATURES OF TYPICAL
HOMOCHAIN POLYMERS[a]

$+CH_2-CRR'+_n$ Made by free radical polymerization		T_g (°C)
R =	R' =	
H	H	-36
H	CH_3	-12
CH_3	CH_3	-68
H	C_6H_5	100
CH_3	CO_2CH_3	105
CH_3	C_6H_5	180

[a] Wrasidlo (1974).

The effect of structure on the glass transition temperature of heterochain polymers is illustrated by considering polyesters and polyamides. In polyesters of oxalic and terephthalic acids with aliphatic diols, the T_g increases and decreases, respectively, with the increase of the chain aliphatic content as shown in Table IV.

TABLE IV

GLASS TRANSITION TEMPERATURES OF TYPICAL POLYESTERS

Oxalate polymers[a] $HO+CH_2+_nOH$	T_g (°C)	Terephthalate polymers[b] of $HO+CH_2+_nOH$	T_g (°C)
n = 3	-19	$n = 2$	90
4	-4.5	4	45
6	29	6	8
		10	-9

[a] Shalaby and Jamiolkowski (1979a).
[b] Yip and Williams (1976a,b).

It should be pointed out that the polyalkylene oxalates of the C_4 and C_6 diols display an anomolous glass transition, as compared with more conventional polyesters. This is in part ascribed to the unique placement of the ester groups along the polymer main chain.

In aliphatic polyamides, the T_g does not vary considerably with the methylene content (Pearce, 1969). The introduction of rigid aromatic or alicyclic moieties into the chain can lead to a noticeable increase in T_g. In case of aliphatic polyamides the T_g can be lowered considerably as substi-

TABLE V

GLASS TRANSITION OF STRUCTURALLY DIFFERENT POLYAMIDES[a]

Repeat unit	T_g (°C)	Reference
$-(CH_2)_5-CO-NH-$	42	Shalaby and Reimschuessel, 1977
$-OC(CH_2)_4-CO-NH-(CH_2)_6-NH-$	80	Pearce, 1969
$-(CH_2)_{11}-CO-N-$ $\qquad\qquad\quad\; CH_3$	−34	Shalaby et al., 1974a
$-(CH_2)_{11}-CO-N-$ $\qquad\qquad\quad\; Et$	−40	Shalaby et al., 1973b
(structure) $CH_2-NH-CO$ (structure) $CONH-$	172	Shalaby et al., 1976c
$-HN-(CH_2)_6-NH-CO-(CH_2)_2\overset{\overset{\textstyle O}{\|}}{P}-(CH_2)_2-CO-$ $\qquad\qquad\qquad\qquad\qquad\quad\; CH_3$	55	Shalaby et al., 1975
$-HN-(CH_2)_8-NHCO-(CH_2)_2-\overset{\overset{\textstyle O}{\|}}{P}-(CH_2)_2-CO-$ $\qquad\qquad\qquad\qquad\qquad\quad\; CH_3$	49	Shalaby et al., 1975
$-HN-(CH_2)_{10}-NHCO-(CH_2)_2-\overset{\overset{\textstyle O}{\|}}{P}-(CH_2)_2-CO-$ $\qquad\qquad\qquad\qquad\qquad\qquad CH_3$	46	Shalaby et al., 1975
$-NH-(CH_2)_{12}-NHCO-(CH_2)_2-\overset{\overset{\textstyle O}{\|}}{P}-CH_2-CO-$ $\qquad\qquad\qquad\qquad\qquad\qquad CH_3$	42	Shalaby et al., 1975
$\overset{\overset{\textstyle O}{\|}}{-C}CH_2CH_2-N$ (ring) $-CH_2NH-$	69	Shalaby et al., 1978a

[a] All samples were quenched from the liquid state prior to measuring the T_g.

tution in the amide group increases. Typical T_g values of different polyamides are shown in Table V.

   b. Molecular Weight. It is well known that T_g increases with increase of molecular weight. This is expressed in the following equation, given by Fox and Flory (1950):

$$T_g = T_g(\infty) - K_g M^{-1} \tag{1}$$

where $T_g(\infty)$ is the limiting T_g at high molecular weight and K_g is a constant.

   It was later noted by Boyer (1974) and Kumler *et al.* (1977) that for polystyrene a plot of molecular weight versus T_g data according to Eq. (1) could be represented by three intersecting straight line regions (Fig. 3),

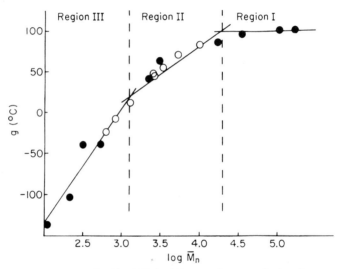

Fig. 3 T_g data as determined by ESR for ($\bullet$) monodisperse anionic polystyrene and ($\bigcirc$) polystyrene binary blends, (—) linear response in each of the three regions, (--) boundaries of the three regions drawn at the intersections of the straight lines. (Reprinted from Kumler *et al.*, 1977, p. 641, by courtesy of Marcel Dekker, Inc.)

each region exhibiting a different value of K_g. Boyer's proposal for polystyrene was then expanded by Cowie (1975), who suggested that a wide variety of polymers exhibit similar three-region behavior for the dependence of T_g on molecular weight. From examination of the literature data for a variety of polymers, Cowie concluded that this three-region behavior is a general feature of T_g–$\overline{M}_n$ plots and suggested the following empirical equation for defining the boundaries between the three regions:

$$T_g(\infty) = 372.6 \log \chi_c - 595 \tag{2}$$

where χ_c is the number of chain segments. Equation (2) defines the boundary between region I (in which the asymptotic value of T_g, i.e., $T_g(\infty)$, has been attained and the observed T_g does not vary with molecular weight) and region II (in which T_g increases steadily with increasing molecular weight). If the range of molecular weights is large enough, a third region (region III) can also be observed in which the increase in T_g with increasing molecular weight is even more pronounced. The equation describing the boundary between regions II and III, the transition between the oligomeric and polymeric regions, is given by

$$T_g(\infty) = 761.5 \log \chi_c - 681.8 \tag{3}$$

The symbol χ_c in Eq. (3) is the number of chain segments at this boundary.

In a recent report on the use of electron spin resonance (ESR) for studying glass transition (Kumler *et al.*, 1977), the ESR data of the T_g dependence on molecular weight were consistent with the previously mentioned three-region interpretation, as can be illustrated by Fig. 3. In this figure, the T_g as determined by ESR for the anionic polystyrenes and binary blends are plotted against log $\overline{M}_n$ and three straight lines have been drawn representing the three regions. The dotted lines separating the three regions were drawn at the intersections of the straight lines. Equations (2) and (3) predict that these divisions should have occurred at log $\overline{M}_n$ values of 3.4 and 4.6, respectively, for polystyrene.

Typical T_g data, as measured by the spin-probe technique using a nitroxide   [2,2,6,6-tetramethyl-4-hydroxy-piperidin-1-oxyl   benzoate (BxONO)] as a probe, of several samples of "monodisperse" polystyrene

TABLE VI

T_g Data for Styrene Oligomers
and Polymers[a]

System	$\overline{M}_n$	T_g (°C)
Styrene monomer	104	−138
Styrene dimer	208	−104
Styrene trimer	312	−40
Polystyrene-600	524	−40
Polystyrene-2100	2,210	40
Polystyrene-4000	3,100	62
Polystyrene-17,400	15,100	86
Polystyrene-37,000	36,000	94
Polystyrene-110,000	111,000	100
Polystyrene-NBS-705	170,900	100

[a] From Kumler *et al.* (1977) and Wall *et al.* (1974).

as well as those of the monomer, dimer, and trimer are shown in Table VI. According to the data in the table, $T_g(\infty)$ for polystyrene is 100°C.

The ESR T_g data reported by these authors (Kumler *et al.*, 1977) agreed in most cases with the T_g data based on other techniques, including dilatometry (Ueberreiter and Kanig, 1952; Fox, 1956; Fox and Flory, 1950, 1954) and differential scanning calorimetry (Cowie, 1975).

The effect of molecular weight on the T_g of heterochain synthetic polymers has received relatively less attention than for homochain polymers. In some of the studies on polyamides reported by this author and his coworkers, T_g was shown to increase with the increase in degree of polymerization (DP) (measured in terms of reduced viscosity). Typical examples of the effect of DP on T_g for several polyamides are shown in Table VII.

TABLE VII

EFFECT OF DP ON T_g OF POLYAMIDES[a]

Polymer	Reduced viscosity	T_g (°C)	Reference
Nylon 6	0.25	25	Shalaby *et al.*, 1974b
	0.45	35	S. W. Shalaby and E. A. Turi, unpublished work, 1973
	0.98	40	Shalaby and Reimschuessel, 1977
	1.80	42	Shalaby *et al.*, 1975
	3.26	42	Shalaby and Reimschuessel, 1977
Polyphosphacaprolactam	0.19	84	Shalaby *et al.*, 1974b
	0.33	90	Shalaby *et al.*, 1974b

[a] The T_g was determined on reheating the quenched melts of the polymer.

Typical differential scanning calorimetry data, illustrating the effect of $\overline{M}_w$ on the T_g of two polymethacrylates, are shown in Fig. 4.

c. T_g Elevation by Chain Entanglement. There have been a few hints in the literature that chain entanglements may influence the T_g of a polymer (Bueche, 1962; Boyer, 1963; Beatty, 1977). These hints took the oblique form of a warning that data obtained above some critical molecular weight should be treated with caution. In a report by Turner in 1978, a direct attempt has been made to analyze this matter on the working hypothesis that, by analogy with chemical cross-links, entanglements may elevate T_g. This hypothesis was evaluated by examination of previously reported data on the dependence of T_g on molecular weight M in order to judge whether there is evidence for a critical-molecular-weight phenomenon that might be attributed to entanglements.

Turner reported that for some polymers, a plot of glass transition temperature T_g versus reciprocal molecular weight can be taken to define two

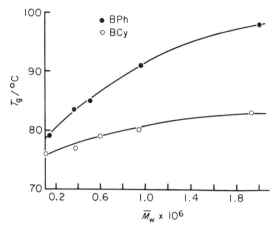

Fig. 4 Plot of vitreous transition temperature T_g as a function of weight-average molecular weight $\overline{M}_w$ for several fractions poly(*p-tert*-butylphenyl methacrylate) (BPh, ●) and poly(*p-tert*-butylcyclohexyl methacrylate) (BCy, ○). Heating rate: 16°C/min. (From Gargallo and Russo, 1975, by courtesy of Hüthig and Wepf Verlag, Basel.)

lines that intersect at a molecular weight designated as M_g (Fig. 5). The value of M_g agrees, within a factor of 2, with critical values of molecular weight reported for other properties that are generally attributed to incipient formation of a network of entanglements. Therefore, it is suggested that an increased elevation of T_g at molecular weights greater than M_g is due to an increasing concentration of entanglements.

A more detailed analysis was made by extending the Fox–Flory theory of the glass transition to include negative contributions to the free volume from entanglements. This extension leads to revised estimates of the free volume per chain end that are much smaller (6–19 Å³ near T_g) than previous estimates that took no account of entanglements. These smaller

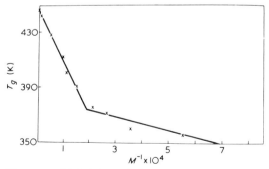

Fig. 5 Plot of T_g versus M^{-1}: Poly(α-methylstyrene), using Cowie and Toporowski DTA data (1968). (From Turner, 1978, by permission of the publishers, IPC Business Press Ltd., ©.)

values are interpreted to mean that the jump units are correspondingly small at T_g, such as envisaged in the Gibbs–DiMarzio theory of the glass transition.

d. Molecular-Weight Distribution. The dependence of T_g of polymers and particularly polystyrene on the number-average molecular weight is well documented (Boyer, 1970). In a recent report (Kumler *et al.*, 1977), two series of polyestyrene blends having a wide range of dispersity were prepared in order to examine the effect of molecular-weight distribution on T_g using the ESR spin-probe technique. The T_g data for these blends and polydisperse ($\overline{M}_w/\overline{M}_n = 20.6$) thermal polystyrene are summarized in Table VIII.

TABLE VIII

EFFECT OF MOLECULAR-WEIGHT DISTRIBUTION ON $T_g{}^a$

Polystyrene system	$\overline{M}_w$	$\overline{M}_n$	T_g (°C)
A	117,700	111,000	100
B	95,000	10,400	83
C	71,800	5,400	70
D	48,300	3,630	53
E	25,100	2,740	43
F	31,600	1,530	43
G	23,300	650	−25

[a] From Kumler *et al.* (1977).

The data in Table VIII clearly demonstrate that T_g values are closer to a linear function of $\overline{M}_n$ rather than of $\overline{M}_w$. This is illustrated in Fig. 6; the solid line is a smooth curve drawn through the experimental points when T_g is plotted versus $\overline{M}_n$ for the series of "monodisperse" polystyrenes. Because of the low polydispersity indices for these samples, essentially the same curve would result if log $\overline{M}_w$ was plotted against T_g. On positioning the data points for the eight blends on the graph as a function of $\overline{M}_n$ and $\overline{M}_w$, it becomes apparent that the T_g values determined by the ESR method are a function of $\overline{M}_n$ rather than $\overline{M}_w$.

The T_g values determined by the ESR method were also shown to agree with those obtained by DSC. The DSC results for the blends and the ESR data for anionic polystyrenes and the binary blends are expressed as a function of $\overline{M}_n$ in Fig. 7. It is obvious from Fig. 7 that the ESR results are internally consistent and are in good agreement with the T_g values determined by DSC; i.e., both techniques measure the same transition.

e. Change in Molecular-Weight Distribution (MWD) in the Amorphous Phase. The presence of certain low-molecular-weight fractions in

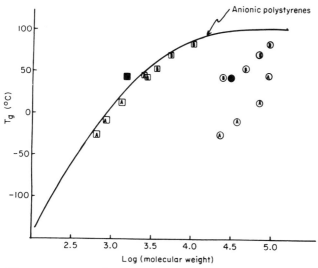

Fig. 6 Glass temperature of polystyrene binary blends as a function of molecular weight: (—) data for "monodisperse" polystyrene samples: (○) T_g plotted versus $\overline{M}_w$; (□) T_g plotted versus $\overline{M}_n$, where A and B indicate the two different series of blends; (●, ■) are data points for the thermal polystyrene PS-1. (Reprinted from Kumler *et al.*, 1977, p. 637, by courtesy of Marcel Dekker, Inc.)

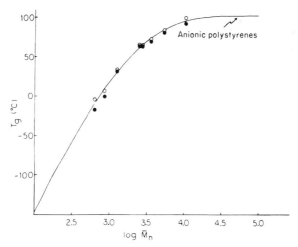

Fig. 7 Comparison of T_g data for polystyrene blends as determined by (●) ESR and (○) DSC; (—) data for "monodisperse" polystyrenes. (Reprinted from Kumler *et al.*, 1977, p. 639, by courtesy of Marcel Dekker, Inc.)

the amorphous region of a semicrystalline polymer can cause its plasticization and lead to a decrease in T_g. The MWD of the amorphous region of a semicrystalline polymer may be dependent on (a) the rate of crystallization from the melt, (b) annealing time and temperature, and (c) type of morphological changes associated with orientation. A decrease in T_g following orientation and/or crystallization, discussed later, may be attributed to changes in MWD of the amorphous phase during the orientation and/or crystallization of these polymers. This is consistent with an earlier proposal suggesting that as a spherulite grows, lower-molecular-weight chains and impurities tend to be rejected from the structure as crystallization proceeds; there may be local accumulation of such matter in the amorphous regions that will depress T_g (Keith and Padden, 1964).

    f. Mode of Crystallization from the Melt.　The molecular chain of ω-amino acid polyamides with odd number of carbons are known to pack in the parallel mode upon crystallization and generally display lower ΔS (fusion) and higher T_m than the ω-amino acid polyamides with even number of carbons. This behavior was extrapolated by Lord (1974) to indicate that "odd"-type polyamides can crystallize at faster rates than their "even" analogs. It was concluded by this investigator from DSC measurements that neither a glass transition nor cold crystallization occurs in quenched samples of "odd"-type polyamides. This conclusion was contradicted by Illers (1977) on the basis of some elegant DSC studies of "odd"- and "even"-type polyamides. It was shown that quenched samples of some "odd"-type polyamides can undergo both a glass transition and a cold crystallization. Furthermore, it was shown that certain "odd" and "even" polyamides having 11 or more carbons in their sequences cannot be prevented from undergoing primary crystallization upon quenching their melts. Primary crystallization can occur at a fast rate on cooling a polymer melt and will be followed by a slow continuous secondary crystallization (Illers, 1975). Now, on quenching a polymer, if crystallinity develops mainly by primary crystallization, the high crystallinity of the quenched material will tend to mask the glass transition upon reheating the sample.

    g. Crystallinity.　According to an early description of crystallites in semicrystalline polymers as fillers, one may expect that T_g will increase with the increase of crystallinity. This is indeed the case in many polymers including poly(ethylene terephthalate) (Woods, 1954), poly-ε-caprolactone (Koleske and Lundberg, 1969), isotactic polystyrene (Newman and Cox, 1960), and isotactic poly(methyl methacrylate) (O'Reilly *et al.*, 1964). However, the effect of crystallinity on T_g is negligible in isotactic polypropylene (Newman and Cox, 1960) and nonexistent in polychlorotrifluoroethylene (Hoffman and Weeks, 1958). On the other hand, the T_g of poly-4-methyl-1-pentene was shown to decrease with the increase in crystallinity (Griffith and Ranby, 1960; Ranby *et al.*, 1962). A few attempts were made to explain the various modes in which crystallinity affects T_g. For poly(ethylene terephthalate), the rate of increase in the value of T_g with the increase in crystallinity is more pronounced for samples containing small crystallites compared to those having larger crystallites (Uematsu and Uematsu, 1959). This suggests that the T_g of polymers having large crystallites should show a minimum or no dependence on the degree of crystallinity. The observed effect of crystallinity on T_g in poly-4-methyl-1-pentene was suggested to be a direct effect of tacticity rather than crystallinity (Griffith and Ranby, 1960). It was thus believed that the increase in crystallinity is associated with an increase of a "low T_g" isotactic fraction and decrease in the "high T_g" syndiotactic fraction. Other views on the "apparent" role of crystallinity on T_g are treated in later sections of this chapter.

Determination of T_g in semicrystalline polymers can be achieved for low-crystallinity samples. As the degree of crystallinity increases, the glass transition will become hardly detectable by direct methods. Indirect techniques such as extrapolation of data for blends or copolymers can be used for determining the T_g of highly crystalline polymers. Measurements of the linear thermal expansion of polyethylene samples having different degrees of crystallinity were made. The temperature at which this parameter becomes independent of crystallinity was denoted as T_g (Stehling and Mandelkern, 1969). More recently, a sensitive calorimetric method for detection of the glass transition in highly crystalline polymers has been described (Chang, 1972). Thus the existence of a glass transition was demonstrated in a linear polyethylene containing less than 5% amorphous phase. The method used in these studies is based on measurements of the kinetic thermal behavior associated with the tendency of the glass to relax to the liquid state in the T_g region. It is interesting to note that both methods give widely different T_gs. This problem was discussed, elegantly, in a recent report by Gaur and Wunderlich (1980), who have given a convincing argument for the glass transition of polyethylene at 237 K.

h. Orientation. It is well documented that orientation of a semicrystalline polymer, by simple drawing of a film or fiber, is associated with noticeable changes in physical properties of the polymer (Bell, 1972). These changes are attributed partly to the orientation of the amorphous phase as suggested by many investigators (Adams, 1971; Bonart, 1969; Buchanan and Walters, 1977; Ito, 1974; Ke, 1964; Nakamura *et al.*, 1972; Nose, 1973; Stein, 1969; Stolting and Muller, 1970; Uejo, 1970; Van Krevelen, 1972; Ward, 1971; Wrasidlo, 1974). However, one of these properties, namely, the glass transition, which can be affected by orientation, received relatively modest attention by a few authors (Adams, 1971; Bonart, 1969; Buchanan and Walters, 1977; Ito, 1974; Nose, 1973; Stolting and Muller, 1970; Uejo, 1970.) For poly(ethylene terephthalate) (PET) fibers, T_g was shown to decrease to a minimum at a draw ratio of 1.5, increase to a maximum at a draw ratio of about 2.0, and then continue to decrease at draw ratios over 2.0 (Ito, 1974). The decrease of T_g at beginning of drawing was suggested to depend on the increase of configurational entropy, whereas at draw ratios above 2.0, the decrease in T_g was proposed to depend on the increase of entropy associated with intermolecular interaction. Hence it was concluded that the change of T_g is determined by local oscillations in the amorphous region. In extensive studies on the effect of orientation on the T_g of aliphatic polyamides, it was shown that (a) the T_gs of these polymers increase steadily with the increase of the draw ratio and (b) the T_gs of oriented fibers measured at 0% RH by an essentially zero deformation frequency technique such as dilatometry or calorimetry may differ by as much as 50°C from those determined by a dynamic method (Rheovibron) with a deformation frequency of 110 Hz (Buchanan and Walters, 1977).

i. Thermal History. The effect of cooling on T_g has been the subject of several investigations in the last 15 years. Wunderlich *et al.* (1964) have shown that one can normally obtain a straight-line fit in a plot of the logarithm of the cooling rate versus T_g, as the latter increases with the increase in cooling rate. Quantitative relationships between T_g and cooling rate (CR) have been given by other authors (Ferry, 1961; Kovacs, 1963), which involved a relaxation activation energy

$$E_a = -2.3R \ d \ \log(CR)/d(1/T_g) \tag{4}$$

and

$$E_a = -2.3R(C_1/C_2)T_g^2 \tag{5}$$

where

$$d \ \log(CR)/dT_g = C_1/C_2 \tag{6}$$

R is the gas constant, and C_1/C_2 are WLF constants. Equations (4) and (5) predict a nonlinear relationship between CR and $1/T_g$. However, for a short T_g range, the relationship may appear linear and give a constant E_a value. Equation (6) predicts a straight line for CR versus T_g with a slope equal to C_1/C_2.

In recent studies of the effect of thermal history on the glass transition of several commercial nylons, both thermal and mechanical techniques were used for the measurement of T_g (Greco and Nicolais, 1976). Results of these studies indicated that during thermal measurements, transitions detected during the initial heating cycle disappeared in the subsequent cooling cycle and appeared again only after a sufficient sample "rest period" and at temperatures different from those initially measured. This behavior was attributed to the structure of the amorphous regions where the hydrogen-bonding groups form an irregular network. The delay in reforming the preceding network was the main cause for the dependency of the observed transition on the thermal history imposed on the samples. On the other hand, it was indicated that mechanical measurements give results that are quite insensitive to thermal treatment of the materials and thus provide reproducible values of the transitions. The effect of thermal history on the T_g of several types of polyamides was described by this author and co-workers over the past 6 to 7 years (Shalaby and Reimschuessel, 1977; Shalaby and Turi, 1977; Shalaby *et al.*, 1973a, b, 1974a, b, 1975, 1976a, 1978a). Polymers that lack hydrogen bonding, such as polystyrene, have been reported to show a substantial increase in T_g upon storing for two years (Spencer and Boyer, 1946). This may be due to less-volatile impurities in the polymer or relaxation. After annealing at low temperature, linear polyethylene displays an unusual feature in its specific heat curve (Sakaguchi *et al.*, 1976). On heating, a maximum is observed just above the annealing temperature. The magnitude of this excess specific heat is dependent on the initial level of crystallinity and the temperature and time of annealing.

   j.   Effect of Heating Rate on the Value of T_g.   The effect of the heating rate is best illustrated by the case of poly(neopentyl methacrylate) having an $\overline{M}_w = 1.93 \times 10^6$. Using the DSC technique, T_g values of 26, 29, 34, and 39°C were recorded according to the midpoint method for heating rates of 4, 8, 16, and 32°C min^{-1}, respectively (Gargallo and Russo, 1975). Obviously, the determined T_g values increased with increasing heating rate. This is not surprising since T_g measured by the midpoint method will often increase with increasing heating rate. On the other hand, Flynn (1974) reported that the "thermodynamic" T_g, defined as the point of discontinuity at the intersection of the glass and liquid enthalpy curves, should not be a function of heating rate and, if measured properly, is not. It is to be noted here that although T_g is heating rate dependent, the enthalpy of a once-cooled glass in not. The enthalpy extrapolation eliminates the heating-rate effect and allows one to measure the T_g that corresponds to the rate used for cooling.

   The combined effect of the number average molecular weight ($\overline{M}_n$) and heating rate (q) on T_g of polystyrene is illustrated in Table IX (Blanchard *et al.*, 1974). The data in this table indicate that T_g increases with the increase in both the polymer DP and the heating rate.

TABLE IX

INFLUENCE OF HEATING RATE q AND MOLECULAR WEIGHT
ON T_g OF POLYSTYRENE[a,b]

$\overline{M}_n$	$q = 80$	$q = 40$	$q = 20$	$q = 10$	$q = 5$
1.8×10^6	115.5	114.0	113.0	—	—
8.6×10^5	114.0	113.0	110.5	109.5	106.5
5.0×10^5	113.5	112.5	110.5	110.0	—
1.6×10^5	—	109.5	109.0	108.5	—
5.1×10^4	111.0	109.0	108.0	106.0	100.0
1.98×10^4	—	104.0	102.0	100.5	—
1.03×10^4	—	97.0	95.0	93.5	—
4.08×10^3	78.0	74.0	71.0	69.0	—
2.03×10^3	—	57.0	55.5	54.0	—
0.90×10^3	—	36.0	34.0	32.0	31.0

[a] From Blanchard *et al.* (1974).
[b] q in degrees Celsius per minute; T_g in degrees Celsius.

k. Combined Effect of Filler and Cooling Rate. The effect of fillers
on T_g is a rather controversial issue, as reviewed by Toussaint (1973-
1974). Many authors proposed that incorporation of fillers can lead to significant changes in T_g of the polymers, whereas others indicated that the
glass transition is virtually unaffected by the incorporation of reinforcing
fillers at technologically important loadings (Kraus and Gruver, 1970). In
his review, Toussaint noted that the fillers can result in an increase, a decrease or no effect on T_g of the polymers. Nevertheless, the most commonly accepted effect of fillers is to increase T_g, which is often attributed
to long-range forces emanating from the surface of the filler which binds
and lowers the mobility of the polymer (Kwei, 1964). On the other hand,
in spite of this increase in T_g, the density of the filled polymer was reported to decrease, indicating a looser packing (Kumins *et al.*, 1963; Lipatov and Geller, 1967). These two concepts of decreased long-range mobility (up to 1500 Å as calculated by Kwie in 1964) and looser packing were
considered as difficult to combine into a single theory and the long-range
aspect of the surface forces was not well accepted (Peyser and Bascom,
1977).

The change of T_g in filled polymers was attributed to the thermal stress
below T_g induced on cooling by Manabe *et al.* (1971); as the thermal stress
relaxes with time, one would expect an annealed sample to display a
much lower filler effect than an unannealed sample. Recent results on
filled polystyrene were more demonstrative of the filler effect, when the
combined effect of filler and cooling rate was examined (Peyser and Bascom, 1977). The effect of cooling rate was described as reminiscent to
the filler effect, so that a quickly cooled polymer will have a higher T_g and

lower density than a slowly cooled one. Thus in the presence of a filler, one may expect a change in the slope of the straight-line fit of the logarithm of cooling rate and T_g documented earlier (Wunderlich *et al.*, 1964). This was indeed the case for silica-filled polystyrene when the filled polymer was shown to have a larger negative slope for the plot of log cooling rate versus T_g^{-1} than the unfilled one. Subsequently, after fast cooling the filled material exhibited a lower T_g than unfilled material. This is illustrated in Fig. 8. However, when the cooling rate was lowered, the T_g of filled and unfilled materials approached each other, and for very slow cooling (annealed samples), the filled material had a higher T_g than unfilled material (see Fig. 8).

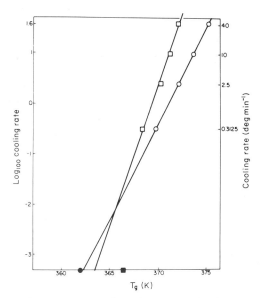

Fig. 8   Plot of cooling rate versus T_g for polystyrene and polystyrene-silica; (□) filled; (○) unfilled; (■) filled, annealed; (●) unfilled, annealed. (Reprinted from Peyser and Bascom, 1977, p. 597, by courtesy of Marcel Dekker, Inc.)

In conclusion, it is this author's opinion that T_g of filled polymer can differ from (usually be higher) that of an unfilled one with identical history and the filler effect varies with the type of interaction.

1. Effect of Stereoregularity on the T_g of Vinyl Polymers. This subject was treated by Karasz and MacKnight (1968) when a correlation between the available T_g data of mono- and disubstituted vinyl polymers $+CH_2-CXY+_n$ led to the conclusion that steric configuration effects T_g only when $X \neq Y$ and neither X nor Y is hydrogen. Conversely, T_g was reported to be independent of configuration when hydrogen is one of the substituents. A basis for this observation was developed in terms of the Gibbs–DiMarzio theory of the glass transition with the assumption that (a) the effect of configuration in disubstituted polymers is intramolecular and is brought about by changes in the flex energy of the isomers and (b) changes in T_g due to side-chain modification are strictly intermolecular. The results of this treatment can be summarized as

$$T_g(\text{syndiotactic}) - T_g(\text{isotactic}) = 0.59 \, \Delta t / k \qquad (7)$$

Equation (7) was considered generally for any pair of stereoisomers, where $T_g(\text{syndiotactic})$ is the T_g of syndiotactic isomer, and $T_g(\text{isotactic})$ is that for isotactic isomer, Δt is the difference in Gibbs–DiMarzio flex energy between the two isomers, and k is Boltzmann's constant. A knowledge of Δt was thus sufficient to determine the difference in T_g between pairs of stereoisomers. Karasz and MacKnight applied Eq. (7) to methacrylate polymers, as illustrated by the data in Table VIII. Based on their results, it was deduced that

$$T_g(\text{syndiotactic}) - T_g(\text{isotactic}) = 112°C \qquad (8)$$

for any isomeric methacrylate polymer pair. This led Wesslen *et al.* (1971) to predict the applicability of Eq. (8) to α-chloroacrylate polymers, since the chloro group is isomorphic with the methyl group of the methacrylate moiety and Δt is simply a function of the α-substituent. This was indeed the case and the T_g data of stereoregular poly(ethyl α-chloroacrylates) were consistent with the developed theory. The effect of stereoregularity on T_g for typical systems is shown in Table X

m. Effect of Electrostatic Interaction on Glass Transition of Ionic Polymers. Intermolecular forces, flexibility of chains, and chain geometry are the three principal variables that govern the glass transition of polymers (Hayes, 1961; Boyer, 1963; Lee and Sewell, 1968; Privalko and Lipatov, 1974). For ionic polymer systems where intermolecular forces predominate, the introduction of ionic moieties into the chain can lead to drastic increases in T_g (Moacanin and Cuddihy, 1966; Otocka and Kwei, 1968). The T_g of some ionic polymers was treated in terms of the parameter q/a, the ratio of the cation charge q to the distance between the centers of cations and anions a (Eisenberg, 1971; Eisenberg *et al.*, 1971). The dependence of T_g on these two parameters was expressed as

$$T_g = 730(q/a) - 67 \qquad (9)$$

This treatment, however, did not fully explain the dependence on T_g on ion concentration or give a satisfactory rationale for the applicability of

TABLE X

EFFECT OF TACTICITY AND OF SIDE-CHAIN LENGTH ON $T_g{}^a$

A. T_g (°C) of Polymethacrylate and Polyacrylate Series

	Methacrylates		Acrylates	
−R	Atactic	Isotactic	Atactic	Isotactic
Methyl	105	43	8	10
Ethyl	65	8	−24	−25
Propyl	35	—	−44	—
Isopropyl	81	27	−6	−11
Butyl	20	−24	−49	—
Isobutyl	53	8	−24	—
Sec-Butyl	60	—	−22	−23
Cyclohexyl	104	51	19	12

B. T_gs (°C) of Stereoregular Poly(Alkyl Methacrylate)

Polymer	Syndiotactic	Isotactic
Methyl	160	43
Ethyl	120	8
Isopropyl	139	27
Butyl	88	−24
Isobutyl	120	8
Cyclohexyl	163	51

[a] From Karasz and MacKnight (1968).

q/a to all molecules. On the other hand, the effect of counterion concentration on the T_g of various ionene polymers was reported later (Tsutsui *et al.*, 1975; Tsutsui and Tanaka, 1972). However, the interpretation of these data on the basis of molecular structure has been discussed by Tsutsui and Tanaka (1977).

The glass transition temperatures of ionic polymers was correlated with cohesive energy densities (CED). The CEDs of ionic polymers, such as ionene polymers and polyacrylates, were calculated with the assumption that they could be approximated to the electrostatic energy of the system. Accordingly, it was shown that the T_g of ionic polymers can be conveniently expressed by

$$T_g = K_1(\text{CED})^{1/2} \tag{10}$$

with different values of K_1, depending on the structure of the polymer.

2. Analysis of Polymeric Glasses at Elevated Pressures

The increase in transition temperature and density with pressure in glassy materials was first reported about 50 years ago (Tammann and Jenckel, 1929). In more recent reports (Wunderlich and Weitz, 1974;

Weitz and Wunderlich, 1974), it was noted that cooling a polymeric melt under elevated pressures through T_g leads to densified glasses. It was shown that densified PS and PMMA glasses have, at sufficiently high pressures, a substantially higher enthalpy than normal glasses; one would have expected the densified glasses to have lower enthalpy than normal ones.

3. Relaxation Phenomena above T_g

It has been recognized that the stepwise change of the specific heat of polymers at the glass transition is sometimes accompanied by a small endothermic peak, depending on the thermal history of the sample and the heating rate (Panke and Wunderlich, 1974). A less-known effect is a second endothermic peak observed in DSC measurements of high DP PMMA, free of cross-links. It was believed that this peak can be attributed to chain disentanglements, which need sufficiently large free volume to proceed at a measurable rate (Panke and Wunderlich, 1974). Thus, at temperatures higher than T_g, the chains of high DP PMMA assume their equilibrium conformation only after disentanglement. Compared to relaxation at glass transition, this process involves longer relaxation times. In a series of PMMA samples with variable DPs, the endothermic peak above T_g becomes more prominent with the increase in DP as shown in Fig. 9.

The variation of the peak temperature with heating rate resulted in activation energy of 91 kJ mole^{-1} and a relaxation time of 4 min at 423 K. The proposed model was further supported by the pressure dependence of the observed effect; it was shown that with a pressure of 70 bar, the relaxation process shifts from 150 to 158°C.

NMR measurements of the nuclear spin–lattice relaxation time T_1 in *cis*-1,4-polybutadiene at room temperature as a function pressure were reported by Geissler (1975). Up to 2000 bar, the relaxation curves revealed a single T_1, the pressure variation of which indicated in the Arrhenius model an activation volume of 14.5 cm^2 mole^{-1}. Above 2000 bar the effects of strains due to partial crystallization become evident and multiple relaxation was observed. By measuring the compressibility of this material up to 10 kbar, an estimate of the free volume (V_t) is made with P in kbar, giving

$$V_t = 1.58(10 - P) \quad \text{cm}^2 \text{ mole}^{-1} \tag{11}$$

4. Double Glass Transition in Semicrystalline Polymers

Double glass transition was discussed in a review by Boyer (1973c), who reported that thermal expansion, specific heat, and/or dynamic mechanical loss data indicate the presence of two glass transitions in bulk-

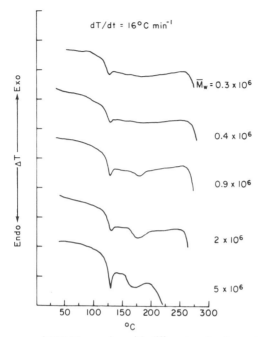

Fig. 9 DSC curves of PMMA samples with different molecular weight. (From Panke and Wunderlich, 1974.)

crystallized polyethylene, polypropylene, and several other polymers. The lower of these transitions was denoted as $T_g(L)$, which appears identical with the conventional T_g at zero crystallinity. The higher transition was designated as $T_g(U)$ and becomes more detectable as the crystallinity increases. The difference $\Delta T_g = T_g(U) - T_g(L)$ tends to approach zero as the fractional crystallinity χ approaches zero. The presence of $T_g(L)$ and $T_g(U)$ in polyethylene was suggested to arise from the morphology of melt-crystallized polymer (Boyer, 1973b), consisting of chain-folded crystals, tight and loose loops, tie molecules, and interlamellar material (Geil, 1963; Keller, 1968). Boyer (1973c) associated $T_g(U)$ with loose loops and tie molecules (no free ends) and $T_g(L)$ with cilia (one free end). It was also noted that amorphous materials rejected by the crystallites might participate in $T_g(L)$ and quantitative heat capacity measurements show no evidence of two T_gs for PE. The relations of $T_g(L)$ and $T_g(U)$ to other transitions in semicrystalline polymer are summarized in Table XI.

The dependence of ΔT_g on crystallinity was well illustrated for polyethylene and polypropylene, as shown in Figs. 10 and 11. For an X of 0.5 ± 0.1, ΔT_g is about 50°C and $T_g(U)/T_g(L)$ is about 1.2 with temperatures in degrees Kelvin. The increases in coefficient of thermal expansion

TABLE XI

CHARACTERISTIC TRANSITIONS FOR SEMICRYSTALLINE POLYMERS ARRANGED
IN ORDER OF ASCENDING TEMPERATURE[a]

Designation	Comments
$T < [T < T_g (L)]$	Various crystalline and amorphous transitions arising from side groups, crystal defects, etc
$T < T_g (L)$	This is an in-chain motion usually occurring at 0.75 T_g (L) and is probably a precursor of T_g (L)
$T_g (L)$	This is the lower of the two amorphous glass transitions. In completely amorphous polymers, T_g (L) $\equiv T_g$, where T_g is the classical glass temperature
$T_g (U)$	This is the upper of the two amorphous glass transitions. It should disappear as crystallinity approaches zero
T_c	This is a crystalline transition that appears to be a premelting phenomenon. It is seen with the least ambiguity in crystals grown from solution under equilibrium conditions. Its value depends on fold length. This definition of T_c, as used by Boyer (1973a), will be limited to this section of the chapter. In the rest of this chapter, T_c is generally used to symbolize crystallization temperature
T_m	Crystalline melting temperature

[a] From Boyer (1973c).

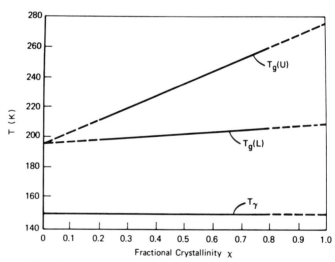

Fig. 10 Schematic representation of the three amorphous transitions in bulk-crystallized semicrystalline polyethylene and their dependence on crystallinity. T_g(U) is the upper glass transition, T_g(L) the lower glass transition, and T_γ the local mode or crankshaft process. (Reprinted from Boyer, 1973c, p. 505, by courtesy of Marcel Dekker, Inc.)

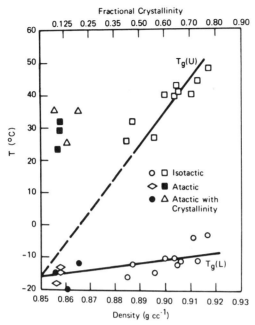

Fig. 11   The double glass transition in polypropylenes of varying tacticity and crystallinity plotted against density. $T_g(L)$ and $T_g(U)$ are based on thermal expansion data. (Reprinted from Boyer, 1973c, p. 512, by courtesy of Marcel Dekker, Inc.)

$(\Delta\alpha)_L$, $(\Delta\alpha)_U$ at those two transitions seem to depend on crystallinity and morphology in the expected manner for polyethylene and polypropylene; for $X = 0.5 - 0.7$, $(\Delta\alpha)_U$ is stronger than $(\Delta\alpha)_L$, and for $X \to 0$, $(\Delta\alpha)_L$ is stronger than $(\Delta\alpha)_U$. Such data are not available for other polymers that display two glass transitions, e.g., poly-1-butene; poly-1-pentene; *cis*- and *trans*-polyisoprene (natural) poly-4-methyl-1-pentene; isotactic polystyrene; poly(vinyl alcohol); oxide polymers of —CH_2—0— with $n = 1$ to 4; poly(ethylene terephthalate); poly(vinylidene fluoride); polyacrylonitrile; poly(vinylidene chloride); and Nylon 6 (Boyer, 1973c).

Some atactic polymers, poly-4-methyl-1-pentene and polystyrene, also seem to have a double T_g, the upper of which was tentatively ascribed (Boyer, 1973c) to the presence of Geil–Yeh type of local order. Since polyethylene, poly(vinylidene chloride), and poly(vinylidene fluoride) exhibit the apparent double T_g, tacticity per se was not considered necessary to cause it. It was also cautioned that special care must be exercised to distinguish $T_g(U)$ from crystalline phase α_c relaxation occurring at T_c. It was shown that T_c for well-annealed crystalline material tends to occur at $0.83–0.85$ T_m, where T_m is given in degrees Kelvin.

5. T_g Relationship for Monomers and Their Polymers

Although a direct relationship between the T_g of monomers and their polymers has not been explored, an empirical relationship was reported for aliphatic lactones and their polymers (Koleske and Lundberg, 1972). Lactones of β-, γ-, δ-, and higher aliphatic hydroxy acids were polymerized and the T_g of the resulting polymers were compared with those of starting lactones. T_g values were based on the mechanical loss data obtained on a recording torsion braid analyzer. All polymers were heated well above their T_m and then quenched in liquid nitrogen prior to measuring their T_g. With the exception of the C_7 and C_8 lactones, which were vitrified by cooling from the liquid state, the T_g of the lactones was determined for their solutions in methyl isobutyl ketone. Extrapolation of the T_g data for those solutions was used to obtain the T_g of the pure monomer. T_g data for a homologous series of unsubstituted lactones, varying in ring size from four to sixteen atoms, and their polymers indicated that (a) examination of these transitions as a function of ring size reveals a maximum in T_g for the seven-atom ϵ-caprolactone ring, which was attributed to a conformational change that occurs in rings containing seven to nine atoms; (b) with the exception of the anomalous ϵ-caprolactone and the strained four-membered lactone, an apparently constant difference of about 70°C between the monomer and polymer T_gs can be observed; and (c) poly-ϵ-caprolactone exhibits the lowest T_g of the polylactones. A summary of these data is shown in Table XII.

TABLE XII

GLASS TRANSITION OF LACTONES AND THEIR POLYMERS[a]

Polylactone of	Ring size of monomer	Reduced viscosity of polymer (dl/g)	$T_{g(P)}$ of polymer (°C)	$T_{g(m)}$ of monomer (°C)	$\Delta G = T_{g(P)} - T_{g(m)}$ (°C)
β-Propiolactone	4	1.36	−28	−131	103
δ-Valerolactone	6	0.75	−57	−124	67
ϵ-Caprolactone	7	0.70	−70	−103	33
ζ-Enantholactone	8	0.61	−45	−112	67
η-Caprylolactone	9	0.83	−38	−111	73
λ-Laurolactone	13	0.35	−25	−98	73
ξ-Pentadodecanolactone	16	2.30	−22	−95	73

[a] From Koleske and Lundberg (1972).

6. Surface Tension of Dilute Solutions for Measuring T_g

The surface tension of polymers when plotted against temperature displays a reversible drop in value at about the glass transition, as was demonstrated for polystyrene (Ferroni, 1964). In a more recent study, T_g of polymethacrylates was determined by both the surface tension and DSC techniques (Gargallo and Russo, 1975). T_g values of poly(p-*tert*-butyl-phenyl methacrylate) (BPh), poly(4-*tert*-butylcyclohexyl methacrylate) (BCy), and poly(neopentyl methacrylate) (NPe), as determined by the surface tension (γ) and DSC techniques are summarized in Table XIII.

TABLE XIII

T_g DATA AS MEASURED BY SURFACE TENSION
(γ) AND DSC TECHNIQUES[a]

Polymer	T_g (°C) by DSC	T_g (°C) from γ
BPh	98	89
BCy	83	79
NPe	34	38

[a] From Gargallo and Russo (1975).

7. *Specific Heat Increment at the Glass Transition* ($\Delta C_p T_g$) *of High Polymers*

The quantity $\Delta C_p T_g$ for 30 different polymers was shown to have an average value of 27.5 cal g^{-1} (Boyer, 1973a). A more accurate value for $\Delta C_p T_g$ was described by the equation

$$\Delta C_p T_g = 15 + 4 \times 10^{-2} T_g \qquad (12)$$

Boyer reported that there is a good correlation between ΔC_p and the difference $\Delta\beta$ in thermal expansion dV/dT above and below T_g. This could be explained empirically since both ΔC_p and $\Delta\beta$ are inversely proportional to T_g. It could also be explained by the fact that T_g is directly proportional to the energy of the hole (related to the flip of the chain statistical segment during the segmental motion at T_g) formation ϵ_h. It was shown that the coefficient of expansion in the liquid state α_l varies inversely with ϵ_h. Finally, a similarity of the hole volume V_l as calculated from the hole theory and the free volume quantity $\Delta\alpha T_g$ from Simha and Boyer was noted (Boyer, 1973a).

8. *The Interplay of Different Parameters Affecting* T_g

It has been shown earlier by Boyer (1963, 1977) that in general the T_g of homopolymers increases with the increase in (a) intermolecular forces as measured by cohesive energy density and (b) interchain steric hindrance imposed by bulky, stiff side groups. On the other hand, the following factors were associated with the decrease in T_g: in-chain groups which promote flexibility such as ether linkage, flexible side groups, and symmetrical substitution of the main chain. Unfortunately, in studying these parameters, it is difficult to identify the effect of each parameter individually and for most polymers there can be a complex interplay between various factors. For instance, polyacrylonitrile is much more polar than polystyrene and yet both have comparable T_g values, as if bulkiness of the phenyl ring offsets the polarity of the nitrile groups. This and other similar cases led to the critical study of the influence of cohesive forces on T_g by Lee and Sewell (1968). These investigators concluded that (a) for polymers with $T_g > 25°C$, there is no correlation between published or calculated values of cohesive energy (CE) in calories or cohesive energy density (CED) in cal cm^{-3} [and hence, on solubility parameters, $\delta = (CED)^{1/2}$]; (b) for polymers with $T_g > 25°C$, there is a good correlation between T_g and CED; (c) the value of 25°C is chosen since most of the CED values are determined at this temperature; (d) better correlation is obtained between T_g and CED than between T_g and CE; and (e) there are certain basic doubts about applying the concepts of CE or CED to polymers in the glassy state, since CE is usually measured on swollen or dissolved polymers. More extensive analysis of this subject was given by Boyer (1977).

The interplay of different parameters affecting the T_g of copolymers is relatively more complex to analyze than for homopolymers. A brief discussion of this subject was given by Boyer (1977) and is alluded to in Chapter 4 on block copolymers and polymer blends.

B. CRYSTALLIZATION AND MELTING

This section will be limited to certain recent and relatively ignored aspects of crystallization and melting, since these two related processes have been treated in Chapter 2 and several reviews (Brydson, 1972; Bueche, 1962; Deanin, 1972; Mandelkern, 1964; Mears, 1965, Van Krevelen, 1972; Wrasidlo, 1974; Wunderlich, 1973, 1976; Wunderlich and Baur, 1970)

1. Effect of Crystallization Conditions on T_m

The well-established effect of crystallization conditions on the melting temperature of semicrystalline polymers has been discussed in the reviews by Brydson (1972), Geil, (1963), and Van Krevelen (1972).

Studies on the crystallization of isotactic poly(methyl methacrylate), indicated that the T_m increases with the increase in the crystallization temperature (De Boer *et al.*, 1975). Figure 12 shows the T_m–T_c relationship for i-PMMA. Melting temperatures were recorded with a DSC using different heating rates. To correct for superheating, T_m values were extrapolated to zero scan speed. T_m values were also estimated on a hot-stage optical microscope. The middle of the melting range (4–8°C) of the crys-

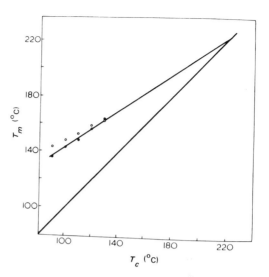

Fig. 12 Melting temperatures T_m of i-PMMA as a function of crystallization temperature T_c. (●) Recorded by DSC extrapolated values with zero scan speed; (○) estimated data by light microscopy with a heating rate of 0.2°C min^{-1}. (From DeBoer *et al.* 1975, by permission of the publishers, IPC Business Press Ltd., ©.)

tals was chosen as T_m. Since it is very time consuming to crystallize i-PMMA at temperatures above 130°C, the line in the T_m–T_c diagram had to be extrapolated over a large distance to obtain T_m°. This means that the extrapolated melting temperatures T_m° can only be determined approximately, about 220°C.

2. Effect of Stereoregularity on T_m

The effect of stereoregularity on the crystallinity and T_m of isotactic polypropylene was reported earlier (Natta, 1959), when polypropylene was fractionated to obtain samples of differing crystallinities and T_ms. These data are shown in Table XIV.

TABLE XIV

Effect of Stereoregularity on Crystallinity and T_m of Polypropylene[a]

Fraction no.[b]	Soluble in	Insoluble in	Crystallinity (%)	T_m (°C)
1	—	Trichloroethylene	75–85	176
2	—	*n*-Octane	64–68	174–175
3	*n*-Octane	2-Ethylhexane	60–66	174–175
4	Ethyl hexane	*n*-Heptane	52–64	168–170
5	*n*-Heptane	*n*-Hexane	41–54	147–159
6	*n*-Hexane	*n*-Pentane	25–37	110–135
7	*n*-Pentane	Ethyl ether	15–27	106–114

[a] From Natta (1959).

[b] Samples 1 and 7 have the highest and lowest isotactic contents, respectively.

In recent studies (Pavan *et al.*, 1977), fractions of predominantly iso-tactic polypropylene with varying stereoregularity were examined by NMR and hot-stage optical microscopy, and both the stereochemical composition and melting temperature of each fraction were determined. The NMR data indicated the presence of six types of tetrads, with the *"mmm"* tetrad corresponding to isotactic fractions of the chain. The melting temperature was shown to be quite dependent on the isotactic content of the chain, as determined by the fraction of the *"mmm"* tetrad. Naturally, maximum T_m was characteristic of the fraction having the high-est content of the *"mmm"* tetrads, as shown in Table XV.

TABLE XV

NMR AND T_m DATA OF POLYPROPYLENE FRACTIONS[a]

Fraction no.[b]	T_{cry}[c]	*"mmm"* fraction by NMR	T_m (°C)
1	93	0.90	167.5
2	88	0.85	163
3	72.5	0.84	158
4	50.5	0.80	148
5	10	0.70	127

[a] From Pavan *et al.* (1977).
[b] Stereoregularity decreases from fraction 1 to 5.
[c] Temperature of isothermal crystallization from octane solution.

The effect of stereoregularity on the basic thermal transitions of poly(*tert*-butyl ethylene oxide) was studied, using DSC measurements (Doddi *et al.*, 1971). The isotactic polymer was shown to have an equilib-rium melting of 135°C and T_g of 58°C. On the other hand, the base-cata-lyzed polymer, which is likely to be made of stereoblocks of isotactic and syndiotactic placements exhibited an equilibrium T_m of 63°C and a T_g of -9°C. These data are summarized in Table XVI.

TABLE XVI

THERMAL CHARACTERISTIC OF ISOTACTIC AND
BASE-CATALYZED POLY(tert-BuEO)[a]

	Isotactic	Base catalyzed
Heat of fusion, ΔH_f, cal g^{-1}	7.30	6.83
Entropy of fusion, cal g^{-1}, deg^{-1}	0.0179	0.0202
T_g, °C	58 ± 1	-9 ± 1
	40 ± 1[b]	
T_m, °C	135	63

[a] From Doddi *et al.* (1971).
[b] For quench-cooled sample.

3. *Effect of Drawing on the Melting Temperature and Heat of Fusion*

Polymer crystals usually melt at temperatures below their thermodynamic equilibrium melting temperatures because of excess of free energies of the crystalline and amorphous phases. It has been quite common to consider that the thickness of crystals plays the most important role in determining T_m (Hoffman and Weeks, 1962). However, it has been shown by Sakurai *et al.* (1974, 1976) that on heating, the thickness of the lamellae increases rapidly, reaching an essentially constant value, depending on the temperature, which is independent of the rate of heating and the thermal history of the sample. This led to the suggestion that T_m is independent of the initial thickness. Meanwhile, it was to disprove, unequivocally, the effect of the initial thickness of the lamellae on T_m. The drawing of semicrystalline polymer is known to cause changes in molecular orientation, crystallite size, distribution of crystallites, and the state of aggregation of the chains in the amorphous regions. In other studies, geared toward determining the relation between T_m and ΔH_f, the changes in these two parameters were measured experimentally for high-density polyethylene as a function of drawing (Sumita *et al.*, 1977). It was shown that both T_m and ΔH_f, as determined from DSC data, increase with an increase in draw ratio. The dependence of T_m and ΔH_f on the draw ratio is illustrated in Figs. 13 and 14. The recorded increases in both the heat fusion and melting temperature of polyethylene terephthalate were ascribed

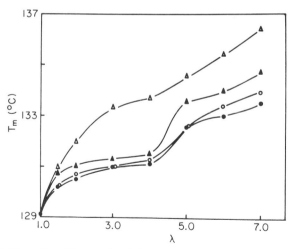

Fig. 13 Melting point of drawn polyethylene samples T_m as a function of draw ratio λ at the indicated rates of heating. (-△-△-) 64°C min⁻¹, (-▲-▲-) 32°C min⁻¹, (-○-○-) 16°C min⁻¹, (-●-●-) 8°C min⁻¹. (From Sumita *et al.*, 1977.)

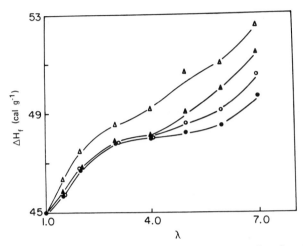

Fig. 14   Heat of fusion of drawn polyethylene samples ΔH_f as a function of draw ratio λ at the indicated rates of heating. Symbols same as for Fig. 13. (From Sumita *et al.*, 1977.)

to orientation of the amorphous phase, as a result of drawing. The excess free energy of the amorphous phase derived from orientation was suggested to increase the T_m, as the amorphous region absorbs heat for its randomization at the T_m. Hence, it was concluded that in samples having an oriented amorphous phase, both the amorphous and crystalline regions contribute to their T_m.

4. *More Accurate Determination of Polymer Melting Temperatures*

In practice, "true" or equilibrium T_m values are usually difficult to obtain. This was ascribed to relatively slow sample heating rates ($<60°C$ min^{-1}), usually used in common thermal techniques (DSC, DTA, thermooptical analysis, and dilatometry), which "will not allow" for sufficient accuracy (Predechi and Karr, 1978). These heating rates were suggested to be slow enough that structural changes such as lamellar thickening or crystallization associated with increasing sample temperature can take place. Thus the measured T_m is no longer characteristic of the starting sample. Furthermore, the dependence of T_m on the sample heating rate is not simple and is strongly affected by prior thermal history. In an attempt to correct this situation, a modified DTA system was constructed. It was capable of providing heating rates between 100 and 600°C min^{-1}. (Wolpert *et al.*, 1971). Similarly, Hager (1972, 1973) has devised a high-speed thin-foil calorimeter for detecting thermal transitions in polymers at rates up to 36,000°C min^{-1}. However, using high heating rates does not solve the problem, for there is a possibility that many polymer structures, particularly those with large crystal thickness in the chain direction, will superheat and lead to erroneously high T_m.

In a recent report, Predecki and Karr (1978) described an improved method for determining T_m that requires only minor modification of conventional thermal optical analysis (TOA). The method makes use of a small, highly conductive substrate and a preselected T_m procedure that results in rapid heating rates ($\sim 1500°C$ min^{-1}) without superheating. Accuracies to within $\pm 0.15°C$ were readily obtained. Application of this method to isothermally crystallized poly(oxymethylene) showed that T_m increases with both time and temperature of crystallization as well as increasing molecular weight. The method was reported to have the advantage that samples with a well-defined thermal history can be easily prepared *in situ*. The potential of this method for characterizing polymer thermal history and molecular weight was also pointed out.

5. Comparison of Crystallization Rates as Measured by Thermal and Other Techniques

Hot-stage microscopy has been used by several investigators for measuring isothermal crystallization rate of thermoplastic polymers (Magill, 1962; Jackson and Longman, 1969; Binsberger and deLange, 1970). In this technique, the light transmission (initial, I_0; final, I_c; and at time t, I_t) through cross-polarizers is monitored as a function of time t with a photocell and chart recorder. Crystallization half-life can be interpolated directly from the I versus t profiles. Rate constants can be calculated from

$$\theta = e^{-kt^n} \tag{13}$$

where θ is the fraction of untransformed material, k is a constant, and n is an integer depending on the type of nucleation and growth (Avrami, 1939; Morgan, 1954). With $\theta = (I_c - I_t)/(I_c - I_0)$, n can be obtained directly from $I_n (-I_n\theta)$ versus $I_n t$ plots. Although this technique is widely used, it is not well established that the depolarized light–time traces should provide a correct measure of polymer crystallization rates or should give the proper Avrami exponent. Binsbergen (1970) has shown, for instance, that only when it is assumed that the birefringent entities in the crystallizing aggregates increase in number and length with time is there a linear relationship between the birefringence and volume of crystalline material. Other techniques may show a similar sensitivity to details of the crystallization process. Godovsky and Slonimsky showed (1974) that for some polymers, calorimetry may give $n = 2$ while dilatometry gives $n = 3$ for the exponent of the Avrami equation. It was suggested that the former technique is more sensitive to the development of two-dimensional lamellae whereas dilatometry reflects the formation of three-dimensional spherulites. Gilbert and Hybart (1974) have noted a similar but smaller difference in DTA and dilatometric measurements of aliphatic polyesters crystallization rates.

Comparative studies of the crystallization rates of polypropylene (PP), poly(butylene terephthalate) (PBT), and poly(ethylene terephthalate) (PET) as measured by DSC and depolarization microscopy were reported recently (Pratt and Hobbs, 1976). Markedly slower crystallization half-times for PBT and PET were recorded by DSC than by depolarizing microscopy (DM). However, similar values of crystallization half-times were determined by DSC and DM for isotactic polypropylene. In both the microscopic and calorimetric experiments, an Avrami analysis of data on each polymer gave $n \sim 3$ (predetermined nucleation and growth) over the range of crystallization temperatures used. The discrepancies between the DM and DSC data for the polyesters were explained in terms of a difference in sensitivity of the two techniques to the different crystallization processes. It was postulated that the rate of primary spherulitic development is given accurately by depolarization measurements, whereas the calorimetric data more accurately reflect the overall rate of crystallization. The latter was suggested to include contributions from small numerous crystals that mirror the development of more well-developed spherulites but do not contribute to increases in birefringence.

6. Side-Chain Crystallinity

Side-chain crystallinity is usually present in atactic vinyl homopolymers having linear side groups in excess of 10–12 carbon atoms (Nielsen, 1962). This was demonstrated by the first-order melting transitions obtained for largely atactic homologs selected from the poly(n-alkyl acrylates) and polymethacrylates and their copolymers (Wiley and Brauer, 1948; Greenberg and Alfrey, 1954), the poly-2-n-alkyl-1,3-butadienes (Overberger *et al.*, 1951), polyvinyl esters (Port *et al.*, 1951), the poly-n-alkylstyrenes (Overberger *et al.*, 1953), the poly(N-n-alkyl acrylamides) (Jordan *et al.*, 1969), and the poly(fluoro-n-alkyl acrylates) (Pittman and Ludwig, 1969). Crystallinity was also reported to develop only by the side chains of isotactic poly(n-alkyl acrylates) (Platé *et al.*, 1968).

7. On the Actual Crystallization of Polymers and Its
   Relevance to the Melting Phenomenon

In simple compounds such as water and simple organic compounds the ordered three-dimensional lattice of the crystalline phase can coexist in equilibrium with the less-ordered liquid phase at a distinct temperature normally referred to as melting temperature T_m or crystallization temperature T_c. Obviously under typical equilibrium conditions (a) $T_m = T_c$; (b) it is not necessary to supercool the molten compound to initiate its crystallization, i.e., the liquid compound need not be cooled well below its T_c to form the nuclei necessary for the crystal growth; and (c) the heat transfer for the crystalline compound is sufficient to allow its melting without superheating, i.e., the sample temperature need not exceed its T_m prior to observing the phase transition. In complex organic systems such as long-chain polymers where relaxation times are longer than those of simple compounds, supercooling and superheating may characterize the crystallization and melting of these materials, respectively. [See Chapter 2 for a comprehensive discussion of melting and Wunderlich (1980).]

Long-range, three-dimensional order in polymers requires parallel alignment of either whole molecules as in extended-chain crystals or relatively long-chain segments, as in periodically folded-chain structures. The most important parameters determining the crystallizability of polymers are chemical and conformational regularity. The latter pertains to ability of chains because of their chain constituents to produce regularly repeating sequence; the chain may conform to a stable helix, planar zigzag, etc. Without such constraints a polymer cannot crystallize. In systems capable of stereoisomerism, additional configurational requirements (tacticity) must be satisfied to attain conformational order. Because of the statistical nature of macromolecules, Wrasidlo (1974) expressed his doubt that perfect conformations will ever exist and consequently that perfect polymer crystals may experimentally ever be achieved. Flory (1949) utilized a lattice model, based on thermodynamic consideration, to predict that polymers with chains satisfying the preceding configurational and conformational requirements for crystallization should attain high degrees of crystallinity. Experimentally this is not generally the case due to (a) the presence of various types of lattice defects and (b) the kinetically controlled component of the crystallization process. This led to many controversial views on polymer crystallization which are discussed by many authors, including Calvert and Uhlmann (1972), Mandelkern (1964), Sharples (1966), Teitel'baum (1974), Wrasidlo (1974), and Wunderlich (1973, 1976).

Polymer crystallization involves the formation of an ordered state from a disordered one. The fact that this process may be persuaded to take different routes, leading to correspondingly different end states, makes it of interest to those concerned with controlling product properties. On the other hand, melting, which may be viewed as the converse process of crystallization, has for its end state the common disorder of the melt, and so might seem at first sight to lack practical relevance to crystallization (Sharples, 1966). However, a good understanding of the course of melting is quite often revealing in that it may provide information on the nature of the crystalline structure being destroyed. Meanwhile, the process of partial melting is followed in some cases by recrystallization, as in annealing. This is of great practical importance in determining the ease with which mechanical properties of a polymer are impaired when temperatures above normal are involved during end use. Although a more extensive treatment of this subject is beyond the scope of the present chapter, it is felt appropriate to outline below the main points of a few basic articles pertinent to the relation of crystallization process of polymers to their melting. Additional treatments can also be found at certain sections of this chapter, in Chapter 2 and in Wunderlich (1980).

It is quite certain that T_m of polymers is not the unambiguous parameter that can be derived from studies of low-molecular-weight compounds. It is equally certain that even with slow crystallization and heating rates, observed T_ms of polymers are well below the true thermodynamic values. Nevertheless, the difference between the observed $T_{m(obs)}$ and equilibrium $T_{m(true)}$ melting temperature can be minimized by increasing the crystallization temperature. An excellent approach to studying the dependence of T_m on T_c was devised by Hoffman and Weeks (1962). If $T_{m(obs)}$ is plotted as a function of T_c, as in Fig. 15, a straight line is obtained. It can then be

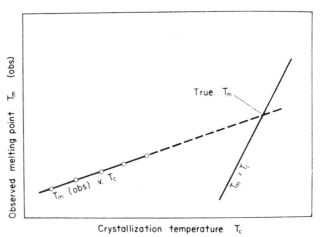

Fig. 15 Hypothetical plot of observed melting point, as a function of crystallization temperature. Its intersection with the $T_m = T_c$ plot is the true melting point, according to the method of Hoffman and Weeks. (From Hoffman and Weeks, 1962.)

argued that the minimum value of $T_{m(obs)}$ is T_c (as the sample cannot melt below its crystallization temperature), so that for the line $T_{m(obs)} = T_c$, the true value of T_m must lie above. The true value of T_m is also likely to lie on the line for $T_{m(obs)}$ versus T_c (Fig. 15) and will be located at the point where the sample crystallizes infinitely slowly. (Crystallization rate is zero when

$\Delta T = 0$ and $T_c = T_m$.) Thus the intersection of these two lines represents the true value of T_m. Using this approach with polyethylene, Geil (1963) has found a value of $T_m = 143 \pm 2°C$.

From the arguments just presented, it is apparent that the observed effects during melting are dependent on both the previous crystallization history of the sample and the conditions under which the melting itself took place. This can be further complicated by the fact that at an intermediate stage of melting, when the temperature has been raised sufficiently to partially reduce the amount of crystalline material, a further process of recrystallization may subsequently take place. This case is illustrated by the data in Fig. 16 reported by Gubler *et al.* (1963). At temperatures just

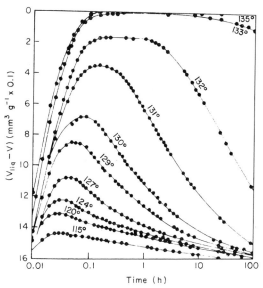

Fig. 16 Annealing of crystallized polyethylene at various temperatures below the melting point ($T_m = 138°C$). (From Gubler *et al.*, 1963.)

below T_m the specific volume of the crystallized sample increases with time; i.e., it melts partially, and only at very small supercooling does it remain constant. At lower temperature, after a period of constant volume, the specific volume starts to decrease, indicating recrystallization. The form of the curves in Fig. 16 depends on the previous crystallization history and reflects the range of crystallite stabilities present. The process of partial melting, when it is followed by subsequent recrystallization steps, is often falsely termed annealing. The observed effects of "annealing" can be considerably enhanced if the sample has been previously quenched from the melt.

In his review of the actual crystallization of polymer, Teitel'baum (1974) indicated that (a) the well-known difference between melting and crystallization temperatures, even for conditions excluding recrystallization and "reorganization" (i.e., for metastable melting) for many polymers can be explained by the nonidentity of nominal and actual crystallization temperatures; (b) the actual crystallization temperature must be equal to the metastable melting temperature; (c) establishing any relationship between T_m and T_c is identical in meaning to stating of the difference

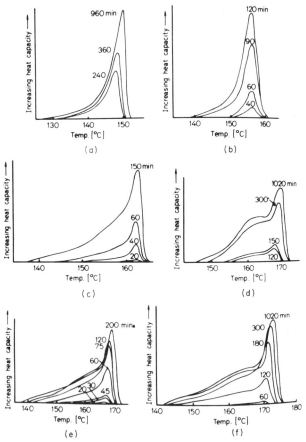

Fig. 17 Differential scanning calorimetry (DSC) thermograms of the crystals grown during isothermal crystallization process. (a) $T_c = 125°C$ for a whole polymer $\overline{M}_v = 5.7 \times 10^3$; (b) $T_c = 135°C$ for a fraction of $\overline{M}_v = 1.88 \times 10^4$; (c) $T_c = 135°C$ for a whole polymer of $\overline{M}_v = 6.25 \times 10^4$; (d) $T_c = 140°C$ for a whole polymer $\overline{M}_v = 1.93 \times 10^4$; (e) $T_c = 135°C$ for a fraction of $\overline{M}_v = 3.21 \times 10^5$; (f) $T_c = 140°C$ for a fraction with $\overline{M}_v = 3.21 \times 10^5$. (From Kamide and Yamaguchi, 1972, by courtesy of Hüthig and Wepf Verlag, Basel.)

between the actual and nominal crystallization temperature; and (d) the value of the melting temperatures (having not been affected by recrystallization) could be considered as a rough approximation of the actual crystallization temperature, as one recognizes from the fact that in a true solid–liquid equilibrium, $T_c = T_m = T_m^0$, where T_m^0 is the equilibrium melting temperature.

Molecular weight is an important parameter that can affect the recrystallization process discussed previously. This is well illustrated by data obtained by Kamide and Yamaguchi (1972) and shown in Fig. 17. The data in Fig. 17 indicate that isotactic polypropylene shows one or two melting endotherms at different temperatures, depending on the (a) molecular weight of the polymer; (b) crystallization temperature; and (c) crystallization time. It should also be noted that two endotherms are due to two different crystallite sizes and/or lattice imperfection. The effect of molecular weight on the heat of crystallization of polytetrafluoroethylene (PTFE) was studied by Suwa *et al.* (1973). These investigators showed that the increase in PTFE molecular weight lowers the heat of crystallization and hence degree of crystallinity considerably. This is illustrated by the data in Fig. 18.

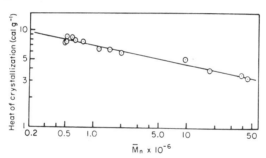

Fig. 18   Relation between number-average molecular weight and heat of crystallization for polytetrafluoroethylene. (From Suwa *et al.*, 1973.)

The crystallization kinetics of polyglycolic acid (PGA) at high undercooling and the effect of T_c on the level of agreement between experimental and calculated crystallization isotherm were recently reported by Chu (1980). In this study the temperature of crystallization T_c ranged from 201 to 207°C. A series of crystallization isotherms at different T_cs were obtained by the usual form of Avrami's equation $\ln \theta = Kt^n$. The half-time increased from 2.7 to 11 min as T_c increased from 201 to 207°C. The calculated and experimental plots of the crystallization isotherms were superimposed by shifting them horizontally along the time axis in order to examine the agreement between Avrami's theory and experimental data involving this particular polymer. It was found that better adherence of the experimental isotherms to the calculated ones was achieved as the T_c increased. At $T_c = 207°C$, the agreement between calculated and experimental values was about 63% relative to the degree of transformation. The Avrami exponent n was 4. However, the experimental isotherms did not correspond to the calculated ones at T_cs lower than 205°C. This suggests that nonequilibrium crystallization had occurred at such high undercoolings.

As with all high-melting polyesters, it is possible to anneal poly(ethylene terephthalate) (PET) not only by physical rearrangement of the macroconformation and improvement of the crystal perfection, but also by chemical changes in the backbone of the molecule (Miyagi and Wunderlich, 1972). In addition, extended-chain crystals of PET have been reported following crystallization under elevated pressure (Siegmann and Harget, 1973). However, no detailed direct thermometric, calorimetric, or dilatometric data on equilibrium crystals have been published as yet, so that it is still necessary to make use of extrapolated data. Until 1978, the accepted values for equilibrium T_m, volume change in melting, heat of fusion, and entropy of fusion were 523–532 K, 11.9 cm³ mole⁻¹, 2.17–2.76 kJ mole⁻¹, and 42.7 J deg⁻¹ mole⁻¹, respectively (Ringwald and Lawton, 1975). Mehta *et al.* (1978) presented a new analysis that led to higher equilibrium melting parameters. Thus the equilibrium melting temperature, volume, and enthalpy and entropy changes on melting of poly(ethylene terephthalate) have been analyzed and heats of fusion have been newly measured with an automated scanning calorimeter to yield the following data: 553 K, 16.9 cm³ mole⁻¹, 2.69 kJ mole⁻¹, and 48.6 J deg⁻¹ mole⁻¹, respectively.

IV. Thermal Behavior of Thermoplastics in Interactive and Degrading Environments

This is a rather broad area that has been constantly surveyed and updated by several authors (Jellinek, 1978; Manche and Carroll, 1972; Madorsky, 1964; Pearce *et al.*, 1975; Reich and Levi, 1967; Reich and Stivala, 1971; Sawada, 1976). Therefore, it is felt unnecessary to prepare a survey of TG profiles of different polymers. Subsequently, this section will cover only new developments in correlating chemical and physical parameters with thermal and thermooxidative stabilities as well as new applications of thermogravimetric and complimentary techniques in selected problem-solving schemes.

A. Effect of Film Thickness on the Oxidation of Polyolefins

Thermooxidative degradation of polyethylene (PE) and polypropylene (PP) is known to be an autocatalytic process, similar to liquid phase oxidation in being a free radical reaction with degenerate chain branching. Of the several factors affecting the oxidation rate, the physical state, structure, and crystallinity of the polymer are most important. Oxidation in both the molten phase and solid state are heterogeneous reactions with rates depending on kinetic parameters as well as transport parameters. Some of the latter are oxygen diffusion to the polymer and diffusion of volatile products from the polymer. Thus the escape of volatile degradation products from molten or solid phase is usually a function of diffusion coefficients in addition to the reaction rate constant and hence depends on sample thickness at constant temperature and pressure. This led Iring *et al.* (1974) to determine the range of sample thickness at which the oxidation of the polymer becomes exclusively kinetically controlled. Extensive TG and DTA studies of PE and PP films indicated that a critical layer thickness of 120 μm could be determined in the initial stage of thermooxidative degradation of both PE and PP. Thus with film thickness less than 120 μm, the polymer degradation becomes exclusively kinetically controlled. Beyond a 120 μm thickness, the thermooxidative process will be affected by diffusion. While the maximum rate of oxygen absorption hardly depends on the film thickness to about 120 μm, the maximum rate of weight loss in the whole range of thickness 0.04–1 mm was shown to be diffusion controlled.

B. Pyrolysis–Molecular-Weight Chromatography–Vapor Phase Infrared Spectrophotometry as a New On-Line System for Polymer Analysis

A new on-line system consisting of a programmable pyrolyzer, a thermal conductivity detector, a mass chromatograph, and a vapor-phase infrared spectrophotometer has been developed for the analysis of polymer degradation (Kiran and Gillham, 1976a, b). Polysulfones were used as an example of the systematic variation of molecular structure as an aid to characterization of thermal degradation using this system (Kiran and Gillham, 1974, 1976a). Application of the pyrolysis system to thermal degradation of polyethylene, polypropylene, and polyisobutylene was described in the literature (Kiran and Gillham, 1976b). The thermal decomposition of polysulfones such as poly(butene 1-sulfone) (PBS), poly(pentene 1-sulfone) (PPS), poly(hexane 1-sulfone) (PHS) and poly(styrene sulfone) (PSS) was also investigated (Kiran *et al.*, 1977). The decomposition was conducted in a helium atmosphere at a heating rate of $20°C$ min^{-1} using the experimental system referred to previously. The PDS, PPS, and PHS samples displayed two-step decomposition; the primary products of decomposition at both steps were the comonomers olefin and SO_2. For PSS, in addition to styrene and SO_2, products with molecular weight corresponding to dimers of styrene were observed. Thus the decomposition of PSS is different from that of polystyrene, which forms mostly styrene on degradation.

C. EFFECT OF END GROUP ON THERMAL STABILITY OF THERMOPLASTICS

Many authors consider thermoplastics to degrade by three basic mechanisms, random degradation, depolymerization, and degradation involving thermally labile defects. Hence, thermoplastics can be discussed separately in these three groups. It must be noted that this is an oversimplification. Some polymers degrade by combinations of several mechanisms and others degrade by side-group stripping, ring closure, inter- and intramolecular chain transfer mechanisms, and even cross-linking and chain formation or diffusion-controlled mechanisms.

1. Polymers That Undergo Random Degradation

When these polymers degrade, bonds in chains may be broken at random. Thus all bonds still intact at any one stage of the degradation have the same probability of being broken. Polyethylene is a typical homochain polymer that undergoes random degradation. Polyethylene terephthalate is a typical heterochain polymer that undergoes random degradation and the presence of carboxy end groups accelerates its degradation.

2. Polymers with Depolymerizable Chains

When a polymer undergoes a depolymerization reaction, the chain scission is essentially the reverse of addition polymerization or ring opening polymerization. This involves the successive generation of monomer units from the chain end. This is one of the most serious types of degradation and it is characteristic of several homochain and heterochain polymers, where the effect of end groups on the chain stability is quite prominent. Poly(methyl methacrylate) undergoes an almost complete thermal depolymerization to its monomer. Similarly, poly-α-methylstyrene undergoes chain unzipping quite readily. The tendency of a polymer to unzip is related to its thermodynamic instability as reflected in the

TABLE XVII

CEILING TEMPERATURE (T_{ce}) OF
TYPICAL POLYMERS[a]

Polymer of	T_{ce} (°C)
Formaldehyde (unstabilized)	10–58
Isobutylene	50
α-Methylstyrene	61[b]
Methacrylonitrile	177
Methyl methacrylate	220
Styrene	300

[a] From Wall (1960).
[b] From Kilroe and Weale (1960).

value of the ceiling temperature. This can be appreciated as one examines the data in Table XVII. At this point, it should be noted that (a) polyisobutylene, with its low ceiling temperature, may undergo partial degradation through chain unzipping, although it is known to cross-link prior to any large-scale degradation; and (b) polystyrene degrades by a combination of at least three mechanisms, including chain unzipping. Naturally, the lower the DP, the higher the concentration of the end groups and hence the less stable the chain becomes. Polyoxymethylene is a classical example of a thermally unstable heterochain polymer. If it is unstabilized it undergoes depolymerization to the trimer at 10–58°C, as shown in Table XVII. Low-DP polymers will be expected to undergo a more facile depolymerization than high-DP ones, because of the high concentration of hydroxy end groups in the former. Capping the hydroxyl end groups of polyoxymethylene through acetal formation (Delrin®) and incorporating ethylene oxide moieties in the chain through copolymerization (Celcon®) do help in improving polymer stability. Nylon 6 is another good example of a depolymerizable heterochain polymer that can be stabilized by end-group modification (Riemschuessel *et al.*, 1973; Pearce *et al.*, 1975).

3. Polymers with Thermolabile Chemical Defects

Polymers with thermolabile chemical defects contain a few or several chemical defects characterized by covalent bonds that are weaker than those of the main chain. On heating these polymers, dissociation occurs first at those weaker bonds. Polystyrene prepared without complete exclusion of oxygen during its free radical formation is known to have peroxide moieties that undergo dissociation before any other part of the chain. The chemical defects can also be an unexpected branch in hydrocarbon chain that destabilizes the chain through the facile radical formation at the methine carbon.

D. EFFECT OF STRUCTURE ON THE THERMAL STABILITY
OF AA–BB-TYPE POLYESTERS

Polyesters based on aliphatic diols and aliphatic or aromatic diacids are dealt with in this section. Minor structural modifications can result in substantial changes in thermal stability.

Madorsky (1964) has reviewed the thermal degradation of poly(ethylene terephthalate) (PET). It was indicated that PET does not unzip but slowly decomposes at 300°C. The gaseous products of pyrolysis under nitrogen at 288°C were reported to be 80% acetaldehyde. The other 20% of the gaseous products contained CO, CO_2, H_2O, C_2H_4, CH_4, C_6H_6, and 2-methyldioxolan.

It is known that under the conditions of PET formation from ethylene glycol and terephthalic acid, the former can self-condense to form diethylene glycol (DEG) (Hovenkamp and Munting, 1970). Once diethylene glycol is formed, it is preferentially incorporated into the polymer, since it has comparable reactivity to ethylene glycol (EG) but a lower volatility. A minimum of 1–3% DEG units can be detected in commercial PET. In an effort to determine the effect of DEG units on the thermal stability of PET, copolymers of terephthalic acid with ethylene glycol and diethylene glycols were prepared to contain 1–24% of DEG units, and their thermal degradation was compared with those of PET and poly(diethylene glycol terephthalate) (PDEGT) (Hergenrother, 1974). The energy and entropy of activation for the thermal degradation were measured for the copolymers, PET, and PDEGT. The activation energy and entropy were found to decrease steadily with increasing diethylene glycol content, as shown in Table XVIII. From these measurements the mechanism of PDEGT degradation was found to differ from that of PET. Fibers prepared from seven

TABLE XVIII

ENERGY (E_a) AND ENTROPY (S) OF ACTIVATION
FOR THE THERMAL DECOMPOSITION OF PET,
PDEGT, AND THE EG, DEG COPOLYMERS[a]

DEG in polymer (%)	E_a (kcal mole^{-1})	S (at 300°C) (cal mole^{-1} deg^{-1})
100.0 (PDEGT)	24.6	−26.4
23.8	32.9	−18.2
18.9	36.6	−11.4
8.0	43.5	0.6
1.0	46.2	5.1
0[b] (PET)	47.3	7.0

[a] From Hergenrother (1974).
[b] Extrapolated from rate versus DEG Concentration plots.

different copolymeric compositions were heat-aged at 121 and 204°C for 24 h. From the changes observed in intrinsic viscosity, percent ether, hydroxyl, and carboxyl end groups during heat aging, it was shown that the mechanisms for decomposition are operative below melt temperatures and can readily destroy such copolymers.

At processing temperatures of 250–280°C, various degradation effects (thermal, oxidative, and hydrolytic) may occur in terephthalic polyesters. Although these effects have been well investigated for PET (Goodings, 1961; Buxbaum, 1968), only recently the degradation of poly(butylene terephthalate) at 240–280°C was examined (Passalacqua *et al.*, 1976). The degradation was monitored by measurements of intrinsic viscosity, carboxyl end groups, and weight loss. A first-order mechanism for chain scission followed by elimination of butadiene was proposed. The values of kinetic constants and activation energy were in good agreement with those of simple esters and PET. Values of the kinetic constant of PBT are shown in Table XIX.

TABLE XIX

Values of the Kinetic Constant for PBT at Various
Temperatures $K_1 \times 10^7$ s^{-1} from Measurements[a]

Temp. (°C)	Viscosity measurements	COOH end groups	Weight loss
240	3.3	2.0[b]	—
250	4.8	4.0	—
260	8.4	8.4	8.7
280	37.1	34.5	39.1

[a] From Passalacqua *et al.* (1976).
[b] Measured at 239°C.

It was concluded that (a) in the range of processing temperature, the thermal degradation of PBT takes place by ester-linkage fission, as is the case of simple carboxylic esters and for the first step in PET degradation; and (b) the chain scissions of PBT are followed by evolution of butadiene, whereas carboxyl end groups increase and unsaturated terminal groups disappear.

The principal mode of thermal breakdown of polyesters prepared from aliphatic diols and aliphatic dibasic acid, reported about 50 years ago, occurs by alkyl–oxygen fission via a cyclic transition state and involving a hydrogen abstraction (Hurd and Blunck, 1938). In order to study the effect of blocking the abstractable β-hydrogen, characteristic of the cyclic transition chain, poly(hexafluoropentamethylene adipate) (PHFPA) was prepared and its thermal stability was compared with that of the unsubstituted poly(2-ethylhexylene adipate) (PEHA) (Trischler and Hollander, 1969). It was shown that the fluorinated backbone is more stable than the unsubstituted one; PEHA and PHFPA decomposed at 280 and 380°C, respectively.

E. DEGRADATION OF LACTONE POLYESTERS

Thermal- and electron-impact-induced degradation reactions of a simple lactone polyester, such as poly-ε-caprolactone, were investigated using a mass spectrometer (Lüderwald, 1977). It was shown that the favored thermal degradation reaction is the cleavage of the ester linkage and formation of ω-hydroxy and ketene end groups, and with a lower intensity the cleavage of the O—CH₂ bond and formation of carboxyl and pentenyl end groups. After electron impact and fragmentation of the pyrolysis products to carboxonium ions, an elimination of caprolactone via an unzipping mechanism was recorded.

The degradation of polyesters and copolyesters based on glycolide and lactide was investigated by direct pyrolysis in the ion source of a mass spectrometer (Jacobi *et al.*, 1978). At 10^{-6} mbar these polyesters were shown to yield cyclic oligomers upon thermal degradation. The oligomers were further degraded by an electron impact mechanism with the elimination of formaldehyde (from the glycolyl moieties) or acetylaldehyde (from the lactyl moieties) and CO_2 giving characteristic series of linear ions. In the pyrolysis mass spectrum of a 50:50 copolyester of lactide and glycolide, intact sequences up to eight monomeric units were observed.

F. PERTINENCE OF STRUCTURE TO THERMAL STABILITY
OF POLYAMIDES

The pertinence of structure to thermal stability of polyamides was reviewed by Pearce *et al.* (1975). Of the many systems examined in the past few years, Nylon 6 is selected as a model system of polylactams and will be treated in this section. Substituting the β-methylene in Nylon 6 with a

$$
\begin{array}{c}
\overset{\displaystyle O}{\overset{\displaystyle \|}{-\text{P}-}} \\
\underset{\displaystyle \text{CH}_3}{\overset{\displaystyle |}{}}
\end{array}
$$

group led to a noticeable depreciation in thermal stability; at 400°C, Nylon 6 and poly(phosphacaprolactam) of similar DP lost 24 and 42% of their original mass, respectively (Shalaby *et al.*, 1974b, 1976b). Introduction of a small fraction of *m*-xylylene isophthalamide units into Nylon 6 backbone led to no detectable stabilization against thermal degradation (Shalaby *et al.*, 1976b). However, copolymers of caprolactam containing

5 and 10% of benzimidazole or 4-phenylene propionamide units were shown to be more stable thermally than Nylon 6 itself (Shalaby *et al.*, 1976a; Shalaby and Turi, 1977).

The thermal stability of a series of poly(alkylene oxamides) was shown to decrease with the decrease in the methylene content of the chain (Shalaby *et al.*, 1973b). The stability of poly(alkylene oxamides) was noted as comparable to or inferior to other more conventional AA–BB aliphatic polyamides.

G. Thermal and Thermooxidative Degradation of Polyolefins

Certain aspects of this subject were reviewed by Shalaby and Pearce in 1974 (1974b) and earlier by Wall in 1960. Selected new developments pertinent to polystyrene, substituted polystyrene, and polypropylene will be discussed in the next few paragraphs.

Differential scanning calorimetry has been used in studying the kinetics of polystyrene (PS) degradation and to determine an exact value for the activation energy of this process, for which a variety of values have been proposed (Kishore *et al.*, 1976). A value of 30 kcal mole^{-1} has been obtained (Kishore *et al.*, 1976) for the activation energy of PS degradation and was shown to be technique dependent. The activation energies were found to be independent of the atmosphere in which the degradation took place; a value of 30 kcal mole^{-1} was obtained when the degradation was conducted in air, nitrogen, or oxygen (see Table XX). This contradicts commonly accepted activation energy values for the PS degradation in oxygen and nitrogen (or vacuum) of about 21.5 and 40–50 kcal mole^{-1}, respectively (J. H. Flynn, private communication, 1979). In a review on the thermal degradation of anionic and thermal (radical) PS at 300 and

TABLE XX

Activation Energy Dependence on Technique
for Polystyrene[a]

Technique	Temp. range (°C)	E_a (kcal mole^{-1})
DSC	370–380	32 ± 3
TGA in air	320–390	29 ± 2
in N$_2$	290–390	32 ± 2
in O$_2$	340–375	28 ± 2
Mass spectrometry for		
formation of styrene	360–430	31 ± 4
formation of benzene	360–430	44 ± 2
formation of ethylene	360–430	30 ± 2

[a] From Kishore *et al.* (1976).

350°C, Cameron *et al.* (1978) indicated that neither random nor chain end initiation of volatile formation occurs in the initial stages of decomposition of the former, but both processes contribute to the early stages of volatilization of the latter. From this difference in behavior it was concluded that the undegraded thermal polymers contain a proportion of labile chain ends in addition to a small number of randomly distributed weak links within the chains.

Studies on the effect of substitution on the thermal and thermooxidative stabilities of polystyrene were reported earlier by Glacoleva and Fratkina (1971). On the basis of their results, polymers of styrene were classified into two groups: (a) those that predominantly undergo degradation [e.g., polystyrene (PS) and styrene–α-methylstyrene copolymers] and (b) those capable of cross-linking (structure formation) [e.g., poly(vinyl toluene) (PVT) and poly-2,4-dimethylstyrene) (PDMS)]. It was established that the replacement of a hydrogen atom in the benzene ring of styrene-type polymers by one or two methyl groups leads to a considerable increase in the concentration of aromatic ketones and aromatic aldehyde in the oxidized specimen. The oxidation and cross-linking of PVT and PDMS were shown to reduce the activation for bond cleavage from 50–60 to approximately 30 kcal mole^{-1} during their thermal oxidation.

A pyrolysis–gas chromatography–mass spectrometer system was used for studying the thermal degradation of polypropylene samples with various isotactic contents (Hosaka *et al.*, 1977). Peaks of the pyrograms obtained were reported to be composed of different homologous series of hydrocarbons. It was found that the isomerization during thermal decomposition was dependent on the stereoregularity.

H. CORRELATION BETWEEN HEATS OF DEPOLYMERIZATION AND ACTIVATION ENERGIES IN THE DEGRADATION OF POLYMERS

The correlation of heats of polymerization and activation energies for the degradation processes leading only to gaseous products was explored in studies by Kishore and Pai Verneker (1976). These studies were limited to the low-temperature region where the degradation leading to volatile formation starts taking place. On the basis of the assumption that the enthalpy change in the formation of solid polymer from gaseous monomer is equal to the enthalpy change during the breaking of the solid polymer into a gaseous monomer, the values for the heat of polymerization of various polymers have been calculated (Brandrup and Immergut, 1966) as presented in Table XXI. In those cases where the data for heats of polymerization for only liquid monomer to solid polymer are available, the heats

TABLE XXI

HEATS OF POLYMERIZATION/DEPOLYMERIZATION AND
ACTIVATION ENERGIES OF POLYMERS[a]

Polymer	ΔH_1 (25°C)[b] kcal mole^{-1} of monomer	ΔH_2 (kcal/mole at 25°C)[c]		ΔH_3 (kcal/mole 25°C)[d] monomer		Energy of activation E_a (kcal mole^{-1})
		Calc. from Cohesive energy data	Re-ported	Calc. as $\Delta H_1 + \Delta H_2$	Re-ported	
Polystyrene	16.7	10.3	9.52	27.0	—	28–30
Poly(methyl methacrylate)	13.6	8.2	8.0	21.8	—	23–30
Polypropylene	20.1	3.4	3.2[e]	23.5	24.9	22.6–24.9
Polyethylene	—	1.5	0.9	—	25.6	26
Polyacrylonitrile	18.3	7.3	8.2	25.6	—	31
Poly(vinyl chloride)	26.7	3.8	—	30.5	31.0	29–32

[a] From Brandrup and Immergut (1966).

[b] Heats of polymerization from liquid monomer to solid polymer.

[c] Heat of volatilization liquid monomer.

[d] Heat of polymerization of gaseous monomer to solid polymer; also equivalent to the heat of depolymerization of solid polymer to gaseous monomer.

[e] Mean value of ethylene and 1-butene.

of vaporization of liquid monomer (Timmermans, 1965) to gaseous monomer have been added. Heats of polymerization–depolymerization and activation energies of several polymers were collected from the literature as summarized in Table XXI. The data in this table show that heats of depolymerization of the cited polymers are almost the same as their respective activation energies. This led Kishore and Pai Verneker (1976) to suggest that for polymers that decompose to only gaseous products, degradation takes place in the following manner:

$$\text{polymer (solid)} \xrightarrow[E_a]{\text{primary reactions}} \text{monomer (gas)} \xrightarrow{\text{secondary reactions}} \text{other products}$$

In agreement with J. H. Flynn (private communication, 1979) this proposed degradation scheme appears to contradict a multitude of studies on molecular-weight change and product analysis, energetics, thermodynamics and the principle of least motion. Flynn also noted that this scheme cannot, for instance, account for the degradation of PVC, which occurs at about 200°C with the stripping off of HCl, formation of double bonds, some cross-linking , and an increase in molecular weight. He then questioned the accuracy of the heats of polymerization–depolymerization values presented by Kishore and Pai Verneker and the validity of their attempted correlation of these parameters with the activation energy of depolymerization.

I. LIFETIME PREDICTION OF POLYMERS FROM DEGRADATION KINETICS

This subject has been discussed in an excellent review by Flynn (1978) and highlights of this review will be summarized in this section.

When one is dealing with a complex polymeric substance, the processes taking place in the solid state over a long period of time in the condensed phase are difficult to measure and their kinetics are too complex to allow a reliable extrapolation from accelerated test conditions to end-use conditions. On the other hand, accelerated aging tests invariably lead to what appears to be reasonable lifetime predictions when extrapolated to service conditions. The bond energies in most organic polymers encompass a fairly narrow range of values so that extrapolation of the rate constants for their degradation reaction will often result in lifetime prediction within the range of 3 to 300 years.

A number of variables or factors may bring about the deterioration of the properties of a polymer. These include temperature, pressure, radiant flux, partial pressures of continuous gases and vapors such as oxygen and water, both external and internal catalysts, mechanical stress, and so on. One or more of these factors is intensified in an accelerated aging test to bring about deterioration in a much shorter period. Usually the temperature and often only the temperature is the intensified variable. If the temperature is chosen, then rate constant or shift factors obtained from two or more isothermal experiments are plugged into a model—the Arrhenius equation—which relates temperature with time. A simple coupling of the process measured and the process responsible for the deterioration was assessed and three techniques for obtaining and analyzing thermal analytical data for accelerated aging predictions were discussed by Flynn in his review. The three techniques involve measurement at constant heating rate, which is advantageous over other isothermal measurements (Flynn, 1977). These advantages include that kinetic parameters may be determined at low conversion and that the entire conversion range from 0 to 100% may be investigated in each experiment.

Specific problems encountered in the attempt to apply thermogravimetric methods to the prediction of aging characteristics of polymers involve the mode of kinetic coupling of the weight loss kinetics to the particular aging process of interest. A new technique in which the entire kinetic spectra are compared among experiments performed at different heating rates in the range from 2°C min⁻¹ to 4°C h⁻¹ was described by Flynn. The comparison of slow and fast heating rates helps to uncouple competing kinetic processes and allows a better prediction of the rate-limiting processes at ambient conditions. Reliable methods for kinetic analysis of data for the first 5% weight loss were described and several examples of application to polyurethane degradation were presented (Flynn, 1978). Flynn concluded that only such a detailed analysis of measurement was a viable method for the prediction of the service behavior of a polymeric material.

J. Effect of Molecular Weight and End Groups

The effect of molecular weight on the thermal degradation of 2.2-bis (4-hydroxyphenyl) propane polycarbonate was studied by Adam *et al.* (1976). It was shown that the amount of thermal degradation, measured thermogravimetrically, increases with decrease in molecular weight. It was also shown that acetylation of the hydroxylic end groups improves the thermal stability of the polycarbonate. Some of these results are illustrated in Fig. 19.

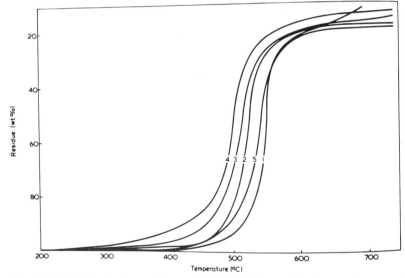

Fig. 19   Weight loss by thermal decomposition of different polycarbonates heated at increasing temperatures: (1) commercial material, $\overline{M}_v = 71,000$; (2) experimental polycarbonate, $\overline{M}_v = 80,000$; (3) experimental polycarbonate, $\overline{M}_v = 16,000$; (4) experimental polycarbonate, $\overline{M}_v = 6000$; (5) acetylated polymer, $\overline{M}_v = 62,200$. Rate of heating: 20°C min⁻¹. (From Adam *et al.*, 1976, by permission of the publisher, IPC Business Press Ltd., ©.)

K. STEADY-STATE PARAMETER JUMP METHODS AND RELAXATION METHODS IN THERMOGRAVIMETRY

Weight loss experiments measure processes that are heterogeneous by necessity. The weight loss step—matter leaving the condensed phase and entering the gaseous phase—involves simple vaporization, sublimation or desorption phenomena. However, the overall weight loss kinetics are often complicated by chemical and physical processes that take place in the condensed phase. Many methods of kinetic analysis require the comparison of data from several separate experiments. In these cases, the correct interpretation of the data is obfuscated further by disparities in the physical and chemical states of the samples, either present initially or induced by the action of diverse values of the rate-forcing variables. All these factors render defining, modeling, and analyzing weight-loss processes an elusive goal.

In a paper by Flynn and Dickens (1976), the inadequacies of conventional techniques of kinetic analysis in coping with these complexities were stressed. Meanwhile, these investigators described new procedures that avoid many of the previous drawbacks and then suggested applications of relaxation techniques to these systems that may point out some new facts of the kinetic processes. In the technique noted by Flynn and Dickens, the magnitude of a rate-forcing variable such as temperature, pressure, gaseous composition, etc., is jumped by discrete steps. This method can be used to determine kinetic relationships between the rate of weight loss and the jumped variable, as illustrated by Flynn and Dickens (1976) with examples from the oxidation of polystyrene.

V. Typical Applications of Thermoanalytical Techniques

This section deals with examples of recent typical applications of different thermoanalytical techniques. Some of these applications were selected for their general nature in order to allow the extrapolation to the use for other polymeric materials. Examples are also given for the measurement of a particular property by more than one technique and analysis of the data. The applications are outlined.

A. DETERMINATION OF HEAT CAPACITY OF COPOLYMERS

Although many studies on heat capacities of linear homopolymers have been reported (Wunderlich and Baur, 1970), data on copolymer heat capacities have been relatively scarce. Examples of the few available studies of the latter subject are reviewed briefly in this section. The molar heat capacity of a 45/55 copolymer of ethylene and tetrafluoroethylene has been measured from 80 to 340 K using an adiabatic calorimeter with an accuracy of 0.3% (Wong *et al.*, 1975). The results were found to be in close agreement with values calculated from the known optical lines of related polymers and the Tarasov model (Tarasov, 1950). The experimental data were also analyzed, together with the available data on PE and four other fluoro-polymers, showing that the principle of additivity for heat capacity (Wunderlich and Jones, 1969) is generally valid within 2%. Typical experimental and calculated heat capacity values are summarized in Table XXII.

TABLE XXII

HEAT CAPACITY (cal/deg-mole) OF HOMOPOLYMERS
AND COPOLYMERS[a,b]

Polymer	300 K		420 K	
	Obs.	Calc.	Obs.	Calc.
Polystyrene	28.8	—	50.2	—
Poly(hexyl methacrylate)	77.0	—	90.2	—
67.4/32.6 Styrene/hexyl methacrylate copolymer	41.7	41.7	62.7	63.2
56.4/43.6 Styrene/hexyl methacrylate copolymer	52.0	52.0	70.5	67.7
26.7/72.3 Styrene/glycidyl methacrylate copolymer	46.0	47.5	69.3	—
34.9/65.1 Styrene/glycidyl methacrylate copolymer	45.0	45.9	65.3	—
Poly(glycidyl methacrylate)	55.0	—	—	—

[a] From Wunderlich and Jones (1969).

[b] Observed experimental data obtained by Lai (1976) or calculated by adding fractional heat capacities of homopolymers.

B. GLASS TRANSITION TEMPERATURES OF POLY(ALKYL-α-CYANOACRYLATES) BY DTA AND DILATOMETRY

The T_gs of several poly(alkyl-α-cyanoacrylates) were determined by dilatometry and DTA (Kulkarni *et al.*, 1973). Some of the obtained data are summarized in Table XXIII and indicate that (1) the T_g decreases with the increase in size of the alkyl group for a given molecular weight; (2) T_g increases with decrease in polymerization temperature; and (3) the T_g of poly(methyl α-cyanoacrylate) is quite dependent on which polymerization technique is employed. The latter two observations may be related to differences in effect of molecular weight and/or stereoregularity among

TABLE XXIII

Glass Transitions of Poly(alkyl α-cyanoacrylates)[a]

Type of alkyl group	Method of polymerization	$\overline{M}_n$	T_g (°C) By dilatometry	By DTA
Methyl	Anionic	2204	100	—
Methyl	Free radical in solution	—	125	135
Methyl	Free radical in bulk	—	160	—
n-Butyl	Anionic	3820	56	—
n-Butyl	Free radical in bulk	—	123	111
Isobutyl	Anionic	2027	51	—
Ethyl	Anionic	1533	115	—
Isopropyl	Anionic	5960	66	—
n-Hexyl	Anionic	2650	86	—
n-Heptyl	Anionic at 25°C	—	—	88
	Anionic at −40°C	—	—	116

[a] From Kulkarni *et al.* (1973).

these samples. If one considered that the poly-*n*-heptyl-α-cyanoacrylate prepared at −40°C has a higher molecular weight than that prepared at 25°C, it becomes apparent that the T_g increases with the increase in molecular weight.

C. DSC for Studying Polymer Crystallization

The limitations of a differential scanning calorimeter, DSC-1B, in measuring directly isothermal crystallization data have been noted by Booth and Hay (1969). These authors concluded that the DSC was not as sensitive as precision dilatometry in detecting low rates of crystallization but that, nevertheless, the time dependence of crystallization could be obtained at higher rates with sufficient accuracy to test the validity of rate equations. Care, however, was necessary in setting and calibrating the temperature, as rate variation between repeated isothermal crystallizations were consistent with an error of ±0.3 K. The use of DSC to determine accurate crystallization parameters was considered by direct comparison with dilatometry (Hay *et al.*, 1976). Two calorimeters were compared, Perkin-Elmer DSC-1B and −2. Fractionated polyethylene was studied, for the narrow-molecular-weight fractions have highly temperature-dependent rate constants. This enables the accuracy of setting and calibrating the temperature of the calorimeters to be gauged. The results of these studies indicated that the DSC-2 instrument can be applied to the study of isothermal crystallization kinetics provided the rate is sufficiently large. [Rates can be measured to lower values (×0.1) than can be measured by DSC-1B.] The isothermal crystallization–times curves were noted as more reproducible with the DSC-2 because of the improved temperature setting (to 0.1 K) and the maximum variation in temperature setting was ± 0.025 K. It was also reported that despite the improved sensitivity of DSC-2, it is more restricted than dilatometry in measuring a low rate of crystallization. This severely limits its use in studying secondary crystallization processes. Its main advantages, however, were indicated to include the speed and accuracy of use and also the small samples required (1–2 mg).

The DSC has been used for studying the crystallization kinetics of poly(vinylidene fluoride) (PVF_2) (Mancarella and Martuscelli, 1977). Specimens of PVF_2 were crystallized isothermally at a series of temperatures in the vicinity of T_m to produce the α-form crystals. The half-time $t_{1/2}$ for the conversion to the α-form of PVF_2 at various crystallization temperatures (T_c) was determined from the crystallization isotherms. The latter curves were used to determine the dependence of $t_{1/2}$ on crystallization temperature as shown in Fig. 20. As observed in all polymeric materials, $t_{1/2}$ increases with T_c for low undercoolings.

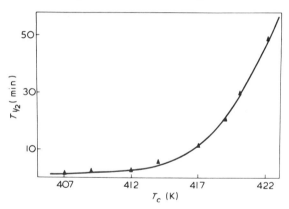

Fig. 20 Variation of $t_{1/2}$ with the crystallization temperature. (From Mancarella and Martuscelli, 1977, by permission of the publishers, IPC Business Press Ltd., ©.)

So far we have discussed only the crystallization of polymer melts; hence this section will be dedicated to the crystallization from polymer solutions. Solution crystallization is associated with phase separation of the solid polymer from the solvent. In one case DSC was used for studying the phase separation of poly(2,6-dimethyl-1,4-phenyl ether) (PPO) solutions in decalin (Janeczek *et al.*, 1978). Three phase separation temperatures were observed for PPO–decalin system. On cooling (at a rate of 20 K min^{-1}) a homogeneous solution of PPO in decalin from a certain tem-

perature T_1, an exotherm appeared over a wide temperature range. The
same sample was then heated at a rate of 10 K min⁻¹ starting at a temperature below T_1. Two endothermic phase transition phenomena were observed on heating the sample. A first, rather wide temperature range corresponding to an initial endothermic effect was found at temperature T_2.
A second endotherm was found at temperature T_3. A typical thermogram
obtained for a 13.6% solution is shown in Fig. 21. For samples of solutions

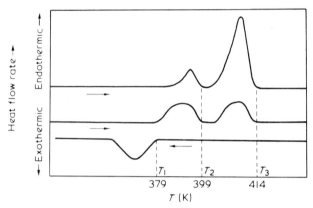

Fig. 21   Phase separation temperature and heat effect determined with DSC for 13.6 wt
% poly(2,6-dimethyl-1,4-phenylene oxide)–decalin solution. (From Janeczek *et al.*, 1978,
by permission of the publishers, IPC Business Press Ltd., ©.)

allowed to stand for longer periods of time (20 min) below T_1, the area of
the second endotherm peak increased while that of the first decreased.
The values of T_2 and T_3 were nevertheless the same. Temperature T_3 was
taken as the melting temperature T_m of the polymer crystals in solutions.
The peak ending at T_2 was thought to represent a decrease of the order in
the system corresponding to the dissolution of the amorphous phase. The
peak ending at T_3 was assumed to represent a decrease of the degree of
order in the system due to melting to the crystalline phase present in it.
Results of these measurements indicated that for 13% solution of PPO in
decalin, the T_m of the crystals is 414 K. The heat of melting of crystals
was also found to be 1285 cal mole⁻¹.

D. MEASUREMENT OF CRYSTALLINITY

It is well documented that the degree of crystallinity of thermoplastic
polymers can be determined by DSC; one divides the measured heat fu-
sion ΔH_f of the sample by ΔH_f°, the heat of fusion for a 100% crystalline
polymer. This practice is almost standard (Dole, 1967) and need not be
discussed any further here. However, for polymers that can have more

than one crystalline structure, the determination of crystallinity from ΔH_f data becomes somewhat complex. To illustrate this point, the reader is referred to a report on the measurement of crystallinity by DSC for Nylon 6, which can be found in both the α- and γ-crystalline forms (Cappola *et al.*, 1975). This report dealt with the construction of a calibration plot for routine measurement of the degree of crystallinity of commercial Nylon 6 through enthalpy of fusion ΔH_f derived by DSC.

Several years ago Yagfarov (1968, 1969) proposed a general method for determining the absolute crystallinities of polymer samples without isolation or prior characterization of either the crystalline or amorphous phase. It was assumed that a given polymer can be described by a two-phase model in which heat capacities of the amorphous and crystalline phases are additive and in which measured enthalpies are proportional to the mass fraction of the crystalline material. However, in a report by Hobbs and Mankin (1971), it was concluded, on the basis of their results on PET and other literature data, that the technique for determining polymer crystallinity by the combination of heat capacity and enthalpy data is not generally applicable and can give large errors in some polymer systems.

E. DSC for Studying the Thermodynamic Properties of Semicrystalline and Amorphous Aromatic Polyethers

The heat capacity of amorphous poly(2,6-dimethoxy-1,4-phenylene ether) PPO was measured as a function of temperature between 280 and 560 K (Savolainen, 1974). Linear correlation between heat capacity and temperature of the semicrystalline PPO was observed between 380 K and the first melting peak (523 K). The second melting peak was observed at 549 K and the heat of melting ΔH_f was found to be 14.8 J g^{-1}. The heat capacity of the amorphous PPO was identical to that of the semicrystalline sample between 380 and 425 K, indicating that crystallinity and chain conformation did not contribute to C_p. A characteristic glass transition of amorphous PPO was recorded at 435–437 K, as indicated by a jump in C_p ($\Delta C_p = 10 \pm 2$ cal mole^{-1} K^{-1}). No melting endotherm could be observed on heating the amorphous sample from 437 to 560 K. It was also emphasized that no manifestation of further crystallization in the semicrystalline PPO sample below T_m was observed. A very low heat of fusion for the semicrystalline PPO was also determined and was suggested to indicate the presence of a long-range order similar to that of the semicrystalline material and to indicate that only minor conformational changes take place during melting.

F. EFFECT OF CHEMICAL STRUCTURE ON THERMAL
TRANSITIONS

1. Effect on T_m by DSC

The effect of chemical structure on T_g and T_m of a homologous series of poly(alkylene terephthalates), poly(alkylene isophthalates), and equimolar random copolyesters of isophthalic and terephthalic acids has been reported (Yip and Williams, 1976a, b). The three series, except for $+CH_2+_5$, were limited to even numbers of methylene groups in the glycol portion of the polyesters. The DSC thermograms for the terephthalate series were similar, differing only in the location of T_m. The T_m of poly(tetramethylene terephthalate) (4GT), poly(hexamethylene terephthalate) (6GT), and poly(decamethylene terephthalate) (10GT) were recorded at 226, 147, and 127°C, respectively. The thermograms of the isophthalate series differed from those of the terephthalates. A small distinct peak was observed 30–40°C lower than the main melting transition. The T_m values of the terephthalate series and T_ms of the major melting endotherms for the isophthalate series are summarized in Table XXIV. The data in this table indicate that for terephthalate polymers made from glycols with an even number of carbons, as the length of the methylene chain increases, T_g and T_m decrease, but crystallinity and heat capacity increase. The T_m, T_g, and crystallinity of the isophthalate series were lower than those for

TABLE XXIV

DSC DATA OF POLY(ALKYLENE ISOPHTHALATES) AND
POLY(ALKYLENE TEREPHTHALATE)[a]

Polymer code[b]	T_g (°C)	T_m (°C)	Cp (cal mole^{-1}) °C at T_g	% Crystallinity
2GT	—	264[c]	—	—
4GT	45	226	72.5	30
5GT	—	130–135	—	—
6GT	8	147, 152	77	41
10GT	−9	127	86	41
2GI	—	103, 108, 240[c]	—	—
4GI	34	141	56	23
5GI	—	Amorphous	—	—
6GI	−1.5	95, 140	66	22, 13
10GI	−15	44	125	16, 20

[a] From Yip and Williams (1976a).

[b] XG = a glycol with X number of methylene groups; XGI = a polymer made of a glycol with X methylene groups and isophthalic acid; XGT = the same as XGI except using terephthalic instead of isophthalic acid.

[c] Based on literature data.

the corresponding terephthalate series. Poly(pentamethylene isophthalate) was amorphous. The values for odd-numbered methylene sequence were fewer than those for the adjacent even-numbered members. A model was used to calculate the heat of fusion of theoretically 100% crystalline poly(hexamethylene isophthalate) (6GI) and poly(decamethylene isophthalate) (10GI).

2. *Effect on T_g and the γ-Relaxation by Dynamic Mechanical Measurements*

The β-relaxation at T_g of the terephthalate and isophthalate polymers is due to the motion of the backbone chains in the amorphous region but differs for the terephthalates in that the p-phenylene group does not exhibit the free rotation possible for the m-phenylene groups. Consequently, the relaxation times of the terephthalates were found to be longer than those for the isophthalate series (Yip and Williams, 1976b). The γ-relaxation temperature T_γ for the higher homologs of the terephthalate series could not be explained in terms of the poly(ethylene terephthalate) analogy. For poly(ethylene terephthalate) and poly(tetramethylene terephthalate), an induced cooperative type of motion of all the moieties was described as being possible, whereby overlapping processes caused by "rocking vibrations" were observed as one γ-peak. The resolution of the γ-loss peak for the two polyesters into components was not possible at the experimental frequency 110 Hz. For poly(hexamethylene terephthalate)

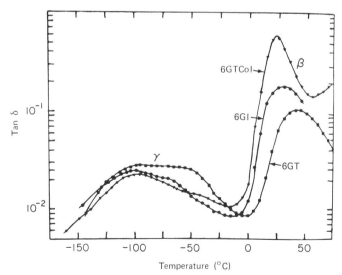

Fig. 22   Tan δ versus temperature for polyhexamethylene polyesters: (■) 6GT, (●) 6GI, (▼) 6GTCoI. (From Yip and Williams, 1976b.)

and poly(decamethylene terephthalate), the "rocking vibrations" between the moieties of the skeletal chain were reduced so that even at a test frequency of 110 Hz the γ-loss peaks could be resolved into two or three components. For poly(decamethylene terephthalate), three components were resolved, the lowest temperature peak γ_1 was attributed to hindered motions of the methylene portions, the γ_2-peak was attributed to motions of the carbonyl group in the gauche conformation, and the γ_3-peak was attributed to the carbonyl group in the trans conformation of the skeletal chain in the amorphous phase. As the length of the methylene chain increased, T_g decreased. As crystallinity increased, the β-relaxation moved to higher temperatures and the damping peak was smaller and broader. The damping peak moved to lower temperatures and increased in size as the length of the methylene chain increased. The damping peak was larger for the isophthalate homologs than for the corresponding terephthalate polyester. Some of these results are illustrated in Figs. 22 and 23.

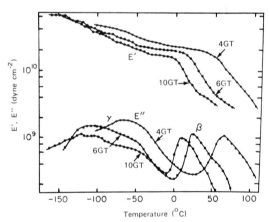

Fig. 23 Dynamic and loss moduli versus temperature for terephthalate series of polyesters: (▼) 4GT, (■) 6GT, (●) 10GT. (From Yip and Williams, 1976b.)

G. DSC FOR THE DETERMINATION OF HEAT OF VOLATILIZATION

The heat of volatilization for a number of polymers was measured by DSC, where the area under the degradation endothermic peak was shown to be directly proportional to the heat of volatilization (Frederick and Mentzer, 1975). Values measured for PMMA, which yield monomer quantitatively in the temperature range investigated, agreed well with predicted values. Reproducibility of the method was shown by an average standard deviation of ±10% for polyethylene, polypropylene, poly(methyl methacrylate), polystyrene and Nylon 6, and their determined heats of volatilization (ΔH_{vol}) in cal/g were 159 ± 13, 151 ± 10, 192 ± 8, 196 ± 8, 188 ± 8, and 135 ± 11, respectively. It is to be pointed out that the ΔH_{vol} measurements should have been stated by Frederick and Mentzer to include the following thermal events:

$$\text{Polymer} \rightarrow \text{condensed phase products} + \Delta H_1 \rightarrow \text{volatile products} + \Delta H_2$$

H. DTA for Comparing the Stability of Ram and Screw Extruded Polypropylene

Talc-filled polypropylene (PP) samples were ram or screw extruded and their oxidative stability with or without aging was compared with unextruded samples (Hampson and Manley, 1976). Figure 24 shows the

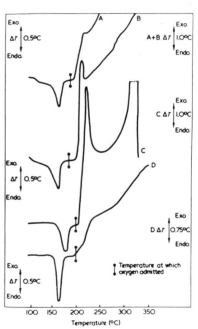

Fig. 24 DTA on polypropylene + talc, in nitrogen with subsequent admission of oxygen: A, Powdered polypropylene, not extruded; B, powdered polypropylene + talc, not extruded; C, ram-extruded polypropylene, aged 21 d at 155°C; D, screw-extruded polypropylene, aged 21 d at 155°C. Note changes of scale. (From Hampson and Manley, 1976, by permission of the publishers, IPC Business Press Ltd., ©.)

DTA results of the original PP (PP + talc) and samples that were extruded on ram and screw machines and subsequently aged for 21 days at 155°C. These samples were initially heated in a static nitrogen atmosphere and then oxygen was introduced at the point indicated in the graph to obtain the decomposition exotherm. This technique has been widely used for studying the effect of inhibitors on the oxidation of polyolefins (Marshal *et al.*, 1973). The change in scale of the ΔT after the introduction of oxygen should be noted. In addition, after such a vigorous reaction, the base line did not remain that of the original PP. It is clearly seen that sample C in Fig. 24, which had been ram extruded, still retains more material for oxidation than the screw-extruded sample (curve *D*). For the latter sample the reaction was almost complete before the oxygen was introduced. Thus filled PP undergoes more degradation during screw as compared with ram extrusion. This technique of introducing oxygen into the DTA conducted under nitrogen was described as a very sensitive and rapid method for determining the amount of oxidation that a sample has undergone and could well be the preferred way of quality control testing by DTA of extruded materials (Hampson and Manley, 1976).

I. DTA, TGA, and Pyrolysis Mass Spectometry for Studying Thermal and/or Thermooxidative Degradation of Polyesters

Zimmermann *et al.* (1971) and Zimmermann and Becker (1973), in their earlier studies of the thermooxidative degradation of PET, found that this process is dependent on the type and concentration of transesterification and polycondensation catalysts, inhibitors, and other modifying additives used during the polymer formation. This was shown by the position and area of the DTA exothermic peak caused by the oxidation of the sample prior to melting. Besides the recorded DTA data, the examined samples also showed weight loss values of 0.1–1%.

The degradation kinetics of PET has been the subject of more studies by Zimmermann and Chu (1973), who proposed that as in the degradation of simple low-molecular-weight ester, a cyclic transition state can be formed, which may lead to the formation of a vinyl ester group and a free carboxylic group. By consecutive reactions of the vinyl ester, acetaldehyde was shown to form and was released as the main degradation product. The carboxylic group was retained and could be determined by titration. The rate constants at 301°C calculated from the increase in carboxylic group concentration were $k = 1.1 \times 10^{-6}$ to 1.4×10^{-6} J K^{-1} mol^{-1} s^{-1}, the energy of activation $E_a = 42$–46 kcal mole^{-1}, and the entropy of activation $\Delta S_A = -4$ to -20 cal deg^{-1} mole^{-1}. These values agreed very well with those calculated from the weight loss values. The fact that the activation paramaters for the pyrolysis of carboxylic esters (e.g., ethyl formate) were $E_a = 44.1$ kcal mole^{-1}, $\Delta S_A = -11$ cal deg^{-1} mole^{-1} were similar to those noted previously for PET led to the suggestion that the thermal scission of the ester bonds is the rate-determining step of the weight loss process caused by the consecutive reactions in the thermal degradation of PET (Zimmermann and Chu, 1973; Zimmermann and Schaaf, 1974).

The reaction in the thermooxidative degradation of PET was proposed to proceed similar to that of hydrocarbons, presumably by a radical chain mechanism with the formation peroxide radical, and hydroperoxides, which decompose in different manners (Zimmermann and Schaaf, 1974). The activation energy for the thermooxidative degradation of liquid hydrocarbon reaches $12-30$ kcal mole^{-1}, a range within which the activation energy for the thermooxidative degradation of molten PET lies, as calculated from the weight loss data. The high negative value of the activation entropy, as compared with that of thermal degradation, led to the suggestion that a transition state with higher steric requirements is the rate-determining step in thermooxidative degradation.

The degradation reactions of poly(ethylene terephthalate), poly(tetramethylene terephthalate), poly(trimethylene terephthalate), and poly(tetradeuteroethylene terephthalate) have been investigated by direct pyrolysis in the ion source of a mass spectrometer, where thermally formed degradation products were ionized and further degraded by electron impact (Lüderwald and Urrutia, 1976a). The thermal and electron-impact-induced degradation reactions could be established. The terephthalate polymers examined were shown to follow the same mechanism in both degradation steps and could be differentiated by their pyrolysis mass spectra.

Pyrolysis mass spectrometry has been used to study the degradation reactions of adipic acid polyesters with ethylene glycol, tetradeuteroethylene glycol, 1,3-propanediol and 2,2'-oxydiethanol, and the polyesters of succinic acid with ethylene glycol, 1,4-butanediol and 2,2'-dimethyl-1,3-propanediol (Lüderwald and Urrutia 1976b). The knowledge of the mechanism of both degradation steps allowed differentiation between isomeric polyesters.

J. Stepwise Thermal Degradation Using Infrared Image Furnace for Studying Effect of Stereoregularity on Polymer Degradation of Acrylic Polymers

A new thermal characterization technique was described in a recent report (Tsuge *et al.*, 1977) to be based on a combination of infrared image furnace, a thermal balance, and a gas chromatograph. This technique was applied for the study of poly(methyl methacrylate) (PMMA) and methyl methacrylate–styrene copolymers (MMA-ST). The pyrolysis data indicated that although both PMMA made anionically (i-PMMA) and free radically (r-PMMA) yield only monomer in the thermal degradation at relatively low temperatures, they degrade differently; more monomer is formed from r-PMMA as compared with i-PMMA. The latter was stable up to about 300°C. This difference was attributed to the existence of much more thermally unstable bonds and end groups in r-PMMA. Thus terminal double bonds, initiator fragment in the polymer chain, and head-to-head or tail-to-tail structures were described as sources of the thermal instability.

K. Solid-State Relaxation in Poly(Ethylene Oxide) by DSC

Poly(ethylene oxide) (PEO) displays multiple relaxation processes, as do many crystalline polymers. These transitions were studied earlier by several athermal techniques. Recently, DSC measurements were made on PEO over a wide range of temperatures from 105 to 365 K (Lang *et al.*, 1977a). The observed and reported transitions were classified into five groups, the $T < T_g(T_\gamma)$, $T_g(L)$, $T_g(U)$, T_α, and T_m peaks in order of increasing temperature. The DSC thermograms prepared by Lang *et al.* revealed three transition regions below the T_m of PEO. The effect of annealing on the intensity and temperature of these transitions made it possible to locate $T < T_g(T_\gamma)$, T_g and T_α at about 130–140, 190–240, and 263–313 K, respectively. The DSC results were used to propose the presence of $T_g(L)$ at 190–200 K with a second $T_g(U)$ above 233 K. [The $T_g(L)$ and $T_g(U)$ were discussed earlier in Section III. A.] Up to three melting endotherms could be observed, one of which was related to the normal primary crystallization process. The peak temperature of the major melting endotherm increased linearly with the annealing temperature, yielding an extrapolated value for the equilibrium melting, T_m° of 347 K. The presence of β- and γ-processes in solid PEO was also suggested on the basis of spin-labeling studies (Lang *et al.*, 1977b).

L. Thermal Analysis for Studying Isomerism and Isomorphism in Polymers

Thermal transition in homopolymers of sebacic acid and isomeric hexanediols were examined by DSC (O'Malley and Stauffer, 1974). Among the homopolymers, only the polyester derived from 1,6-hexanediol was shown to be crystalline. The poly(hexamethylene sebacate) revealed a T_m of 65–70°C (depending on DP) and a T_g of -62°C. Other isomeric homopolymers derived from the branched diols 2-methyl-2-ethyl-1,3-propanediol and 2,5-hexanediol were amorphous and underwent glass transition at -69 and -66°C, respectively.

The phenomenon of isomorphism and the associated presence of mixed crystals on cocrystallization in several polymers was reviewed by a few authors (Allegra and Bassi, 1969, 1975; Natta *et al.*, 1969; Wunderlich, 1973, 1976). The two basic requirements for isomorphism are that (1) the different types of monomer units have approximately the same shape and occupy the same volume, and (2) the same chain conformation be compatible with either of them. Different aspects of isomorphism were examined in polyolefins (Allegra and Bassi, 1969; Benedetti *et al.*, 1975), polyethers (Natta *et al.*, 1969), polyesters (Allegra and Bassi, 1969, 1975; Ishibashi, 1964), and polyamide (Prince *et al.*, 1970; Prince and Fredericks, 1972).

Isomorphism in polyamides will be used to illustrate the use of DTA for monitoring isomorphism. In an effort to study the ability of 4-aminomethylcyclohexanecarboxylic acid (AMCC) to "isomorphously" replace ε-aminocaproic acid residues in Nylon 6, a series of copolymers were prepared from ε-caprolactam (CL) and AMCC having high trans or cis contents. The DTA of CL/AMCC copolymers showed that the AMCC moieties generally led to the increase of T_g, T_c, and T_m of the polyamides. This is illustrated by the data in Table XXV.

TABLE XXV

DTA DATA OF NYLON 6 AND CL/AMCC COMPOLYAMIDES[a]

Mole % CL	Mole % AMCC	AMCC configuration	T_g (°C)	T_c (°C)	T_m (°C)
100	—	—	45	75	221
90	10	High trans	58	93	218
70	30	High trans	90	135	240
50	50	High trans	115	165	355
90	10	High cis	60	95	216
70	30	High cis	80	130	238
50	50	High cis	105	160	350

[a] From Prince *et al.* (1970).

The display of isomorphism by a series of oxalate copolyesters based on *trans*-1,4-cyclohexanedimethanol and 1,6-hexanediol was demonstrated using DSC and X-ray techniques (Shalaby and Jamiolkowski, 1979b). All copolymers having between 5 and 90% alicyclic sequences were shown to be crystalline. Isomorphism in copolyesters of ε-caprolactone and its ether analog (1,5-dioxepan-2-one) was proposed to be manifested in certain copolymeric compositions (Shalaby, 1980).

Properties of solution-grown crystals of isotactic propylene/1-butene copolymers made from different ratios of the comonomers were studied by Cavallo *et al.* (1977). Measurement of the apparent enthalpy of fusion of copolymer crystals showed a eutectic point corresponding to composition of approximately 0.48 1-butene.

M. Thermal Analysis for Studying Formation and
   Chemical Reactions of Polymers

1. Polymer Formation

The use of isothermal DSC to follow the course of free radical bulk polymerization has been demonstrated by Horie *et al*. (1968, 1969). They studied the diffusion-controlled polymerization of both methyl methacrylate and styrene. Barrett (1970) has used DSC for studying emulsion polymerization, and earlier (1967) he used this technique for determining rates of thermal decomposition of polymerization initiators. The kinetics of the free radical polymerization of methyl methacrylate in *n*-dodecane to produce dispersions of polymer, stabilized with a stearic acid barrier of soluble polymer chains, have been determined by Barrett and Thomas (1969) using thermal analysis. Haas *et al*. (1970) have measured overall activation energies for the bulk polymerization of several monomers by an analogous DTA method, but some of the values they obtained were not in good agreement with those obtained by conventional methods. Ebdon and Hunt (1973) have used isothermal DSC for examining some kinetic features of free radical polymerization of styrene. They have demonstrated the usefulness of this technique for obtaining reliable quantitative information about free radical polymerization kinetics.

2. Reactions of Polymers

Ke (1963) has used DTA to study the oxidation of polymers, such as isotactic poly-4-methyl-1-pentene at elevated temperature and the efficiency of antioxidants. The application of DTA to the study of thermal degradation has been quite common. Paciorek *et al*. (1962) have used DTA to compare the thermal stability of polyethylene with those of poly(vinyl fluoride), poly(vinylidene fluoride), and polytetrafluoroethylene. Fock (1967) studied the effects of ionizing radiation on polytetrafluoroethylene.

N. Thermomechanical Analysis (TMA) for Studying
   Glass Transition

In DSC and DTA measurements of glass transition temperature (T_g) of polymers, it has been noted earlier in this chapter that the value of T_g is dependent on the rate of heating. An increase in the heating rate leads to an increase in the recorded T_g. The use of TMA to study the influence of heating and cooling rates and sample thickness on the expansion coefficients and T_g of carboxy-terminated polybutadiene (PB) was reported by Schwartz in 1978. The detection of T_g by TMA in expansion and penetra-

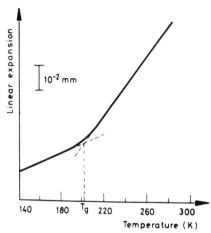

Fig. 25   Linear expension of PB. Specimen thickness = 2.47 mm. Heating rate = 4.16 K min^{-1}. (From Schwartz, 1978.)

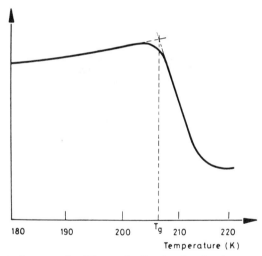

Fig. 26   Penetration test of a PB sample. Load = 7 g, heating rate = 4.16 K min^{-1}, specimen thickness = 2.52 mm. (From Schwartz, 1978.)

tion experiments is illustrated in Figs. 25 and 26. The results on the effect of heating and cooling rates on T_g are shown in Figs. 27 and 28. These indicate that the value of T_g is dependent on both the heating and cooling rates. Extrapolation to zero rates gives the same T_g for both cooling and heating. Schwartz also demonstrated that the expansion coefficient is not influenced by the rates of heating and cooling or by the sample thickness.

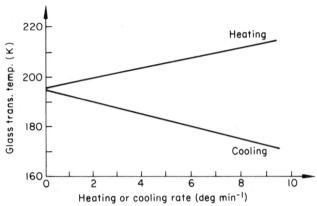

Fig. 27 Glass transition temperature of PB as a function of the heating and cooling rates in expansion measurements. (From Schwartz, 1978.)

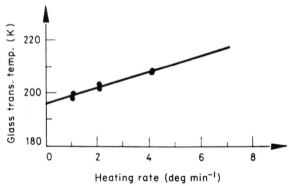

Fig. 28 Glass transition temperature of PB as a function of the heating rate in a penetration test. (From Schwartz, 1978.)

O. Study of PVC Degradation and Stabilization

Thermal degradation of PVC is characterized by two major types of processes. Those pertinent to the dehydrohalogenation to form polyenes are referred to as primary processes. Reactions associated with the polyenes constitute the second type of thermal degradation processes. These include cross-linking and gel formation, cyclization, chain scission, benzene formation, and Diels–Alder reaction with additives as well as oxidation of polyenes.

In their studies of the primary degradation processes, Abbas and Sorvik (1976) examined the influence of branch sites as triggers for the dehydrohalogenation of PVC. The conclusion was made that branch sites with tertiary hydrogens play a lesser role than do the more predominant linkages with allylic chlorines in the thermal degradation of PVC. The detection of unsaturation in PVC has been achieved by Caraculacu and Bezdadea (1969) using Fourier transform ¹H NMR. Ahlstrom *et al.* (1976) used pyrolysis GC, IR, and ¹³C NMR techniques for studying the effect of the microstructure of PVC on its degradation behavior. From the results obtained by these authors, the branch content in a series of reduced PVC samples was estimated at four to eight branches per 1000 carbon along the polymer chain. In a report by Liebman *et al.* (1978), these authors reported the application of "time-resolved" pyrolysis gas chromatography (PGC) in addition to derivative thermogravimetric analysis (DTGA) to a series of PVC homopolymers with differing branch content and to a model copolymer series with low amounts of propylene in an otherwise vinyl chloride chain. Benzene and toluene generation and decay envelopes were determined during the controlled thermal degradation and related to the derivative TGA experiments (Figs. 29 and 30). These data allowed in-

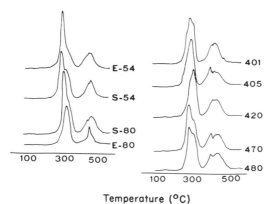

Fig. 29   Derivative TG in nitrogen for (left) PVC Nordforsk series and (right) Airco copolymers. Heating rate = 10°C min⁻¹. (From Liebman *et al.*, 1978.)

terpretation as to the microstructure of the respective polymers and its effect on the degradative pathways. It was concluded that the branch content was not a dominant factor in the initiation of thermal degradation, although it was a factor in other stages of the complex mechanism. A unique fragmentation step was noted only under oxidative thermal exposures, which gave additional support for the theory of low-level unsaturation sites as being significant triggers in the decomposition mechanism.

Among the most important secondary processes in the thermal degradation of PVC are cross-linking and gel formation. The early stages of these reactions have been monitored by Kelen (1978) through measuring the increase in the number-average molecular weight ($\overline{M}_n$), using both

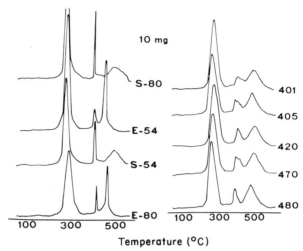

Fig. 30 Derivative TG in air for (left) PVC Nordforsk series and (right) Airco copolymers. Heating rate = 10°C min⁻¹. (From Liebman *et al.*, 1978.)

membrane osmometry (OSM) and gel permeation chromatography (GPC). The change in $\overline{M}_n$ with thermal degradation of PVC is shown in Figure 31.

A method for studying the thermal stability of PVC compounded with different stabilizers using DTA has been described by Dick and Westerberg (1978). In these studies dioctyl phthalate (DOP), tribasic lead sulfate (TBLS), and china clay were used for compounding PVC to impart thermal stabilization. Samples of the compounded PVC were heated in nitro-

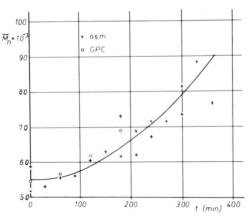

Fig. 31 Change in molecular weight with thermal degradation of PVC (180°C in air). (Reprinted from Kelen, 1978, p. 349, by courtesy of Marcel Dekker, Inc.)

gen using a heating rate of 5°C min⁻¹. The resulting DTA thermograms are shown in Fig. 32. Prior to the endothermal breakdown of the compounded PVC, all systems containing stabilizers show an exothermic maximum. This maximum appears to shift to higher temperatures as the amount of stabilizer is increased.

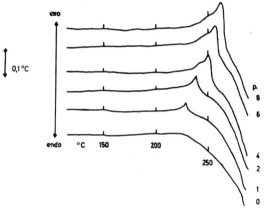

Fig. 32　Scanning DTA of formulations containing 100 parts PVC, 50 phr DOP, 8 phr china clay, and 0, 1, 2, 4, 6, or 8 phr TBLS, 5°C min⁻¹ N₂ atmosphere. (Reprinted from Dick and Westerberg, 1978, p. 455, by courtesy of Marcel Dekker, Inc.)

VI. Thermal and Complementary Techniques for Investigative Studies

Thermal analysis has been used extensively by itself. In many cases the effectiveness of this increased when used with one or several other analytical techniques in investigative types of studies. Typical cases of these studies are discussed in this section.

A. Thermal Development of Crystallinity in Branched Nylon 3

Typically, an initial heating thermogram of linear Nylon 3 shows a single melting exotherm at about 350°C (Shalaby and Turi, 1973). However, heating of branched Nylon 3 was shown to increase its crystallinity through two successive crystallizations that were associated with two exotherms at temperatures well below the melting endotherm (Camino and Guaita, 1977). The two exotherms were detected by DTA at 115 and 260°C. The first exotherm was interpreted as a "cold crystallization" and the second as a crystallization occurring when the most imperfect crystalline regions begin to melt.

B. UNDERSTANDING THE EFFECT OF ANNEALING BELOW
T_g ON MECHANICAL PROPERTIES

Polymeric structures that depend on thermal mechanical history can be quantitatively determined from the relative enthalpy of the system. It was demonstrated earlier that enthalpy decreases on annealing below the T_g and increases under tensile stress (Matsuoka *et al.*, 1973, 1974). An increase in enthalpy reduces the material's relaxation time and thereby enables a local brittle-to-ductile transition to be incuded by stress concentration. This transition is more difficult to induce in well-annealed materials. Studies made by a few investigators on the effect of excess enthalpy and presumably volume on molecular mobility and on the mode of failure of glassy PET revealed a correlation between the transition from ductile behavior to brittle fracture and the loss of excess enthalpy with annealing below T_g (Petrie, 1974; Mininni *et al.*, 1973). Meanwhile, those property changes were not related to the development of crystallinity nor to enhancement of any secondary loss mechanism, such as the γ-relaxation process.

In some studies on the semicrystalline poly(butylene terephthalate) (PBT), the effect of annealing below T_g on both the polymer thermal behavior and tensile properties was investigated (Bair *et al.*, 1976). A sample of PBT was cooled from the melt at 250°C to 23°C at 40°C min^{-1} and the heat capacity was measured from 0 to 250°C and then plotted against temperature, as shown by the solid line in Fig. 33. From 0 to 40°C, C_p increases smoothly, but near 50°C, a rise in C_p was recorded which is in-

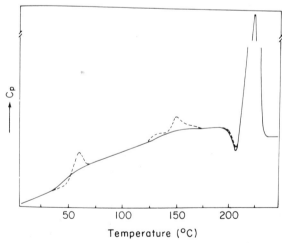

Temperature (°C)

Fig. 33 Comparative specific heat behavior of PBT after quenching from the melt (—) and following 7 weeks annealing at 29°C (---). (From Bair *et al.*, 1976.)

dicative of T_g. Between 180 and 200°C, C_p is depressed because of the heat of crystallization of metastable crystals, formed on rapid cooling. Under these conditions the inception of crystallization was detected at 191°C. On slow cooling (0.31°C min^{-1}), with the sample from the melt followed by remelting, there was no sign of exotherm between 180 and 210°C. When a sample of the polymer was cooled from 250 to 23°C at a 40°C min^{-1} rate and then annealed at 29°C (20°C below T_g and 200°C below T_m) for 7 weeks, the C_p change with temperature was different from that of the unannealed sample, and a noticeable change was recorded near the T_g, as can be observed in the dotted line of Fig. 33. The heat capacity change in the T_g region is given on an expanded scale in Fig. 34. With the exception of the samples that were run immediately after grinding (dotted line in Fig. 34), all other samples showed an absorption of thermal energy that superimposed on the normal increase of C_p at T_g. The magnitude of this endotherm increased with increased annealing period. Since the integral of C_p with respect to temperature is a change in enthalpy, the size of the endotherm is a measure of the enthalpy change from the glassy to the rubbery state. Hence the enthalpy of the sample annealed at 29°C per 1000 h was lowered by 0.32 cal g^{-1} below that of the initial quenched state. In addition, during this annealing process, the T_g shifted from 48 to 55°C. Annealing PBT below T_g not only decreased the excess thermodynamic energy but also led to the development of a few low-melting crystals, as shown by the dotted line about 150°C in Fig. 34.

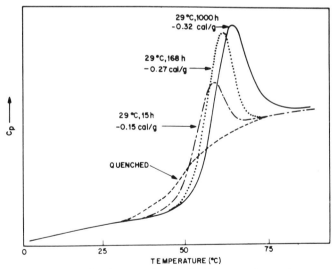

Fig. 34   DSC scans at a heating rate of 40°C min^{-1} for PBT after various annealing periods at 29°C after cooling from 260°C. (From Bair *et al.*, 1976.)

So far, we have dealt with the effect of annealing in reducing the excess thermodynamic energy. The question is, How does this correlate with tensile properties of PBT? It is well demonstrated by Bair *et al.* (1976) that annealing has a noticeable effect on the tensile properties parallel to that recorded for excess enthalpy. For instance, the elongation at yield (EAY) of PBT tensile bars stored at 23°C for 18 months increased by 60% after heating to 100°C in a circulating air oven, followed by quenching (Fig. 35). Heating at a temperature well below the detectable

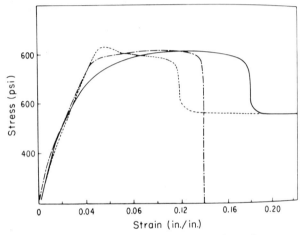

Fig. 35   Stress versus strain for PBT tensile bars (…) 18 months at room temperature; (—) 1 h at 100°C, quenched; (–·–) aged 233 h at 29°C. (From Bair *et al.*, 1976.)

melting range of PBT crystals should have restored the glassy region to its state of highest enthalpy. This is consistent with the fact that when the quenched polymer melt was annealed at 29°C, its EAY decreased progressively, until after 9 d it brittle-fractured (Fig. 35).

Reading the preceding discussion, one may be convinced that annealing below T_g does produce significant changes in molecular mobility and thus influences the polymer thermal behavior and its drawing characteristics in a tensile test.

C. NATURE OF THE SECONDARY LOSS PEAKS IN GLASSY AMORPHOUS POLYMERS

The dependence of modulus and damping of an amorphous polymer on temperature is well known. At T_g the modulus drops by a factor of at least 1000. This transition is accompanied by a high loss peak, with the maximum of tan δ generally lying between 1 and 3; the corresponding peak is usually referred to as the α-peak. In addition to the main transition at T_g, which has a tremendous effect on mechanical properties, there almost always are minor transitions in the glassy regions at which the modulus usually decreases by a factor of 2 or less. These minor or secondary transitions may have a marked effect in the mechanical behavior of polymers; sometimes a material is brittle below and tough above the secondary transitions. When there is more than one secondary transition, they are labeled from higher to lower temperatures with the successive letters β, γ, δ, and so on.

It is usually preferable to plot the tan δ on a logarithmic scale, for the small loss peaks due to secondary transitions are more prominent, their maxima are more distinct, and slopes are more straight than in a linear plot. In a recent review (Heijboer, 1977) the behavior of the secondary maxima in glassy amorphous polymers was argued to be different from that of the glass transition in the following manner:

(1) They obey an Arrhenius relationship over a broad frequency range.

(2) For other than local main-chain motions, the activation energy E_a of the molecular process underlying the loss maximum can be calculated from $E_a = 0.060 \, T_m$ with an accuracy of 10%. For local main-chain motions, E_a is usually slightly higher, say 20%.

(3) For many secondary maxima the origin of E_a is mainly intramolecular.

(4) Plasticization, in most cases, does not affect the temperature location of a secondary peak.

(5) Sufficiently strong intermolecular hindrance usually decreases the height of the loss maximum, but does not increase temperature.

The use of thermomechanical analysis (TMA) to estimate the tensile compliance D for polymeric materials has been reported by Gillen (1978). Values of D for polymers ranging from hard plastics to rubbers have been estimated using a commercial thermomechanical analyzer in a novel manner. The method was described as being capable of quickly and easily estimating values of D^{-1} ranging from 10^6 to 5×10^9 Pa.

D. RECOGNITION OF THE TWO CRYSTALLINE FORMS OF
   NYLON 6

Nylon 6 is known to have two principal, well-marked crystalline forms: the monoclinic α-modification and the hexagonal γ-modification. The melting behavior and the fraction of each of these forms in a Nylon 6 sample depends to a great extent on the thermal history of the sample. The DSC thermograms of samples with α-γ mixed structures showed up to five distinct melting endotherms (Weigel *et al.*, 1974). By variation of the temperature and duration of crystallization, it was possible to relate the peaks to the crystalline structure. Peak 1 occurs (at about 220°C) in almost all samples and is caused by the melting of the monoclinic fraction. Peak 2 occurs (at about 215°C) only with samples containing the hexagonal modification. Peak 3 appears about 30°C above T_c and is associated with the annealing of the γ-form. Peak 4 is observed at about 35°C above T_c and is related to the α-form in samples crystallized between 170 and 190°C. Peak 5 occurs in all samples and appears about 18–20°C above T_c when crystallization is conducted at low temperature. Thus peaks 2 and 3 are related to the hexagonal fraction, peaks 1 and 4 are related to the monoclinic form, but peak 5 is independent of the crystalline structure.

E. Cause of Superheating in Semicrystalline Polymers

When the heating rate of crystals exceeds their melting rate, resulting in an erroneously high T_m, this phenomenon is referred to as superheating. For recognition of superheating, it is to be realized that the recording of a high T_m must not be due to heat transport lags, which leads only to erroneous temperature measurement. The actual temperature of the substance must be above the normal melting temperature for true superheating to occur. At very high molecular weights, it was proposed that the entanglements of the chain in the amorphous phase slow the melting rate and cause an increase in the T_m characterizing the crystalline fraction, and this was related to superheating. Over the last few years, thermal analysis has been used for investigating the phenomenon of superheating in semicrystalline polymers (Hellmuth *et al.*, 1966; Kamide and Yamaguchi, 1972; Chansy *et al.*, 1974; and Chapter 2 of this book).

Superheating in a high DP polymethylene having an average molecular weight of $1–9 \times 10^5$ and 80–90% crystallinity was investigated by Mucha (1974). This was prepared by the catalyzed decomposition of diazomethane. Fiberlike polymethylene samples were shown to exhibit superheating on thermal analysis. The amount of superheating was shown to depend on DP, as judged by the melting peak temperature change as a function of heating rates, less than $5–100°C$ min^{-1}. High melting temperatures caused by superheating were linked to tie molecules between the crystalline regions in the polymer. This was consistent with the results of etching experiments. Thus it was demonstrated that breaking the ties by oxidation, during etching of amorphous phase, causes the disappearance of the high T_m values associated with superheating.

F. ASSESSMENT OF THE EFFECT OF END GROUPS ON THE CRYSTALLINITY AND FUSION OF POLY(ETHYLENE OXIDE) (PEO)

In extended-chain crystals the end groups of PEO chains are concentrated in the surface layers of the lamellar crystals. It has been argued (Kovacs and Gonthier, 1972) that the formation of hydroxyl–hydroxyl hydrogen bonds in the surface layers of the crystalline lamellae is an important factor in determining the mode of crystallization and stability of the crystals. However, melting temperature data of well-characterized fractions of PEO having hydroxy, methoxy, ethoxy, acetoxy, phenoxy, and chloro groups were presented in a recent report as evidence against the unique effect of the hydroxy end groups in PEO (Fraser *et al.*, 1977). Based on the data presented in Table XXVI, it was concluded that the stability of crystalline PEO relative to its melt can be substantially affected by the nature of the end groups and that the stabilizing effect of the hydroxy end group is not unique.

TABLE XXVI

DSC MELTING TEMPERATURES (°C) OF POLY(ETHYLENE OXIDE) FRACTIONS[a]

	End group					
$\overline{M}_n$	—OH	—OCH$_3$	—O—C$_2$H$_5$	—OCOCH$_3$	O—C$_6$H$_5$[b]	Cl—[b]
1000 ± 100	38.4	37.4	—	—	31	30
1500 ± 100	47.8	47.5	47.9	45.1	43	43.6
2000 ± 100	53.6	53.2	53.8	51.9		

[a] From Fraser *et al.* (1977).
[b] Measured by polarizing microscopy to ±1°C.

G. EFFECT OF MOLECULAR WEIGHT ON CRYSTALLIZATION RATE AND CRYSTALLINITY IN HIGH -MOLECULAR-WEIGHT POLY(ETHYLENE OXIDE) (PEO)

Dilatometric crystallization isotherms have been determined for a set of PEO fractions ranging in molecular weight from 2×10^4 to 1.6×10^6 (Maclaine and Booth, 1975). For a given fraction the isotherms obtained for different crystallization temperatures could be superimposed over most of the crystallization. For a given crystallization temperature, the degree of crystallinity obtained in the primary stage of crystallization varied greatly with molecular weight, and superimposition of the isotherms was not possible. This was associated with the possibility that the nature of the spherulite initiation process varies somewhat within the range of molecular weights investigated. Secondary crystallization processes were described as pronounced when the molecular weight $(\overline{M}_v)$ exceeds 10^5.

H. DSC for Predicting the Spinning Performance of PET

Polymer pellets manufactured under supposedly identical conditions were reported to give yarns of either "good" or "poor" spinning performance (Turi and Sibilia, 1978). The "poor" spinning performance was characterized by frequent filament breakage during melt-spinning and subsequent yarn processing. Differential scanning calorimetry was used successfully to identify differences between the two PET batches of resins. The polymer with "good" spinning performance was shown to have short and broad crystallization exotherms and a cold crystallization temperature, $T_{c_{(c)}}$ at 160–165°C. On the other hand, the polymer with "bad" spinning performance was characterized by a long and narrow crystallization exotherm and its cold crystallization temperature $[T_{c_{(c)}}]$ was recorded at 174–185°C.

I. Variation of Nylon 6 T_g with Its Water Content by Dilatometry

On examining the effect of water content on the properties of Nylon 6, one can find that the T_g of Nylon varied considerably. This led Kettle (1977) to measure the T_g of carefully prepared samples of Nylon 6 with variable water contents. The measurement of the T_g values of the samples were made using a dilatometric technique. The results obtained are summarized in Table XXVII, in which the environment RH and water content for each sample are listed. From these data, a graph (Fig. 36) was constructed to show the effect of water on T_g. This graph did not show the expected steady decrease in T_g with increasing water content but levels off between 2.5 and 5.0% (Wt. %) water content. This was in agreement with the findings of Kaimin et al. (1975), who ascribed this fall in T_g of Nylon 6 to the plasticizing effect of water. However, Fox (1956) has defined a plasticizer in terms of the following equation:

$$1/T_g^s = (W/T_g) + (W^p/T_g^p) \qquad (14)$$

where W and W^p are the weight fraction of pure polymer and pure plasticizer, respectively, and T_g, T_g^p, and T_g^s are the glass transition temperatures of the pure polymer, pure plasticizer, and plasticized polymer, respectively.

If the Fox equation applied, a plot of $1/T_g^s$ against W^p would be a straight line. However, when a plot of $1/T_g^s$ against the water content was constructed (Fig. 37), it was found to be nonlinear. Therefore, it is ap-

TABLE XXVII

Effect of Environment on the
Equilibrium Water Contents
and T_g of Nylon 6[a]

RH (%)	Water content (%)	$T_g \pm 1$ (°C)
[b]	0.35	94
[c]	0.70	84
12	1.17	71
33	1.99	56
44	2.70	45
55	3.47	43
66	4.45	40
86	6.61	23
97	10.33	6

[a] From Kettle (1977).
[b] Dried over silica gel.
[c] Tested as received.

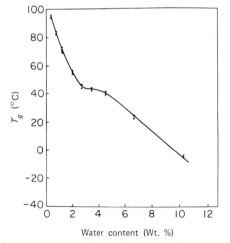

Fig. 36 Effect of water content on the T_g of Nylon 6. (From Kettle, 1977, by permission of the publishers, IPC Business Press Ltd., ©.)

parent that the water is not behaving just as a plasticizer in Nylon 6. This more complex situation tends to support the mechanism proposed by Puffer and Sebenda (1976) and Kawasaki and Sekita (1964) for the inter-action of water with Nylon 6.

It is interesting to note that by extrapolating the initial linear portion of the T_g/water content, a value of 100 ± 2°C was obtained for T_g of dry Nylon 6 granules.

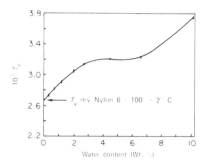

Fig. 37 Fox plot for Nylon 6. (From Kettle, 1977, by permission of the publishers, IPC Business Press, Ltd., ©.)

J. Pertinence of Chain Conformation to Thermal Stability of Nylon 1 Polymers and Copolymers

Organic isocyanates can be converted, with suitable catalysts, to an N-substituted polyamide of the Nylon 1 type as

$$n\,N{=}C{=}O \longrightarrow \left(\!\!\begin{array}{c} O \\ \| \\ N{-}C \\ | \\ R \end{array}\!\!\right)_{\!n}$$

The structure of Nylon 1 was shown to be associated with a highly extended chain conformation in solution, as described by Berger (1973) and Bur and Fetters (1976). Hence, Nylon 1 molecules are also highly extended with a nearly rodlike conformation in the solid state. This particular property, as will be seen later in this section, has a direct effect on the thermal stability of Nylon 1 type polymers.

Generally, amorphous polymers in the glassy state are regarded as being frozen into one particular assembly of all possible chain conformations. On heating above T_g, all conformations of the molecules become accessible and interchangeable and this leads to the many well-known characteristics of glass transition and the phenomenon of rubber elasticity. However, if the polymer molecules have a nearly rodlike conformation, there may be no possibility of substantial changes in conformational state above T_g and some features of the glass transition could be affected. In an effort to check this possibility, Owadh *et al.* (1978) prepared polymers and copolymers of alkyl isocyanates, and their properties, including their thermal behavior, were investigated. All the homopolymers were shown to have a significant decomposition rate at 150°C, but the copolymers of butyl isocyanate and ethyl isocyanate were somewhat more stable, thermally, than the butyl homopolymer. The copolymers generally showed lower crystallinities than the homopolymers. Dynamic mechanical measurements gave damping peaks with a drop in the modulus in the temperature range of 50–150°C, but the peaks were broader and the decline in modulus was noted to be less steep than for conventional vinyl-type polymers. Over the same temperature range, the DSC thermograms were nearly linear and showed no observable discontinuity. In mechanical tests, both modulus and yield stress correlated with the NCO–hydrocarbon ratio. It is interesting to note that the stress–strain curve in tension was similar to those of cellulose derivatives known to have molecules with linear conformation; they did not show a maximum and exhibited uniform extension. The highly linear conformation of Nylon 1 polymers was suggested to be responsible for its stress–strain properties. This conclusion would be in accordance with any model in which yielding and strain hardening were associated with the straightening of molecules in a random configuration as in rubber elasticity (Thackray and Haward, 1968).

K. NATURE OF THREE THERMAL TRANSITIONS IN AN ALTERNATING COPOLYMER OF ETHYLENE AND CHLOROTRIFLUOROETHYLENE

Three transitions have been reported to exist in a 1:1 alternating copolymer (PE-CTFE) of ethylene and chlorotrifluoroethylene (Sibilia *et al.*, 1976). Mechanical relaxation techniques have been used for recording these transitions on a quick-quenched sample. These transitions were recorded with peaks at 130, 88, and −65°C and were assigned the symbols α, β, and γ, respectively. The intensity of the α-transition was also recorded in the mechanical relaxation spectra of an annealed sample. X-ray diffraction and dichroic infrared measurements were conducted on oriented specimens as a function of temperature. The results showed that the α-transition was accompanied by a change of lateral binding distances in the crystalline phase and a conformational change attributed to an ''unkinking'' of the molecular chain. The β-transition was found to be related to the amorphous phase and the onset of the β-peak corresponded to the transition at 35°C found by infrared techniques and previously by a torsional modulus method. Hence, the β-transition was assigned to the glass transition. A more definitive assignment of the β-transition was contingent on a detailed structural analysis of the γ-transition, which was not explored to any great extent by Sibilia *et al.*(1976).

L. T_g OF AMORPHOUS AND SEMICRYSTALLINE POLYETHYLENE (PE)

Polyethylene is one of the most well-studied polymers. It has been used as a model system and reference frame whenever scientists investigate the properties of new and conventional polymers. However, there is still no consensus on the precise temperature of the glass transition of PE, both linear and branched. This was attributed in part to the inability to obtain linear PE in an amorphous state (Simon *et al.*, 1975). As mentioned earlier in this chapter, the T_g of PE was determined indirectly from linear thermal expansion data (Stehling and Mandelkern, 1969) or through sensitive calorimetric measurements of the kinetic thermal behavior associated with the tendency of the glass to relax to the liquid state in the T_g region (Chang, 1972).

A few years ago, Hendra and his co-workers (1975) announced the significant achievement of reaching an amorphous state of linear PE. Their technique was to quench from the melt a thin film of polymer between two aluminum foils. Infrared spectra near 720 cm^{-1} indicated the absence of crystallinity on quenching and were used to monitor its reappearance on heating. They noted the reappearance of split absorption bands near 730 cm^{-1}, on heating the quenched film to 180 K and above. Since recrystallization is known to occur only about 20–30 K above the glass transition temperature, they concluded that T_g must be less than 180 K. However, this conclusion was questioned in the following year (Boyer, 1976) for the following reasons, among others:

(a) Many lines of evidence pointed to an extrapolated T_g for linear PE T_g of 195 ± 5 K (Boyer, 1973b).

(b) The specimen employed had a broad molecular-weight distribution and the low-molecular-weight fraction could effectively lower T_g over that reported by Stehling and Mendelkern (1969).

(c) Three factors might favor recrystallization below nominal T_g: (i) local structure present in the melt and frozen into the quenched specimen, (ii) high degree of supercooling, and (iii) molecular motion in the glass at secondary relaxation.

These and related arguments led Boyer and Snyder (1977) to reexamine the IR spectra reported by Hendra *et al.* (1975) and the conclusion that the T_g of quenched linear PE is not equal to 180 K as reported by the latter investigators. The IR spectra of the quenched sample was noted to differ from that of the melt. Whereas the melt appeared truly amorphous, the quenched solid showed evidence of local structure, partly in the form of trans sequences. Hence, recrystallization is a process of going from a somewhat ordered solid to a semicrystalline solid. These arguments, along with some new ESR data, were used to substantiate earlier conclusions that T_g of amorphous PE is about 193 K. Boyer and Snyder then emphasized that IR results of Hendra *et al.* are compatible with a T_g around 195 K and do not require the existence of a T_g below 180 K. It was also suggested that recrystallization in the glassy state is facilitated by order in the quenched specimen and/or the strong γ-relaxation around 148 K.

In a slightly different approach from those just discussed, molecular relaxations in partially hydrogenated *cis*-1,4-polybutadienes were used as a guide to T_g of amorphous PE (Cowie and McEwen, 1977). Thus a series was prepared of partially hydrogenated polybutadienes that were regarded as random copolymers of ethylene and acetylene. The T_g of amorphous PE was estimated by plotting T_gs of the copolymers, derived from DSC measurements as a function of composition and extrapolated to 100% ethylene content. A value in the range 190–200 K was obtained using two extrapolation methods. The dynamic mechanical spectra, using torsional braid analysis (TBA), exhibited a number of damping maxima. These results were used to suggest that the γ-relaxation in PE is not a T_g but is more likely to result from a crankshaft motion of the short methylene sequences in the chain.

M. Thermal Degradation of Head-to-Head (H–H) and Head-to-Tail (H–T) Polymers

Undoubtedly, polystyrene (PS) is one of the most important commercial polymers and its thermal degradation has been the subject of many early studies. It had been observed that the molecular weight of polystyrene decreases rapidly in the initial stages of thermal degradation and some investigators have explained this behavior by suggesting "weak links" to various degrees in the polymer (Cameron and Grassie, 1961). It was then proposed that some H–H linkages exist in the normally obtained, radically initiated styrene polymer and that the initial degradation pattern may be related to this type of link. In later investigations, styrene–stilbene copolymers were prepared and examined as model systems of commercial PS with variable amounts of H–H linkages (Cameron and Grassie, 1962). These systems were used since pure H–H polystyrene was not available at that time. It was, however, found that the presence of stilbene units in those stilbene–styrene copolymers did not affect significantly the thermal degradation behavior as compared with normal radically polymerized styrene polymer.

In 1977, Inoue *et al.* reported the successful formation of a high-molecular-weight pure H–H PS. Subsequent investigations of the thermal degradation behavior of H–H and H–T polystyrene by DTG and TG showed no significant differences in their thermal stability, and it was concluded that the vinyl end groups obtained from the radical disproportionation reaction were the weak links of radically polymerized polystyrene (Strazielle *et al.*, 1978). In more recent studies, Lüderwald and Vogl (1979) examined the thermal behavior of H–H and H–T polystyrenes and polyvinylcyclohexanes by direct pyrolysis in a mass spectrometer. Both H–H and H–T isomers show only small differences in their initial temperatures of decomposition but remarkably different degradation processes. Whereas H–T polystyrene decomposes in accordance with earlier investigations mainly by a radical depolymerization into the monomer and yields only a small amount of dimer and trimer, the H–H polystyrene shows no unzipping and only a statistic degradation into oligomeric styrenes. The formation of stilbene is a diagnostic reaction of H–H polystyrene. The pyrolysis-mass spectra of H–H and H–T polyvinylcyclohexanes are more similar than those of the polystyrenes but the favored thermal cleavage between the two tertiary carbon atoms or next to the tertiary carbon atoms of the polymer backbone chain leads to some characteristic fragments, e.g., 1,3-dicyclohexylpropene from the H–T and 1,2-dicyclohexylethane from the H–H polymer.

N. THERMAL DEGRADATION OF POLY(BUTYLENE TEREPHTHALATE) (PBT)

During melt processing of polyesters, including PBT, high melt temperatures and moisture can result in polymer degradation. Although investigation of the initial weight loss behavior of PBT was reported a few years ago (Passalacqua *et al.*, 1976), it is only recently that the volatile products of thermal degradation have been characterized (Lum, 1979). Laser microprobe analysis and dynamic mass spectrometric techniques were used to identify the primary volatile degradation products and initial pyrolysis reactions that control polymer degradation. A complex multistage decomposition mechanism was proposed that involves two major reaction pathways. Initial degradation occurs by an ionic decomposition process that results in the evolution of tetrahydrofuran. This is followed by concerted ester pyrolysis reactions that involve an intermediate cyclic transition state and yield 1,3-butadiene. Simultaneous decarboxylation reactions occur in both decomposition regimes. Finally, the latter stages of polymer decomposition were characterized by evolution of CO and complex aromatic species such as toluene, benzoic acid, and terephthalic acid. Activation energies of formation for the main pyrolysis products were determined from the dynamic measurements of the major ion species and indicate values of $E = 27.9$ kcal mole^{-1} for the production of tetrahydrofuran and $E = 49.7$ kcal mole^{-1} for the production of butadiene.

O. THERMAL DEGRADATION OF POLY-α-ESTERS

The influence of substituents (R_1 and R_2) on the degradation of poly-α-esters, $+O—CR_1R_2—CO+_n$ has been studied by several investigators (Cooper *et al.*, 1973; Lovett *et al.*, 1973; Patterson *et al.*, 1973; Sutton *et al.*, 1973; Sutton and Tighe, 1973). It has been shown by these authors that in the case of the simplest poly-α-esters ($R_1 = R_2 = $ —H or —CH_3) the thermal decomposition in the temperature range 250–400°C is best interpreted in terms of an intramolecular ester interchange process, resulting mainly in the production of the relevant glycolide and substituted glycolides. These reactions are characterized by first-order kinetics, with respect to both formation of volatile material and decrease in molecular weight. Furthermore, in each case a large and negative entropy of activation is found with a low value for both the frequency factor and activation energy. The magnitudes of these parameters are indicative of a degradation process involving a fairly high degree of steric order in the transition state of the rate-controlling step.

Norton *et al.* (1978) has used TGA and pyrolysis GC to study the thermal degradation of poly(cyclopentylidene carboxylate) and poly(cyclohexylidene carboxylate) over the temperature range 200–500°C by using a combination of kinetic and analytical techniques. The results obtained indicate the simultaneous operation of mechanisms previously observed with different compounds in the series. Thus kinetic studies by thermogravimetric analysis and characterization of the products formed suggest the occurrence of an essentially first-order hydrogen abstraction process with superimposed random scission reactions in the early stages. The main products are cyclopentene-1-carboxylic acid and cyclohexene-1-carboxylic acid, respectively. This corresponds to the behavior previously observed in the degradation of poly-3-pentylidene carboxylate. On the other hand, in-line pyrolysis directly linked to a gas chromatograph clearly indicates the production of cyclopentanone and cyclohexanone and, more importantly, in the case of poly(cyclopentylidene carboxylate), the formation of dicyclopentyl glycolide. The appearance of these products in addition to the unsaturated acids suggests that some degradation occurs by the intramolecular ester interchange process previously postulated for poly(methylene carboxylate) and poly(isopropylidene carboxylate). The associated thermodynamic parameters and relative quantities of the different products observed by gas-chromatographic methods indicate that the hydrogen abstraction process predominates.

P. THE BRILL TRANSITION IN NYLON 66

The Brill transition in Nylon 66 represents a change from a triclinic unit cell to a pseudohexagonal form. This transition was reported to take place between 100 and 240°C, and appears to be sensitive to crystallization, annealing, and deformation conditions (Itoh, 1976). Wilhoit and Dole (1953) reported earlier that the Brill transition is associated with a small latent heat, but the thermodynamic properties of this transition have not been well defined. In recent studies by Starkweather and Jones (1981), powders of Nylon 66 crystallized from methanol were shown to exhibit a latent heat of about 4.5 cal g^{-1} at the Brill transition near 200°C, where the unit cell changes from a triclinic to a pseudohexagonal form. The dimensions of the hydrogen-bonded sheets remain almost unchanged up to 240°C, but the separation between the sheets increases with increasing temperature. Both DSC and x-ray techniques were used in these studies.

VII. Special Topics

The last section of this chapter deals with topics that are of general interest, relatively new, or controversial. Parts of this section may be looked on as introductions to detailed reviews of the treated topics a few years from now.

A. ORDER IN THE AMORPHOUS PHASE

It is well recognized that the morphology of the amorphous phase is one of the most important factors influencing the physical properties of a polymeric material, and the presence of "order" in the amorphous phase will definitely affect its morphology. The controversial issue on the presence of order in the amorphous phase has attracted the attention of many investigators in recent years (Tonelli, 1971). One of the most investigated polymers is poly(ethylene terephthalate) (PET), which will be treated in this section as a model case. Furthermore, "unconventional" thermal transitions recorded for PET will also be discussed.

Poly(ethylene terephthalate) exhibits a number of transition temperatures at which property changes have been shown to occur. It has been noted earlier that the T_g for amorphous PET occurs at 70°C by DTA (Ke, 1964). Changes in loss modulus were recorded at −65 and 70°C (Illers and Breuer, 1963). Thermal mechanical results have shown changes in tension and expansion at −130, −62, and 77°C (Miller, 1969). For samples of PET drawn at 60 and 85°C, it was concluded that at both temperatures molecules are held by tie points from ball-like structures, and some alignment was caused on drawing (Koenig and Mele, 1969). The effect of orientation on the transition near −130°C has been demonstrated for amorphous and drawn films, showing that alignment or orientation influences the amount of elongation at this temperature; the transition temperature remained independent of the drawing of the film (Miller, 1969). The transitions at −130 and −62°C have not been assigned conclusive molecular origin.

It has been reported that the amorphous phase of PET contains no crystallinity that is detectable by x-rays (Lawton and Cates, 1969; Illers and Breuer, 1963). However, molecular order, which is not measurable by x-rays, was suggested to be present in PET amorphous phase at ambient temperatures (Stuart, 1967). Additional evidence for the presence of some order in the amorphous phase was contained in a detailed report by Yeh and Geil (1967), indicating that at ambient temperatures PET has ball-like paracrystalline structures of 75Å diameter, which merge to fibrillar ribbons when heated above 70°C.

The presence of some order in the PET amorphous phase was associated with the unusually low difference (16°C) between the lowest T_c (86°C) and T_g (70°C) (Miller, 1975). It has been shown that some x-ray crystallinity would be detected only after periods of annealing at 86°C, which is adopted as the minimum T_c. The unusually small value of T_c-T_g of PET can be appreciated as one reviews the T_c and T_g values of the other thermoplastics shown in Table XXVIII.

TABLE XXVIII

T_c-T_g VALUES OF TYPICAL THERMOPLASTICS

Polymer	T_c (°C)	T_g (°C)	T_g-T_c, (°C)	Reference
Polyethylene	−50	−125	75	Stehling and Mandelkern, 1970
Polycarbonate	−50	−88	38	Boyer and Snyder, 1977
	180	−137	317	Miller, 1971
				Kampf, 1962
PET	86	70	16	Miller, 1975

In their studies of the morphology and physical properties of polycarbonates, Neki and Geil (1973) reported that the changes in morphology and physical properties caused by annealing below T_g are, in general, closely related. They also noted that this relationship cannot be explained only by changes in free volume, and proposed that changes in the degree and type of order (nodular structure) also play a role.

In dilatometric TMA measurements of PET near 70°C, an abrupt change was recorded and proposed to indicate that the transition may be first order (Miller, 1975). The dilatometric results for three forms of PET are illustrated in Fig. 38. Forms I and II are two highly crystalline forms (Bell and Dumbleton, 1969); form III is an amorphous cast film. Figure 38 shows the volume expansivity of the amorphous form of PET, also two transitions at 69 and 78°C, followed by the volume shrinkage accompanying the recrystallization of the amorphous form at 127°C, with a final fusion noted at 256°C. The onset of the exotherm by DTA began near 127°C. Form I showed the occurrence of two first-order phenomena at 72 and 82°C, followed by volume shrinkage at 124°C and final fusion at 256°C. Form II showed no volumetric change in the 70–80°C transition interval

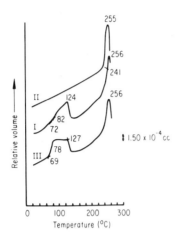

Fig. 38 Dilatometric response of PET. (From Miller, 1975.)

but did illustrate the first-order fusion beginning at 241°C and ceasing at 255°C.

The DSC and DTA data have shown the normal heat capacity change associated earlier with T_g at 70°C, whereas annealing for 30 min at 110°C induced further molecular order, as shown in Fig. 39. Rapid quenching of

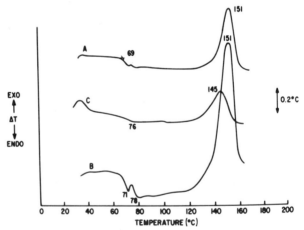

Fig. 39 Differential thermal response of amorphous PET (10°C min⁻¹). A. DSC; B. DTA; C. DTA after 30 min at 110°C. (From Miller, 1975.)

the amorphous PET after heating at 120°C yielded an amorphous sample, and measurement by depolarized light transmission as the sample was re-heated revealed a rapid crystallization occurring near 80°C (Fig. 40). The shrinkage detected by tensile measurements near 68°C and the apparent

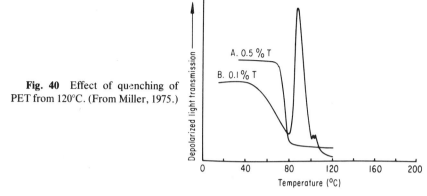

Fig. 40   Effect of quenching of
PET from 120°C. (From Miller, 1975.)

recrystallization at slow heating rates based on depolarized light analyses
were considered as supporting evidence for the proposed first-order character of the phase transition at about 70°C. Isothermal annealing for short
periods at temperatures between 45 and 80°C led to further increase in
molecular order, as detected by thermal depolarization analysis (TDA)
(Fig. 41).

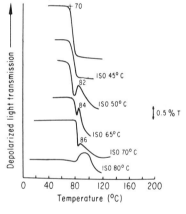

Fig. 41   Effect of annealing from 45 to
80°C by TDA. (From Miller, 1975.)

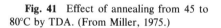

In the course of a study on amorphous multifilaments of PET with a
varying degree of preorientation and variable annealing time at room temperature (during storage), Hagege (1977) was able to report evidence,
especially for oriented polymer, to supplement in part Miller's proposal
(1975) that the transition near 70°C of amorphous PET film is of a "crystalline character." The peak near 70°C was given the name "mobile peak of
the third kind" by Hagege (1977), without excluding the possibility that
this endotherm is related to water desorption phenomenon. Hagege (1977)
showed that the endothermic peak for preoriented PET yarn could be re-

corded at 59, 61, 64, and 69°C upon storage for 3, 4, and 6 h, and 2 months at room temperature. For a sample annealed for 13 d under standard conditions and then for 36 h at 40°C, the endothermic peak was recorded at 76°C. After considering Miller's data (1969) as well as his own results and those of others, Hagege proposed that this endothermic peak is due to a fusion of a mesomorphic phase—not a truly crystalline one—as reported by Miller for a nonfibrous "amorphous" PET probably of a nematic order. This fusion was considered a prerequisite for the onset of thermal motion of the molecular segments (i.e., the properly termed glassy transition). In addition to being of fundamental interest, it was noted by Hagege that this peak might be of some practical value in characterizing preoriented PET yarn instead of the exothermic crystallization peak which is also well known to be orientation and heating rate dependent (Maclean, 1974; Quynn, 1972).

The effect of annealing on the thermal behavior of PET, as examined by DSC, was studied by several investigators (Roberts, 1969; Lawton and Cates, 1969; Holdsworth and Turner-Jones, 1971; Roberts, 1970; Miyagi and Wunderlich, 1972; Sweet and Bell, 1972). These studies have been conducted on oriented and unoriented samples. It was concluded that for PET the double melting endotherms are caused by crystallites of different sizes and/or perfection. In a recent report (Oswald *et al.*, 1977) a middle endothermic peak (MEP), in an intermediate range well below the PET melting endotherm, was reported to develop when drawn PET fibers were thermally treated. The DSC thermograms in Fig. 42 illustrate the development of an endothermic peak between 130 and 234°C upon treating the PET fibers thermally. The shape and position of the MEP, depend on the temperature and duration of the thermal treatment. The reported thermal, x-ray, and tensile data suggested strongly that the MEP is the result of melting of very small crystallites or perhaps nuclei, located in the highly extended anisotropic amorphous phase of the oriented fibers. These studies also indicated that the temperature of the MEP increases with the increase in the size of nuclei or crystallites. The energy associated with MEP was reported to reach a value of 2 cal g^{-1} compared with 17 cal g^{-1} for the major melting endotherm. The "small crystallites" or nuclei were proposed to form in highly strained, anisotropic amorphous fraction of the fibers as a result of two opposing and competitive processes; one operates under the driving force to increase the entropy of the system and the second one is governed by the tendency to decrease the enthalpy. The formation of nuclei was suggested to prevent the complete disorientation of the amorphous phase and control the amount of shrinkage the yarn will undergo at temperatures above the T_g.

Using the preceding data for PET and other literature data pertinent to

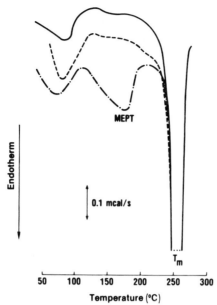

Fig. 42 DSC thermograms for PET yarn: (—) control, as is; (--) submerged in silicone oil at room temperature (RT); (––) heat-treated in silicone oil at 175°C for 15 s (under zero load). (Reprinted from Oswald *et al.*, 1977, p. 231, by courtesy of Marcel Dekker, Inc.)

the orientation of the amorphous, a structural model was proposed in an attempt to explain the relevance of the amorphous phase morphology to physical and thermal behavior of polymers (Harget and Oswald, 1978).

B. CRYSTALLIZATION AND MELTING UNDER PRESSURE

1. Polyethylene

Using high-pressure DTA and dilatometry, an additional transition for polyethylene was observed. This transition takes place, reversibly and at relatively high temperature during cooling and heating, under elevated pressures, above 4000 kg cm^{-2}. Much attention is being devoted to this phenomenon, as it relates closely to the so-called extended-chain crystals (ECC) of polyethylene. From some optical and x-ray observations (Bassett *et al.*, 1974; Yasuniwa and Takermura, 1974) of polyethylene under high pressures and temperatures, a new crystalline form, different from the usual orthorhombic form, has been recognized in a small temperature region just below the T_m. It was determined that the new form is most probably of hexagonal structure, with the molecules in a disordered helical conformation (Bassett *et al.*, 1974). It was contended further that the chain-extended growth is the result of crystallization from the melt via the hexagonal phase, whereas chain-folded growth is the familiar melt crystallization into the orthorhombic phase. On the other hand, the additional transition just discussed was later believed to represent the crystallization and melting of two kinds of ECC crystals with different thermal stabilities (Maeda and Kanetsuna, 1974). These crystals were reported to form during the first and second stages of volume shrinkage from the melt under elevated pressure and were designated as highly ECC and ordinary ECC, respectively. The highly ECC was suggested to correspond partially to the hexagonal, high-pressure case determined by Bassett *et al.* (1974). The coexistence of ordinary ECC and highly ECC in the pressurized product conflicts with the generally accepted concept of the morphological and thermal properties of the so-called ECC, in which the T_m is at about 140°C and consists of extended-chain lamellae, having an average thickness of about 2000–3000 Å, and in which the lamellar size distribution of ECC in the c-axis direction varies from 1000 Å and several micrometers (Prime and Wunderlich, 1969; Prime *et al.*, 1969; Rees and Bassett, 1971).

In an effort to confirm the existence of ordinary ECC and highly ECC in polyethylene under atmospheric and elevated pressures, a thorough investigation of the effect of the crystallization process and crystallization conditions on the formation of ECC was recently completed (Maeda and Kanetsuna, 1974, 1975, 1976). Thus the thermal behavior of extended-chain crystals of polyethylene formed during various crystallization processes and conditions under about 5000 kg cm^{-2} were studied by high-pressure dilatometry and DSC. Results of these experiments indicate that by isothermal crystallization at small undercoolings for prolonged periods, the products show two endothermic peaks in the melting region of the usual extended-chain crystals. This meant the presence of bimodal lamellar thickness distribution in ECC. A phase diagram has been made for pressures up to 5000 kg cm^{-2}. The experimental results confirm the existence of two kinds of ECC, i.e., ordinary extended-chain and highly extended-chain crystals. Typical illustrations of the effect of crystallization conditions on the thermal transitions are shown in Fig. 43.

The effects of molecular weight on high-pressure crystallization of linear polyethylene and the properties of the resulting specimens were reported by Hoehn *et al.* (1978) and Ferguson and Hoehn (1978).

The high-pressure crystallized specimens were characterized by x-ray diffraction, density, and differential scanning calorimetry. Nitric acid etching followed by gel permeation chromatography and DSC on the residues provided further characterization of the morphology. The crystallinity decreased from 100 to 80% with increasing molecular weight, over the range $M_v = 4.9 \times 10^4$ to 4.6×10^6. The crystallites were predominately in ex-

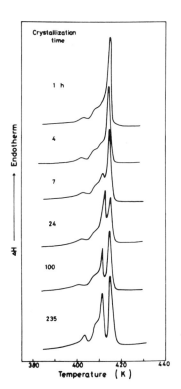

Fig. 43 DSC curves of melting of the unfractionated samples crystallized through process A, isothermally at 234°C and 5130 kg cm⁻² for indicated times. Melting curves at a rate of 5°C min⁻¹ were not corrected for instrumental lag. (From Maeda and Kanetsuna, 1976.)

tended chain morphology, but the percentage in extended-chain form decreased from 98 to 85% over the indicated molecular-weight range. In all cases, there was a wide, bimodal or multimodal, distribution of lamellar thickness. The physical characterization results were consistent with and complementary to the results from electron microscopy. The large differences in crystallinity and gross morphology readily account for observed molecular-weight-dependent differences in mechanical properties.

2. Nylon 6

Studies on the crystallization of Nylon 6 from the melt under elevated pressure indicated that crystallization induced by pressures up to 8 kbar at temperatures between 270 and 310°C did not lead to a significant increase in T_m of a polymer sample containing 8% caprolactam (Gogolewski and Pennings, 1973, 1975). However, the melting peak temperature increased from 220 to 250°C for Nylon 6 free of caprolactam, which was crystallized under pressure exceeding 5 kbar for 50 h. Meanwhile, the heat of fusion of a Nylon 6 specimen crystallized under these conditions increased from 14 to 37 cal g⁻¹. It was reported recently that pressure-

crystallized Nylon 6 fractures very easily, pointing to a considerable reduction of intercrystalline links and tie molecules (Gogolewski and Pennings, 1977).

3. Nylon 12

The influence of pressure on the crystallization and annealing of Nylon
12 has been investigated recently (Stamhuis and Pennings, 1977). The increase of the final melting temperature of this polymer with pressure
amounted to 20°C kbar^{-1} as determined by high-pressure dilatometry,
which exceeds the increase in T_m noted for Nylon 6. Thus the crystallization of Nylon 12 is more sensitive to pressure than that of Nylon 6. Nylon
12 samples crystallized at a pressure of 4.9 kbar displayed multiple melting behavior, whereas annealing under pressure gave rise to one melting
peak as shown by DSC. Thermograms of Nylon 12 samples crystallized
under different conditions are shown in Fig. 44. The heat of fusion could
be enhanced from 16 to 32 cal g^{-1} and the melting peak temperature could
be increased from 179 to 209°C by annealing under 4.9 kbar and 260°C for
336 h.

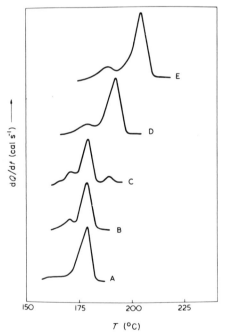

Fig. 44 DSC melting thermograms for various Nylon 12 samples: A. original sample; B.
sample crystallized at 260°C and 4.9 kbar for 16 h; C. sample crystallized at 250°C and 4.9
kbar for 16 h; D. sample crystallized at 240°C and 4.9 kbar for 16 h; E. sample crystallized at
240°C and 4.9 kbar for 48 h. (From Stamhuis and Pennings, 1977, by permission of the publishers, IPC Business Press Ltd., ©.)

Crystallization as well as annealing under pressure led to a partial transformation of the pseudohexagonal or monoclinic crystal structure to the alpha modification. The pressure-treated Nylon 12 consisting of well-developed spherulites could be fractured very easily along the interspherulitic and transspherulitic planes.

C. THERMAL PROPERTIES OF HIGH-PRESSURE EXTRUDED POLYMERS

Perhaps the most successful combination of flow, temperature, and pressure factors has been achieved in the Instron capillary rheometer with extrusion procedure devised by Southern and Porter (1970a,b). Polyethylene prepared by their technique surpassed all conventionally processed PE in orientation (Desper *et al.*, 1974), chain extension (Weeks and Porter, 1975), and tensile modulus and strength (Capiati and Porter, 1975). However, the thermal properties of these polymers received relatively minor attention. This section is dedicated to discussion of the thermal properties of high-pressure-extruded thermoplastics.

1. High-Pressure-Extruded Poly(Ethylene Terephthalate) (PET)

The procedure of Southern and Porter was used to prepare PET segments by high-pressure extrusion in an Instron capillary rheometer at temperatures from 245 to 265°C (Griswold and Cuculo, 1978). During the initial stage of extrusion, highly oriented, translucent segments were generated by flow-induced crystallization. The melting behavior of translucent segments prepared at different temperatures was examined by DSC. Typical shapes of the melting endotherms of short-growth (30 min) segments are shown in Fig. 45. It is apparent from these endotherms that the T_m increases with the increase of the extrusion temperature. It was believed that these single-peak endotherms are directly related to the state of translucent segment before the melting experiment. This was consistent with several reports in which the mechanisms involved in the structural reorganization of PET samples during calorimetric measurements were generally clarified (Holdsworth and Turner-Jones, 1971; Sweet and Bell, 1972; Miller, 1974). Usually, PET samples must be annealed at high temperatures for sufficient times before the double endotherms indicative of recrystallization phenomena give way to a single endotherm, which is characteristic of the original crystallization conditions. The single endotherm, which is commonly referred to as a form-II peak, was associated with annealed samples that solely contain larger and more perfect crystals. These are unable to recrystallize after melting during the heating scan (Sweet and Bell, 1972). Similarly, it was believed that the DSC

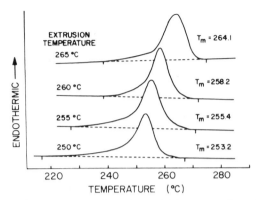

Fig. 45 DSC melting endotherms of translucent PET segments formed during high-pressure extrusion at 250, 255, 260, and 265°C for 30 min. DSC heating rate was 20°C min⁻¹. (From Griswold and Cucula, 1978.)

curves in Fig. 45 are form-II endotherms, because the time, temperature, and pressure conditions of the extrusion experiment should have been sufficient to produce noncrystallizable material (Griswold and Cuculo, 1978). The densities of the translucent segments were consistent with the proposed perfection of the crystalline components. In addition, the high scan rate should have suppressed the recrystallization phenomenon (Sweet and Bell, 1972). At this point it is to be noted that the melting temperatures recorded by Griswold and Cuculo for the translucent PET samples extruded under high pressure are not much higher than that of ordinary PET of 256°C, particularly if the onset of melting is taken as the T_m.

2. Thermal Transitions of Hydrostatically Extruded PMMA and HIPS

There have been very few studies on the thermal properties of hydrostatically extruded poly(methyl methacrylate) (PMMA) and high-impact polystyrene (HIPS). Recent DTA data (Ariyama et al., 1977) on the T_g and T_m of PMMA indicated that (a) the T_g of a sample with 50% area reduction is 10°C lower (91°C) than that of virgin PMMA and samples with other values of reduction in area; (b) the T_m (initial melting) values of hydrostatically extruded samples ranged from 207 to 231°C, which are higher than the 202°C characteristic of the virgin sample; and (c) the T_m (the peaking melting temperature) of all samples, with the exception of the polymer with 70% area reduction, was higher (249 to 252°C) than that of the virgin sample (228°C). A similar effect, that is, a decrease in T_g and an increase in the initial melting temperature, was recorded for hydrostatically extruded high-impact polystyrene; the T_g decreased from 97 to 93°C and the T_m (initial melting) increased from 188 to 207°C.

Karasz *et al.* (1965) reported that isotacticity in PMMA lowers its T_g; T_g for atactic PMMA is 103°C, whereas T_g for isotactic PMMA is 42. On the other hand, Ariyama *et al.* (1977) associated the change in PMMA transition in hydrostatically extruded samples with considerable orientation of the chains in the direction of extrusion. However, if Tonelli's (1971) views on the presence of order in the amorphous phase are considered, the recorded decrease in T_g of PMMA may be related to a difference in cooling rate between the virgin sample and that prepared by hydrostatic extrusion and associated reduction in the number of rotational states that the chains can adopt upon reheating.

D. LIQUID–LIQUID TRANSITION IN AMORPHOUS POLYMERS AT T_{ll}

The existence of a relaxation above the T_g of amorphous polymers is a subject of controversy. Evidence of the existence of such a "transition" and its basic features were reported as early as 1966 and then recently in 1977 (Boyer, 1966; Gillham and Boyer, 1977a,b). The T_{ll} was detectable by thermomechanical techniques (TMA), where experiments above T_g for amorphous polymers were made possible by using supported samples, employing a mixture of low DP polymer in a high DP matrix of the same polymer. Using this approach, the dependency of T_{ll} on molecular weight has been studied for 1,4-polybutadiene (Sidorovich *et al.*, 1971) and the dynamic mechanical experiment (~ 50 Hz) revealed the transition by a loss peak. A second approach depends on the use of a composite specimen consisting of a polymer and an inert glass substrate in a torsional pendulum experiment. This technique, torsional braid analysis (TBA) (Gillham, 1974), was used by several investigators for studying T_{ll} (Stadnicki *et al.*, 1976; Glandt *et al.*, 1976; Gillham *et al.*, 1976; Gillham and Boyer, 1977a,b; Gillham, 1979). A typical illustration for the use of TBA in detecting T_{ll} was contained in a recent review of the T_{ll} of polystyrene and block copolymers (Gillham and Boyer, 1978). For the most part, the T_{ll} relaxation behaves like the glass transition in its dependence on molecular weight, on number-average molecular weight in binary polystyrene blends, and on composition in a polystyrene, homogeneously plasticized throughout the range of composition. Diblock and triblock copolymers were reported to display a $T > T_g$ relaxation above the T_g of the polystyrene phase.

Two results, in particular, were noted (Gillham and Boyer, 1978), to suggest that the T_{ll} relaxation is molecularly based: (1) The temperature T_{ll} is determined by the number-average molecular weight for binary blends of polystyrene when both components have molecular weights below M_c (critical molecular weight for chain entanglement); (2) homopolymers and diblock and triblock copolymers of styrene have a $T > T_g$ relaxation at approximately the same temperature when molecular weight of the styrene block is equal to that of the homopolymer.

In recent studies, Smith *et al.* (1979) used the spin-probe technique to show the presence of a transition above the T_g that was ascribed to T_{ll} in amorphous polymers. In these studies, samples of 50% plasticized polystyrene were examined at temperatures larger than T_g by the nitroxide spin-probe BzONO (Tempol benzoate). Correlation times τ_c were calculated by line-width analyses. Arrhenius plots showed a break in slope and this break was associated with the liquid–liquid transition T_{ll}, observed by other methods (with due allowance for increased frequency of the spin-probe measurements) (Fig. 46). The transition was observed in atactic polypropylene, but not in isotactic poly-1-butene (Fig. 47).

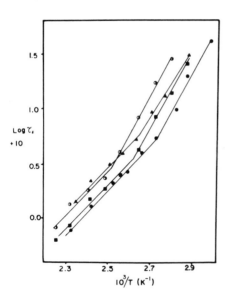

Fig. 46 Arrhenius plots for BzONO in plasticized polystyrene: (●) PS 4000 + 50% plasticizer, (■) PS 17,500 + 50% plasticizer, (▲) PS 110,000 + 50% plasticizer, and (◐) PS 4000 + 30% plasticizer. (From Smith *et al.,* 1979, reprinted with permission from Macromolecules, **12**, 61 (1979) American Chemical Society.)

So far, arguments have been presented in support of the presence of a thermal transition above T_g at T_{ll} and it is now appropriate to outline a few opposing views and counterarguments as to the presence of a liquid–liquid transition. It was reported by Nielson (1977) that the liquid–liquid transition at T_{ll} in polymer melts need not be a true polymer transition but can be an artifact of the torsional braid analysis technique. The shift in T_{ll} with molecular weight was also attributed to the variation in viscosity with molecular weight. Nevertheless, Nielson did rule out the presence of transitions in polymer melts and indicated that an apparatus which measures the true dynamic properties of polymer melts should be used to de-

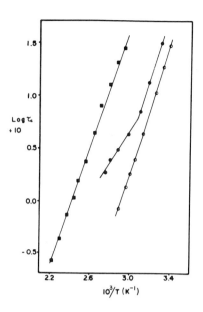

Fig. 47 Arrhenius plots for polypropylene and poly-1-butane: (●) PP + BzONO, (◐) PP + Tempol, and (■) PB-1 + BzONO. (From Smith *et al.*, 1979, reprinted with permission from Macromolecules, **12**, 61 (1979) American Chemical Society.)

tect those transitions rather than the torsional braid apparatus which measures the properties of a composite system. A year before Nielsen's report, Patterson *et al.* (1976) had reported that T_{ll} is a kinetic flow phenomenon, on the basis of experimental results obtained by techniques other than the torsional braid analysis. These authors pointed out that there is no quantitative agreement between T_{ll} values reported by Stadnicki *et al.* (1976) and the T_f values of Kargin and Malinskii (1950), but Glandt *et al.* (1976) have noted that the location of T_{ll} is sensitive to changes in molecular weights and polydispersity. It has also been suggested that the glass braid support used by Gillham introduces the transition artificially by reflecting either polymer flow within the support matrix or changes in capillary action or surface tension in the polymer–matrix system when the polymer is above T_g.

Such criticisms could be countered if the support medium is varied or eliminated altogether and T_{ll} is still observed and if the phenomenon is demonstrated to be of a more general nature. To this end Cowie and McEwen (1979a) have examined the dynamic mechanical behavior of a number of polymers using a variety of supports and presented evidence for the existence of a superglass transition in amorphous polymers. Using cellulose or glass fiber mats or braids as support systems, these authors obtained reproducible results irrespective of the supporting medium. In addition, the existence of the superglass transition in unsupported copolymer systems has been demonstrated by Cowie and McEwen (1979a) in order to remove the doubt concerning the part played by the inert support medium (including surface tension effects or capillary action, which were believed to have resulted in spurious damping above T_g).

In Nielsen's argument (1977) against T_{ll}, he observed that in a polymer–glass braid composite, one could expect to see two effects; above T_g the damping in the polymer will increase with temperatures as viscous flow becomes important, but the braid, with a relatively constant modulus, will show a decrease in damping when impregnated with viscous polymer. The superposition of these two effects will result in a damping maximum in the system that arises from the nature of the composite and not the polymer. This is difficult to counter, although it assumes that polymer and braid impregnated with polymer may act independently above T_g, which is not necessarily the case. Cowie and McEwen's work (1979a) with the unsupported copolymer sample eliminates this possibility. If Nielsen is correct, then Cowie and McEwen would have seen a constant increase in the damping above the polystyrene T_g, instead of which a damping maximum in the expected temperature region has again been observed.

Neumann *et al.* (1978) have analyzed data for polystyrene supported in a steel spring. They were unable to locate a glass transition, but reported a substantial damping peak in the T_{ll} region, which they interpreted as an artifact of the spring–polymer composite system. Their arguments are based on a spring–dashpot model that predicts a maximum in tan δ corresponding to T_{ll}. This suggests T_{ll} corresponds to an isoviscous state and hence will be a function of the composite characteristics. Cowie and McEwen (1979a) appear to believe that for this particular support, Neumann's analysis may be partly correct, but that a spring will respond differently in a composite compared with the supports used by Cowie and McEwen. Above T_g the spring will tend to move through the viscous polymer rather than with it, and damping will occur because of this out-of-phase motion. This would be an artifact of the composite, but similar behavior would not be experienced with support media such as cellulose and glass fiber mats. In the latter, the fibers of the support are randomly distributed and associated with a thin film of the polymer rather than with an independent bulk phase. Thus the behavior of the composite will not result in the out-of-phase movement of polymer and support, as may be experienced with the spring. The observed damping then reflects a real process in the polymer rather than a spurious interaction of polymer and support. Again the data from the unsupported systems (Cowie and McEwen, 1979a) answers this comment most effectively.

Other techniques have revealed the existence of a superglass transition process: infrared (Enns *et al.*, 1978), thermally stimulated discharge studies (Marchal *et al.*, 1978), thermal diffusivity (Ueberreiter and Naghizaden, 1972), and elastic recovery studies (Maxwell and Nguyen, 1978) all support the dynamic mechanical work and reduce doubt about it being an artifact of the system.

Additional evidence for the existence of T_{ll} was given by Cowie and McEwen (1979b) based on their observation of a superglass transition event in the damping spectra of amorphous block copolymers. These authors indicated that T_g and T_{ll} appear to have similar origins.

E. On the Nature of Multiple Melting Endotherms

Multiple melting endotherms are frequently observed in DSC traces of polymers, the bulk or single-crystal form, and various explanations have been proposed to account for particular cases. The shape of the DSC trace is determined in general by two competing processes: melting of the sample according to type, size, and perfection of the crystallites and reorganization of the sample as it is heated up during the scan. The relative importance of those processes in a particular instance will depend on the reorganization rate in relation to the DSC scan rate and, hence, on the thermal stability of the sample as determined by its nature and previous thermal history.

Single crystals often produce two endotherms, sometimes with an exotherm between them, indicating that the higher melting peak is due to ultimate melting of materials that have been reorganized during the experiment (Mandelkern and Allan, 1966; Bair *et al.*, 1967). This is also borne out by studying the effect of irradiation, which prevents refolding by cross-linking the material and suppresses formation of the higher-melting species (Bair *et al.*, 1967, 1968). Double-melting endotherms in bulk Nylon 66, polystyrene, and poly(ethylene terephthalate) have been interpreted as a result of conversion of kinetically favored lamellar crystal to thermodynamically favored extended-chain crystals (Bell and Dumbleton, 1969; Bell *et al.*, 1968; Nealy *et al.*, 1970). However, later investigations indicated that this behavior can be explained by reorganization during the DSC scan (Holdsworth and Turner-Jones, 1971; Sweet and Bell, 1972), in agreement with the results of Roberts (1970). Multiple melting of poly-*N*-methyllaurolactam was also attributed to reorganization during the DSC scan (Shalaby *et al.*, 1974a). The lower melting peak referred to so far is generally found to decrease in size and move up the temperature scale with increased annealing time. This behavior can be explained by the crystals becoming more stable so that they melt at higher temperatures and therefore recrystallize less readily.

Recent studies on low-density polyethene samples, which have been treated to develop a range of well-defined lamellar textures, revealed one, two, or more melting endotherms associated with these samples, depending on the annealing conditions (Pope, 1976). Before testing, these samples have been drawn, rolled, and annealed at different temperatures. The melting behavior of the samples was examined as a function of annealing temperature and time, subsequent heating, and irradiation dose (to prevent refolding by cross-linking). Three types of endotherms were observed. One of these was proposed to correspond to the primary melting of the lamellae seen by x-rays. The other two endotherms were attributed to the melting of the products of reorganization during scan and to melting of the material recrystallized during cooling from the original annealing temperature. Irradiation was found to suppress the refolding that normally occurs when a sample is heated above its annealing temperature. Irradiation also caused an increase in T_m.

In the case of isotactic polystyrene (i-PS), Boon et al. (1968) suggested that a recrystallization process may take place during the scan of melt-crystallized samples of i-PS in a DSC apparatus. A similar conclusion was drawn by Pelzbauer and Manley (1970) from their study on the influence of the heating rate on melting peak area. Besides the two melting peaks, as observed in other polymers too, Pelzbauer and Manley found a small melting endotherm just above the T_c. They suggested that this peak originates from the melting of "impurities" such as stereoblocks or other sterically inhomogeneous molecular species. Lemstra et al. (1972) examined the melt behavior of melt-crystallized and solution-crystallized i-PS in order to obtain further evidence of the origin of the different melting endotherms. At large supercooling, crystallization from the melt produced a small melting endotherm, just above T_c. This peak was reported to originate from secondary crystallization of the melt trapped within the spherulites. Next to this endotherm, a second melting endotherm, related to the normal crystallization process, was recorded and its peak temperature increased linearly with T_c, yielding an extrapolated value for the equilibrium melting temperature T_m° of $242 \pm 1°C$. For normal values of T_c and heating rate, a third endotherm appeared on the second (major) melting endotherm. From the effects of heating rate and partial scanning on the ratio of the peak areas and of previous heat treatment on dissolution temperature, it was concluded that the third endotherm arises from the second one by continuous melting and recrystallization during the scan.

F. Thermal Behavior of Comblike Polymers and Their Monomers

1. Phase Transitions in Polymers

Polymers in which each repeat unit has a long side chain display transition characteristics of both linear and branched systems and are usually denoted as comblike polymers. Those having alkyl side groups are the most common type of comblike polymers and hence will be covered in this chapter. This type of comblike polymer can be represented as

$$\left[\begin{array}{c} S \\ | \\ (CH_2)_n-H \end{array} \right]_P$$

where S is the chain sequence. The comblike polymers have an intrinsic capacity of packing through the main chain or the side groups. This was recognized earlier by Greenberg and Alfrey in 1954, who reported the ability of atactic branched polymers to form crystalline phase where the corresponding three-dimensional structure was due entirely to the side-chain packing. The glass transition temperature (T_g) of these polymers was shown to decrease with the number of methylene groups in the side chain until a critical length is reached; thereafter, the T_g started to increase with the chain length.

The increase in T_g with increase in chain length beyond a critical number of carbon atoms (N_c) was ascribed by many authors to the crystallization of these side chains and progressive increase in T_m and hence T_g; the crystallites act as fillers in the amorphous phase and hence raise the Tg (Greenberg and Alfrey, 1954; Platé and Shibaev, 1974). Effect of structure in crystalline phase morphology was treated in an excellent review by Platé and Shibaev in 1974 and was examined earlier by Turner-Jones in 1965. Most authors advocate the thesis that the side chains crystallize in the same crystal structure as n-paraffins and polyethylene.

So far we have dealt with transitions beyond N_c and avoided the more difficult effects prevailing below this critical length of the side chain. In a recent report (Reimschuessel, 1979), the monotonic decrease in T_g toward a critical value T_{g_c} at N_c was described as a predictable change on the basis of the equations

$$N_c = 1 + 2.8 \times 10^{-2}T_{g_1} \tag{15}$$

$$T_{g_c} = T_{g_1} - N_c (N_{c-1}) \tag{16}$$

relating N_c to the glass transition at $N = 1$ (T_{g_1}). Combining these equations yields

$$T_{g_c} = 0.97 \, T_{g_1} - 8 \times 10^{-4}T_{g_c}^2 \tag{17}$$

which relates T_{g_c} solely to T_{g_1}. These equations were derived with the assumption that the T_g of the comblike polymers is, for a given backbone chain structure, directly related to the molecular weight of the monomer unit. Reimschuessel illustrated the applicability of these equations to the calculation of N_c and T_{g_c} for polybutadienes (PB), polystyrenes (PS), poly(vinyl ethers) (PVE), polyacrylates (PA), poly(methyl methacrylate) (PMMA), and other polymers. The calculated and experimental N_c and T_{g_c} values of some of these polymers are shown in Table XXIX.

TABLE XXIX

CALCULATED AND EXPERIMENTAL N_c AND T_{g_c} VALUES OF COMBLIKE POLYMERS[a]

	N_c		T_{g_c} (K)		
Polymer	Exp.	Calc'd.[b]	Exp.	Calc'd.[c]	Calc'd.[d]
Poly-2-alkyl-1,3-butadienes	7	7	190	171	170
Poly(alkyl styrenes)	10	11	208	233	238
Poly(alkyl vinyl ether)	8	8	193	185	187
Poly(alkyl acrylates)	9	9	187	201	205
Poly(alkyl methacrylates)	12	12	240	246	252

  [a] From Reimschuessel (1979).
  [b] Using Eq. (15).
  [c] Using Eq. (16) and using calculated N_c values.
  [d] Using Eq. (17).

The effect of side-chain crystallinity in copolymers, made from n-octadecyl acrylate or vinyl stearate, on their heat of fusion, T_g and T_m was reported by Jordan (1971) and Jordan *et al.* (1971a,b). Thus the effect of interrupting the long ordered 18-carbon side chains by randomly interspersed amorphous side chains of various lengths was examined using a series of copolymers of n-octadecyl acrylate and its lower acrylate homologs (mostly C_1 through C_8). It was found that simple dilution of the crystalline component (from monomer with C_{18} side chain) by the amorphous component (from monomers with short side chains) governed the decline in the heat of fusion values and the fraction of crystallinity present. High crystallization rates were encountered because equilibrium crystallinity was nearly achieved for most copolymers. Melting temperature depression was less than theoretically expected in copolymers having short amorphous comonomer side chains but approached the theoretical depression as these side chains became very long. The decline in the glass transition temperature was linear with increasing weight fraction of n-octadecyl acrylate for all systems in the composition range where the copolymers were amorphous. Beyond a critical fraction of n-octadecyl acrylate (0.3–0.5), developing side-chain crystallinity due to this comonomer raised the T_g steadily for all systems. The effect of composition of these copolymers on their mechanical properties was studied by Jordan *et al.* (1972), who discussed these properties in terms of the thermal transitions of the systems.

2. Phase Transition and Effect on Polymerizability of Monomers

Phase transition of long-chain vinyl monomers such as octadecyl methacrylate (OMA), octadecyl acrylate (OA), and vinyl stearate (VS) have been detected and examined by DTA and other analytical techniques such as x-ray diffraction and infrared spectroscopy. It was shown by DTA that OMA and OA can be present in three crystalline modifications, which were denoted as the α-, β-, and β'-forms (Shibasaki and Fukuda, 1977). Based on the x-ray diffraction data, the crystal structures of the α-form and β-form were assigned to monoclinic and hexagonal structures, respectively. No crystal structure was assigned to the unstable β'-form.

The polymerizability of long-chain vinyl monomers in the solid state was shown by a few investigators to depend markedly on the geometric disposition and the mobility of the monomer molecules in the crystal lattice (Shibasaki and Fukuda, 1977). Monomers in the β-form crystals exhibited very high polymerizability, but the polymerization of monomers in their α-form was strongly hindered. The β'-form was too unstable to study its polymerizability. The β-form of highly purified OMA was a metastable system and completely changed to the more stable α-form upon storage for a few hours at 0°C.

Heats of fusion and melting temperatures of monomeric n-alkylacrylates, N-n-alkylacrylamides, and vinyl esters were obtained by Jordan (1972). The α-hexagonal crystal modification was recorded near T_m for the higher n-alkyl acrylates but a β-form was stable at low temperatures for the entire series. The magnitude of the heats of fusion indicated β-polymorphs for vinyl esters.

G. THERMAL CONDUCTIVITY OF POLYMERS

Thermal conductivity of polymers is a property of considerable importance, both scientifically and technologically, and yet has received comparatively minor attention by a few investigators. Most of the investigations on this subject were reviewed in the recent literature (Knappe, 1971; Kline and Hansen, 1970; Choy, 1977).

In his review, Choy (1977) concentrated on the interpretation of three aspects essential to the thermal conductivity K (mW cm K^{-1}) of polymers: the temperature (T, K) dependence, the crystallinity dependence, and the orientation effect. It was reported that K for all amorphous polymers is approximately equal in magnitude and characterized by a T^2 dependence below 0.5 K, but between 5 and 15 K it becomes independent of T (plateau region). At higher temperatures, K again increases, but above 60 K it becomes proportional to specific heat. For a semicrystalline polymer, the plateau region is absent and K normally exhibits a T and T^3 dependence between 0.1 and 20 K. At higher temperatures, K increases more slowly up to the T_g, except for highly crystalline polymers (e.g., polyethylene and polyoxymethylene for which K reaches a peak value of 100 and then decreases slowly with temperature. As exemplified by poly(ethylene terephthalate) (PET), K also depends strongly on the degree of crystallinity. Although K for PET at 30 K and above increases with increasing crystallinity, values below 10 K show the opposite trend. At 1.5 K, K of a 50% crystalline sample is an order of magnitude lower than the amorphous material. Thus, for semicrystalline polymers, K shows both strong crystallinity and temperature dependence, with a distinctive cross-over point at about 10 K. These marked features were proposed to be the result of the interplay between two competing factors, the intrinsically high conductivity in the crystalline regions and the reduction in K due to an additional phonon scattering mechanism, which becomes important at low temperatures. This scattering could arise from either the correlation in the spatial fluctuation of the sound velocity in the polymer or the acoustic mismatch at the interfaces between the crystallites and the amorphous matrix.

Orientation was reported (Choy, 1977) to produce a very large anisotropy in semicrystalline polymers, which decreases at low temperature and becomes insignificant below 10 K. This feature could also be explained in terms of the same competing mechanism, discussed earlier, if one realizes that the molecular chains in the crystallites are essentially lined up along the direction of orientation, thus offering very little resistance along this direction. For polyethylene, with an extension ratio of 25, the thermal conductivity at 100 K along the extrusion direction is 91 mW cm^{-1} K^{-1}, a value that is extremely high for organic polymers and close to that of stainless steel. At this temperature, the anisotropy was only about 20, yet because of the different temperature dependence of the thermal conductivity along and perpendicular to the extrusion direction, an anisotropy as high as 60 at room temperature was predicted.

At this point, it was felt appropriate to review the results of studies other than those covered by Choy (1977). Thus several aspects of thermal conductivity will be treated in the next paragraphs.

Qualitatively, the thermal conductivity of amorphous polymers is lower than that of crystalline polymers; highly crystalline polymers are most conductive, thermally (Van Krevelen, 1972). It was shown that in an amorphous polymer, thermal conductivity displays a mild maximum about the T_g which decays beyond this temperature. For highly crystalline polymers, the thermal conductivity decreases smoothly with the rise in temperature prior to melting. Meanwhile, thermal conductivity of partially crystalline polymers is a hybrid of the crystalline and amorphous phases, and the net behavior of the material depends on the level of crystallinity. Unfortunately, no adequate theory exists that may be used to predict accurately the thermal conductivity of polymeric melt or solids. However, Van Krevelen (1972) has demonstrated that thermal conductivity of amorphous polymers and polymer melts can be calculated by means of additive quantities (Rao function, molar heat capacity, and molar volume). He also noted that empirical rules can be used to estimate the thermal conductivity of crystalline and semicrystalline polymers.

It was shown earlier that the thermal conductivity of a polymer is quite dependent on its morphology. However, in a polymer melt, morphological factors are essentially nonexistent and the thermal conductivity of a polymer was shown to be quite dependent on its molecular weight and molecular weight distribution. Hanson and Ho (1965) developed a theory that predicts that thermal conductivity increases proportionately with the increase in molecular-weight to the two-third power and that the effect of molecular weight on thermal conductivity is negligible at high molecular weights. Fuller and Fricke (1971) have shown experimentally that the thermal conductivity of a polyethylene system varied proportionately with molecular weight to the one-half power and that the effect of molecular weight was negligible above 90,000 daltons. However, the deviation of the polyethylene system from the theory of Hanson and Ho was later associated with the difference in molecular weight distribution among the studied samples (Ramsey *et al.*, 1973). The effect of molecular-weight distribution on the thermal conductivity of a series of polypropylene and polyethylene samples was verified by Ramsey *et al.* in 1973. These authors have shown that thermal conductivity in the studied polymer melts depends on the molecular-weight distribution and degree of branching of the polymer chains.

In recent studies by Fesciyan and Frisch (1978) on the thermal conductivity of polymer melts, the authors reported that the method of time correlation functions was used to calculate the thermal conductivity for high molecular weight melts. The results were shown to be independent of molecular weight, in agreement with experimental data, and had qualitatively the right high-temperature behavior.

H. Charge Transfer Complexes

Thermal techniques such as DTA and TGA have been used for studying charge transfer complex of donor polymers and acceptor monomeric compounds (Barrales-Rienda and Gonzales-Ramos, 1975; Chu and

Stolka, 1975; Pielichowski, 1972; Pielichowski and Morawiec, 1977). Poly(vinyl carbazole) (PVCZ), poly(vinyl chlorocarbazole) (PVCCZ) and poly(vinyl dichlorocarbazole) are typical examples of the donor polymers used for the formation of charge transfer complexes with tetracyanoethylene or hydrogen halides. Most of these complexes undergo a series of thermally induced physical changes as is shown for PVCZ and PVCCZ in the accompanying tabulation.

	PVCZ (°C)	PVCCZ (°C)
Temperature of minimal endothermic effect for complex	149	169
Temperature of minimal endothermic effect for polymer	450	428
Temperature of 1% weight loss	353	381
Initial temperature of intensive degradation of complex	75	97
Initial temperature of intensive degradation of polymer	350	378

For PVCCZ, all the changes below 181°C and with 35% weight loss were associated with the dissociation of tetracyanoethylene of the charge transfer complex. From the weight loss of a split acceptor, it was possible to determine, accurately, the composition of a complex (Pielichowski and Morawiec, 1977).

I. A METHOD FOR MOLECULAR-WEIGHT CALCULATION USING DSC DATA

For insoluble polymers, the usual methods for molecular-weight measurement are not applicable. Sperati and Starkweather (1961) presented a practical method for molecular-weight calculation of polytetrafluoroethylene (PTFE) using specific gravity data. However, the test method of the standard specific gravity is rather complicated and sometimes the sample prepared has voids that result in low specific gravity. In the course of their studies on the melting and crystallization behaviors of PTFE, Suwa *et al.* (1973) found that the heat of crystallization of the polymer melt is quite dependent on its molecular weight. Using heat of crystallization data obtained by DSC, these investigators related the number-average molecular weight ($\overline{M}_n$) of PTFE to its heat of crystallization (ΔH_c, cal g^{-1}) according to the following relationship in molecular-weight range of 5.2×10^5 to 4.5×10^7 daltons.

$$\overline{M}_n = 2.1 \times 10^{10} \, \Delta H_c^{-5.16} \qquad (18)$$

The heat of crystallization was noted by Suwa *et al.*, to be independent of the cooling rate in the range of 4–32°C min^{-1}. It is interesting to note that earlier in the chapter, dependence of crystallization rate and crystallinity on the molecular weight was documented in more than one instance.

J. STUDY OF THERMAL TRANSITIONS BY GC METHODS

Gas chromatography has been used to study polymer phase transitions. The GC retention volume (Young, 1968; Guillet, 1973; Braun *et al.*, 1975; Braun and Guillet, 1975) reflects the interaction of a "probe" vapor with the polymer; changes in the polymer that occur at the melting point or the glass transition temperature should therefore exert a marked effect on the retention volumes. Sharp discontinuities have been detected in the variation of retention polymers and polymer melts can be calculated by means of additive quantities (Rao function, molar heat capacity, and molar volume). He also noted that empirical rules can be used to estimate the thermal conductivity of crystalline and semicrystalline polymers.

It was shown earlier that the thermal conductivity of a polymer is quite dependent on its morphology. However, in a polymer melt, morphological factors are essentially nonexistent and the thermal conductivity of a polymer was shown to be quite dependent on its molecular weight and molecular-weight distribution. Hanson and Ho (1965) developed a theory which predicts that thermal conductivity increases proportionately with the increase in molecular weight to the two-third power and that the effect of molecular weight on thermal conductivity is negligible at high molecular weights. Fuller and Fricke (1971) have shown experimentally that the thermal conductivity of a polyethylene system varied proportionately with molecular weight to the one-half power and that the effect of molecular weight was negligible above 90,000 daltons. However, the deviation of the polyethylene system from the theory of Hanson and Ho was later associated with the difference in molecular-weight distribution among the studied samples (Ramsey *et al.*, 1973). The effect of molecular-weight distribution on the thermal conductivity of a series of polypropylene and polyethylene samples was verified by Ramsey *et al.* in 1973. These authors have shown that thermal conductivity in the studied polymer melts depends on the molecular-weight distribution and degree of branching of the polymer chains.

In recent studies by Fesciyan and Frisch (1978) on the thermal conductivity of polymer melts, the authors reported that the method of time correlation functions was used to calculate the thermal conductivity for high molecular volumes with temperature for hydrocarbons on columns containing polyethylene and polypropylene near the polymer melting points (Alishoev *et al.*, 1965). This work was extended by Guillet and

Stein (1970) and Gray and Guillet (1971), who determined melting points, degrees of crystallinity, and crystallization kinetics for polyolefins by GC. Glass transition temperatures of polystyrene, poly(vinyl chloride), poly(methyl methacrylate) (Lavoie and Guillet, 1969), and polycarbonates (Yamamoto *et al.*, 1971) were also detected by this method. Gas chromatographic evidence has been presented for a second-order transition other than the glass transition in cellulose acetate (Nakamura *et al.*, 1972).

Early experiments by Smidsrod and Guillet (1969) and by Lavoie and Guillet (1969) showed that inverse gas chromatography could be used to study glass transition behavior in polymers. Liebman *et al.* (1972) reported the detection of the glass transition T_g and of a secondary transition T_β in poly(vinyl chloride) (PVC) by recording the chromatographic separation of cis-trans isomers as a function of temperature. For plasticized PVC, however, their experiments did not show the expected behavior. On this basis, Liebman *et al.* concluded that the gas–liquid chromatography (GLC) technique was not applicable to the study of plasticized polymers. This prompted Braun and co-workers (1975) to reinvestigate the use of the so-called molecular probe method with particular reference to the study of glass transitions in plasticized PVC and polystyrene. Results obtained by these authors suggest that other experimental factors may have led to the failure of the Liebman technique, since they observed the expected behavior for both plasticized and unplasticized samples. The studies made by Braun *et al.* (1975) on the gas chromatographic behavior of a variety of polymeric stationary phases with alkane or alkanol probes showed that the occurrence of the typical z-shaped retention diagram is indeed a general feature that occurs just above the glass transition for all the polymers studied. Under appropriate conditions of probe solubility and coating thickness, such behavior was considered as diagnostic for T_g, but the actual value of T_g corresponds most nearly with the onset of bulk sorption as estimated from the first deviation from linearity of the portion of the retention diagram corresponding to surface adsorption. The same general behavior was observed with both internally and externally plasticized samples, as well as with those whose glass transition is altered by changes in molecular weight or stereoregularity. It was concluded that the gas chromatographic method may be used with some confidence in the detection and estimation of the glass transition in polymers.

K. Thermal Transitions in Mesomorphic Systems

Thermotropic liquid crystals offer uniquely ordered media for polymerization. The liquid crystalline state permits full two-dimensional (nematic and cholesteric) and one-dimensional (smectic) movement. Various

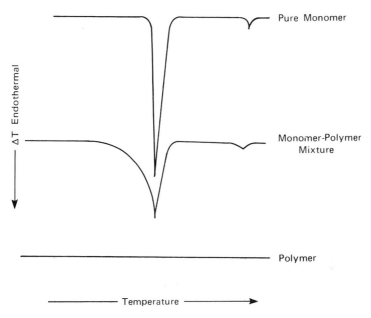

Fig. 48 Thermograms of *N*-(*p*-cyanobenzylidene)-*p*-aminostyrene. (From Hsu and Blumstein, 1977.)

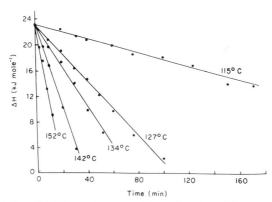

Fig. 49 Variation of ΔH (crystal $\rightarrow$ nematic) as a function of time at various temperatures. (From Hsu and Blumstein, 1977.)

degrees of translational freedom are also attainable depending on the class of mesogen. Such freedom of choice, plus the ability to orient nematic and cholesteric mesogens on a molecular scale in an electric or magnetic field or on certain surfaces, makes the liquid crystal state an attractive polymerization medium. This subject has been treated thoroughly in two excellent books (Blumstein, 1978a, b) and a review (Barrall and Johnson,

1979). This section deals only with specific aspects of thermal transitions in mesomorphic monomers and/or polymer systems.

Hsu and Blumstein (1977) have reported the polymerization of N-(p-cyanobenzylidene)-p-aminostyrene in the nematic and isotropic liquid phases. The monomer forms the nematic mesophase at 113.8°C, which forms the isotropic liquid at 140.5°C. The crystal mesophase transition requires 5480 cal mole^{-1} (14.2 cal mole K^{-1}). The monomer polymerizes readily on heating in both the nematic and isotropic liquid by a free radical mechanism. Since the polymer is almost insoluble in the solid phase, more soluble in the nematic, and most soluble in the isotropic liquid, the thermogram of a partially polymerized monomer is substantially different from that of the pure compound (Fig. 48). Hsu et al. (1976) used the decrease in the solid to nematic transition heat with increased polymerization to follow the progress of the polymerization. The plots of the apparent solid heat of fusion versus time after treatment at various temperatures are quite linear (Fig. 49). This not only supports the validity of the thermoanalytical approach, but also is strong evidence that the polymer does not form cocrystals with the solid. It is to be noted that the nematic to isotropic liquid transition becomes broader, smaller, and shifted to lower temperatures as polymerization advances. This indicates that the polymer is more soluble in the isotropic than in the nematic phase.

Polyesters with mesogenic elements and flexible spacers in the main chain were prepared and their properties were examined by Blumstein et al. (1979). Contrary to their expectation, these investigators reported that DSC and x-ray data of most of these polyesters are due to layered (smectic) organization above the crystal–liquid transition. The existence of nematic order in these polymers was noted as being less frequent than the smectic type.

Jackson and Kuhfuss (1976) have shown that injection molding of a 40:60 (molar) copolyester of ethylene terephthalate (ET)/p-oxybenzoate gives articles having mechanical properties that are highly anisotropic and strongly dependent on processing conditions. McFarlane et al. (1977) then suggested that this copolyester forms a nematic melt.

In their investigation of a [poly(ethylene terephthalate)-co-p-oxybenzoate] containing 30 mole % oxybenzoate units (referred to as T2/30), Krigbaum and Salaris (1978) identified the endotherm at t_{III} = 244°C as a nematic → isotropic transition. Subsequent investigation revealed that the t_{III} endotherm disappeared if the polymer was heated to the isotropic melt or dissolved and reprecipitated. The loss of the t_{III} transition might be due to molecular-weight reduction, sequence randomization, or erroneous identification of the transition. Viscosity and high-field NMR data obtained by Lader and Krigbaum (1979) were suggested to eliminate the first two explanations. Annealing studies at temperatures higher than those of the earlier work led those authors to conclude that the t_{III} transition represents the melting of crystallites formed during a high-temperature annealing operation.

L. FERROELECTRIC PROPERTIES OF FLUOROCARBON POLYMERS

Since their discovery, fluorocarbon polymers have been known for their excellent chemical and thermal stability and a few of them display some rather unique properties. Poly(vinylidene fluoride) (PVDF) has attracted considerable attention as a transducer material because of its piezoelectric properties (Kawai, 1969; Fukada and Takahashi, 1969). It has been reported that uniaxially stretched PVDF has a polar form I (β-phase) crystal structure that behaves as a ferroelectric material (Tamura *et al.*, 1977; Naegele and Yoon, 1978; Furukawa *et al.*, 1980). Lando and Doll (1968) have shown that the introduction of trifluorethylene (TrFE) into PVDF enhances crystallization into the polar form I phase. Yagi *et al.* (1980) has synthesized copolymers of vinylidene fluoride and trifluoroethylene and observed DSC peaks indicating that a phase transition has occurred below T_m. Furukawa *et al.* (1981) has studied this ferroelectric phase transition near 70°C in a 55%/45% VDF/TrFE copolymer by dielectric and specific heat measurements.

VIII. Conclusion

Thermoplastic polymers are the most important group of commercial polymeric materials and their broad range of properties and applications makes it impossible to review them in a manner satisfactory to all or almost all readers. The author must admit an obvious tendency in certain parts of this review to dwell on subjects that are more pertinent to polymer scientists than to thermal analysts. On the other hand, attempts were made in Sections VI and VII to impress thermal analysts with (a) the relevance of athermal techniques, such as chromatography, spectroscopy, and x-ray to investigative studies using conventional and less conventional thermal methods; (b) their important role in contributing to the solution, or, at least, understanding of controversial problems, including the question of order in the amorphous phase and the liquid–liquid transition above T_g; (c) their present and potential contributions to the development of emerging technologies such as high-pressure extrusion; and (d) the pertinence of thermal techniques to the study of some intriguing polymeric materials that are not necessarily typical examples of thermoplastics, e.g., charge transfer complexes, comblike polymers, and mesomorphic systems.

ACKNOWLEDGMENTS

The author wishes to express his deepest appreciation to those colleagues who have reviewed this chapter and offered excellent suggestions that led to some basic changes in the revised manuscript. Thanks are due to Dr. E. M. Barrall II of IBM; Dr. R. F. Boyer of Midland Macromolecular Institute; Dr. J. H. Flynn of the National Bureau of Standards; Mr. R. L. Fyans of Perkin-Elmer; Dr. L. G. Roldan of J. P. Stevens and Co., Inc.; Dr. E. A. Turi of Allied Chemical Corporation; and Professor B. Wunderlich of Rensselaer Polytechnic Institute.

References

Abbas, K. B., and Sorvik, E. M. (1976). *J. Appl. Polym. Sci.* **20**, 2395.

Adam, G. A., Hay, J. N., Parsons, I. W. , and Haward, R. N. (1976). *Polymer* **17**, 51.

Adams, G. C. (1971). *J. Polym. Sci., Polym. Phys. Ed.* **9**, 1235.

Ahlstrom, D. H., Liebman, S. A., and Abbas, K. B. (1976). *J. Polym. Sci., Polym. Chem. Ed.* **14**, 2479.

Alishoev, V. R., Berezkin, V. G., and Mel'Nikova, Yu. Z. (1965). *Zh. Fiz. Khim.* **39**, 200.

Allegra, G., and Bassi, I. W. (1969). *Adv. Polym. Sci.* **6**, 549.

Allegra, G., and Bassi, I. W. (1975). In "Polymer Handbook" (J. Brandrup and E. H. Immergut, eds.), 2nd ed., p. II-205. Wiley (Interscience), New York.

Ariyama, T., Nakayama, T., and Inoue, N. (1977). *J. Polym. Sci., Polym. Lett. Ed.* **15**, 427.

Avrami, M. J. (1939). *J. Chem. Phys.* **7**, 1103.

Bair, H. E., Salovey, R., and Huseby, T. W. (1967). *Polymer* **8**, 9.

Bair, H. E., Huseby, T. W., and Salovey, R. (1968). *Anal. Calorium.* **1**, 31.

Bair, H. E., Bebbington, G. H., and Kelleher, P. G. (1976). *J. Polym. Sci., Polym. Phys. Ed.* **14**, 2113.

Barrall, E. M., II, and Johnson, J. F. (1979). *J. Macromol. Sci., Rev. Macromol. Chem.* **C17**, 137.

Barrales-Rienda, J. M., and Gonzales-Ramos, J. (1975). *Angew. Makromol. Chem.* **43**, 105.

Barrett, K. E. J. (1967). *J. Appl. Polym. Sci.* **11**, 1617.

Barrett, K. E. J. (1970). *Br. Polym. J.* **2**, 45.

Barrett, K. E. J., and Thomas, H. R. (1969). *J. Polym. Sci., Polym. Chem. Ed.* **7**, 2621.

Bassett, D. C., Block, S., and Piermarini, G. J. (1974). *J. Appl. Phys.* **45**, 4146.

Beatty, C. L. (1977). *Am. Chem. Soc., Div. Org. Coat. Plast. Chem. Prepr.* **37**, 344.

Bell, J. P. (1972). *Text. Res. J.* **42**, 292.

Bell, J. P., and Dumbleton, J. H. (1969). *J. Polym. Sci., Polym. Phys. Ed.* **7**, 1033, 1950.

Bell, J. P., Slade, P. E., and Dumbleton, J. H. (1968). *J. Polym. Sci., Polym. Phys. Ed.* **6**, 1773.

Benedetti, E., Corradini, P., and Pedone, C. (1975). *Eur. Polym. J.* **11**, 585.

Berger, M. N. (1973). *J. Macromol. Sci., Rev. Macromol. Chem.* **C9**, 269.

Billmeyer, F. W., Jr. (1971). "Textbook of Polymer Science," 2nd ed., pp. 208, 288. Wiley (Interscience), New York.

Binsberger, F. L. (1970). *J. Macromol. Sci., Phys.* **B4**, 837.

Binsberger, F. L., and deLange, B. G. M. (1970). *Polymer* **11**, 309.

Blanchard, L. P., Hesse, J., and Malhotra, S. L. (1974). *Can. J. Chem.* **52**, 3170.

Blumstein, A., ed. (1978a). "Liquid Crystal and Ordered Polymers," Academic Press, New York.

Blumstein, A., ed. (1978). "Mesomorphic Order in Polymers and Polymerization in Liquid Crystalline Media," ACS Symp. Ser., Vol. 74. Am. Chem. Soc., Washington, D.C.

Blumstein, A., Sivaramakrishnan, S. N., Sohn, W. H., Blumstein, R. B., and Clough, S. B. (1979). *Polym. Prepr., Am. Chem. Soc., Div. Polym. Chem.* **20,** 638.

Bonart, R. (1969). *Kolloid-Z.* **231,** 4381.

Boon, J., Challa, G., and Van Krevelen, D. W. (1968). *J. Polym. Sci., Polym. Phys. Ed.* **6,** 1791.

Booth, A., and Hay, J. N. (1969). *Polymer* **10,** 95.

Boyer, R. F. (1963). *Rubber Chem. Technol.* **36,** 1303.

Boyer, R. F. (1966). *J. Polym. Sci., Part* **C14,** 267.

Boyer, R. F. (1970). *Encycl. Polym. Sci. Technol.* **10,** 313.

Boyer, R. F. (1973a). *J. Macromol. Sci., Phys.* **B7,** 467.

Boyer, R. F. (1973b). *Macromolecules* **6,** 288.

Boyer, R. F. (1973c). *J. Macromol. Sci., Phys.* **B8,** 503.

Boyer, R. F. (1974). *Macromolecules* **7,** 142.

Boyer, R. F. (1976). *Polymer* **17,** 997.

Boyer, R. F. (1977). *Encycl. Polym. Sci. Technol., Suppl.* 2 (H. F. Mark, N. G. Gaylord, and N. M. Bikales, eds.), p. 745. Wiley, New York.

Boyer, R. F., and Snyder, R. G. (1977). *J. Polym. Sci., Polym. Lett. Ed.* **15,** 315.

Bandrup, J., and Immergut, E. H., eds. (1966). "Polymer Handbook," Wiley (Interscience), New York.

Braun, J. M., and Guillet, J. E. (1975). *J. Polym. Sci., Polym. Chem. Ed.* **13,** 1119.

Braun, J. M., Lavoie, A., and Guillet, J. E. (1975). *Macromolecules* **8,** 311.

Brydson, J. A. (1972). *Polym. Sci.* **2,** 193.

Buchanan, D. R., and Walters, J. P. (1977). *Text. Res. J.* **47,** 398.

Bueche, F. (1962). "Physical Properties of Polymers," Wiley (Interscience), New York.

Bur, A. J., and Fetters, L. F. (1976). *Chem. Rev.* **76,** 727.

Buxbaum, L. H. (1968). *Agnew. Chem., Int. Ed. Engl.* **7,** 182.

Calvert, P. D., and Uhlmann, D. R. (1972). *J. Appl. Phys.* **43,** 944.

Cameron, G. G., and Grassie, N. (1961). *Polymer* **2,** 367.

Cameron, G. G., and Grassie, N. (1962). *Makromol. Chem.* **51,** 130.

Cameron, G. G., Meyer, J. M., and McWalter, I. T. (1978). *Macromolecules* **11,** 676.

Camino, G., and Guaita, M. (1977). *Eur. Polym. J.* **13,** 903.

Capiati, N. J., and Porter, R. S. (1975). *J. Polym. Sci., Polym. Phys. Ed.* **13,** 1177.

Cappola, G., Filippini, R., and Pallesi, B. (1975). *Polymer* **16,** 546.

Caraculacu, A., and Bezdadea, E. (1969). *J. Polym. Sci., Polym. Chem. Ed.* **15,** 611.

Cavallo, P., Martuscelli, E., and Pracella, M. (1977). *Polymer* **18,** 42.

Champetier, G., and Pied, J. P. (1961). *Makromol. Chem.* **44,** 64.

Chang, S. S. (1972). *Polym. Prepr., Am. Chem. Soc., Div. Polym. Chem.* **13,** 322.

Chansy, H. D., Bonyour, E., and Marchessault, R. H. (1974). *Colloid Polym. Sci.* **252,** 8.

Choy, C. L. (1977). *Polymer* **18,** 984.

Chu, C. C. (1980). *Am. Chem. Soc., 14th Middle Atlantic Regional Meeting* Abstracts, p. 121.

Chu, I. Y. C., and Stolka, M. (1975). *J. Polym. Sci., Polym. Chem. Ed.* **13,** 2867.

Cooper, D. R., Sutton, G. J., and Tighe, B. J. (1973). *J. Polym. Sci., Polym. Chem. Ed.* **11,** 2045.

Cowie, J. M. G. (1975). *Eur. Polym. J.* **11,** 297.

Cowie, J. M. G., and McEwen, I. J. (1977). *Macromolecules* **10,** 1124.

Cowie, J. M. G., and McEwen, I. J. (1979a). *Polymer* **20,** 719.

Cowie, J. M. G., and McEwen, I. J. (1979b). *Macromolecules* **12**, 56.

Cowie, J. M. G., and Toporowski, P. M. (1968). *Eur. Polym. J.* **4**, 621.

Deanin, R. D. (1972). "Polymer Structure, Properties and Applications," Cahners Books, Boston, Massachusetts.

DeBoer, G. O. R., Van Ekenstein, A., and Challa, C. (1975). *Polymer* **16**, 930.

Desper, C. R., Southern, J. H., Ulrich, R. D., and Porter, R. S. (1974). *J. Appl. Phys.* **41**, 4284.

Dick, W., and Westerberg, C. (1978). *J. Macromol. Sci., Chem.* **A12**, 455.

Doddi, N., Forsman, W. C., and Price, C. C. (1971). *Macromolecules* **4**, 648.

Dole, M. (1967). *J. Polym. Sci., Part C* **18**, 57.

Ebdon, J. R., and Hunt, B. J. (1973). *Anal. Chem.* **45**, 804.

Eisenberg, A. (1971). *Macromolecules* **4**, 125.

Eisenberg, A., Matsura, H., and Yokoyama, T. (1971). *J. Polym. Sci., Polym. Phys. Ed.* **9**, 2131.

Enns, J. B., Boyer, R. F., Ishida, H., and Koenig, J. L. (1978). *Org. Coat. Plast.* **38**, 373.

Ferguson, R. C., and Hoehn, H. H. (1978). *Polym. Eng. Sci.* **18**, 446.

Ferroni, E. (1964). *J. Polym. Sci., Part* **B2**, 51.

Ferry, J. D. (1961). "Viscoelastic Properties of Polymers," p. 212. Wiley, New York.

Fesciyan, S., and Frisch, H. L. (1978). *J. Chem. Phys.* **67**, 5691.

Flory, P. J. (1949). *J. Chem. Phys.* **17**, 223.

Flynn, J. H. (1974). *Thermochim. Acta.* **8**, 69.

Flynn, J. H. (1977). In "Aspects of Degradation and Stabilization of Polymers" (H. G. Jellink, ed.), Chapter 12. Elsevier, Amsterdam.

Flynn, J. H. (1978). In "Thermal Methods in Polymer Analysis (S. W. Shalaby, ed.), p. 163. Franklin Inst. Press, Philadelphia, Pennsylvania.

Flynn, J. H. (1979). Private Communication.

Flynn, J. H., and Dickens, B. (1976). *Thermochim. Acta* **15**, 1.

Fock, J. (1967). *Polym. Lett.* **5**, 635.

Fox, T. G. (1956). *Bull. Am. Phys. Soc.* [2] **1**, 123.

Fox, T. G., and Flory, P. J. (1950). *J. Appl. Phys.* **21**, 581.

Fox, T. G., and Flory, P. J. (1954). *J. Polym. Sci.* **14**, 315.

Fraser, M. J., Cooper, D. R., and Booth, C. (1977). *Polymer* **18**, 852.

Frederick, W. J., Jr., and Mentzer, C. C. (1975). *J. Appl. Polym. Sci.* **19**, 1799.

Fukada, E., and Takahashi, S. (1969). *Jpn. J. Appl. Phys.* **8**, 1960.

Fuller, T. R., and Fricke, A. L. (1971). *J. Appl. Polym. Sci.* **15**, 1729.

Furukawa, T., Date, M., and Fukada, E. (1980a). *J. Appl. Phys.* **51**, 1135.

Furukawa, T., Johnson, G. E. Bair, H. E., Tajitsu, Y., and Fukada, E. (1981). *Ferroelectrics* **34**, 61.

Gargallo, L., and Russo, M. (1975). *Makromol. Chem.* **176**, 2735.

Gaur, U., and Wunderlich, B. (1980). *Macromolecules* **13**, 445.

Geil, P. H., ed. (1963). "Polymer Single Crystals." Wiley, New York.

Geissler, E. (1975). *J. Polym. Sci., Polym. Phys. Ed.* **13**, 1301.

Gilbert, M., and Hybart, F. J. (1974). *Polymer* **15**, 407.

Gillen, K. T. (1978). *J. Appl. Polym. Sci.* **22**, 1291.

Gillham, J. K. (1974). *AI Ch E J.* **20** (6), 1066.

Gillham, J. K. (1979). *Polym. Eng. Sci.* **19**, 749.

Gillham, J. K., and Boyer, R. F. (1977a). *Polym. Prepr., Am. Chem. Soc., Div. Polym. Chem.* **18**, 468.

Gillham, J. K., and Boyer, R. F. (1977b). *J. Macromol. Sci., Phys.* **B13**, 497.

Gillham, J. K., and Boyer, R. F. (1978). In "Thermal Methods in Polymer Analysis" (S. W. Shalaby, ed.), p. 5. Franklin Inst. Press, Philadelphia, Pennsylvania.

Gillham, J. K., Benci, J. A., and Boyer, R. F. (1976). *Polym. Eng. Sci.* **16**, 357.

Glagoleva, Yu. A., and Fratkina, G. P. (1971). *J. Polym. Sci. USSR* **12** (9), 2263.

Glandt, C. A., Toh, H. K., Gillham, J. K., and Boyer, R. F. (1976). *J. Appl. Polym. Sci.* **20**, 1277, 2009.

Godovsky, Yu. K., and Slonimsky, G. L (1974). *J. Polym. Sci., Polym. Phys. Ed.* **12**, 1053.

Gogolewski, S., and Pennings, A. J. (1973). *Polymer* **14**, 463.

Gogolewski, S., and Pennings, A. J. (1975). *Polymer* **16**, 673.

Gogolewski, S., and Pennings, A. J. (1977). *Polymer* **18**, 647.

Goodings, E. P. (1961). *Soc. Chem. Ind.* **13**, 211.

Gordon, G. A. (1971). *J. Polym. Sci., Polym. Phys. Ed.* **9**, 1693.

Gray, D. G., and Guillet, J. E. (1971). *Macromolecules* **4**, 129.

Greco, R., and Nicolais, L. (1976). *Polymer* **17**, 1049.

Greenberg, S. A., and Alfrey, T. (1954). *J. Am. Chem. Soc.* **76**, 6280.

Griffith, H., and Ranby, B. G. (1960). *J. Polym. Sci.* **44**, 369.

Griswold, D., and Cuculo, J. A. (1978). *J. Appl. Polym. Sci.* **22**, 163.

Gubler, M., Rabesiaka, J., and Kovacs, A. J. (1963). *In* "Polymer Single Crystals" (P. H. Geil, ed.), Wiley, New York.

Guillet, J. E. (1973). *In* "New Developments in Gas Chromatography" (J. H. Purnell, ed.), Wiley, New York.

Guillet, J. E., and Stein, A. N. (1970). *Macromolecules* **3**, 103.

Haas, H. C., Manning, M. J., and Hollander, S. A. (1970). *J. Polym. Sci., Polym. Chem. Ed.* **8**, 3657.

Hagege, R. (1977). *Text. Res. J.* **47**, 229.

Hager, N. E. (1972). *Rev. Sci. Instrum.* **43**, 1116.

Hager, N. E. (1973). *J. Polym. Sci., Polym. Symp.* **43**, 77.

Hampson, F. W., and Manley, T. R. (1976). *Polymer* **17**, 723.

Hansen, D., and Ho, C. C. (1965). *J. Polym. Sci., Part A* **3**, 651.

Harget, P. J., and Oswald, H. J. (1978). *In* "Thermal Methods for Polymer Analysis" (S. W. Shalaby, ed.), p. 23. Franklin Inst. Press, Philadelphia, Pennsylvania.

Hay, J. N., Filtzgerald, P. A., and Wiles, M. (1976). *Polymer* **17**, 1015.

Hayes, R. A. (1961). *J. Appl. Polym. Sci.* **5**, 318.

Heijboer, J. (1977). *Int. J. Polym. Mater.* **6**, 11.

Hellmuth, E., Wunderlich, B., and Rankin, J. M. (1966). *Appl. Polym. Symp.* **2**, 101.

Hendra, P. J., Jobic, H. P., and Holland-Moritz, K. (1975). *J. Polym. Sci., Polym. Lett. Ed.* **13**, 365.

Hergenrother, W. L. (1974). *J. Polym. Sci., Polym. Chem. Ed.* **12**, 875.

Hobbs, S. Y., and Mankin, G. I. (1971). *J. Polym. Sci., Polym. Phys. Ed.* **9**, 1907.

Hoehn, H. H., Ferguson, R. C., and Herbert, R. R. (1978). *Polym. Eng. Sci.* **18**, 457.

Hoffman, J. D., and Weeks, J. J. (1958). *J. Res. Natl. Bur. Stand. (U. S.)* **60**, 465.

Hoffman, J. D., and Weeks, J. J. (1962). *J. Res. Natl. Bur. Stand. Sect. A.* **66**, 13.

Holdsworth, P. J., and Turner-Jones, A. (1971). *Polymer* **12**, 195.

Horie, K., Mita, I., and Kambe, H. (1968). *J. Polym. Sci., Polym. Chem. Ed.* **6**, 2663.

Horie, K., Mita, I., and Kambe, H. (1969). *J. Polym. Sci., Polym. Chem. Ed.* **7**, 2561.

Hosaka, Y., Kojima, T., Noguchi, T., and Fujii, T. (1977). In "Thermal Analysis" (H. Chihara, ed.), p. 293. Heyden, London.

Hovenkamp, S. G., and Munting, J. P. (1970). *J. Polym. Sci., Polym. Chem. Ed.* **8**, 679.

Hsu, E. C., and Blumstein, A. (1977). *J. Polym. Sci., Polym. Lett. Ed.* **15**, 129.

Hsu, E. C., Lim, L. K., Blumstein, R. B., and Blumstein, A. (1976). *Mol. Cryst. Liq. Cryst.* **33**, 35.

Hurd, C. D., and Blunck, C. H. (1938). *J. Am. Chem. Soc.* **60**, 2419.

Illers, K. H. (1975). *Prog. Colloid Polym. Sci.* **58**, 61.

Illers, K. H. (1977). *Polymer* **18**, 551.
Illers, K. H., and Breuer, H. (1963). *J. Colloid Sci.* **18**, 1.
Inoue, H., Helbig, M., and Vogl, O. (1977). *Macromolecules* **10**, 1331.
Iring, M., Hedvig-Laszlo, Zs., Kelen, T., and Tudos, F. (1974). *In* "Thermal Analysis" (I. Buzas, ed.), Vol. 2, p. 127. Heyden, London.
Ishibashi, M. (1964). *Polymer* **5**, 103.
Ito, E. (1974). *J. Polym. Sci., Polym. Phys. Ed.* **12**, 1477.
Itoh, T. (1976). *Jpn. J. Appl. Phys.* **15**, 2295.
Jackson, J. B., and Longman, G. W. (1969). *Polymer* **10**, 873.
Jackson, W. J., Jr., and Kuhfuss, H. F. (1976). *J. Polym. Sci., Polym. Chem. Ed.* **14**, 2043.
Jacobi, E., Luderwald, J., and Schulz, R. C. (1978). *Makromol. Chem.* **179**, 429.
Janeczek, H., Turska, E., Szekely, T., Lengyel, M., and Till, F. (1978). *Polymer* **19**, 85.
Jellinek, H. H. G., ed. (1978). "Aspects of Degradation and Stabilization of Polymers." Am. Elsevier, New York.
Johnson, N. W. (1976). *J. Macromol. Sci., Rev. Macromol. Chem.* **C14**, 215.
Jordan, E. F., Jr. (1971). *J. Polym. Sci., Polym. Chem. Ed.* **9**, 3367.
Jordan, E. F., Jr. (1972). *J. Polym. Sci., Polym. Chem. Ed.* **10**, 3347.
Jordan, E. F., Jr., Riser, G. R., Artymyshyn, B., Parker, W. E., Pensabene, J. W., and Wrigley, A. N. (1969). *J. Appl. Polym. Sci.* **13**, 1777.
Jordan, E. F., Jr., Artymyshyn, B., Speca, A., and Wrigley, A. N. (1971a). *J. Polym. Sci., Polym. Chem. Ed.* **9**, 3349.
Jordan, E. F., Jr., Feldeisen, D. W., and Wrigley, A. N. (1971b). *J. Polym. Sci., Polym. Chem. Ed.* **9**, 1835.
Jordan, E. F., Jr., Riser, G. R., Artymshyn, B., and Pensabene, J. W. (1972). *J. Polym. Sci., Polym. Phys. Ed.* **10**, 1657.
Kaimin, I. F., Apinis, A. P., and Galvenowskii, A. Ya. (1975). *Polym. Sci. USSR* **17**, 46.
Kamide, K., and Yamaguchi, K. (1972). *Makromol. Chem.* **162**, 205, 219.
Kampf, G. (1962). *Kolliod Z. Z. Polym.* **185**, 6.
Karasz, F. E., and MacKnight, J. W. (1968). *Macromolecules* **1**, 537.
Karasz, F. E., Bair, H. E., and O'Reilly, J. M. (1965). *J. Phys. Chem.* **69**, 2657.
Kargin, V. A., and Malinskii, U. M. (1950). *Dokl. Akad. Nauk SSSR* **72**, 725, 915.
Kawai, H. (1969). *Jpn. J. Appl. Phys.* **8**, 975.
Kawasaki, K., and Sekita, Y. (1964). *J. Polym. Sci., Part A-2* **2**, 2437.
Ke, B. (1963). *J. Polym. Sci., Part A* **1**, 1453.
Ke, B. (1964). "Newer Methods of Polymer Characterization," p. 389. Wiley (Interscience), New York.
Keith, H. D., and Padden, F. J., Jr. (1964). *J. Appl. Phys.* **35**, 1270.
Kelen, T. (1978). *J. Macromol. Sci., Chem.* **A12**, 349.
Keller, A. (1968). Review article: *Rep. Prog. Phys.* **31**, (2), 623.
Kettle, G. J. (1977). *Polymer* **18**, 742.
Kilroe, J. G., and Weale, K. E. (1960). *J. Chem. Soc.* p. 3849.
Kiran, E., and Gillham, J. K. (1974). *J. Macromol. Sci., Chem.* **A8**, 211.
Kiran, E., and Gillham, J. K. (1976a). *J. Appl. Polym. Sci.* **20**, 931.
Kiran, E., and Gillham, J. K. (1976b). *J. Appl. Polym. Sci.* **20**, 2045.
Kiran, E., Gillham, J. K., and Gipstein, E. (1977). *J. Appl. Polym. Sci.* **21**, 1159.
Kishore, K., and Pai Verneker, V. R. (1976). *J. Polym. Sci., Polym. Sci. Ed.* **14**, 761.
Kishore, K., Pai Verneker, V. R., and Nair, N. R. (1976). *J. Appl. Polym. Sci.* **30**, 2355.
Kline, D. E., and Hansen, D. (1970). *In* "Thermal Characterization Techniques" (P. E. Slade and L. J. Jenkins, eds.), p. 247. Dekker, New York.
Knappe, W. (1971). *Adv. Polym. Sci.* **7**, 477.

Koenig, J. L., and Mele, M. D. (1969). *Anal. Calorim.* **2**, 83.

Koleske, J. V., and Lundberg, R. D. (1969). *J. Polym. Sci., Polym. Phys. Ed.* **7**, 795.

Koleske, J. V., and Lundberg, R. D. (1972). *J. Polym. Sci., Polym. Phys. Ed.* **10**, 323.

Kovacs, A. J. (1963). *Adv. Polym. Sci.* **3**, 394.

Kovacs, A. J., and Gonthier, A. (1972). *Kolliod-Z.* **250**, 530.

Kraus, G., and Gruver, J. T. (1970). *J. Polym. Sci., Polym. Phys. Ed.* **8**, 571.

Krigbaum, W. R., and Salaris, F. (1978). *J. Polym. Sci., Polym. Phys. Ed.* **16**, 883.

Kulkarni, R. K., Porter, H. J., and Leonard, F. (1973). *J. Appl. Polym. Sci.* **17**, 3509.

Kumins, C. A., Roteman, J., and Ralle, C. J. (1963). *J. Polym. Sci., Part A* **1**, 541.

Kumler, P. L., Keinath, S. E., and Boyer, R. F. (1977). *J. Macromol. Sci., Phys.* **B13**, 631.

Kwei, T. K. (1964). *J. Polym. Sci., Part A* **3**, 3299.

Lader, H. J., and Krigbaum, W. R. (1979). *J. Polym. Sci. Polym. Phys. Ed.* **17**, 1661.

Lai, J. H. (1976). *J. Appl. Polym. Sci.* **20**, 1059.

Landa, J. B., and Doll, W. W. (1968). *J. Macromol. Sci. Phys.* **B2**, 205.

Lang, M. C., Noel, C., and Legrand, A. P. (1977a). *J. Polym. Sci., Polym. Phys. Ed.* **15**, 1319.

Lang, M. C., Noel, C., and Legrand, A. P. (1977b). *J. Polym. Sci., Polym. Phys. Ed.* **15**, 1329.

Lavoie, A., and Guillet, J. E. (1969). *Macromolecules* **2**, 443.

Lawton, E. L., and Cates, D. M. (1969). *Anal. Calorim.* **2**, 89.

Lee, W. A., and Sewell, J. H. (1968). *J. Appl. Polym. Sci.* **12**, 1397.

Lemstra, P. J., Kooistra, T., and Challa, G. (1972). *J. Polym. Sci., Polym. Phys. Ed.* **10**, 823.

Liebman, S. A., Ahlstrom, D. H., and Foltz, C. R. (1972). *J. Chromatogr.* **67**, 153.

Liebman, S. A., Ahlstrom, D. H., and Foltz, C. R. (1978). *J. Polym. Sci., Polym. Chem. Ed.* **16**, 3139.

Lipatov, Yu. S., and Geller, T. E. (1967). *Vysokomol. Soedin., Ser. A* **9**, 222.

Lord, F. W. (1974). *Polymer* **15**, 42.

Lovett, A. J., O'Donnell, W. G., Sutton, G. J., and Tighe, B. J. (1973). *J. Polym. Sci., Polym. Chem. Ed.* **11**, 2031.

Lüderwald, I. (1977). *Makromol. Chem.* **178**, 2603.

Lüderwald, I., and Urrutia, H. (1976a). *Makromol. Chem.* **177**, 2079.

Lüderwald, I., and Urrutia, H. (1976b). *Makromol. Chem.* **177**, 2093.

Lüderwald, I., and Vogl, O. (1979). *Makromol. Chem.* **180**, 2295.

Lum, R. M. (1979). *J. Polym. Sci., Polym. Chem. Ed.* **17**, 203.

McFarlane, F. E., Nicely, V. A., and Davis, T. G. (1977). *In* "Contemporary Topics in Polymer Science" (E. M. Pearce and J. R. Schaefgen, eds.), p. 109. Plenum, New York.

Maclaine, J. Q. G., and Booth, C. (1975). *Polymer* **16**, 680.

MacLean, D. L. (1974). *J. Appl. Polym. Sci.* **18**, 625.

Madorsky, S. L. (1964). "Thermal Analysis of Organic Polymers." Wiley, New York.

Maeda, Y., and Kanetsuna, H. (1974). *J. Polym. Sci., Polym. Phys. Ed.* **12**, 2551.

Maeda, Y., and Kanetsuna, H. (1975). *J. Polym. Sci., Polym. Phys. Ed.* **13**, 637.

Maeda, Y., and Kanetsuna, H. (1976). *J. Polym. Sci., Polym. Phys. Ed.* **14**, 2057.

Magill, J. H. (1962). *Polymer* **3**, 35.

Manabe, S., *et al.* (1971). *Int. J. Polym. Mater.* **1**, 47.

Mancarella, C., and Martuscelli, E. (1977). *Polymer* **12**, 1240.

Manche, E. P., and Carroll, B. (1972). *Phys. Methods Macromol. Chem.* **2**, 240.

Mandelkern, L. (1964). "Crystallization of Polymers." McGraw-Hill, New York.

Mandelkern, L., and Allan, A. L. (1966). *J. Polym. Sci.* **4**, 447.

Marchal, E., Benoit, H., and Vogl, O. (1978). *J. Polym. Sci., Polym. Phys. Ed.* **16**, 949.
Marshal, D. I., George, E. G., and Turnipseed, J. M. (1973). *Polym. Eng. Sci.* **13**, 415.
Matsuoka, S., Aloisio, C. J., and Bair, H. E., (1973). *J. Appl. Phys.* **44**, 4265.
Matsuoka, S., Bair, H. E., and Aloisio, C. J. (1974). *J. Polym. Sci. Polym. Sump.* **46**, 415.
Maxwell, B., and Nguyen, My. (1978). *Soc. Plast. Eng. Tech. Paper* p. 540.
Mears, P. (1965). "Polymers—Structure and Bulk Properties." Van Nostrand-Reinhold, Princeton, New Jersey.
Mehta, S., Gaur, U., and Wunderlich, B. (1978). *J. Polym. Sci., Polym. Chem. Ed.* **16**, 289.
Miller, G. W. (1969). *Anal. Calorim.* **1**, 71.
Miller, G. W. (1971). *J. Appl. Polym. Sci.* **15**, 2335.
Miller, G. W. (1974). *Thermochim. Acta* **8**, 129.
Miller, G. W. (1975). *J. Polym. Sci., Polym. Phys. Ed.* **13**, 1831.
Minnini, R. M., Moore, R. S., Flick, J. R., and Petrie, S. E. B. (1973). *J. Macromol. Sci., Phys.* **B8**, 343.
Miyagi, A., and Wunderlich, B. (1972). *J. Polym. Sci., Polym. Phys. Ed.* **10**, 2073, 2085.
Moacanin, J., and Cuddihy, E. F. (1966). *J. Polym. Sci., Part C* **14**, 313.
Morgan, L. B. (1954). *J. Appl. Chem.* **4**, 160.
Mucha, M. (1974). *In* "Thermal Analysis" (I Buzas, ed.), Vol. 2, p. 25. Heyden, London.
Naegele, D., and Yoon, D. Y. (1978). *Appl. Phys. Lett.* **33**, 132.
Nakamura, K., Watanabe, T., Katayama, K., and Amano, T. (1972). *J. Appl. Polym. Sci.* **16**, 1077.
Natta, G. (1959). *J. Polym. Sci.* **34**, 531.
Natta, N., Allegra, G., Bassi, I. W., Carlini, C., Chiellini, E., and Montagnoli, G. (1969). *Macromolecules* **2**, 311.
Nealy, D. L., Davis, T. G., and Kibler, C. J. (1970). *J. Polym. Sci., Polym. Phys. Ed.* **8**, 2141.
Neki, K., and Geil, P. H. (1973). *J. Macromol. Sci. Phys.* **B8**, 295.
Neumann, R. M., Senich, G. A., and MacKnight, W. J. (1978). *Polym. Eng. Sci.* **18**, 624.
Newman, S., and Cox, W. P. (1960). *J. Polym. Sci.* **46**, 29.
Nielsen, L. E. (1962). "Mechanical Properties of Polymers," p. 23. Van Nostrand-Reinhold, Princeton, New Jersey.
Nielsen, L. E. (1977). *Polym. Eng. Sci.* **17**, 713.
Northolt, M. G., Tabor, B. J., and Van Aartsen, J. J. (1975). *Colliod Polym. Sci.* **57**, 225.
Norton, G. P., Tighe, B. J., and Molloy, R. (1978). *J. Polym. Sci., Polym. Chem. Ed.* **16**, 283.
Nose, T. (1973). *Polym. J.* **4**, 217.
O'Malley, J. J., and Stauffer, W. J. (1974). *J. Polym. Sci., Polym. Chem. Ed.* **12**, 865.
O'Reilly, J. M., Karasz, F. E., and Bair, E. H. (1964). *Bull. Am. Phys. Soc.* [2] **9**, (11), 285.
Oswald, H. J., Turi, E. A., Harget, P. J., and Khanna, Y. P. (1977). *J. Macromol. Sci., Phys.* **B13** (2), 231.
Otocka, E. P., and Kwei, T. K. (1968). *Macromolecules* **1**, 401.
Overberger, C. G., Arond, L. H., Wiley, R. H., and Garrett, R. R. (1951). *J. Polym. Sci.* **7**, 431.
Overberger, C. G., Frazier, C., Mandelman, J., and Smith, H. F. (1953). *J. Am. Chem. Soc.* **75**, 3326.
Owadh, A. A., Parsons, I. W., Hay, J. N., and Haward, R. N. (1978). *Polymer* **18**, 386.
Paciorek, K. L., Lajiness, W. G., Spain, R. G., and Lenk, C. T. (1962). *J. Polym. Sci.* **61**, S-41.
Panke, D., and Wunderlich, B. W. (1974). *In* "Thermal Analysis" (I. Buzas, ed.), Vol. 2, p. 35. Heyden, London.

Passalacqua, V., Pilati, F., Zamboni, V., Fortunato, B., and Manaresi, P. (1976). *Polymer* **17**, 1044.

Patterson, A., Sutton, G. J., and Tighe, B. J. (1973). *J. Polym. Sci., Polym. Chem. Ed.* **11**, 2343.

Patterson, G. C., Bair, H. E., and Tonelli, A. E. (1976). *J. Polym. Sci., Part C* **17**, 171.

Pavan, A., Provasoli, A., Moraglio, G., and Zambelli, A. (1977). *Makromol. Chem.* **178**, 1099.

Pearce, E. M. (1969). *Trans. N.Y. Acad. Sci.* **31**, 629.

Pearce, E. M., Shalaby, S. W., and Barker, R. H. (1975). *In* "Flame Retardant Polymeric Materials" (M. Lewin, S. M. Atlas, and E. M. Pearce, eds.), Chapter 6. Plenum, New York.

Peebles, L. H., Jr. (1976). *Encycl. Polym. Sci. Technol. Suppl.* **1**, 2.

Pelzbauer, Z., and Manley, R. S. J. (1970). *J. Polym. Sci., Polym. Phys. Ed.* **8**, 649.

Petrie, S. E. B. (1974). *Polym. Prepr., Am. Chem. Soc., Div. Polym. Chem.* **15**, 336.

Peyser, P., and Bascom, W. D. (1977). *J. Macromol. Sci., Phys.* **B13**, (4), 597.

Pielichowski, J. (1972). *J. Therm. Anal.* **4**, 339.

Pielichowski, J., and Morawiec, E. (1977). *In* Thermal Analysis" (H. Chihara, ed.), p. 220. Heyden, London.

Pittman, A. G., and Ludwig, B. A. (1969). *J. Polym. Sci., Polym. Chem. Ed.* **7**, 3053.

Platé, N. A., and Shibaev, V. P. (1974). *J. Polym. Sci., Macromol. Rev.* **8**, 117.

Platé, N. A., Shibaev, V. P., Petrukhin, B. S., and Kargin, V. A. (1968). *J. Polym. Sci., Part C* **23**, 37.

Pope, D. P. (1976). *J. Polym. Sci., Polym. Phys. Ed.* **14**, 811.

Port, W. S., Hansen, J. E., Jordan, E. F., Jr., Dietz, T. J., and Swern, D. (1951). *J. Polym. Sci.* **7**, 207.

Pratt, C. F., and Hobbs, S. Y. (1976). *Polymer* **17**, 12.

Predecki, P., and Karr, P. H. (1978). *Polym. Eng. Sci.* **18**, 1.

Prime, R. B., and Wunderlich, B. (1969). *J. Polym. Sci., Polym. Phys. Ed.* **7**, 2061.

Prime, R. B., Wunderlich, B., and Melillo, L. (1969). *J. Polym. Sci., Polym. Phys. Ed.* **7**, 2091.

Prince, F. R., and Fredericks, R. J. (1972). *Macromolecules* **5**, 168.

Prince, F. R., Pearce, E. M., and Fredericks, R. J. (1970). *J. Polym. Sci., Polym. Chem. Ed.* **8**, 3533.

Privalko, V. P., and Lipatov, Yu. S. (1974). *J. Macromol. Sci., Phys.* **B9**, 551.

Puffer, R., and Sebenda, J. (1976). *J. Polym. Sci., Part C* **16**, 77.

Quynn, R. G. (1972). *J. Appl. Polym. Sci.* **16**, 3393.

Ramsey, J. C., III, Fricke, A. L., and Caskey, J. A. (1973). *J. Appl. Polym. Sci.* **17**, 1597.

Ranby, B. G., and Cahn, K. S., and Brumberger, H. (1962). *J. Polym. Sci.* **58**, 545.

Rees, D. V., and Bassett, D. C. (1971). *J. Polym. Sci., Polym. Phys. Ed.* **9**, 385.

Reich, L., and Levi, W. (1967). *Macromol. Rev.* **1**, 173.

Reich, L., and Stivala, S. S. (1971). "Elements of Degradation." McGraw-Hill, New York.

Reimschuessel, H. K. (1979). *J. Polym. Sci., Polym. Chem. Ed.* **17**, 2447.

Reimschuessel, H. K., Shalaby, S. W., and Pearce, E. M. (1973). *J. Fire Flammability* **4**, 299.

Ringwald, E. L., and Lawton, E. L. (1975). *In* "Polymer Handbook" (J. Brandrup and E. H. Immergut, eds.), 2nd ed., p. V–71. Wiley (Interscience), New York.

Roberts, R. C. (1969). *Polymer* **10**, 113, 117.

Roberts, R. C. (1970). *J. Polym. Sci., Part B* **8**, 381.

Sakaguchi, F., Mandelkern, L., and Maxfield, J. (1976). *J. Polym. Sci., Polym. Phys. Ed.* **14**, 2137.

Sakurai, K., Miyasaka, K., and Ishikawa, K. (1974). *J. Polym. Sci., Polym. Phys. Ed.* **12,** 1587.

Sakurai, K., Oota, T., Miyasaka, K., and Ishikawa, K. (1976). *J. Polym. Sci., Polym. Phys. Ed.* **14,** 1527.

Savolainen, A. (1974). *Eur. Polym. J.* **10,** 9.

Sawada, H. (1976). "Thermodynamics of Polymerization." Dekker, New York.

Schwartz, A. (1978). *J. Therm. Anal.* **13,** 489.

Shalaby, S. W. (1980). U.S. Patent 4,205,399 (to Ethicon, Inc.).

Shalaby, S. W., and Jamiolkowski, D. D. (1979a). U.S. Patent 4,140,678 (to Ethicon, Inc.).

Shalaby, S. W., and Jamiolkowski, D. D. (1979b). U.S. Patent 4,141,087 (to Ethicon, Inc.).

Shalaby, S. W., and Pearce, E. M. (1974a). "Chemistry of Macromolecules." Am. Chem. Soc., Washington, D.C.

Shalaby, S. W., and Pearce, E. M. (1974b). *Int. J. Polym. Mater.* **3,** 81.

Shalaby, S. W., and Reimschuessel, H. K. (1977). *J. Polym. Sci., Polym. Chem. Ed.* **15,** 1349.

Shalaby, S. W., and Turi, E. A. (1977). *ACS Symp. Ser.* **59,** 251.

Shalaby, S. W., and Turi, E. A. (1973). Unpublished work.

Shalaby, S. W., Fredericks, R. J., and Pearce, E. M. (1972). *J. Polym. Sci., Polym. Chem. Ed.* **10,** 1699.

Shalaby, S. W., Pearce, E. M., Fredericks, R. J., and Turi, E. A. (1973a). *J. Polym. Sci., Polym. Phys. Ed.* **11,** 1.

Shalaby, S. W., Fredericks, R. J., and Pearce, E. M. (1973b). *J. Polym. Sci., Polym. Phys. Ed.* **11,** 939.

Shalaby, S. W., Fredericks, R. J., Pearce, E. M., and Wenner, W. M. (1974a). *J. Polym. Sci., Polym. Phys. Ed.* **12,** 223.

Shalaby, S. W., Sifniades, S., Klein, K. P., and Sheehan, D. (1974b). *J. Polym. Sci., Polym. Chem. Ed.* **12,** 2917.

Shalaby, S. W., Turi, E. A., Riggi, M. H., and Harget, P. J. (1975). *J. Polym. Sci., Polym. Chem. Ed.* **13,** 669.

Shalaby, S. W., Turi, E. A., and Harget, P. J. (1976a). *J. Polym. Sci., Polym. Chem. Ed.* **14,** 2407.

Shalaby, S. W., Sifniades, S., and Sheehan, D. (1976b). *J. Polym. Sci., Polym. Chem. Ed.* **14,** 2675.

Shalaby, S. W., Turi, E. A., and Pearce, E. M. (1976c). *J. Appl. Polym. Sci.* **20,** 3185.

Shalaby, S. W., Turi, E. A., and Reimschuessel, H. K. (1978a). *Polym. Prepr., Am. Chem. Soc., Div. Polym. Chem.* **19,** 654.

Shalaby, S. W., Turi, E. A., and Wenner, W. M. (1978b) *In* "Thermal Methods in Polymer Analysis" (S. W. Shalaby, ed.), p. 35. Franklin Inst. Press, Philadelphia, Pennsylvania.

Sharples, A. (1966). "Introduction to Polymer Crystallization." Arnold, London.

Shibasaki, Y., and Fukuda, K. (1977). *In* "Thermal Analysis" (H. Chihara, ed.), Vol. 2, p. 34. Heyden, London.

Shiono, T., Niki, E., and Kamiya, Y. (1977). *J. Appl. Polym. Sci.* **21,** 1635.

Sibilia, J. P., Schaffhauser, R. J., and Roldan, L. G. (1976). *J. Polym. Sci., Polym. Phys. Ed.* **14,** 1021.

Sidorovich, E. A., Marci, A. I., and Gashthol'd, N. S. (1971). *Rubber Chem. Technol.* **44,** 166.

Siegmann, A., and Harget, P. J. (1973). *J. Polym. Sci., Polym. Phys. Ed.* **18,** 346.

Simon, J., Beatty, C. L., and Karasz, F. E. (1975). *J. Therm. Anal.* **7,** 187.

Smidsrod, O., and Guillet, J. E. (1969). *Macromolecules* **2,** 272.

Smith, P. M., Boyer, R. F., and Kumler, P. L. (1979). *Macromolecules* **12,** 61.

Southern, J. H., and Porter, R. S. (1970a). *J. Macromol. Sci., Phys.* **B4,** 541.

Southern, J. H., and Porter, R. S. (1970b). *J. Appl. Polym. Sci.* **14,** 2305.

Spencer, R. S., and Boyer, R. F. (1946). *J. Appl. Phys.* **17,** 398.

Sperati, C. A., and Starkweather, H. W., Jr. (1961). *Fortschr. Hochpolym.-Forsch.* **2,** 465.

Stadnicki, S. J., Gillham, J. K., and Boyer, R. F. (1976). *J. Appl. Polym. Sci.* **20,** 1245.

Stamhuis, J. E., and Pennings, A. J. (1977). *Polymer* **18,** 667.

Starkweather, H. W., Jr., and Jones, G. A. (1981). *J. Polym. Sci. Polym. Phys. Ed.* **19,** 467.

Stehling, F. C., and Mandelkern, L. (1969). *Polym. Lett.* **7,** 255.

Stehling, F. C., and Mandelkern, L. (1970). *Macromolecules* **3,** 242.

Stein, R. S. (1969). *Ann. N.Y. Acad. Sci.* **155,** 566.

Stolting, J., and Muller, F. H. (1970). *Kolloid-Z.* **238,** 459.

Strazielle, C., Benoit, H., and Vogl, O. (1978). *Eur. Polym. J.* **14,** 331.

Stuart, H. A. (1967). *Angew. Chem.* **6,** 844.

Sumita, M., Miyasaka, K., and Ishikawa, K. (1977). *J. Polym. Sci., Polym. Chem. Ed.* **15,** 837.

Suwa, T., Takehisa, M., and Machi, S. (1973). *J. Appl. Polym. Sci.* **17,** 3253.

Sutton, G. J., and Tighe, B. J. (1973). *J. Polym. Sci., Polym. Chem. Ed.* **11,** 1069.

Sutton, G. J., Tighe, B. J., and Roberts, M. (1973). *J. Polym. Sci., Polym. Chem. Ed.* **11,** 1079.

Sweet, G. E., and Bell, J. P. (1972). *J. Polym. Sci., Polym. Phys. Ed.* **10,** 1273.

Tammann, G., and Jenckel, E. (1929). *Z. Anorg. Allg. Chem.* **184,** 416.

Tamura, M., Hagiwara, S., Matsumoto, S., and Ono, N. (1977). *J. Appl. Phys.* **48,** 513.

Tarasov, V. V. (1950). *Zh. Fiz. Khim.* **24,** 111.

Teitel'baum, B. Ya. (1974). *In* "Thermal Analysis" (I. Buzas, ed.), Vol. 2, p. 13. Heyden, London.

Thackray, G., and Haward, R. N. (1968). *Proc. R. Soc. London, Ser. A* **302,** 453.

Timmermans, J. (1965). "Physico-Chemical Constants of Pure Organic Compounds," Vol. 2. Elsevier, Amsterdam.

Tonelli, A. E. (1971). *Macromolecules* **4,** 653.

Toussaint, A. (1973-1974). *Prog. Org. Coat.* **2,** 273.

Trischler, F. D., and Hollander, J. (1969). *J. Polym. Sci., Polym. Chem. Ed.* **7,** 971.

Tsuge, S., Murakami, K., Esaki, M., and Takeuchi, T. (1977). *In* "Thermal Analysis" (H. Chihara, ed.), p. 289. Heyden, London.

Tsutsui, T., and Tanaka, T. (1972). *Polym. J.* **5,** 332.

Tsutsui, T., and Tanaka, T. (1977). *Polymer* **18,** 817.

Tsutsui, T., Tanaka, R., and Tanaka, T. (1975). *J. Polym. Sci., Polym. Phys. Ed.* **13,** 2091.

Turi, E. A., and Sibilia, J. P. (1978). *In* "Thermal Methods in Polymer Analysis" (S. W. Shalaby, ed.), p. 77. Franklin Inst. Press, Philadelphia, Pennsylvania.

Turner, D. T. (1978). *Polymer* **19,** 789.

Turner-Jones, A. (1965). *Makromol. Chem.* **71,** (6), 249.

Ueberreiter, K., and Kanig, G. (1952). *J. Colloid Sci.* **7,** 569.

Ueberreiter, K., and Naghizadeh, J. (1972). *J. Kolloid Z.Z. Polym.* **250,** 927.

Uejo, H. S. (1970). *J. Appl. Polym. Sci.* **14,** 317.

Uematsu, Y., and Uematsu, I. (1959). *Rep. Prog. Polym. Phys. Jpn.* **2,** 27.

Van Krevelen, D. W. (1972). "Properties of Polymers." Am. Elsevier, New York.

Wall, L. A. (1960). *SPE J.* **16,** (8), 810, 1031.

Wall, L. A., Roestamsjah, and Aldridge, H. (1974). *J. Res. Natl. Bur. Stand., Sect. A* **78,** 447.

Ward, I. M. (1971). *J. Mater. Sci.* **6,** (11), 1397.

Weeks, N. E., and Porter, R. S. (1975). *J. Polym. Sci., Polym. Phys. Ed.* **13,** 2049.

Weigel, P., Hirte, R., and Ruscher, H. (1974). *In* "Thermal Analysis" (I. Buzas, ed.), Vol. 2, p. 43. Heyden, London.

Weitz, A., and Wunderlich, B. (1974). *J. Polym. Sci., Polym. Phys. Ed.* **12**, 2473.

Wesslen, B., Lenz, R. W., MacKnight, W. J., and Karasz, F. E. (1971). *Macromolecules* **4**, 24.

Wiley, R. H., and Brauer, G. M. (1948). *J. Polym. Sci.* **3**, 647.

Wilhoit, R. C., and Dole, M. (1953). *J. Phys. Chem.* **57**, 14.

Wolpert, S. M., Weitz, A., and Wunderlich, B. (1971). *J. Polym. Sci., Polym. Phys. Ed.* **9**, 1887.

Wong, K. C., Chen, F. C., and Choy, C. L. (1975). *Polymer* **16**, 649.

Woods, D. W. (1954). *Nature (London)* **174**, 753.

Wrasidlo, W. (1974). *Adv. Polym. Sci.* **13**, 5.

Wunderlich, B. (1973). "Macromolecular Physics," Vol. 1, Sect. 2.4.8. Academic Press, New York.

Wunderlich, B. (1976). "Macromolecular Physics," Vol. 2, pp. 259, 266. Academic Press, New York.

Wunderlich, B. (1980). "Macromolecular Physics," Vol. 3. Academic Press, New York.

Wunderlich, B., and Baur, H. (1970). *Adv. Polym. Sci.* **7**, 151, 281.

Wunderlich, B., and Jones, L. D. (1969). *J. Macromol. Sci., Phys.* **B3**, 67.

Wunderlich, B., and Weitz, A. (1974). *In* "Thermal Analysis" (I Buzas, ed.), Vol. 2, p. 3. Heyden, London.

Wunderlich, B., Bodily, D. M., and Kaplan, M. H. (1964). *J. Appl. Phys.* **35**, 95.

Yagfarov, M. Sh. (1968). *Vysokomol. Soedin., Ser. A* **10**, 1267.

Yagfarov, M. Sh. (1969). *Vysokomol. Soedin., Ser. A* **11**, 1195.

Yagi, T., Tatemoto, M., and Sako, J. (1980). *Polym. J.* **12**, 209.

Yamamoto, Y., Tsuge, S., and Takeuchi, T. (1971). *Bull. Chem. Soc. Jpn.* **44**, 1145.

Yasuniwa, M., and Takermura, T. (1974). *Polymer* **15**, 661.

Yeh, G. S. Y., and Geil, P. H. (1967). *J. Macromol. Sci., Phys.* **B1** (4), 235, 251.

Yip, H. K., and Williams, H. L. (1976a). *J. Appl. Polym. Sci.* **20**, 1209.

Yip, H. K., and Williams, H. L. (1976b). *J. Appl. Polym. Sci.* **20**, 1217.

Young, C. L. (1968). *Chromatogr. Rev.* **10**, 129.

Zimmermann, H., and Becker, D. (1973). *Faserforsch. Textiltech.* **24**, 479.

Zimmermann, H., and Chu, D. (1973). *Faserforsch. Textiltech.* **24**, 445.

Zimmermann, H., and Schaaf, E. (1974). *In* "Thermal Analysis" (I. Buzas, ed.), Vol. 2, p. 137. Heyden, London.

Zimmermann, H., Schaaf, E., and Seganowa, A. (1971). *Faserforsch. Textiltech.* **22**, 255.

CHAPTER 4

Block Copolymers and Polyblends

SHALABY W. SHALABY

Research and Development Division
Ethicon, Inc.
Somerville, New Jersey

HARVEY E. BAIR

Bell Laboratories
Murray Hill, New Jersey

I. Introduction

As growth in polymer applications increases, we are almost always faced with a parallel increase in demands for new systems. To synthetic polymer scientists (who may be called macromolecular engineers) and polymer technologists, new systems have meant (a) totally

ISBN 0-12-703780-2

new polymers, (b) new "simple" copolymers of known monomers, (c) melt-blends, (d) composites of thermoplastic polymers and inorganic or organic fillers, (e) chemically modified known, or old, polymers, and (f) block and graft copolymers.

New thermoplastic polymers as well as the so-called new "simple" copolymers, which are in effect random and alternating copolymers, are discussed in Chapter 3. Unfortunately, with the exception of the very few known alternating copolymers, most random copolymers have less-than-desirable properties as thermoplastic systems and accordingly will be given limited attention in this chapter. Plasticized polymers are treated in Chapter 9 and will not be dealt with here to any significant extent. Some chemically modified polymers were covered in Chapter 3, but one large segment of this type of polymer (graft copolymers) was not, and will be treated here. Therefore, Chapter 4 deals primarily with (a) block and graft copolymers that are either commercial or experimental, (b) melt-blends of two or more polymers, at least one of these polymers being a thermoplastic material, and (c) composites based on thermoplastic polymers and organic or inorganic fillers.

The last part of this chapter will review a special case of composite polymers—impact-modified plastics. Typically, these commercially important systems consist of a rigid matrix and a dispersed rubber phase. The former phase is often composed of a polymer or copolymer and the latter an unsaturated rubber. In addition, graft and block copolymerization methods are used commonly to enhance the adhesion between the rubber and the matrix. The thermodynamic criteria for compatibility for these and other polyblends will be presented and thermal analysis techniques for determining the miscibility of polymers, in general, will be reviewed. Lastly, the utilization of quantitative C_p measurements for assaying the composition of polyblends will be described.

II. Block and Graft Copolymers

A. INTRODUCTION

Block copolymers are made of two or more monomeric moieties that are present as long sequences in the chain. When the blocks are present end to end in a linear structure, the copolymer is referred to as a *block copolymer*. On the other hand, if the copolymer is made of several grafts on the main chain, the copolymer is normally called a *graft copolymer*. Special types of block copolymers have a starlike arrangement of the blocks; these are given the name *star copolymers.* When the copolymers can be made totally of homochain or heterochain blocks, they are referred to as *homoblock* or *heteroblock copolymers,* respectively. If the copolymer is composed of both homochain and heterochain blocks, it will be referred to in this chapter as a *homo-heteroblock copolymer.*

Synthetic block and graft copolymers constitute a group of relatively new materials, a few of which are now commercial products. The polyether polyesters (Hytrel), polyether urethanes (Lycra and Biomer), and polystyrene polydienes (Kraton) are the most important members of the commercial block copolymers which behave both as thermoplastics and elastomers and hence are commonly referred to as elastoplastics or thermoplastic elastomers.

This section will be limited to the use of thermal analysis in studying the effect of important chemical and physical parameters on the thermal and mechanical properties of commercial and experimental block and graft copolymers.

B. BLOCK AND GRAFT COPOLYMERS AS MULTICOMPONENT SYSTEMS

Less than 20 years ago, it was rather common to use the term *multicomponent systems* synonymously with plasticized polymers, polymer blends, and composites of polymers and inorganic fillers where the components of each system could be recognized microscopically as two or more well-defined phases or, better called, macrophases. In graft and block copolymers with long sequences, one can normally recognize the consistent components as separate phases only by electron microscopy. The size of the phases in block and graft copolymers ranges from less than 100 to several hundred angstroms. Hence, these are usually called microphases. A two-component microphase system consists of a uniformly dispersed phase or domains and a matrix. Inversion of phases is dependent on the copolymer compositions and the conditions under which the microphases are allowed to reach ambient conditions. The reader must be cautioned that *not* all block and graft copolymers are capable of domain formation. Only those systems having blocks of certain large dimensions and a reasonable level of incompatibility between the blocks are capable of domain formation. For a detailed discussion of the chemical aspects of block and graft copolymers, the reader is referred to several reviews that have appeared over the last 15 years (Aggarwal, 1970; Allport, 1973; Battaerd and Tregear, 1967; Ceresa, 1973; Estes *et al.,* 1970; Noshay and McGrath, 1977).

C. Typical and Atypical Applications of Thermal
and Complementary Techniques

Examples for the investigative use of thermal and athermal analytical
techniques for studying commercial and experimental block and graft copolymers are discussed in the rest of this section.

1. Detection of Block Formation in "Randomly Formed" Heterochain Copolymers

During the random copolymerization of two comonomers that are
greatly different in reactivity, block formation may take place instead of
or along with random copolymerization. In copolyamides of m-xylylene
diammonium isophthalate (MXD—I) and caprolactam (CL) having 31–
91% of MXD–I, block formation was detected in certain compositions
(Shalaby *et al.*, 1976c). Using DSC, the T_g of those copolyamides was
shown to decrease with the decrease in their aromatic (MXD–I) content.
The T_g of a typical copolyamide with an MXD–I mole fraction of 0.34 was
about 50°C higher than that of Nylon 6. A few of the copolyamides exhibited noticeable deviation from the rest of the copolymers when solution
viscosity data were plotted against composition. Similar deviation was
displayed by a few copolymers (I–C, I–D, I–F, and I–G) when a linear
plot of the T_g was constructed, as shown in Fig. 1. These deviations were
associated with the presence of block structures in the respective copolymers.

Block formation was reported to occur upon copolymerizing, "randomly," a mixture of caprolactam (CL) and β-(4-aminophenyl)propionic
acid (APP) (Shalaby *et al.*, 1976a). When a series of copolymers having
between 5 and 90% of the APP moieties was prepared, the 50/50 CL/APP

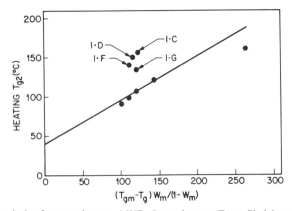

Fig. 1 Wood plot for caprolactam–MXD–I copolymer. (From Shalaby *et al.*, 1976c.)

copolymer displayed two melting endotherms. One of these endotherms (at 225°C) was attributed to Nylon 6 moieties, whereas the second one was ascribed to a copolymeric block made mostly of APP units. This was suggested to indicate the formation of blocks during the copolymerization of two comonomers (CL and APP) with different reactivities, a situation that can be most expected to result in blocks at the 50/50 comonomer composition. The DSC data supporting this conclusion are summarized in Table I. Note that even in copolymers having up to 15% APP units, the

TABLE I

DSC Data of CL and APP Copolymers
and the Corresponding Homopolymers[a]

	Polymer number						
	I	II	III	IV	V	VI	VII
% CL	100	95	90	85	50	10	0
% APP	0	5	10	15	50	90	100
DSC data:							
Initial heating T_m (°C)	219	221	222	218	225	360	360
					335		
Reheating T_g (°C)	40	42	43	44	96	[b]	[b]
T_c (°C)	69	69	69	72	128	[b]	[b]
T_m (°C)	220	220	220	216	225	[b]	[b]
					335		

[a] From Shalaby *et al.* (1976b).
[b] Could not be detected upon reheating the quenched melt.

T_ms of the copolymers were essentially the same as that of Nylon 6 itself. This is another support for the block formation proposal. Similarly, one can attribute to block formation the undetectable difference in T_m between the 10/90 CL/APP copolymer and the APP homopolymer. It must also be added that a similar conclusion could as easily be drawn on the basis of the x-ray diffraction data of these systems.

2. Block Formation through Interchange Reactions

Melt-blending of two heterochain polymers may lead in some instances to block formation. The extent of this reaction is dependent on both reaction temperature and time. Excessive interchange reaction may first lead to formation of short blocks and then random copolymeric chains. Both TGA and DSC have been used for studying block formation under these conditions, as illustrated in the next few paragraphs.

Block formation was detected when Nylon 6 granules coated with poly(methylphosphacaprolactam) (PMPC) were heated at 260°C for 20 min in the capillary of an Instron rheometer (Shalaby *et al.*, 1976b). The blend was extruded and the resulting fibers were extracted with water or methanol to remove unreacted PMPC. The block formation by amide interchange reaction in the melt was verified by (a) the presence of phosphorous moieties in the extracted polymer, as determined by elemental analysis; (b) the decrease in melt viscosity during extrusion (Nylon 6 itself has a stable melt viscosity at 260°C for the noted residence time); (c) the decrease in solution viscosity with the increase of the low degree of polymerization (DP) PMPC fractions used for blending; and (d) the unaffected T_m of Nylon 6. Some of the experimental data supporting these conclusions are summarized in Table II.

TABLE II

THERMAL AND OTHER PROPERTIES OF NYLON 6-PMPC MELT BLENDS
AND CORRESPONDING HOMOPOLYMERS[a]

	Polymer number				
	I	II	III	IV	V
Initial blend data:					
% Nylon, η_{red}	100/1.75	95/1.75	95/1.75	75/1.75	—
% PMPC, η_{red}	0/0.33	5/0.19	5/0.33	25/0.33	100/0.33
Blend reduced viscosity	[b]	1.36	1.46	1.35	[b]
Melt rheology data:					
Melt viscosity (poise × 10^{-3},					
After 15 min flow time	5.2	1.4	1.8	1.3	—
After 25 min flow time	5.9	1.1	1.4	0.8	—
Extrudate properties after extraction:					
η_{red}	1.85	1.43	1.65	1.59	—
DSC initial T_g (°C)	30	—	—	—	52
T_m (°C)	219	220	220	—	—
Reheating, T_g (°C)	42	42	43	43	90
T_c (°C)	69	71	69	70	—
T_m (°C)	219	219	220	219	—
Approx. % interchange based % P	[b]	20	81	38	[b]

[a] From Shalaby *et al.* (1976b).
[b] Homopolymer.

Melt-blending of three parts Nylon 6 (η_{red} = 3.36) and one part poly[β-(4-methylpiperidyl)propionamide] (PAMP) (η_{red} = 0.5) in a Brabender extruder was shown to lead to different levels of block formation through amide interchange reaction, depending on the melt residence time in the extruder (Shalaby *et al.*, 1978). Blending for 3 and 9 min produced systems having 13 and 8% extractables and reduced viscosities of 3.13 and 2.65, respectively. The extraction of less than 25%, which represents the initial content of PAMP in the system, and the drop in solution viscosities were used to propose block formation upon melt-blending. Additional data used to support the proposed block formation included the higher T_g and T_c of the "block copolymer" (made by melt-blending for 9 min) as compared with those of Nylon 6, as shown in Table III.

TABLE III

DSC AND DTA DATA OF NYLON 6, PAMP, AND THEIR "BLOCK COPOLYMER"[a]

	Nylon 6	Nylon 6–PAMP	PAMP
DTA data:			
Initial heating T_g (°C)	—	62	—
T_m (°C)	—	—	212
DSC data:			
Reheating T_g (°C)	35	55	48
T_c (°C)	67	—	85
T_m (°C)	220	—	223

[a] From Shalaby et al. (1978).

Recently, a series of elastomeric polyether polyester copolymers were prepared by ester interchange reaction in the melt, and their thermal behavior, among other properties, was correlated with their chemical compositions (Wolfe, 1978). Thus block copolymers made by melt-blending mixtures of 30/70, 40/60, 50/50, and 82/18 poly(tetramethylene terephthalate) and poly(polyoxybutylene terephthalate) were shown to have T_m values of 152, 172, 189, and 212°C, respectively. By changing the diol of poly(alkylene terephthalates) melt-blended with equal amounts of poly(polyoxybutylene terephthalate), the resulting block copolymers exhibited T_m values of 224, 198, 189, 106, 122, and 106°C when diols having 2, 3, 4, 5, 6, and 10 methylene, respectively, were used.

3. The Study of Multiple Glass Transitions in Copolymers

Multiple glass transitions of polymers have been reported previously by Angelo et al. (1965), in a treatment of the glass transition in block copolymers and mixtures of incompatible homopolymers. Krause and Roman (1965) reported that mixtures of compatible homopolymers, on the other hand, exhibit a single glass transition. Accordingly, if phase mixing takes place in block copolymers, the system will be expected to display one T_g. Extending this argument to random copolymers, these systems ought not to have more than a single glass transition. However,

Chandler and Collins (1969) found that copolymers of butadiene and acrylonitrile having more than 36% acrylonitrile had a single T_g whereas those with less than 36% acrylonitrile were shown by DTA to have two glass transition temperatures. Typical T_g data for different copolymer compositions are summarized in Table IV as a function of comonomer ratios. The

TABLE IV

DTA T_g VALUES OF DIFFERENT COPOLYMERS (°C)[a]

Polymer no.	% Acrylonitrile unit in the copolymer	First heat		Reheating	
1	63.0	—	− 7	—	− 4
2	51.0	—	−14	—	−12
3	35.0	−32	−27	−31	−23
4	29.3	−54	−30	−52	−29
5	20.0	−69	−35	−69	−33
6	11.1	−77	—	−77	—
7	0.0	−85	—	−85	—

[a] From Chandler and Collins (1969).

presence of two T_gs was considered to be unrelated to simple physical separation of two incompatible components but the exact cause of the two T_gs was not clearly defined. It is our opinion that the display of two T_gs is related to the presence of two types of blocks, one rich in acrylonitrile units and the other rich in butadiene units. This is consistent with the vast difference in the reactivity ratios of the two monomers, a situation that can very easily lead to block formation.

4. Monitoring the Effect of Hydrogenation on the T_g of Unsaturated Systems

Triblock copolymers of poly(styrene-*b*-butadiene-*b*-styrene) (S–B–S) and poly(styrene-*b*-isoprene-*b*-styrene) (S–I–S), widely used as elastoplastics, suffer severe drawbacks in their applications because of poor aging properties of the unsaturated soft phase. In principle, these properties could be improved by introducing saturated soft phases and hard phases with higher T_g or T_m. This was skillfully achieved by hydrogenating the S–B–S and S–I–S block copolymers (Zotteri and Giuliani, 1978). Thus, by hydrogenation, the soft phase becomes a saturated rubber and the hard phase a polyvinylcyclohexane block with a T_g nearly 50°C higher than that of polystyrene. Although the effect of hydrogenation of the soft phase on the aging properties is well known, only recently was its effect on the elastic properties of the copolymer examined (Zotteri and Giulianai, 1978). The dynamic mechanical spectra of hydrogenated S–B–S and S–I–S revealed an increase in the T_g of the hard segments (formerly polystyrene) by about 40–50°C in both polymers. The effects of T_g increase on the long-term elastic properties as a function of temperature were different for the two copolymers.

5. Melting and Glass Transition of Radiation-Induced Graft Polyethylene

It is well documented that the physical and mechanical properties of a polymer can be changed by radiation-induced grafting of side chains. Correlation between these changes with changes in polymer morphology has attracted the attention of several investigators, including those who have selected styrene-grafted high-density polyethylene (PE) for their studies (Toi *et al.*, 1977). Melting and glass transition data obtained from DSC measurements were reported for styrene-grafted high-density PE. The grafting was achieved in the presence of γ radiation. Judging by the T_m and heat of fusion data, it was concluded that the polystyrene grafts have no effect on the PE crystallites. However, the half-width of the melting endotherm was observed to increase slightly in the grafted polymer, an effect that was attributed to the increase in the crystallite size distribution. It was also shown that the polystyrene grafts do not affect the PE amorphous phase and hence its T_g remained unaffected. Thus it was suggested that the free volume or segmental mobility does not decrease by radiation-induced grafting. Note that the independence of T_g on the PS grafts is questionable, particularly if we acknowledge the literature arguments on the value of T_g of polyethylene. For instance, Beatty and Karasz (1979) advocated a T_g of 145 K, while Gaur and Wunderlich (1980) stressed that polyethylene undergoes a glass transition at 237 K.

6. The Study of Microphase Separation

Styrene–butadiene–styrene (S–B–S) triblock copolymers have been commonly used for studying microphase separation in block copolymers. The phase separation in these copolymers has been studied extensively by x-ray diffraction and electron microscopy (Hendus *et al.*, 1967; Dluglasz *et al.*, 1970). Solvent cast films were found on annealing to demonstrate a variety of morphologies ranging from spheres to rods to lamellae, depending on the ratio of styrene to butadiene and on the solvent used in casting (Fisher, 1968; Lewis and Price, 1971). Well-developed morphologies could be obtained either by extrusion (Dluglasz *et al.*, 1970) or by compression molding, followed by high temperature annealing (Pedemonte *et al.*, 1976). Although the viscoelastic relaxation of such materials has been studied extensively (Smith and Dickie, 1969; Shen and Kaelble, 1970; Lin *et al.*, 1971; Cohen and Tschoegl, 1972), dielectric (Pochan and

Crystal, 1972) and ultrasonic relaxation studies (Shen *et al.*, 1973) have been relatively few. In general, the relaxation spectrum of the styrene–butadiene block copolymers is composed of two principal features: the high-temperature loss occurring at about 393–413 K, depending on the frequency of observation and characteristic of the glass transition process of the polystyrene domains, and the low-temperature loss occurring at approximately 223 K associated with the relaxation of polybutadiene blocks. Nuclear magnetic relaxation studies (Wardell *et al.*, 1976) have recently been reported on SBS macroscopic single crystals, formed by extrusion. The sample had a unique morphology composed of polystyrene cylinders arranged in a hexagonal array parallel to the extrusion axis within a continuous polybutadiene matrix. The effects of magnetic spin diffusion were quite evident in these studies and were noted to lead to an increase in the magnitude of the butadiene spin-lattice relaxation time (T_1) minima at the expense of the styrene process. Studies of T_1 and the spin–spin relaxation time (T_2) indicated minima at 233 and 373 K, whereas the rotating-frame experiment (T_{1_ρ}) led to the observation of minima at 233 K. Consideration of these data in terms of various models indicated the absence of molecular orientation in either component phases but did not reveal the effects that morphology may have on the observed relaxation process. This led Datta and Pethrick (1977) to investigate the effect of morphology on the relaxation of these systems. Thus acoustic absorption and velocity measurements were conducted over a temperature range from 213 to 293 K on a series of linear and four-arm styrene–butadiene–styrene triblock copolymers. The loss peak associated with the segmental motion of the butadiene block was shown to be sensitive to the composition (styrene-butadiene ratio) and structure of the copolymer. The results of these studies also showed that (a) the effect on the chain dynamics of chemical "cross-linking" (i.e., in the star copolymer) can clearly be observed by the appearance of an additional loss peak; (b) anisotropy as a result of the development of a high degree of morphological perfections can be observed by the velocity measurements; and (c) distinct effects of morphology and "cross-linking" can be vividly demonstrated.

7. Effect of Branching on Properties of Block Copolymers

Studies on the effect of branching on the properties of block copolymers were limited essentially to those of polystyrene (S) and polybutadiene (B). The rheological behavior of butadiene–styrene block copolymers was considered to be determined by their two-phase domain structure, which persists in the melt (Kraus *et al.*, 1967; Arnold and Meier, 1970; Bianchi *et al.*, 1970), i.e., at temperatures exceeding the T_g of the polystyrene domains. Since the association of polystyrene blocks into domains represents a type of physical branching (in B–S and B–S–B block copolymers) or cross-linking [in S–B–S and (S–B)$_n$ block copolymers], the effects of additional chemical branching were considered to be not predictable (Kraus *et al.*, 1971). This led Kraus *et al.* (1971) to study the effects of symmetrical star-branching on the viscoelastic behavior of butadiene–sytrene block copolymers with narrow molecular-weight distribution and having the following structures: B–S–B, (BS–)$_3$, (S–B–)$_3$ and (SB–)$_4$. At constant molecular weight and total styrene content, viscosities were greater for polymers terminating in styrene blocks, irrespective of branching. Branching was found to decrease the viscosity of either polybutadiene-terminated or polystyrene-terminated systems, compared at equal $\bar{M}_w$. However, comparison at equal block lengths showed that the length of the terminal block, not the total molecular weight, governs the viscoelastic behavior of these polymers to a surprisingly good approximation. This unusual result was rationalized in terms of two-phase domain structure of these polymers that persists to a significant degree in the melt. The persistence of the domain structure in the melt was clearly demonstrated by the loss tangent data. The dynamic mechanical measurements indicated that at 130°C, even short polystyrene blocks were suitable for producing networks with relatively low loss values, but not at 160°C. When the block length was large, tan δ was smaller at the higher temperature, but this did not indicate a higher storage modulus. Thus the network deteriorates with rising temperature at all block lengths, but does so more rapidly when the blocks are short. At 160°C, where the network becomes increasingly imperfect, the effect of branch points becomes apparent in somewhat lower values of tan δ.

8. The Study of Crystallization-Induced Block Formation

Solid-state isomerization of poly(ester acetals) and the related double-bond isomerization reaction of 1,4-polybutadiene were reported several years ago by Lenz and his co-workers (1970, 1973a,b). It was reported by these investigators that any reversible isomerization or reorganization reaction that can be carried out on the repeating units of semicrystalline copolymers should be capable of being forced in an antiequilibrium direction by permitting expansion of existing crystalline regions through incorporation of newly reacted units. If the units in the crystalline regions are inaccessible and unreactive, the copolymer will either increase in composition of the crystallizable units beyond the equilibrium amount for an isomerization reaction or be converted from a random to a block structure without compositional change for a reorganization reaction. The principle of reversible reorganization reaction was then applied, successfully, to the ester–ester interchange reaction of polyesters (Lenz and Go, 1973).

Thus random copolymers of *cis-* and *trans*-1,4-cyclohexylenedimethylene terephthalate (PCHDT) were prepared and permitted to undergo ester-interchange reactions at temperatures just below their T_ms. As predicted from the principle of crystalline-induced reactions of semicrystalline copolymers, the terephthalate copolymers were shown to undergo reorganization through the ester-interchange reactions. This was shown by the observed changes in physical properties (including thermal ones) that are associated with the conversion of a random to a block copolymer. Supporting data for the formation of block structures was obtained by monitoring the solubility, crystallinity, and changes in T_c and T_m as a function of copolymer composition. Some of these data are shown in Table V and Figs. 2 and 3.

TABLE V

MELTING AND RECRYSTALLIZATION TEMPERATURES
OF CIS/TRANS PCHDT OF DIFFERENT COMPOSITIONS[a]

Trans content (%)	Initial copolymer		Crystallization-induced reaction products	
	T_m (°C)	T_c (°C)	T_m (°C)	T_c (°C)
5	242	218	245	220
50	269	240	279	246
70	280	256	293	263

[a] Lenz and Go (1973).

The effects of time and temperature on the extent of this crystallization-induced reaction (or reorganization) were carefully examined and typical results are shown in Table VI. It was believed that the reorganization is responsible for the direct preparation of block copolymers by the solid-state polycondensation reaction used in these studies.

As can be realized from the data presented previously, the formation of block copolymers from random ones was based on the observed increase in T_m and crystallinity. The decrease in T_m and crystallinity of the reorganized sample by heating in the presence of an active catalyst was considered (Lenz and Go, 1974) as additional evidence for the block formation through reorganization.

The application of ester-interchange reactions to the crystallization-induced reorganization of block copolymers discussed previously was further supported by quantitative studies of this reaction (Lenz and Go, 1974). Structural changes based on dyad compositions in copolyesters of terephthalic and 2-methylsuccinic acids with ethylene glycol were deter-

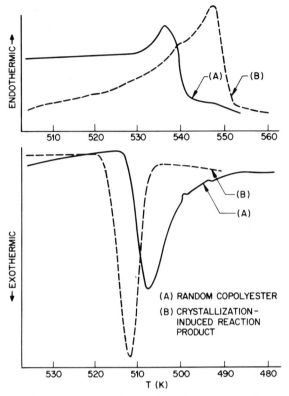

Fig. 2 Melting (above) and recrystallization (below) thermograms from DSC analysis of 50% PCHDT copolymer. (From Lenz and Go, 1973).

mined by high-resolution NMR. Calculated sequence length distributions were related to the effects of copolymer composition, catalyst concentration, and reaction temperature on the rate of the ordering (or reorganization) and to the effect of structure on crystalline properties. Results of these studies were quite consistent with earlier conclusions based on the thermal analysis data.

9. Thermal Stability of Block Copolymers

So far we have discussed the effect of several structural and physical parameters on the thermal transitions of block copolymers. On the other hand, one must not forget that the effects of those parameters on the thermal stability of block copolymers are quite relevant to their performance as elastoplastics, especially if they are to be fabricated by melt-processing. To illustrate these effects, the thermal stability of segmented polyurethanes will be treated in this section.

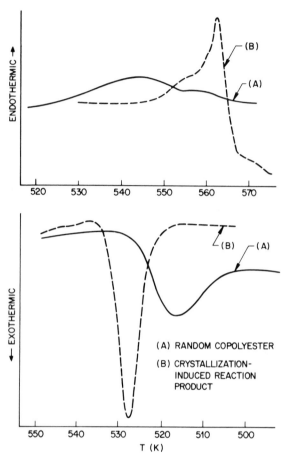

Fig. 3 Melting (above) and recrystallization (below) thermograms from DSC analysis of 70% PCHDT copolymer. (From Lenz and Go, 1973.)

During early studies on chemical-structure–physical-property relationships (Ferguson and Patsavoudis, 1972, 1974; Ferguson *et al.*, 1972) in melt-spun fibers of segmented polyurethanes [$(AB)_n$ type of block copolymers], it was noticed that these polymers were relatively less stable thermally than the more conventional fiber-forming polymers. In an effort to assess, systematically, to what extent the thermal stability is structure dependent, two series of segmented polyurethanes were prepared and their thermal stabilities were examined by both DTA and TGA (Ferguson and Petrovic, 1976). The polyurethanes were synthesized from trimethylene diamine; 4,4'-diphenylmethane diisocyanate and two polyether-based (polytetrahydrofuran, PTHF) macrodiisocyanates. The soft segment con-

TABLE VI

Progress of Crystallization-Induced Reaction
as a Function of Temperature and Time[a]

Polymer	Temp. (°C)	Time (h)	T_m (°C)	Relative crystallinity
70 % *trans*-PCHDT	285	0	298	1.00
	285	1	321	1.76
	285	1.5	322	1.93
	325	1.5	329	0.96
50% *trans*-PCHDT	250	0	282	1.00
	250	1	309	3.42
	250	2	313	3.59
	250	10	320	5.05

[a] From Lenz and Go (1973).

centration was varied from 0 to 70.8% for the first series and kept at 82.8% for the second series. By comparing the changes in weight loss and DTA peaks with chemical structure, it was possible to distinguish between soft segment and hard segment degradation. In the initial stages of degradation, the stability was shown to increase as the soft segment concentration increases, whereas the reverse effect was found to be true at the later stages of degradation. The 100% soft segment polymer (PTFH) and the 100% hard segment polymers appeared to behave dissimilarly to their corresponding blocks in the copolymers. It was suggested that the hard segment has a stabilizing effect on the degradation of the soft segment.

10. Comparison of Block Copolymers with Random Copolymers

One may consider block copolymers as intermediates between random copolymers and polymer blends in that long sequences are covalently linked to form block copolymers. In terms of thermal transitions and other thermal properties, the block copolymers are relatively more similar to polymer blends than to random copolymers. The thermal properties of some random and block copolymers were discussed sporadically in different contexts in this and the previous chapters. Some recent comparative studies between random and block copolymers are outlined in this section. This is intended to help the reader develop an appreciation for the differences between these systems and select a proper approach for the detection of random copolymers as impurities in block copolymers or vice versa.

Whereas volumetric properties (VP) have been reported for a variety of homopolymers, relatively few data have been published for random copolymers. Meanwhile, the VP data for block copolymers are extremely rare (Brandrup and Immergut, 1975). In a communication by Renuncio and Prausnitz (1977), experimental density measurements for copolymers of butadiene and styrene at 75 and 100°C and at pressures up to 1 kbar were reported. Measurements have been made for various copolymers from completely random to completely block copolymers. The experimental results were well represented by the Tait equation

$$v/v_0 = 1 - C \ln(1 + P/B) \tag{1}$$

where v/v_0 is the ratio of specific volumes at pressure P to that at 1 atm and B and C are constants. Renuncio and Prausnitz defined compressibility (β) by

$$\beta = (1/v_0)(\partial v/\partial p)_T \tag{2}$$

This compressibility is readily found from Eq. (1) to be

$$\beta = C(B + P)^{-1} \tag{3}$$

To a first approximation, specific volumes and compressibilities were linear functions of weight percent styrene. For a fixed composition, the effect of structure on volumetric properties was small. However, it was noted that when compared at the same overall composition, the specific volume and compressibility declines slightly when the styrene was in a block rather than random form.

11. Analysis of Block Copolymer Blends

Blends of poly-2,6-dimethyl-1,4-phenylene oxide (PPO) and the triblock styrene–butadiene–styrene (S–B–S) copolymers were selected as typical systems for illustrating the effect of a high-T_g polymer on the properties of an elastoplastic in a series of melt-blends. Kambour (1972) found that the blending of PPO with an S–B–S copolymer markedly elevated the elastomer's use temperature. The introduction of rather small amounts of PPO to the S–B–S copolymer raised the yield temperature of its "network"-stabilizing polystyrene domains to technologically advantageous levels. The enhancement of mechanical properties exceeded that based solely on the expected effects of T_g elevation. Shultz and Beach (1974, 1977) have studied in some detail the degree of correspondence between the thermooptical transition temperatures (T_{T0A}) in S–B–S/PPO blends and the T_{T0A} of polystyrene/PPO blends having the same styrene/aromatic ether unit composition ratios. It was shown that blending PPO with the S–B–S copolymer increases, monotonically, the softening temperature of the latter as measured by T_{T0A}. The T_{T0A} transition temperatures of the S–B–S/PPO phases were shown to approximate closely those of polystyrene/PPO blends having the same styrene/aromatic ether unit composition ratios.

D. THERMAL AND ATHERMAL ANALYSIS OF SPECIFIC TYPES OF COPOLYMERS

This section deals with the characterization of several types of block and graft copolymers grouped on the basis of their chemical structure. It was felt necessary not to limit ourselves to thermoanalytical methods in discussing the characterization of these copolymers, whose complexity makes it imperative to rely on more than one analytical technique in studying their properties most effectively.

1. Heterochain Block Copolymers

a. Polyamide Systems. Monitoring the isothermal degradation of *trans*-2,5-dimethyl piperazine-based copolyamides, synthesized by a two-stage solution, polycondensation was used to study the presence of block structures in these systems (Bruck and Levi, 1967; Bruck, 1969; Bruck and Thadani, 1970). Comparison of the isothermal weight loss curves of random with those of block copolyamides showed a clear difference in behavior in that random copolymers were much less stable than block ones, as shown in Figs. 4 and 5.

The structure and properties of aliphatic copolyamides, obtained by interfacial polycondensation, with emphasis on their T_ms, were studied by Ke and Sisko (1961). The DTA curves of poly(hexamethylene adipamide-*co*-sebacamides) revealed a shift in the position of the melting peaks toward lower T_m than that of Nylon 66; the peaks were reported to widen and their area decreased as the 50/50 copolymer composition was approached.

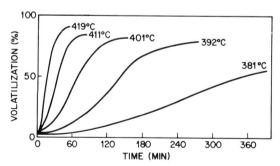

Fig. 4 Isothermal degradation of 50/50 mole % *random* copolymer of terephthaloyl *trans*-2,5-dimethylpiperazine/isophthaloyl *trans*-2,5-dimethylpiperazine. (From Bruck and Levi, 1967; Bollinger, 1977–1978. Reprinted by courtesy of Marcel Dekker, Inc.)

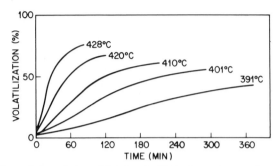

Fig. 5   Isothermal degradation of 50/50 mole % *block* copolymer of terephthaloyl *trans*-2,5-dimethylpiperazine/isophthaloyl *trans*-2,5-dimethylpiperazine. (From Bruck and Levi, 1967; Bollinger, 1977–1978. Reprinted by courtesy of Marcel Dekker, Inc.)

The DTA curves of copolyamides containing 50 and 60 mole % of se-bacamides exhibited rather broad melting endotherms. This was associated with the formation of block copolymeric chains. Block formation in melt-blends of poly(*p*-diphenylene adipamide) and poly(*p*-diphenylene terephthalamide) could not be ascertained from their TGA data and those of the homopolymers prior to mixing (Bollinger, 1977–1978).

Thermal techniques have been used by Marupov *et al.* (1977) in studying the heat resistance of Nylon 6–polyacrylonitrile graft copolymers. The DTA curves of Nylon 6 (N-6) with 9.8% polyacrylonitrile (PAN) grafts indicated no exothermal effect of cyclization, typical for pure PAN. However, there was a strongly expressed exothermal effect of cyclization, with a maximum below 330°C, in the graft copolymer modified with sodium sulfide. It was also noted that in PAN, increased thermal effects of cyclization may be observed below 260°C. The thermal behavior of N-6/PAN copolymer, as shown by DTA, was suggested to indicate that intramolecular cyclization of the PAN moieties increased in the presence of Na_2S. Isothermal TG data obtained at 150°C indicated that thermal stability decreased in the order PAN, N-6/PAN and N-6/PAN + Na_2S. The low thermal stability of the latter system was ascribed to a large intramolecular cyclization.

Nylon 6–polyoxybutylene, prepared by end-grafting caprolactam onto poly(oxybutylene diamine), was analyzed by DSC for the presence of a macromolecular species other than the expected A–B–A block structures (Shalaby *et al.*, 1973a). The DSC thermograms of these copolymers were characterized by a maximum of three melting endotherms (X, Y, and Z), depending on the nylon-to-polyether ratio in the system, as summarized in Table VII. The endotherm near 220°C (endotherm Z) was ascribed to Nylon 6 melting. The X and Y endotherms were attributed to the polyether blocks in the copolymers. The authors used the heat fusion

TABLE VII

DSC DATA OF NYLON 6–POLYOXYBUTYLENE END-GRAFTS[a]

Polymer no.	I	II	III	IV	V	VI
Molar % of nylon in the system	74	65	54	50	45	44
Initial heating data T_m, °C/ΔH_f cal g^{-1}						
Endotherm X	15.5/2.4	14/1.2	15/0.24	17.5/2.7	16/2.2	18/5
Endotherm Y	0/0	39/1.3	38/1.7	42/1.1	44/3.9	41/1.0
Endotherm Z	223/11.9	219.5/104	222/8.4	221/10	219/9.3	215/9.9
Reheating data T_m, °C/ΔH_f cal g^{-1}						
Endotherm XY[b]	20/2.3	21/2.6	21/5.2	22/4.8	21/6.4	23/6.7
Endotherm Z	220/12.3	218.5/9.8	218/8.4	217/9.4	218/8.0	216/8.3

[a] From Shalaby *et al.* (1973a).

[b] The X,Y endotherms displaced by the copolymer upon reheating form one common endotherm (XY) during reheating.

and chemical composition data to show that the X endotherm is due to the central polyether block in an A–B–A arrangement. The higher-T_m Y endotherm was associated with the polyether in a diblock system. Thus it was noted that the polyether blocks in a triblock structure do not crystallize as freely as their counterparts in the diblock system. The presence of two types of polyether blocks was further confirmed by dynamic mechanical properties data obtained on a rheovibron (see Fig. 6). The data indicated the presence of two types of blocks in copolymer V, which contains 55 mole *g* polyether. The two components α_c and α_d of Nylon 6 loss peak

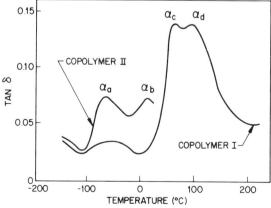

Fig. 6 Temperature dependence of tan δ of two polyoxybutylene–Nylon 6 block copolymers. (From Shalaby *et al.*, 1973b.)

were proposed to be due to nylon moieties in A–B–A and A–B (or nylon homopolymer) block structures, respectively.

Diblocks and triblocks were reported to be present in end-grafted polyoxybutylene with laurolactam (Shalaby *et al.*, 1973b). Arguments similar to those presented previously for Nylon 6–polyoxybutylene copolymers were used to support the proposed structure of the Nylon 12 polybutylene copolymer. Typical DSC data for the latter system are shown in Table VIII. Endotherms X and Y were related to polyether blocks in A–

TABLE VIII

DSC Data of Nylon 12–Polyoxybutylene End-Grafts[a]

Polymer no.	I	II	III	IV	V
Molar % of Nylon 12	68	60	23	19	17
Initial heating data: T_m, °C/ΔH_f cal g^{-1}					
Endotherm X	13.5/1.5	15/4.1	15/6.8	16.5/8.1	16.5/7.6
Endotherm Y	37/1.0	39/0.3	37.5/0.4	38/0.4	38/0.1
Endotherm Z	183,(171)[b]/7.8	181/8.9	177,(172)[b]/6.4	180.5,171.5/5.2	182,(173)[b]/4.3

[a] From Shalaby *et al.* (1973b).
[b] Small peak.

B–A and A–B structures. Endotherm Z was attributed to Nylon 12 blocks. The presence of a two-component endotherm for Nylon 12, near 170 and 180°C, was associated with Nylon 12 blocks in A–B–A and A–B (or Nylon 12 homopolymer) block structures, respectively.

Block copolymers of Nylon 66 ($\overline{M}_n \simeq 1000$) and poly-ε-caprolactone were prepared and their thermal properties were examined (Huet and Marechal, 1974). The T_g dependence on the block copolymer composition was studied. It was also noted that complete segregation (or microphase separation) can be complete.

b. Polyurethane Systems. Among the most common members of this class of block copolymers are those made of 2,4 and 2,6-toluene diisocyanate (TDI), 1,4-butanediol (BD), and polyoxybutylene (POB). The polyether and polyurethane long sequences are referred to as the soft and hard segments, respectively. In recent dynamic mechanical property studies of 2,4-TDI polyurethanes, it was reported that those made from the 2,4-TDI with POB having a molecular weight of 1000 (2,4-T-1P) display extensive phase mixing (Senich and MacKnight, 1978). This conclusion was based on the observed dependence of the position of the T_g relaxation on the hard segment content and absence of modulus enhancement at temperatures greater than T_g. On the other hand, polyurethanes made with POB having a molecular weight of 2000 (2,4-T-

2P) were shown to phase-separate. The latter copolymers displayed a low temperature and relatively composition-insensitive soft-segments glass transition as well as a considerable degree of modulus enhancement above the soft T_g. The level of modulus enhancement increased as more hard segments were incorporated in the copolymer. The 2,4-T-2P samples exhibited an increasingly well-defined hard segment T_g as the hard segment content was increased. The soft segment of 2,6-TDI copolymer, with POB having a molecular weight of 2000, gave evidence of crystalline melting, indicating that minimal phase mixing occurred in these systems.

The development of crystallinity in the polyoxybutylene soft segment of polyurethane block copolymers containing hard segments of piperazine extended with the bischloroformate of 1,4-butanediol, was monitored by DSC (Allegrezza *et al.*, 1974). For polyether segments having an average molecular weight of 1744, T_m was observed at 8 and $-10°C$ for samples having narrow and wide molecular-weight distribution, respectively. The polyether–urethane systems are treated in detail in Section II.E of this chapter.

Polyurethane block copolymers when used as elastomers do not have to be thermoplastic in all instances. This means that they can be cross-linked and used as conventional elastomers. It is rather difficult to assess with certainty the effect of cross-linking on the properties of polyurethane block copolymers due to the complexity of their structure. Therefore, it was felt appropriate to introduce the reader to this subject by reviewing some recent studies on the dependence of crystallization capability of somewhat simple polyurethanes on the degree of cross-linking (Slowikowska and Daniewska, 1975). Investigation of crystallization capability of polyurethanes has been carried out on two series of polymers in which the degree of cross-linking was systematically changed. The first series was obtained by the reaction of poly(oxyethylene glycols) (POE), having molecular weights of 4000 and 6000 and a triisocyanate. The molar ratio of triisocyanate to the POE was varied from 1 to 3. The second series was prepared by the polyaddition of poly(ethylene sebacate) to triisocyanate. Samples of poly(ethylene sebacate) (PS-2) with various molecular weights have been used in preparing different members of this series. The crystallization capability of the different polymers was monitored by the depolarized light method, optical microscopy and DSC. The analytical data indicated that there is definite dependence of crystallizability on the degree of cross-linking. The cross-linked polyurethanes crystallized only within defined ranges of chain length between the cross-links (or entanglements). In any of these ranges, the capability of crystallization and T_m were shown to decrease with the increase in the degree of cross-linking. This was attributed to the difficulties for crystal lattice formation by shorter PS-2 or

POE chains between the cross-links. Polyurethanes having very short POE or PS-2 chains did not crystallize at all. We feel that chain length is not the only factor affecting crystallization; crystallization will be retarded or inhibited if the chains are restricted from their motion to align for nucleation.

   c. Polyesters. Some of the polyester block copolymers are discussed in the chapter under polyamides or polysulfones since the second blocks of their chains are made from either of these chemical entities. Certain types of polyester block copolymers were discussed under interchange reactions for block formation and crystallization-induced reactions. Other polyester block copolymers, which belong strictly to this section, are discussed later.

   Block copolymers of polycaprolactone and bisphenol-A-based polycarbonates have $\overline{M}_n$ values of 1000–4500 were prepared and their thermal properties were studied (Huet and Marechal, 1974). The T_g dependence on the composition of the block copolymers was recorded. Only partial segregation (or microphase separation) was observed in these systems.

   Measurements of the dynamic mechanical properties were used along with other analytical techniques for the morphological characterization of polyester-based elastoplastics (Seymour *et al.*, 1975). A series of elastoplastic materials based on a semicrystalline poly(tetramethylene terephthalate hard segment and a polyoxytetramethylene (E-4) soft segment was prepared and used in these studies. Results from low-angle laser light scattering and polarized light microscopy revealed the presence of spherulitic superstructures over a wide range of compositions. Evidence for chain folding within the lamellae has been obtained by the chemical etching technique. Mechanical relaxation measurements were reported to suggest that the noncrystalline soft phase is a mixed phase of polyether soft segment and uncrystallized polyester hard segments. The response of these copolymers to uniaxial extension, as determined using low-angle laser scattering, was associated with deformation of the spherulites into ellipsoids having their major axis in the stretch direction. This deformation was only partially recoverable on release of load. These and other data led to the conclusion that these elastoplastics are more analogous, both in morphology and properties, to common semicrystalline thermoplastics than to block copolymer systems that undergo microphase phase separation into hard and soft domains. In the polyesters subject of these studies, the spherulitic crystallization persisted even in samples having very short number average, crystallizable block lengths. Although incompatibility of the hard and soft segments did not lead to hard–soft-phase separation in these systems, it was proposed to serve as an added driving force for hard-segment association and subsequent crystallization.

Interesting types of polyester block copolymers are those based on amorphous soft and crystalline hard heterochain blocks. Typical examples of these include isomeric block copolyesters based on crystalline and amorphous sebacate long sequences (O'Malley, 1975) and polyester–siloxane block copolymers (O'Malley and Stauffer, 1977). The isomeric polyester block copolymers were synthesized by coupling hydroxyl-terminated poly(hexamethylene sebacate) (HMS) and poly(2-methyl-2-ethyl-1,3-propylene sebacate) (MEPS) with hexamethylene diisocyanate. Poly(HMS) and poly(MEPS) represented the crystalline and amorphous blocks in the copolymer, respectively. Differences in thermal transitions between the block copolymer and mixtures of the homopolymers were studied using DSC. These mixtures were prepared by melt-blending, and rheological properties of the resulting blends were noted to indicate that the two isomeric homopolymers are not grossly phase separated above T_m, and thus the mixture was considered as a useful reference for judging the effect of copolymerization on the thermal transitions. The data in Fig. 7 show that both the T_ms and T_cs change very slightly ($\Delta T < 4°C$) as the crystallizable component, poly(HMS), in the mixture is decreased from 100 to 25%. This was reported to indicate that the P(HMS) phase separates and crystallizes readily in the mixture and that the amorphous component, poly(MEPS), does not substantially interfere with this process.

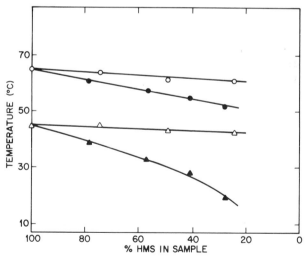

Fig. 7 The effect of sample composition on the melting and crystallization temperatures of poly(HMS) and poly(MEPS) homopolymer mixtures and block copolymers. Scan speed $10°$ min^{-1}. Open circles and triangles are melting and crystallization temperatures data, respectively, for mixtures and the corresponding filled points are for the block copolymers. (From O'Malley, 1975.)

However, we suggest the decrease in T_m, associated with melt-blending, may be due to partial ester-interchange reaction between the amorphous and crystalline phase, especially since they do not phase-separate grossly in the melt. Alternatively, the T_m drop could have been caused by the formation of less perfect crystallites in the blend, as compared with those of the homopolymer. For the block copolymer (Fig. 7), a T_m depression of ~12°C was observed as the crystallizable block concentration decreased from 100 to 28%. Again, we feel that this decrease in T_m can be related to the decrease in the block size, increased amount of imperfection in the crystallites, and/or ester-interchange reaction (leading to increased sequence heterogeneity of the blocks), as the concentration of the crystallizable block in the copolymer decreases. On the other hand, the 30°C decrease in T_c over the same concentration range was proposed to reflect the difficulty that poly(HMS) has in crystallizing in the block system, compared to crystallizing in the mixture. A similar observation was recorded earlier for polystyrene–polyethylene oxide block copolymers (O'Malley *et al.*, 1970). A comparison of the T_m data obtained by microscopy and DSC of sample crystallized isothermally is shown in Fig. 8. Note that the T_m values of the copolymers and not the homopolymers determined by microscopy are higher than those obtained by DSC, although the same heating rate (10°C min^{-1}) was used in both cases. This was attributed to the fact that optical T_ms correspond to the melting of the most perfect crystallites, whereas the calorimetric T_m is the temperature at which the rate of sample melting is maximum. If this argument is valid, one must expect the homopolymer to have "most perfect" crystals since the T_ms determined by both techniques are almost identical. This could very well be true if one realizes that the difference in T_ms as measured by the two techniques increases with the decrease in the concentration of the crystalline block and the subsequent decrease in the cystallite imperfection.

Heat-of-fusion data, as determined by DSC, were used in calculating the percent crystallinity of the poly(HMS) and poly(MEPS) block copolymers. In Fig. 9 the heat-of-fusion data of isothermally crystallized block copolyesters and two different-molecular-weight poly(HMS) samples are shown. As the concentration of HMS segments in the copolymer increased, the measured heat of fusion and hence degree of crystallinity increased. The crystallinity values as determined from the DSC heat-of-fusion data and those based on density measurements are compared for several copolymers in Fig. 10. It is obvious from these data in Fig. 10 that the DSC-determined values of crystallinity are lower than those based on the density measurements. To explain this, it was postulated that the measured areas under the endothermic DSC peaks do not adequately take into

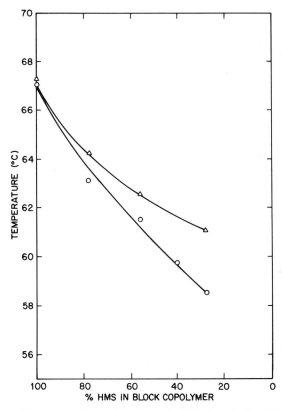

Fig. 8   The melting temperatures of isothermally (47°C) crystallized block copolymers determined by optical microscopy (open triangles) and DSC (open circles). (From O'Malley, 1975.)

account the areas associated with premelting below the major melting peak.

For studying the properties of the polyester–siloxane type of block copolymers, a new series of crystalline $(AB)_n$-type multiblock copolymers based on crystalline poly(hexamethylene sebacate) (HMS) and amorphous poly(dimethyl siloxane) (DMS) has been recently prepared, and their solid-state properties have been examined (O'Malley and Stauffer, 1977). The composition of the copolymers ranged from 0 to 69% DMS. The copolymers crystallized in spherulitic textures when cast as films from solution or the melt. As the DMS concentration in the copolymers increased, the spherulite sizes decreased, but only a very small T_m depression was observed. This is well illustrated in Fig. 11. In spite of the

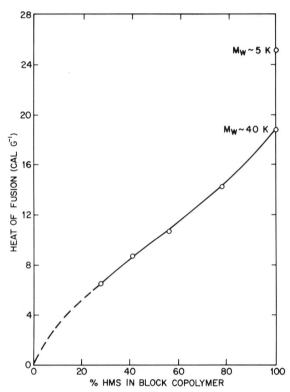

Fig. 9   The effect of composition on the heat of fusion of isothermally (47°C) crystallized block copolymers. (From O'Malley, 1975.)

incorporation of large amounts of DMS in the copolymer spherulites, the T_m data in Fig. 11 show only a 3.5°C depression over the entire copolymer composition range. This remarkably small T_m depression is close to the ~ 2°C value calculated from the Flory theory (Mandelkern, 1964), assuming the DMS block to be a comonomer unit occurring every 11 hexamethylene sebacate repeat units and taking the heat of fusion of HMS to be 32 cal g^{-1} (O'Malley and Stauffer, 1974). By comparison with the DMS systems, the T_m depression for the HMS/MEPS multiblock copolymers over a comparable composition range was 6°C (O'Malley, 1975). It was reported that the implication of these results is that the HMS and DMS blocks in the copolymers microphase-separate at the molecular level and that the siloxane blocks perturb only to a very minor extent the size and perfection of the HMS crystallites. In addition, it was suggested that phase separation is more complete in HMS/DMS systems than in the HMS/MEPS block copolymers (described earlier in the section), as a

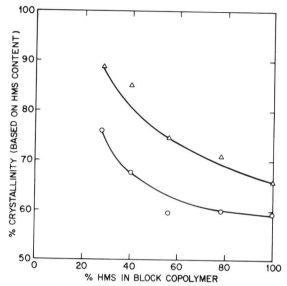

Fig. 10 The % crystallinity of the HMS component in the block copolymers as a function of copolymer composition. The samples were isothermally (47°C) crystallized and the crystallinity data are derived from DSC (open circles) and density (open triangles) data. The filled data points for 100% HMS are for a $\bar{M}_w \sim 5$ K sample, whereas the corresponding open points are for a $\bar{M}_w \sim 40$ K homopolymer. (From O'Malley, 1975.)

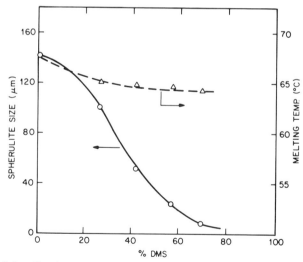

Fig. 11 Spherulite size (open circle) and melting temperature (open triangle) as a function of composition in HMS/DMS block copolymer films isothermally crystallized from the melt at 47 ± 1°C. (From O'Malley and Stauffer, 1977.)

result of the much larger solubility parameter differences of the components in the siloxane copolymers. This is a reasonable interpretation of the reported results, since the efficiency of phase separation in both these multiblock copolymers should be enhanced by having the enthalpy of crystallization available to drive the phase separation process (Meier, 1974; Matzner *et al.*, 1973).

The commercially important class of polyester block copolymers, namely, the polyether-esters, will not be discussed in this section, for it will be treated in great detail in Section II.E of this chapter.

d. Polysulfone Systems.   Most studies on block copolymers have been directed toward thermoplastic elastomers (Noshay and McGrath, 1977). However, interest in studying block copolymers other than elasto-plastics was demonstrated in a recent report on bisphenol-A-based block copolymers of poly(arylene ether sulfones) and poly(aryl carbonate) (McGrath *et al.*, 1977a,b). Results of these investigations indicated that these block copolymers are essentially homogeneous, whereas the corresponding homopolymer blends of equivalent or even lower molecular weights were definitely heterogeneous. McGrath *et al.* (1977b) reported that the dynamic mechanical spectra of the essentially homogeneous block copolymers revealed one major loss peak and a second minor loss peak that was attributed to residual homopolymer. However, in more recent studies (Ward *et al.*, 1978), DSC measurement revealed the presence of two glass transitions caused by the two block structures in systems similar to those noted earlier as "essentially homogeneous." The DSC trace of polysulfone ($\overline{M}_n$ = 16,000)–polycarbonate ($\overline{M}_n$ = 17,000) block copolymer revealed two T_gs at 437 and 458 K, indicating a microphase separation. Thermomechanical data recorded for other samples were consistent with these DSC data.

e. Polyester Systems.   This section pertains to heterochain block copolymers that are made strictly of two different types of polyether blocks. Typical examples of this class of copolymers are those made of poly(ethylene oxide) (PEO) and poly(propylene oxide) (PPO). The thermodynamic properties pertinent to thermal behavior of these polymers have been studied by a few investigators (Booth and Pickles, 1973; Ashman and Booth, 1975, 1976; Ashman *et al.*, 1975).

The PEO–PPO block copolymers were selected by Ashman and Booth (1976) for their thermodynamic studies, for the short-component blocks are compatible in the melt and only one type of block is crystallizable. This made it possible to undo the complexity in the molecular interpretation of the thermodynamics of block copolymers. Multiblock copolymers of PEO and PPO type (PEO–PPO)$_n$ were prepared by condensation of hydroxy- and chlorocarboxy-terminated polymers. Small-angle x-ray scattering and dilatometry were used to determine lamellae spacings and T_ms of the copolymers, respectively. The T_ms of these copolymers were reported to be predictable from the T_ms of corresponding triblock (PPO–PEO–PPO) copolymers studied earlier by Booth and Pickles (1973).

Ashman and Booth (1975) have described the crystallinity and fusion of predominantly extended-chain crystals of type PEO–PPO block copolymers and interpreted the results in terms of a simple stacked-lamellae model with alternating amorphous and crystalline layers. Earlier, Booth and Pickles (1973) described the melting of predominantly folded-chain crystals of triblock type PPO–PEO–PPO copolymers, with PEO block having an average DP of 70–100. Recently, Ashman *et al.* (1975) studied crystallinity and fusion of PPO–PEO–PPO triblock copolymers having relatively short blocks; DP of PEO was 48, whereas that of PPO ranged from 1 to 7. Lamellae spacings, specific volumes, and T_ms have been determined for a series of these block copolymers, and the obtained data were also interpreted in terms of a stacked-lamellae model with alternating amorphous and crystalline layers. Both extended-chain and once-folded chain crystalline lamellae were found, the former with thickness of 32 ethylene oxide (EO) chain units and the latter with thickness of about 21 EO chain units. On the other hand, the specific volume of the polymer in the amorphous layer, which is predominantly formed of chains emerging from the end interfaces of the crystalline lamellae, was lower than that of the corresponding supercooled melt having the same composition. The T_ms of the copolymers were low compared to that of perfectly crystalline PEO (i.e., 37–55°C, compared with 70°C). This was attributed to the large positive free energy of formation from the melt of the crystalline/amorphous end interfaces (σ_0) and the amorphous layer (σ_a). For extended polycrystals, values for σ_0 and σ_a were found to be ~6 and 2 kJ mole^{-1}, respectively. It was also found that $\sigma_{0,x} = 2.5$ kJ mole^{-1} for a completely extended chain end interface and $\sigma_{0,f} = 10$ kJ mole^{-1} for a completely folded chain end interface.

2. Homochain Block and Graft Copolymers

The use of thermal and related techniques for studying copolymers containing at least one block of thermoplastic nature will be dealt with in this section. These include copolymers of polyolefins such as polystyrene and acrylic polymers such as poly(methyl methacrylate).

a. Effect of Block Size on T_g in Vinyl Systems. Domain size dependence of T_g was treated theoretically in an excellent report by Couchman and Karasz (1977). Experimentally, it is well documented that as in homopolymers, the T_g of the block copolymers increases with the increase in block length at a certain DP range. For example, in dynamic mechanical measurements of styrene–butadiene block copolymers, Kraus et al. (1967) recorded that the temperature of loss peak due to the styrene block decreased from 104 to 60°C when the styrene block length decreased by a factor of about 4. Shortly thereafter Childers and Kraus (1967) postulated that the lowering of T_g with decreasing block length may be attributed to increased compatibility and some mixing of the styrene and butadiene blocks. In a relatively more recent investigation, Kraus and Rollman (1976) have used dynamic mechanical measurement for studying T_g of styrene–butadiene triblock copolymers. It was demonstrated in these studies that the T_g of the styrene block decreased from 120 to 20°C as the block molecular weight decreased from 20,000 to 5000 daltons. On the other hand, the T_g of the butadiene block remained unaffected by the block length until it reached a low value of 10^4, below which a minor decrease was recorded. However, other earlier investigators (Robinson and White, 1970) used DSC to show a definite T_g dependence on block length for both the styrene and isoprene blocks in a triblock copolymer. Recently, DSC data on the T_g of styrene–dimethylsiloxane block copolymers were used to dispute the generally accepted role of mixing in lowering the T_g in block copolymers with short blocks (Krause and Iskandar, 1978). Thus it was shown that the T_g of the styrene block in styrene–dimethylsiloxane block copolymers was lower than that in a styrene–isoprene block copolymer. This is contrary to what one would expect in terms of compatibility (i.e., the styrene blocks are more compatible with the isoprene blocks than with the dimethylsiloxane blocks and one would expect the styrene–dimethylsiloxane block copolymer to have a styrene block with a higher T_g than that in the styrene–isoprene system). It was thus concluded by these authors that molecular weight alone cannot be responsible for the T_g dependence on the block length.

b. Multiplicity of Glass Transition in Vinyl Systems. Reports in the literature on the T_g of block and graft copolymers show frequently that these systems have two separate T_gs (Baer, 1964; Cooper and Tobolsky, 1966) and each one is close to the T_g of the parent homopolymer. Block and graft copolymers with T_g values at intermediate temperatures have also been reported (Beevers and White, 1960; Kenney, 1968). Until the effect of phase mixing on T_g was recognized, the T_gs of block copolymers appeared rather unpredictable. On the other hand, one must acknowledge the fact that the occurrence of mutual solubility is rare (Krause and Roman, 1965; Stoelting et al., 1969) and the block copolymers of such pairs would be expected to show only one T_g. For other pairs to show only one T_g in their block copolymers, when each block is of reasonable size, the single link between blocks would have to be unrealistically sufficient to overcome their mutual incompatibility. Hence the two phases will be incompatible and the system will display two T_gs.

Unfortunately, most reports on T_g deal with one or a few polymer systems and different methods are used in different reports that cause an inevitable variability in the reported T_g values for any particular system. However, this situation was corrected (at least in part) by Black and Worsfold (1974), who determined the T_gs of a large series of block and graft copolymers using only one technique—thermal expansion. The block and graft copolymers were prepared from several vinyl monomers to form polar, nonpolar, and polar–nonpolar block systems. Typical T_g values of several copolymers are shown in Table IX. It was found in all

TABLE IX

GLASS TRANSITIONS OF VINYL-TYPE BLOCK AND GRAFT COPOLYMERS[a]

M_1	M_2	$\%M_1$	Total MW	Lower T_g	Upper T_g
α-Methylstyrene	Vinyl acetate	18	103,000	35	182
α-Methylstyrene	Vinyl chloride	67	39,000	−8	182
α-Methylstyrene	Styrene	45	61,000	127	—
Styrene	Methyl methacrylate	40	70,000	—	98
Styrene	Butyl acrylate	46	104,000	−55	99
Styrene	Ethylene oxide	50	40,000	−72	100
Styrene	Isoprene	50	1,000,000	−75	101
Styrene	Isobutylene	40	141,000	−69	102
Methyl Methacrylate	Ethyl acrylate	56	162,000	−23	115
Methyl Methacrylate	Vinyl acetate	50	96,000	38	98
Methyl Methacrylate	Ethyl methacrylate	50	104,000	69	106

[a] From Black and Worsfold (1974).

examined systems, except one, that the T_gs were close to those found in the individual homopolymers. The exception was a block copolymer of styrene and α-methylstyrene, whose homopolymers would give clear, mixed solutions in a mutual solvent. The block copolymer of styrene and α-methylstyrene showed one glass transition at 127°C between those of the corresponding homopolymers.

3. Homochain–Heterochain Block and Graft Copolymers

In homochain–heterochain block and graft copolymers the blocks are made of both homochain and heterochain long sequences. Examples of these polymers are not as numerous as other types of block copolymers. The use of thermal and athermal techniques to study the physical properties of some of these copolymers is illustrated in the next few paragraphs.

a. Polystyrene–Polycaprolactone Systems (S–CL).   Polystyrene–polycaprolactone systems were first prepared by Teyssie *et al.* (1977) by a combination of an anionic and a coordination polymerization catalyst. These copolymers consist of amorphous polystyrene blocks (PS) and crystallizable poly-ϵ-caprolactone (PCL) blocks. Accordingly, they were expected (Herman *et al.*, 1978) to exhibit mesophases in the presence of a selective solvent for a given block, as, for example, polystyrene–poly(ethylene oxide) copolymers do (Gervais *et al.*, 1971). In order to preserve the crystalline character of the PCL blocks, Herman *et al.* (1978) have swelled the S–CL copolymers with diethyl phthalate (DEP), a selective solvent for PS, prior to examining its morphology and thermal properties. It was found that the selectively swollen S–CL copolymer adopts a lamellar crystalline structure at room temperature, as in the case of polystyrene–poly(ethylene oxide) (S–EO) block copolymers (Gervais and Gallot, 1973). However, above the T_m of the CL blocks, no mesophases were observed in the S–CL/DEP system (Herman *et al.*, 1978), contrary to the S–EO/DEP system, which adopts a hexagonal, lamellar, or even inversed hexagonal structure above the EO T_m, depending on the copolymer composition (Gervais and Gallot, 1973). The T_m and degree of crystallinity of the S–CL copolymer were determined using DSC.

b. Polystyrene–Polycaprolactam Systems (S–CLM).   Block copolymers of polycaprolactam can in principle be prepared by two routes —by bonding prepolymers with reactive end groups in condensation and addition reactions or by building up the polyamide chain at the end groups of the second prepolymer. The latter route has been employed in the preparation of AB- and ABA-type block copolymers where A is the polyamide segment formed by the polymerization of lactam. The hydrolytic polymerization of lactams can take place at growth centers consisting of amine end groups on a desired prepolymer (Shalaby *et al.*, 1973a,b). Similarly, the anionic polymerization can take place on active groups attached to one or both ends of a prepolymer selected for end grafting (Hergenrother and Ambrose, 1974; Korshak *et al.*, 1976; Owen and Thompson, 1972; Yamashita *et al.*, 1972,1973). The polymerization conditions used by various investigators differed considerably; the copolymers, when characterized, contained major (Hergenrother and Ambrose, 1974) or minor (Yamashita *et al.*, 1972) fractions of the corresponding homopolymers.

Compared with the hydrolytic polymerization, the use of the anionic polymerization of lactams in the preparation of block copolymers has the advantage of allowing lower polymerization temperatures and shorter reaction times where functional exchange reactions of the polyamide end-graft leading to homopolymer formation is negligible. Anionic end-grafting has been used effectively for preparing ABA block copolymers, where *A* and *B* represent polyamide and polystyrene segments, respectively (Stehlicek and Sebenda, 1977a). Thus caprolactam was end-grafted anionically onto polystyrene-bisacryllactams under mild conditions (in a solvent at 120°C) to produce rather pure copolymers; traces of polystyrene and $\leq 2.5\%$ polycaprolactam represented the minor impurities in these block copolymers. The length of the nylon block was affected by reaction time and by the amount and type of the solvent used. Typical DTA data of several block copolymers are shown in Table X.

TABLE X

DTA Data of PS–Nylon 6 Block Copolymers[a]

Polymer	$\bar{P}_B{}^b$	Solvent type	Reaction time (h)	% Conversion[d]	$\bar{P}_A{}^b$	Specific viscosity	T_m (°C)[e]
1	43	$C_6H_5CH_3$	1	3.4	11	0.21	None
2	43	$C_6H_5CH_3$	6	14.9	26	0.91	215
3	146	$C_6H_5CH_3$	1	3.2	8	0.25	199
4	81	$C_6H_5CH_3$	6	15.2	30	1.01	222
5	174	$C_6H_5CH_3$	6	17.8	35	1.00	213
6	174	$C_6H_5CH_3$	24	15.8	30	1.23	213
7	81	$C_6H_5CH_3{}^c$	6	26.5	53	1.68	223
8	43	None	6	97.0	–	7.12	222
9	43	DMAC	1	5.6	12	0.45	211
10	181	DMAC[c]	1	28.9	68	1.06	223

[a] From Stehlicek and Sebenda (1977a).
[b] $\bar{P}_A$ and $\bar{P}_B$ = number average DP of Nylon 6 and PS blocks.
[c] Used half the normal amount of solvent.
[d] % conversion for caprolactam.
[e] T_m of Nylon 6 block.

It is interesting to note the excellent correlation between the T_m of Nylon 6 and number average DP of the nylon block. It is also worth mentioning that the T_m data of samples 7 and 10 indicate that the reaction solvent has no effect on the T_m of the nylon block.

c. Copolymers of Polystyrene with Nylon 8 and Nylon 12. Copolymers consisting of triblocks, namely, Nylon 8–polystyrene–Nylon 8 and Nylon 12–polystyrene–Nylon 12 were prepared recently (Stehlicek and Sebenda, 1977b) by anionic end-grafting of the proper lactam (L-9 and L-3, respectively) onto polystyrene-bisacryllactam at 120°C. Both cross-linked and soluble products were obtained with 8-octane-lactam (L-9) ir toluene, whereas polymerization in dimethylacetamide gave rise to soluble copolymers. This illustrates the role of solvent in affecting side condensation reactions leading to cross-linking.

Softening and melting of the block copolymers were investigated by DTA and penetrometry without attempting to interpret the observed effect. Figure 12 shows the DTA curves of melting of chosen copolymers

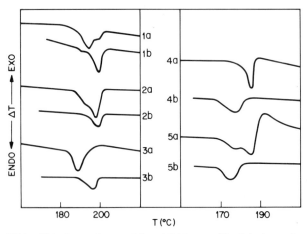

Fig. 12 DTA of block copolymers L9 and L13 crystallized during polymerization (a) and subjected to remelting (b), heating rate 10 K min^{-1}. (From Stehlicek and Sebenda, 1977b.)

crystallized during polymerization and after remelting. For Nylon 8 copolymers prepared in toluene, the maximum of the endotherm was raised on remelting from 191–194 to 198–200°C. This was attributed to better ordering in the crystalline domain. It is our opinion that the increased order in the crystalline phase is associated with the slower cooling rate of the melted polymer as compared with the rate of precipitation (fast cooling) during polymer formation. This is consistent with the fact that when the crystallization–melting cycle was repeated, no additional rise in T_m was recorded. Additional evidence for our hypothesis is that the copolymer made from DMAC, a better solvent for the nylon block than toluene and hence low precipitation rate, revealed a T_m of 199–200°C, which is comparable to that of the melt-crystallized samples. On the other hand, the T_m of Nylon 12 block was lowered on melting and recrystallization from 175–177 to 167–170°C and the endotherm broadened. This was attributed to the possibility of partial "dissolution" of the polystyrene phase in the polyamide phase or "thermal degradation" of the copolymer. Again, it is our opinion that partial phase-mixing may have contributed to plasticization of the Nylon 12 phase and hence lowered its T_m—a situation that is not likely to occur with a fast precipitating block copolymer during formation from a selective solvent for either of the blocks. The proposed partial phase-mixing is consistent with the recorded dependence of the softening temperature (T_s) of the block copolymer on composition (see Fig. 13). It is apparent from the T_s data that whereas a radical decrease in T_s for Nylon 8 copolymers occurred at 30% PS content, the T_s of Nylon 12 copolymers showed a gradual decrease, reaching a minimum at 70% PS and then rose slowly.

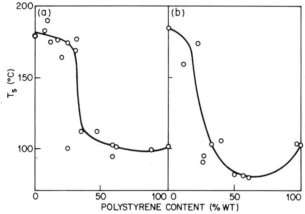

Fig. 13 Dependence of the softening temperature (T_s) on the weight content of polystyrene segments in copolymers of L9(a) and L13(b). (From Stehlicek and Sebenda, 1976b.)

d. Graft Copolymer of Polypivalolactone on Ethylene Terpolymers. The anionic ring-opening polymerization of pivalolactone (2,2-dimethyl-propiolactone, PVL) was used to graft side chains on poly(ethylene-co-vinyl acetate-co-methacrylic acid) and similar terpolymers (Sundet, 1978). Grafting was effected in a homogeneous solution in tetrahydrofuran, using tetralkylammonium salts as initiators. Graft frequency was determined by the concentration of copolymerized methacrylic acid and the graft degree of polymerization. Under the employed grafting conditions, grafting of pivalolactone was noted to lead to the growth of polyester side chains with strong tendency to crystallize from the ethylene copolymer substrate (Sundet, 1978). Differential scanning calorimetry curves showed an endotherm marking the T_m of the polypivalolactone side chain, which was distinct from those of the terpolymer substrate appearing at lower temperature (Fig. 14). The crystallization exotherm of the polyester grafts was recorded on cooling the graft copolymer melt. The presence of crystalline microdomains in polypivalolactone copolymers has been demonstrated earlier using wide-angle x-ray and electron

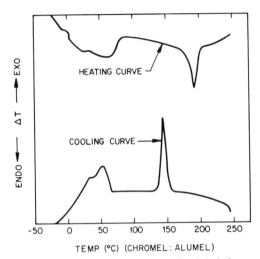

Fig. 14 Differential scanning calorimetry curves of poly(ethylene-*co*-vinyl acetate-*co*-methacrylic acid-*g*-pivalolactone). Base resin composition E7 3.5%, VOAc 25%, MAA 1.5%; $\overline{M}_w = 13,000$; PVL grafted; 18%, DP 12. [Reprinted with the permission from S. A. Sundet (1978), *Macromolecules* **11**, 148. Copyright 1978 by American Chemical Society.]

microscopy (Buck, 1977); hence it was expected to form in the present graft copolymers of polypivalolactone and the ethylene copolymers.

To the extent that these side chains form crystalline microdomains of polypivalolactone (PPVL), they develop a thermoplastic network within the backbone substrate. For example, the polymer whose DSC curve is shown in Fig. 14 flows only when the temperature of the PPVL is reached

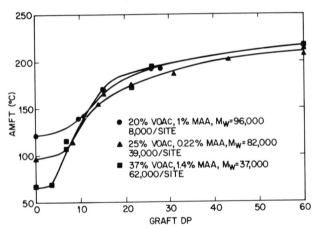

Fig. 15 Effect of graft length on automated minimum flow temperatures. [Reprinted with permission from S. A. Sundet (1978), *Macromolecules* **11**, 148. Copyright 1978 by American Chemical Society.]

and the crystalline regions melt. Since the T_m of the PPVL side chains increase with their degree of polymerization, the minimum flow temperatures were shown to depend on the chain length. This effect is illustrated with three different substrate compositions in Fig. 15. Torsion modulus loss curves (Fig. 16) revealed some evidence for the T_g of PPVL reported to be $-10°C$ (Oosterhof, 1974). The T_m of the ethylene terpolymer at 60°C was detectable because of the structural support of the grafted side chains.

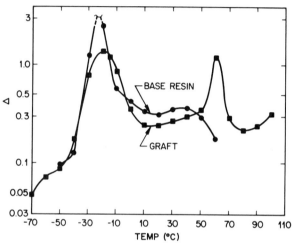

Fig. 16   Loss curve from torsion modulus data. Base resin E 61.3%, VOAc 37.3%, MAA 1.4%; $\overline{M}_w = 37,000$; graft 10% PVL, for a DP of 7. [Reprinted with permission from S. A. Sundet (1978), *Macromolecules* **11**, 148. Copyright 1978 by American Chemical Society.]

E. A Survey of Athermal Techniques for Characterizing Block and Graft Copolymers

The use of thermal and other techniques for characterizing block and graft copolymers was covered, partly, in a recent review by Bollinger (1977–1978). Although this review dealt with heterochain systems, most of the techniques discussed can also be used for homochain systems. Infrared spectroscopy, ultraviolet spectroscopy, nuclear magnetic resonance, x-ray diffraction, DTA, and TGA were the basic techniques described by Bollinger (1977–1978). The use of light scattering for characterizing homochain block copolymers and its potential use for heterochain copolymers was discussed by a few authors (Arpin and Strazielle, 1975; Bollinger, 1977–1978; Burke, 1973; Bushuk and Benoit, 1958). Attempts to apply gel permeation chromatography to the study of block copolymers were also made (Arpin and Strazielle, 1976; O'Malley, 1975; Panaris and Pallas, 1970).

In an attempt to illustrate the use of thermal and athermal techniques for the characterization of block copolymers, some recent studies on two important types of block copolymers are discussed in this section. The two types of polymers subject of this discussion are the polyether–esters and polyether–urethanes.

1. Polyether–Ester Block Copolymers

Less than 10 years ago, segmented polyether–ester copolymers were developed as thermoplastic elastomers (Cella, 1973; Witsiepe, 1973). Typical representatives of this group of polymers are poly(tetramethylene oxide)–poly(tetramethylene terephthalate) (PTMO–4GT), which have become important commercial elastoplastics. These exhibit the high elongation characteristics of elastomers coupled with the properties of plastics such as high modulus and strength and the ability to be processed by conventional thermoplastic technology. In 1976 Lilaonitkul *et al.* studied the physical properties including the thermal behavior of these polymers using (a) small-angle light scattering, (b) dynamic mechanical testing, (c) DSC, (d) stress–strain measurements, and (e) dichroism experiments. Discussion of these studies will be limited to the reported thermal properties of PTMO–4GT. For a better appreciation of this discussion some background information on the chemical structure and basic morphology of these polymers is given briefly in the next paragraph.

At typical service temperatures, the PTMO–4GT copolymers behave like chemically cross-linked elastomers, but at elevated temperatures, they soften and flow like conventional thermoplastics. Chemically, the copolymers are based on blocks of poly(tetramethylene oxide) (PTMO) and poly(tetramethylene terephthalate) (4GT). The PTMO component is characterized as a rubber or "soft" segment, since it possesses a relatively low T_g. The 4GT block represents the reinforcing, rigid, "hard" segments, for it has a relatively high T_g, well above room temperature. It is the intermolecular association of the hard blocks that provides the physical cross-linking for the system. On heating above the T_m of 4GT, these quasi-cross-links will no longer exist and the liquid copolymer flows as a thermoplastic material. On cooling the molten polymer, crystallization of the hard segments takes place to provide a superstructure for the elastomer phase consisting of uncrystallized hard and soft segments.

The copolymers studied by Lilaonitkul *et al.* (1976) were prepared by melt transesterification of dimethylterephthalate, 1,4-butanediol, and PTMO (having a $\overline{M}_n$ of about 1000). The overall molecular weight of these copolymers was approximately 25,000–30,000. The thermal behavior of a series of polyether–esters consisting of 33–84% by weight 4GT hard segments was studied using DSC and dynamic mechanical techniques. The experimental results are summarized in Table XI, where the samples are numbered according to the weight percent of the 4GT hard segment (e.g., H-57 represents the copolymer containing 57% hard segment by weight). The data in Table XI indicate that the T_g is due to an amorphous phase that is made largely of soft blocks, since its value increases from -68 to $-9°C$ as the hard content increases from 33 to 84%. On the other hand, the T_m, ΔH_f, and degree of crystallinity were shown to increase with the increase in the hard segment content. This is quite consistent with an earlier conclusion (Cella, 1973), indicating that the crystalline phase in the PTMO–4GT system is formed primarily by 4GT segments. The detection of one T_g in this series of segmented copolymers, both by the DSC and dynamic mechanical measurement, was noted by Lilaonitkul *et al.* (1976) to be associated with an amorphous phase made of a nearly compatible mixture of soft PTMO segments and uncrystallized 4GT hard segments. As expected, the T_g increases with the increase in the overall hard content of the copolymers. The DSC and dynamic mechanical measurement data reported by Lilaonitkul *et al.* (1976) are given in Table XI.

TABLE XI

THERMAL PROPERTIES OF PFMO–4GT COPOLYMERS[a]

	H33	H50	H57	H63	H76	H84
4 GT (wt %)	33	50	57	63	76	84
Average 4 GT block length, L	2.64	4.95	6.43	8.14	14.8	24.26
DSC results[a]						
T_g (°C)	-68	-59	-55	-51	-33	-9
T_m (°C)	163	189	196	200	209	214
ΔH_f (cal g^{-1})	3.9	7.8	9.8	11.4	13.9	14.6
Crystallinity (%)	11.5	22.9	28.6	33.3	40.7	42.8
Rheovibron results[b]						
E'' peak (°C)[b]	-63	-58	-53	-48	-30	-4
Tan δ peak (°C)[c]	-51	-42	-34	-27	10	30

[a] From Lilaonitkul *et al.* (1976).
[b] Compression-molded samples.
[c] For β relaxation.

Lilaonitkul *et al.* (1976) applied the Gordon–Taylor equation (Gordon and Taylor, 1952), as rearranged by Wood (1958), for the effect of composition on the T_g in compatible copolymers, to the polyether–ester system. The Wood equation can be represented by the following equation, using the notations associated with the PTMO–4GT system for convenience:

$$T_g = [k(T_{g_2} - T_g)W_2/(1 - W_2)] + T_{g_1} \qquad (4)$$

where T_g is the glass transition temperature of a block copolymer containing weight fractions W_1 and W_2 of PTMO and 4GT blocks; T_{g1} and T_{g2} are the glass transition temperatures of the PTMO and 4GT blocks, respectively; and k is a constant.

Since the 4GT hard segment of the copolyester is crystallizable, Eq. (1) was applied using W_2, defined by two different means. First, W_2 was taken as the weight fraction of 4GT. Second, it was assumed that only the noncrystallized 4GT contributes to the soft phase T_g. The normalized 4GT weight fraction W_2' was defined by the expression $W_2/(W_2 + W_1)$, where W_2 is the weight fraction of amorphous 4GT and W_1 is the weight fraction of PTMO. A plot of Eq. (1) for different definitions of W_2 is shown in Fig. 17, T_{g2} was fixed at 50°C [the T_g of poly(tetramethylene terephthalate)]. A least-square analysis of the data gave k values of 0.5 and 0.22, and T_g values of -89°C and -84°C for the normalized and unnormalized

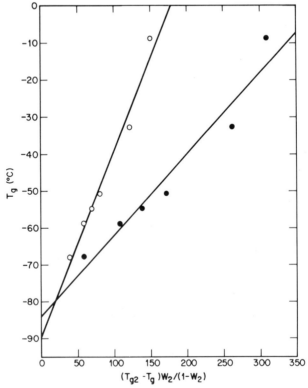

Fig. 17 T_g data plotted according to Gordon-Taylor equation: (open circle) W_2 defined as weight fraction of amorphous 4GT; and (solid circle) W_2 as weight fraction 4GT. (From Lilaonitkul *et al.*, 1976.)

weight fractions, respectively. The predicted $-89°C$ T_g for PTMO was in good agreement with the literature value of $-88°C$ (Yoshida *et al.*, 1973). The fit of the data for the total 4GT weight fraction W_2 was not as good. The data fit in this latter case was aided by the availability of the adjustable parameter k. The experimental data and predicted values of T_g versus normalized 4GT weight fraction are shown in Fig. 18 for $k = 0.5$, $T_{g_1} = -88°C$ and $T_{g_2} = 50°C$. Although these copolymers displayed the glass transition behavior characteristic of compatible systems, Lilaonitkul *et al.* presented some arguments in favor of describing the PTMO–4GT system as partially heterogeneous.

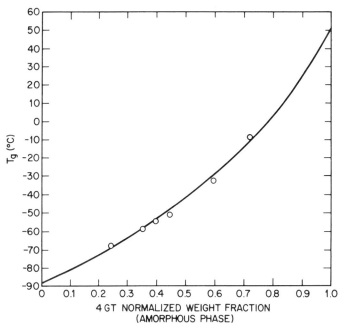

Fig. 18   Plot of T_g versus normalized 4GT content. Curve is Gordon–Taylor equation with $K = 0.5$. (From Lilaonitkul *et al.*, 1976.)

   Other studies (West *et al.*, 1974; Seymour *et al.*, 1975; Lilaonitkul *et al.*, 1976; Lilaonitkul and Cooper, 1977) have shown that the basic structure of polyether–esters is spherulitic and that three different types of spherulites could be prepared (Lilaonitkul *et al.*, 1976; Lilaonitkul and Cooper, 1977), depending on the sample composition and fabrication procedure. Specifically, positive birefringence spherulites (type III), negative birefringence spherulites (type II), and spherulites that have their optical axes tilted at 45° to the spherulite radius (type I) have been observed. Samples showing no spherulitic structure (type NO) have also been reported and concluded to have continuous and interpenetrating crystalline and amorphous regions. In a more recent investigation, a 56–44 PTMO–4GT sample was selected for extensive studies (Lilaonitkul *et al.* 1977). From this particular composition the four types of spherulitic textures have been prepared by varying the solvent and casting temperature at a fixed solvent evaporation rate. Although the exact mechanisms that lead to the formation of each structure remained undefined, the data obtained by Lilaonitkul *et al.* 1977) suggest that they may relate to the mobility of the polymer chains in the system when crystal nucleation and growth occur. The DSC and mechanical properties data of the 56–44 PTMO–4GT system containing the spherulitic structures are summarized in Table XII and show that (a) the spherulitic structure results in changes in both

TABLE XII

Characterization of a 56–44 PTMO–4GT Copolymer[a]

	NO	I	II	III
DSC: T_g (°C)	−66	−65	−66	−67
T_m (°C)	177	176	175	173
ΔH_f (cal/g)	8.2	7.1	7.5	8.0
% crystallinity	24.1	20.8	22.1	23.5
Mechanical properties				
True stress, dynes cm²) $\times$ 10^{-7}				
at 100% ϵ	15.1	16.3	17.8	14.7
at 500% ϵ	71.2	76.0	86.3	62.2
Elongation at break (%)	760	740	625	580

[a] From Lilaonitkul *et al.* (1977).

T_g and T_m; (b) the different types of morphologies are associated with some differences in ΔH_f and percent crystallinity; (c) samples exhibiting type II spherulitic structure are superior to others, mechanically, and also produce the strongest-oriented samples; (d) samples containing type I spherulites are somewhat inferior to type II spherulites in terms of tensile strength; and (e) samples based on type I and II spherulites are much stronger, mechanically, than the nonspherulitic systems.

2. Polyether–Urethanes

Over the past decade, segmented polyurethane elastomers have been studied by Cooper *et al.* (1976), who demonstrated that the mechanical and thermal properties are markedly influenced by composition and morphological detail (e.g., degree of phase separation and the perfection of order in the domains). Based on these studies and earlier investigations using infrared dichroism (Estes *et al.*, 1971), it has become understood that the hard (urethane) segments and soft (usually polyether) segments that comprise the backbone of polyurethane elastomers segregate in the solid state into hard and soft domains and hence produce a two-phase morphology. Based on their studies of the effect of thermal treatment on the morphology of polyether–urethanes (PEU), Wilkes and co-workers (1975) prepared the following model. Even in well-aged samples, the two phases exist in these systems, first, hard domains that are reasonably pure and, second, soft domains comprising soft segments. On heating, the extended soft segments contract or relax, causing additional hard segments to be pulled out of the hard domains and phase mixing to occur. Subsequent rapid cooling creates a thermodynamic driving force for the two phases to separate again, but the high viscosity requires the demixing (and domain purification) to occur slowly over a period of time. The new equilibrium state is, however, eventually achieved. Because the soft segment T_g is sensitive to the soft-segment domain purity (high T_g is associated with high purity), it can be used to monitor the mixing–demixing process. This later process was then followed using DSC measurements (Hesketh and Cooper, 1977), as shown in the next paragraph.

The DSC data obtained by Hesketh and Cooper (1977) on series of segmented PEU allowed those authors to suggest that phase mixing occurs on annealing and that phase demixing (or soft-domain purification) occurs after quenching from the annealing condition. This conclusion was in agreement with the model proposed earlier by Wilkes and co-workers (1975). The major observed effect of the phase mixing–demixing processes was an increase and decrease in T_g on mixing and demixing, respectively. The ability of the hard segment to crystallize was reported to affect directly the magnitude of the T_g shifts. It was also noted that DSC events arising from the hard-segment domains appear to be affected to a lesser degree by phase mixing–demixing than those arising from the soft-segment domains (e.g., the soft-segment glass transition).

In earlier studies by Seymour and Cooper (1973) on PEU, the obtained DSC and infrared spectroscopy data for a series of copolymers indicated that (a) the DSC endotherms observed are due to short- and long-range ordering of the hard segments rather than to hydrogen-bond dissociation as previously thought; (b) hydrogen-bond disruption as monitored by IR spectroscopy is influenced primarily by the T_g of the hard segments; and (c) differences in domain morphology of these materials, which are sensitive to sample thermal history, are not revealed by changes in the extent of thermal stability of the hydrogen bond. The IR analysis revealed that hydrogen bond dissociation does occur, increasing steadily as the temperature of the polymer is raised above the T_g of the hard segment. However, significant hydrogen bonding did persist at 200°C and the thermal behavior of the hydrogen bonds is insensitive to the morphological details. The DSC thermograms and not the IR spectra of the polymers were affected noticeably by annealing. The lack of correlation between the IR and DSC data and insensitivity of the DSC experiment to the hydrogen bond can be explained as follows. Since hydrogen bond dissociation is accompanied by an enthalpy change, one would expect to observe a response in a calorimetric experiment. However, such response is prevented, for the DSC experiment did not allow sufficient time for maximum hydrogen bond dissociation effects to occur. Even if the scan rate were sufficiently slow, the contribution of hydrogen bond dissociation to the observed DSC curve would be very slight.

III. Polyblends

A. Introduction

At least three major stages can be noted in the development of modern plastics to meet an ever-expanding and seemingly endless number of industrial applications. First, new families of polymers were produced from an abundant supply of monomers. Second, random, and then block and graft, copolymerization techniques have been developed to modify a resin's chemical and physical properties. The third and most recent industrial venture into producing new plastics deals with physical blending of two or more polymers. This latter method has begun to attract interest and commercial development.

The largest commercial exploitation of polymer blending has been connected with the attempt to improve the impact strength of plastics by the inclusion of elastomeric materials. Early work was aimed almost exclusively at modifying polystyrene and has since expanded to include poly(vinyl chloride), poly(phenylene oxide), poly(methyl methacrylate) as well as other thermoplastic and some thermosets. Undoubtedly, blending will play an important role in the future recycling of plastic scrap.

In this section the use of TA techniques for studying polymer compatibility is reviewed. In some cases, these thermal methods may be used to determine the composition of polyblends both qualitatively and quantitatively. The use of DSC measurements to study effects of chemical aging by heat and sunlight on ABS (acrylonitrile butadiene styrene) resins is also discussed.

B. Compatibility

1. Thermal Criterion

The thermodynamic criterion for a stable mixture of two polymers is that the second derivative of the free energy with respect to concentration is positive. However, the actual determination of when this condition has been met is usually decided on somewhat inexact definitions of compatibility based on experimentally measured properties of the polymer systems. Some of the most common methods include light transmission microscopy, thermal expansion, heat capacity, NMR, dynamic mechanical and dielectric techniques. Each of these approaches has its strengths and weaknesses in terms of resolution and interpretation. Thus, whenever possible, one should apply a number of these techniques in a complimentary fashion to characterize polymer blends.

Krause (1972, 1978) has made a comprehensive literature survey of the compatibility of blends of over 500 polymer pairs. This review deals with data collected from a variety of experimental techniques and indicates that about two dozen polymer pairs are compatible in all proportions and a slightly greater number that appear to be at least conditionally compatible at room temperature.

Once two polymers are determined to be compatible, it is important to examine their stability as a function of temperature. It has been found that when the temperature of an initially homogeneous mixture is lowered below a point on the phase boundary called the upper critical solution temperature (UCST), phase separation begins to occur. In addition, in some polyblends a similar phenomenon occurs when the temperature is raised. This critical point is called the lower critical solution temperature (LCST).

Bank *et al.* (1972) made the first experimental determination of a homogeneous polymer system that has a LCST. In their study of polystyrene (PS) and poly(vinylmethyl ether) (PVME) it was found by DSC that polyblends of PS–PVME yielded a single glass temperature (T_g) that was composition dependent for polymer mixtures cast from toluene. However, heating the samples to 125°C induced phase separation (Fig. 19).

Vapor sorption measurements (Kwei *et al.*, 1974) by thermogravimetry and gas–liquid chromatography experiments (Su and Patterson, 1977) have been used to calculate the interaction parameter χ'_{23} for polymer binary mixtures as a function of composition and temperature. In the former technique, the parameters χ_{12} and χ_{13} (where the subscript 1 refers to the solvent and subscripts 2 and 3 to the two polymers) are determined in separate experiments from the relationship

$$\ln a_1 = \ln \phi_1 + \phi_2 + \chi_{12}\phi_2^2 \tag{5}$$

where a_1 is the activity of the solvent and ϕ refers to the volume fraction of each component. Once x_{12} and x_{13} have been determined, x'_{23} for the polymer pair can be calculated from

$$\ln a_1 = \ln \phi_1 + (1 - \phi_1) + (\chi_{12}\phi_2 + \chi_{13}\phi_3)(1 - \phi_1) - \chi'_{23}\phi_2\phi_3 \tag{6}$$

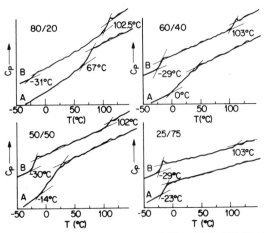

Fig. 19 Effect of heat treatment on the compatibility of PS–PVME mixtures cast from toluene. Samples were heated at 20° min⁻¹in a DSC from 25 to 150°C and quenched-cooled in liquid nitrogen after phase separation occurred: (A) before phase separation; (B) after phase separation (From M. Bank, J. Leefingwell, and C. Thies, *J. Polymer. Sci., Polym. Phys. Ed.* **10**, 1087–1109 (1972). Copyright © 1972. Reprinted by permission of John Wiley & Sons, Inc.)

The calculated values of χ'_{23} for PS–PVME polyblends are shown in Fig. 20. The negative values of χ'_{23} meet the thermodynamic criterion for stability in a binary mixture. In addition, extrapolation of the data in Fig. 20 indicates χ'_{23} would become positive at about 70% PVME and suggests the presence of a LCST. Further, NMR measurements and microscopic observations of PS–PVME polyblends show that phase separation has occurred (Nishi *et al.*, 1975b). A plot of the inception and completion temperatures for phase separation is shown in Fig. 21. It should be noted that in the study of a blend of crystalline and amorphous polymers the interaction parameter cannot be obtained by the vapor sorption method. However, an alternative approach based on the depression of T_m will be reviewed in a later section.

One of the best-documented cases of compatible polymer blends is that of polystyrene (PS) and poly(2,6-methyl-1,4-phenylene oxide) (PPO). A single T_g has been detected for all blend compositions by thermal optical analysis (TOA) (Shultz and Gendron, 1973), DSC (Stoelting *et al.*, 1970; Bair, 1970a,b; Shultz and Gendron, 1972), and dielectric measurements (MacKnight *et al.*, 1971), thermomechanical analysis (Prest and Porter, 1972), and dynamic mechanical techniques (Schultz and Beach, 1974). The modification of PS–PPO blending by including random copolymers of styrene and 4-chlorostyrene has been investigated in an attempt to examine the effects of partial miscibility on thermal transitions (Fried *et al.*, 1978).

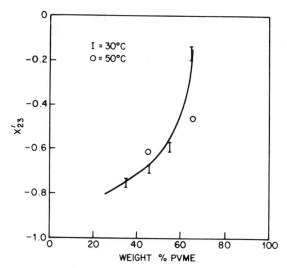

Fig. 20 Interaction parameter x'_{23} as a function of PVME concentration in PVME–PS blends [From T. K. Kwei, T. Nishi, and R. F. Roberts (1974). *Macromolecules* **7**, 672. Copyright by the American Chemical Society.]

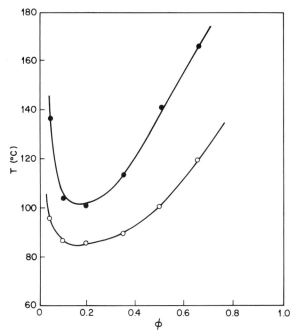

Fig. 21 Initiation (open circle) and completion temperatures (solid circle) for phase separation of several PS concentrations ϕ of PS–PVME blends at a heating rate of 0.2° min⁻¹ [From T. Nishi, T. T. Wang, and T. K. Kwei (1975b), *Macromolecules* **10**, 160. Copyright by the American Chemical Society.]

Lastly, Karasz (1977) has reported the phase diagram for a blend of *p*-chlorostyrene/*o*-chlorostyrene copolymer with PPO as a function of PPO concentration (Fig. 22). Thus if one knows the phase boundaries, the processing conditions can be controlled to yield either a homogeneous polyblend at room temperature or a mixture containing two discrete phases. Obviously, additional tests must be carried out to determine the most desirable physical state.

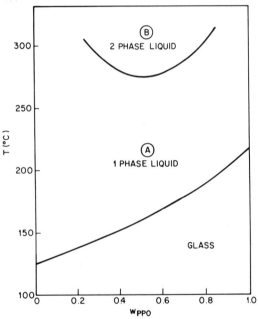

Fig. 22   Phase diagram for a blend of *p*-chlorostyrene/*o*-chlorostyrene copolymer with PPO as a function of weight fraction PPO(w_{PPO}) From Karasz and MacKnight, in "Contemporary Topics in Polymer Science," Vol. 2 (E. M. Pearce and J. R. Schaefgen, eds.). Plenum, New York.

2. Solid-State Transitions

a. Amorphous Polyblends.   Probably the least ambiguous measure of polymer compatibility is the detection of a single glass transition whose temperature falls somewhere between those corresponding to the two-component polymers. DSC measurements of incompatible blends of poly(vinyl chloride) and a copolymer of styrene and acrylonitrile (SAN) whose respective T_gs are 87 and 102°C can resolve the two individual T_gs at heating rates as fast as 40°C/min^{-1} (Bair, 1970a,b). However, if the difference between T_gs of two incompatible polymers is less than 10°, the resolution of the T_gs in a blend of the two will be difficult if not impossible unless special techniques are employed. This may entail annealing one phase selectively or plotting the derivative of the heat capacity C_p curve.

Additional difficulties can arise in detecting minor components in blends that are near the edge of compatibility. From quantitative DSC investigations of incompatible blends of PPO and random copolymers of styrene and p-chlorostyrene with compositions around the compatibility–incompatibility transition, Fried *et al.* (1978) has found that the width of the T_g interval increased as blend composition approached the phase separation point, and as much as 40% of the total sample is undetected in terms of C_p discontinuities. These authors hypothesize that the loss in ΔC_p could be caused by large, diffuse interfacial regions or the formation of minute dispersed phases that cannot be detected by DSC. However, the latter effect does not seem plausible in light of the detection of C_p discontinuities in plasticized PVC associated with domains as small as 50 Å (Bair and Warren, 1979). Clarification of these effects await more definitive studies.

Numerous attempts have been made to relate the T_g of a compatible blend to blend composition, as is often done with random copolymers. A few compatible blends such as butadiene–acrylonitrile copolymers with PVC (Zakrzewski, 1973) and PVC in ethylene–vinylacetate copolymers (Hammer, 1971) have been found to exhibit T_g-composition dependences, which can be predicted by the simple Fox (1956) equation

$$1/T_g = W_1/T_{g1} + W_2/T_{g2} \tag{7}$$

where W_1 and W_2 represents the mass fraction of the components and T_g, T_{g1}, and T_{g2} are the T_gs of the blend, component 1 and component 2, respectively. However, the T_g-compositional dependence of a number of compatible blends cannot be correlated by any of the well-known expressions such as the Kelley–Bueche (1961), Gordon–Taylor (1952), Gibbs–Dimarzio (1959), and Couchman (1978) (Chapter 9, III.A.2) equations.

b. Partially Crystalline Polyblends. Until recently it was believed that compatible blends of a crystalline polymer with any other polymer was unlikely. However, during the past decade several groups have shown that compatible polymer blends can be formulated in systems that have one polymer component that is crystalline. Examples of five such systems are blends of poly(vinylidene fluoride) (PVF_2) and poly(methyl methacrylate) (PMMA) by Noland *et al.* (1971), Nishi and Wang (1975), and Paul and Altamirano (1974); PVF_2 and poly(ethyl methacrylate) (PEMA) Noland *et al.* 1971), Imken *et al.* 1976), and Kwei *et al.* 1976); blends of PVC and polyϵ-caprolactone) (PCL) by Koleske and Lundberg (1969), Robeson (1973), and Khambatta *et al.* (1976); mixtures of PPO and isotactic polystyrene (PS) by Wenig *et al.* (1975, 1976); blends of PVC and an ethylene terpolymer (TP) by Bair *et al.* (1977); and mixtures of PVC with a polyester (Robeson, 1978) or a copolyester (Nishi *et al.*, 1975a).

All blends that include a crystallizable component have several common characteristics. First, they exhibit a single T_g that is composition dependent. Second, the noncrystalline polymer acts as a diluent that lowers the crystallization temperature T_c of the crystalline phase while T_g is increasing. In this manner crystallization can be inhibited when $T_c - T_g$ is near or below 0°. This latter effect leads to an increasing depression of T_m and the apparent heat of fusion ΔH_f^* as the concentration of polymeric diluent is increased. This behavior is illustrated in Figs. 23 and 24 for a

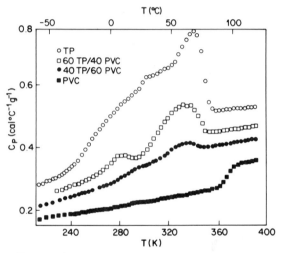

Fig. 23 Specific heat as a function of temperature for TP, 40 TP/60 PVC, 60TP/40PVC and PVC. (From Anderson *et al.*, 1979.)

polyblend of PVC and a terpolymer (TP) of ethylene vinyl acetate and carbon monoxide (Anderson *et al.*, 1979). In Fig. 23 the melting temperature is lowered by 6°C and T_g increased by 28°C when 60 wt % PVC is mixed with TP. In addition, the area under the melting endotherm is reduced significantly. If the apparent heats of fusion ΔH_f^* of the blends are normalized by comparing them to the apparent heat of fusion of slow-cooled, pure TP, the rapid decrease in blend crystallinity can be noted (Fig. 25). Here the broken line represents the expected ΔH_f^* if TP were free to crystallize with no interference from PVC. Obviously, the apparent heats of fusion were lower for all blends than would be predicted based on the blend composition. Thus it appears that the mobility of the crystallizable TP is severely restricted as T_c begins to approach T_g as the concentration of PVC

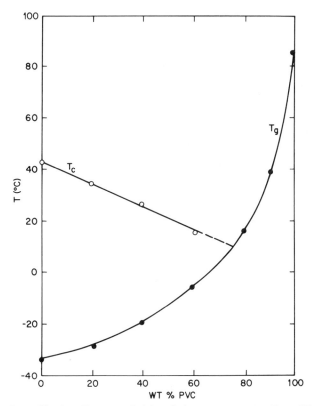

Fig. 24 Crystallization (T_c, open circle) and glass temperatures (T_g, solid circle) as a function of PVC concentration in blends of TP and PVC. (From Anderson *et al.*, 1979.)

in the blend is increased. Similar observations have been made for blends of PVF_2 and PMMA by Paul and Altamirano (1975) and Nishi and Wang (1975) and PVF_2 and PEMA by Kwei *et al.* (1976).

Recently, Nishi and Wang (1975) have derived an analytical expression for polymeric diluent systems from Scott's equation (Scott, 1949) for the thermodynamic mixing of two polymers. Their equation for the melting-point depression is

$$T_m^0 - T_m = -T_m^0 (V_{2u}/\Delta H_{2u}) B \phi_1^2 \tag{8}$$

where T_m^0 and T_m are the equilibrium and observed melting temperatures, respectively, V_{2u} is the molar volume of the crystalline polymer, ΔH_{2u} is the heat of fusion of the unblended crystalline component, ϕ is the volume fraction of the polymeric diluent, and B is related to the interaction parameter x by the expression

$$\chi = BV_1/RT \tag{9}$$

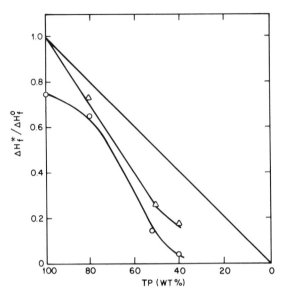

Fig. 25 Ratio of $\Delta H_f^* / \Delta H_f^0$ plotted against weight percent of TP in blends of TP and PVC: (open triangle) slow cooled than annealed at 60°C; (open circle) cooled from melt at 40° min^{-1}. (From Bair *et al.*, 1977.)

A plot of $\Delta T_m / \phi_1 T_m^0$ versus ϕ_1 should result in a straight line and the value of the interaction parameter for the polymer–polymer blend can be obtained from the slope of the plot.

Melting temperature of PVF$_2$ in the presence of PMMA (Nishi and Wang, 1975) or PEMA (Kwei *et al.*, 1976) have yielded x values of -0.30 and -0.34 at 160°C. The T_m depression of PVF$_2$ is shown in Fig. 26. The negative value of x, which is sufficient for thermodynamic stability, confirms the compatibility of the two pairs of polymer blends. In a recent extension of this work, it was found that two incompatible polymers, PMMA and PEMA, can be brought together by PVF$_2$ to form ternary mixtures that are compatible (Kwei *et al.*, 1977). The lowering of PVF$_2$s T_m in the ternary system can be calculated from Scott's equation. Ternary mixtures that contain 40–70 wt % of PVF$_2$ are amorphous when quenched from the melt and each mixture consists of a single phase whose T_g is equal to the volume fraction average of the T_g of the component polymers:

$$T_g = \Sigma_i \phi_i T_{gi} \tag{10}$$

The agreement between experimental and calculated values as shown in Fig. 27 is excellent.

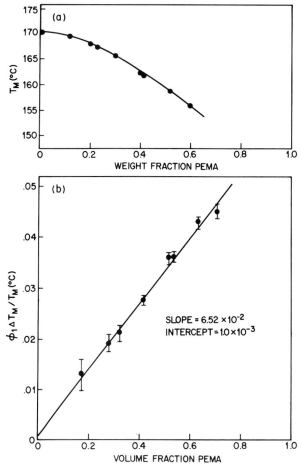

Fig. 26 Melting point depression of PVF_2 in mixtures of PVF_2–PEMA. [Reprinted with permission from T. K. Kwei, G. D. Patterson, and T. T. Wang, (1976), *Macromolecules* **9**, 783. Copyright by the American Chemical Society.]

C. Quantitative Analysis of Polyblends

Most polymer mixtures are not compatible and thus form phases that can be quantitatively identified by their transition temperatures. In some cases microphases of about 100 Å or smaller have been detected by DSC (Bair *et al.*, 1972; Anderson *et al.*, 1979; Bair and Warren, 1979, 1980, 1981). In addition, the concentration of a glassy polymeric component in a blend can be determined from the magnitude of increase in C_p at T_g. The importance of these measurements is obvious whether one is interested in

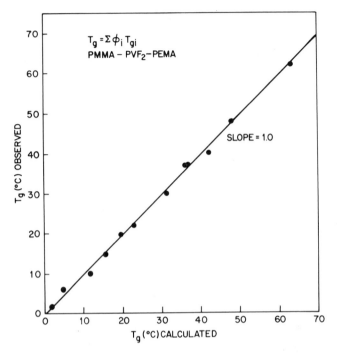

$$T_g = \Sigma \, \phi_i \, T_{gi}$$
PMMA – PVF$_2$–PEMA

SLOPE = 1.0

Fig. 27   Observed (solid circle) and calculated (line) T_gs of a ternary mixture of PMMA–PVF$_2$–PEMA. [Reprinted from T. K. Kwei, H. L. Frisch, W. Radigan, and S. Vogel (1977), *Macromolecules* **10,** 157. Copyright by the American Chemical Society.]

relating the impact strength of a rubber-modified plastic to the amount of rubber in the system or checking the level of components in a polyblend prior to production. In some cases, these methods may be used to monitor the effects of chemical degradation due to processing (Congdon *et al.,* 1979; Bair *et al.,* 1980a).

1. Glassy Components

The glass temperature may be taken as either the onset or halfway point of the transition interval between the C_p of the glass and liquid. The latter point agrees closely with the break in enthalpy versus temperature plot that occurs at T_g and thus corresponds to a dilatometrically determined glass temperature, if prior thermal histories and heating rates are the same.

Although the C_p of an unknown sample can be measured by comparison to a standard reference material, the magnitude of the increase in the specific heat ΔC_p at the glass transition may be determined directly by

DSC without reference to any standard according to the relationship

$$\Delta C_p = \Delta S / MR \qquad (11)$$

where Δ is fractional increase in the ordinate at the transition, S is full-scale value of the power (ordinate) in millicalories per second, M is the mass of the sample, and R is the heating rate of the calorimeter in degrees per second.

This technique affords a rapid and accurate measurement of ΔC_p to within 1%. However, a few precautionary rules should be remembered for an accurate measurement of ΔC_p to be obtained: First, a heating rate should be selected that coincides with the effective cooling rate for the formation of the glass; or the glass should be cooled at a rate that will be greater than the subsequent heating rate; second, any stored energy due to orientation effects must be released by heating prior to evaluation of the material; third, if any diluent or plasticizer exists in the material, it will not only lower T_g but also spread the transition over a wide temperature interval and make the measurement of ΔC_p more difficult. Thus the diluent should be extracted and identified, and, if possible, a concentration series should be evaluated in order to understand the role of the diluent within the polyblend.

The fraction x of a component in a polyblend is the ratio of the observed change in C_p for a particular component in the polyblend ΔC_p^{blend} to the known increase in C_p of the pure parent of the component in the blend ΔC_p^{parent}

$$x = \Delta C_p^{blend} / \Delta C_p^{parent} \qquad (12)$$

An alternative thermal method of determining x without calibrating the DSC ordinate is by comparing the transition height, TH^{blend}, at T_g for the particular component in the blend to the transition height, TH^{parent}, of a known pure amount of the same material as in the polyblend:

$$x = (TH^{blend} / TH^{parent})(M^{parent} / M^{blend}) \qquad (13)$$

where M is the weight of the pure parent material and the weight of the polyblend, respectively.

2. Partially Crystalline Components

Crystalline phases in incompatible polyblends can be identified by the position of T_m and quantified x from the ratio of the observed heat of fusion ΔQ_f for a particular component in the polyblend ΔQ_f^{blend} to the known apparent heat of fusion ΔQ_f^* of the pure parent of one component in the blend $\Delta Q_f^{*parent}$:

$$x = \Delta Q_f^{blend} / \Delta Q_f^{*parent} \qquad (14)$$

In cases where one polymer acts as a diluent in a crystalline polyblend, Eq. (11) does not hold. However, if the effect of polymeric diluent on ΔQ_f is measured as a function of blend composition, subsequent polyblends of these two components can be analyzed (see Section III.B.2).

3. Experimental Difficulties

The techniques for the detection and quantification of T_g and T_m have been discussed earlier in the chapter. In addition to the anomalies that can be associated with diluent effects, several other limitations on the use of T_g for quantitative analysis in polyblends should be noted. Temperature shifts and the broadening of T_g can occur if one of the components in a blend is nearly compatible. Also, if one component is present in a very small quantity (< 5 wt %), it may not be detected and may lead to an erroneous assumption of compatibility. In these cases it is advisable to anneal the questioned component overnight (~ 15 h) by holding it 10 or 20° below its T_g prior to rescanning from $\sim 50°$ below T_g and looking for an endotherm astride T_g, which should help detect the minor phase if it is present (Bair and Warren, 1980, 1981). This technique can also be employed to study polyblends whose T_g are 20° or less apart.

Another complication can arise in using ΔC_p at T_g for assaying the concentration of a component in a polyblend and that is the attenuation or apparent absence of T_g in blends of one or more crystalline polymers. For certain polymers with sharp and distinct crystalline-amorphous regions, it is possible to detect both a melting endotherm and a glass transition by standard calorimetric procedures. Under these conditions Karasz *et al.* (1965) have shown that the degree of crystallinity x may be calculated from either the area under the endotherm or from ΔC_p by the relation

$$x = (\Delta Q_f/\Delta H_f^\circ) \approx 1 - (\Delta C_p^{obsd}/\Delta C_p^a) \tag{15}$$

where ΔH_f is the heat of fusion of the completely crystalline polymer, ΔQ_f is the experimentally determined heat of fusion for the semicrystalline sample, ΔC_p^{obsd} is the observed height of T_g for the partially crystalline sample, and ΔC_p^a is the magnitude of the increase in C_p at T_g for a totally amorphous sample. Recently, Bair and Warren (1980) have used this technique to estimate the level of crystallinity in poly(vinyl choride) samples prepared at different temperatures. In this way the thermal degradation that normally accompanies the melting of PVC was avoided. Many semicrystalline polymers, such as polycarbonate (O'Reilly *et al.*, 1964) or PPO (Karasz *et al.*, 1968), behave differently, as crystallinity in the sample is increased, a disproportionate reduction in ΔC_p occurs. This effect in blends can obscure T_g and make quantitative analysis difficult, if not impossible.

D. MULTICOMPONENT POLYBLENDS

The production of multiphase polymer systems is reportedly the fastest-growing field in the entire chemical industry. One type of these multicomponent materials is the rubber-reinforced plastics, where small, micron-sized rubber particles are blended or covalent bonded to a rigid polymer matrix. By this technique a brittle copolymer such as styrene–acrylonitrile (SAN) may be toughened by the addition of a rubberlike polybutadiene (PBD). In this case, the resulting combination of plastic and rubber is an acrylonitrile–butadiene–styrene (ABS) resin with high impact strength. The elucidation of the composition and structure of these polyblends may be accomplished by a combination of phase separation methods, spectroscopic analysis and electron microscopy (Gesner, 1968; Matsuo, 1969). In addition to these techniques rapid thermal analysis methods can be used to characterize these types of polyblends (Bair, 1970a,b). The quantitative thermal analysis of several of these impact-modified polyblends—namely, ABS, PVC, and Noryl—are reviewed later.

1. Incompatible Blends

In Fig. 28 the low-temperature specific heat C_p of *trans*-1,4-polybutadiene (BD) and an ABS resin are plotted against absolute temperature (Bair, 1974). ΔC_p was measured to be 0.088 cal °C⁻¹ g⁻¹ for the BD sample. The ABS sample has lower specific heat values as compared with the BD polymer; and, more importantly, it has a comparatively smaller ΔC_p

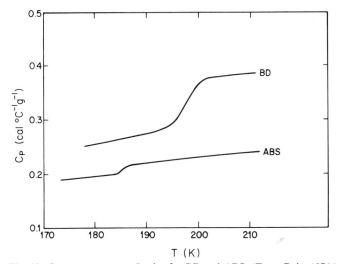

Fig. 28   Low temperature C_p plot for BD and ABS. (From Bair, 1974.)

at 184 K with a value of 0.011 cal $K^{-1} g^{-1}$. This T_g at 184 K is attributed to the BD phase of the ABS. Thus the concentration of BD in the ABS resin is equal to the ratio of 0.011 (ΔC_p of the BD in ABS) to 0.087 (ΔC_p of pure BD), or about 0.13. The estimated 13 wt % BD in the ABS agrees well with value determined by other chemical techniques. The T_g of BD polymer in the ABS is about 10° lower than T_g of the pure polymer. Subsequent studies have shown the lowering of the BD polymer's T_g is a diluent effect resulting from the presence of about 1 wt % styrene (Reed *et al.*, 1974).

The preceding room-temperature C_p behavior of a commercial resin is shown in Fig. 29. In addition to the expected second-order transition at 93°C due to SAN, two first-order transitions at 70 and 150°C were observed. The T_g at 93°C with ΔC_p equal to 0.070 cal °C^{-1} g^{-1} is due to 76 wt % SAN, the matrix of the ABS blend. One should note that this technique does not give any measure of the level of grafting in the polyblend. The small melt at 70°C is believed to be the melting of about 0.24 wt % fatty acid residue. The larger ΔQ_f at 150°C is due to the fusion of a molding lubricant (see Chapter 9).

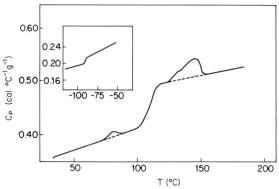

Fig. 29 Specific heat versus temperature for a commercial ABS resin. (From Bair, 1974.)

These same thermal methods have been used to study impact modifiers such as methacylate–butadiene–styrene (MBS) and methacrylate–acrylonitrite–butadiene–styrene (MABS) (Bair, 1970b). The amount of rubber in these systems was relatively high and varied from 45 to 56 wt %. This is far greater than the amount of rubber found in any commercial ABS resins. The T_gs of -57 to -52°C indicate the presence of copolymers of BD. These materials should be viewed as impact-modifying additives.

The C_p of an impact-modified PVC resin is compared to C_p for PVC

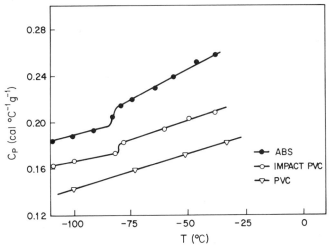

Fig. 30 Comparative C_p behavior of PVC, ABS, and an impact-modified PVC between −110 and −25°C. (From Bair, 1970b.)

and ABS and shown in Figs. 30 and 31. Three T_gs were observed at −80, 80, and 104°C. The low-temperature transition is attributed to polybutadiene; the intermediate T_g at 80° is ascribed to PVC, and the highest T_g is due to SAN. Therefore, it is inferred that this impact-modified PVC resin has been produced by blending about 48 wt % ABS with 45 wt % PVC

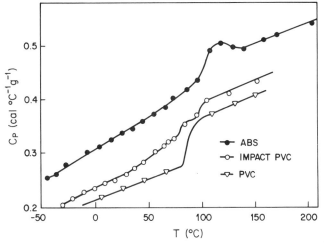

Fig. 31 Specific heat comparison between PVC, ABS, and an impact-modified PVC from −25 to 200°C. (From Bair, 1970b.)

(the remaining 7 wt % is believed to be due to additives). The composition of this resin was based on ΔC_p measurements (Bair, 1970b).

Since the role of rubber particles in a rubber-toughened plastic is to modify the deformational behavior of the matrix polymer, the composition of the matrix-to-rubber graft in this type of material should have a dominant effect in determining physical properties, such as impact strength and tensile elongation. Classically, the characterization of the level of grafting in these toughened polymers has been made by solvent–phase separation techniques. Bair *et al.* (1981) have found that the magnitude of ΔC_p at the rubber's T_g diminished in direct proportion to increasing graft content in ABS resins. These findings indicate that in order to maintain a given level of impact strength the concentration of BD in a rubber-toughened plastic must be raised as the amount of grafting (SAN to BD) was increased.

The structural relaxation in glassy polymers can be studied by monitoring the time dependence of the enthalpy using DSC techniques (Petrie, 1972; Matsuoka and Bair 1977; Bair *et al.*, 1980, 1981). Several aspects of nonlinear viscoelastic response in these materials can be interpreted in terms of the excess enthalpy associated with dilatation under strain (Matsuoka *et al.*, 1973, 1974). In fact the onset of brittle behavior following annealing below T_g can be explained by the fact that it would take a greater magnitude of strain to regain a critical amount of excess enthalpy through stress-induced dilation than the glassy polymer can tolerate without failure (Matsuoka *et al.*, 1974). Tant and Wilkes (1981) have followed the physical aging in styrene–butadiene block copolymers by DSC experiments. It was found that annealing these copolymers below T_g not only lowered their excess enthalpies but also increased their mechanical response time.

2. Compatible Blends

One of the most studied compatible blends is that of PS and PPO (MacKnight *et al.*, 1978). PS and PPO are compatible in all proportions and films formed from their blends are clear. In addition, only single glass transitions intermediate in temperature between the T_gs of PPO and PS have been reported (Bair, 1970a,b; Prest and Porter, 1972; Shultz and Gendron, 1972) (Fig. 32).

From quantitative DSC measurements Bair (1970a,b) deduced that commercial Noryl resins are actually blends of PPO and high-impact polystyrene (HIPS). In Fig. 33 C_p versus T for a Noryl resin is plotted. Two T_gs were observed at -61 and $141°C$ with ΔC_ps equal to 0.005 and 0.044 cal $°C^{-1}$ g^{-1}, respectively. The low T_g is due to about 5 wt % styrene–butadiene rubber (SBR). The $141°C$ T_g indicates that the resin is composed

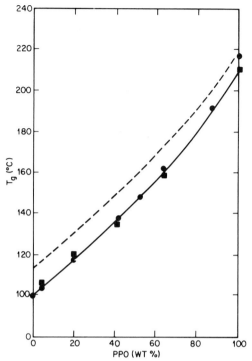

Fig. 32 T_g plotted against concentration of PPO in blends of PS–PPO. (Solid circle) DSC measurements (Prest and Porter, 1972). (Solid square) Probe penetration (Prest and Porter, 1972). Solid and broken curves were drawn from equations given by Shultz and Gendron (1972) for their DSC and thermal optical analysis (TOA) data, respectively. (From Fried, 1976.)

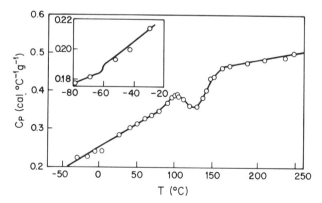

Fig. 33 C_p versus T for a Noryl® resin. (From Bair, 1970b.)

of approximately 47 wt % of both PS and PPO. In addition, a small melting was detected between 83 and 103°C, with ΔQ_f of 0.2 cal g^{-1}. Based on the crystallization and subsequent melting behavior of this unknown component, it was presumed to be about 1 wt % of low-density polyethylene acting as a mold lubricant.

It has been found that thermodynamically incompatible polymer pairs can be captured kinetically in an intimate admixture (Miyata and Hata, 1970). In their initial work poly(methyl methacrylate)–poly(vinyl acetate) blends freeze-dried from cosolutions in benzene were reported to be homogeneous and dynamic mechanical and dilatometric measurements revealed a single T_g intermediate between those of the pure polymers. Further dilatometric and calorimetric data corroborated the original observations of homogeneity in these blends (Ichihara *et al.*, 1971). Recently, Shultz and Young (1980) have reported that homogeneous blends of noncompatible poly(methyl methacrylate) and polystyrene were prepared by a rapid freezing of their cosolutions in naphthalene followed by room-temperature sublimation of the solvent. In this work the test for homogeneity was based on DSC data that showed single T_gs for these blends intermediate between the glass temperature of the component polymers.

E. Effects of Aging on Rubber-Toughened Polyblends by DSC

DSC analysis of ABS samples either weathered outdoors for up to 10 yr or aged at 71°C has shown that these plastics deteriorate most readily through the polybutadiene component (Bair *et al.*, 1980a). Oxidation of the BD in ABS leads to an increasing and broadening T_g as well as a decrease in ΔC_p at T_g (Fig. 34). The latter quantity can be used to determine quantitatively the amount of unoxidized rubber in a photo- or thermo-oxidized sample. The loss of impact strength of these resins with chemical aging can be correlated with the oxidation of the polybutadiene phase. Studies of rubber content from ΔC_p measurements as a function of sample thickness have revealed that an embrittled layer containing oxidized BD develops at the surface of an aged, 100 mil ABS plaque (natural) and grows to a thickness of about 10 mil after 3 yr exposure to sunlight (Fig. 35). ABS containing carbon black is effectively screened from this photoinduced degradation and consequently has much better retention of its initial impact strength values. Last, it appears possible to conduct accelerated aging studies at elevated temperatures to evaluate the relative stabilities of various ABS formulations (Bair *et al.*, 1980a).

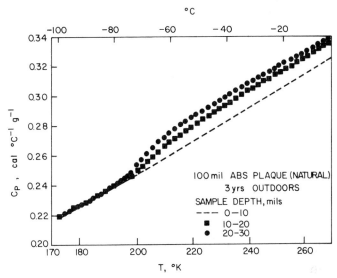

Fig. 34 C_p curves for a 100 mil, plaque of ABS (natural) aged outdoors for three years as a function of sample depth (H. E. Bair, unpublished work, 1980.)

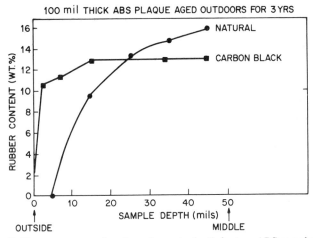

Fig. 35 Rubber content as a function of sample depth for two ABS samples aged for 3 years outdoors. (Open circle) ABS, natural (Solid circle) ABS, with about 2 wt % carbon black.

IV. Conclusion

In this chapter we have tried to cover some of the most recent studies on block and graft copolymers as well as polymer blends (polyblends). It is obvious that the thermal behavior and other technologically important physical properties of commercial polymeric composites constituted a major part of this review. This was done without excluding selected (a) theoretical treatments pertinent to the subject of multicomponent systems; (b) several discussions of experimental multicomponent systems; and (c) occasional discussion of athermal analytical techniques that have been used for the characterization of some multicomponent systems as major or complementary techniques to several thermoanalytical methods.

The presence of macromolecular impurities in block copolymers was noted in this review as a distinct cause of improper assessment of the thermal and physical properties of these systems. Although a few investigators have recognized and properly treated the presence of macromolecular impurities in block copolymers, it is felt that more studies need to be dedicated to this problem.

Our discussion of polymer blends has indicated explicitly the important role of thermal methods in characterizing and predicting the end use performance of such complex systems in a more satifactory manner than most other analytical techniques.

ACKNOWLEDGMENTS

The authors are indebted to Dr. F. E. Karasz of the University of Massachusetts, Dr. T. K. Kwei of Industrial Technology Research Institute, Hsin-Chu, Taiwan, and Dr. E. A. Turi of Allied Corporation for their excellent critical review of the chapter and for their constructive comments and excellent suggestions.

References

Aggarwal, S. L., ed. (1970). "Block Copolymers." Plenum, New York.
Allegrezza, A. E., Jr., Seymour, R. W., Ng, H. N., and Cooper, S. L. (1974). *Polymer* **15,** 433.
Allport, D. C. (1973). "Block Copolymers." Appl. Sci. Publ., London.
Anderson, E. W., Bair, H. E., Johnson, G. E., Kwei, T. K., Padden, R. J., and Williams, D. (1979). *Adv. Chem. Ser.* **176,** 413.
Angelo, R. J., Ikeda, R. M., and Wallach, M. L. (1965). *Polymer* **6,** 141.
Arnold, K. R., and Meier, D. J. (1970). *J. Appl. Polym. Sci.* **14,** 427.
Arpin, M., and Strazielle, C. (1975). *C. R. Hebd. Seances Acad. Sci., Ser. C* **28,** 1293.
Arpin, M., and Strazielle, C. (1976). *Makromol. Chem.* **177,** 293.
Ashman, P. C., and Booth, C. (1975). *Polymer* **16,** 889.
Ashman, P. C., and Booth, C. (1976). *Polymer* **15,** 105.
Ashman, P. C., Booth, C., Cooper, C. R., and Price, C. (1975). *Polymer* **16,** 897.
Baer, M. (1964). *J. Polym. Sci., Part A* **2,** 217.
Bair, H. E. (1970a). *Polym. Eng. Sci.* **10,** 247.
Bair, H. E. (1970b). *Anal. Calorim.* **2,** 51.
Bair, H. E. (1974). *Polym. Eng. Sci.* **14,** 202–205.

Bair, H. E., and Warren, P. C. (1979). *Bull. Am. Phys. Soc.* [2] **24**, 288.

Bair, H. E., and Warren, P. C. (1980). *Abstr. Int. PVC Symp. 3rd, 1980* pp. 101–104.

Bair, H. E., and Warren, P. C. (1981). *J. Macromol. Sci. Phys.* (in press).

Bair, H. E., Matsuo, M., Salmon, W. A., and Kwei, T. K. (1972). *Macromolecules* **5**, 114.

Bair, H. E., Kwei, T. K., Williams, D., and Padden, F. J., Jr. (1977). *Org. Coat. Plast. Chem.*, **38**(1), 240.

Bair, H. E. Anderson, E. W., Johnson, J. E., and Kwei, T. K. (1978). *Polym. Prepr. Am. Chem. Soc., Div. Polym. Chem.* **19**, 143.

Bair, H. E., Boyle, D. J., and Kelleher, P. G. (1980a). *Polym. Eng. Sci.* **20**, 995.

Bair, H. E., Johnson, G. E., Anderson, E. W., and Matsuoka, S. (1980b). *ACS Polym. Preprints* **21**(2), 21–23.

Bair, H. E., Boyle, D. J., and Shepherd, L. (1981). *Bull. Am. Phys. Soc.* **26**(3), 398. Also in "Thermal Analysis in Polymer Characterization," (E. Turi, ed.). Heyden, Philadelphia, Pennsylvania (to be published).

Bank, M., Loffingwell, J., and Thies, C. (1972). *J. Polym. Sci., Polym. Phys. Ed.* **10**, 1097.

Battaerd, H. A. J., and Tregear, G. W. (1967). "Polymer Reviews," Vol. 16. Wiley (Interscience), New York.

Beatty, C. L., and Karasz, F. E. (1979). *J. Macromol. Sci., Rev. Macromol. Chem.* **C17**, 37.

Beevers, R. B., and White, E. F. T. (1960). *Trans Faraday Soc.* **56**, 1529.

Bianchi, U., Pedemonte, E., and Turturro, A. (1970). *Polymer* **11**, 268.

Black, P., and Worsfold, D. J. (1974). *J. Appl. Polym. Sci.* **18**, 2307.

Bohn, L. (1968). *Rubber Chem. Technol.* **41**, 495.

Bollinger, J. C. I. (1977). *J. Macromol. Sci., Rev. Macromol. Chem.* **C16**, 23.

Booth, C., and Pickles, C. J. (1973). *J. Polym. Sci., Polym. Phys. Ed.* **11**, 249.

Brandrup, J., and Immergut, E. H., Eds. (1975). "Polymer Handbook," 2nd ed. Wiley, New York.

Bruck, S. D. (1969). *Polymer* **10**, 939.

Bruck, S. D., and Levi, A. A. (1967). *J. Macromol. Sci., Chem.* **A1**, 1095.

Bruck, S. D., and Thadani, A. (1970). *In* "Block Copolymers" (S. L. Aggarwal, ed.), p. 321. Plenum, New York.

Buck, W. H. (1977). *Rubber Chem. Technol.* **50**, 109.

Burke, J. J., (1973). *J. Macromol. Sci., Chem.* **A7**, 187.

Bushuk, W., and Benoit, H. (1958). *Can. J. Chem.* **36**, 1616.

Cella, R. J. (1973). *J. Polym. Sci., Polym. Symp.* **42**, 727.

Ceresa, R. J. (1973). "Block and Graft Copolymers," Vol. 1. Wiley, New York.

Chandler, L. A., and Collins, T. E. (1969). *J. Appl. Polym. Sci.* **13**, 1585.

Childers, C. W., and Kraus, G. (1967). *Rubber Chem. Technol.* **40**, 1183.

Cohen, R. E., and Tschoegl, N. W. (1972). *Int. J. Polym. Mater.* **2**, 49.

Congdon, W. I., Bair, H. E., and Khanna, S. K. (1979). *Org. Coat. Plast. Prepr.* **40**, 739.

Cooper, S. L., and Tobolsky, V. A. (1966). *Text. Res. J.* **36**, 800.

Cooper, S. L., West, J. C., and Seymour, R. W. (1976). *Encycl. Polym. Sci. Technol. Suppl.* **1**, 521.

Couchman, P. R. (1978). *Macromolecules* **11**, 1156.

Couchman, P. R., and Karasz, F. E. (1977). *J. Polym. Sci., Polym. Phys. Ed.* **15**, 1037.

Datta, P. K., and Pethrick. R. A. (1977). *Polymer* **18**, 919.

Dluglasz, J., Keller, A., and Pedemonte, E. (1970). *Kolloid Z. Z. Polym.* **242**, 1125.

Estes, G. M., Cooper, S. L., and Tobolsky, V. A. (1970). *J. Macromol. Sci., Rev. Macromol. Chem.* **C4**, 313.

Estes, G. M., Seymour, R. W., and Cooper, S. L. (1971). *Macromolecules* **4**, 452.

Ferguson, J., and Petrovic, Z. (1976). *Eur. Polym. J.* **12**, 177.

Ferguson, J., and Patsavoudis, D. (1972). *Eur. Polym. J.* **8**, 385.

Ferguson, J., and Patsavoudis, D. (1974). *Rheol. Acta* **13**, 72.

Ferguson, J., Hourston, D. J., Meredith, R., and Patsavoudis, D. (1972). *Eur. Polym. J.* **8**, 369.

Fisher, E. (1968). *J. Macromol. Sci., Chem.* **A2**, 1285.

Fox, T. G. (1956). *Bull. Am. Phys. Soc.* [2] **1**, 123.

Fried, J. R. (1976). Ph. D. Thesis, Univ. of Massachusetts.

Fried, J. R., Karasz, F. E., and MacKnight, W. J. (1978). *Macromolecules* **11**, 150.

Gaur, U., and Wunderlich, B. (1980). *Macromolecules* **13**, 445.

Gervais, M., and Gallot, B. (1973). *Makromol. Chem.* **171**, 157.

Gervais, M., Douy, A., and Gallot, B. (1971). *Mol. Cryst. Lig. Cryst.* **13**, 289.

Gesner, B. D. (1968). *Polym. Sci. USSR (Engl. Transl.)* **3A**, 3825.

Gibbs, J., and Dimarzio, E. (1959). *J. Polym. Sci.* **40**, 121.

Gordon, M., and Taylor, J. S. (1952). *J. Appl. Chem.* **2**, 493.

Hammer, C. F. (1971). *Macromolecules* **4**(1), 69.

Hendus, H., Illers, K. I., and Ropte, E. (1967). *Kolloid Z. Z. Polym.* **216**, 110.

Hergenrother, W. L., and Ambrose, R. J. (1974). *J. Polym. Sci., Polym. Chem. Ed.* **12**, 2613.

Herman, J., Jerome, R., Teyssie, P., Gervais, M., and Gallot, B. (1978). *Makromol. Chem.* **179**, 1111.

Hesketh, T. R., and Cooper, S. L. (1977). *Org. Coat. Plast. Prepr.* **37**(2), 509.

Huet, J. M., and Marechal, E. (1974). *Eur. Polym. J.* **10**, 771.

Ichihara, S., Komatsu, A., and Hara, T. (1971). *Polym. J.* **2**, 640.

Imken, R. L., Paul, D. R., and Barlow, J. W. (1976). *Polym. Eng. Sci.* **16**, 593–601.

Kambour, R. P. (1972). U. S. Patent (to General Electric Co.) 3,639,508.

Karasz, F. E. (1977). *Contemp. Top. Polym. Sci.* **2**, 143.

Karasz, F. E., Bair, H. E., and O'Reilly, J. M. (1965). *J. Phys. Chem.* **69**, 2657.

Karasz, F. E., Bair, H. E., and O'Reilly, J. M. (1968). *J. Polym. Sci., Polym. Phys. Ed.* **6**, 1141.

Ke, B., and Sisko, A. W. (1961). *J. Polym. Sci.* **50**, 87.

Kelley, F. N., and Bueche, F. (1961). *J. Polym. Sci.* **50**, 549.

Kenney, J. F. (1968). *Polym. Eng. Sci.* **8**, 216.

Khambatta, F. B., Warner, F., Russel, T., and Stein, R. S. (1976). *J. Polym. Sci., Polym. Phys. Ed.* **14**, 1391.

Koleske, J. V., and Lundberg, R. D. (1969). *J. Polym. Sci., Polym. Phys. Ed.* **7**, 795.

Korshak, V. V., Frunze, T. M., Kurashev, V. V., and Danilevskaya, L. B. (1976). *Vysoko-mol. Soedin., Ser. A* **18**, 848.

Kraus, G., and Rollmann, K. W. (1976). *J. Polym. Sci., Polym. Phys. Ed.* **14**, 1133.

Kraus, G., Childers, C. W., and Gruver, J. T. (1967). *J. Appl. Polym. Sci.* **11**, 1581.

Kraus, G., Naylor, F. E., and Rollmann, K. W. (1971). *J. Polym. Sci., Polym. Phys. Ed.* **9**, 1839.

Krause, S. (1972). *J. Macromol. Sci., Rev. Macromol. Chem.* **C7**(2), 251.

Krause, S. (1978). *In* "Polymer Blends" (D. R. Paul and S. Newman, eds.), Vol. 1, pp. 16–106. Academic Press, New York.

Krause, S., and Iskandar, M. (1978). *Polym. Prepr.*, **19**, 44.

Krause, S., and Roman, N. (1965). *J. Polym. Sci.* **3**, 1631.

Kwei, T. K., Nishi, T., and Roberts, R. F. (1974). *Macromolecules* **7**, 667.

Kwei, T. K., Patterson, G. D., and Wang, T. T. (1976). *Macromolecules* **9**, 78–784.

Kwei, T. K., Frisch, H. L., Radigan, W., and Vogel, S. (1977). *Macromolecules* **10**, 157.

Lenz, R. W., and Go, S. (1973). *J. Polym. Sci., Polym. Chem. Ed.* **11**, 2927.

Lenz, R. W., and Go, S. (1974). *J. Polym. Sci., Polym. Chem. Ed.* **12**, 1.
Lenz, R. W., Awl, R. A., Miller, W. R., and Pryde, E. H. (1970). *J. Polym. Sci., Polym. Chem. Ed.* **8**, 429.
Lenz, R. W., Martin, E., and Schuler, A. N. (1973a). *J. Polym. Sci., Polym. Chem. Ed.* **11**, 2265.
Lenz, R. W., Ohata, K., and Funt, J. (1973b). *J. Polym. Sci., Polym. Chem. Ed.* **11**, 2273.
Lewis, P. R., and Price, (1971). *Polymer* **13**, 20.
Lilanonitkul, A., and Cooper, S. L. (1977). *Rubber Chem. Technol.* **50**(1), 1.
Lilaonitkul, A., West, J. C., and Cooper, S. L. (1976). *J. Macromol. Sci., Phys.* **B12**, 563.
Lilaonitkul, A., Estes, G. M., and Cooper, S. L. (1977). *Polym. Prepr., Am. Chem. Soc., Div. Polym. Chem.* **18**(2), 500.
Lin, C. K., Chin, R. E., and Tschoegl, N. W. (1971). *Adv. Chem. Ser.* **99**, 397.
McGrath, J. E., Ward, T. C., Schori, E., and Wnuk, A. J. (1977a). *Polym. Prepr., Am. Chem. Soc. Div. Polym. Chem.* **18**, 346.
McGrath, J. E., Ward, T. C., Schori, E., and Wnuk, A. J. (1977b). *Polym. Eng. Sci.* **17**, 647.
MacKnight, W. J., Stoelting, J., and Karasz, F. E. (1971). *Adv. Chem. Ser.* **99**, 29.
MacKnight, W. J., Karasz, F. E., and Fried, J. R. (1978). *In* "Polymer Blends" (D. R. Paul and S. Newman, eds.), Vol. 1, p. 242. Academic Press, New York.
Mandelkern, L. (1964). "Crystallization of Polymers. McGraw-Hill, New York.
Marupov, R., Kalontarov, I. Ya., Asroıov, Y., and Mavyanov, A. M. (1977). *J. Polym. Sci., Polym. Chem. Ed.* **15**, 2835.
Matsuo, M. (1969). *Polym. Eng. Sci.* **9**, 226.
Matsuoka, S., and Bair, H. E. (1977). *J. Appl. Phys.* **48**(10), 4058.
Matsuoka, S., Bair, H. E., and Aloisio, C. J. (1973). *J. Appl. Phys.* **44**, 4265–4268.
Matsuoka, S., Bair, H. E., and Aloisio, C. J. (1974). J. Polym. Sci.: Symposium No. 46, 115–126.
Matzner, M., Noshay, A., and McGrath, J. E. (1973). *Polym. Prepr., Am. Chem. Soc., Div. Polym. Chem.* **14**, 68.
Meier, D. J. (1974). *Polym. Prepr.* **15**, 171.
Miyata, S., and Hata, T. (1970). *Proc. Int. Congr. Rheol., 5th, 1968* Vol. 3, pp. 71–81.
Nishi, T., and Wang, T. T. (1975). *Macromolecules* **8** (6), 909.
Nishi, T., Kwei, T. K., and Wang, T. T. (1975a). *J. Appl. Phys.* **46**, 4175.
Nishi, T., Wang, T. T., and Kwei, T. K. (1975b). *Macromolecules* **8**, 227.
Noland, J. S., Hsu, N.-C., Saxon, R., and Schmitt, J. M. (1971). *Adv. Chem. Ser.* **99**, 15.
Noshay, A., and McGrath, J. E. (1977). "Block Copolymers: Overview and Critical Survey." Academic Press, New York.
O'Malley, J. J. (1975). *J. Polym. Sci., Polym. Phys. Ed.* **13**, 1353.
O'Malley, J. J., Crystal, R. G., and Erhardt, P. F. (1970). *In* "Block Copolymers" (S. L. Aggarwal, ed.), p. 163. Plenum, New York.
O'Malley, J. J., and Stauffer, W. J. (1974). *J. Polym. Sci., Polym. Chem. Ed.* **12**, 865.
O'Malley, J. J., and Stauffer, W. J. (1977). *Polym. Eng. Sci.* **17**, 510.
O'Reilly, J. M., Karasz, F. E., and Bair, H. E. (1964). *J. Polym. Sci., Part C* **6**, 109.
Ossterhof, H. A. (1974). *Polymer* **15**, 49.
Owen, M. J., and Thompson, J. (1972). *Br. Polym. J.* **4**, 297.
Panaris, R., and Pallas, G. (1970). *J. Polym. Sci., Part B* **8**, 441.
Paul, D. R., and Altamirano, J. O. (1974). *Polym. Prepr., Am. Chem. Soc., Div. Polym. Chem.* **15**(1), 409.
Paul, D. R., and Altamirano, J. O. (1975). *Adv. Chem. Ser.* **142**, 371.
Pedemonte, E., Dondeo, G., de Candta, F., and Romano, G. (1976). *Polymer* **17**, 73.

Petersen, R. J., Cornelussen, R. D., and Rozelle, L. T. (1969). *Polym. Prepr., Am. Chem. Soc., Div. Polym. Chem.* **10**, 385.

Petrie, S. E. B. (1972). *J. Polym. Sci. A 2*, **10**, 1255–1272.

Pochan, J. M., and Crystal, R. G. (1972). *In* "Dielectric Properties of Polymers" (F. E. Karasz, ed.), p. 313. Plenum, New York.

Prest, W. M., and Porter, R. S. (1972). *J. Polym. Sci., Polym. Phys. Ed.* **10**, 1639.

Reed, T. F., Bair, H. E., and Vadimsky, R. G. (1974). *In* "Recent Advances in Polymer Blends, Grafts and Blocks" (L. H. Sperling, ed.), p. 359. Plenum, New York.

Renuncio, J. A. R., and Prausnitz, J. M. (1977). *J. Appl. Polym. Sci.* **21**, 2867.

Robeson, L. M. (1973). *J. Appl. Polym. Sci.* **17**, 3607.

Robeson, L. M. (1978). *J. Polym. Sci., Polym. Lett. Ed.* **16**, 261.

Robinson, R. A., and White, E. F. T. (1970). *In* "Block Copolymers" (S. L. Aggarwal, ed.), p. 123. Plenum, New York.

Scott, R. L. (1949). *J. Chem. Phys.* **17**(3), 279.

Senich, G. A., and MacKnight, W. J. (1978). *Polym. Prepr., Am. Chem. Soc., Div. Polym. Chem.* **19**, 11.

Seymour, R. W., and Cooper, S. L. (1973). *Macromolecules* **6**, 48.

Seymour, R. W., Overton, J. R., and Corley, L. S. (1975). *Macromolecules* **8**, 331.

Shalaby, S. W., Pearce, E. M., and Reimschuessel, H. K. (1973a). *Ind. Eng. Chem., Prod. Res. Dev.* **12**, 128.

Shalaby, S. W., Reimschuessel, H. K., and Pearce, E. M. (1973b). *Polym. Eng. Sci.* **13**, 88.

Shalaby, S. W., Turi, E. A., and Harget, P. J. (1976a). *J. Polym. Sci., Polym. Chem. Ed.* **14**, 2407.

Shalaby, S. W., Sifniades, S., and Sheehan, D. (1976b), *J. Polym. Sci., Polym. Chem. Ed.* **14**, 2675.

Shalaby, S. W., Turi, E. A., and Pearce, E. M. (1976c). *J. Appl. Polym. Sci.* **20**, 3185.

Shalaby, S. W., Turi, E. A., and Reimschuessel, H. K. (1978). *Polym. Prepr., Am. Chem. Soc., Div. Polym. Chem.* **19**, 654.

Shen, M., and Kaelble, D. H. (1970). *J. Polym. Sci., Part B* **8**, 149.

Shen, M., Kamiskia, V. A., Biliyar, K., and Boyd, R. H. (1973). *J. Polym. Sci., Polym. Phys. Ed.* **11**, 2261.

Shultz, A. R., and Beach, B. M. (1974). *Macromolecules* **7**, 902.

Shultz, A. R., and Beach, B. M. (1977). *J. Appl. Polym. Sci.* **21**, 2305.

Shultz, A. R., and Gendron, B. M. (1972). *J. Appl. Polym. Sci.* **16**, 461.

Schultz, A. R., and Gendron, B. M. (1973). *Polym. Prepr. Amer. Chem. Soc. Div. Polym. Chem.* **14**(1), 571.

Shultz, A. R., and Young, A. L. (1980). *Macromolecules* **13**, 663–668.

Slowikowska, I., and Daniewska, I. (1975). *J. Polym. Sci., Polym. Symp.* **53**, 187.

Smith, T. L., and Dickie, R. A. (1969). *J. Polym. Sci., Part C* **26**, 163.

Stehlicek, J., and Sebenda, J. (1977a), *Eur. Polym. J.* **13**, 949.

Stehlicek, J., and Sebenda, J. (1977b). *Eur. Polym. J.* **13**, 955.

Stoelting, J., Karasz, F. E., and MacKnight, W. J. (1969). *Polym. Prepr., Am. Chem. Soc., Div. Polym. Chem.* **10**, 628.

Stoelting, J., Karasz, F. E., and MacKnight, W. J. (1970). *Polym. Eng. Sci.* **10**(3), 133.

Su, C. S., and Patterson, D. (1977). *Macromolecules* **10**, 708.

Sundet, S. A. (1978). *Macromolecules* **11**, 146.

Tant, M. R., and Wilkes, G. L. (1981) *Polym. Eng. Sci.* **21**, 325–330.

Teyssie, P., Bioul, J. P., Hamitou, A., Heuschen, J., Hock, L., and Jerome, R. (1977). *ACS Symp. Ser.* **59**, 165.

Toi, K., Kikuchi, M., and Tokuda, T. (1977). *J. Appl. Polym. Sci.* **21**, 535.

Ward, T. C., Wnuk, A. J., Henn, A. R., Tang, S., and McGrath, J. E. (1978). *Polym. Prepr.,*
    Am. Chem. Soc., Div. Polym. Chem. **19,** 115.
Wardell, G. E., Douglass, D. C., and McBrietry, V. (1976). *Polymer* **17,** 41.
Wenig, W., Karasz, F. E., and MacKnight, W. J. (1975). *J. Appl. Phys.* **46,** (10), 4194.
Wenig, W., Hammel, R., MacKnight, W. J., and Karasz, F. E. (1976). *Macromolecules*
    9(2), 253.
West, J. C., Lilaonitkul, A., Cooper, S. L., Mehra, U., and Shen, M. (1974). *Polym. Prepr.,*
    Am. Chem. Soc., Div. Polym. Chem. **15**(2), 191.
Wilkes, G. L., Bagrodia, S., Humphries, W., and Wildnauer, R. (1975). *J. Polym. Sci.,*
    Polym. Lett. Ed. **13,** 321.
Witsiepe, W. K. (1973). *Adv. Chem. Ser.* **129,** 39.
Wolfe, J. R., Jr. (1978). *Polym. Prepr., Am. Chem. Soc., Div. Polym. Chem.* **19,** 5.
Wood, L. A. (1958). *J. Polym. Sci.* **28,** 319.
Yamashita, Y., Matsui, H., and Ito, K. (1972). *J. Polym. Sci., Polym. Chem. Ed.* **10,** 3577.
Yamashita, Y., Murase, Y., and Ito, K. (1973). *J. Polym. Sci., Polym. Chem. Ed.* **11,** 435.
Yoshida, S., Suga, H., and Seki, S. (1973). *Polym. J.* **5,** 25.
Zakrzewski, G. A. (1973). *Polymer* **14,** 347.
Zotteri, L., and Giuliani, G. P. (1978). *Polymer* **19,** 476.

CHAPTER 5

Thermosets

R. BRUCE PRIME

International Business Machines Corporation
San Jose, California

I. Introduction

This chapter covers the application of thermal analysis to rigid thermosetting polymers. Thermal analysis techniques covered are differential scanning calorimetry (DSC), differential thermal analysis (DTA), thermogravimetry (TG), thermomechanical analysis (TMA), torsional braid analysis (TBA), and dynamic mechanical analysis (DMA). Instrumentation is described in Chapter 1 and the general response of materials to the addition or withdrawal of heat in Chapter 2. Flexible thermo setting polymers, or elastomers, are treated in Chapter 6. These thermal analysis techniques are often complemented by chemical identity measurements such as infrared (IR), gas chromatography (GC), liquid chromatography (LC), and mass spectrometry (MS); molecular size measurements such as gel permeation chromatography (GPC); and rheological measurements. One such application area is quality control (May *et al.*, 1978; Chen, 1978; Thomas *et al.*, 1979; Penn, 1979; Zucconi, 1979). Another is the study of mechanisms and kinetics of curing (epoxies: Acitelli *et al.*, 1971; Schneider *et al.*, 1979; phenolics: Katovic, 1967a,b; Kurachenkov and Igonin, 1971; King *et al.*, 1974; Westwood, 1975). Results of these measurement techniques are discussed only where they have been used to supplement thermal analysis methods; TMA rheological techniques are described.

As the name suggests, thermosetting resins become set, i.e., infusible and insoluble, as a consequence of the chemical cross-linking reactions accompanying cure. Most formulations require heat for curing; sometimes pressure is used to enhance flow. The resulting polymer, if properly processed, is a highly cross-linked, three-dimensional infinite network.

The majority of thermosets are used in filled or reinforced form to reduce cost, to modify physical properties, to act as a binder for particles, to reduce shrinkage during cure, or to provide or enhance flame retardance. In general, thermosets possess good dimensional stability, thermal stability, chemical resistance, and electrical properties. Because of these properties, they find widespread use in several fields, for example,

(1) Aerospace: radomes, ablative materials for atmospheric reentry, wing and fuselage structural members, helicopter rotor blades, missile components;

(2) Appliance: appliance housings, knobs, pot handles, corn poppers;

(3) Automotive: distributor caps, integrated front-end panels, brake pads, paint, as a lightweight replacement for metal;

(4) Building and housing: plywood paneling, floor tiles, roofing and siding, counter tops, mobile home construction, foamed-in-place insulation;

(5) Clothing: textile treating, pearlescent shirt buttons;

(6) Electrical: wiring boards, conductive polymeric elements, pro-

tective coatings, potting and encapsulation from minute silicon chips to large switches and coils;

(7)   Furniture: simulated wood door and drawer fronts, lamp shades, pedestals, dinnerware;

(8)   Medical and dental: dental fillings, orthopedic implant material;

(9)   Radio and television: housings, switches, implosion barriers for television tubes;

(10)   Recreation: bowling balls, boat hulls, marine finishes, dune buggies, fishing rods, tennis racquets;

(11)   Tooling: abrasive grinding wheels, sandpaper.

Because it is crucial to the utilization of all thermosets, we begin with a discussion of the general nature of curing. The curing of a thermoset is complex in that several steps are involved. As illustrated in Fig. 1, the chemistry of cure begins by formation and linear growth of the chain that soon begins to branch, and then to cross-link. As the reaction proceeds, the molecular weight increases rapidly and eventually several chains

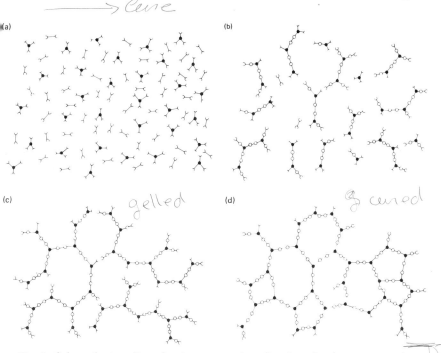

Fig. 1   Schematic, two-dimensional representation of curing of a thermoset, starting with A-stage monomer(s) (a); proceeding via simultaneous linear growth and branching to a B-stage material below the gel point (b); continuing with formation of a gelled but incompletely cross-linked network (c); and leading finally to the fully cured, C-stage thermoset (d).

become linked together into networks of infinite molecular weight. This sudden and irreversible transformation from a viscous liquid to an elastic gel, which marks the first appearance of the infinite network, is called the gel point.

Gelation is characteristic of thermosets, and it is of foremost significance. From a processing standpoint, gelation is critical since the polymer does not flow and is no longer processable beyond this point. Gelation occurs at a well-defined and calculable stage in the course of the chemical reaction and is dependent on the functionality, reactivity, and stoichiometry of the reactants. Means to detect gelation include the abrupt inability of bubbles to rise in the thermosetting mass, the rapid approach to infinite viscosity, and the back-extrapolations to zero modulus or zero gel fraction. Gelation typically occurs between 55 and 80% conversion (degree of cure $\alpha = 0.55-0.80$). For example, the gel point of a difunctional epoxy cross-linked with a stoichiometric amount of tetrafunctional amine in which all hydrogens are of equal reactivity is calculated to be $\alpha = 0.577$ (Flory, 1953). Gelation does not inhibit the curing process (e.g., the reaction rate remains unchanged) and cannot be detected by techniques sensitive only to the chemical reaction, such as DSC and TG. Beyond the gel point, reaction proceeds toward the formation of one infinite network possessing the dimensions of the reaction vessel and with substantial increase in cross-link density, glass transition temperature, and ultimate physical properties.

Another phenomenon, distinct from gelation, that may occur at any stage during cure is vitrification of the growing chains or network. This transformation from a viscous liquid or elastic gel to a glass begins to occur as the glass transition temperature of these growing chains or network becomes coincidental with the cure temperature. Further curing in the glassy state is extremely slow and, for all practical purposes, vitrification brings an abrupt halt to curing. Vitrification is a reversible transition, and cure may be resumed by heating to devitrify the partially cured thermoset. The onset of vitrification causes a shift from chemical control to diffusion control of the reaction and may be observed by a gradual decay of the reaction rate.

A time–temperature-transformation (TTT) diagram of curing, such as Fig. 2, is a useful tool for illustrating these phenomena. The TTT diagram is taken after Gillham (1980) and Enns *et al.* (1981) and can be generated by either TBA or DMA measurements. On the diagram the times to gelation and the times to vitrification are plotted as a function of the isothermal cure temperature. At temperatures below the glass transition temperature of the unreacted resin or resin mixture (T_{c0}, see Table I and Fig. 2), reaction is confined to the solid state and is therefore very slow to occur. T_{c0} serves to define storage temperatures for unreacted

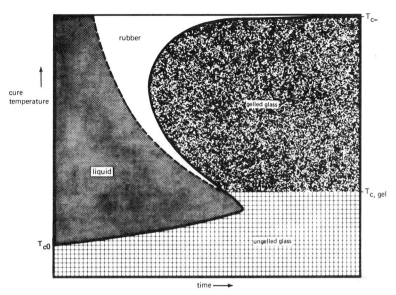

Fig. 2 Time–temperature-transformation (TTT) diagram, showing the four material states encountered during cure: liquid, rubber, ungelled glass, and gelled glass. T_{c0}, $T_{c,gel}$, and $T_c\infty$ (see Table I) are critical temperatures in the curing process. TTT diagrams are obtained by measuring times to gel (dashed line) and times to vitrify (solid line) versus isothermal cure temperature (e.g., by TBA or DMA). (After Gillman, 1980, and Enns *et al.*, 1981.)

TABLE I

GLOSSARY OF CURING TERMS AND CHARACTERISTIC CURING PARAMETERS

α = extent of conversion (e.g., of epoxide groups), fraction reacted, degree of cure
α_{gel} = α at the gel point
α_{ult} = maximum achievable extent of conversion ≤ 1

t_{gel} = time to gelation (gel time)
t_{vit} = time to vitrification

T_c = cure temperature
T_{c0} = T_c below which no significant reaction of the uncured resin mixture occurs, $\cong T_{g0}$
  (cf. storage temperature for uncured resin mixture)
$T_{c,gel}$ = T_c at which vitrification and gelation occur simultaneously
$T_{c\infty}$ = minimum T_c at which ultimate conversion occurs

T_g = glass transition temperature
T_{g0} = T_g for thermoset with degree of conversion $\alpha = 0$
$T_{g,gel}$ = T_g for thermoset with degree of conversion α_{gel}; $T_{g,gel} \geq T_{c,gel}$
$T_{g,ult}$ = T_g for thermoset with degree of conversion α_{ult}
$T_{g\infty}$ = T_g for thermoset with degree of conversion $\alpha = 1$

B (stoichiometric parameter) = ratio of equivalents of comonomer 1 (e.g., amine hydrogens) to comonomer 2 (e.g., epoxide)

---

resins. Between T_{c0} and $T_{c,gel}$ the liquid resin will react until its continuously rising glass transition temperature becomes coincidental with the cure temperature, at which stage vitrification will commence, and the reaction becomes diffusion controlled and is eventually quenched when vitrification is complete. Note that $T_{c,gel}$ is the cure temperature at which vitrification and gelation occur simultaneously.

Between $T_{c,gel}$ and $T_{c\infty}$, gelation precedes vitrification, and a cross-linked rubbery network forms and grows until its glass transition temperature coincides with the cure temperature and the reaction becomes quenched. Above $T_{c\infty}$, the minimum cure temperature required to achieve full and complete cure, the thermoset remains in the rubbery state after gelation unless other reactions occur, such as oxidative cross-linking or chain scission. When combined with DSC or TG, degree of conversion can be added to the TTT diagram. These latter measurements permit reaction rates and kinetics to be measured and applied to the processing of thermosets.

The handling, processing, and development of ultimate properties are very much dependent on gelation and vitrification. For example, thermosets are often identified at three conditions or "stages" of cure: A, B, and C stages. A stage refers to an unreacted resin; B stage to a partially reacted and usually vitrified system, *below the gel point*, requiring only heat to cure; C stage to the completely cured material. Thus to safely B-stage a thermoset requires vitrification prior to gelation, and this can be accomplished by maintaining the reaction temperature below $T_{c,gel}$. B staging is desirable because it provides systems that are easy to handle and process. To achieve complete cure, and thereby develop ultimate properties, necessitates the avoidance of vitrification during cure. To cross-link a thermoset completely, it is necessary to cure above the cure temperature of the fully cured polymer network ($T_{c\infty}$). Both of these latter concepts have originated with or been strongly supported by thermal analysis studies [see, for example, Babayevsky and Gillham (1973), Enns *et al.* (1981), Gillham (1980), Sourour and Kamal (1976)].

To take proper advantage of thermosets requires detailed knowledge of the handling, processing, and curing of the base resins, including gel points, glass transition temperatures, and reaction rates and kinetics; ability to measure the properties of the finished product, especially extent of cure; ability to determine the effects of fillers, catalysts, and other constituents including atmospheric moisture on the curing and physical properties; ability to determine or estimate the useful life under various stresses and environments; ability to measure degradation and decomposition, including the nature of decomposition products and analysis of failed parts; and ability to identify and/or quantify the polymeric resins, fillers, and other constituents, especially for quality control purposes.

This chapter focuses on the application of thermal analysis techniques to each of these areas. In addition, major emphasis is placed on the study by thermal analysis of curing kinetics and mechanisms. The old but still not settled question of the equivalency of cure kinetics derived from dynamic experiments with accepted isothermal data is also treated.

II. Thermophysical Property Measurements

A. CURING AND THE PROPERTIES OF UNCURED RESINS

This section covers various aspects of curing: mechanisms and kinetic equations, gelation, vitrification, and the experimental determination of cure kinetics. The objective in characterizing the cure of a thermoset is to provide a complete description of curing that is commensurate with its ultimate use. The generation of accurate time–temperature–degree of conversion curves is of great practical utility for establishing optimum cure schedules (B staging or approach to complete cure) and determining shelf lives (approach to gelation). The methods discussed are also capable of contributing to the determination of curing mechanisms. The primary objective of this section will be to demonstrate how thermal analysis techniques can provide such information. It should be noted that it is the complementary nature of several techniques, including, but not limited to, thermoanalytical, that provides a complete description of thermoset cure.

1. Mechanisms and Kinetic Equations

All kinetic studies start with the basic rate equation that relates the rate of conversion $d\alpha/dt$ at constant temperature to some function of the concentration of reactants $f(\alpha)$ through a rate constant k:

$$d\alpha/dt = kf(\alpha) \tag{1}$$

Mechanistically, thermoset curing can be divided into two general categories: nth order and autocatalyzed. It should also be pointed out that curing is not necessarily limited to one chemical reaction, and the kinetics may be those of an overall process when the chemical reactions occur simultaneously. Curing by two or more consecutive reactions is also possible.

For thermosets that follow nth-order kinetics, the rate of conversion is proportional to the concentration of material which has yet to react, for example,

$$f(\alpha) = (1 - \alpha)^n \tag{2}$$

where α is the fractional concentration of reactants consumed after time t (α is also the degree of conversion) and n is a reaction order. Thermoset cure reactions that are autocatalyzed are characterized by a maximum reaction rate at 30–40% of the reaction. Kinetics of autocatalyzed reactions are described by relations such as

$$f(\alpha) = \alpha^m(1 - \alpha)^n \tag{3}$$

where m is also a reaction order.

Diligence must be exercised to ensure that the correct equation is used to interpret thermoset curing data.

In the usual manner, the temperature dependence is assumed to reside in the rate constant through an Arrhenius relationship given by

$$k = A \exp(-E/RT) \tag{4}$$

where E is an activation energy, R the gas constant, T the absolute temperature, and A the preexponential or frequency factor. In practice, the course of the curing reaction of most thermosets can be adquately described in terms of these simple chemical models. That this is true suggests that the chemical reactions are controlling the rate of cure.

The following is a detailed development of kinetic equations based on known chemistry of the amine–epoxy reaction, probably the most frequently studied thermoset. The reaction most extensively characterized by thermoanalytical techniques is that between the diglycidyl ether of bisphenol-A (DGEBA) and m-phenylenediamine (mPDA), and it is recommended as a reference system. A unique feature of the amine–epoxy reaction is that at its extremes, it can follow either nth-order or autocatalyzed kinetics. The objectives here are to put the general kinetic equations into perspective, to describe the necessary simplifying assumptions, and to show that, depending on the particular reactive system, one equation may be more representative than another. Shechter *et al.* (1956) studied the chemistry of cure of epoxy resins by amine hardeners using model compound reactions and wet chemical analytical procedures. They found that combination of epoxide and primary amine lead to two principal reactions:

They found that hydroxyl groups generated during the reaction or provided by the addition of solvent (plus other catalysts, including water) accelerated the amine–glycidyl ether reaction markedly. In all cases, the hydroxyl groups served only as a catalyst for the reaction and not as a serious contender for epoxide in competition with amine. Dusek and Bleha (1977) again showed this to be true when amine is present in excess or at the stoichiometric concentration; only when epoxide was present in excess did the secondary hydroxyl groups formed in the preceding reaction add to the epoxide ring, Reaction 59. Byrne *et al.* (1980), using reverse-phase high-performance liquid chromatography and Fourier transform infrared applied to model compounds, again found only Reactions (5) and (6). However, introduction of a substituted urea accelerator caused a dramatic change in the nature of the reaction. Finally, it was concluded that conversion of both primary amine to secondary amine [Reaction (5)] and secondary amine to tertiary amine [Reaction (6)] proceeded more or less at random; i.e., the reaction rates were indistinguishable. Charlesworth (1979a,b) reached the same conclusion based on agreement between probability theory and GPC studies. Studies by Horie *et al.* (1970a) and Lunak and Dusek (1975) suggest that these reaction rates are close but not identical. It can be safely stated that the reactions of primary and secondary hydrogens with epoxide occur simultaneously, and only one overall reaction is detected. This is manifested by a single DSC peak, as in Fig. 73. Although a single overall reaction is the general rule for thermoset curing, there are some notable exceptions. For example, the curing of epoxides with polyamides (Prime and Sacher, 1972), a complex amine–expoxide containing an excess of epoxide and dicyandiamide (dicy) as one of the amines (Schneider *et al.*, 1979), some phenol–formaldehyde reactions (Fig. 5), and some unsaturated polyester and diallyl phthalate (DAP) curing reactions (Fig. 74).

A mechanism was proposed in which hydroxyl aids in the opening of the epoxide ring by hydrogen bonding in the transition state. Smith (1961) assumed this mechanism to be the rate-controlling step for the reaction between a secondary amine and epoxide and proposed third-order kinetics consistent with the results of Schechter *et al.* (1956). Horie *et al.* (1970a) extended this reaction scheme to include epoxide reaction with primary amine. The assumption that external catalyst or impurity initially present in the system $(HX)_0$ and reaction products having hydroxyl groups $(HX)_A$ act as true catalysts and are not consumed in any side reactions led to the following reaction scheme:

$$A_1 + E + (HX)_A \xrightarrow{\kappa_1} A_2 + (HX)_A \tag{7}$$

$$A_1 + E + (HX)_0 \xrightarrow{\kappa_1'} A_2 + (HX)_0 \tag{8}$$

$$A_2 + E + (HX)_A \xrightarrow{\kappa_2} A_3 + (HX)_A \tag{9}$$

$$A_2 + E + (HX)_0 \xrightarrow{\kappa_2'} A_3 + (HX)_0 \tag{10}$$

E, A_1, A_2, and A_3 represent epoxide, primary amine, secondary amine produced by addition of epoxide to primary amine, and tertiary amine as a final product, respectively. The corresponding e, a_1, and a_2 are the molar concentrations of E, A_1, and A_2 at time t; e_0 and a_0 the initial concentrations of E and A_1; c_0 the concentration of $(HX)_0$; and x the epoxide consumed, and also the concentration of $(HX)_A$, at time t. The rate of consumption of epoxide is given by

$$dx/dt = \kappa_1 a_1 ex + \kappa_1' a_1 e c_0 + \kappa_2 a_2 ex + \kappa_2' a_2 e c_0 \tag{11}$$

Some simplifying assumptions, including $\kappa_2/\kappa_1 = \kappa_2'/\kappa_1' + 1/2$ when reactivity of all amine hydrogens is equal, led to

$$dx/dt \approx (\kappa_1 x + \kappa_1' c_0)(e_0 - x)(a_0 - x/2) \tag{12}$$

In isothermal DSC studies of model reactions between phenyl glycidyl ether and n-butyl amine catalyzed with n-butyl alcohol, Horie *et al.* (1970a) observed the following:

(a)   The uncatalyzed reaction behaved according to the anticipated third-order, autocatalyzed kinetics with κ_1 (primary amine reaction) showing an activation energy of 58.2 kJ mole^{-1} (13.9 kcal mole^{-1}). The secondary amine reaction (κ_2) showed on activation energy of 56.1 kJ mole^{-1} (13.4 kcal mole^{-1}). Values of κ_2/κ_1 ranged from 0.61 to 0.65.

(b)   The reaction catalyzed with n-butyl alcohol behaved as predicted. At high catalyst concentrations, pseudo-second-order (nth-order) kinetics were observed. Data suggested the existence of trace impurity in the starting materials that accelerated the amine–epoxide reaction (compare, for example, the finite intercept $= k_1'$ in Fig. 31).

The same authors also studied three amine–epoxide thermosetting systems. They observed κ_1 values equal to or somewhat greater than those measured for the model reaction. Again, activation energies of 54–59 kJ mole^{-1} (13–14 kcal mole^{-1}) were observed.

The agreement between these activation energies for the thermosets and those for elementary rate constants (model reactions) suggests that there may be little difference between activation energies of the four basic chemical rate constants and explains why the rate-controlling mechanism for thermoset curing may be the same as for the model system.

Equation (12) is easily converted to a fractional concentration basis, where $\alpha = x/e_0$.

$$d\alpha/dt = (k_1\alpha + k_1')(1 - \alpha)(B - \alpha) \tag{13}$$

where $k_1 = \frac{1}{2}k_1 e_0^2$, $k_1' = \frac{1}{2}k_1' e_0 c_0$, and B ($= 2a_0/e_0$) is the ratio of amine hydrogen equivalents to epoxide equivalents initially; $B = 1$ when stoichiometric quantities of reactants are mixed. Equations (12) and (13) have been used by Horie *et al.* (1970a) and Sourour and Kamal (1976) in the study of amine–epoxide reactions by DSC. A very similar equation was used by Ryan and Dutta (1979) and Dutta and Ryan (1979).

Thermosetting systems obeying Eq. (13) would be expected to operate between two extremes. At the one $c_0 = 0$, and the cure would be autocatalyzed, exhibiting an induction time. Examples include Figs. 8, 9, 29, 30, 76, and 77. At the other extreme, where $k_1' \gg k_1\alpha$ curing would behave according to nth-order kinetics, with the maximum rate at $t = 0$. Figures 3, 4, 18, 39, 69, and 70 are examples of nth-order reactions.

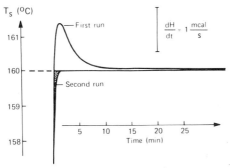

Fig. 3   Isothermal DSC method 1 curing of epoxy resin. Such behavior is indicative of an nth-order reaction. Sample placed in heated furnace at 160°C. Hatched area between first and second run used to measure unrecorded heat. DGEBA–MTHPA ($B = 0.9$), benzyldimethylamine catalyst. Mettler TA2000 System. (From Widmann, 1975b.)

2. Experimental Determination of Cure Kinetics

The generation of time–temperature–degree of cure curves is of great practical utility for establishing optimum cure schedules and determining shelf lives. It is also an integral part of mechanistic studies into the nature of curing, as discussed in the previous section. In this section we treat thermal analysis techniques for measuring the time dependence of curing at constant temperature. Dynamic kinetic studies in which time and temperature vary simultaneously are the subject of a later section. Dynamic method B, which uses the dependence of the peak reaction temperature with heating rate, is quick and accurate and is valuable in itself, as a precursor to isothermal studies and especially where isothermal measurements are inaccurate or unreliable. Examples of the latter include reac-

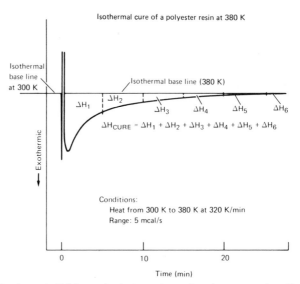

Fig. 4 Isothermal DSC method 1 curing of polyester resin. For $T_c > T_{c,\infty}$, $\Delta H_{CURE} = \Delta H_{RXN}$. Perkin-Elmer DSC-2. (From Cassel, 1977a.)

tions with multiple exotherms (Fig. 74), reactions with difficult-to-resolve baselines (Fig. 6), partially reacted systems, and solvent effects (see Fig. 65). This highly recommended technique is described in Section IV.B. It is the opinion of this author that, except for dynamic method B, isothermal techniques are more accurate and informative in characterizing the kinetics of curing of thermosetting systems.

    a. Degree of Conversion from Heat of Reaction.   A commonality of all thermosetting systems is the liberation of heat accompanying cure.

$$\text{reactants} \xrightarrow{-\Delta H_{RXN}} \text{products} \tag{14}$$

where ΔH_{RXN} is the exothermic heat per mole of reacting groups. For some systems the heat of reaction is well known and can be used to quantify the extent to which complete reaction takes place or as a reference for isothermal kinetic studies. It is assumed that the reaction rate $d\alpha/dt$ is directly proportional to the rate of heat generation dH/dt, which is true, providing the cure reaction is the only thermal event. Note that dH/dt is the ordinate of a DSC trace (see Fig. 3).

$$\frac{d\alpha}{dt} = \frac{dH/dt}{\Delta H_{RXN}} \tag{15}$$

Although both DTA and DSC have been employed to measure heats of thermosetting reactions, the method of choice is clearly DSC, which, because it is a heat flow measurement, yields a quantitative measure of both the heat and rate of reaction. For this reason, and because of the ready availability of DSC instrumentation, this topic will be discussed in terms of DSC only.

Kinetic measurements involving the heat of reaction fall into two categories: method 1, in which the rate and extent of reaction are continuously monitored in the calorimeter, and method 2, in which the heat evolved during the completion of cross-linking of partially cured samples is measured. Fava (1968) published a classic paper on these methods, which are described in detail later in this section.

Procedure.   Care must be exercised in sample preparation to ensure that no reaction occurs prior to the experiment and to prevent degradation of ingredients. A recommended procedure (Acitelli *et al.*, 1971) is to mix solutions of ingredients at low temperature (e.g., 0°C), freeze-dry, and store at very low temperatures, e.g., $-20°C$ ($<T_{c0}$, see Tables I and V). Even at these conditions, some reaction may occur over long periods, necessitating periodic discarding of material (e.g., after one week). Amines tend to be unstable and should be stored in a dark, cool, dry atmosphere. Sample size is a compromise between detectability of reaction ($\rightarrow$large samples) and minimizing thermal gradients and contribution of sample exotherm to sample heating ($\rightarrow$small samples); 5–20 mg is a reasonable sample size. Heat capacity of the reference cell should approximate that of the empty sample cell. Partial volatilization of the sample can occur during cure; it is absolutely necessary to prevent any loss of sample, since this will contribute to the measured heat. Normally, volatile sample pans are adequate. In extreme cases special, sealed pans (Freeberg and Alleman, 1966) or a high-pressure cell (Levy *et al.*, 1970) are necessary.

For some thermosetting resins the exothermic cure reaction may be altered or masked by competitive thermal events, for example, vaporization of water of condensation or solvent or even volatilization of reactants. Levy *et al.* (1970) demonstrated the utility of pressure DSC in such situations (Fig. 5). At ambient pressure the exothermic cure reaction and endothermic vaporization of condensation products (water, formaldehyde) in the phenolic system are competitive in the 100–200°C range, and the net energy change detected by DSC does not yield usable information. By raising the pressure to 800 psig, vaporization is shifted out of the phenolic cure region, and a two-stage curing exotherm is observed. The large exotherm centered at $\sim$340°C is attributable to polymer decomposition. Manley (1974a,b) employed this technique in studying the cure of amino resins. He found one potential drawback: that retention of formaldehyde in the pressure DSC experiment can lead to a different product than that obtained if molding is done at ambient pressure. When elevated pressure

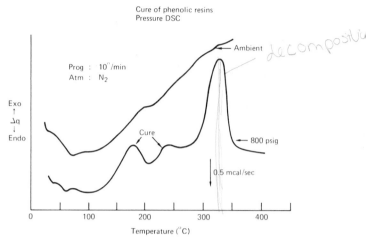

Fig. 5 Comparison of ambient and high-pressure DSC of phenolic resin. Endothermic volatilization of condensation products, which occurs simultaneously with the exothermic curing at ambient pressure, is shifted to higher temperatures in the pressure DSC experiment. DuPont 900, pressure DSC cell. (From Levy *et al.*, 1970.)

is part of the process, then pressure DSC can simulate the processing environment (Chow and Steiner, 1979). Pressure DSC data of Levy *et al.* (1970) for DAP curing shows an additional exotherm, which develops at 800 psi. A hermetically sealed stainless steel capsule, which contains the water of curing in the condensed state, may also be used when curing takes place below 200°C (Freeberg and Alleman, 1966). Sebenik *et al.* (1974) used this type of capsule to study addition and condensation reactions in phenol–formaldehyde resins. Westwood (1975), using similar pressure capsules to study cure of phenolic resole resins via isothermal method 2, found that the predicted cure level (pressure DSC) was consistently less than actually measured (atmospheric oven). Westwood also employed IR to follow the decrease in methylol concentration with cure. He found a significant concentration of residual methylol groups, suggesting that $\alpha_{ultimate} < 1$ (see Section II.A.2.c). Other calorimetric studies of condensation cure include phenolic resole resins by ambient pressure DTA and IR (Katovic, 1967a,b), phenolic and amino resins by pressure DTA (Ezrin and Claver, 1969), phenolic novolac and resole resins by pressure DTA and IR (Kurachenkov and Igonin, 1971), phenolic resole resins by DSC, GPC, and NMR (King *et al.*, 1974), phenolic resole and melamine–formaldehyde resins by sealed capsule DSC (Kay and Westwood (1975), commercial injection molding phenolic compounds by ambient pressure DSC (Siegmann and Narkis, 1977), phenolic resole resins by ambient-pressure DSC (Koutsky and Ebewele, 1977), and phenolic resole and no-volac systems by ambient and high-pressure DSC (Chow and Steiner, 1979).

ΔH_{RXN}. Application of isothermal methods 1 and 2 and dynamic methods A and C required an accurate value of ΔH_{RXN} or a reliable method for its determination. For reasons to be discussed, it is recommended that ΔH_{RXN} be determined from scanning experiments, preferably at more than one heating rate. Precisions of $\pm 4\%$ and accuracies of $\pm 6\%$ or better are achievable. Scanning rates between 2 and 20°C min^{-1} are recommended.

For method 1 isothermal experiments above the cure temperature of the fully cured network ($T_{c\infty}$), ΔH_{RXN} can be measured directly (Gray, 1972; see Fig. 4). However, this method is not recommended since care must be taken that any unrecorded heat of reaction liberated during heating and establishment of instrumental equilibrium is either small or corrected for, that the correct base line is found, and that complete cure is achieved, i.e., heat liberated slowly over long times near complete cure is not neglected (see discussion on method 1). Accurate measurement of ΔH_{RXN} from scanning experiments requires that a baseline can be drawn between a discernible beginning and end of the cure reaction. As Fig. 6 illustrates, such is not always the case. Multiple exotherms will also render the determination of ΔH_{RXN} more difficult.

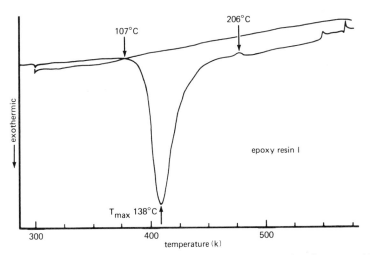

Fig. 6 Dynamic DSC scan of epoxy resin that does not return to baseline (second heating) at apparent end of reaction. Small endotherm at 206°C in first heating corresponds to melting of hardener. Inflection near 107°C in second heat is the glass transition. DGEBA–dicy ($B \approx 1$). Perkin-Elmer DSC-2. (From Schneider *et al.*, 1979.)

Fava (1968) measured the heat of reaction of one epoxy–anhydride at several heating rates (Table II). His results suggest that an optimum range

TABLE II

Apparent Heats of Reaction Measured by Isothermal Method 1 and from Scanning Experiments[a]

Isothermal method 1[b]		Scanning experiments[c]		
		Scan rate		Temperature at thermogram peak
T (K)	ΔH (cal g^{-1})	(K min^{-1})	ΔH (cal g^{-1})	(K)
360	54			
380	68	0.5	61	379
390	65	1	83	390
400	81	2	93	401
420	87	4	93	415
430	86	8	88	425
435	80	16	68	441
440	76	32	60	509
445	62	64	51	522
450	64			

[a] From Fava (1968), epoxy–anhydride, see Fig. 7. Reproduced by permission of the publishers, IPC Business Press Ltd.©.

[b] Re isothermal method 1: curing was incomplete below $T_{g\infty}$ (~400 K); above ~435 K a significant portion of ΔH occurred during instrumental equilibration and was not recorded (see Widmann, 1975a, also Fig. 3).

[c] Re scanning experiments: at slowest heating rate ΔH occurring below instrument sensitivity may not have been recorded; at two fastest heating rates thermal decomposition probably overlapped the curing reaction.

exists for the heating rate. At slow heating rates (≤ 1 K min^{-1}), some of the initial reaction and final reaction went unrecorded because of lack of instrument sensitivity. At fast heating rates (≤ 16 K min^{-1}), thermal decomposition probably interfered with the later stages of cure, again diminishing the exotherm. The progressive decrease in T_g with increasing cure temperature above 420 K for this material (Fig. 7) suggests that thermal degradation can occur simultaneously with curing at high temperatures (also see Fig. 62b). Several authors have found ΔH_{RXN} to be independent of heating rate between heating rates of 2 and 20°C min^{-1} (Prime, 1973; Sourour and Kamal, 1976; Hagnauer et al., 1978).

Ideally, ΔH_{RXN} is the total heat liberated when an uncured material is taken to complete cure, and this value is a constant for a particular thermosetting resin. For several DGEBA*–primary amine systems and model

* Diglycidyl ether of bisphenol-A; see Section V.A.

reactions (Klute and Viehmann, 1961; Horie *et al.*, 1970a), ΔH_{RXN} has been measured and found to be constant between 103 + 3 (24.5 ± 0.6) and 109 ± 3 kJ mole^{-1} (26.0 ± 0.7 kcal mole^{-1}) of epoxide, respectively. Where such values are known, they can be used in Eqs. (15) and (16) or as a check on experimental results. For several systems, ΔH_{RXN} is not known and must be measured.

In several instances (Horie *et al.*, 1970a; Prime, 1973; Barton, 1974a,b; Sourour and Kamal, 1976; Dutta and Ryan, 1979), ΔH_{RXN} was found to agree well with published values for the type of reaction studied, demonstrating that, within experimental uncertainties, complete conversion was achieved (i.e., $\alpha_{ult} = 1$) (see Table V). However, such is not always the case. For short-chain amine hardeners, Horie *et al.* found the heat of reaction to be less than ΔH_{RXN}, indicating that $\alpha_{ult} < 1$ for these systems (see Table III). This determination required prior knowledge of ΔH_{RXN}.

TABLE III

ILLUSTRATION OF α_{ult} AND $T_{g,ult}$ FOR SYSTEMS FOR WHICH
COMPLETE CURE IS SHORT OF 100% CONVERSION[a]

Diamine curing agent	α_{ult}	$T_{g,ult}$ (°C)	$T_{g,\infty}$ (°C, extrapolated)
Ethylenediamine (EDA)	84 ± 3%	77 ± 3	92 ± 4
Trimethylenediamine (TMDA)	93 ± 1	81 ± 3	87 ± 4
Hexamethylenediamine (HMDA)	99 ± 1	72 ± 2	72 ± 3

[a] Data of Horie *et al.* (1970a) for DGEBA–diamine systems cured isothermally at 50, 60, and 70°C followed by postcure (4°C m^{-1} to 200°C). $\alpha_{ult} = (\Delta H_{ISO} + \Delta H_{postcure})/\Delta H_{RXN}$, ΔH_{RXN} from 100% cure of model system. Extrapolation to $T_{g\infty}$ performed by this author.

Westwood (1975) showed by combined DSC and IR measurements that less than 100% conversion took place in the phenolic resole resins studied. If it is not possible to obtain or measure a value for ΔH_{RXN}, then ΔH_{ult} should be used. In these cases a relative degree of cure is measured and any kinetic evaluation may be affected. For example, in the preceding phenolic resoles, it can be shown that absolute degrees of cure (IR) follow first-order kinetics whereas relative degrees of cure (DSC) follow second-order kinetics.

Method 1.   By far the more versatile technique, this method capitalizes on the ability of DSC to monitor, simultaneously, the rate of reaction $d\alpha/dt$ and extent of reaction α over the entire course of reaction. This allows direct use of derivative forms of rate equations, which are far easier to use than the integrated forms. Method 1 is applicable between some minimum rate of heat generation, dependent primarily on calorimeter sensitivity and some maximum rate of heat production, dependent on the time to heat and establish instrumental equilibrium relative to the reaction rate. As an example, at high temperatures a significant portion of the reaction can take place before the calorimeter equilibrates and data are being recorded. Widmann (1975a) has proposed a means to correct for such unrecorded heat (Fig. 3); for the sample studied, he reports that 5% of ΔH_{RXN} goes unrecorded at 150°C and 20% at 170°C. It is expected that this problem will be more accute with nth-order reactions in which the reaction rate is a maximum at $t = 0$. For autocatalyzed reactions where the stoichiometry is known, an alternate method exists for estimating the unrecorded heat; see further discussion in this section.

Two method 1 experimental techniques exist. In the first technique the calorimeter is preheated to the desired reaction temperature prior to placing the unreacted sample in the calorimeter cell. In the second technique the sample is placed in the calorimeter cell at a temperature at which essentially no reaction can take place, and the temperature is then raised as rapidly as possible to a predetermined reaction temperature. With both techniques the rate of heat generation is followed with time. The criterion in choosing between these techniques is the establishment of instrument equilibrium (e.g., sample temperature = constant = preselected cure temperature) relative to the amount of unrecorded reaction that has taken place. Miller and Oebser (1980) found the time to establish equilibrium to be shorter for the first method, but observed no significant differences between results obtained by either method for several epoxy resin systems.

The degree of conversion is given by

$$\alpha_t = \Delta H_t / \Delta H_{RXN} \tag{16}$$

where α_t is the extent of reaction and ΔH_t the heat generated up to time t. Analogously, the rate of reaction is given by Eq. (15).

Fava (1968) was the first to demonstrate the benefits and potential deficiencies of this method in a study of one epoxy–anhydride system. His results indicated that ΔH_{RXN} from isothermal experiments was not constant but varied systematically with temperature as shown in Table II. Below $T_{g\infty}$, ~ 400 K $\equiv 127$°C, dynamic scans subsequent to the isothermal cures showed that incomplete cure contributed to the low values of ΔH_{RXN}, as illustrated in Fig. 7. Similar increases in ΔH_{RXN} with temperature from 75 to 96°C were observed by Cernee et al. (1977) for the free radical polymerization of diallyl fumarate. Near the apparent completion of the reaction at long times, heat will be unrecorded if the reaction rate falls below the sensitivity of the calorimeter. At high temperatures, the unrecorded portion of the reaction occurring during heating of the sample and establishment of instrumental equilibrium (see Fig. 3) became more significant, again contributing to the low values observed.

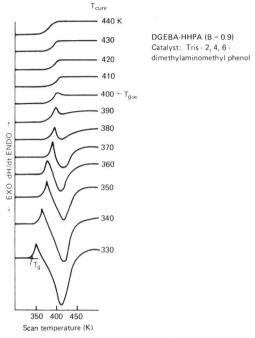

Fig. 7 DSC scans at 8°C/min on epoxy–anhydride samples previously cured to apparent completion at indicated temperatures. Note increasing T_g followed by residual exotherm for samples cured below $T_{g\infty}$. Diminishing T_g at highest cure temperatures is suggestive of degradation (see Fig. 62). Perkin-Elmer DSC-1. (From Fava, 1968. Reproduced by permission of the publishers, IPC Business Press Ltd. ©.)

Sourour and Kamal (1976) noted that, for autocatalyzed reactions, the extent of reaction up to the peak or maximum rate (compare Fig. 8) was constant and independent of temperature for a given thermosetting system. The measured values for an amine–epoxy system were $\alpha_{max} = 0.33$ for $B = 1.0$ and $\alpha_{max} = 0.42$ for $B = 1.5$; subscript max refers to values at the maximum reaction rate. When the rate is a maximum, $d^2\alpha/dt^2 = 0$ which, utilizing Eq. (13), allows α_{max} to be calculated from

$$3\alpha_{max}^2 - 2\alpha_{max}(B + 1) + B \cong 0 \qquad (17)$$

when $k_1\alpha \gg k_1'$. Calculated values of α_{max} are 0.33 ($B = 1.0$) and 0.39 ($B = 1.5$). For the same amine–epoxy system with $B = 1.3$ (see Figs. 29 and 30), this author observed $\alpha_{max} = 0.37 \pm 0.03$ between 23 and 157°C; the calculated value is 0.37. The constancy of α_{max} is important for three reasons. First, the reaction rate $(d\alpha/dt)_{max}$ can be used as a kinetic parameter to determine an overall activation energy (Sourour and Kamal, 1976, Section II.A.3.b). Second, Sourour and Kamal showed that when the

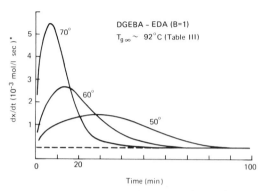

Fig. 8 Isothermal DSC method 1 curing of epoxy–amine at three temperatures. Maximum reaction rate at 30–40% of reaction is characteristic of autocatalyzed reactions.* Reaction rate (see Eqs. (11) and (12)). Perkin-Elmer DSC-1. (From Horie *et al.*, 1970a.)

stoichiometry is known, α_{max} could be used to estimate the unrecorded ΔH_{RXN} previously discussed. The unrecorded heat will be the difference between the measured heat up to the peak and $\alpha_{max} \times \Delta H_{RXN}$. And third, since the extent of reaction at the exotherm peak is constant, the time to reach α_{max} versus temperature can be used to measure the activation energy [compare Eq. (29)]. See Section II.A.3.b and Table V.

Method 2. This method becomes necessary when the rate of heat evolution is too small for detection by method 1, for example, at low temperatures. Several samples are cured isothermally, in an oven or the calorimeter itself, for various times until no additional curing can be detected. Samples are scanned in the calorimeter at heating rates between 2 and 20°C min⁻¹ from which the residual heat of reaction, i.e., the heat evolved during completion of cross-linking $\Delta H_{t,resid}$, is measured. The degree of conversion for method 2 is calculated from

$$\alpha_t = \frac{\Delta H_{RXN} - \Delta H_{t,resid}}{\Delta H_{RXN}} \equiv \frac{\Delta H_t}{\Delta H_{RXN}} \tag{18}$$

where ΔH_{RXN} and ΔH_t are as previously defined. A shortcoming of this method is the lack of a direct measure of the reaction rate; however, reaction rates can be estimated from tangents to the α versus time curves. Fava (1968) showed the equivalency of method 1 and method 2 (Fig. 9). This method loses sensitivity as the residual exotherm diminishes; this problem is especially acute when the glass transition immediately precedes the small residual exotherm ($T_{cure} < T_{c\infty}$; see Fig. 7, 380 and 390 K scans).

Photocuring. The photocuring of coatings and inks is gaining interest, especially with the need for systems free of volatile solvents. Papers by

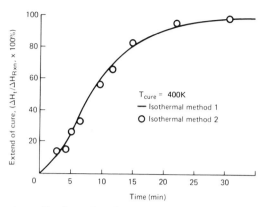

Fig. 9   Comparison of isothermal method 1 with isothermal method 2. Same epoxy–anhydride described in Fig. 7. Perkin-Elmer DSC-1. (From Fava, 1968. Reproduced by permission of the publishers, IPC Business Press Ltd. ©.)

Wight and Hicks (1978), Evans *et al.* (1978), and Tryson and Schultz (1979) describe the modification of differential scanning calorimeters to accept an ultraviolet (UV) light source. See Figure 10. Experiments similar to both isothermal methods were reported, where cure was initiated by unshuttering the lamp. Evans *et al.* studied the photoinitiated curing of polyester–styrene systems between 0 and 70°C (Fig. 11) and, with the aid of neutral density filters, showed that the initial rate of cure followed the expected square-root dependency on the incident light intensity. Wight and Hicks studied effects of oxygen, reactive diluents, and photoinitiator concentration on the UV curing of acrylated urethane and acrylated epoxy compounds. Tryson and Schultz measured the kinetics of photoinitiated conversion of multifunctional acrylates.

   b. T_{c0}, T_{g0}, $T_{c\infty}$, $T_{g\infty}$, α_{ult}, *and* $T_{g,ult}$ *by DSC.*   A brief description of these terms is given in Table I. The glass transition is a transition between a metastable glassy state and a supercooled liquid or rubbery state. Chapter 2 contains a thorough discussion of this phenomenon, and Flynn (1974) presents an excellent discussion for measurement of the glass transition temperature. Table V contains measured values for one amine–epoxy system at three stoichiometric ratios. For thermosets that cure via a condensation reaction, thermogravimetry may be used to measure extents of reaction (see Section II.A.2.d). T_{c0} is the temperature below which curing of the unreacted resin or resin mixture must take place in the glassy state. Since no significant reaction rate occurs below this temperature, T_{c0} has significance as a storage temperature. Sourour and Kamal (1976) showed that a semilogarithmic plot of isothermal cure temperature T_c against $1 - \alpha_{T_c}$, the fraction unreacted because of quenching of the

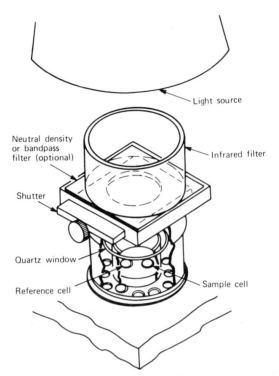

Fig. 10 Sample chamber of DSC for differential photocalorimetry experiment. Perkin-Elmer DSC-1b. (From Wight and Hicks, 1978.)

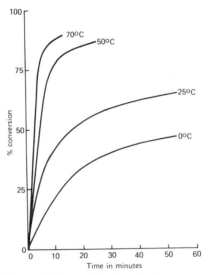

Fig. 11 Curing by differential photocalorimetry of polyester–styrene thermoset via iso-thermal DSC method 1. Perkin-Elmer DSC-1b. (From Evans *et al.*, 1978.)

reaction on vitrification, could be extrapolated to T_{c0}:

$$1 - \alpha_{T_c} = (\Delta H_{RXN} - \Delta H_{T_c})/\Delta H_{RXN} \qquad (19)$$

where ΔH_{T_c} is the heat liberated at T_c prior to vitrification (see Fig. 12).

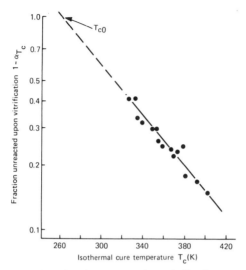

Fig. 12 Dependence of the fraction unreacted on vitrification, $1 - \alpha_{Tc}$, on isothermal cure temperature for DGEBA–mPDA ($B = 1.0$). Extrapolation to $1 - \alpha_{Tc} = 1$ gives $T_{c0} = 259$ K ($-14°C$). Perkin-Elmer DSC-1. (From Sourour and Kamal, 1976.)

Measurement of the glass transition temperature of the unreacted co-monomer mixture T_{g0} should yield a value close to T_{c0}.

$T_{c\infty}$ is the minimum cure temperature at which total ($\alpha = 1$) or ultimate ($\alpha = \alpha_{ult}$) conversion takes place. Extrapolation of a T_c versus ΔH_{T_c} plot (Fig. 13) gives $T_{c\infty}$ as the temperature at which $\Delta H_{T_c} = \Delta H_{RXN}$. Also, the relation between T_g of a vitrified thermoset and its cure temperature will yield $T_{c\infty}$ at $T_g = T_{g\infty}$ (see Enns *et al.*, 1981). $T_{c\infty}$ may be close to $T_{g\infty}$, and these two terms are sometimes used interchangeably. However, data of Sourour and Kamal (1976, Table V) suggest that $T_{c\infty}$ may be 20–25°C higher than $T_{g\infty}$ for some thermosets, while data of Enns *et al.* (1981) imply that $T_{c\infty}$ can be 20°C lower than $T_{g\infty}$ for others.

$T_{g\infty}$ is the glass transition temperature of the totally cured thermoset, i.e., at 100% conversion of reactive groups or $\alpha = 1$.* For some thermo-

---

* In actual practice it is improbable that conversion of every group will occur, especially when stoichiometric amounts react. A practical definition of total conversion is $\alpha \geq 0.99$, which can be achieved for several thermosets. This definition is also in accord with the limits of detection of calorimetry and thermogravimetry.

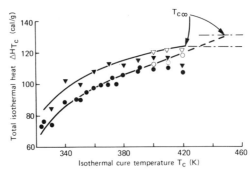

Fig. 13 Total heat of reaction, ΔH_{Tc}, at different isothermal cure temperatures T_c for two DGEBA–mPDA systems. Experimental data: (●) $B = 1.0$; (▼) $B = 1.5$. Data corrected for conversion spent during transient heating to T_c: (○) $B = 1.0$; (▽) $B = 1.5$. $T_{c\infty}$ is the minimum temperature at which ultimate conversion occurs. Perkin-Elmer DSC-1. (From Sourour and Kamal, 1976.)

sets the final or ultimate extent of conversion is considerably less than 100%, even after postcuring. $T_{g,ult}$ is the glass transition temperature for such a system. The ultimate extent of conversion is expressed as

$$\alpha_{ult} = \Delta H_{ult}/\Delta H_{RXN} \qquad (20)$$

where ΔH_{ult} is the total heat a thermosetting system is capable of liberating, whereas ΔH_{RXN} is the heat corresponding to total conversion of all reactive groups. Note that an independent determination of ΔH_{RXN}, e.g., from model reactions, is required for calculation of α_{ult}. Independent measure of reactive group concentration, e.g., by spectroscopy, may help to establish the value of α_{ult} and ΔH_{RXN}. Measurement of residual methylol concentration in phenolic resole curing is one example (Westwood, 1975). These parameters can be very important for characterizing cure. All but one of the thermal analysis techniques (TG) are capable of measuring glass transition temperatures and are discussed in appropriate later sections. Here we discuss how T_{g0}, $T_{g\infty}$, $T_{g,ult}$ and α_{ult}, plus T_gs of uncured and partially cured thermosets, can be measured as part of the DSC cure study.

In reporting $T_{g\infty}$ or $T_{g,ult}$, care must also be taken that the sample is indeed fully or ultimately cured, and it is recommended that the sample be given a known thermal history, e.g., by cooling from above T_g at the same rate as the subsequent heating. See also Chapter 2.

To achieve complete cure and develop ultimate physical properties, it is necessary to cure a resin above $T_{c\infty}$. One criterion for the achievement of complete or ultimate cure is the absence of a residual exotherm [for example, isothermal cure temperatures ≥ 400 K in Fig. 7 (Fava, 1968)].

Horie *et al.* (1970a) showed that, based on values of ΔH_{RXN} for a model amine–epoxy reaction, for which they demonstrated 100% conversion, some systems never reached 100% conversion, even with postcuring and in the absence of a residual exotherm (Fig. 14). Their data for systems

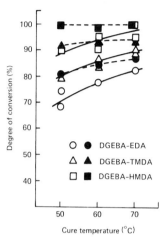

Fig. 14   Final conversion of isothermal cure ($\bigcirc$, $\triangle$, $\square$) and ultimate conversion added by temperature scanning ($\bullet$, $\blacktriangle$, $\blacksquare$) for the cure of DGEBA with diamines ($B = 1$) against the temperature of isothermal cure. Extrapolated values of $T_{g\infty}$ in Table III. Perkin-Elmer DSC-1. (From Horie *et al.*, 1970a.)

where $\alpha_{ult} < 1$ suggest that plots of degree of conversion against T_g, including α_{ult} and $T_{g,ult}$, will extrapolate to $T_{g\infty}$ (i.e., T_g at $\alpha = 1$). Data for three epoxy–diamine systems are shown in Table III. The extrapolations were performed by this author. The data are internally consistent and show an increasing $T_{g\infty}$ with decreasing diamine chain length ($\rightarrow$ increasing cross-link density). The data also suggest that some mobility in the network is required for achievement of 100% conversion. Similar results were obtained by Cernee *et al.* (1977) for the free radical polymerization of diallyl fumarate.

DSC has been used successfully to measure unblocking temperatures of blocked isocyanate compounds used in the formulation of stable, one part polyurethane systems (Anagnostou and Jaul, 1981). Effects of additives and pigments were measured, and changes were observed in the temperature range in which unblocking occurs in the presence of a compound containing primary hydroxyl groups. TG was not successful in studying this phenomenon. DSC can also be useful in studying physical aging below the glass transition temperature. See Section II.C.

c. $T_{c,gel}$, $T_{g,gel}$, and α_{gel}: DSC plus Complementary Technique and/or Gelation Theory. See Table I for a brief description of terms, Chapter 2 and Flynn (1974) for measurement of the glass transition temperature, and Table V for some measured values. For thermosets that cure via a condensation reaction α_{gel} is measured by thermogravimetry.

The gel point signifies the irreversible transition from a viscous liquid to a cross-linked gel. Beyond the gel point, a thermosetting resin is no longer capable of flow, and processability is difficult to impossible. Thus gelation is the most common mechanism involved in shelf life. For this reason, the degree of conversion at the gel point (α_{gel}) and the time–temperature dependencies of gelation are important aspects of the curing of thermosets.

As has been demonstrated, DSC is fully capable of measuring the degree of conversion. But it is totally insensitive to the physical changes occurring at the gel point. TBA and DMA and possibly TMA, which are discussed next, are sensitive to gelation, but are not capable of measuring the degree of conversion. Therefore, complementary experiments are needed; for example, determination of gel times at a series of constant temperatures by TBA, DMA, or some other technique sensitive to gelation, plus the measurement of the extent of conversion at these times and temperatures by DSC. Schneider et al. (1979) report such complementary TBA and DSC experiments.

Acitelli et al. (1971) studied the isothermal curing of an epoxy–amine by means of DSC, IR, and dc conductivity. The latter technique is relatively insensitive prior to gelation, but subsequent to gelation the conductivity decreases by several orders of magnitude because of inhibition that the growing three-dimensional network offers to migrating charge carriers. A conductivity–time plot, Fig. 15, shows the measurement of gel time by this technique. Comparison with α versus time measurements (isothermal method 2), using the measured activation energy to correct for small temperature differences between experiments, gave the gelation data in Tables IV and V and Fig. 29 and 30. Values of α_{gel} for a difunctional monomer (e.g., epoxy) reacting via step-growth polymerization with a tetrafunctional monomer (e.g., primary diamine) can be calculated from (Flory, 1953)

$$\alpha_{gel} = \sqrt{B/3} \qquad (21)$$

A value of $\alpha_{gel} = 0.66$ can be calculated for the preceding system ($B = 1.3$). The complementary nature of these methods was instrumental in showing that the same reaction occurs both prior to the gel point and subsequent to it. Choi (1970) compared DSC isothermal method 2 data with relative viscosity data to establish α_{gel} as 24–30% of the residual cure

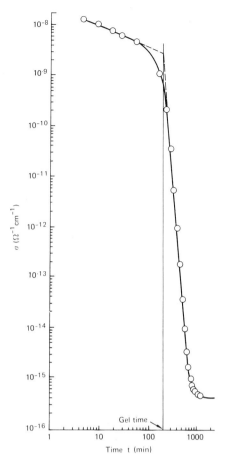

Fig. 15 Conductivity-time plot during cure of epoxy–amine showing measurement of gel time. Correlation with isothermal DSC method 2 (Fig. 29) gives degree of conversion at the gel point (α_{gel}, Table IV). (From Acitelli *et al.*, 1971. Reproduced by permission of the publishers, IPC Business Press Ltd. ©.)

TABLE IV

MEASUREMENT OF DEGREE OF CONVERSION AT THE GEL POINT (α_{gel})
BY COMBINED DSC AND DC CONDUCTIVITY, DGEBA–mPDA $(B = 1.3)^a$

T_c (°C)	Gel time (min)[b]	$\alpha_{gel}{}^c$ (%)
23	1860	70
56	155	68
75	70	72

[a] Unpublished data of R. B. Prime and E. Sacher.

[b] From dc conductivity–time plots (Fig. 15), measured activation energy used to convert conductivity gel times (52.5 and 72.5°C) to DSC gel times at 56 and 75°C.

[c] From gel times plus measured α versus time by DSC method 2 (see Fig. 29).

TABLE V

KINETIC, THERMODYNAMIC, AND CHARACTERISTIC PARAMETERS FOR CURE OF DGEBA–m-PHENYLENE DIAMINE EPOXY SYSTEM

Parameter	$B = 1.0$[a]	$B = 1.3$[b]	$B = 1.5$[a]
ΔH_{RXN}	130 ± 5 cal^{-1} g $\equiv 26.1 \pm 1.0$ kcal mole^{-1} epoxide	120 ± 4 cal^{-1} g $\equiv 25.0 \pm 0.8$ kcal mole^{-1} epoxide	$123 \pm$ cal^{-1} g $\equiv 26.4 \pm 1.1$ kcal mole^{-1} epoxide
T_{c0} (°C)	-14		
k_1 (min^{-1})	$8.06 \times 10^5 \exp(-11.4$ kcal mole$^{-1}/RT)$	$7.50 \times 10^5 \exp(-11.2$ kcal mole$^{-1}/RT)$	$3.70 \times 10^5 \exp(-11.0$ kcal mole$^{-1}/RT)$
k_1' (min^{-1})	$5.53 \times 10^8 \exp(-19.4$ kcal mole$^{-1}/RT)$		$9.65 \times 10^9 \exp(-21.1$ kcal mole$^{-1}/RT)$
α_{max}	0.33 (measured)	0.37 ± 0.03 (measured)	0.42 (measured)
	0.33 (calculated), Eq. 17	0.37 (calculated), Eq. 17	0.39 (calculated), Eq. 17
k_1 (min^{-1})[c]	$31 \times 10^5 \exp(-12.4$ kcal mole$^{-1}/RT)$	$1.8 \times 10^5 \exp(-10.1$ kcal mole$^{-1}/RT)$	$17 \times 10^5 \exp(-11.9$ kcal mole$^{-1}/RT)$
t_{max} (min)		$9.6 \times 10^{-7} \exp(12.4$ kcal mole$^{-1}/RT)$	
α_{gel}	0.58 (calculated, Eq. 21)	0.66 (calculated, Eq. 21)	0.70 (calculated, Eq. 21)
		0.70 ± 0.02 (measured)	
t_{gel}	$1.0 \times 10^{-6} \exp(12.8$ kcal mole$^{-1}/RT)$		$1.0 \times 10^{-6} \exp(12.5$ kcal mole$^{-1}/RT)$
$T_{c,gel}$ (°C)	56		63
$T_{g,gel}$ (°C)	69		75
$T_{c\infty}$ (°C)	178		147
$T_{g\infty}$ (°C)	158	130	122

[a] Data of Sourcour and Kamal (1976).

[b] Unpublished data of R. B. Prime.

[c] $k_1 \cong (d\alpha/dt)_{max}/\alpha_{max} (1 - \alpha_{max})(B - \alpha_{max})$, Eq. 30.

for a partially reacted diallyl phthalate (DAP) compound. The gel point of DAP occurs near 75% conversion of all double bonds (Sundstrom and English, 1978). The point at which the rate of viscosity increase approached infinity was chosen as the gel point.

Sourour and Kamal (1976) used calculated values of α_{gel} [Eq. (21)] to measure gelation times (t_{gel}) via isothermal method 1 for the same epoxy–amine studied by Acitelli *et al.*, but at different stoichiometries. They measured the times to reach that fraction of the total heat of reaction corresponding to the extent of reaction at the gel point [ΔH_{gel}, cf. Eq. (16)], at several isothermal temperatures. Agreement with rheologically measured gel times (cone-and-plate viscometer) and TBA gel times on the same system [Babayevsky and Gillham (1973); see also Section II.A.2.e, Figure 34] was excellent. Sourour and Kamal also measured $T_{g,gel}$, the glass transition temperature of material taken to degree of conversion α_{gel} at $T_{c,gel}$. $T_{c,gel}$ is the cure temperature at which the total heat liberated is ΔH_{gel}, the reaction ceasing because of vitrification. Note that $T_{c,gel}$ and $T_{g,gel}$ are characteristic parameters of a thermoset. $T_{g,gel}$ was found to be 10–13°C higher than $T_{c,gel}$. $T_{c,gel}$ has significance in terms of shelf lives and B staging; it is the isothermal cure temperature at which vitrification and gelation occur simultaneously. See Table V for data of one amine–epoxy system at three stoichiometric ratios. Comparison with Figs. 29 and 30 suggests that the $T_{c,gel}$ data of Table V correspond to the *onset* of vitrification or diffusion control and that gelation can occur at longer times below these $T_{c,gel}$ temperatures until the vitrification process is complete.

d. Degree of Conversion from Loss of Condensation Products: TG. For several thermosets [phenolic, amino, poly(acrylic acid)], an integral part of the cure is the formation of condensation products, such as water or formaldehyde.

$$\text{reactants} \xrightarrow{-\Delta H_{RXN}} \text{thermoset} + \text{condensation products} \qquad (22)$$

Under ideal conditions the weight loss is equivalent to the enthalpy loss as a measure of the degree of conversion. However, the reaction weight loss can be contaminated, e.g., by loss of absorbed water or other volatile components. If weight loss is to be used as a quantitative measure of degree of conversion, the source and extent of any contaminants must be identified (e.g., by TG–MS, TG–GC, or GC–MS). When utilizing the weight loss rate (DTG), diffusion must also be considered. For example, if the water of condensation must first diffuse to the surface, and if this is the rate-controlling step, then the measured rate (diffusion-controlled) will not be the desired reaction rate (chemically controlled). Combination of weight loss measurements with spectroscopic (IR), calorimetric (DSC), and dynamic mechanical (DMA, TBA) measurements would be very informative. An example of condensation curing followed by TG is shown in Figs. 16 and 17 (Cassel, 1976). Figure 16 shows isothermal curing at

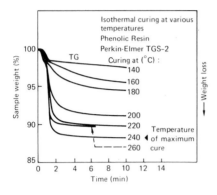

Fig. 16 Isothermal weight loss during cure of phenolic resin heated to several isothermal cure temperatures at 160°C/min. Small initial weight loss may be absorbed water. Perkin-Elmer TGS-2. (From Cassel, 1976, 1977a.)

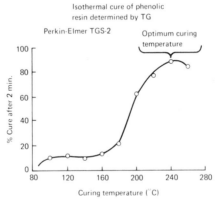

Fig. 17 Isochronal degree of conversion against cure temperature from data of Fig. 16; 100% conversion identified with maximum weight loss at 240°C. Perkin-Elmer TGS-2. (From Cassel, 1976.)

seven temperatures. In each case an initial, rapid, and constant weight loss, representing approximately 15% of the complete curing weight loss, was observed. The author attributed this to the first stage of curing; a similar phenomenon in cross-linking via anhydride formation in poly(acrylic acid) was attributed to removal of bound water (Greenberg and Kamel, 1977). The increase in the ultimate weight loss, and hence extent of cure, with increasing temperature was attributed to vitrification occurring at successively higher cure levels with increasing temperature; unfortunately, no T_g data were reported. The lower weight loss at the highest temperature was ascribed to degradation. It may be attributable to oxidative cross-linking. Based on observations by this author, it is probable that oxidative cross-linking causes vitrification and quenching of further condensation reaction prior to complete or ultimate conversion. In Fig. 17 is plotted the relative percent cure after 2 min, showing the plateau between the two stages of weight loss, the most efficient cure temperature as defined by the author, and the apparent onset of degradation. Greenberg and Kamel (1977) used a similar approach to study the kinetics of anhydride formation in poly(acrylic acid) composites (Fig. 18).

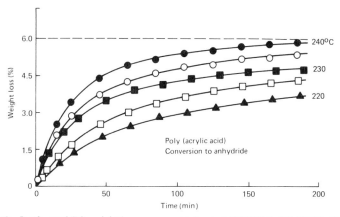

Fig. 18 Isothermal TG weight loss curves obtained at (●) 240°C; (○) 235°C; (■) 230°C; (□) 225°C; (▲) 220°C. Sample is poly(acrylic acid)–alumina composite containing 48% polymer by weight. The dotted line represents the maximum calculated weight loss due to anhydride formation (see Section V.F). DuPont 950 TGA. (From Greenberg and Kamel, 1977.)

When using this method quantitatively, it is necessary to establish that the desired weight loss is indeed being monitored. By a mass spectroscopic technique, Greenberg and Kamel showed that volatile reaction products in a poly(acrylic acid) composite consisted of water and carbon dioxide in a ratio of approximately 12:1 by weight. Manley (1974a,b) showed by evolved gas analysis that both absorbed water and formaldehyde accompany the loss of water of condensation during cure of phenol–formaldehyde resins.

e. Mechanical Spectroscopy and Rheology During Cure: TBA, DMA, TMA. Two of the thermal techniques are capable of measuring dynamic mechanical spectra of curing and cured thermosets. A third can monitor viscosity during cure into the region of gelation or vitrification. Torsional braid analysis (TBA) subjects a resin–glass braid, resin–thread, or resin–

film composite to free torsional oscillations at approximately 1 cps ≡ 1 Hz (Gillham, 1974). Dynamic mechanical analysis (DMA) subjects a sample specimen of known geometry, or resin supported on a glass fabric or other support, to flexural deformation and monitors its natural resonant frequency from approximately 2–70 Hz (DuPont Instruments, a). In the DMA experiment, energy is supplied to maintain the sample in oscillation at constant amplitude. TBA and DMA provide information on the elastic modulus (stored energy) and mechanical damping (loss energy) of the sample. Both have a lower limit of detectability (in terms of modulus) that renders them relatively insensitive below the gel point.

Thus TBA and DMA are sensitive to the large changes in modulus that occur beyond the gel point. Since the elastic modulus in the rubbery state at constant temperature is a measure of the cross-link density (Flory, 1953), these methods are theoretically capable of following the cure reaction from gelation to the completion of the reaction. Both gelation and vitrification are manifested as mechanical damping peaks accompanying large changes in rigidity or modulus (see Figs. 19 and 20). The thermome-

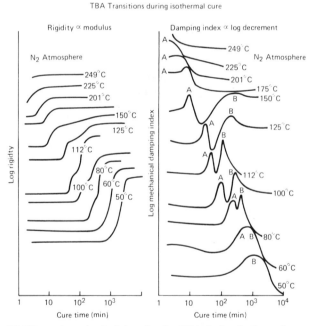

Fig. 19 Rigidity and mechanical damping by TBA during isothermal cure of cycloaliphatic epoxy. 3,4-Epoxycyclohexylmethyl-3,4-epoxycyclohexane carboxylate (Table XIV) + HHPA ($B = 0.9$), ethylene glycol (initiator), benzyldimethylamine (catalyst). Mechanical damping peak $A \Rightarrow$ gelation, peak B $\Rightarrow$ vitrification. (From Gillham et al., 1974.)

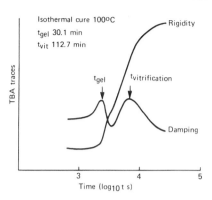

Fig. 20 Determination gelation and vitrification times from TBA experiment. Epoxy–dicy ($B \sim 1$). (From Schneider *et al.*, 1979.)

chanical analyzer (TMA), when modified to function as a microparallel plate rheometer (PPR), has been used to generate viscosity data during the early stages of curing (Cessna and Jabloner, 1974). Using this method, it is possible with small samples to investigate several phenomena associated with curing, including melting or devitrification of powdered resins and the early advancement of cure, including gelation. This technique is limited to initially solid resins. A similar TMA characterization of powdered coating resins, but using the spherical indentor accessory, was reported by Zicherman and Holsworth (1974). However, this technique does not yield quantitative data, and the PPR method is strongly recommended for this type of study. Another potential application of TMA to the cure of thermosets involves use of the volume dilatometer to follow volume changes during cure. A volume measurement could be important when trying to minimize stresses by minimizing shrinkage during cure and when characterizing thermosets, such as those reported by Bailey (1975), that undergo negligible or even positive volume change on curing.

TBA, DMA, and TMA are complementary to DSC and TG in the study of thermosets. DSC and TG directly measure the extent and rate of chemical conversion, are insensitive to gelation, and become less sensitive beyond the gel point as the reaction rate diminishes. Mechanical spectroscopy (TBA, DMA) is little affected prior to the gel point, but at the gel point the rigidity of the nascent network begins to build and subsequent changes in rigidity as a consequence of curing are detected. These methods are not able to measure the degree of conversion directly, but when coupled with a method that can, such as DSC, they are able to monitor the chemical conversion reaction in the rubbery state between the gel point and either vitrification or the end of the reaction. TMA–PPR is complementary to TBA and DMA since it is most sensitive to rheological changes occurring at and below the gel point.

Torsional Braid Analysis (TBA). Torsional braid analysis (TBA) was introduced by Lewis and Gillham (1962) [also see Gillham (1974)]. In the study of thermosets, a multifilamented glass braid is impregnated with a solution of uncured reactants. During cure, the composite specimen is subjected to free torsional oscillations from which the relative rigidity $1/p^2$ (p = period of oscillation in seconds) and mechanical damping index $1/n$ (n = number of cycles for the damping peak amplitude to decrease by a fixed amount) are obtained intermittently through the isothermal cure. The relative rigidity is proportional to the shear modulus (G') and the mechanical damping index is a measure of the logarithmic decrement, which is proportional to the ratio of the out-of-phase shear modulus (G'') to the storage shear modulus (G'). $G''/G' = \tan \delta$. The logarithmic decrement (Δ) is equal to $\ln(A_i/A_{i+1})$, where A_i is the amplitude of the ith oscillation. Papers by Babayevsky and Gillham (1973), Gillham et al. (1974, 1977), Gillham (1976, 1979a,b, 1980), Schneider et al. (1979), Lewis et al. (1979), Doyle et al. (1979), Enns and Gillham (1980), and Enns et al. (1981) exemplify the TBA technique in the study of isothermal curing of thermosets. Schneider and Gillham (1980) used single-ply composite strips in place of the glass braid to study curing of prepreg materials.

In these studies three different types of dynamic mechanical behavior were observed, depending on the temperature of cure relative to the characteristic temperatures of the thermoset (see Fig. 2). The first type of behavior is observed at temperatures below $T_{c,gel}$ where the sigmoidal increase in rigidity on vitrification is accompanied by a single damping peak B, and vitrification is attributed to the increase in molecular weight of the polymer below the gel point (Fig. 19, 50°C curves). A shoulder on the damping curve preceding peak B is evidence that gelatin is occurring in the vicinity of vitrification, i.e., the temperature of cure is close to $T_{c,gel}$. When the temperature of cure exceeds $T_{c,gel}$, the reactive system first passes through gelation giving damping peak A and subsequently passes through vitrification yielding damping peak B (Fig. 19, 80–125°C curves). This corresponds to the second type of behavior. In this case vitrification is attributed primarily to the increasing crosslink density of the network after gelation. This type of curing behavior is manifested by a two-step increase in rigidity: the first (and smaller) during gelation and the second during vitrification. Note from Fig. 19 that no reaction can be detected in the glassy state. When the cure temperature exceeds $T_{c\infty}$ (third type of behavior), vitrification cannot occur and peak B on the damping curve disappears (Fig. 19, 175°C and higher).

Figure 20 shows how times to gelation and vitrification are determined from the relative rigidity and mechanical damping index curves. Use of strips of prepreg materials in place of the glass braid results in a marked weakening of the gelation mechanical damping peak (Schneider and Gillham, 1980). Independent measurements of the gel fraction with cure times may be useful in identifying the TBA gel peak (Enns and Gillham, 1980). When TBA and DSC measurements are combined, degree of conversion at the gel point can be determined (Schneider *et al.*, 1979). In Fig. 2 these times are plotted against temperature of isothermal cure. $T_{c,gel}$ is determined as the intercept of the gelation and vitrification curves, i.e., the cure temperature at which the two phenomena are coincident. $T_{c\infty}$ corresponds to the temperature at which the time to vitrify becomes infinite.

Babayevsky and Gillham (1973) derived an expression for the extent of reaction *from the gel point,* based on TBA values of relative rigidity. The derivation stemmed from work of others, showing that the equation of rubber elasticity could be applied to epoxy–diamine networks in the rubbery state, where the elastic modulus and shear modulus are directly proportional to cross-link density at constant temperature. The same athors describe an approximate means of measuring the relative degree of cure that occurs in the glassy state.

Note that the kinetic rate equations described in this chapter may not apply to conversion in the glassy state when the rate is diffusion controlled. It should also be noted that degrees of cure in the rubbery and glassy states are only relative measures of the degree of conversion. Thus the TBA data obtained in this manner can be greatly enhanced by complementary measurements of the actual degree of conversion, e.g., by DSC or by knowledge of the degree of conversion at the gel point. In a study on a similar epoxy–diamine system, Acitelli *et al.* (1971) demonstrated the complementary nature of DSC, IR, and dc conductivity measurement; the latter measurement yields information that is very similar to TBA. Schneider *et al.* (1979) combined TBA, DSC, and IR to unravel the complex curing behavior of two commercial dicy-containing epoxy resins.

TBA has also been used in a scanning mode to characterize the physical changes that occur during cure (Cessna and Jabloner, 1974; Zicherman and Holsworth, 1974). Figure 21 illustrates the type of information obtained.

Manzione *et al.* (1981a,b) used TBA, TMA, electron microscopy, and mechanical property testing to study the development of a variety of morphologies in rubber modified epoxies. These morphologies parallel a spectrum of phase-separated and dissolved rubber, which could be developed through control of rubber–epoxy compatibility and cure conditions. The volume fraction, domain size, and the number of particles of phase-separated ruber were shown to be determined by the competing effects of incompatibility, rate of nucleation and domain growth, and the quenching of morphological development by gelation. Optimum materials contained a

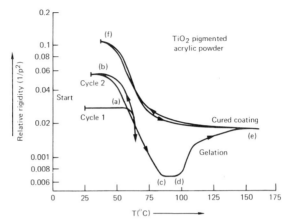

Fig. 21   Relative rigidity versus temperature at 2°C/min. (a) T_g/T_{melt} of powder = 62°C; (b) T_g, uncured film; (c) end of melting; (d) onset of cure, gelation (~95°C); (e) cure effected; (f) T_g, cured film. Compare T_g/T_{melt} by TMA penetration (63°C) and DSC (56°C). Chemical Instruments Corp. Model 100-B1 TBA. (From Zicherman and Holsworth, 1974.)

relatively large amount of dissolved rubber and a low volume fraction of phase-separated rubber, which combined high elongations at low strain rates with improved impact properties.

Dynamic Mechanical Analysis (DMA).   DMA is a form of dynamic me-chanical spectroscopy that is a new member of the thermal analysis fam-ily. To date detailed studies of curing, similar to those using TBA, have not been reported. However, initial studies reported by Hassel (a, 1978) Reed (1979, 1980), Gill (1980a), Strauss (1980), and Grentzer *et al.* (1981a, b) illustrate the vast application potential of this technique. DMA appears potentially capable of the same quantitative curing studies that have been performed by TBA.

Unsupported samples that have gelled or samples supported on a glass fabric or other substrate can be studied. An example of the former experi-ment, where the sample is run in the verticle mode, is shown in Fig. 42. Figure 22 illustrates the horizontal mode for studying samples supported on a glass fabric. This technique allows liquid samples, such as uncured resins, to be studied. Prepregs can also be run in this mode. Deformation in the horizontal mode is complex, involving shearing, bending, and com-pressive forces, and quantitative modulus data cannot be obtained. The damping information (tan δ, the ratio of energy lost to energy stored in a cyclic deformation), however, can be made quantitative since this is not deformation or geometry dependent, and the contribution of the fabric can be subtracted from the quantitative damping measurement.

Reed (1979, 1980) observed significant influences of silane-finished,

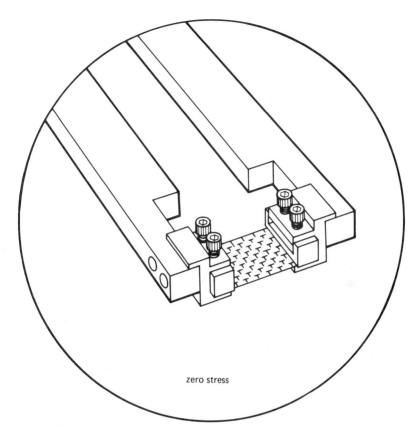

zero stress

Fig. 22 Horizontal mode for studying samples supported on glass fabric by DMA. Du-Pont 981 DMA. (From Gill, 1980.)

glass fiber reinforcement on the damping behavior of unidirectional epoxy composites run in the horizontal mode. Broadening of the main glass transition and the appearance of an additional relaxation above T_g were observed as the fiber orientation changed from transverse (perpendicular to the direction of displacement) to longitudinal and as the sample thickness (number of plies) decreased. The damping behavior of transverse composite and neat resin were very similar. Softening temperatures measured by TMA also exhibited the high-temperature transition. No transition above T_g was observed by DSC. It was suggested that the high-temperature transition may be characteristic of resin in the interfacial region between fiber and resin. Since glass fabrics are employed as "inert" supports for DMA resin studies, these results warrant further investigation. Of course, coupling agents should not be used on inert supports.

Figure 23 illustrates the isothermal curing of an epoxy–carbon fiber prepreg by DMA. Shown is the buildup in relative modulus beyond the gel

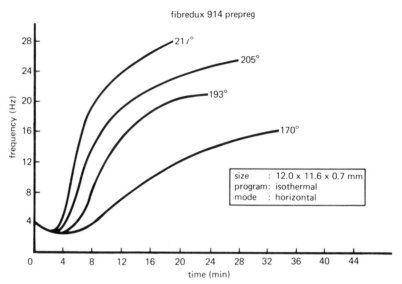

Fig. 23 Isothermal curing of epoxy–carbon fiber prepreg by DMA. Back extrapolation to zero modulus (2–3 Hz) gives gel times. Approximate cure rates subsequent to gelation may be obtained from the slopes of the frequency–time curves. DuPont 1090/981 DMA. (From Gill, 1980.)

point. The time to gel can be measured by back extrapolation to zero modulus. This figure can be compared to the rigidity curves of Fig. 19. Strauss (1980) used this technique, combined with mass spectroscopy and ^{13}C NMR, to study curing of a liquid phenol–formaldehyde resin. DMA of uncured resins can also be run in the scanning mode, as illustrated by Fig. 24a. The frequency or relative modulus response can be compared to the TBA traces in Fig. 21. By comparison of the tan delta trace with the DSC scan on the same resin, Fig. 24b, several aspects of the cure become evident. The first peak at ~80°C is associated with the glass transition of the epoxy powder coating. As can be seen, the break in the frequency curve is in better agreement with the DSC value than is the damping peak. The effect of frequency on T_g (~8°C/decade of frequency) accounts for this discrepancy; DMA damping peak temperatures adjusted to 1 Hz should be in good agreement with T_gs from DSC and TMA (Blaine, 1981). The next three peaks are associated with the cure. Because of the increase in modulus, the highest temperature peak can be identified as gelation (cf. Fig. 21). The relatively low modulus at the completion of curing at high temperatures is evidence that vitrification has not occurred and that $T_{g\infty}$ is below 180°C (again compare Fig. 21). This is not the case for the phenolic system in Fig. 43, where it is clear that $T_{g\infty}$ must be above 250°C.

DMA appears to be an attractive technique and the applications to thermoset curing are obvious. DMA of cured and partially cured thermo-

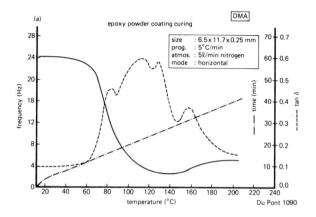

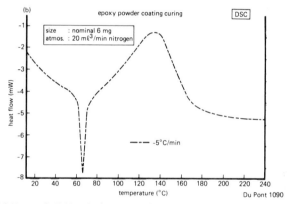

Fig. 24   (a) Dynamic DMA during cure of epoxy powder coating. (b) DSC trace. The effect of frequency on T_g accounts for the damping peak occurring at a higher temperature than T_g by DSC. Because of the increase in modulus, the highest-temperature damping peak can be identified as gelation. DuPont 1090/981 DMA and 1090/910 DSC. (From Gill, 1980.)

sets is treated in Section II.B.2.a. Because it is a new technique, many of the quantitative studies done by TBA remain to be done by DMA. These include precise measurements of times to gelation and vitrification.

Thermomechanical Analysis (TMA–PPR).   Time–temperature–viscosity data on thermosets can be generated during the early stages of cure when the TMA is employed as a parallel plate rheometer. Cessna and Jabloner (1974) pioneered in this application of the TMA, and Dienes and Klemm (1946) and DuPont Instruments (b) are good additional references for physical and experimental detail. Bartlett (1978) and Bloechle (1978) have applied the PPR technique to characterization of fiber-reinforced

Parallel plate rheometer

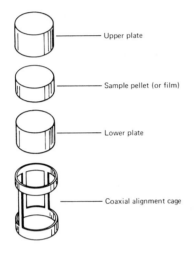

Upper plate

Sample pellet (or film)

Lower plate

Coaxial alignment cage

Fig. 25 TMA microparallel plate assembly for type (a) experiment, where sample radius (R) = plate radius = constant. (From DuPont Instruments, (b).

resins. Two types of experiments are possible: (a) when the sample fills the space between the plates and exudes from between them after the load is applied (constant sample radius, equal to radius of plates (Fig. 25) and (b) when the radius of the test plates is larger than that of the sample throughout the test and the sample volume is constant (Fig. 26). The Wallace plastimeter is representative of the first (a) and the Williams plastimeter of the second (b). Both methods require a flat-ended cylindrical sample. Viscosities from 10^2 to 10^8 poise can be measured with low shear rates of 10^{-3} to 10^0 sec^{-1} being generated. For method a experiments, the viscosity η is calculated as

$$\eta = (4F/3\pi R^2)/[d(1/h^2)/dt] \qquad (23)$$

where F is the applied force, R is the sample radius, h is the distance between the plates = sample thickness, and t is time. Cessna and Jabloner also show an equation relating shear rate to these parameters. For method b experiments, the relationship is

$$\eta = (8\pi F/3V^2)/[d(1/h^4)/dt] \qquad (24)$$

where V is the sample volume. In actuality, the parallel plate methods measure the creep compliance, which has elastic, time-dependent elastic (viscoelastic), and viscous components. The measurement of viscosity requires that the viscous component dominate the deformation, and the criterion is the constancy of the respective isothermal slopes, $d(1/h^2)/dt$

Fig. 26   TMA microparallel plate assembly for a type (b) experiment, where the sample radius is smaller than the plate radius and sample volume (*V*) is constant. (a) Before experiment. (b) After experiment. Thermoset cure would require some means to prevent adhesion to assembly shown (e.g., disposable film or plates). Perkin-Elmer TMS-1. (From R. B. Prime, unpublished.)

and $d(1/h^4)/dt$, with time, See Dienes and Klemm (1946), who described the physics of these experiments. The complexity of the calculations make on-line data reduction desirable (Blaine a). Note that corrections for thermal expansion are necessary for programmed temperature experiments. This method is limited to samples that can meet the sample geometry requirements.

The first of the TMA–PPR methods (a) has been used to study thermoset cure (Cessna and Jabloner, 1974; Blain and Lofthouse, a). Figure

Fig. 26 (*continued*)

27 compares several aspects of the TMA–PPR experiments with DSC on a powdered polyaromatic thermosetting resin (Blaine and Lofthouse, a). Figure 27a shows the deformation (i.e., reduction in sample thickness) on heating.

Figure 27b displays the viscosity computed from the TMA experiment. The apparent increase in the viscosity near 80°C is probably due to the volume increase on melting (cf. DSC trace, Fig. 27d). Thus judgment must be exercised in interpreting viscosity data near a phase change. The increase in viscosity beginning near 115°C is real and, as can be seen by comparison with the DSC trace, is brought about by initiation of the curing reaction. Note that gelation can occur shortly after initiation of the

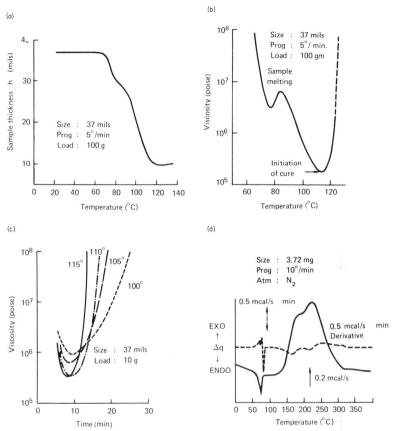

Fig. 27 TMA-PPR method (a) study of a powdered polyaromatic thermosetting resin. (a) TMA thermogram showing motion of probe as sample is compressed and exudes through alignment cage; (b) viscosity computed from thermogram a; (c) family of isothermal viscosity curves; (d) DSC trace showing melting and curing. DuPont 943 TMA and PPR Accessory. (From Blaine and Lofthouse, a.)

curing reaction for B-staged thermosets. Figure 27c shows a family of curves, representing the initiation of curing and consequent increase in viscosity for the same resin at several isothermal temperatures.

Cessna and Jabloner (1974) used this method to investigate curing of several polyaromatic thermosetting resins, including the one described earlier. They were able to distinguish between low melt viscosity resins suitable for low-pressure molding of highly filled parts or preparation of fiber-reinforced composites, and intermediate and high melt viscosity resins designed for compression and injection molding of less highly filled parts. Their data, Fig. 28, show time–temperature relationships for one

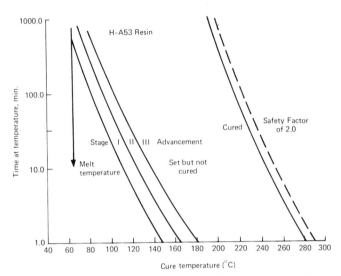

Fig. 28 Time–temperature curves for polyaromatic resin generated from TMA-PPR method (a) experiments. Stage I: initial viscosity increase; stage II: advancement to where melt processing becomes difficult (shelf life); stage III: material set but not cured. DuPont 943 TMA. (From Cessna and Jabloner, 1974. Reprint courtesy of *The Journal of Elastomers & Plastics*, Technomic Publishing Company, 265 Post Road West, Westport, Ct. 06880.)

resin at several stages of cure from the initial significant increase in viscosity (stage I) to advancement where melt processing may be difficult (stage II) to advancement where the material is set (stage III, gelation) to final cure. Final cure was not determined by the TMA–PPR method. Shelf life was correlated with stage II advancement.

3. Rate Constants and Activation Energies

In a practical sense the activation energy has a dualistic nature. On the one hand, it can give valuable information on barriers to reaction and chemical mechanisms, and its constancy or change can be a valuable tool in understanding molecular aspects of resin and catalyst chemistry or addition of filler. For these purposes measurement of the activation energy alone may be sufficient. In a more pragmatic sense the activation energy is a time–temperature shift factor and as such it is useful in predicting behavior of thermosets in real processes. Figure 28 is a good example of the latter, even though an activation energy per se was not determined. Barton (1979) describes time–temperature superposition of DSC degree of cure data of an epoxy resin system. When measured, the activation energy can predict how much slower or quicker the same event will occur, be it gelation or 95% of cure, at a lower or higher temperature, respectively. The form of the rate equation, i.e., $f(\alpha)$ in Eq. (1), is necessary for predicting time–temperature behavior and is also useful in studying cure mechanisms.

Figures 29 and 30 demonstrate the general relation between the activa-

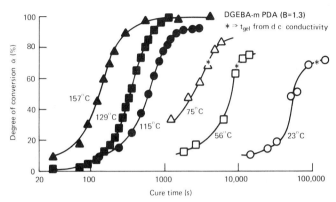

Fig. 29 Isothermal cure curves for epoxy–amine from DSC method 1 (solid symbols) and method 2 (open symbols) experiments. Perkin-Elmer DSC-1. (From R. B. Prime, unpublished data; results in Acitelli *et al.*, 1971.)

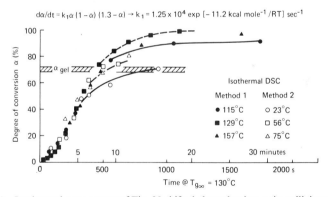

Fig. 30 Isothermal cure curves of Fig. 29 shifted along the time axis, utilizing the activation energy, to cure times at $T_{g\infty} = 130°C$. See Figs. 15 and 29 and Table IV for determination of α_{gel}. Perkin-Elmer DSC-1. (From R. B. Prime, unpublished data; results in Acitelli *et al.*, 1971.)

tion energy and time–temperature superposition. Figure 29 shows a description of the curing of an epoxy–diamine as measured by DSC methods 1 and 2. The gel points were determined by combined DSC-dc conductivity measurements at three temperatures (Fig. 15, Table IV). $T_{g\infty}$

was found to be 130°C as measured by DSC at 20°C min^{-1} heating rate (onset temperature) on samples freely cooled in the calorimeter from the rubbery state. Figure 30 shows these same data, shifted along the time axis using the measured activation energy of 46.8 kJ mole^{-1} (11.2 kcal mole^{-1}); note that the activation energy could actually be determined by time–temperature shifting of the curves in Fig. 29. All data have been shifted to the times corresponding to curing at $T_{g\infty}$ by means of

$$t_2/t_1 = \exp\{[E(T_1 - T_2)]/RT_1T_2\} \tag{25}$$

Figure 30 also illustrates the shift from chemical control to diffusion control of the reaction at the onset of vitrification at temperatures below $T_{c\infty}$. This shift occurs at approximately 40% conversion at 23°C, 55% conversion at 56°C, and 75% conversion at 115°C.

   a. The Rate Constant.   The rate constant is a measure of the velocity of the reaction. As defined in Eq. (1), it is measured as the slope of the reaction rate $d\alpha/dt$ versus $f(\alpha)$ curve. Thus to obtain the rate constant, one must first establish the form of the rate equation. As previously discussed, all thermosetting cure reactions can be represented by either autocatalyzed or nth-order rate equations. The distinguishing characteristic of autocatalyzed reactions is the occurrence of the maximum rate at 30–40% completion of the reaction (see Figs. 8, 9, 29, and 30). In an nth-order reaction the maximum rate occurs at $t = 0$ (see Figs. 3, 18, 69, and 70). Experimentally, autocatalyzed reactions can best be distinguished from nth-order reactions from plots of α versus time, measured from either isothermal method 1 or method 2 experiments. Compare, for example, Fig. 9 and 18. Once the general form of the rate equation has been established, the specific form (e.g., reaction order) must be determined or verified. By reference to Eq. (13), for reactions following autocatalyzed kinetics, a linear variation of $(d\alpha/dt)/(1 - \alpha)(B - \alpha)$, a reduced rate, with extent of reaction α will occur. Sourour and Kamal (1976) demonstrated that such a linear relationship exists up to the onset of vitrification where the cure reaction becomes diffusion controlled (Fig. 31). Note that the slope is k_1 and the intercept k_1'. Similar behavior was observed by Horie *et al.* (1970a) for different epoxy–amine systems. They attributed the small values of k_1 to the presence of trace impurities that catalyze the reaction.

   For an nth-order reaction, the value of n must be established. Although this must be done experimentally, knowledge of the chemistry can be very useful. By reference to Eq. (1) and (2), an nth-order reaction will show a linear relationship between $\ln(d\alpha/dt)$ and $\ln(1 - \alpha)$, where n is the slope and $\ln k$ is the intercept. Since the majority of reactions are either first or second order, the recommended practice is to utilize the preceding

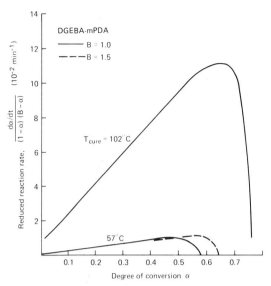

Fig. 31 Reduced isothermal cure rate against degree of conversion for epoxy–amine. The falloff in reaction rate is due to the onset of vitrification, where the reaction becomes diffusion controlled. Initial slope $= k_1$, intercept $= k_1'$. Perkin-Elmer DSC-1. (From Sourour and Kamal, 1976.)

relationship with data at one temperature to distinguish between these two possibilities. Then it should be verified that data at all reaction temperatures follow the particular rate equation. If this fails, a fractional reaction order may be employed. An nth-order reaction where reactants are not present in stoichiometric amounts will require a modified rate equation. For example, a second-order reaction where the equivalent ratio of reactants is B will follow $f(\alpha) = (1 - \alpha)(B - \alpha)$; cf. Eqs. (3) and (13).

In the previous discussion the reaction rate $d\alpha/dt$ was utilized, which is measured directly in an isothermal method 1 experiment (DSC and DTG). In a method 2 experiment, which does not measure the reaction rate directly, the experimenter has several options. First, values of $d\alpha/dt$ can be obtained from tangents to the α versus time curve, either graphically or numerically. This approach was used by Acitelli *et al.* (1971). Another approach, used by Choi (1970) for an nth-order reaction, is to plot the data according to integrated forms of the rate equations. For a first-order reaction,

$$-\ln(1 - \alpha) = kt \tag{26}$$

and for a second-order reaction,

$$1/(1 - \alpha) = 1 + kt \tag{27}$$

Choi found that the curing of a DAP model compound followed second-order kinetics. Greenberg and Kamel (1977) used a third approach to analyze the nth-order anhydride formation in poly(acrylic acid). Their method, developed by Wilkinson (1961), utilizes the relationship

$$t/\alpha \cong 1/k + nt/2 \tag{28}$$

which is accurate up to $\alpha \sim 0.4$. Analysis of their thermogravimetric data according to Eq. (28) gave values of $n = 1.8–2.3$, strongly suggestive of a second-order reaction. The kinetic analyses of the TG data (Fig. 18), utilizing Eq. (27) verified second-order kinetics (Fig. 32).

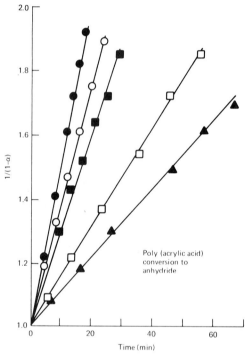

Fig. 32   Plot of isothermal TG degree of conversion data, up to $\alpha \approx 0.4$, according to second-order rate equation. DuPont 950 TGA. (From Greenberg and Kamel, 1977.)

b. The Activation Energy.   As previously discussed, in the most pragmatic sense the activation energy is a time–temperature shift factor. Activation energies are normally obtained from an Arrhenius plot.* Such a plot is shown in Fig. 33 for the autocatalyzed rate constants k_1 of an

---

* Activation energies and preexponential constants can also be obtained by dynamic method B. See Section IV.B.

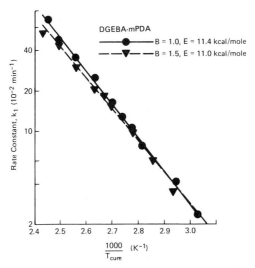

Fig. 33   Arrhenius plot of autocatalyzed reaction rate constants k_1 for epoxy–amine at two hardener levels. Perkin-Elmer DSC-1. (From Sourour and Kamal, 1976.)

amine–epoxy system at two stoichiometric levels (Sourour and Kamal, 1976). Activation energies of approximately 46 kJ mole^{-1} (11 kcal mole^{-1}) were obtained; activation energies for the much smaller k_1's (catalyzed rate constants) were close to 84 kJ mole^{-1} (20 kcal mole^{-1}). An Arrhenius plot of the second-order rate constants for anhydride formation in poly(acrylic acid) gave an activation energy of 159 kJ mole^{-1} (38 kcal mole^{-1}, Greenberg and Kamel, 1977).

The activation energy can also be measured from the temperature dependence of the time to reach a constant extent of conversion. The time to gelation satisfies this requirement.* Consider the integration of Eq. (1) up to the gel point:

$$\int_{\alpha=0}^{\alpha_{\mathrm{gel}}} \frac{d\alpha}{f(\alpha)} = \mathrm{const} = k \int_{t=0}^{t_{\mathrm{gel}}} dt = kt_{\mathrm{gel}} \tag{29}$$

demonstrating that the time to gelation is inversely proportional to the rate constant, independent of $f(\alpha)$. Figure 34 shows an Arrhenius plot of gel times for one of the amine–epoxy systems of Fig. 33 ($B = 1.0$). Times to gelation were obtained by DSC using calculated values of α_{gel} and cone-

---

* Caution must be exercised to distinguish between gel times that occur during chemical control and diffusion control of the curing process. Figure 30 demonstrates that gelation at 23°C occurs during diffusion control, gelation at 56 and 75°C occurs close to onset of diffusion control, and gelation at higher temperatures occurs during chemical control of the reaction.

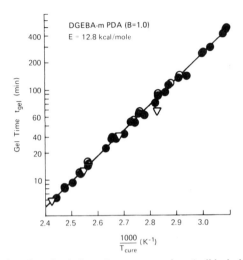

Fig. 34 Arrhenius plot of gel times for epoxy–amine: (solid circle) kinetic gel times from DSC–gelation theory; (open circle) rheological gel times from viscosity measurements; (▽) dynamic mechanical gel times by TBA. (From Babayevsky and Gillham, 1973. Sourour and Kamal, 1976.)

and-plate viscosity measurements (Sourour and Kamal, 1977), and by TBA (Babayevsky and Gillham, 1973). The activation energy of 53.5 kJ mole^{-1} (12.8 kcal mole^{-1}) was in close agreement with the value found for k_1, 47.8 kJ mole^{-1} (11.4 kcal mole^{-1}). Agreement was expected since k_1 was observed to be several orders of magnitude larger than k_1' [see Eq. (13)]. The larger activation energy from times to gelation may be due to the onset of diffusion control of the reaction prior to gelation at the lower temperatures. See footnote on p. 483. Barton (1979) found excellent agreement between activation energies from gel times and from DSC isothermal method 1 data.

Equation 29 can be adapted to the times to any constant degree of conversion. For example, as previously discussed, the occurrence of a maximum in the reaction rate at a constant extent of conversion for autocatalyzed systems can be utilized to obtain kinetic information. In isothermal method 1 experiments, the times t_{max} to reach the maximum reaction rate can be used in the same manner as gel times to obtain the activation energy. The maximum reaction rate versus isothermal cure temperature will also yield kinetic parameters, since from Eq. (13)

$$\frac{(d\alpha/dt)_{max}}{\alpha_{max}(1 - \alpha_{max})(B - \alpha_{max})} \cong k_1 = A \exp[-E/RT] \qquad (30)$$

when $k_1 \gg k_1'$. Results in Table V suggest that both methods give reasonable approximations (E and $\ln A \pm 10\%$) to results obtained over the entire course of the reaction.

One of the major applications of kinetic data is the prediction of the time–temperature–degree of conversion relationships for real systems. An example of the expected accuracy of such predictions is illustratd by Fig. 35. Calculated and experimental curves agree precisely during the

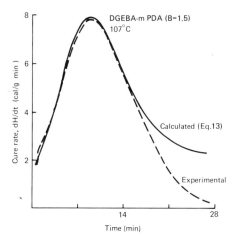

Fig. 35   Comparison between experimental and calculated cure rates for an autocatalyzed epoxy–amine. Falloff in experimental rate at 15–16 min due to onset of vitrification. Perkin-Elmer DSC-1. (From Sourour and Kamal, 1976.)

early reaction, but the experimental rate falls off at the onset of vitrification where the reaction becomes diffusion controlled. When coupled with times to vitrification, as determined by TBA, for example (Fig. 20), this approach should possess excellent predictive utility.

B.   PROPERTIES OF CURED THERMOSETS

In this section the physical properties of cured thermosets are treated. Properties that relate to the utility and quality control* of materials are emphasized, and these can be categorized under degree of cure, physical properties, and composition. All physical properties, including stability, ultimately depend on the extent to which a material is fully cured. Some of these properties are excellent indicators of the degree of cure, as will be discussed. Properties that are important to the end use of a material

---

* In several studies thermal analysis, combined with chromatography and spectroscopy techniques, have been applied to quality control of thermosets (May *et al.,* 1978; Chen, 1978; Thomas *et al.,* 1979; Penn, 1979; Zucconi, 1979).

include linear and volumetric coefficients of thermal expansion, dimensional stability, softening temperature, cross-link density and modulus,
and composition. The latter includes not only the amounts of resins and
fillers but also their identification.

1. Residual Reaction and Degree of Cure

a. ΔH_{resid}: DSC.   This method of measuring degree of cure is an
extension of isothermal method 2, Section II.A.2.a) The method can be
used routinely to monitor cure at the 95% conversion level (residual cure
>5%) and with care at the 98% conversion level. The residual exotherm
or heat evolved during completion of cross-linking is a measure of the
uncured fraction, and thus the degree of cure calcuated from Eq. (18).
Prior knowledge of ΔH_{RXN} is required to obtain the absolute degree of
cure. Typically, quantitative measurement of the residual exotherm for a
material that is between 95 and 100% cured is difficult, as can be pictured
by reference to Fig. 7 and 36. This is especially true when the glass

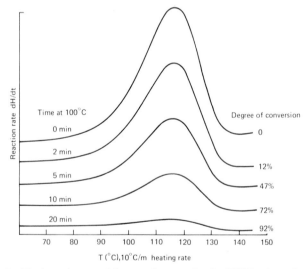

Fig. 36   Residual exotherm and degree of conversion by DSC isothermal method 2 of a
free radical initiator (AIBN). DuPont 990/910 DSC. (From Swarin and Wims, 1976.)

transition with accompanying endotherm immediately precedes the
exotherm, as in Fig. 7. With modern high-precision calorimeters,
especially those with computerized data analysis, it should be possible to
improve greatly the sensitivity of this method, to perhaps the final 1–2%
of cure, by measuring the area between two successive heating cycles.

Brett (1976) used this method to correlate bond strengths with degree of cure for an epoxy–amine adhesive. His bond strength data show a good correlation with the residual exotherm for both isothermal and isochronal curing. Storage of assembled adhesive joints at room temperature prior to elevated temperature cure caused deterioration of the bond strength and concommitant reduction of the residual exotherm, indicating that premature curing had a deleterious effect on the adhesive bond. Based on his DSC results, Brett could conclude that bond failure was cohesive in the range of 70–90% cure and became adhesive above 90% cure.

Parker and Smith (1979) also found that the lap shear strength of an epoxy film adhesive decreased as the residual exotherm decreased. A minimum in ΔH_{resid}, corresponding to a minimum bond strength, was used as a criterion, in conjunction with DSC-measured cure kinetics, to evaluate shelf lives at various storage temperatures. It was also shown that film adhesive lots could be accepted or rejected on the basis of residual exotherm as well as lap shear strength. The same authors correlated the residual exotherm of epoxy prepregs with the percentage of resin flow, which is used as a material specification. Again, shelf-life storage conditions were evaluated on the basis of a minimum in ΔH_{resid}. In a similar study Jackson *et al.* (1975) correlated the residual exotherm, spiral flow, and epoxide equivalent weight of an epoxy–novolac molding powder.

b. ΔW_{resid}: *TG.*   Where the curing chemistry involves loss of condensation products, e.g., phenolic and amino resins, a measure of the degree of cure is possible from detection of the weight loss upon completion of cross-linking. Basically, measurement of the residual weight has the same sensitivity as measurement of the residual enthalpy. However, sorbed water, residual solvent, and other volatile components can interfere; desiccated samples, and the use of DTG to separate curing from other processes will improve accuracy. An illustrative example of a brake lining material containing a phenolic binding resin is shown in Fig. 37 (Cassel, 1976). Note that full scale represents only 5% weight loss. Curves A were run on the uncured composite and provide a basis for calculating degree of cure; note apparent loss of absorbed water from the first portion of the DTG trace plus similarity of the thermogram to the pressure DSC experiment of a similar phenolic (Fig. 5). Sample B, previously cured for 1 min at 160°C, shows a residual weight loss of 1.8%, indicating by comparison with sample A that approximately 40% of cure had been achieved. Sample C was cured at 180°C and shows a small residual weight indicating greater than 90% cure. Since undercured brake linings can soften and seize at lower temperatures than those properly cured, a rapid test such as this has practical utility, for example, in quality control.

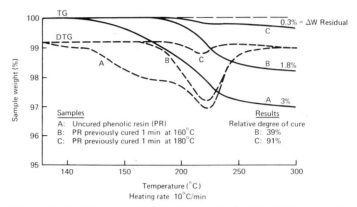

Fig. 37   Residual weight loss and degree of conversion by TG/DTG isothermal method 2. Phenolic bonding resin. Perkin-Elmer TGS-2. (From Cassel, 1976, 1977a.)

Figure 38 shows measurement of the residual cure and amount of polymer directly on a poly(amide–imide) coated wire (Cassel, 1976). Strictly speaking, poly(amide–imide) is not a thermoset, but the curing reaction is similar to several thermosets. Scans were run on two separate samples at different sensitivities. In the lower curve (box) more than 99% of the sample weight was suppressed. From this analysis, a weight loss of 0.022% of the original sample was attributed to residual curing. From the stoichiometry of the curing reaction, it was calculated that had all the polymer present been uncured a weight loss of 0.27% would have been observed. Thus the residual cure calculates to 8% and the degree of cure is therefore 92%.

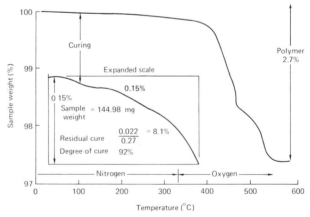

Fig. 38   Residual weight loss and degree of conversion plus percentage polymer of poly(amide-imide) coated wire by TG. A weight loss of 0.27% was calculated for 100% conversion. Perkin-Elmer TGS-2. (From Cassel, 1976.)

An undercured wire coating will tend to soften and lose its insulation properties at a lower temperature.

2. Physical Properties and Degree of Cure

a. Glass Transition and Softening Temperatures: DSC, DMA, TBA, TMA. See Chapter 2 and Flynn (1974) for interpretation and measurement of the glass transition. The glass transition and concomitant effects, such as softening, not only provide a means of characterizing the state of cure, but often define an upper use temperature. Enthalpy relaxation (enthalpy recovery) accompanying the glass transition can give information about the thermal history or physical aging of thermosets. See Section II.C.1 and Ophir *et al.* (1978). Fava (1968) showed that the glass transition temperature increases smoothly as the reaction proceeds to completion (Fig. 7). Gray (1972) established an empirical correlation between relative degree of cure and glass transition temperature, both measured by DSC. The subject of this study was a B-staged epoxy–fiber glass laminate. The relative degree of cure was determined by DSC method 1 (Fig. 39). Glass transition temperatures as a function of time at the same isothermal cure temperature were determined on samples that

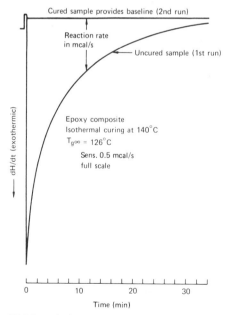

Fig. 39   Isothermal DSC method 1 curing of B-stage epoxy–fiber glass laminate. Sample heated from 50 to 140°C at 320°C/min. (~ 17 sec). Area between first and second run used to measure ΔH_{RXN}. Perkin-Elmer DSC-2. (From Gray, 1972.)

had been programmed to temperature at 320°C min⁻¹, held for the desired
time, and cooled at 320°C min⁻¹. T_g was taken as the temperature at the
midpoint between the initial and final heat capacity curves on samples
heated at 10°C m⁻¹ (Fig. 40). A "blank," curve 2 in Fig. 40, showed

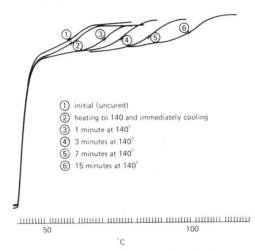

Fig. 40 Glass transition temperatures of epoxy–glass laminate as a function of curing
times at 140°C. Perkin-Elmer DSC-2. (From Gray, 1972.)

that an approximately 1°C increase in T_g resulted from the process of
heating and cooling. From glass transition temperatures and corresponding
degrees of cure, Gray established the correlation in Fig. 41. If the T_g
can be determined to ±1°C, as claimed by the author, then relative
degree of cure can be measured to within ±2% by this technique. On
an absolute degree of cure basis, data of Gray,* Horie *et al.* (1970a),
and Barton (1974a,b) suggest that beyond the gel point a ± 1% change in
degree of cure will be reflected by a ± 1–4°C change in T_g. This suggests
that this technique has the potential (e.g., in a quality control application)
to ensure >98% cure. However, other factors, such as effects of physical
aging or relief of stress, may alter the glass transition temperature or
render its evaluation more difficult. In actual practice, it is reasonable to
expect this method to ensure that >95% of cure has taken place.

The α-transition (highest temperature transition, T_g) in the dynamic
mechanical spectra of thermosetting resins is synonymous with the glass
transition. As such the peak temperature in the α-transition is a measure
of the degree of cure. This is illustrated in Fig. 42 for an epoxy laminate

* Assuming the initial degree of conversion of the B-staged laminate to be approximately
50%.

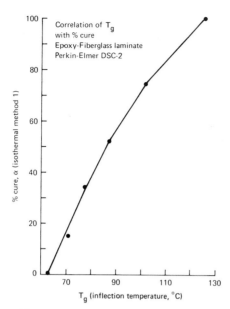

Fig. 41 Glass transition temperature as a measure of degree of cure for epoxy–glass laminate. T_g from Fig. 40, degree of cure by isothermal method 1. Perkin-Elmer DSC-2. (From Gray, 1972.)

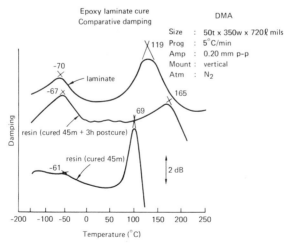

Fig. 42 DMA damping curves show the laminate to be intermediate in cure between the undercured resin and the postcured resin. Extent of cure is related to the α-transition temperature (69–165°C) and magnitude of the γ-transition at ~ -70°C. DuPont 980 DMA. (From Blaine and Lofthouse, a).

(Blaine and Lofthouse, a; Hassel, 1978). The α-transition peak temperature of the "laminate" (119°C) relative to the "postcured resin" (165°C) demonstrates that the laminate is considerably undercured. The γ-transition, at ~-70°C, was attributed to a crankshaft rotation of the cross-linking segments. In addition to the shift in the α-transition peak temperature with increasing conversion, the γ-transition is seen to increase in intensity with the advancement of cure. Thus the magnitude of the γ-transition damping peak may also be a measure of degree of cure. A β-transition at 50–60°C has been observed by Hartmann and Lee (1977) and Babayevsky and Gillham (1973) for epoxy–amine thermosets by TBA. The former authors attribute this to an undercured network. Figure 43 shows the rela-

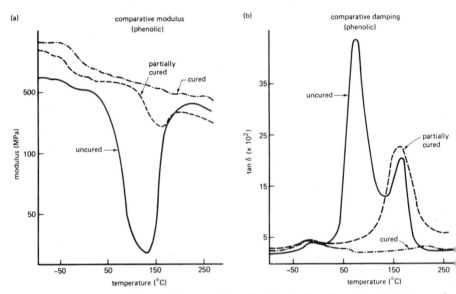

Fig. 43 Comparative modulus (a) and damping (b) of a phenolic resin at three stages of cure. DuPont 981 DMA. (P. S. Gill personal communication, 1980.)

tive modulus and damping of a phenolic at three stages of cure (P. S. Gill, personal communication, 1980). The uncured sample shows a large damping peak at T_g and a smaller, well-separated damping peak associated with cure. The partially cured sample exhibits a broad, single damping peak that is actually a combination of the peaks associated with the glass transition and cure. The third sample shows essentially no damping at all, demonstrating that complete cure has taken place and that $T_g > 250$°C (cf. modulus curves). The small damping peak at ~-10°C was attributed to plasticizer. Reed (1979, 1980) observed broadening of the α-transition and an additional damping peak above T_g in a silane-finished, glass-fiber, reinforced epoxy composite. This high-temperature relaxation was highly dependent on fiber orientation in the unidirectional composite, and it may be indicative of a fiber–resin interfacial region distinct from the bulk resin. Grentzer *et al.* (1981a,b) studied the damping behavior of can coatings on the metal can stock by DMA. Damping behavior was related to can fabrication performance. They also related modulus and damping characteristics of gel coatings to their environmental stress cracking behavior.

TBA gives information similar to that given by DMA, but it has limited application to cured thermosets. In the normal mode of operation, the braid must be coated from solution preventing study of already gelled samples. Such application of TBA was described in Section II.A.2.e. However, Sykes *et al.* (1977) describes a novel use of TBA (Fig. 44) in which a 200 × 2 mm composite specimen replaces the braid. Using this technique, they showed that absorption of moisture caused a decrease in rigidity, a broadening and decrease in the α-transition, and an increase in intensity of the γ-transition. Their results led them to conclude that the polymer is plasticized by absorbed water, which is, in part, reversible. Schneider and Gillham (1980) used a similar technique to study curing of prepreg materials.

TMA can be used in the penetration mode to detect relative states of cure, as demonstrated by Cassell (1977b, Figs. 45 and 46). Typically high loadings (10–100 g) are employed with small-area probes producing contact pressures up to 50 psi. Neet *et al.* (1978) used a light load on the expansion probe to measure softening temperatures of rigid polyurethane foams. Their results were more consistent when the probe was placed on the skin or molded surface. The same sample preparation and conditioning procedures described for thermal expansion measurements (Section II.B.2.b) may be applicable to measurement of softening temperatures. Lamoureux (1978) found the derivative trace to be as useful as the expansion curve for measuring softening or penetration temperatures.

In this mode the glass transition temperature is not measured directly, but the softening associated with the large-scale decrease in modulus at the glass transition. These "softening temperatures," although not identical to T_g, have been shown to correlate well with glass transition temperature (Angeloni, 1967; Fava, 1968; Zicherman and Holsworth, 1974; Lamoureux, 1978; Brewis *et al.*, 1979). Figure 45 demonstrates use of the hemispherical probe where less penetration at higher temperatures indicates a harder material corresponding to higher cross-link density. As illustrated by this example, high loadings are necessary to effect a sufficient amount of indentation. In Fig. 46 the knife-edged flexure probe is used to accentuate softening at the glass transition temperature. Note that the softenings of both the outer polyester coating and inner poly(amide–imide) coating were detected. Sykes *et al.* (1977) describe a fixture (Fig. 47) for measuring heat distortion temperatures (HDT) in a manner similar

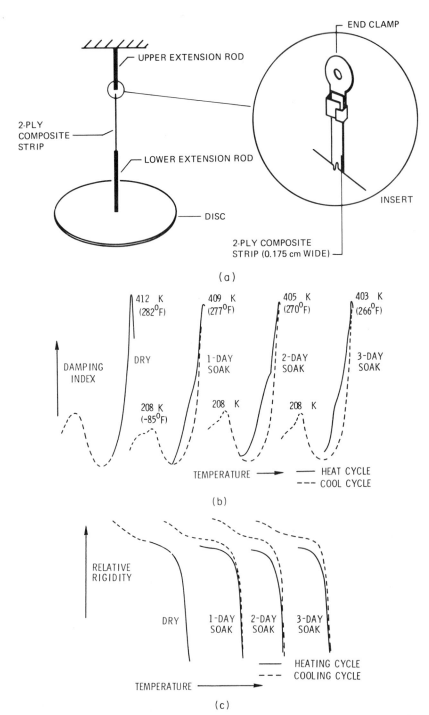

Fig. 44 TBA of unsupported sample. (a) Sketch of the TBA pendulum. Insert shows method used to attach end clamps to composite specimen. (b) Damping curves at different times for two-ply unidirectional, graphite–epoxy composite cured at 400 K. (c) Relative rigidity curves for same composite. (From Sykes *et al.,* 1977, by permission of the Society for the Advancement of Material and Process Engineering.)

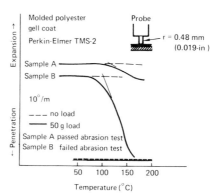

Fig. 45   TMA penetration test to detect relative states of cure. The more highly cross-linked material softens at a higher temperature, undergoes less penetration, and is more abrasion resistant. Perkin-Elmer TMS-2. (From Cassel, 1977a.)

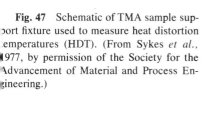

Fig. 46   Use of the TMA flexure probe to accentuate softening of both the outer and inner coating of a motor-winding wire. Perkin-Elmer TMS-2. (From Cassel, 1976.)

Fig. 47   Schematic of TMA sample support fixture used to measure heat distortion temperatures (HDT). (From Sykes *et al.*, 1977, by permission of the Society for the Advancement of Material and Process Engineering.)

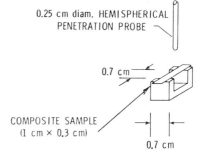

to the ASTM D648 technique. Reproducibility was reported as better than $\pm 2°C$. The effect of water sorption, from immersion in distilled water, on HDT is shown in Fig. 48. In comparison to TBA results on the same samples (Fig. 44), sorption corresponding to a weight gain of 1.0% resulted in a much larger decrease in HDT (49°C) than in T_g (12°C). This suggests

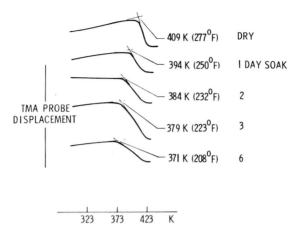

Fig. 48 TMA probe displacement curves and extrapolated heat distortion temperature (HDT) values for two-ply unidirectional graphite–epoxy composite. (From Sykes *et al.,* 1977, by permission of the Society for the Advancement of Material and Process Engineering.)

that HDT shows the combined effects of moisture on both T_g and modulus. For dry specimens HDT was about 3°C lower than T_g. McKague *et al.* (1973) describe similar TMA tests using the knife-edged flexure probe and give a detailed comparison with the ASTM D648 test. The effects of exposure of very thin (0.005–0.020 in.) epoxy and phenolic samples to high humidity environments were studied.

Brewis *et al.* (1977a) showed close agreement between increasing joint strength and softening temperatures of an epoxide–isoprene copolymer adhesive. Neet *et al.* (1978) observed the influence of the cure temperature and isocyanate index (stoichiometry) on softening temperatures of rigid polyether and polyester urethane foams. An increasing cure temperature produced an increase in the softening temperature until a maximum value was achieved. An increase in the isocyanate index from 1.0 to 1.4 also caused the softening temperature to increase. Knowles (1978) observed effects of mix ratio and temperature–humidity aging on the softening temperatures of polyurethane and epoxy encapsulating materials. Brewis *et al.* (1979) found that, on postcuring of amine-hardened epoxide resins, large increases in softening temperatures can occur without reaction of the epoxide groups. Reed (1979, 1980) observed an additional softening temperature above that associated with T_g for a silane-finished, glass-fiber-reinforced epoxy; by DMA this behavior was manifested as an additional damping peak.

Another means of determining degree of cure from T_g involves measurement of thermal expansion coefficients through the glass transition interval by TMA. See discussion in Section II.B.2.b. By definition, the glass transition temperature is the temperature of intersection of the expansion curves of the glassy and supercooled rubbery states (see Chapter 2). Typically, thermal expansion in the rubbery state is about three times that in the glassy state. As illustrated by Figs. 49 and 50, considerable in-

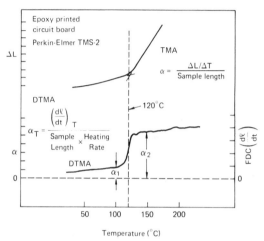

Fig. 49   Linear expansion through the glass transition interval by TMA-DTMA. Linear expansion coefficient α from TMA trace requires measurement of the slope; DTMA allows direct recording of α. Perkin-Elmer TMS-2. (From Cassel, 1977a,b.)

Fig. 50   DTMA traces showing that a more completely cured thermoset is characterized by (a) higher T_g, (b) lower expansion rate in both the glassy (α_1) and rubbery (α_2) states, and (c) smaller change in expansion rate. Also see Table VI. Perkin-Elmer TMS-2. (From Cassel, 1977a.)

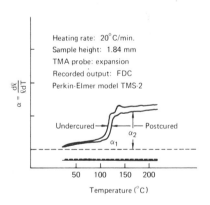

formation in addition to T_g is provided by the TMA measurement. The calibrated derivative trace (DTMA) provides a direct reading of the thermal expansion coefficients in the glassy (α_1) and rubbery (α_2) states. As

TABLE VI

THERMAL EXPANSIONS OF EPOXY–ANHYDRIDE RESIN
BY TMA (DUPONT 940)

Sample	T_g (°C)[b]	Coefficient of linear expansion α (10^{-6} in. in.$^{-1}$ °C^{-1})		
		α_1, below T_g	α_2, above T_g	$\alpha_2 - \alpha_1$
1	137	70	125	55
2	133	73	153	80
3	122	76	187	111
4[c]	119	80	238	158

[a] From Angeloni (1967).
[b] T_g as in Fig. 49 (TMA).
[c] Sample 4 has lowest cross-link density.

demonstrated by Fig. 50 and Table VI, a more completely cured material
is characterized by a higher glass transition temperature, lower thermal
expansion coefficients in both the glassy and rubbery states, and a smaller
change in expansion coefficients ($\alpha_2 - \alpha$). However, as discussed in the
next section, before measuring thermal expansion, it is first necessary to
condition the sample by heating through the glass transition and cooling
with the loaded probe in place. This was done prior to the measurements
in Figs. 49 and 50. It must be established, as Gray (1972) did (see Fig. 40),
that the state of cure is not significantly altered. This method appears to
be highly sensitive to degree of cure, as suggested by Angeloni (1967) for a
series of epoxy–anhydride resins (Table VI). Unfortunately, values for
degree of cure were not reported, although sample 4 was reported to have
the lowest degree of cross-linking. Angeloni also found a direct proportionality between the deflection temperature under load (ASTM D648)
and T_g by TMA in both the expansion and penetration modes. Lamoureux
(1978) correlated T_g by TMA expansion with softening temperatures by
TMA penetration for printed circuit board materials. He found the penetration method to give more reproducible results that were a few degrees
lower than T_g.

   b. *Thermal Expansion: TMA.*   TMA is also capable of providing
quantitative values of linear coefficients of thermal expansion. However,
certain precautions must be taken in order to obtain accurate data. Accuracies of $\pm 5\%$, which include uncertainties in the individual measurement and the instrument calibration, are attainable (Prime *et al.*, 1974;
Gaskill and Barrall, 1975). Appropriate thermal expansion standards

(Prime *et al.*, 1974; Thomas and Scharr, 1975; Gaskill and Barrall, 1975) and correct expansion values at the particular calibration temperatures (Fig. 51) must be employed. For precise work, especially with fiber-rein-

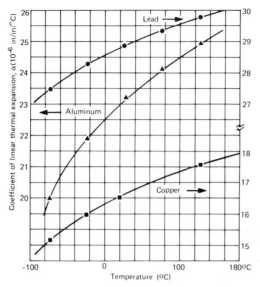

Fig. 51 Coefficients of linear thermal expansion for three calibration standards. Values for lead and aluminum taken from American Institute of Physics Handbook (Gray, 1963). Values for copper from Hahn (1970).

forced composites, the expansion coefficient of fused quartz, 0.55×10^{-6} in. in.$^{-1}$ °C^{-1} (20–320°C) (Weast, 1978–1979), should be subtracted from the value for the calibration standard, since the measured expansion is the difference between that of the standard and quartz.

$$\alpha(\text{measured}) = \alpha(\text{sample}) - \alpha(\text{quartz}) \tag{31}$$

See further discussions in this section. Accessories for volume dilatometry are available for TMA instruments, but little information has been published on their accuracy and use.

Sample preparation and conditioning are integral parts of the experimental procedure for measuring linear expansion. Samples should have flat surfaces, which can be achieved by molding, machining, or melt-pressing; a motorized microtome is useful. A room temperature pellet press can also be used. If the sample is punched from a sheet, the burred edge must be removed unless the sample can be measured burred edge up with the probe well within the burrs. Reliable TMA expansion data require that the sample first be conditioned in the apparatus by heating through the glass transition and cooling with the loaded probe in place. Relief of internal stress and conformation of the sample to the apparatus take place during conditioning. The latter, which includes penetration of the probe, accommodation of probe misalignment, and accommodation of surface roughness of the probe, sample holder, and sample, dictates that the conditioning take place in the TMA apparatus. Ennis and Williams (1977) have an excellent discussion on sample conditioning. The necessity for conditioning erases thermal history; if this is of interest, it can be observed by DSC.

The sample height should be measured with a micrometer or with the TMA; accuracy must be commensurate with that expected for the expansion coefficient. Sample thicknesses < 2.5 mm are recommended in order to keep thermal gradients in the sample < 1°C. A 2°C gradient is expected in a 5 mm thick sample (unpublished data of this author). Loading on the sample should be light, just sufficient to maintain the probe in contact with the sample on cooling, e.g., 1 g on the flat-tipped expansion probe (~ 2.5 mm diameter, 0.3 psi) or 0.1 g on the hemispherical probe (2.5 mm diameter). Ennis and Williams (1977) claim better accommodation of probe misalignment with the hemispherical probe. They also showed that both probes give equivalent results, provided the hemispherical probe is positioned away from the sample edge. Thomas and Scharr (1975) observed significant effects of heating rate, loading, and sample diameter on the measured expansion rate.

Figure 49 shows the calculation of the expansion coefficient from the slope of the TMA trace and illustrates use of the calibrated derivative trace (DTMA) to record directly the linear expansion coefficient (Cassel, 1977a,b). When measuring materials with small expansion coefficients (e.g., $<50 \times 10^{-6}$ in. in.$^{-1}$ °C^{-1}, a correction for the expansion of quartz should be made (Prime *et al.*, 1974) [see Eq. (31)]. Since the sample or calibration standard displaces the quartz probe, the actual expansion measured is the difference between the expansions of the specimen and quartz.

Thermosetting polymers, especially those that are highly filled, can absorb sufficient water from the atmosphere to affect thermal expansion. Brand (1977) showed that absorbed water in epoxy–glass circuits boards plasticized the polymer, resulting in significant increases in thermal expansion coefficients (TMA) and lowering of glass transition temperatures (TMA, DSC). The high thermal expansion contributed to a processing problem. However, the effect was found to be highly reversible, i.e., desorption resulted in the lower expansion coefficients and higher glass transition temperatures of the original dry materials. Mark and Findley (1978) showed that, over longer time periods, even unfilled thermosets will undergo reversible dimensional changes on sorption and desorption of water. Prime *et al.* (1974) showed that times for thin (<0.3 mm) composite samples to equilibrate with the environment are short and that these samples needed to be dry before meaningful thermal expansion data could be obtained. Although drying of thicker samples may not be required to obtain reproducible expansion data, absorbed water will nonetheless affect the results (see, for example, Brand, 1977; Adamson, 1979, 1980). Many materials, especially those with filler, can be anisotropic and inhomogeneous with respect to thermal expansion. In some cases, "warping" can be misinterpreted as expansion. With fiber-reinforced composites, thermal expansion is dominated by the high-modulus, low-expansion fibers (see Section III.C.). Thus both fiber loading and orientation must be known when interpreting thermal expansion data. Some reasons for these precautions are exemplified by the linear expansion studies of Ritchie *et al.* (1972a,b) on several molding compounds. Figure 52 shows some

ELECTRONIC COMPONENT ENCAPSULATION

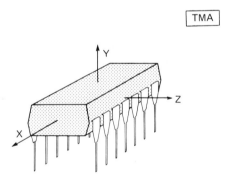

Compound	$\alpha_{150\,°C}$ (10^{-6} in. in.$^{-1}$ °C)		
	X	Y	Z
Phenolic A	23.6	27.5	23.5
DAP A	37.1	72.5	73.9
Silicone A	26.0	33.4	42.2
Epoxy D	35.4	56.2	32.5

Fig. 52   Directional dependence of linear expansion of semiconductor device molded from four encapsulants. The *x* axis is the direction of flow. DuPont 941 TMA. (From Ritchie *et al.,* 1972a,b.)

dependencies on materials and directional orientation. These authors found thermal expansion, both the average and the deviation from the average, to have a very large dependence on molding condition. The repeatability within a given molding cavity and molding condition was much better (15–20%) than repeatability among molding cavities. There are several factors that can account for such variations in thermal expansion, including degree of cure, filler loading (see Table VIII), and orientation of polymer and/or filler, absorbed water, or solvent (Prime *et al.*, 1974; Brand, 1977; Mark and Findley, 1978), and relief of local stress induced during cooling in a mold (Ennis and Williams, 1977).

Expansion measurements on thin samples, e.g., films and sheets, present certain challenges. In the thin z direction, it is often necessary to load the probe heavily to prevent buckling or warping and assure that only thermal expansion is being measured. In-plane measurements (*xy* direction) on samples too thin to support themselves require special techniques, such as those described by Prime *et al.* (1974). The directional dependence of thermal expansion of random and pseudoisotropic composites was measured by TMA. The very thin samples, 0.2 mm ≡ 7.5 mils, were suspended between special TMA quartz members by means of Invar chucks (Fig. 53). The thermal expansion rate of Invar is very close to that of quartz. Figure 54 illustrates some of the experimental details, including the sample mounting template and copper calibration standards, that helped to establish the effective sample length in addition to calibrating the measurement.

The composite samples were found to equilibrate rapidly with atmospheric moisture, and the associated hygroscopic expansion β was found to be of the same order as thermal expansion, necessitating measurement on dry samples in a dry environment:

$$\beta_T = (\Delta L/L_0 \, \Delta RH)_T \tag{33}$$

The desire to estimate hygroscopic expansion, plus the experimentally observed need for one heating and cooling cycle to condition the sample and assembly, was incorporated into the experimental procedure (see Table VII). Figure 55 is an actual TMA trace and demonstrates calculation of the thermal expansion coefficient. Because of the very small ΔLs measured, it was necessary to operate the TMA module on a floating air table. Results for random and pseudoisotropic samples at two orientations are given in Table VIII; much of the difference between the random composites can be attributed to differences in fiber loading as measured by TG.

c. Specific Heat (C_p): *DSC.* There is little specific heat data reported for thermosets. Specific heat measurements are straightforward, and

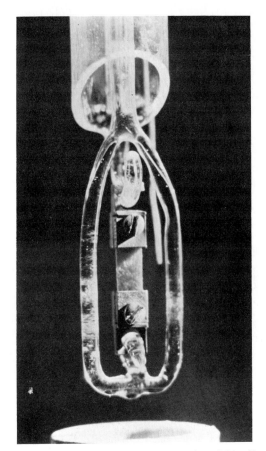

Fig. 53   TMA technique used to measure linear expansion of thin, fiber-reinforced composites. Quartz hooks are separated by 0.75 in. Loading was 10 g $\cong$ 30 psi. Perking-Elmer TMS-1. (From Prime *et al.*, 1974.)

those interested in making these measurements are referred to other chapters in this book, especially Chapter 2. The study of a phenolic resole polymer by Warfield and Kayser (1975) is a good example of the capabilities of DSC in this area, especially the capability of estimating T_g when it is not accessible to direct determination. The specific heat at constant pressure C_p between 47 and 197°C was determined with a Perkin-Elmer DSC-1b to a stated accuracy of $\pm 2\%$. Extrapolation to 25°C gave $C_p = 0.353$ cal g^{-1} °C^{-1}. A secondary transition, decrease in slope dC_p/dT, was observed at 125–130°C. Employing a previously developed corresponding states relationship and the specific heat at constant volume C_v, obtained from C_p and other measurable physical parameters, Warfield and Kayser estimated T_g for this polymer to occur at about 325°C. Since

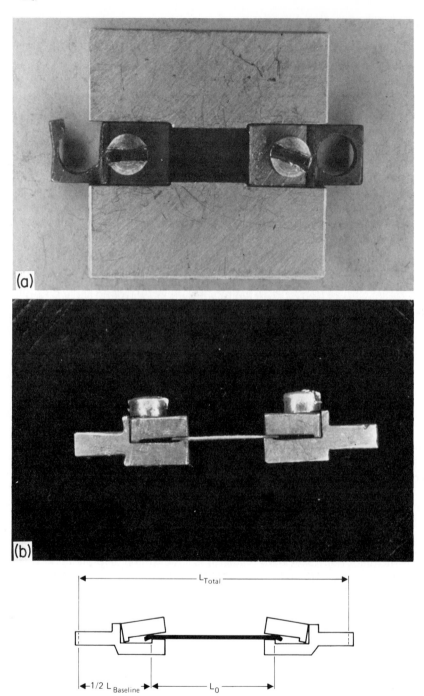

Fig. 54 Experimental details for TMA technique of Fig. 53. (a) Mounting template to yield aligned specimen at reproducible Invar chuck separation; (b) actual side view and exaggerated schematic demonstrating how chucks grip composite; (c) NBS and OF copper standards for calibration. Perkin-Elmer TMS-1. (From Prime *et al.*, 1974.)

Fig. 54 (*continued*)

TABLE VII

TMA Procedure for Measurement of Thermal Expansion of Dry
Composites at 0% Relative Humidity and Estimation
of Hygroscopic Expansion[a]

---

1. Equilibrate sample at lab ambient for 24 h, record RH.
2. Cut and mount sample specimen in Invar chucks, using template (Fig. 54a).
3. Clamp specimen in TMA at 18°C, ambient RH. Record pen position.
4. Dry N_2 at 50 cm^3 min^{-1}, heat at 5° min^{-1} to 75°C, and hold until length stable (~2 h). Cool at 5°C min^{-1} to 18°C. Estimate β from length change re step 3 (desorption).
5. Heat/cool at 1.25°C m^{-1} from 18 to 75 to 18°C, two cycles. Measure α (50°C, 0% RH) from 50° slope of heating cycles.
6. N_2 off, reequilibrate with ambient RH overnight. Estimate β from length change (resorption).

---

[a] Prime *et al.* (1974).

degradation starts at about 200°C, this transition is not easily accessible to direct determination. Era *et al.* (1977) measured specific heats of a series of resole model compounds at varying phenol–formaldehyde molar ratios. C_p values were all lower than those reported by Warfield and Kayser, and the secondary transitions were much weaker. C_p and $dC_p dT$ data of Era *et al.* were typical of amorphous polymers. Chang (1981) measured the specific heat of the same resole-type phenolic studied by

$$\alpha_{50^\circ} = \frac{S \cdot \Delta mV}{L_0 \Delta T} - \frac{L_{BL}}{L_0} \Delta \alpha + \alpha_{Quartz} \tag{32}$$

$$\alpha_{50^\circ} = \frac{(61.9 \ \mu in./mV) \ (0.558 \ mV)}{(0.242 \ in.) \ (50^\circ C)} - \frac{0.509 \ in.}{0.242 \ in.} \times 0.37 \ ppm/^\circ C + 0.55 \ ppm/^\circ C$$

$$\alpha_{50^\circ} = 2.62 \ ppm/^\circ C$$

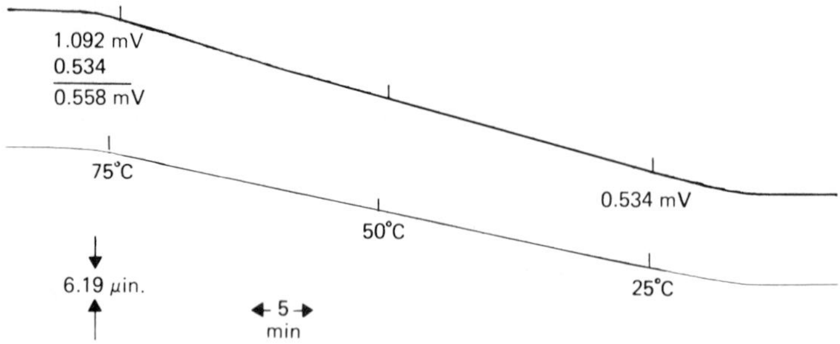

Fig. 55 Actual TMA trace at 1.25°C min^{-1} and thermal expansion calculation for thin, fiber-reinforced composite (90° pseudoisotropic, Table VII). Perkin-Elmer TMS-1. (From Prime *et al.*, 1974.)

TABLE VIII

THERMAL AND HYGROSCOPIC EXPANSION RESULTS OF THIN,
FIBER-REINFORCED COMPOSITES

Composite	Thermal expansion (10^{-6} in. in.$^{-1}$ °C^{-1})		Hygroscopic expansion (10^{-6} in. in.$^{-1}$ %RH)	
	α_{0°	α_{90°	β_{0°	β_{90°
Random[b]	21.4 ± 0.1	14.9 ± 0.5	29 ± 7	17 ± 5
Pseudoisotropic[c]	2.59 ± 0.01	2.60 ± 0.03	2.7 ± 0.1	2.8 ± 0.5

[a] From Prime *et al.* (1974).

[b] DuPont Kevlar® high-modulus organic fibers randomly distributed in epoxy matrix (USPolymeric E-702), 7.5 mils thick. Fiber content by TG (DuPont 950) in N$_2$ at 20°C m^{-1}. 0° sample: 36.3 wt % fiber ~28 vol %. 90° sample: 43.0 wt % fiber ~35 vol %.

[c] Courtaulds HMS high-modulus graphite fibers in epoxy matrix (Shell DX210/BF$_3$400), 6 plys, (0 ± 60)$_s$, 7.5 mils thick. 45–50 vol % chopped fibers (manufacturer's data).

Warfield and Kayser between 4 and 450 K by adiabatic calorimetry and DSC. Meaningful specific heat data required extensive postcuring and desorption of absorbed water. For a wide temperature range specific heat of the phenolic resin was represented by a simple proportionality to temperature: $C_p = 0.0042T$ J g^{-1} K$^{-1} = T$ mcal g^{-1} T^{-1}. These data were in close agreement with the results of Era *et al.* Chang also summarized specific heat data of thermosetting resins, cross-linked polymers, and varnishes between 0.1 and 500 K.

d. Cross-link Density by Solvent Swell: TMA, TG.   Solvent swell is a traditional measure of the cross-link density v_e/V_0, which is the number of moles of elastically effective cross-links per unit volume of polymer. For unfilled, cross-linked polymers, the equation of Flory and Rehner (1943) relates cross-link density to v_2, the measured volume fraction of polymer in a sample that has reached its equilibrium swollen state:

$$v_e/V_0 = -[\ln(1 - v_2) + v_2 + \chi_1 v_2^2]/V_1(v_0^{2/3} v_2^{1/3} - 2v_2/f) \qquad (34)$$

where χ_1 is the polymer–solvent interaction parameter, V_1 the molar volume of solvent, v_0 the volume fraction polymer in the diluent–polymer mixture at the time of cross-linking, and f the functionality of cross-links. Because curing involves the formation of cross-links, cross-link density is a sensitive measure of degree of cure. It is also a measure of the molecular weight between cross-links and is proportional to Young's modulus. For filled, cross-linked polymers, the calculations are inexact because of assumptions that must be made and uncertainties in the polymer–solvent interaction parameter. However, when swelling data are directly proportional to Young's modulus, which is an unambiguous measure of cross-link density, they are very precise measures of degree of cure (see Prime, 1978). Two of the thermal analysis techniques have been employed as micromethods to measure solvent swell of elastomers, and both should be applicable to thermosets. Examples of solvent swell measurements of thermosets are provided by Kwei (1963) and Bell (1970). Greenberg and Kamel (1977) measured solvent swell (by means other than described here) and loss of water of condensation by TG in a study of the kinetics of anhydride formation in poly(acrylic acid). See Section II.A.2.c. For epoxies and poly(acrylic acid), volume swelling of cured thermosets was 5–20%.

Machin and Rogers (1970) measured swelling in one dimension via a thermomechanical analyzer. The samples were thermoplastics and cross-linked elastomers. To convert to volume swelling, it was necessary to assume homogeneity and isotropic swelling. The volume fraction v_2 is related to the experiment by

$$v_2 = V_0/V_s = (l_0/l_s)^3 \qquad (35)$$

where V_0 and l_0 represent the dry sample dimensions, free of cross-linked material, and V_s and l_s are the swollen volume and corresponding swollen thickness. Prime (1978) used the microbalance of a thermogravimetric analyzer to measure swelling of silica-filled siloxane elastomers. The volume fraction v_2 is related to the TG experiment by

$$v_2 = (1/\rho_p)/(1/\rho_p + SR/\rho_s) \qquad (36)$$

where ρ_p is the polymer density and ρ_s the solvent density. The swell ratio SR is defined as

$$SR = W_s/W_0 - 1 \tag{37}$$

where W_s is the swollen weight and W_0 the weight of dry sample free of uncross-linked material. The measured swell ratios were shown to have a direct correlation with Young's modulus.

Both techniques give a measure of the volume fraction v_2 that is directly relatable to cross-link density. TG is the preferred method, because it is independent of sample geometry, homogeneity, and isotropicity, and because an extractable fraction (uncross-linked fraction, placticizer) is also measured. An advantage of the TMA technique is the ability to measure swelling of a coating without removing it from the substrate. In the measurement of degree of cure or quality control, it is first necessary to establish the swell ratio (TG) or linear swelling (TMA) corresponding to complete cure. For filled systems an independent measure of filler content is necessary. Because it is a microtechnique and sample geometry is irrelevant, the TG method can detect gradients and local variabilities in cross-link density.

e. Thermal Stability: TG. Thermal degradative stability has been related to degree of cure of an epoxy resin (Lee and Levi, 1969). Overall activation energies for thermal decomposition were determined as a function of curing temperature from both isothermal and dynamic thermogravimetry. The data showed an orderly increase in activation energy with increasing cure temperature. Even more dramatic were the decomposition temperatures, which increased in direct proportion to the cure temperatures. These results are consistent with the view that higher cure temperatures result in a higher degree of cross-linking causing an increase in the energy to degrade a given mass (but one containing more chemical bonds) into its volatile products. Although the hypothesis is logical and this type of test appears potentially attractive as an indicator of completeness of cure (and/or effects of postcuring), the data available are not sufficiently convincing and need to be correlated with more absolute measurements of degree of cure.

3. Composition

In the majority of cases the desired compositional information is a quantitative measure of the percentage of filler(s) and the method is thermogravimetry. Quantification of the amount of coating, for example, on wire or particles, and the ratio of two polymers, often an elastomer or thermoplastic blended with a thermoset, are other compositional applications of TG. In these instances the nature of the resins and fillers is usually known. There are other circumstances where quantitative and/or qualitative analyses are needed. In these cases environment and method of decomposition, derivative thermogravimetry (DTG), DSC, DTA, and even DMA may contribute to the analyses. Combination of mass spectroscopy or gas chromatography with thermal techniques (i.e., TG–MS and TG–GC) is especially useful in identification of the polymer phase and elucidation of decomposition mechanisms.

a. Quantitative/Qualitative Analyses: TG, DTG. In any quantitative measure involving the pyrolysis of polymer, it must be established that the polymer phase decomposes without residue or with a known amount of residue. For example, some polymers will carbonize in nitrogen, and it may be necessary to finish the analysis in air, or to do the complete measurement in air providing the filler is inert. In materials with oxidizable (carbon) or thermally reactive (calcium carbonate, aluminum hydroxide) fillers, or a mixture of fillers, more sophisticated procedures such as nitrogen-followed-by-air pyrolyses or parameter-jump procedures (Flynn and Dickens, 1976) are necessary (see also Cassel, 1976).

Figure 56 shows the determination of the glass concentration in an epoxy–fiber glass prepreg by TG in nitrogen (Blaine, 1974). Epoxies generally decompose without residue in nitrogen, but this must be established

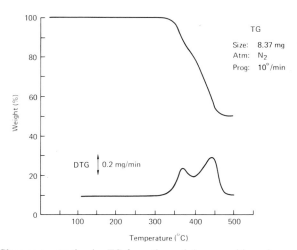

Fig. 56 Glass concentration by TG from thermal decomposition of epoxy–fiber glass prepreg in N_2. Characteristic DTG trace suggests potential as identification for resin. DuPont 951 TGA. (From Blaine, 1974.)

for each thermoset. With a glass filler, visual inspection can readily identify carbonaceous residues. Such carbonaceous residues will usually show a distinct decomposition in air in the 500–700°C region. Although the percent filler can be read from the chart, the recommended procedure is to weigh the residue by electronically suppressing its weight for greater accuracy. The characteristic nature of the derivative trace suggests that it has the potential to identify the resin in addition to giving a relative measure of thermal stability. Figure 57 demonstrates the power of TG in measuring uniformity of filler content in a molded motor housing (Cassel, 1977a). Figure 58 illustrates the determination of carbonate filler in a polyester from its thermal conversion to the oxide. Figure 59 demonstrates the

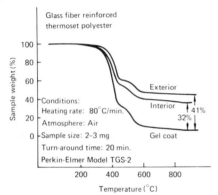

Fig. 57   TG shows a large nonuniformity in filler content for a molded motor housing. A 9% difference was measured between the bulk (INTERIOR) and a narrow inside ridge (EXTERIOR). The outer surface, which contained little or no filler, was presumed to be the GEL COAT. Perkin-Elmer TGS-2. (From Cassel, 1977a.)

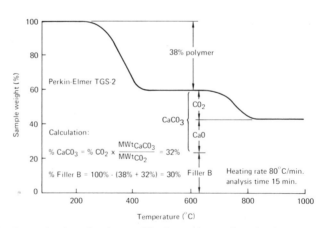

Fig. 58   Determination of carbonate filler in a polyester from its thermal conversion to the oxide. Perkin-Elmer TGS-2. (From Cassel, 1976.)

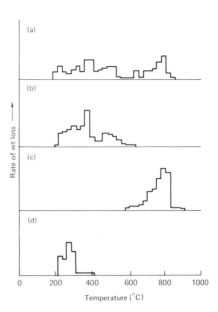

Fig. 59 DTG curves of cured bisphenol-A epoxy filled with (a) $CaCO_3$ and (b) $Al(OH)_3$. Curves (c) and (d) are $CaCO_3$ and $Al(OH)_3$ alone, respectively. (From Manley, 1974a,b, by courtesy of Marcel Dekker, Inc.)

utility of DTG in qualitative identification of fillers, in this case hydrated alumina, widely used for improvement in tracking properties and calcium carbonate, employed where castings have to `be machined (Manley, 1974a,b). Other common fillers, such as silica and asbestos, can also be identified by thermal analysis (Manley, 1967). Measurement of the very small amount of polymer on a polymer-coated wire is exhibited in Fig. 38, suggesting potential for reasonable accuracy in measurement of particulate coatings and paint films. Figure 60 shows the measurement of filler content and relative thermal stability of a series of filled molding compounds. This information was used as an identification tool for various encapsulating compounds. It is clear that TG is not only a quantitative analysis for filler content, but that TG–DTG is a potential qualitative identification of both polymer and filler.

In both qualitative and quantitative thermogravimetry, it is often necessary to identify, and sometimes quantify, the volative products. This is especially true when decomposition reactions overlap. Manley (1974a,b) showed, by evolved gas analysis, that volatilization of absorbed water and formaldehyde accompanied the loss of water of condensation during cure of a melamine formaldehyde resin. Greenberg and Kamel (1977) analyzed the volatile products formed during isothermal anhydride formation in poly(acrylic acid) by mass spectroscopy. Analysis of the effluent from the TG experiment showed the presence of 8% CO_2 (attributed to decar-

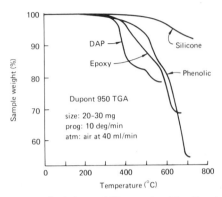

Fig. 60 TG measurement of relative stability, used as identification tool for encapsulating compounds. Filler content is also measured. DuPont 950 TGA. (From Thomas, 1974.)

boxylation) in the water of condensation. Pierron and Bobos (1977) coupled a gas chromatograph with a thermobalance (TG–GC) for identification of decomposition products of encapsulant molding compounds at processing temperatures. They were interested in determining the effect of these contaminants on the extended behavior of integrated circuit devices. TG–GC also permitted rapid evaluation of thermal stability and identification of the various constituents in the compounds. Lum and Feinstein (1980, 1981) did similar studies on novolac–epoxy encapsulant molding compounds by TG, DSC, and dynamic pyrolytic mass spectrometry. The effect of a brominated flame retardant on a corrosion-related failure rate was investigated. Their data showed that the failures were probably due to the presence of chloride ion from the polymerization process and water generated by condensation reactions, and not to the presence of the flame retardant. Isothermal accelerated aging studies were performed by TG at high temperatures. But the presence of a degradation reaction in the same temperature region, present in both TG and DSC scans, precluded reliable extrapolation to operating temperatures.

b. Qualitative Analyses: DSC/DTA. Manley (1974a,b) reports that the DTA curves of common commercial epoxy resins are sufficiently distinctive to be a qualitative identification. Pierron and Bobos (1977) report that the presence of phenolic novolac, cresolic novolac, and brominated DGEBA can be identified from DSC peak exotherm temperatures of molding compounds in the 350–450°C temperature range. Experiments were conducted in an inert atmosphere at 50°C min⁻¹ heating rate. Manley (1974a,b) also reports that DTA can distinguish between the types of bonding present in a melamine–formaldehyde resin (Fig. 61). In this case,

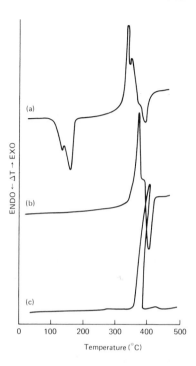

Fig. 61 DSC of melamine formaldehyde resin. (a) Curing in pressure DSC showing curring endotherms (suggestive of vaporization of water at high pressure) followed
by decomposition exotherm above 300°C.
(b) Conventional DSC of resin cured by
typical compression molding; higher decomposition temperatures suggest product is
different from that cured in pressure DSC
(see text). (c) Pressure DSC of same molded
resin; absence of endotherm following exotherm implies it may be due to volatilization of decomposition products. (From
Manley, 1974a,b, by courtesy of Marcel
Dekker, Inc.)

it was hypothesized that in commercial molding conditions, excess formaldehyde escapes, favoring formation of methylene bridges as the cross-
link units (Fig. 61b). However, when cured in a pressure DTA cell, the
contained formaldehyde may react with melamine, producing less stable
methylol and ether groups (Fig. 61a; see also Section V.C). These data
demonstrate that care must be exercised to ensure that the preferred
chemistry takes place in the thermal analysis experiment. In Fig. 61c the
pressure DTA experiment suggests that the endotherm following the decomposition exotherms may represent the volatilization of decomposition
products.

Decomposition profiles by TG and the presence or absence of distinctive low-temperature transitions by DMA or TBA may also be useful as
qualitative analysis tools.

C. STABILITY AND AGING

In this section two primary aspects of stability will be emphasized: the
analysis of aged or failed parts, primarily to determine cause or mechanism of failure, and experimentation allowing valid predictions of the life

of materials. Accelerated aging falls into the latter category. The first of these involves measurements such as the glass transition temperature or solvent swell. The objective is to ascertain why a material failed, for example, from inherent instability or improper curing. The ability to assess and predict stability falls into two main categories. The first is simple tests that rank several materials according to relative thermal stability. The second involves somewhat sophisticated analysis of decomposition data to obtain meaningful kinetic information, which may include combining evolved gas analysis with thermal methods.

For the simpler types of analyses the techniques have already been described, and the intention here is only to show a few specific applications. With regard to accelerated aging, this topic will not be treated in depth because current thermal analysis methods are not sufficiently reliable to replace conventional methods (Brown *et al.*, 1970; Krizanovsky and Mentlik, 1978). However, because thermoanalytical methods offer significant savings in time and cost relative to conventional methods, there is a considerable driving force to improve the techniques so that accurate predictions of life can be made with good reliability. The parameter-jump methods introduced by Flynn and Dickens (1976) and Flynn (1978b) combined with increasing capabilities of instruments, including microprocessors, to analyze data and interact with the experiment, offer promise in this area.

1. Analysis of Aged or Failed Parts

Where aging or failure is a problem, there is often an involvement with the cross-link density. As discussed earlier, failure often results from an originally undercured part, and analyses such as those in Fig. 36–42, 45, 46, 49, and 50 may help to ascertain this situation. Brett (1976) found a dramatic relationship between degree of cure (residual exotherm, e.g., Fig. 36) and bond performance for an epoxy resin adhesive. A transition in the failure mode from cohesive to adhesive was reported to occur at 85–90% cure for single lap adhesion tests. Using DSC, TMA, and IR, Brand (1977) found that the effect of moisture on the properties of a cured epoxy composite played a major role in intermittent electrical problems encountered on multilayer circuit boards. It was concluded that failure, traceable to opens in plated-through holes used for plug-in socket terminals, was the net result of high z-axis thermal expansion stresses acting on marginal plated-through holes. High thermal expansions were traceable to absorbed water. Solutions to the problem included drying of B-stage and C-stage materials prior to lamination and maintaining multilayer boards dry prior to soldering. Sykes *et al.* (1977) showed that moisture sorption had a significant effect on T_g and rigidity (TBA) (Fig. 44) and a major effect on the heat-deflection temperature (HDT TMA) (Fig. 48), of a commercial graphite–epoxy composite. Such composites are being considered for use as primary structural components in aircraft. Moisture sorption effects were shown to be largely, although not 100%, reversible. Desorption restored original values such as thermal expansion and T_g. Sykes *et al.*, however, showed that repeated sorption–desorption cycles resulted in lower HDTs after the second sorption.

Also, the gain or loss of material (e.g., moisture or antioxidant) can cause detrimental chemical or physical changes. In filled systems it is possible to lose polymer preferentially resulting in a relative increase in filler content. Where it is also possible to lose additives, such as plasticizers or antioxidants, solvent swell or decomposition profile and/or products becomes an indicator of the change.

For properly cured, highly cross-linked thermosets, aging is often accompanied by a change in cross-link density. The glass transition temperature has been shown to be a sensitive measure of degree of cure, i.e., cross-link density. In Fig. 62, T_g is also shown to be a sensitive monitor of

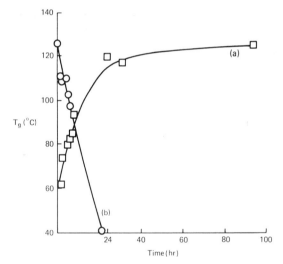

Fig. 62   Glass transition temperatures of a bisphenol-A epoxy cured with phthalic anhydride. (a) After curing at 160°C for indicated times, showing increase in T_g from increasing cross-link density. (b) After heating at 220°C for indicated times; T_g monitors degradation, presumably accompanied by decrease in cross-link density. (From Manley, 1974a,b, by courtesy of Marcel Dekker, Inc.)

degradation, again, presumably, as an indicator of change in cross-link density. T_g trends of Fig. 7 show that degradation can accompany cure at very high cure temperatures. Pierron and Bobos (1977) found that

the magnitude of the decomposition exotherm of epoxy molding compounds diminished in an orderly fashion with time at temperatures between 175 and 245°C. Jones *et al.* (1974) made similar observations for
a boron–epoxy composite. By analogy with isothermal method 2, such
data are a potential measure of the extent of degradation and may have
predictive utility (e.g., life). A useful corollary would be provided by
measurement of commensurate changes in sample weight (e.g., by TG).

Deterioration of properties of thermosets may also occur as a result of
physical aging. When a polymer is rapidly cooled to a temperature below
its glass transition temperature, during processing for example, it is not in
thermodynamic equilibrium. This results from the inability of the polymer
chains to attain their equilibrium conformation due to the rapid decrease
in mobility as the glass transition temperature is approached. One overall
result of such a quenching process is that there is an excess free volume
trapped into the glassy state which is manifested as a lowering of its density. Another result is the trapping of excess enthalpy. The so-called
physical aging phenomenon is the time-dependent approach toward equilibrium and it is typically viewed as a recovery (e.g., volume recovery,
densification) or relaxation (enthalpy relaxation) phenomenon. Enthalpy
relaxation can easily be measured by DSC (Ophir *et al.*, 1978; Kong *et al.*,
1979), for example by superimposing the first and second DSC scans for a
sample.

Associated with physical aging are significant effects on mechanical
and dielectric properties. Changes corresponding to measured enthalpy
relaxation in epoxies include decreases in strain to break and level of
stress relaxation (Ophir *et al.*, 1978) and decrease in solvent uptake (Kong
et al., 1979).

Failure mechanisms have also been investigated with the aid of thermal analysis. Huneke and Dorey (1980) used DSC and TBA to study electrical failure mechanisms in printed wiring boards. Failure occurred in the
presence of heat and high humidity and was attributed to loss of adhesion
between the glass and epoxy. Silane coupling agents were shown to be
effective in reducing the failure rate and also the amount of water absorption. A lowering of both the glass transition temperature and degree of
enthalpy relaxation on aging below T_g by the coupling agents was interpreted as evidence for a tougher, more homogeneous interface between
glass and resin, and one that retards polymer densification during aging.

2. Assessment of Stability and Life

a. Relative Thermal Stability: TG, DSC. Comparison of decomposition profiles, such as those in Fig. 60 (Thomas, 1974), is a common means

of classifying materials according to their *relative* thermal stability. Based on these data, diallyl phthalate appears to be the least stable, silicone the most stable, and epoxy and phenolic of comparable intermediate stability. By TG and DTA in air Antal and Csillag (1975) studied the relative thermal stabilities of epoxy resins in which the epoxide and the hardener were varied. Their results indicated greater thermal stability of anhydride-cured relative to amine-cured resins. The degradation processes differed, the amine-cured resins exhibiting entirely exothermic behavior whereas the anhydride-cured systems showed an initial exothermic reaction followed by a series of endothermic processes. Diglycidyl ether resins were observed to be more thermally stable than diglycidyl ester resins.

Several studies have been reported on the thermal stability of thermoset resins and resin composites in relation to their fire and burning properties. See also Chapter 8. Fire-resistant materials are characterized by high anaerobic thermal stability (TG in N_2), high char yield (residue at 700–800°C from N_2 TG), and high oxygen index. Increased char formation limits the production of combustible and/or toxic gases, decreases the heat produced, and lowers smoke emission. One series of studies (Pearce *et al.*, 1978; Lin and Pearce, 1979a,b; Arada *et al.*, 1979) has focused on the search for inherently flame-retardant epoxy thermosets. Copolymer structure, as determined by the epoxy resins, curing agents and degree of cure, was shown to have a significant effect. Similar studies by Kourtides and Parker (1979) and Kourtides *et al.* (1979) were directed at materials for aircraft interiors. Kubin's (1979) work addressed the ability of composites for use as aircraft structural components to survive a fire. TG was used to estimate activation energies (dynamic method A) that ranged between 20 and 25 kcal mole^{-1}. Exothermic heats of reaction by DSC were 18–20 cal g^{-1}. Because of the exothermic nature of the decompositions, it was concluded that effective fire fighting would have to reduce bulk temperatures well below 300°C to prevent the fire from continuing or regenerating. The work of Wentworth *et al.* (1979, 1980) and Wentworth (1981) was directed at concern over the accidental release to the environment of conductive graphite and carbon fibers that could occur as the result of fire. The primary concern was for unprotected electrical circuits. By dynamic and isothermal TG in air, they measured the relative thermal stabilities of several resin–fiber systems. The "worst-case" system consisted of an unstable resin (epoxy) but very stable fibers that could survive a fire and be released. This situation is aggravated by the most stable, highly graphitized fibers also being the most conductive. The "best-case" system contained the most stable resin, a polyquinoxaline. Systems having intermediate potential for fiber release contained either resins of intermediate stability, such as polyimides, or fibers of poor thermal stability. Serafino *et al.* (1979) addressed the problem of release of carbon fibers by utilizing bis(imide–amine) curing agents to increase the char yield by 50–100%.

Absolute classification of thermal stability is difficult because of the myriad of chemical and physical processes that can complicate the *overall* kinetics to be measured (Flynn and Dickens, 1976). Decomposition mechanisms, especially oxidative degradations, are often diffusion controlled so that sample geometry and fillers can alter the decomposition process. As discussed by Wendlandt (1974), reactions occurring at elevated temperatures in a thermobalance may not be the same as those occurring at lower temperatures and over longer time periods in real situations. Also, relative stabilities in an inert atmosphere may differ from those in air. Rather than being an absolute measure, this type of information should be used as a guide for further studies on the decomposition mechanisms. In all instances care must be taken when ascertaining relative stability to ensure as much as possible identical experimental conditions, including sample size and geometry, atmosphere, flow rate, and instrument configuration. Parameter-jump methods introduced by Flynn and Dickens (1976) and Flynn (1978b) can eliminate these variables for a particular sample, where, for example, temperature can be jumped between two or more levels in the same experiment.

b. Kinetic Analyses and Lifetime Predictions: TG, DSC. Life predictions require an experimentally determined relationship between time to failure and appropriate variable(s) that force the failure (e.g., temperature or environment). The most common accelerating variable is temperature, which is related to time by the activation energy. For the accelerated test to be valid, the same activation energy must control both the accelerated test and the failure process. Thus valid life predictions require, first, measurement of the correct relationship between time and a forcing variable (activation energy for temperature), and, second, identification of a failure time from the ensemble of times. Neither of these criteria is simple to satisfy, and at the current state of the art the success of thermoanalytical techniques has been marginal; for example, they have not been successful at replacing the more conventional, but also more costly, methods (Brown *et al.,* 1970). As stated previously, success in this area requires both better experimental techniques, such as the parameter-jump method, and better analytical capabilities, such as those provided by microprocessors. Following is an example that demonstrates the current state of the art.

Brown *et al.* (1970) employed DSC and isothermal and dynamic TG methods for estimation of the thermal life of magnet wires. The thermal life rating, the extrapolated temperature at which electrical failure occurs in 20,000 h, according to ASTM D2307, was used as a reference for these techniques. All thermal analysis measurements were performed in dry oxygen. In dynamic TG a correlation was observed between the temperature at 5% weight loss and the ASTM thermal life ratings. In DSC the extrapolated onset of the first major decomposition peak was found to have a similar correlation with thermal life ratings. Isothermal rates of weight loss were measured at 5% weight loss between 150 and 350°C. Assuming zero-order kinetics allowed calculation of activation energies; extrapolation of Arrhenius plots to rates of approximately 10^{-7} mg mg^{-1} min^{-1} yielded temperatures close to the ASTM temperatures for the majority of samples. In all cases two distinct correlations were observed, one for low-temperature-rating enamels (primarily aliphatic), and the other for high-temperature-rating enamels (mostly aromatic). As the authors point out, such correlations are useful for predicting probable values of thermal life ratings for new wire enamels but, because of exceptions to the correlations that are not understood, they cannot at this time supplant the ASTM procedure. The recent literature (Krizanovsky and Mentlik, 1978) suggests that the conclusions of Brown *et al.* are still current. To provide good predictive utility to the TG and DSC methods requires elucidation of failure mechanisms. Polymer transitions, interactions between substrate surface and polymer, and the influence of metals and metal oxides on decomposition need to be investigated.

III. Effects of Catalysts, Fillers, Water, and Other Constituents

In commercial applications thermosets are rarely used without the incorporation of some other material. As the names suggest, catalysts and accelerators are employed to increase the speed of the reaction. Hardeners, such as diamines in epoxies, become incorporated in the polymer chain; thus the relative amount of hardener to resin can affect not only the reaction rate, but also the ultimate properties of the thermoset, such as $T_{g\infty}$ and dynamic mechanical properties. Fillers are used both to enhance or provide physical properties, such as modulus, thermal expansion, thermal and electrical conductivity, and magnetic recording properties, and to reduce cost. Additives are commonly employed to enhance the thermal stability or burning characteristics of materials. Fillers and additives alike can have an effect, sometimes detrimental and sometimes beneficial, on the ability of a thermoset to cure properly. Solvents that are part of the process and water that is an unavoidable constituent of the environment can become involved in the chemistry of curing as well as plasticizing the cured thermoset. This section focuses on the application of thermal analysis in these areas.

A. CURE

Since the vast majority of commercial thermosets are used in conjunction with other materials (fillers, decorative papers, stabilizing additives), it is imperative to know or be able to measure their effects on curing. Variations in the amount of catalyst can affect both the rate of cure and ultimate extent of cure. Residual solvents and absorbed moisture can have a similar effect.

To define quantitatively the effect of catalysts, fillers, etc., on curing requires measurement of the heat of reaction ΔH_{RXN} or weight loss of reaction ΔW_{RXN}, activation energy E, reaction order n, and rate constant k. Decreases in ΔH_{RXN} greater than expected from the additive content reflect on the ability of the "composite system" to cure completely. A change in k shows a change in the reaction rate, a change in n reflects a change in mechanism, and a change in E may indicate a catalytic effect. For fillers, additives, moisture, etc., measurement is made relative to the "pure" or neat thermosetting system. ΔH_{RXN} and E can easily be determined from three dynamic DSC scans at differing heating rates between 2 and 20°C min^{-1} (dynamic method B, see Section IV.B). Measurement of changes in the rate constant and reaction order requires the appropriate isothermal method and rate equation as discussed in Section II. However, an estimation of the rate may be obtained from changes in the DSC peak exotherm temperature.

Willard (1974a,b) used DSC to evaluate the effects of peroxide initiators and fillers on the heats of reaction and reaction rate constants of a single diallyl phthalate (DAP) prepolymer. The catalysts are described in Table XVII. Rate constants were calculated from the dynamic DSC scans by dynamic method A (see Section IV.A). Earlier work (Willard, 1973) (see Fig. 69) showed good agreement between rate constants obtained in this manner and isothermal data by infrared measurements. The data of Fig. 63a show the rate constant to be sensitive to both catalyst type and level. At three parts per hundred (phr) t-butyl perbenzoate (TBP), the reaction rate was twice that at the 1 phr level. Even more dramatic, the reaction catalyzed by TBP proceeded approximately 50 times faster than that catalyzed by t-butyl hydroperoxide (TBHP), both catalysts at the 3

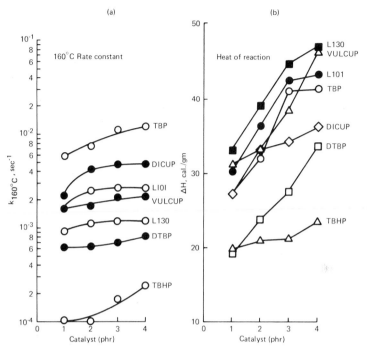

Fig. 63 DSC data showing effect of catalyst type and level on curing of DAP pre-polymer. (a) Reaction rate via dynamic method A, Eq. (45). (b) Heat of reaction. See Table XVII for description of catalysts. DuPont 900 DSC. (From Willard, 1974a,b, by courtesy of Marcel Dekker, Inc.)

phr level. Effects of initiator type and level on heat of reaction (proportional to extent of cure) are shown in Fig. 63b. The steep concentration dependence on heat of reaction can be important to the manufacturer of commercial molding compounds since minor variations in catalyst concentration may result in significant variation in degree of cure of the finished product. For commercial applications, the combination of high heat of reaction and fast curing rate is desirable; *t*-butyl perbenzoate (TBP) and dicumyl peroxide (Di-Cup) are in general usage at the 3 phr level. Similar behavior was observed by Cernee *et al.* (1977) for the free radical polymerization of diallyl fumarate and related compounds. Kubota (1975) studied the effects of reactive diluent, pressure, low-profile agent, and filler on the curing of a polyester resin by pressure DSC. Belani and Spork (1978) characterized catalyst contents in semiconductor device molding compounds by isothermal DSC.

Widmann (1975b) found by DSC isothermal method 1 that diminishing catalyst levels led to reduced reaction rates and higher activation energies for an epoxy–anyhdride system catalyzed with a tertiary amine. Dutta and Ryan (1979) studied the effects of two fillers on the cure kinetics of the mPDA–DGEBA epoxy system, again by DSC isothermal method 1. They evaluated the effects of a carbon black filler and a silica filler on ΔH_{RXN}, activation energies, rate constants, and reaction orders. Equation (13) was utilized, but the reaction orders were considered to be variables, and expressions for the filler content were included. These kinetic expressions were found to yield results that were in good agreement with the experimental data for the two fillers. Carbon black was found to have no effect on the heat of reaction or activation energies and only a small effect on reaction rates. The silica filler caused an increase in ΔH_{RXN} of $\sim 13\%$ and a reduction in the activation energy for the autocatalyzed reaction [k_1 in Eq. (13)] from 15.4 to 14.4 kcal mole^{-1}.

Miller and Oebser (1980), using a version of DSC isothermal method 1, showed that the activation energy of the uncatalyzed reaction of DGEBA epoxy resins with various curing agents depends only on the functionality of the curing agent. See Table IX. They also showed that accelerators

TABLE IX

SUMMARY OF ACTIVATION ENERGIES
FOR CURE OF DGEBA RESINS
WITH
VARIOUS CLASSES OF CURING AGENTS

Curing agent	E (kcal mole^{-1})
Primary and secondary amines	11–14
Aromatic	11–12
Aliphatic	12–14
Tertiary amines	26
Anhydrides	12–15
Dicy	15–17
Imidazoles	19–22
Alcohols	14–15

[a] From Miller and Oebser (1980).

used in dicy-cured* systems can be divided into two types. Co-curing agents, such as acetoguanidine and the imidazoles, shorten cure times but do not affect the activation energy. Catalysts that promote ring opening shorten the cure time and substantially lower the activation energy—for example, alcohols to 11–12 kcal mole^{-1}, chlorophenyl ureas to 7 kcal

* Dicyandiamide; crystalline particles dispered in epoxide resin to give one-component epoxy system.

mole^{-1}, and Shell D type catalysts to 8–9 kcal mole^{-1}. Activation energies for the uncatalyzed reaction were 15–17 kcal mole^{-1}.

Figure 64 shows the effects of four types of commonly used DAP fill-

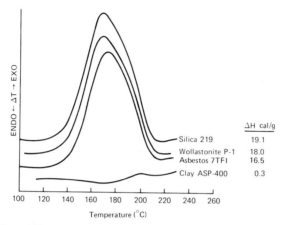

Fig. 64   Effect of fillers (50 wt. %) on the curing of a DAP prepolymer catalyzed by 3 phr dicumyl peroxide. Heat of reaction of 34 cal g^{-1} measured for unfilled system. Acidic clay surface postulated to have decomposed the initiator. DuPont 900 DSC. (From Willard, 1974a,b, by courtesy of Marcel Dekker, Inc.)

ers on the heat of reaction of the dicumyl peroxide–initiated reaction. Three of the fillers show little effect to slight enhancement of the heat of reaction, whereas the kaolinite clay virtually eliminates curing. The latter effect was attributed to decomposition of the initiator by the acidic clay surface. Similar but less dramatic effects were observed for the *t*-butyl perbenzoate initiated cures.

Kaelble and Cirlin (1971) employed DSC in conjunction with Instron thermomechanical analysis to show that an aluminum flake filler in an epoxy–phenolic adhesive was inert. Apparent activation energies, calculated from the variation in peak reaction temperatures with heating rate, dynamic method B, Section IV.B, were identical for the filled and unfilled systems. The heats of reaction for the two systems were in proportion to the filler loading. At fast heating rates a cavitation process in the unfilled adhesive was observed just above the boiling point of water (Fig. 65). This phenomenon, which results in a void structure in the adhesive, was not observed in the metal-filled adhesive. In addition to reducing void formation, the metal filler provided higher modulus and better fracture toughness in the cured adhesive.

The curing chemistry of commercially important dicy–epoxy thermosets is very complex (see Section V.A); and water, dicy particle size, and

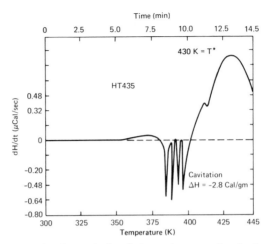

Fig. 65 DSC curve showing cavitation during early stages of curing in an epoxy–phenolic adhesive. Cavitation, which begins immediately above the boiling point of water, results in a cavity and void structure in the adhesive. Perkin-Elmer DSC-1b. (From Kaelble and Cirlin, 1971.)

epoxide properties have been shown to play important roles. Using DSC, Brand (1977) showed that curing B-staged samples in a saturated water environment at $\sim 150°C$ caused an irreversible lowering of $T_{g,ult}$ by approximately $10°C$ relative to dry samples cured in ambient air. Samples cured in a trichloroethylene atmosphere exhibited an initially very low T_g, but removal of the solvent raised T_g to the level of samples cured in ambient. Based on this information, it was concluded that absorbed water altered the cure mechanism and that B-stage and C-stage materials should be dried prior to circuit board lamination. Noll (1977) also studied effects of moisture on the curing of B-stage composites; he used DSC to observe both T_g and ΔH of curing and correlated these results with gel time and peel strength measurements. Peel strength was taken as the ultimate test of B-stage quality. Noll observed the same effect as Brand for samples exposed to high humidity prior to curing. But his further results suggested that there were more complex effects of water on curing and that some optimal water content greater than zero exists for achieving maximum adhesion. In fact, his data suggest that commonly practiced drying procedures prior to lamination could lead to reduced peel strengths. Sacher (1973) observed effects of dicy particle size, surfactant, and epoxide molecular weight on curing by DSC and TMA. It was found that fine particles ($< 125 \ \mu m$) impart longer shelf life and higher activation energy of curing than coarse particles; that surfactant allows the reaction to proceed at lower temperatures but also at diminished rates and with an apparent change in mechanism; and that with a higher-molecular-weight epoxide, reaction subsequent to the primary cure resulted in a more rigid epoxy than expected. Schneider and Gillham (1978) and Schneider *et al.* (1979) employed DSC and TBA in studying curing of two commercial dicy resins. They found that cure of one system, containing difunctional epoxide (DGEBA) and an approximate stoichiometric amount of amine hydrogens, was simple in that curing exhibited one DSC exotherm (5°C min^{-1} in N$_2$, Fig. 6) and additional curing beyond that recommended did not significantly alter the thermoset properties (T_g). Curing of the other system, which contained a tetrafunctional epoxide that was present in excess relative to dicy, was complex. The dynamic DSC trace showed three exothermic peaks, and TBA studies showed that additional curing beyond that recommended significantly increased T_g from 138 to 228°C. These results implied that at least the first two DSC exotherms contribute to curing, and it was hypothesized that etherification of epoxide groups [reaction (59)] is part of the overall cure. Lunak *et al.* (1978) investigated the curing of an epoxide resin with derivatives of dicy.

Bank *et al.* (1972, 1973) used DSC to demonstrate that microencapsulation of epoxy curing agents does not significantly alter the curing process. The shapes of curing curves were similar, with encapsulation of the curing agents shifting peak temperatures higher by 3–10°C. Microencapsulation allows mixing of resin and curing agent without immediate initiation of curing. Heating or rupture of the capsule by pressure releases the curing agent. Kaelble *et al.* (1974) employed similar DSC techniques to show that differing surface treatments of a graphite fiber did not substantially alter the curing kinetics or extent of reaction of an epoxy resin.

Anagnostou and Jaul (1981) used DSC to measure unblocking temperatures of three blocking agents used in the formulation of stable one-part polyurethane systems. The unblocking reactions were found to be endothermic. On addition of a polyol containing only primary hydroxyl groups, only exothermic behavior, attributed to reaction of unblocked isocyanate with the hydroxyl groups, was observed. In each case the exotherm occurred at a lower temperature than the original unblocking reaction. The addition of several pigments to the blocked isocyanate polyol mixture had no effect on one blocking agent, raised the unblocking temperature of another by 10°C, and lowered the unblocking temperature of the third by approximately 20°C.

Sebenik *et al.* (1974) investigated the complex reaction between phenol and formaldehyde in the presence of sodium hydroxide by DSC (sealed pan) and IR. They observed two dominant reactions: the second-order addition of formaldehyde to phenol with predominant formation of *o*-hydroxymethyl phenol and subsequent first-order condensation of the latter with the formation of methylene bridges (see Section V.B for chemistry). At a heating rate of 4 K min^{-1} peak temperatures for the latter reaction were 428–440 K (155–167°C). Increasing the catalyst concentration from 0.25 to 1.0% produced an orderly increase in reaction rate and decrease in activation energy; no effect on reaction order was observed. King *et al.* (1974) studied the effects of formaldehyde–phenol (F–P) ratio and catalyst type on the curing of phenolic resole resins by DSC (sealed pan), GPC, and NMR. Their work showed that sodium hydroxide as catalyst favors methylene bridge formation [cf. reactions (65) and (68)] whereas triethylamine favors methylene ether bridges [cf. reaction (69)]. It was important that the catalyst was present during cure and not just during synthesis of resins. Ether formation became relatively more probable with increasing methylol content of the resin (increasing FP). At a heating rate of 8°C min^{-1}, peak exotherm temperatures for methylene bridge formation were ~ 155°C and for methylene ether bridges ~ 185°C. Exothermic peaks at ~ 220°C have been attributed to conversion of methylene ether bridges to methylene bridges and formaldelyde (Kurachenkov and Igonin, 1971).

B. STABILITY

Antal and Csillag (1975) showed by TG and DTA in air that anhydride-cured epoxide resins were more stable than their amine-cured counterparts. And the degradation processes were different; the amine-cured resins exhibited entirely exothermic behavior whereas the anhydride-cured materials showed an initial exothermic reaction followed by a series of endothermic processes.

To improve their fire and safety performance, thermosets often contain fire and flame retardants. These additives prevent the open flaming of a material as well as encouraging the decomposition of the polymer in such a way as to prolong its physical integrity. TG can be a useful tool to study the effects of flame retardants, as illustrated in Fig. 66. This illustration compares the thermogravimetry in air and oxygen of an epoxy molding compound, first without and then with flame retardant. Comparison of the air versus oxygen environments is a good indicator of where burning begins; the considerable heat evolution in oxygen causes the sample temperature, monitored on the x axis, to leap ahead of the program temperature, closely represented by the air trace. The sample without retardant begins to burn near 420°C, where the sample with retardant does not start to burn until temperatures above 450°C are reached. It is interesting that in air the sample with flame retardant begins to decompose at a lower temperature, but does so at a more protracted rate. Weight loss of the sample without flame retardant accelerates rap-

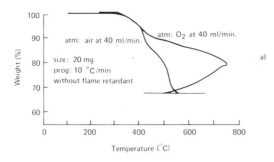

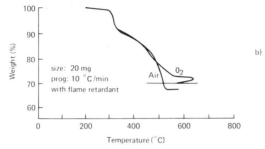

Fig. 66 Thermogravimetry in air and oxygen of epoxy molding compound (a) without and (b) with flame retardant. Compound without retardant begins to burn at ~420°C (in O_2 temperature leaps ahead of program). Compound with retardant starts to burn above 450°C and burning is less eventful. DuPont 950 TGA. (From Thomas, 1974.)

idly as decomposition progresses. Wentworth *et al.* (1979) showed by TG in air that the presence of graphite fibers had no influence on the decomposition profile of an epoxy resin.

Pierron and Bobos (1977) studied the effects of flame retardants on uncured epoxy molding compounds in an inert environment by TG and TG–GC. They found an appreciable *decrease* in the initial decomposition temperature (1% weight loss) when flame retardancy was induced by brominated DGEBA alone. Optimum performance, in terms of both decomposition temperatures and number of decomposition products, was achieved when the flame retardant system consisted of a brominated resin plus antimony oxide. Lum and Feinstein (1980, 1981) studied the effect of a brominated flame retardant in epoxy encapsulant molding compounds on a corrosion-related failure rate. TG, DSC, and pyrolytic mass spectrometry data suggested that the failure was probably due to the presence of chloride ion from the resin polymerization and water generated by condensation reactions, and not to the presence of the flame retardant.

Silane coupling agents have been shown to affect the stability of printed wiring boards (Huneke and Dorey, 1980). DSC and TBA were used to investigate an electrical failure process attributed to loss of adhesion between glass fibers and epoxy resin in high temperature and high humidity environments. The effect of the coupling agents was to decrease the water sorption and lower both the glass transition temperature and degree of enthalpy relaxation on aging below T_g. The data were interpreted as evidence of a tougher, more homogeneous interface between glass and resin, and one that retards polymer densification during aging.

C. THERMAL EXPANSION

The proper interpretation and use of thermoset thermal expansion data requires some background in the thermal expansion of filled materials. The following discussion presumes that the thermoset is stable during the measurement. It is assumed that the polymer is fully cured and that no postcuring occurs, that no transitions occur, no enthalpy or volume relaxations occur, and that no weight changes occur. The polymer is assumed to be below the glass transition temperature, although the equations are valid for temperatures above T_g. Thermal expansion above T_g has been found to be particle size dependent (Pinheiro and Rosenberg, 1980), with the expansion rate decreasing with decreasing particle size. The following equations have been found to be *representative* of thermosets, when the resin matrix is bonded to the filler. For more detailed discussions the reader is directed to Schapery (1968), Holliday and Robinson (1973), and Ashton et al. (1969).

First, there is a universal relationship between cubical (γ) and linear (α) expansion that holds for all materials:

$$\gamma = \alpha_x + \alpha_y + \alpha_z \tag{38}$$

See Table X for explanation of terms. Several equations exist that relate cubical expansion to filler and matrix properties (Holliday and Robinson, 1973). It is the opinion of the author that the most representative equation for fiber- and particle-filled thermosets is the equation of Kerner (1956):

$$\gamma = \gamma_f v_f + \gamma_m v_m - (\gamma_m - \gamma_f) v_m v_f \theta \tag{39}$$

where

$$\theta = \frac{(1/K_m) - (1/K_f)}{(v_m/K_m) + (v_f/K_f) + (3/4G_m)} \tag{40}$$

At 40–60 vol % filler, γ calculated from Eq. (39) will be 75–90% of the mean value [rule of mixtures, first two terms of Eq. (39)]. For example, for 50 vol % E glass in a typical epoxy matrix, Eq. (39) yields a value of $\gamma \sim 84 \times 10^{-6}$ in.^{3} in.$^{-3}$ °C^{-1}, the rule of mixtures gives 98×10^{-6} in.^{3} in.$^{-3}$ °C^{-1}.

Linear thermal expansion of thermosets filled with a uniform distribution of *spherical particles* is isotropic:

$$\alpha_x = \alpha_y = \alpha_z = \gamma/3 \tag{41}$$

An epoxy filled with 50 vol % glass spheres should have a linear thermal expansion of $\sim 30 \times 10^{-6}$ in. in.$^{-1}$ °C^{-1}.

For a unidirectional fiber-reinforced composite, thermal expansion parallel to the fibers is given by α_{11} and perpendicular to the fibers by α_{22} (Table X). For this case the relation between volumetric and linear expansion is

$$\gamma = \alpha_{11} + 2\alpha_{22} \tag{42}$$

For a two-dimensional, fiber-reinforced isotropic composite, the in-plane linear expansion is given by

$$\alpha_{\text{ISO}} = \alpha_x = \alpha_y = \frac{1}{2}\left[\alpha_{11} + \alpha_{22} + \frac{(E_{11} - E_{22})(\alpha_{11} - \alpha_{22})}{E_{11} + (1 + 2\nu_{12})(E_{22})}\right]$$

$$\approx \alpha_f + \frac{5v_m E_m \alpha_m}{v_f E_f} \tag{43}$$

TABLE X

GLOSSARY OF THERMAL EXPANSION TERMS

$f \rightarrow$ (filler) $m \rightarrow$ matrix
γ = coefficient of cubical thermal expansion
α = coefficient of linear thermal expansion
$z \rightarrow$ perpendicular to fibers, x and $y \rightarrow$ parallel to fibers
K = bulk modulus = $E/3(1 - 2\nu)$
G = shear modulus = $E/2(1 + \nu)$
E = Young's modulus (tensile modulus)
ν = Poisson's ratio
v = volume fraction
v/o = volume percent, w/o = weight percent
11 $\rightarrow$ parallel to fiber direction in unidirectional layer
22 $\rightarrow$ normal to fiber direction in unidirectional layer

The exact equation is from Ashton *et al.* (1969). The approximation (for "back-of-the-envelope" calculations, good to $\sim 25\%$) has been observed by this author and serves as an indicator of the relative influence of the various fiber and matrix properties. For a pseudoisotropic composite containing 50 vol % E glass in epoxy, Eq. (43) yields an exact value of $\alpha_{\text{ISO}} = 16.4 \times 10^{-6}$ in. in.$^{-1}$ °C^{-1}, identical to copper, and an approximate value of 20×10^{-6} in. in.$^{-1}$ °C^{-1}. Because the fibers in a fabric are not in a straight line, an increase in α_{ISO} of some 10–30% is to be expected (Halpin, 1970). In composites reinforced with short fibers with aspect ratios less than about 250, α_{ISO} will increase in a predictable manner (Halpin and Pagano, 1969). Linear thermal expansion perpendicular to the fiber plane is simply

$$\alpha_z = \gamma - \alpha_x - \alpha_y \tag{44}$$

For the previous E-glass pseudoisotropic composite, $\alpha_z = 52 \times 10^{-6}$ in/in-°C is calculated.

Angeloni (1967) measured linear coefficients of expansion of pultrusion-formed glass-reinforced polyester rods by TMA. Measurements were made in the longitudinal direction ($\alpha_{11} = 5.5 \times 10^{-6}$ in. in.$^{-1}$ °C^{-1}). An independent dilatometric measurement gave $\alpha_{11} = 6.0 \times 10^{-6}$ in. in.$^{-1}$ °C. By reference to Table X and substitution of appropriate numbers,* it can be shown that these values are reasonable at fiber loadings of 50–60 vol %.

The calculated γ from Eq. (42), 92×10^{-6} in.^{3} in.$^{-3}$ °C^{-1}, is quite reasonable. In a similar manner it can be shown that the thermal expansion data of Prime *et al.* (1974) (See Table VIII) is in agreement with the preceding discussion.

Water has been shown to alter significantly the thermal expansion behavior of thermosets. Brand (1977) showed that z-axis thermal expansion (α_{22}) increased by approximately 7×10^{-6} in. in.$^{-1}$ °C for every 0.1 w % water absorbed by 0.062 in. thick epoxy–glass laminates. However, the magnitude of these results must be viewed in light of the fast heating and cooling rate (160°C min^{-1}), inclusion of the glass transition, and lack of sample conditioning. The measurements did, however, provide a temperature range (0–215°C) and heating rate similar to soldering operations. It was concluded that the effects of water on thermal expansion, which were especially large above T_g when samples were heated rapidly, resulted in sufficient stresses during soldering to cause intermittent electrical problems (see Section II.C.1). Prime *et al.* (1974) studied the thermal expansion, below T_g, of very thin composites that equilibrated rapidly with the environment. It was concluded that specimens needed to be dry prior to measurement, but analysis of the data shows that expansion coefficients of "wet" specimens would have been lower than dry specimens, implying that hygroscopic contraction and thermal expansion were occurring simultaneously. Adamson (1979, 1980) showed that expansion coefficients of thicker epoxy resin specimens saturated with water were approximately twice that of the dry specimens; in these measurements no change in moisture concentration was allowed. Hygroscopic expansivi-

---

* $\alpha_f = 4.9 \times 10^{-6}$ in. in.$^{-}$ °C^{-1}, $E_f = 10 \times 10^6$ psi, $\nu_f = 0.2$, $\alpha_m = 60 - 80 \times 10^{-6}$ in. in.$^{-1}$ °C^{-1}, $E_m = 0.3$–0.6×10^6 psi, $\nu_m = 0.3$–0.4.

ties of the thin composites [Eq. (33), estimated from TMA data] were found to be of the same magnitude as thermal expansion. Both studies showed that effects of moisture sorption are reversible.

D. OTHER PHYSICAL PARAMETERS

Several other properties are affected by the presence or amount of catalyst, filler, and other constituents. The glass transition is one such property. Where the curing agent or hardener is a comonomer, in epoxies, for example, the ratio of hardener to resin has a significant effect on T_g, as shown in Table V. Knowles (1978) showed that TMA softening temperatures of epoxy and polyurethane encapsulating materials could increase or decrease when incorrect mix ratios were used. Another factor influencing T_g is the type and amount of filler loading. Depending on the surface properties of the filler, the glass transition relative to the unfilled thermoset may increase, decrease, or remain unchanged. Zicherman and Holsworth (1974) reported on a TiO_2-filled acrylic that showed an orderly decrease in T_g by DSC and TMA penetration with increasing filler content. The penetration rate, however, decreased with increased filler loading.

Droste and Dibenedetto (1969) (DSC) and Pinheiro and Rosenberg (1980) (thermal expansion) found that the glass transition temperature increased with increasing filler concentration and with increasing specific surface area (decreasing particle size) of the filler. It was suggested that the strongly bonding particles increase T_g by changing the nature of the polymer, not only at the particle surface but also for a considerable distance into the polymer itself.

Reed (1979) observed significant influences of silane-finished, glass fiber reinforcement on properties of unidirectional composites by DMA and TMA. Observed were broadening of the α-transition (T_g), an additional damping peak above T_g, and an additional softening temperature above that associated with T_g. It was suggested that these results may be indicative of a fiber–resin interfacial region distinct from the bulk resin.

As reported earlier, Kaeble and Cirlin (1971) observed that an aluminum flake filler in an epoxy–phenolic adhesive provided higher modulus and better fracture toughness than the unfilled adhesive without affecting the cure. In a similar vein, Bank *et al.* (1972, 1973) found that bond strengths for an epoxy cured with a microencapsulated curing agent were substantially higher than the identical system cured with pure curing agent. The authors suggested that the capsules might act in a unique manner as an active filler because of the ability of the curing agent and/or resin to permeate the capsule wall.

Moisture has been shown to act as a plasticizer in cured thermosets, reducing the modulus and glass transition temperature, as well as affecting the thermal expansion coefficient. Both Brand (1977), utilizing DSC and TMA in the expansion mode, and Sykes *et al.* (1977), employing TBA and TMA in the heat-deflection-temperature (HDT) mode, and Lauver (1978) using TMA in the penetration mode showed consistent lowering of T_g with increasing water sorption (see Fig. 44 and 48). Sykes *et al.* also showed a diminishing rigidity of the composite with increasing water sorption (Fig. 44). The major drop in HDT is obviously the combined result of lower T_g and lower modulus. McKague *et al.* (1973) also used TMA in the heat-deflection mode to study effects of humidity exposure on an epoxy–graphite composite and epoxy and phenolic resins. They observed that, in addition to diminishing rigidity, water sorption induced significant creep at room temperature. Although effects of moisture sorption were mostly reversible, Sykes *et al.* showed that, starting with dry composite samples, repeated sorption–desorption cycles resulted in permanently lower HDTs after the second sorption leg. Each leg of the sorption–desorption cycle was 14 days in duration. Other physical properties influenced by catalysts, fillers, and additives amenable to study by thermal analysis methods include modulus (DMA, TBA, TMA flexure), hardness (TMA indentation), and moisture pickup and retention (TG).

IV. Cure Kinetics from Dynamic Experiments

The means to extract kinetic information from dynamic experiments can be divided into three categories. The first (method A) involves analysis of only one exotherm. In theory, a dynamic DSC (also DTA and TG–DTG) trace contains all the kinetic information normally embodied in a series of isothermal experiments, which makes it highly attractive. In practice, these techniques have been successful only for some reactions, especially those that are first order, i.e., $f(\alpha) = 1 - \alpha$. The one deficiency of this method is its lack of consistent accuracy in measuring kinetic parameters. The second technique (method B) capitalizes on the variation in peak exotherm temperature with heating rate. Methods of this type give an accurate measure of the activation energy and frequency factor for all reactions. Method B finds practical application both as a precursor to isothermal studies and when multiple exotherms or unresolvable baselines preclude application of the isothermal methods. This method is even more useful when followed by a check, e.g., measurement of an isothermal half-life (see next paragraph). The third method (method C) requires, as does method B, experiments at more than one heating rate and analyzes the dependence of the temperature to reach a constant conversion on heating rate; in actuality the second technique is a special case of this more general method. The third technique has been applied to autocatalyzed reactions, i.e., $f(\alpha) = \alpha(1 - \alpha)(B - \alpha)$, with moderate success. However, for the amount of effort required, it makes more sense to do three or more method 1 isothermal experiments that yield more information and more accurate results. Another potentially attractive approach, which has not been reported in the literature, is the merger of isothermal method 1 with temperature-jump methods (Flynn and Dickens, 1976; Flynn, 1978b). By jumping the temperature between three levels, the equivalent of three isothermal experiments can be done simultaneously.

The criteria for judging the dynamic experiment must be its ability to describe accurately the known and accepted isothermal course, i.e., the ability to predict degree of conversion at any given time and temperature. An accuracy of $\pm 10\%$ is reasonable to expect. Duswalt (1974) suggests that an isothermal method 2 experiment carried out at the calculated half-life conditions (time–temperature to 50% conversion)* is a reasonable check. For some reactions, the time–temperature to 90–95% conversion* is a better test (see Fig. 77, for example).

A. METHOD A: ANALYSIS OF ONE EXOTHERM

This method is very attractive because of the abundance of information contained, or potentially contained, in a single temperature programmed experiment. There is appeal to this method, to be able to extract from one experiment what would otherwise require three or more experiments. It works reasonably well, but not well enough to be consistently reliable when used to predict the course of a reaction over a wide time–temperature range. As will be shown, method A has yielded accurate kinetic parameters for some reactions, especially the first-order thermal decomposition of polymerization initiators. For the majority of reactions, however, this method consistently overestimates the activation energy and preexponential factor when compared to values obtained from isothermal experiments. For some reactions it has been shown that extrapolation to zero heating rate yields data that are in agreement with isothermal experiments, but this behavior is also inconsistent and eliminates the advantages of the method. The bottom line, in the opinion of this author, is that this method lacks the reliability necessary for routine use in measuring cure kinetics but has sufficient attractiveness and potential to warrant continued study.

Borchardt and Daniels (1956) were the first to describe the application of dynamic DTA and DSC to the study of reaction kinetics. They derived

---

* For first-order reactions, $t_{1/2} = \ln 2/k$, $t_{0.95} = \ln 20/k$; for nth-order reactions, $t_{1/2} = (2^{n-1} - 1)/k(n - 1)$, $t_{0.95} = (20^{n-1} - 1)/k(n - 1)$.

equations relating the shape of a DTA curve to the reaction kinetics giving rise to the curve; the equations were based on reactions occurring in stirred solutions where heat capacities and heat transfer coefficients of sample and reference are very nearly identical. They derived equivalent equations for the DSC experiment. Prime (1973) reviewed their DTA and DSC methods with several other dynamic methods. Assuming that the heat evolved in a small time interval is directly proportional to the number of moles reacting during that time, they arrived at Eq. (15) in Fig. 67. As

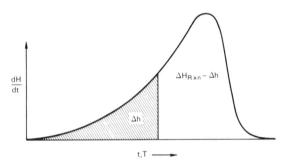

$$\frac{d\alpha}{dt} = \frac{1}{\Delta H_{Rxn}} \frac{dH}{dt} \tag{15}$$

$$\text{1st order: } k_1 = \frac{dH/dt}{\Delta H_{Rxn} - \Delta h} \tag{45}$$

$$\text{nth order: } k_n = \frac{(V\Delta H_{Rxn}/N_0)^{n-1}\, dH/dt}{(\Delta H_{Rxn} - \Delta h)^n} \tag{46}$$

$$\text{Autocatalyzed: } k_A = \frac{(V\Delta H_{Rxn}/N_0)^2\, dH/dt}{\Delta h\,(\Delta H_{Rxn} - \Delta h)\,(B\Delta H_{Rxn} - \Delta h)} \tag{47}$$

Fig. 67 Hypothetical dynamic DSC trace that permits determination of kinetic parameters via dynamic method A and the appropriate listed equations. V is volume, N_0 is number of moles originally available to react.

before, ΔH_{RXN}, ideally, is the total heat liberated when a totally uncured resin is taken to complete cure. Rate constants in terms of the DSC data are shown for first-order, nth-order, and autocatalyzed reactions. Using DTA, Borchardt and Daniels (1956) measured the first-order thermal decomposition of benzenediazonium chloride in aqueous solution at 1°C min^{-1} heating rate, and the pseudo first-order kinetics for the reaction of N,N-dimethylaniline with ethyl iodide using N,N-dimethylaniline as solvent at 2.3°C min.$^{-1}$ The DTA method was accurate for the first-order reaction (E, 5% high), but not for the pseudo-first-order reaction (E, 50% higher than isothermal value) (see Prime, 1973). Barrett (1967) measured the kinetics and heats of reaction for the first-order thermal decomposition of free radical polymerization initiators by DSC using Eq. 45 to calculate rate constants from the dynamic DSC curves. Activation energies and frequency factors were reported for the decomposition of azobisisobu-tyronitrile (AIBN), benzoyl peroxide (BP), and diisopropyl peroxydicar-bonate (DIPP) at 16°C/min. heating rate. The activation energies, all close to 30 kcal/mole, agreed to within ±5% of the literature values. For de-composition of AIBN, which was shown to be an unambiguous first-order reaction, additional measurements were made at 4, 8, and 32°C min^{-1}. Both activation energy and frequency factor increased with increasing heating rate; the best values, relative to literature values, were obtained at a heating rate of 16°C min.$^{-1}$ Swarin and Wims (1976) obtained similar re-sults for AIBN and BP by DSC. By use of Eq. (46), they were able to analyze several potential values for the reaction order. From Ar-rhenius plots of these analyses (Fig. 68), they again demonstrated that these reactions were first order.

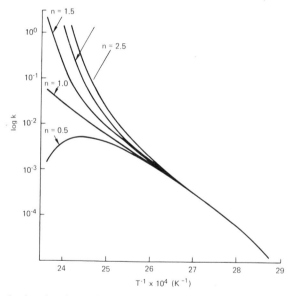

Fig. 68   Arrhenius plots from DSC dynamic method A data utilizing Eq. (46) and several potential values of the reaction order *n*. This technique may be used to determine the order of *n*th-order reactions, equal to one for this example. DuPont 990/910 DSC. (From Swarin and Wims, 1976.)

Willard (1972) studied the polymerization of diallyl phthalate (DAP) catalyzed with ducumyl peroxide. Data analyzed according to Eq. (45) gave a linear Arrhenius plot, demonstrating the reaction to be first order. The measured activation energy of 36.0 kcal mole^{-1} was close to an iso-thermal value of 35.2 kcal mole^{-1} reported by Duswalt (1974) for the ther-mal decomposition of dicumyl peroxide. Figure 69 demonstrates the excellent agreement with isothermal results for this system. As Willard

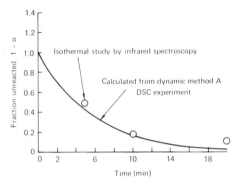

Fig. 69 Isothermal infrared data demonstrate predictive utility of DSC dynamic method A for first-order polymerization of DAP with dicumyl peroxide. DuPont 900 DSC. (From Willard, 1972.)

points out, reliable dynamic DSC data of this sort are useful in practical applications for estimating the extent of conversion that can occur during heating prior to attainment of the isothermal molding temperature of the specimen.

Swarin and Wims (1976) studied, in addition to the thermal decomposition of AIBN and BP, the curing of an epoxy powder paint and an epoxy potting compound. By the Arrhenius plot method previously discussed, they demonstrated all four reactions to be first order. As shown in Table XI, a heating-rate dependence was observed. The zero heating-rate

TABLE XI

KINETIC PARAMETERS AT DIFFERENT HEATING RATES
FOR THE DECOMPOSITION OF FREE RADICAL INITIATORS
AND THE CURING OF EPOXIES (ALL REACTIONS
DETERMINED TO BE FIRST ORDER; SEE FIG. 68;
E IN KILOCALORIES PER MOLE, A IN INVERSE SECONDS)[a]

Heating rate (°C/min)	BP		AIBN		Powder paint[b]		Potting compound[b]	
	E	$A \times 10^{-14}$	E	$A \times 10^{-14}$	E	$A \times 10^{-7}$	E	$A \times 10^{-10}$
20	30.8	3.90	31.3	22.1				
10	30.8	4.10	30.5	9.7	25.3	40,000	25.5	7.7
5	30.5	3.19	31.7	62.9	21.2	49	24.6	3.0
2	30.3	2.47	30.6	16.7	18.6	2.17	24.0	1.4
0	29.6	0.99	32.2	16.2	18.1	1.37	23.4	0.76
	(30.0)[c]		(30.8)	(15.8)				

[a] Dynamic method A, from Swarin and Wims (1976).

[b] Powder paint and potting compound were epichlorohydrin–bisphenol A-type epoxies, the former with an amine catalyst and the latter an anhydride hardener.

[c] Values in parentheses are literature values.

values were obtained first by extrapolating rate constants, from Eq. (45), against heating rate, and then constructing an Arrhenius plot from the extrapolated rate constants. Within experimental error, the extrapolated zero-heating-rate values for the initiators are identical to the actual DSC values. However, the extrapolated values are truly different for the epoxies. The authors showed that the actual dynamic DSC kinetic values did not yield accurate predictions, whereas the extrapolated values did (Fig. 70).

As reported earlier (Section III.A), Sebenik *et al.* studied reactions

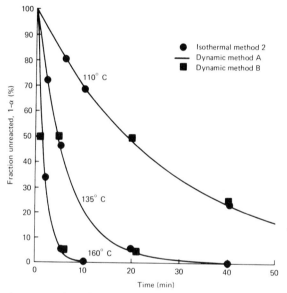

Fig. 70   Curing curves show that dynamic method A (extrapolated to zero heating rate, see text), dynamic method B, and isothermal method 2 yield identical results for this system. Method B calculations performed by this author. DuPont 990/910. (From Swarin and Wims, 1976.)

between phenol and formaldehyde by DSC (dynamic method A). The NaOH-catalyzed reactions proceeded in two steps to yield first *o*-hydroxymethyl phenol and then its condensation product (Fig. 71). The reactions were carried out in stainless steel pans fashioned after Freeberg and Alleman (1966). After it was established that the first exotherm resulted from the formation of *o*-hydroxymethyl phenol [reaction (64)], it became possible to deconvolute the multiple exotherms by comparison with the condensation of the pure monomer. The deconvoluted exotherms were analyzed according to Eq. (46) (Fig. 72), yielding a reaction order of 2 for the addition of formaldehyde to phenol and reaction order of 1 for condensation of the addition product. Activation energies decreased

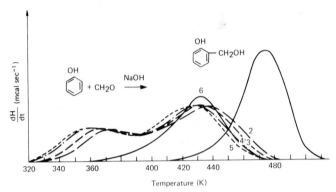

Fig. 71 Thermograms at 4°C min.$^{-1}$ for the reaction between phenol and formaldehyde. 1 (——) no catalyst added; 2 (————), 0.25% NaOH; 3 (———), 0.50% NaOH; 4 (—·—) 0.75% NaOH; 5 (-------), 1% NaOH; 6 (——), condensation of o-hydroxymethyl-phenol, 1% NaOH Perkin-Elmer DSC-lb. [Reprinted with permission from Sebenik *et al.* (1974). Copyright 1974 Pergamon Press, Ltd.]

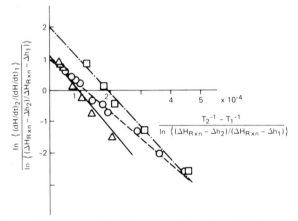

Fig. 72 Plots of dynamic method A data, according to Eq. (46), from which the activation energy E and reaction order n for addition and condensation reactions between phenol and formaldehyde were obtained. Slope is $-E/R$, ordinate intercept is n. Catalyst 0.5% NaOH. □, Addition; ○, condensation; △, condensation of o-hydroxymethyl-phenol (1% NaOH); Perkin-Elmer DSC-1b. [Reprinted with permission from Sebenik *et al.* (1974). Copyright 1974 Pergamon Press, Ltd.]

with increasing catalyst concentration; by comparison with reported isothermal values, they were consistently high by 10–20%. Kay and Westwood (1975) applied this technique to the curing of phenol–formaldehyde and melamine–formaldehyde systems. All reactions exhibited only one cure exotherm. An NaOH-catalyzed phenolic resole system showed decreasing activation energies with increasing heating rate. The average value was ~30% higher than that which can be calculated from peak temperature–heating-rate data (dynamic method B). The effect of an acid catalyst on the melamine system was shown to be a reduction in ΔH_{RXN} and the expected reduction in activation energy and increase in reaction rate.

Other studies of thermosets utilizing these methods have not been successful. Prime (1973) reported on an epoxy–polyamide system that was not autocatalyzed and with reaction order close to 1. At all heating rates, E and A were considerably larger than the isothermal values but did not show the trends observed by Swarin and Wims. However, the curing of this system was complicated by higher-temperature reactions occurring after or overlapping with the primary epoxy–amine reaction (Prime and Sacher, 1972). The dynamic DSC results of Abolafia (1969) on an epoxy–dicyandiamide (dicy) varnish formulation were also considerably larger than isothermal values reported by Sacher (1973), Schneider and Gillham (1978), and Schneider *et al.* (1979). Dynamic DSC studies by Taylor and Watson (1970) again yielded activation energies for curing of epoxy resin with aromatic amines that were significantly larger than those commonly observed. Similar results were obtained by Prime (1970, 1973) for an epoxy–diamine system.

As has been shown, the ability to extract reliable kinetic data from one dynamic DSC trace has been truly successful only for some first-order reactions. For a pair of apparently first-order epoxy cure reactions, data obtained by extrapolation to zero heating rate were of predictive utility. However, the experiments at several heating rates and extrapolation offer no advantage over the isothermal methods. Several examples were noted where the method was unreliable. Additional examples were given by Duswalt (1974), who warned that, "This technique should be used with considerable caution." This is very good advice.

It is the opinion of this author that, at this time, isothermal method 1 offers the best compromise between conservation of time and reliability of results. However, it is also his belief that the time is approaching (with the accelerated technological progress) when *consistently reliable* kinetic data will be obtained from a single scan.

Where discrepancies exist between isothermal and dynamic experiments, there are numerous potential reasons. Swarin and Wims (1976) attribute their observed heating-rate dependence to thermal lag of the instrument and possibly the reaction itself. Much time was spent in Section II.A describing the reaction kinetics of the common epoxy–amine reaction. It was seen that this reaction can vary between third-order, autocatalyzed with induction time, to pseudo-second-order without induction time. Several real systems, including epoxy–amine, epoxy–anhydride and polyester–styrene, have been observed to generally follow the autocatalyzed reaction scheme. Much of the unreliability of dynamic kinetics may be attributed to attempting to fit measured data to a kinetic equation that does not truly describe the course of the reaction. Also, as proposed by Kamal and Sourour (1973), the dynamic DSC trace may contain more than just the desired kinetic data. Perhaps kinetic analyses should be made utilizing a "reaction" base line attributable solely to curing rather than the "experimental" base line attributable to cure plus nonlinear changes in specific heat. Changes in volume that occur during cure may have some influence. Dynamic Arrhenius equations that are different from the isothermal equation have been proposed by Draper (1970), Prime (1970), and MacCallum and Tanner (1970). Prime (1973) had reasonable success in applying this approach to the cure of an epoxy–diamine and an epoxy–polyamide. This approach was also proved successful in relating the dynamic and isothermal cure data for epoxide resin cured with an imidazole (Barton and Shepherd, 1974). Simons and Wendlandt (1972) argue against such equations.

A systematic investigation of these variables and perhaps others appears necessary before *consistently reliable* results can be obtained. It is possible that strict attention to certain experimental and computational details will ultimately be necessary, and finally lead to the general success of this method.

B. Method B: Variation of Peak Exotherm Temperature with Heating Rate

The peak exotherm temperature (T_p) varies in a predictable manner with the heating rate (ϕ). Figure 73 is an example. Several methods to distill the activation energy E and preexponential A have been devised. Kinetic parameters obtained by the method presented here have been very satisfactory with respect to known materials properties (Duswalt, 1974). It is identical to the ASTM Method E698-79 for determining the Arrhenius kinetic constants for thermally unstable materials. For nth-order reactions, this author has observed E, A, and reaction rates from these values to be as accurate as the same parameters from the isothermal methods of Section II. Even for autocatalyzed reactions, dynamic method B gives an accurate measure of the activation energy and preexponential factor. Tables XII and XIII show typical results for some thermosetting systems. This method is valuable as a precursor to isothermal studies and is often the only means to analyze the curing kinetics of systems with multiple exotherms (Fig. 74), solvent effects (Fig. 65), or unreliable baselines (Fig. 5). A simple, usable yet accurate relationship between activation energy E, heating rate ϕ, and peak exotherm temperature T_p is based on work of Ozawa (1965, 1970). Ozawa's is a difference method, starting with a com-

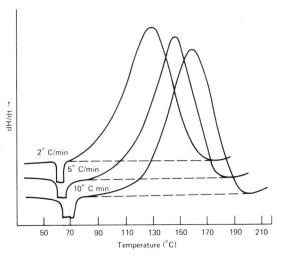

Fig. 73 Dynamic method B: effect of heating rate on the DSC exotherms, especially the peak temperatures T_p, of an epoxy powder paint (epichlorohydrin-bisphenol-A type with amine hardener). DuPont 990/910 DSC. (From Swarin and Wims, 1976.)

TABLE XII

KINETIC PARAMETERS FOR CURING OF EPOXY POWDER PAINT[a,b]

	E (kcal mole^{-1})	A (s^{-1})	$k_{135°C}$ (s^{-1}), First order
Swarin and Wims method	18.1	13.7×10^6	0.00265
Dynamic Method B[c]	18.0	10.8×10^6	0.00249

[a] Method of Swarin and Wims (1976) compared with dynamic method B.
[b] Epichlorohydrin–bisphenol A-type resin with amine catalyst/hardener.
[c] Calculations by this author from data of Swarin and Wims (1976).

bined formed of Eqs. (1) and (4):

$$d\alpha/dt = f(\alpha)A \exp[-E/RT] \tag{48}$$

Integration of this equation proceeds as follows:

$$\int_0^{\alpha_p} \frac{d\alpha}{f(\alpha)} = A \int_{t_0}^{t_p} e^{-E/RT} \, dt = \frac{A}{\phi} \int_{T_0}^{T_p} e^{-E/RT} \, dT$$

$$\approx \frac{A}{\phi} \int_0^{T_p} e^{-E/RT} \, dT \approx \frac{AE}{\phi R} p(E/RT) \tag{49}$$

Values for $p(E/RT)$ were tabulated by Doyle (1961), where

$$\log p(E/RT) \approx -2.315 - 0.4567E/RT \tag{50}$$

for $20 < E/RT < 60$. It has been observed (Horowitz and Metzger, 1963;

TABLE XIII

Combined DSC Dynamic Method B and Isothermal Method 2 Data for Multiple Curing Reactions Occurring in Epoxy (DGEBA)–Polyamide (Versamid® 140)[a]

| | Dynamic method B | | | Isothermal DSC method 2 + Isothermal dc conductivity | | | | |
Reaction	E (kcal m⁻¹)	A (s⁻¹)	$t_{1/2}$ (h)	E (kcal mole⁻¹)	A (s⁻¹)	$t_{1/2}$	ΔH_{RXN} (cal g⁻¹)	ΔH_{RXN} (kcal mole⁻¹ epoxide)
1	14.5	5.9×10^5	4.3 at 40°C	14.5	6.3×10^5	4.0 at 40°C	86.6 ± 3.0	26.3 + 0.9
2	19.0	5.0×10^5	22.4 at 80°C			12.0 at 80°C	17.0 ± 6.3	
3	27.5						14.8 ± 2.4	

[a] From unpublished data of R. B. Prime and E. Sacher.

542

Prime, 1973; Peyser and Bascom, 1974) that the extent of reaction at the peak exotherm α_p is constant and independent of heating rate. The latter two references are specific to thermoset curing. Therefore, the first integral in Eq. (49) is a constant, and the following relationship is obtained:

$$E \approx \frac{-R}{0.4567} \frac{\Delta \log \phi}{\Delta(1/T_p)} \equiv \frac{-R}{1.052} \frac{\Delta \ln \phi}{\Delta(1/T_p)} \tag{51}$$

From the peak reaction temperatures as a function of heating rate, the activation energy can be obtained with a precision of $\pm 3\%$ (Duswalt, 1974) and estimated accuracy of $\pm 10\%$. As the data of Tables XII and XIII suggest, the observed accuracy is usually better. This author has found that more accurate values of $p(E/RT)$ (Doyle, 1961) do not significantly improve the accuracy of Eq. (51) and that this equation may be used with confidence.

Fava (1968) employed an equivalent but more difficult to use approach. By plotting $(d\alpha/dt)_p$ measured at differing heating rates against T_p^{-1}, he observed an activation energy of 17.8 kcal mole^{-1} for an epoxy–anhydride that appeared to follow autocatalyzed kinetics.

Other expressions exist that relate E to ϕ and T_p (Kissinger, 1957; Kaelble and Cirlin, 1971; Crane *et al.*, 1973). These methods either are in error (see Peyser and Bascom, 1975) or yield inaccurate results (see Prime, 1973) and should not be used.

Kissinger derived a useful and accurate expression for the frequency factor for nth-order reactions

$$A = \frac{\phi E \, \exp[E/RT_p]}{RT_p^2[n(1 - \alpha_p)^{n-1}]} \approx \frac{\phi E \, \exp[E/RT_p]}{RT_p^2} \tag{52}$$

Kissinger argued that $n(1 - \alpha_p)^{n-1} \sim 1$ and is independent of heating rate. This is true by definition for first-order reactions, and Prime (1973) showed this quantity to be constant and only 2–4% greater than unity for an nth-order epoxy cure reaction. Using ϕ, T_p, and the resulting E from Eq. (51) in Eq. (52), the preexponential factor can be obtained with a precision of $\pm 2\%$ (Duswalt, 1974). By following Kissinger's method, it can be shown that the frequency factor for autocatalyzed reactions can be approximated from

$$A \cong \frac{\phi E \, \exp[E/RT_p]}{RT_p^2[2\alpha_p + 2B\alpha_p - 3\alpha_p^2 - B]} \tag{53}$$

Duswalt (1974) utilized dynamic method B to examine the cure kinetics of a commercial epoxy resin, obtaining $E = 11.3$ kcal mole^{-1} and $A = 4.03 \times 10^4$ min^{-1}. Employing the first-order equation, he calculated the half-life at 100°C and aged the sample at those conditions. The degree of conversion, measured by DSC method 2, was 51%. The data of Swarin and Wims (1976) (Fig. 73) were analyzed by this author according to Eqs. (51) and (52). As can be seen from Table XII, the results compare very favorably with those of Swarin and Wims. The dynamic method B results are compared with the dynamic DSC and measured isothermal data (DSC method 2) in Fig. 70; the data are indistinguishable. Table XIII compares dynamic method B and isothermal kinetic data for an nth-order epoxy with multiple curing reactions. For the autocatalytic epoxy cure reaction of Fig. 30, dynamic method B yields $E = 12.3$ kcal mole^{-1} and $A = 27 \times 10^5$ min^{-1}.

As discussed by Duswalt (1974), advantages of this method include simplicity, applicability to many types of reaction, and relative insensitivity to solvent effects, baseline problems, and secondary reactions. The peak temperature has also been shown to be relatively insensitive to degree of reaction prior to the experiment (Chow and Steiner, 1979); see also Figs. 7 and 36. Duswalt describes the application of this technique, in combination with isothermal method 2, to study the complex curing of a hypothetical thermosetting resin that produces multiple curing exotherms. Real systems that exhibit this behavior include some unsaturated polyester and DAP molding compounds (Maas, 1978) (Fig. 74), epoxy–

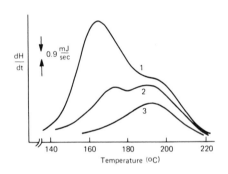

Fig. 74 Complex curing behavior of unsaturated polyester molding compound initially (1, $\alpha = 0$) and after partial curing (2, $\alpha = 0.51$ and 3, $\alpha = 0.82$). Perkin-Elmer DSC-1b. (From Mass, 1978.)

polyamide for which three curing reactions were found (Prime and Sacher, 1972), a dicy–epoxide in which the epoxide was in excess (Schneider and Gillman, 1978; Schneider *et al.*, 1979) and some phenolic resole resins [see, for example, King *et al.* (1974) and Fig. 5]. As discussed by Duswalt, first E and A values are obtained via dynamic method B for as many reactions as can be resolved. From calculated rate constants, a low temperature is identified at which the first reaction proceeds much faster (50–100 times) than reaction 2. Aging at this temperature followed by comparison of isothermal method 2 scans can yield an estimate of ΔH_{RXN} for reaction 1 (see Fig. 74). At appropriate higher temperatures the procedure can be repeated for reactions 2, 3, etc. Kinetic data for unresolvable peaks may be obtained by isothermal aging at two or more isothermal temperatures. Interference from overlapping reactions can be corrected by using rate constants measured by dynamic method B. Also see Flynn (1980). One shortcoming of this method is the requirement of an assumed reaction order. Reference to the detailed procedure of Duswalt is recommended for those wishing to apply this technique. Prime and Sacher (1972) combined this procedure with a curve-resolving capability in studying the complex cure behavior of an epoxy (DGEBA)–polyamide (Versamid®* 140) at two stoichiometric ratios. Table XIII contains the dynamic method B results for one of the systems. From this data it is estimated that at 40°C, reaction 1 will proceed 1600 times faster than reaction 2. At 40°C it was found that only the first reaction occurred. IR identified this as the expected amine–epoxide reaction (reaction 57), and the ΔH_{RXN} of 26.3 kcal mole^{-1} supports this conclusion (Horie *et al.*, 1970a, 26.0 kcal mole^{-1}; Klute and Viehmann, 1961, 24.5 kcal mole^{-1}). At 80°C the second reaction proceeds at a measurable rate; IR suggests it to be an amide–ester interchange. At 110°C a third complex reaction occurs that is accompanied by the loss of amide absorption in the IR.

C. Method C: Variation of Temperature to Constant Conversion with Heating Rate

The third dynamic method is the general case of the second method. It has been applied with reasonable success to the curing of autocatalyzed epoxies (Fava, 1968; Barton, 1974a,b). With this technique, time–temperature–degree of cure predictions can be made *without knowledge* of $f(\alpha)$, which is a distinct advantage over the two previous methods. It may also be used to measure constancy of or changes in apparent activation energy with temperature and degree of conversion (see Flynn, 1978a,b; Barton, 1974a,b). Fava (1968) showed that the area under a plot of reciprocal rate $(d\alpha/dt)^{-1}$ versus degree of conversion α constructed at a temperature T_0 from a series of dynamic DSC scans is equivalent to the time to reach α at isothermal temperature T_0. See Fig. 75 and Eq. 54. Using this technique, Fava constructed the series of isothermal cure curves (dashed lines) in Fig. 76, which are shown with isothermal method 1 curves. Except at the lowest temperature, the two results compare favorably. Had the isothermal method 1 curves been corrected for instrumental

---

* ® Trademark, General Mills Corp.

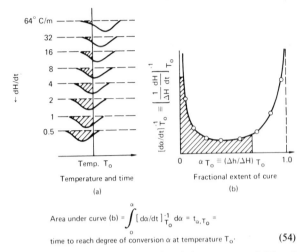

$$\text{Area under curve (b)} = \int_0^\alpha [\, d\alpha/dt\,]^{-1}_{T_0}\, d\alpha = t_{\alpha, T_0} =$$

time to reach degree of conversion α at temperature T_0.   (54)

Fig. 75 Dynamic method C: (a) set of displaced thermograms yielding the rate and extent of reaction at temperature T_0; (b) reciprocal reaction rate versus degree of conversion curve from (a) used to generate isothermal cure curve at T_0 (Figs. 76 and 77, for example). Perkin-Elmer DSC-1. (From Fava, 1968. Reproduced by permission of the publishers, IPC Business Press Ltd. Ⓒ.)

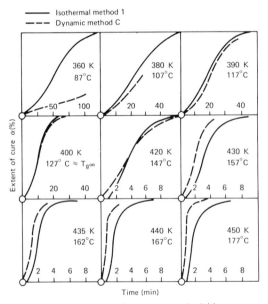

Fig. 76 Comparison of cure curves for epoxy–anhydride generated by isothermal method 1 and dynamic method C. The time for the sample to reach temperature (~1 min) contributes to the deviations at temperatures above 140°C. Perkin-Elmer DSC-1. (From Fava, 1968. Reproduced by permission of the publishers, IPC Business Press Ltd. Ⓒ.

equilibration (Section II.A.2.a; Widmann, 1975a), the comparison might have been considerably better above 400 K.

Barton (1974a,b) extended this technique to include calculation of activation energy that permitted the generation of a family of isothermal cure curves from experiments at two heating rates. He employed the basic Arrhenius equation, Eq. (48), from which it follows that

$$\ln \frac{(d\alpha/dt)_{T_1}}{(d\alpha/dt)_{T_2}} = \frac{E}{R}\left(\frac{1}{T_2} - \frac{1}{T_1}\right) \tag{55}$$

where the Ts are temperatures for a constant degree of conversion at different heating rates. DSC experiments of an epoxy–diamine thermoset at two heating rates generated the necessary rate data and activation energy. Using the measured activation energy and Eq. 55, a 10°C m⁻¹ DSC scan was converted to a 100°C isothermal plot similar to curve b in Fig. 75. Integration of this plot gave the 100°C isothermal cure curve, which, via the activation energy, was transformed into a family of isothermal curves (Fig. 77). It can be seen that the predicted curves agree well with experi-

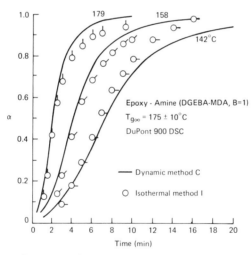

Fig. 77 Isothermal epoxy–amine cure curves predicted from dynamic method C data (solid curve) and corresponding isothermal experimental points. DuPont 900 DSC. (From Barton, 1974a,b, by courtesy of Marcel Dekker, Inc.)

ment up to ~50% conversion, but are in serious disagreement in the 90–95% conversion region. The overprediction at low temperatures and underprediction at high temperatures are identical to Fava's results and correspond to dynamic activation energies that are higher than the true isothermal values.

V. Specific Polymers: Chemistry and Trade Names

The purpose of this section is to support the other sections with information on chemical structures and chemical reactions and to match common trade names with the respective chemical structures. An attempt has been made to cover most of the thermosetting systems referenced in the thermal analysis literature and cited in this chapter. To make the chapter more understandable and more usable, several examples of chemical curing reactions are given. Only the major reactions are presented; possible side reactions and reactions of less frequently encountered thermosets have been omitted. Those interested in more detail are referred to monographs on thermosetting resins, such as those by Lee and Neville (1967) and May and Tanaka (1973) on epoxy resins, Martin (1956) and Knop and Scheib (1979) on phenolic resins, and Boenig (1964) on unsaturated polyester resins.

A. Epoxy Resins

Epoxy resins are a general class of materials characterized by a chemical structure containing the epoxide group or oxirane ring:

$$R-HC\overset{\displaystyle O}{\underset{\displaystyle \diagdown}{\diagup}}CH_2 \tag{56}$$

They can be cured by a variety of cross-linking agents known as hardeners (reactive comonomers) or by catalysts that promote self-polymerization. The more common epoxy resins, hardeners, and catalysts, and associated chemistry, are highlighted below. The reader interested in more detail is referred to general references on this topic, such as Lee and Neville (1967) and May and Tanaka (1973).

1. Resins

Epoxy resins can be divided into two major groups: epichlorohydrin-derived resins and cycloaliphatic resins. The oldest and most widely used epoxy resins are prepared from the reaction of epichlorohydrin with a polyhydroxy compound, such as bisphenol A. Their structural formula is shown in Table XIV. The compound corresponding to $n = 0$ is DGEBA (the diglycidyl ether of bisphenol A) with an equivalent weight of 174 g eq^{-1} of epoxide. In high melting solid epoxies, n can be as high as 20.

Other epichlorohydrin-derived resins are epoxidized novolaks, whose structure is also shown in Table XIV. These resins differ from the latter resins in that phenol–formaldehyde novolaks are substituted for bisphenol A. Epoxy novolaks find application where performance at elevated temperatures is important.

TABLE XIV

SOME COMMON EPOXIDE RESINS

A. Bisphenol A–epichlorohydrin (DGEBA) resins

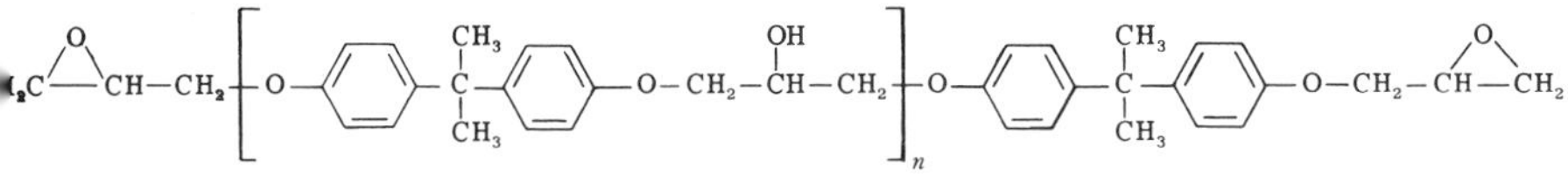

n	*Trade names*
0(DGEBA)	Ciba-Geigy Araldite AY105, 6004, Dow DER 332, Shell Epon 828
2-3	Ciba-Geigy Araldite 6071, Dow DER 661, Shell Epon 1001
5-6	Ciba-Geigy Araldite 6084, Dow DER 664, Shell Epon 1004

B. Epoxy novolak resins

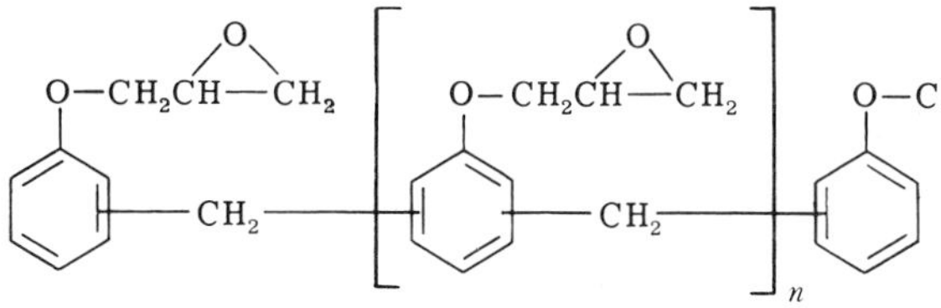

n	*Trade names*
0.2	Ciba-Geigy Araldite EPN 1138, Dow DEN 431, Shell Epon 152
1.6	Ciba-Geigy Araldite EPN 1139, Dow DEN 438, Shell Epon 154

C. Cycloaliphatic epoxides, e.g.

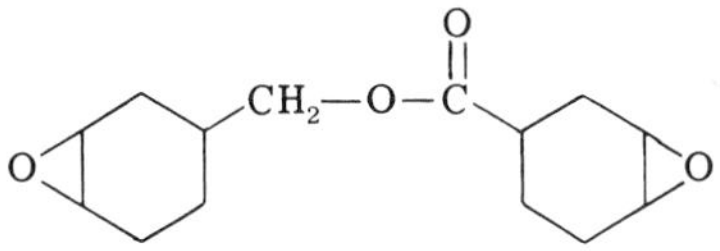

Chemical Structure	*Tradenames*
3,4-Epoxycyclohexylmethyl-3,4-epoxyclohexane carboxylate (above)	Union Carbide Bakelite Resin ERL 4221 Ciba-Geigy Araldite CY 179

D. Tetrafunctional epoxides

Tetraglycidylmethylene Dinaline (TGMDA)

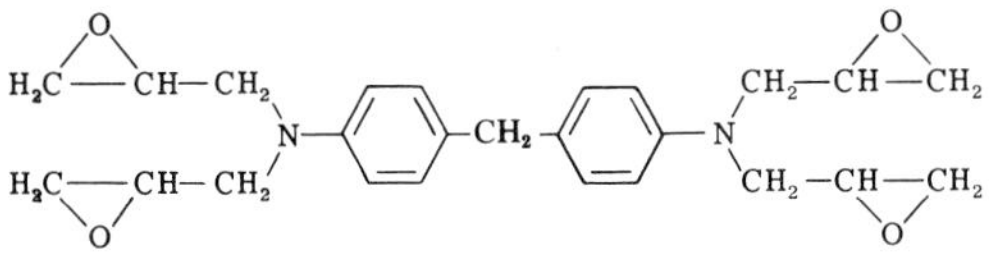

Trade name Ciba-Geigy MY 720

MY 720

A relatively new but important class of resins is the cycloaliphatic epoxides, an example of which is shown in Table XIV. They find application in electronic encapsulation systems, because of the absence of sodium and chlorine impurities, and as exterior high-voltage electrical insulation. Bis(2,3-epoxy cyclopentyl) ether, a cycloaliphatic epoxy that does not contain ester linkages, finds application in the aerospace and hydrospace industries (particularly in graphite and boron composites) because of its high strength. A tetrafunctional epoxide with aerospace applications as a resin for graphite composites is also shown.

2. Hardeners

Numerous amine and anhydride hardeners are available as curing agents for epoxy resins. For simplicity, it can be stated that aliphatic amines are used with bisphenol A resins when room temperature cures are desired and when heat deflection temperatures below 100°C can be tolerated. Examples are given in Table XV; May and Tanaka (1973, Chap. 4) give a more extensive description. Aromatic amines can provide heat deflection temperatures in the 150–170°C range, but require curing at elevated temperatures.

Primary and secondary amines react with epoxide in the following manner (see also Section II.A.1).

$$R_2'-NH \; + \; \overset{H_2C----CH-R}{\underset{O}{\diagdown\diagup}} \; \longrightarrow \; R_2'-N-CH_2-\underset{\underset{OH}{|}}{CH}-R \qquad (57)$$

Anhydride-hardened bisphenol A systems offer long pot life, low exotherm, excellent adhesion and electrical properties, and heat deflection temperatures between 125 and 170°C. Examples are shown in Table XV. Small amounts of a tertiary amine, such as benzyldimethylamine or tris-2,4,6-dimethylaminomethyl phenol, are frequently used to accelerate the cure reaction, which results in the formation of ester linkages:

$$R-HC\underset{O}{\overset{\diagup\diagdown}{----}}CH_2 \qquad O=C\underset{O}{\overset{O}{\diagup\diagdown}}C=O \qquad \longrightarrow \qquad R-\underset{\underset{}{|}}{\overset{O^-}{CH}}-CH_2O-\overset{O}{\overset{\|}{C}} \quad \overset{O}{\overset{\|}{C^+}} \qquad (58)$$

Cycloaliphatic epoxides respond only to acidic hardeners, such as polycarboxylic acid or acid anhydrides, via reaction (58). Because of their compact structure, systems cured with hexahydrophthalic anhydride can have heat deflection temperatures above 190°C.

Another common reaction, sometimes a side reaction, is etherification

TABLE XV

Some Common Epoxide Hardeners and Catalysts

Aliphatic amine hardeners	
Ethylenediamine (EDA)	$H_2N-(CH_2)_2-NH_2$
Diethylenetriamine (DETA)	$H_2N-(CH_2)_2-NH-(CH_2)_2-NH_2$
Triethylenetetramine (TETA)	$H_2N-(CH_2)_2-NH-(CH_2)_2-NH-(CH_2)_2-NH_2$
Hexamethylenediamine (HMDA)	$H_2N-(CH_2)_6-NH_2$
Dicyandiamide (dicy)	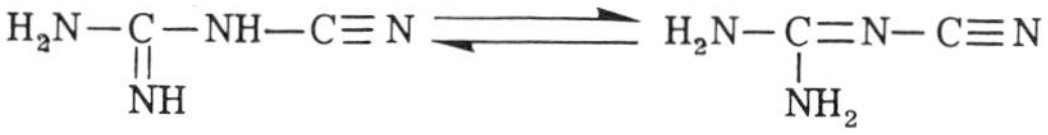
Aromatic amine hardeners	
m-Phenylenediamine (mPDA)	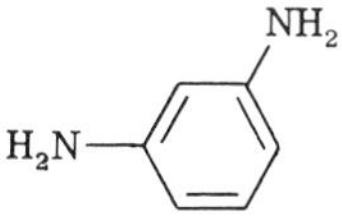
4,4′-Diaminodiphenylmethane (methylenedianiline, MDA)	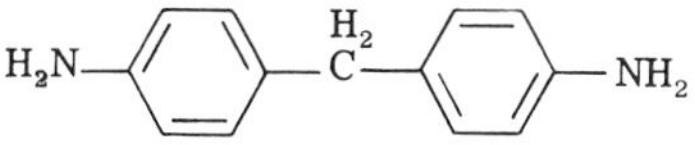
Diaminodiphenylsulfone (DDS)	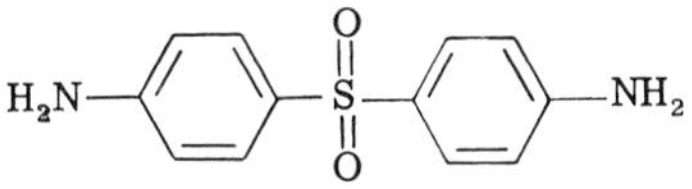
Polyamide hardeners	
General Mills Versamid® resins	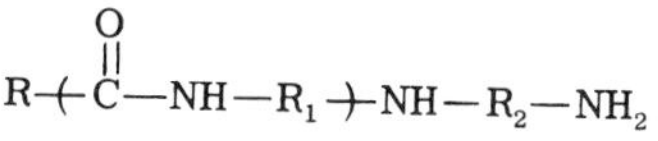
Acid anhydride hardeners	
Hexahydrophthalic anhydride (HHPA)	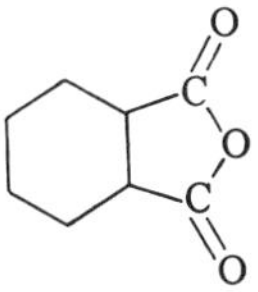
Methyl tetrahydrophthalic anhydride (MTHPA)	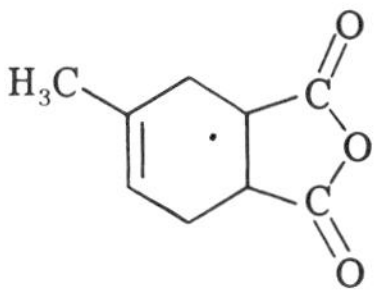
Pyromellitic dianhydride (PMDA)	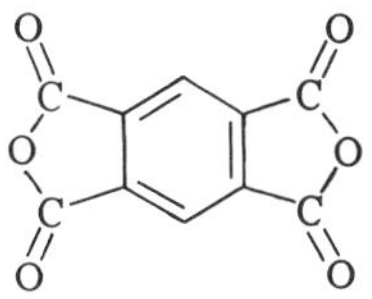
Tertiary amine catalysts	
Benzyldimethylamine	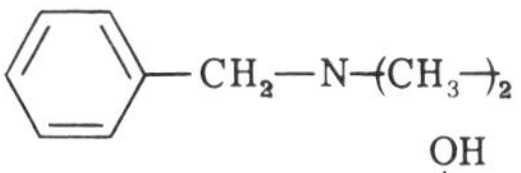
Tris-2,4,6-dimethylaminomethyl phenol	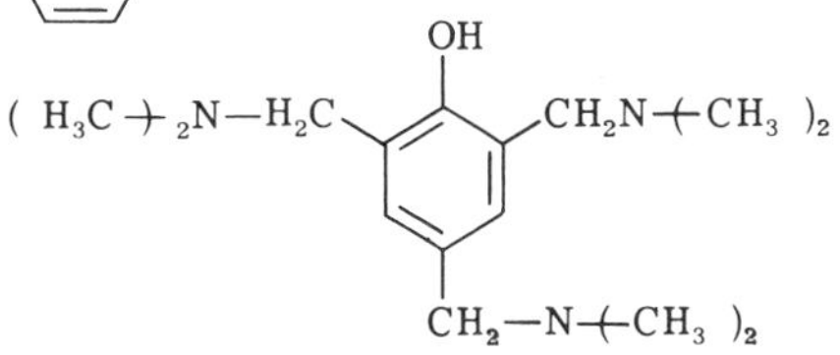

or self-polymerization of epoxide, shown in reaction (59). Many commercially important epoxy resin systems contain dicy (dicyandiamide, cyanoguanidine) as the amine hardener plus a tertiary amine catalyst. Fine crystalline dicy particles are dispersed in the epoxy resin, and because the initiation of curing requires dissolution of dicy at elevated temperatures, this one-part epoxy system offers excellent stability and shelf life in addition to low cost and product uniformity. The curing chemistry is very complex and, as has been shown in Section III.A, includes the participation of water. Saunders *et al.* (1967) give an excellent description of the curing chemistry, which is shown to proceed in two steps. The first step, primary cure, is an exothermic reaction that occurs during B staging and the period prior to gelation. It includes contributions from reactions (59) and (60), both of which proceed until all available epoxy groups have been consumed.

Etherification or self-polymerization, for example,

$$
R'NH-CH_2-\underset{\underset{OH}{|}}{CH}-R \;+\; \underset{\underset{O}{\diagdown\!\diagup}}{H_2C\!-\!CH-R} \;\xrightarrow{\;NR_3''\;}\; R'NH-CH_2-\underset{\underset{\underset{\underset{CHOH}{|}}{\overset{|}{CH_2}}}{\overset{|}{O}}}{CH}-R \tag{59}
$$

N-Alkyl cyanoguanidine formation:

$$
4\;\; \underset{\underset{O}{\diagup\!\diagdown}}{RHC\!-\!CH_2} \;+\; \underset{\underset{NH}{||}}{H_2N-C}-NH-C\!\equiv\!N \;\longrightarrow\; R-CHOH-CH_2-\underset{\underset{\underset{\underset{R}{|}}{CHOH}}{\overset{|}{CH_2}}}{N}-\underset{\underset{N}{|}}{C}-N\!\equiv\!N \tag{60}
$$

The second and final stage of curing, which involves additional crosslinking and occurs almost entirely in the press, can be adequately represented by

$$
-\underset{\underset{OH}{|}}{CH}-CH_2-R'' \;+\; N\!\equiv\!C-\underset{\underset{R'}{|}}{N}-\underset{\underset{N}{||}}{C}-NR_2' \;\xrightarrow{\;NR_3\;}\; -\underset{\underset{NH}{|}}{CH}-CH_2-R'' \tag{61}
$$

where R″ may be

$$-O-CH_2-CH-CH_2-O-C_6H_4- \qquad -N-C-N-C\equiv N$$
$$\qquad\qquad\ \ OH \qquad\qquad\qquad\qquad\quad R' \ \|\ \ R$$
$$\qquad\qquad\qquad\qquad\qquad\qquad\qquad\qquad\quad NR'$$

$$\qquad\qquad NR_2$$
$$-N{=}C-N-C\equiv N \quad or \quad -N-C{=}N-R$$
$$\qquad\ \ |\ \ R' \qquad\qquad\qquad\ C\equiv N$$
$$\qquad NR_2'$$

and R′ may be $-H$ or $-CH_2-CH(OH)-CH_2-O-C_6H_4$.

Saunders *et al.* report that epoxy-dicy formulations commonly contain one-half the stoichiometric equivalent of dicy, i.e., $B = \frac{1}{2}$. They also report that significant traces of ammonia, aminotriazines such as melamine, and isocyanurates are produced during the cure, but do not seem to contribute materially to the overall curing process. Sacher (1973) found that decomposition of dicy into melamine (formulation 66) played a significant role in the cure.

The interaction of moisture with the cure may be attributable to the following proposed reactions of water with dicy (Noll, 1977).

$$H_2N-C-NH-C\equiv N \ + \ H_2O \longrightarrow H_2N-C-NH-C-NH_2$$
$$\qquad\ \ \|\qquad\qquad\qquad\qquad\qquad\qquad\qquad \|\qquad\quad \|$$
$$\qquad\ \ NH \qquad\qquad\qquad\qquad\qquad\qquad\quad NH\qquad O \qquad\qquad (62)$$

$$\qquad Dicy \qquad\qquad\qquad\qquad\qquad Dicarbonimidic\ diamide$$

$$H_2N-C-NH-C\equiv N \ + \ H_2O \longrightarrow H_2N-C{=}O \ + \ H_2N-C\equiv N$$
$$\qquad\ \ \|\qquad\qquad\qquad\qquad\qquad\qquad\qquad\quad |$$
$$\qquad\ \ NH \qquad\qquad\qquad\qquad\qquad\qquad\quad NH_2 \qquad\qquad\qquad (63)$$

$$\qquad\qquad\qquad\qquad\qquad\qquad\qquad Urea \qquad\quad Cyanamid$$

Both reactions increase the number of amine-hydrogens available to react with epoxide. Since reaction (63) results in two compounds capable of cross-linking the epoxy resin, it may account for lowering of the cross-link density which was observed by Brand (1977) and Noll (1977) and discussed in Section III.A.

B. PHENOLIC RESINS

Phenolic resins have been in commercial use longer than any other synthetic polymer except cellulose nitrate. Martin (1956) and Knop and Scheib (1979) are general references on this topic. They are based almost exclusively on the reaction products of phenol and formaldehyde.

The first step is an exothermic addition reaction forming compounds known as methylol derivatives, the reaction taking place at the ortho or

para positions, e.g.,

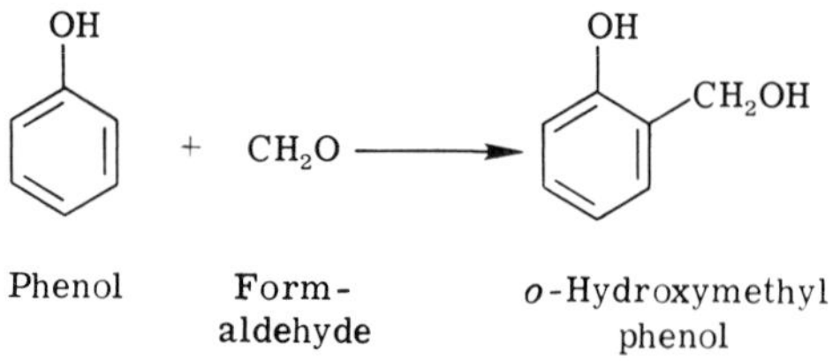

$$\text{Phenol} \quad + \quad CH_2O \longrightarrow \quad o\text{-Hydroxymethyl phenol} \tag{64}$$

Phenol   Form-aldehyde   o-Hydroxymethyl phenol

Complete ortho–para substitution can take place with additional formaldehyde. These products are formed most satisfactorily under neutral or alkaline conditions.

The next step is the formation of either novolak or resole resins, depending on the mole ratio of formaldehyde to phenol (F–P) and catalyst. Novolak resins are formed in the presence of acid catalysts with F–P $<$ 1. The methylol derivatives condense exothermally with phenol to form linear polymers (A-stage resins) with the following structure, where ortho and para links occur randomly:

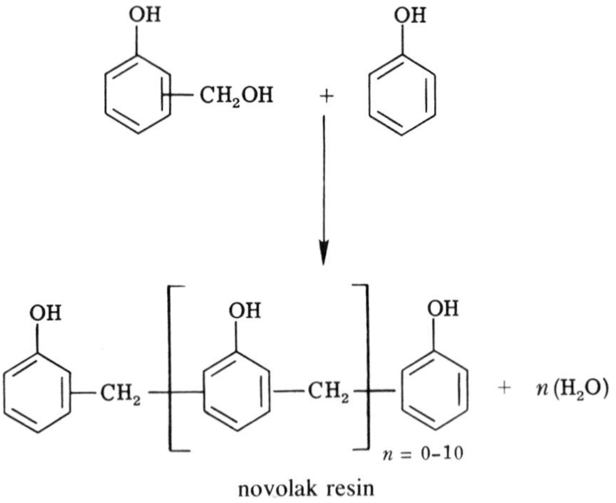

$$\tag{65}$$

$$+ \quad n(H_2O)$$

$$n = 0\text{--}10$$

novolak resin

Resole formation takes place in the presence of alkaline catalysts when F–P $>$ 1, where the methylol phenols can condense either through methylene linkages or through methylene ether linkages [cf. reaction (69)]. In the latter case, subsequent loss of formaldehyde may occur with methylene bridge formation. Resole resins are similar to novolacs but contain reactive methylol groups that can condense to a network structure.

The formation of resoles and novolaks, respectively, leads to the production of one-stage and two-stage phenolic resins. In a two-stage or novolak resin, the additional formaldehyde needed for cross-linking is added as hexamethylenetetramine (HMTA), which decomposes in the presence of heat and moisture to give formaldehyde and ammonia. Ammonia catalyzes the curing. Phenolic molding compounds are B-staged resole or novolac resins.

Because of the chemistry of these resins, the curing of phenolics is amenable to study by DSC and TG. In DSC both the addition and condensation reactions can be studied (Sebenik *et al.*, 1974), but specially sealed pans or pressure DSC is usually required. Note that if the resin is designed for volatilization of excess formaldehyde, this cannot take place in the pressure or sealed-pan DSC experiment, and somewhat altered chemistry and a different product may result. TG can monitor the condensation reaction, but volatilization of absorbed water, formaldehyde, ammonia, or reactive constituents may interfere, suggesting that TG–GC, TG–MS, or evolved gas analysis may be necessary to deconvolute the weight loss curves (see Manley, 1974a,b, and Section V.C).

C. Amino Resins

The principal amino resins are the condensation products of formaldehyde with melamine and/or urea (see following reactions). Both react with formaldehyde, first by addition to form methyol compounds and then by condensation in reactions much like those of phenol and formaldehyde.

$$
\begin{array}{cc}
\text{Melamine} & \text{Urea}
\end{array}
\tag{66}
$$

The addition reaction takes the form

$$
\text{R}-\text{C}-\text{NH}_2 \;+\; \text{H}_2\text{C}{=}\text{O} \longrightarrow \text{R}-\text{C}-\text{NH}-\text{CH}_2\text{OH}
\tag{67}
$$

Two primary reactions lead to cross-linked polymers, methylene bridge formation:

$$
\text{R}_2-\text{N}-\text{CH}_2\text{OH} \;+\; \overset{\overset{\text{R}'}{|}}{\text{H}-\text{N}-\text{CH}_2\text{OH}} \longrightarrow \text{R}_2-\text{N}-\text{CH}_2-\overset{\overset{\text{R}'}{|}}{\text{N}}-\text{CH}_2\text{OH} \;+\; \text{H}_2\text{O}
\tag{68}
$$

and methylene ether formation:

$$
\text{R}_2-\text{N}-\text{CH}_2\text{OH} \;+\; \text{HOCH}_2-\text{N}-\text{R}_2' \longrightarrow \text{R}_2-\text{N}-\text{CH}_2-\text{O}-\text{CH}_2-\text{N}-\text{R}_2' \;+\; \text{H}_2\text{O}
\tag{69}
$$

As with the phenolic resins, the cure of amino resins can be studied by DSC and TG. However, Manley (1974a,b) has shown that escape of formaldehyde, which occurs in the TG experiment and may occur in practice, leads to a different product than is obtained via high-pressure DSC. He also showed, by evolved gas analysis, that volatilization of absorbed water and formaldehyde accompany the loss of water of condensation.

D. ALLYL RESINS

Allyl resins are diallyl ester monomers and B-staged prepolymers that can be cross-linked by free radical polymerization through their double bonds. Allyls in widest use are the diallyl phthalate (DAP) and diallyl isophthalate (DAIP) resins (see Table XVI).

TABLE XVI

SOME COMMON ALLYL RESINS

	Trade names
Diallyl phthalate (DAP) 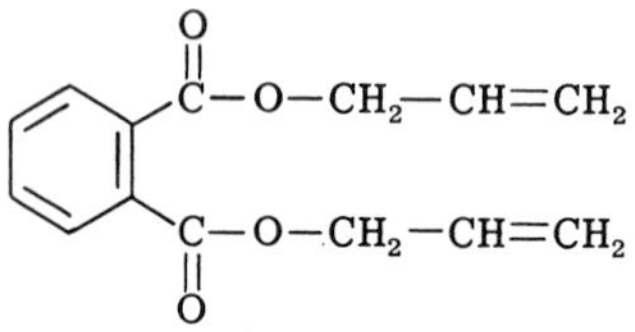	Prepolymer, FMC Corp. Dapon® 35

Diallyl isophthalate (DAIP) 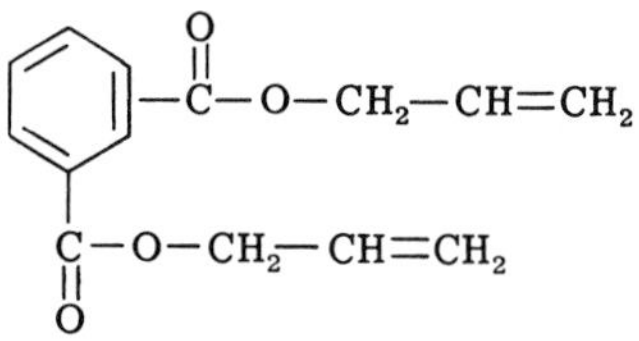	Prepolymer, FMC Corp. Dapon® M

Diethylene glycol bis(allyl carbonate) 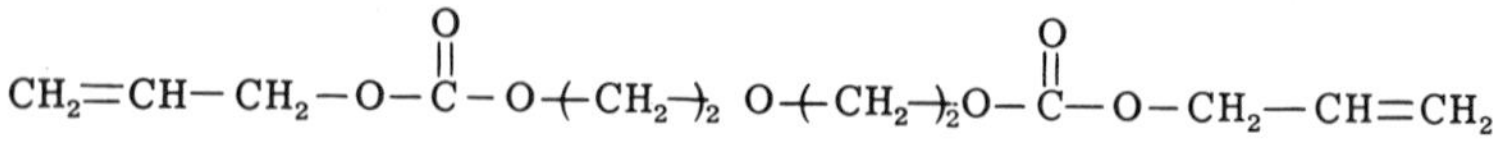	PPG Ind. CR-39 Allyl Diglycol Carbonate®

$$CH_2{=}CH{-}CH_2{-}O{-}\overset{\displaystyle O}{\overset{\|}{C}}{-}O{\left(CH_2\right)_2}O{\left(CH_2\right)_2}O{-}\overset{\displaystyle O}{\overset{\|}{C}}{-}O{-}CH_2{-}CH{=}CH_2$$

Triallyl cyanurate (TAC)

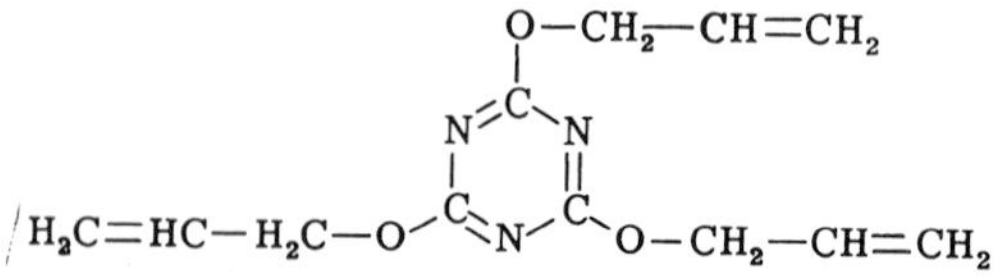

The prepolymers can be used as the sole resin, whereas the monomers are commonly used as cross-linking agents for polyesters. Triallyl cyanurate (TAC), when used in the cross-linking of unsaturated polyesters, results in excellent strength retention at elevated temperatures.

These resins are unsurpassed in their ability to maintain electrical properties under high temperatures and humidity conditions. They are also characterized by excellent dimensional stability, chemical resistance, and a chemical purity that minimizes ionic contamination.

Diethylene glycol bis(allyl carbonate), an allyl diglycol carbonate (Table XVI), is used directly for casting clear, hard glasslike products used for eyeglasses. Properties include impact resistance, light weight, and low grinding costs.

Probable reactions leading to DAP prepolymer are as follows (Slysh *et al.*, 1974):

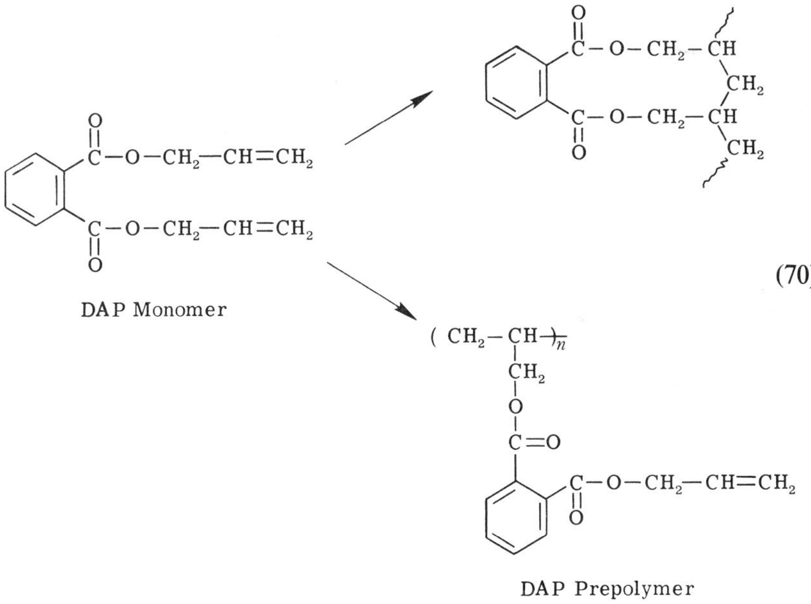

(70)

DAP Monomer

DAP Prepolymer

Some common free radical initiators are shown in Table XVII.

E. Unsaturated Polyester Resins

Commercial resins consist of a linear polyester polymer, a cross-linking monomer, and inhibitors to retard the cross-linking until the resin is to be used. Boenig (1964) is a general reference on this topic. The linear polyester is typically the condensation product of an unsaturated dibasic

TABLE XVII

Some Free Radical Initiators Commonly Used
to Catalyze Cure of Thermosets

Free radical initiator	Abbreviation	Trade names
Azobisisobutyronitrile	AIBN	
Benzoyl peroxide	BP	Lucidol 98[a]
t-Butyl perbenzoate	TBP	
Dicumyl Peroxide	DP	Di-Cup,[b] Luperox 500R[a]
α,α'-Bis(*t*-butyl peroxy) Diisopropyl benzene		Vul-Cup[b]
2,5-Dimethyl-2,5 di(*t*-butyl peroxy) hexane		Lupersol 101[a]
2,5-Dimethyl-2,5 di(*t*-butyl peroxy) hexyne-3		Lupersol 130[a]
Di-*t*-butyl peroxide	DTBP	
t-butyl hydroperoxide	TBHP	

[a] Registered Trademark, Lucidol Division, Pennwalt Corp.
[b] Registered Trademark, Hercules, Inc.

acid (commonly maleic anhydride or fumaric acid), a glycol (e.g., ethylene, propylene, diethylene, or dipropylene glycol) and a saturated dibasic acid (such as phthalic anhydride, isophthalic acid, or adipic acid) to modify the degree of unsaturation. Some common cross-linking monomers are styrene, vinyl toluene, methyl methacrylate, diallyl phthalate, and triallyl cyanurate. Conventional inhibitors are hydroquinone, quinone, and *t*-butyl catechol.

The basic resins can be cross-linked by free radical initiators, usually an organic peroxide (see Table XVII). Free radicals can be formed by thermal decomposition of the catalyst (e.g., *t*-butyl perbenzoate or benzoyl peroxide), ultraviolet light decomposition, or by chemical decomposition in ambient temperature applications (e.g., methyl ethyl ketone peroxide or benzoyl peroxide).

F. Poly(Acrylic Acid)

Both intermolecular and intramolecular anhydride formation can occur in poly(acrylic acid) (PAA) in the 150–300°C range. Intermolecular anhydride formation results in a cross-linked polymer network. The kinetics and type of cross-linking in a composite material consisting of PAA (48% by weight) and alumina have been studied by thermogravimetry (Greenberg and Kamel, 1977). These materials are suitable for use in orthopedic implant applications. Complete conversion of all acid groups to anhydride corresponds to 12.5% loss of water by weight of linear polymer. Complete conversion in the preceding composite would thus result in a 6.0% weight loss. Intermolecular anhydride formation is

$$
\begin{array}{c}
\text{R}-\text{CH}_2-\underset{\displaystyle |}{\text{CH}}-\text{R}' \\
\underset{\displaystyle |}{\text{C}}{=}\text{O} \\
\underset{\displaystyle |}{\text{OH}} \\
\text{OH} \\
\underset{\displaystyle |}{\text{C}}{=}\text{O} \\
\text{R}-\text{CH}_2-\text{CH}-\text{R}'
\end{array}
\xrightarrow{\text{heat}}
\begin{array}{c}
\text{R}-\text{CH}_2-\underset{\displaystyle |}{\text{CH}}-\text{R}' \\
\underset{\displaystyle |}{\text{C}}{=}\text{O} \\
\text{O} \\
\underset{\displaystyle |}{\text{C}}{=}\text{O} \\
\text{R}-\text{CH}_2-\text{CH}-\text{R}'
\end{array}
+ \text{H}_2\text{O} \qquad (71)
$$

VI. Conclusions

The full complement of thermal characterization techniques offers both user and maker of thermosets a truly unique capability. Combination with spectroscopic methods for direct chemical group analysis, such as IR and NMR, and methods for evolved gas analysis, such as GC, MS, and IR, afford complete and total characterization capability. Thermal analysis microtechniques can provide a detailed road map of curing that has predictive utility. They also provide the ability to assess degree of cure and composition in addition to several thermodynamic, mechanical, and rheological parameters. Stability, i.e., the ability to perform a function or maintain properties over an extended period, can often be judged. The ability to measure experimentally the effects of fillers, additives, moisture, and so on, on curing and physical properties is a definite asset since these are often part of real thermosetting systems. In view of such powerful capabilities, it is surprising that use of these thermoanalytical techniques is not more widespread. As an example, very little and very incomplete coverage is provided in recent books, such as "Epoxy Resins" (May and Tanaka, 1973) and "Chemistry and Application of Phenolic Resins (Knop and Scheib, 1979). It can be concluded only that application of thermal analysis to thermosets is in its infancy and that potential for growth is tremendous.

In this chapter the study of thermosets has been divided into three major areas: curing, properties of cured materials, and stability of cured materials. Since the user of thermosets, e.g. semiconductor device manufacturers or the automobile industry, must cure the A- or B-staged resins, and since ultimate properties depend on the state of cure, it is only reasonable that curing is the most critical factor affecting performance. Shelf life, which involves primarily the avoidance of gelation, is also a key parameter.

To make optimum use of thermosets requires accurate time–temperature descriptions of the path to gelation and to some reasonable extent of
cure (e.g., >95%). Figure 30 is a good example. Other key parameters are
$T_{c\infty}$, the minimum temperature at which total cure can be achieved, and
$T_{c,gel}$, the cure temperature at which gelation and vitrification occur simultaneously. $T_{c,gel}$ is an important parameter in B staging and shelf life. It is
also necessary to know or establish values of ΔH_{RXN} or ΔW_{RXN}, the heat
change and weight loss, respectively, corresponding to conversion of all
reactive groups.

Determination of the time–temperature–degree of cure relationship of
a thermoset requires an orderly progression of experiments, which have
been discussed in Section II.A. In summary, one must first determine
whether the reaction follows nth-order or autocatalyzed kinetics, fol-
lowed by evaluation of $T_{c\infty}$ and ΔH_{RXN} or ΔW_{RXN}. It must then be demonstrated that the data are indeed represented by the appropriate equation
over the entire course of reaction (remembering that below $T_{c\infty}$, vitrification will halt the reaction short of 100% conversion) and over the temperature range, catalyst concentration range, etc., of interest. For nth-order
reactions, this involves determination of the reaction order; most curing
reactions are first or second order. For most autocatalyzed reactions,
verification of the most common model [cf. Eq. (13)] is sufficient; occasionally the constants m and n in Eq. (3) must be evaluated. If comonomers are present in nonstoichiometric amounts ($B \neq 1$), this must be
taken into account for both types of reaction. Before embarking on the
kinetic experimentation, some attention to experimental detail is necessary. Precautions should be taken to avoid premature reaction prior to the
experiment and loss of reactants or unrecorded reaction in a DSC experiment. In a TG experiment, it should be verified that the measured weight
loss is indeed due to the expected reaction products. TG–MS and TG–
GC are especially useful combinations. Greater use of thermogravimetry
in the study of condensation reactions should occur with the current availability of high-precision instruments with good derivative capability. If
the rate of weight loss is being utilized, care must be taken to insure that
the chemical reaction kinetics are differentiated from other effects, such
as diffusion and evaporation from surfaces. Isothermal kinetic experimentation is recommended, method 1 being preferred because of simultaneous measurement of extent and rate of reaction. Dynamic method B experiments, which utilize the variation in peak exotherm temperature with
heating rate (Section IV.B), are always valuable. This method quickly
gives an accurate measure of the activation energy, and frequency factor
if the rate equation is known. The data should be verified by comparing
calculated half-lives ($t_{1/2}$) or times to 95% conversion ($t_{0.95}$) with values

measured by one of the isothermal methods. Method B is often the only means to analyze the curing kinetics of complex systems with multiple exotherms, solvent effects or unreliable base lines. Dynamic method A has tremendous potential but is not yet at a level of consistent reliability. Times and extents of reaction at which vitrification occurs at temperatures below $T_{c\infty}$ and independent measurement of gel times (by TBA or DMA, for example), allowing complementary determination of α_{gel}, complete the cure study.

The key property of a cured thermoset, which ultimately determines the level of all physical properties (modulus, thermal expansion, tensile and bond strengths, conductivity, stability, etc.) is the degree of cure. A very sensitive measure of the degree of cure was shown to be the measurement of thermal expansion through the glass transition interval. The transition temperature, the expansion rates (above and below T_g), and the change in expansion rate on devitrification are all measures of the state of cure. Measurement of the softening temperature by TMA can be a good indicator of the glass transition but is also sensitive to independent changes in modulus from, for example, water sorption. Measurement of the modulus (by DMA, for example) or its equivalent, cross-link density, by solvent swell (TMA or TG) are also sensitive indicators of degree of cure. Solvent swell by TG has the advantage of not requiring a specific sample geometry and simultaneously measuring the soluble fraction. Measurements of T_g or residual exotherm by DSC, or residual weight loss by TG, are also good indicators of degree of cure. All of these methods require attention to experimental and theoretical details. When used properly, sensitivity to $>95\%$ to $>98\%$ of cure can be achieved.

Another important parameter of cured thermosets easily addressed by thermal analysis is composition. The most common compositional measurement is filler content by TG. This is very important in quality control. Qualitative identification of fillers (e.g., $CaCO_3$ and $Al(OH)_3$) and resins can be very useful, especially where a thermoset contains multiple fillers and/or polymers.

Stability implies product reliability, i.e., the ability of a material or assembly to maintain its properties or perform a function over an extended period. The tacit assumption of a temperature-activated failure mechanism is commonly made, resulting in an accelerated aging study, which is not unlike a curing study. The major difficulty is elucidation of a failure mechanism that, on the one hand, is amenable to study in an accelerated mode, and on the other hand, is truly representative of the actual failure mode. To date, thermal analysis methods have not been sufficiently reliable to replace the more tedious and costly ''standard'' techniques. The application of new experimental methods, such as parameter jump methods (Flynn and Dickens, 1976; Flynn, 1978b) plus increasing on-line, interactive computational capabilities, should provide the needed reliability. Thermal characterization techniques are commonly used to rank materials according to relative stability (thermal, thermal-oxidative, or burning stability for example). Such rankings are strictly *relative,* and the reliability of such studies is dependent on the relationship between events occurring in a thermobalance or calorimeter and the actual environment in which the thermoset must function.

There are several apparent applications of thermal characterization techniques that have not been fully utilized, or not reported if they have been utilized. One of these is the interaction of thermosets with water, which is just beginning to receive attention. The area of dimensional stability, which often involves the simultaneous response of materials to changes in temperature and humidity, can be studied by TMA. Hydrolytic stability, which can be assessed by TG where decomposition products are volatile or by TMA, TBA, or DMA where cross-link density is affected, can also be studied by these methods. These applications require the maintenance of a constant but variable water vapor pressure in the purge gas, which is not difficult. The general study of stability and degradation in various environments is another area where potential is not being fully utilized. Use of the thermobalance and thermal mechanical analyzer to measure cross-link density by solvent swell on a rapid, microscale has great potential.

ACKNOWLEDGMENTS

Several people have made significant and material contributions to this work. Foremost, I must express my most sincere gratitude to the editor, Dr. Edith A. Turi, without whom this book would not exist. Her always cheerful disposition, her constant and unfaltering encouragement and concern, her uncountable suggestions and help have made the writing of this chapter a real joy and most rewarding experience for me.

I was privileged to have four distinguished colleagues review the various stages of this chapter. A special thanks to Professor Musa Kamal of McGill University, whose painstakingly thorough review, numerous suggestions, and very useful critique was invaluable in the preparation of the final manuscript. Appreciation is also expressed to Professor John Gillham of Princeton University and to Dr. Nathan Schneider and Dr. Robert Sacher of the Army Materials and Mechanics Research Center, all of whose comments and suggestions resulted in significant improvements to this chapter.

Sincere appreciation is expressed to the IBM Corporation for generously providing the resources for this work. I would especially like to recognize Ms. Barbara Blank for meticulous preparation of the first two manuscripts, Mr. Haik Marcar for editing the final manuscript, Mr. John Minick for the illustrations, and the Word Processing Center for the beautiful and speedy preparation of the final manuscript.

Finally, I must acknowledge my family, who patiently accepted the loss of their husband and father for too many weekends.

References

Abolafia, O. R. (1969). *Soc. Plast. Eng. [Tech. Pap.]* **15**, 610–616.

Acitelli, M. A., Prime, R. B., and Sacher, E. (1971). *Polymer* **12**, 333–343.

Adamson, M. J. (1979). *NASA Tech. Memo.* **NASA TM-X-78610.**

Adamson, M. J. (1980). *J. Mater. Sci.* **15**, 1736–1745.

Anagnostou, T., and Jaul, E. (1981). *J. Coatings Tech.* **53**, 35–45.

Anderson, H. C. (1966). *Polymer* **7**, 193–195.

Andrejs, B., Schulz, J. P., and Wappler, E. (1979). *Thermochim. Acta* **29**, 309–314.

Angeloni, F. M. (1967). *Polym. Prepr., Am. Chem. Soc., Div. Polym. Chem.* **8**(2), 950–962.

Antal, I., and Csillag, L. (1975). *In* "Thermal Analysis" (I. Buzas, ed.), Vol. 3, pp. 347–358. Heyden, London.

Arada, B., Lin, S. C., and Pearce, E. M. (1979). *Int. J. Polym. Mater.* **7**, 167–184.

Ashton, J. E., Halpin, J. C., and Petit, R. H. (1969). "Primer on Composite Materials Analysis." Technomic, Stamford, Connecticut.

Babayevsky, P. G., and Gillham, J. K. (1973). *J. Appl. Polym. Sci.* **17**, 2067–2088.

Bailey, W. J. (1975). *J. Macromol. Sci., Chem.* **A9** (5), 849–865.

Bank, M., Bayless, R., Botham, R., and Shank, P. (1972). *Polym. Prepr., Am. Chem. Soc., Div. Polym. Chem.* **13** (2), 1250–1255.

Bank, M., Bayless, R., Botham, R., and Shank, P. (1973). *Mod. Plast.* **50**, 84–86.

Barrett, K. E. J. (1967). *J. Appl. Polym. Sci.* **11**, 1617–1926.

Bartlett, C. J. (1978). *J. Elastomers Plast.* **10**, 369–376.

Barton, J. M. (1973). *Makromol. Chem.* **171**, 247–251.

Barton, J. M. (1974a). *J. Macromol. Sci., Chem.* **A8** (1), 53–64.

Barton, J. M. (1974b). *In* "Polymer Characterization by Thermal Methods of Analysis" (J. Chiu, ed.), pp. 25–32. Dekker, New York.

Barton, J. M. (1979). *Brit. Polym. J.* **11**, 115–119.

Barton, J. M., and Shepherd, P. M. (1975). *Makromol. Chem.* **176**, 919–930.

Barton, J. M., Lee, W. A., and Wright, W. W. (1978). *J. Therm. Anal.* **13**, 85–89.

Batzer, V. H., Lohse, F., and Schmid, R. (1973). *Angew. Makromol. Chem.* **29/30**, 349–411.

Belani, J. G., and Sporck, C. R. (1978). *Soc. Plast. Eng. [Tech. Pap.]* **24**, 577–579.

Bell, J. P. (1970). *J. Polym. Sci., Polym. Phys. Ed.* **6** (6), 417–436.

Blaine, R. L. (a). *DuPont Appl. Brief* **TA-61.**

Blaine, R. L. (1981). Private communication.

Blaine, R. L., and Lofthouse, M. G. (a). *Dupont Appl. Brief* **TA-65.**

Blaine, R. L., Gill, P. S., and Hassel, R. L. (1979). *Soc. Plast. Eng. [Tech. Pap.]* **25**, 822–824.

Bloechle, D. P. (1978). *J. Elastomers Plast.* **10**, 377–385.

Boneig, H. V. (1964). "Unsaturated Polyester: Structure and Properties." Elsevier, Amsterdam.

Borchardt, H. J., and Daniels, F. (1956). *J. Am. Chem. Soc.* **79**, 41–46.

Brand, J. (1977). *Tech. Pap. IPC-TP Inst. Printed Circuits* **IPC-TP-152.**

Brennan, W. P. (1978). *Perkin-Elmer* **TAAS 26.**

Brennan, W. P., and Cassel, B. (1978). *Perkin-Elmer* **TAAS 25.**

Brett, C. L. (1976). *J. Appl. Polym. Sci.* **20**, 1431–1440.

Brewis, D. M., Comyn, J., and Fowler, J. R. (1979). *Polymer* **20**, 1548–1552.

Brown, G. P., Haarr, D. T., and Metlay, M. (1970). *Thermochim. Acta* **1**, 441–449.

Byrne, C. A., Hagnauer, G. L., Schneider, N. S., and Lenz, R. W. (1980). *Polym. Composites* **1**, 71–76.

Carpenter, J. F. (1976). *Natl. SAMPE Symp. Exhib. [Proc.]* **21,** 783–802.

Cassel, B. (1976). *Perkin-Elmer* **MA-29.**

Cassel, B. (1977a). *Perkin-Elmer* **TAAS 19.**

Cassel, B. (1977b). *Perkin-Elmer* **TAAS 20.**

Cassel, B. (1980), *Polym. News* **6**(3), 108–115.

Cassel, B., and Gray, A. P. (1980). *Thermochim. Acta* **36,** 265–277.

Cernee, F., Osredkar, U., Moze, A., Vizovisek, I., and Lapanje, S. (1977). *Makromol. Chem.* **178,** 2197–2203.

Cessna, L. C., Jr., and Jabloner, H. (1974). *J. Elastomers Plast.* **6,** 103–113.

Chang, S. S. (1981). *In* "Thermal Analysis in Polymer Analysis" (E. A. Turi, ed.), Hayden, London.

Charlesworth, J. M. (1979a). *J. Polym. Sci., Polym. Phys. Ed.* **17,** 1557–1569.

Charlesworth, J. M. (1979b). *J. Polym. Sci., Polym. Phys. Ed.* **17,** 1571–1580.

Chen, J. S. (1978). *SME Tech. Pap. [Ser.] EM* **EM78-403.**

Choi, S. V. (1970). *SPE J.* **26,** 51–54.

Chow, S., and Steiner, P. R. (1979). *J. Appl. Polym. Sci.* **23,** 1973–1985.

Crane, L. W., Dynes, P. J., and Kaelble, D. H. (1973). *J. Polym. Sci., Polym. Lett. Ed.* **11,** 533–540.

Creedon, J. P. (1970). *Anal. Calorim.* **2,** 185–199.

Cuthrel, R. E. (1968). *J. Appl. Polym. Sci.* **12,** 955–967.

David, D. J. (1967). *Insulation (Libertyville, Ill.)* **13** (November), 38–45.

Dienes, G. J., and Klemm, H. F. (1946). *J. Appl. Phys.* **17,** 458–471.

Doyle, C. D. (1961). *Anal. Chem.* **33,** 77–79.

Doyle, M. J., Lewis, A. F., and Li, H. M. (1979). *Polym. Eng. Sci.* **19,** 687–691.

Draper, A. L. (1970). *Proc. Toronto Symp. Therm. Anal., 3rd,* pp. 63–69.

Droste, D. H., and Dibenedetto, A. T. (1969). *J. Appl. Polym. Sci.* **13,** 2149–2168.

DuPont Instruments (a). "Thermal Analysis Review: Dynamic Mechanical Analysis." Dupont Instruments, Wilmington, Delaware.

DuPont Instruments (b). Preliminary Sales Information for the Parallel Plate Rheometer TMA Accessory. Dupont Instruments, Wilmington, Delaware.

Dusek, K., and Bleha, M. (1977). *J. Polym. Sci., Polym. Chem. Ed.* **15,** 2393–2400.

Dusek, K., Ilavsky, M., and Lunak, S. (1975). *J. Polym. Sci., Polym. Symp. Ed.* **53,** 29–44.

Duswalt, A. A. (1974). *Thermochim. Acta* **8,** 57–68.

Dutta, A., and Ryan, M. E. (1979). *J. Appl. Polym. Sci.* **24,** 635–649.

Eisenmann, D. E., and Halyard, S. M., Jr. (1976). *Thermochim Acta* **14,** 87–97.

Ennis, B. C., and Williams, J. G. (1977). *Thermochim. Acta* **21,** 355–367.

Enns, J. B., and Gillham, J. K. (1980). *N. Am. Therm. Anal. Soc. [Proc.]* **10,** 303–309.

Enns, J. B., Gillham, J. K., and Small, R. (1981). *Am. Chem. Soc. Div. Polym. Chem. Preprints* **22**(2), 123–124.

Era, V. A. Lindberg, J. J., Mattila, A., Vauhkonen, L., and Linnahalme, T. (1976). *Angew. Makromol. Chem.* **50,** 43–52.

Era, V. A., Mattila, A., and Lindberg, J. J. (1977). *Angew. Makromol. Chem.* **64,** 235–238.

Evans, A. J., Armstrong, C., and Tolman, R. J. (1978). *J. Oil. Colour Chem. Assoc.* **61,** 251–255.

Eyerer, P. (1974). *J. Appl. Polym. Sci.* **18,** 975–992.

Ezrin, M., and Claver, G. C. (1969). *Appl. Polym. Symp.* **8,** 159–170.

Fava, R. A. (1968). *Polymer* **9,** 137–151.

Flory, P. J. (1953). "Principles of Polymer Chemistry." Cornell Univ. Press, Ithaca, New York.

Flory, P. J., and Rehner, J., Jr. (1943). *J. Chem. Phys.* **11**, 521–526.

Flynn, J. H. (1974). *Thermochim. Acta* **8**, 69–81.

Flynn, J. H. (1978a). *In* "Thermal Methods in Polymer Analysis" (S. W. Shalaby, ed.), pp. 163–186. Franklin Inst. Press, Philadelphia, Pennsylvania.

Flynn, J. H. (1978b). *In* "Aspects of Degradation and Stabilization of Polymers" (H. H. G. Jellinek, ed.), pp. 573–603. Elsevier, Amsterdam.

Flynn, J. H. (1980). *Thermochim. Acta* **37**, 225–238.

Flynn, J. H., and Dickens, B. (1976). *Thermochim. Acta* **15**, 1–16.

Flynn, J. H. and Wall, L. A. (1966). *J. Res. Natl. Bur. Stand., Sect. A* **70**, 487–523.

Freeberg, F. E., and Alleman, T. C. (1966). *Anal. Chem.* **38**, 1806–1807.

Gaskill, R., and Barrall, E. M., II (1975). *Thermochim. Acta* **12**, 102–104.

Gill, P. S. (1980), "Thermal Analysis: New Techniques and Applications for Characterization of Thermosetting Polymers," ACS Lab. Prof. Ser.

Gillham, J. K. (1974). *AIChE J.* **20**, 1066–1079.

Gillham, J. K. (1976). *Polym. Eng. Sci.* **16**, 353–356.

Gillham, J. K. (1979a). *Polym. Eng. Sci.* **19**, 319–326.

Gillham, J. K. (1979b). *Polym. Eng. Sci.* **19**, 676–682.

Gillham, J. K. (1980). *Soc. Plast. Eng.* [*Proc. An. Tech. Conf.*] **38**, 268–271.

Gillham, J. K., and Mentzer, C. C. (1972). *Polym. Prepr., Am. Chem. Soc., Div. Polym. Chem.* **13**(1), 247–252.

Gillham, J. K., Benci, J. A., and Noshay, A. (1974). *Polym. Prepr., Am. Chem. Soc., Div. Polym. Chem.* **15**(1), 241–247.

Gillham, J. K., Glandt, C. A., and McPherson, C. A. (1977). *In* "Chemistry and Properties of Crosslinked Polymers" (S. S. Labana, ed.), pp. 491–520. Academic Press, New York.

Gordon, S. E. (1967). *Polym. Prepr., Am. Chem. Soc., Div. Polym. Chem.* **8**(2), 955–962.

Gordon, S. E. (1969). *In* "Thermal Analysis" (R. F. Schwenker, Jr., and P. D. Garn, eds.), Vol. 1, pp. 667–682. Academic Press, New York.

Gray, A. P. (1972). *Perkin-Elmer* **TAAS 2.**

Gray, D. D., ed. (1963). "American Institute of Physics Handbook." McGraw-Hill, New York.

Greenberg, A. R., and Kamel, I. (1977). *J. Polym. Sci., Polym. Chem. Ed.* **15**, 2137–2149.

Grentzer, T. H., Holsworth, R. M., and Prouder, T. (1981a). *Amer. Chem. Soc. Div. Org. Coat. and Plast. Preprints* **44**, 515–519.

Grentzer, T. H., Holsworth, R. M., and Prouder, T. (1981b). *Soc. Plast. Eng.* [*Proc. An. Tech. Conf.*] **39**, 156–157.

Hagnauer, G. L., Sprouse, J. F., Sacher, R. E., Setton, I., and Wood, M. (1978). *U.S. Army Mater. Mech. Res. Cent.* [*Tech. Rep.*] **AMMRC TR 78-8** (AD 054625).

Hahn, T. A. (1970). *J. Appl. Phys.* **41**, 5096–5101.

Halpin, J. C., and Pagano, N. J. (1969). *J. Compos. Mater.* **3**, 720–724.

Hartmann, B. (1975). *J. Appl. Polym. Sci.* **19**, 3241–3255.

Hartmann, B., and Lee, G. F. (1977). *J. Appl. Polym. Sci.* **21**, 1341–1349.

Hassel, R. L. (a). *Dupont Appl. Brief* **TA-75.**

Hassel, R. L. (1978). *Ind. Res. Dev.* **20**, 160–163.

Hassel, R. L., and Blaine, R. L. (1979). *Natl. SAMPE Symp. Exhib.* [*Proc.*] **24**, 342–350 (Book 1).

Heijboer, J. (1979). *Polym. Eng. Sci.* **19**, 664–673.

Higgs, D. A., and Manley, T. R. (1973). *Chem. Ind.* (*London*) **23**, 1112.

Hinrichs, R., and Thuen, J. (1979). *Natl. SAMPE Symp. Exhib.* [*Proc.*] **24,** 404–421 (Book 1).

Holliday, L., and Robinson, J. (1973). *J. Mater. Sci.* **8,** 301–311.

Holsworth, R. M. (1969). *J. Paint Technol.* **41,** 167–168.

Horie, K., Hiura, H., Souvada, M., Mita, I., and Kambe, H. (1970a). *J. Polym. Sci., Polym. Chem. Ed.* **8,** 1357–1372.

Horie, K., Mita, I., and Kambe, H. (1970b). *J. Polym. Sci., Polym. Chem. Ed.* **8,** 2839–2852.

Horowitz, N. H., and Metzger, G. (1963). *Anal. Chem.* **35,** 1464–1468.

Huneke, J. T., and Dorey, J. K., II (1980). *N. Am. Therm. Anal. Soc.* [*Proc.*] **10,** 175–178.

Jackson, J., Thomas, R., Scharr, T., and Zettek, R. (1975). *Soc. Plast. Eng.* [*Proc. An. Tech. Conf.*] **33,** 488–490.

Jones, J. F., Bartels, T. T., and Fountain, R. (1974). *Polym. Eng. Sci.* **14,** 240–245.

Kaelble, D. H., and Cirlin, E. H. (1971). *J. Polym. Sci., Part C* **35,** 79–100.

Kaelble, D. H., Dynes, P. J., and Cirlin, E. H. (1974). *J. Adhes.* **6,** 23–48.

Kamal, M. R. (1974). *Polym. Eng. Sci.* **14,** 231–239.

Kamal, M. R., and Sourour, S. (1973). *Polym. Eng. Sci.* **13,** 59–64.

Katovic, Z. (1967a). *J. Appl. Polym. Sci.* **11,** 85–93.

Katovic, Z. (1967b). *J. Appl. Polym. Sci.* **11,** 95–102.

Kay, R., and Westwood, A. R. (1975). *Eur. Polym. J.* **11,** 25–30.

Kerner, E. H. (1956). *Proc. Phys. Soc., London, Sect. B* **69,** 808–813.

King, P. W., Mitchell, R. H., and Westwood, A. R. (1974). *J. Appl. Polym. Sci.* **18,** 1117–1130.

Kissinger, H. E. (1957). *Anal. Chem.* **29,** 1702–1706.

Klute, C. H., and Viehmann, (1961). *J. Appl. Polym. Sci.* **5,** 86–95.

Knop, A., and Scheib, W. (1979). "Chemistry and Application of Phenolic Resins." Springer-Verlag, Berlin and New York.

Knowles, K. F. (1978). *Tech. Rep* **E-78-23,** NTIS AD-A062-602, National Technical Information Service, U.S. Dept. of Commerce.

Kong, E. S. W., Tant, M. R., Wilkes, G. L., Banthia, A. K., and McGrath, J. E. (1979). *Am. Chem. Soc. Div. Polym. Chem. Preprints* **20,** 531–534.

Kourtides, D. A. and Parker, J. A. (1979). *Soc. Plast. Eng.* [*Proc. An. Pac. Tech. Conf., 4th*] **4,** 51–54.

Kourtides, D. A., Gilwee, W. J., Jr., and Parker, J. A. (1979). *Polym. Eng. Sci.* **19,** 24–29.

Koutsky, J. A., and Ebewele, R. (1977). *In* "Chemistry and Properties of Cross-linked Polymers" (S. S. Labana, ed.), pp. 521–533. Academic Press, New York.

Kreahling, R. P., and Kline, D. E. (1969). *J. Appl. Polym. Sci.* **13,** 2411–2425.

Kreibich, U. T., and Schmid, R. (1975). *J. Polym. Sci., Polym. Symp. Ed.* **53,** 177–185.

Krizanovsky, L., and Mentlik, V. (1978). *J. Therm. Anal.* **13,** 571–580.

Kubin, R. F. (1979). *Tech. Rep.* NWC-TP-6104, Naval Weapons Center, China Lake, California.

Kubota, H. (1975). *J. Appl. Polym. Sci.* **19,** 2279–2297.

Kurachenkov, V. I., and Igonin, L. A. (1971). *J. Polym. Sci., Polym. Chem. Ed.* **9,** 2283–2289.

Kwei, T. K. (1963). *J. Polym. Sci., Part A* **1,** 2977–2988.

Labana, S. S., and Chang, V. F. (1972). *Polym. Prepr., Am. Chem. Soc., Div. Polym. Chem.* **13**(1), 241–246.

Lamoureux, R. T. (1978). *Proc. Tech. Program—Natl. Electron. Packag. Prod. Conf., 1978* pp. 161–168.

Lauver, R. W. (1978). *Proc. Annu. Conf.—Reinf. Plast./Compos. Inst., Soc. Plast. Ind.* **33,** Sect. 15-C.

Lee, H., and Neville, K. (1967). "Handbook of Epoxy Resins." McGraw-Hill, New York.

Lee, H. T., and Levi, D. W. (1969). *J. Appl. Polym. Sci.* **13,** 1703–1705.

Levy, P. F. (1975). *In* "Thermal Analysis" (I. Buzas, ed.), Vol. 3, pp. 3–14. Heyden, London.

Levy, P. F., Nieuweboer, G., and Semanski, L. C. (1970). *Thermochim. Acta* **1,** 429–439.

Lewis, A. F., and Gillham, J. K. (1962). *J. Appl. Polym. Sci.* **6,** 422-nnn.

Lewis, A. F., Doyle, M. J., and Gillham, J. K. (1979). *Polym. Eng. Sci.* **19,** 683–686.

Lin, S. C., and Pearce, E. M. (1978). *Polym. Prepr., Am. Chem. Soc., Div. Polym. Chem.* **19**(2), 17–22.

Lin, S. C. and Pearce, E. M. (1979a). *J. Appl. Polym. Sci.* **23,** 3095–3119.

Lin, S. C. and Pearce, E. M. (1979b). *J. Appl. Polym. Sci.* **23,** 3355–3374.

Lum, R. M., and Feinstein, L. G. (1980). *Elec. Components Conf.* [*Proc.*] **30,** 113–120.

Lum, R. M., and Feinstein, L. G. (1981). *J. Microelec. Reliability* **21**(1), 15–31.

Lunak, S., and Dusek, K. (1975). *J. Polym. Sci., Polym. Symp. Ed.* **53,** 44–55.

Lunak, S., Klaban, J., Vladyka, J., and Smreka, J. (1978). *Congr. FATIPEC* **14,** 405–411.

Maas, T. A. M. M. (1978). *Polym. Eng. Sci.* **18,** 29–32.

MacCallum, J. P., and Tanner, J. (1970). *Nature* **225,** 1127–1128.

Machin, D., and Rogers, C. E. (1970). *Polym. Eng. Sci.* **10,** 300–304.

McKague, E. L., Jr., Reynolds, J. D., and Halkias, J. E. (1973). *J. Test. Eval.* **1,** 468–471.

McPherson, C. A., and Gillham, J. K. (1978). *Org. Coat. Plast. Chem.* **38,** 229–231.

Malavasic, T., Moze, A., Vizovisek, I., and Lapanje, S. (1975). *Angew. Makromol. Chem.* **44,** 89–97.

Manley, T. R. (1967). *Plast. Inst., Trans. J.* **35,** 525–527.

Manley, T. R. (1974a). *J. Macromol. Sci., Chem.* **A8**(1), 53–64.

Manley, T. R. (1974b). *In* "Polymer Characterization by Thermal Methods of Analysis" (J. Chiu, ed.), pp. 53–64. Dekker, New York.

Manz, W., and Creedon, J. P. (1972). *In* "Thermal Analysis" (H. G. Wiedemann, ed.), Vol. 3, pp. 145–157. Birkhaeuser, Basel.

Manzione, L. T., Gillham, J. K., and McPherson, C. A. (1981a). *J. Appl. Polym. Sci.* **26,** 889–905.

Manzione, L. T., Gillham, J. K., and McPherson, C. A. (1981b). *J. Appl. Polym. Sci.* **26,** 907–919.

Mark, R., and Findley, W. N. (1978). *Polym. Eng. Sci.* **18,** 6–15.

Martin, R. W. (1956). In "The Chemistry of Phenolic Resins." Wiley, New York.

May, C. A., and Tanaka, Y. (1973). "Epoxy Resins." Dekker, New York.

May, C. A., Whearty, D. K., and Fritzen, J. S. (1976). *Natl. SAMPE Symp. Exhib.* [*Proc.*] **21,** 803–818.

May, C. A., Hadad, D. K., and Browning, C. E. (1978). *Proc. Annu. Conf.—Reinf. Plast./Compos. Inst., Soc. Plast. Ind.* **33,** Sect. 15-D.

Miller, R. L., and Oebser, M. A. (1980). *Thermochim. Acta* **36,** 121–131.

Murayama, T., and Bell, J. P. (1970). *J. Polym. Sci., Polym. Phys. Ed.* **8,** 437–445.

Neet, T. E., Parker, B. G., and Smith, C. H. (1978), *J. Cell. Plast.* **14**(4), 213–218.

Noll, T. E. (1977). *Tech. Pap. IPC-TP Inst. Printed Circuits* **IPC-TP-140.**

Olcese, T., and Spelta, O. (1975). *J. Polym. Sci., Symp.* **53,** 113–126.

Ophir, Z. H., Emerson, J. A., and Wilkes, G. L. (1978). *J. Appl. Phys.* **49,** 5032–5038.

Orsi, F., Bertalan, R. G., Anna, P. Gyurkovics, I., and Laczko, M. (1975). *In* "Thermal Analysis" (I. Buzas, ed.), Vol. 2, pp. 105–115. Heyden, London.

Ozawa, T. (1965). *Bull. Chem. Soc. Jpn.* **38,** 1881–1886.

Ozawa, T. (1970). *J. Therm. Anal.* **2**, 301–324.

Pappalardo, L. T. (1973). *Soc. Plast. Eng.* [*Tech. Pap.*] **20**, 13–16.

Parker, B. G., and Smith, C. H. (1979). *Mod. Plast.* **56**, 58–61.

Pearce, E. M., Lin, S. C., Lin, M. S., and Lee, S. N. (1978). *In* "Thermal Methods in Polymer Analysis" (S. W. Shalaby, ed.), pp. 187–198. Franklin Inst. Press, Philadelphia, Pennsylvania.

Penn, L., Morra, B., and Mones, E. (1977). *Composites* **8**, 23–26.

Penn, L. S. (1979). *ASTM Spec. Tech. Publ.* **674**, 519–532.

Peyser, P., and Bascom, W. D. (1974). *Anal. Calorim.* **3**, 537–554.

Peyser, P., and Bascom, W. D. (1975). *J. Polym. Sci., Polym. Lett. Ed.* **13**, 129–130.

Peyser, P., and Bascom, W. D. (1977). *J. Polym. Sci.* **21**, 2359–2373.

Pierron, E. D., and Bobos, G. E. (1977). *J. Electron. Mater.* **6**, 333–348.

Pinheiro, M. DeF. F., and Rosenberg, H. M. (1980). *J. Polym. Sci, Polym. Phys. Ed.* **18**, 217–226.

Porowska, E., and Stareczek, T. (1975). *In* "Thermal Analysis" (I. Buzas, ed.), Vol. 3, pp. 401–413. Heyden, London.

Porowska, E., and Tokarzewska, M. (1979). *J. Macromol. Sci., Chem.* **A13**(7), 909–921.

Prime, R. B. (1970). *Anal. Calorim.* **2**, 210–210.

Prime, R. B. (1973). *Polym. Eng. Sci.* **13**, 365–371.

Prime, R. B. (1978). *Thermochim. Acta* **26**, 165–174.

Prime, R. B., and Sacher, E. (1972). *Polymer* **13**, 455–458.

Prime, R. B., Barrall, E. M., II, Logan, J. A., and Duke, P. J. (1974). *AIP Conf. Proc.* **17**, 72–83.

Reed, K. E. (1979). *Proc. Annu. Conf.—Reinf. Plast./Compos. Inst., Soc. Plast. Ind.* **34**, Sect. 22G.

Reed, K. E. (1980). *Polym. Composites* **1**, 44–49.

Ritchie, K. (1976). *Microelectron. Reliab.* **15**, 489–490.

Ritchie, K., Hunter, W. L., and Malkiewicz. (1972a). *In* "Thermal Analysis" (H. G. Wiedemann, ed.), Vol. 3, pp. 179–185. Birkhaeuser, Basel.

Ritchie, K., Hunter, W., Malkiewicz, C., and Maze, C. (1972b). *Soc. Plast. Eng.* [*Tech. Pap.*] **18**(Pt1), 114–116.

Roller, M. B. (1979). *Polym. Eng. Sci.* **19**, 692–698.

Ryan, M. E., and Dutta, A. (1979). *Polymer* **20**, 203–206.

Sacher, E. (1973). *Polymer* **14**, 91–95.

Sanders, C. I. (1970). *J. Paint Technol.* **42**, 405–408.

Saunders, T. F., Levy, M. F., and Serino, J. F. (1967). *J. Polym. Sci., Polym. Chem. Ed.* **5**, 1609–1617.

Schapery, R. A. (1968). *J. Compos. Mater.* **2**, 380–404.

Schindlbauer, H., Henkel, G., Weiss, J., and Eichberger W. (1976). *Agnew. Makromol. Chem.* **49**, 115–128.

Schneider, N. S., and Gillham, J. K. (1978). *Org. Coat. Plast. Chem.* **38**, 491–496.

Schneider, N. S., and Gillham, J. K. (1980). *Polym. Composites* **1**, 97–102.

Schneider, N. S., Sprouse, J. F., Hagmauer, G. L., and Gillham, J. K. (1979). *Polym. Eng. Sci.* **19**, 304–312.

Sebenik, A., Vizovisek, I., and Lapanje, S. (1974). *Eur. Polym. J.* **10**, 273–278.

Senich, G. A., MacKnight, W. J., and Schneider, N. S. (1979). *Polym. Eng. Sci.* **19**, 313–318.

Serafini, T. T., Delvigs, P., and Vannucci, R. D. (1979). *NASA Tech. Memo.* **NASA TM-X-79226.**

Shechter, L., Wynstra, J., and Kurkjy, R. P. (1956). *Ind. Eng. Chem.* **48**, 94–97.

Siegmann, A., and Narkis, M. (1977). *J. Appl. Polym. Sci.* **21**, 2311–2318.
Simmons, E. L., and Wendlandt, W. W. (1972). *Thermochim. Acta* **3**, 498–500.
Slysh, R., and Gayler, K. E. (1978). *Polym. Eng. Sci.* **18**, 607–610.
Slysh, R., Hettinger, A. C., and Gayler, K. E. (1974). *Polym. Eng. Sci.* **14**, 264–272.
Smith, I. T. (1961). *Polymer* **2**, 95–108.
Sourour, S., and Kamal, M. R. (1976). *Thermochim. Acta* **14**, 41–59.
Strauss, C. R. (1980). *N. Am. Therm. Anal. Soc.* [*Proc.*] **10**, 311–319.
Sundstrom, D. W., and English, M. F. (1978). *Polym. Eng. Sci.* **18**, 728–733.
Swarin, S. J., and Wims, A. M. (1976). *Anal. Calorim.* **4**, 155–171.
Sykes, G. F., Burks, H. D., and Nelson, J. B. (1977). *Natl. SAMPE Symp. Exhib.* [*Proc.*] **22**, 350–364.
Taylor, L. J., and Watson, S. W. (1970). *Anal. Chem.* **42**, 297–299.
Thomas, G. R., Halpin, B. M., Sprouse, J. F., Hagnauer, G. L. and Sacher, R. E. (1979). *Natl. SAMPE Exhib.* [*Proc.*] **24**, 458–505.
Thomas, R. E. (1974). Motorola Semiconductor Products Div., Phoenix, AZ 85003 (personal communication to R. L. Blaine, DuPont).
Thomas, R. E., and Scharr, T. (1975). *Soc. Plast. Eng.* [*Proc. An. Tech. Conf.*] **33**, 217–218.
Ting, R. Y., and Nash, H. C. (1979). *Proc. Annu. Conf.—Reinf. Plast./Compos. Inst., Soc. Plast. Ind.* **34**, Sect. 22F.
Tryson, G. R., and Shultz, A. R. (1979). *J. Polym. Sci., Polym. Phys. Ed.* **17**, 2059–2075.
Warfield, R. W., and Kayser, E. G. (1975). *J. Macromol. Sci., Phys.* **B11**(3), 325–328.
Weast, R. C., ed. (1978–1979). "CRC Handbook of Chemistry and Physics," 59th ed. CRC Press, Inc., West Palm Beach, Florida.
Wendlandt, W. W. (1974). "Thermal Methods of Analysis," 2nd ed. Wiley (Interscience), New York.
Wentworth, S. E., King, A. O., and Shuford, R. J. (1979). *U.S. Army Mater. Mech. Res. Cent.* [*Tech. Rep.*] *AMMRC TR* **AMMRC TR 79-1** (AD A065962).
Wentworth, S. E., King, A. O., and Shuford, R. J. (1980). *Polym. Composites* **1**, 103–109.
Wentworth, S. E. (1981). *N. Am. Therm. Anal. Soc.* [*Proc.*] **11**, 733–738.
Westwood, A. R. (1975). *In* "Thermal Analysis" (I. Buzas, ed.), Vol. 3, pp. 337–346. Heyden, London.
White, R. H., and Rust, T. F. (1965). *J. Appl. Polym. Sci.* **9**, 777–784.
Widmann, G. (1975a). *Thermochim. Acta* **11**, 331–333.
Widmann, G. (1975b). *In* "Thermal Analysis" (I. Buzas, ed.), Vol. 3, pp. 359–366. Heyden, London.
Wight, F. R., and Hicks, G. W. (1978). *Polym. Eng. Sci.* **18**, 378–381.
Wilkinson, R. W. (1961). *In* "Kinetics and Mechanism" (A. A. Frost and R. G. Pearson, eds.), p. 47. Wiley, New York.
Willard, P. E. (1972). *Polym. Eng. Sci.* **12**, 120–124.
Willard, P. E. (1973). *SPE J.* **29**, 38–42.
Willard, P. E. (1974a). *J. Macromol. Sci., Chem.* **A8**(1), 33–41.
Willard, P. E. (1974b). *In* "Polymer Characterization by Thermal Methods of Analysis" (J. Chiu, ed.), pp. 33–41.
Winkler, E. L., and Parker, J. A. (1969). *In* "Thermal Analysis" (R. F. Schewenker, Jr. and P. D. Garn, eds.), Vol. 1, pp. 481–492. Academic Press, New York.
Zicherman, J. B., and Holsworth, R. M. (1974). *J. Paint Technol.* **46**, 55–61.
Zucconi, T. D. (1979). *ASTM Spec. Tech. Publ.* **STP 674**, 533–540.

CHAPTER 6

Elastomers

J. J. MAURER

Corporate Research Laboratories
Exxon Research and Engineering Company
Linden, New Jersey

I. Introduction

A. Fundamental Characteristics of Conventional Elastomers

Materials that exhibit "rubbery" or elastomeric behavior have two important primary characteristics in common. One of these is the ability to withstand large deformation without rupturing. The other is the ability to recover spontaneously to the original dimensions on release of the deforming force. These substances are generally composed of long polymer chains that normally take up an irregular, statistically random configuration in the undeformed state. These chains elongate during deformation with an attendant decrease in entropy and return to the randomly coiled state during retraction. The retractive force is primarily associated with the change in entropy due to stretching of the chains. For these processes to occur, polymer chain mobility must be high. Therefore, the polymer must not have a high degree of crystallinity in the unstretched state. Further, for conventional elastomers, the glass transition temperature should be well below the normal use temperature range or be capable of being lowered significantly by the addition of plasticizers. In addition, the polymer chains are usually connected by some type of cross-linking process in order to form a permanent network. This leads to elastic recovery, rather than plastic flow, following deformation.

B. Definition of Elastomer Systems

For most important commercial applications, elastomers are used in complex formulations that are here termed elastomer systems. The major components of such systems are shown in Table I. Each of the compo-

TABLE I

Components of Practical Rubber Formulations

1. Elastomers
2. Fillers (carbon blacks, mineral fillers)
3. Process oils, plasticizers
4. Additives (stabilizers, process aids, etc.)
5. Curatives (ZnO, S, stearic acid, accelerators, peroxides, sulfur donor systems, etc.)

nents in a particular elastomer formulation contributes in some way toward meeting the cost versus performance requirements associated with both the manufacturing process and the end-use application.

A comprehensive description of the science and technology associated with the different components shown in Table I and the specific choice of these ingredients for a given end-use formulation is beyond the scope of this review. Detailed discussions of many of these areas are available in the following sources: *Vulcanization of Elastomers,* Alliger and Sjothun (1964); *Reinforcement of Elastomers,* Kraus (1965); *The Chemistry and Physics of Rubber-like Substances,* Bateman (1963); *Rubber Technology,* Morton (1973); *The Science and Technology of Rubber,* Eirich (1978), *Polymer Chemistry of Synthetic Elastomers,* Kennedy and Tornqvist (1969); and *Rubber Chemistry and Technology,* the *Journal of the Rubber Division of the American Chemical Society.*

A general description of the role of the components listed in Table I may, however, prove useful for those less familiar with elastomer systems, and it is presented for these readers. A primary basis for choosing a particular elastomer or elastomer blend is, of course, that it meets key performance criteria of the intended end use. In addition, it may be chosen to enhance processing in factory equipment or to facilitate construction (e.g., adhesion or building tack requirements for tires). Process oils and plasticizers can also serve a number of functions, ranging from modification of end-use characteristics (such as hardness and low-temperature properties) to improving factory processability and reducing compound costs via diluent action. Reinforcing fillers are used to improve such properties as tensile strength, abrasion resistance, and stiffness. Carbon black is the major filler used for this purpose. A wide range of properties can be achieved through variation in carbon black type (particle size, surface activity, structure, etc.) and concentration or by using blends of different blacks. Mineral fillers, singly or in blends with carbon blacks, are also used in some types of elastomer systems. The cure system used to form the network in an elastomer vulcanizate is commonly based on zinc oxide, sulfur, and stearic acid, plus organic accelerators, whose function is mainly to increase the rate of the curing process. Manipulation of cure system type and cure conditions enables wide variation in the type and degree of cross-linking in the system. Both factors can significantly influence vulcanizate properties. Finally, a wide variety of miscellaneous additives may also be present in a practical elastomer system. Their function ranges from modifying cure behavior to stabilizing the system (thermal or oxidative) in manufacturing and/or end-use environments, improving factory processing, reducing cost, etc. Overall, a dominant consideration in the development of a practical rubber formulation is to meet the desired production cost and end-use performance criteria.

C. Characteristics of Less Conventional Elastomer Systems

Two elastomeric systems have evolved whose general properties require some modification in the description of elastomers given previously. One of these, known as thermoplastic elastomers or thermoelastics, is a class of block copolymers composed of a rubbery soft block and a glassy, crystalline, or ionic domain hard block. The soft blocks give the material the general elastomeric properties described earlier. The hard blocks serve to provide the junction points in the system, thus filling the role of the chemical cross-links obtained with the various cure systems used in conventional elastomers. At the same time, the hard domains also serve as reinforcing filler, similar to the effect of carbon black in conventional elastomers. A key property of this type of copolymer is that it is thermoplastic in nature; i.e., the material can be processed like a plastic above T_g, T_m, or the "dissociation" point of the hard block; below these temperatures the material acts like a cross-linked elastomer system. Comparison of the properties of thermoelastic versus conventionally cured vulcanizates is given by Noshay and McGrath (1977).

A second class of less-conventional elastomer systems is composed of *blends* of elastomers and thermoplastics that exhibit thermoplastic behavior similar to block copolymers. In the blends, of course, the elastomers and thermoplastics function as the rubbery "soft block" and "hard block,'' respectively. Both the thermoplastic systems and the blends of elastomers and plastics are briefly reviewed in this chapter and treated in detail in Chapters 3 and 4.

D. Representative Types of Commercial Elastomers

Commercial elastomers encompass a wide variety of chemical compositions and, accordingly, a wide range of properties. Representative types of conventional elastomers that will be illustrated in this chapter are shown in Table II, which also provides the ASTM abbreviation code used

TABLE II

Nomenclature and Abbreviation Code for Selected Elastomers[a]

BIIR	Bromo–isobutene–isoprene
BR	Butadiene
CIIR	Chloro–isobutene–isoprene
CR	Chloroprene
EPM	Ethylene–propylene (saturated)
EPDM	Ethylene–propylene–diene
IIR	Isobutene–isoprene
IR	Isoprene (synthetic)
NBR	Nitrile–butadiene
NCR	Nitrile–chloroprene
NIR	Nitrile–isoprene
NR	Natural rubber
SBR	Styrene–butadiene

[a] American Society for Testing and Materials (1979c). Reprinted with permission of ASTM, 1916 Race Street, Philadelphia, PA 19103, Copyright.

for these materials. There are numerous additional types of both conventional and less-conventional elastomers. The former include polysulfides, polyacrylates, silicones, urethanes, and polyisobutylenes. The latter, the thermoelastic materials, are represented by segmented polyurethanes, polyether–polyesters, and ionic-domain-forming systems (e.g., carboxylate or sulfonate-based systems). A general comparison of the properties of conventional elastomers is given by Billmeyer (1971). Such comparisons, as he noted, should be considered as general guidelines in view of the degree to which properties may be varied by manipulation of cure system and condition, filler type and level, polymer and/or filler blends, etc.

E. End-Use Applications for Elastomers

The major end use for elastomers is in tires for passenger cars, trucks, buses, aircraft, and off-the-road equipment. The transportation industry uses substantial additional quantities of elastomers for a variety of purposes, including window channels, door seals, bumpers, shock absorption, damping, electrical wiring, mats, hoses, and belts. A random listing of other end-use areas includes clothing, protective equipment, sporting goods, electrical insulation, sound and shock absorption systems, and a wide array of mechanical goods (seals, O-rings, belts, hoses, washers, gaskets, etc.), caulks, sealants, adhesives, reservoir liners, etc.

F. Applications for Thermal Analysis in Elastomer Science and Technology

Complexity is a distinguishing feature of both elastomer systems and their applications. Not only is a given formulation complex, but there are also a wide variety of formulations, depending on the particular end-use requirements. This situation creates a need for analytical procedures that can provide guidance to both manufacturers and purchasers of rubber goods and in both the research and development aspects of elastomer science and technology. A variety of procedures for analyzing vulcanizates are reported, but all of them are too lengthy for use as routine quality control procedures. Moreover, a cross-linked elastomer vulcanizate is an intractable solid, insoluble in most solvents, and thus not easily amenable to classical chemical methods. A summary of many of the wet chemical procedures is presented in ASTM D297-79 (American Society for Testing and Materials, 1979b). Similarly, although instrumental procedures such as pyrolysis-infrared spectrometry (MacKillop, 1974) and pyrolysis-gas chromatography (Krishen, 1972) have been utilized, they too require considerable time and effort. It will be shown in this review that the basic thermal analytical procedures (DTA, DSC, TG/DTG, TMA and DMA) have broad applicability for evaluation of elastomer systems. This will be illustrated for a variety of elastomer systems, including the polymers themselves (the so-called gum elastomers), uncured compounds composed of these polymers plus various combinations of the additives described in Section I.B, and cured gum elastomers or compounds.

G. Objectives of This Review

A comprehensive treatment of even the major opportunities for thermal analysis of elastomer systems is beyond the scope of this discussion. Rather, the objective is to *illustrate* the important opportunities for thermal analysis using representative systems and literature references as available and/or appropriate. Key references have been provided regarding these areas and should be consulted for details of the investigations. Overall, the objective is to provide information regarding proven applications for these techniques and to stimulate thinking regarding additional potential applications.

II. Nomenclature

Elastomer nomenclature is based on the procedure recommended by the American Society for Testing and Materials (1979c). This follows the practice of *Rubber Chemistry and Technology*, the *Journal of the Division of* Rubber Chemistry, of the American Chemical Society. The ASTM system is illustrated in Table II for the elastomer types most frequently mentioned in this chapter.

Two systems of carbon black nomenclature are encountered in the cited references. Both have been retained to facilitate consideration of these references and of the general literature of elastomer science and technology. A comparison of these systems, by Donnet and Voet (1976), is presented in Table III. This comparison is based solely on carbon black

TABLE III

ASTM DESIGNATION AND SURFACE AREA LIMITS FOR COMMERCIAL BLACKS[a]

ASTM designation	Older type	SA limits $(m^2\ g^{-1})$	ASTM designation	Older type	SA limits $(m^2\ g^{-1})$
N110	SAF	125–155	S301	MPC	105–125
N219	ISAF-LS[b]	105–135	N440	FF	43–69
N220	ISAF	110–140	N550	FEF	36–52
N242	ISAF-HS[c]	110–140	N601	HMF	26–42
N285	IISAF-HS[c,d]	100–130	N660	GPF	26–42
N326	HAF-LS[b]	75–105	N770	SRF	17–33
N330	HAF	70–90	N774	SRF-NS[f]	17–33
N347	HAF-HS[c]	80–100	N880	FT	13–17
S300	EPC	95–115	N990	MT	6–9
N339	HAF-HS(NT)[c,d]	90–105			

[a] Donnet and Voet (1976). Reprinted by courtesy of Marcel Dekker, Inc.
[b] LS = low structure.
[c] HS = high structure.
[d] IISAF = "Intermediate" intermediate super abrasion black.
[e] NT = new technology.
[f] NS = nonstaining.

surface area, as determined by the standard BET method. Additional properties that determine the selection of carbon black type(s) for particular end-use applications are discussed by Donnet and Voet (1976).

A general listing of abbreviations, symbols, definitions, tradenames, etc., is presented in Table IV. Common abbreviations are used for the basic types of thermal analysis. These same abbreviations have also been used to indicate a given type of thermal analysis equipment, or a type of analysis that was conducted with a particular type of thermal analysis equipment. For example, *TMA* stands for *thermomechanical analysis,* a

TABLE IV

IDENTIFICATION OF SYMBOLS, ABBREVIATIONS,
TEST PROCEDURES, AND MATERIALS[a]

ADL Rebound Test	ADL Ball Rebound Tester. Transistor Automation Corp., Cambridge, Mass.
ASTM	American Society for Testing and Materials
DMA	Dynamic mechanical analysis
DSC	Differential scanning calorimetry
DTA	Differential thermal analysis
DTG	Derivative thermogravimetry
TG	Thermogravimetric analysis
TMA	Thermomechanical analysis
E_a	Energy of activation
GC	Gas chromatography
Gehman rigidity modulus	A measure of the low-temperature stiffening of elastomer systems as determined by the Gehman torsional wire apparatus (ASTM D1053)
ΔH_f	Heat of fusion
ΔH_v	Heat of vulcanization
IR analysis	Infrared analysis
phr	Parts by weight of an ingredient in a rubber compound of formulation per hundred parts of rubber
Probability plot	A graphical procedure for testing cumulative frequency data for fit to the normal or log normal distribution
®	® appearing as an "exponent" (e.g., Rheovibron®) indicates a registered trademark
Rheovibron®	Direct reading, dynamic viscoelastometer, DDV-II Toyo Measuring Instruments Co., Ltd.
Polygard®	Tri(nonylated phenyl)phosphite, a stabilizer-antioxidant (Uniroyal Chemical Co.)

Polymers

BR-1220	Butadiene rubber, 97% cis-content; Cariflex®; Ghijsels (1977)
Epon 1002®	Epoxy resin (Shell Chemical Co.), Gillham *et al.* (1977)
Estane®	Polyether polyurethane thermoplastic elastomer, B. F. Goodrich Co. Additional description in this chapter and Illinger *et al.* (1972)
Hypalon 20®	Chlorosulfonated polyethylene (E. I. DuPont & Co.)
Hytrel®	Polyester thermoplastic elastomer (E. I. DuPont & Co.). Additional description in this chapter and Nishi *et al.* (1975)
Kapton®	Polyimide film (E. I. DuPont & Co.), Gillham *et al.* (1977)
Kraton 1101®	A linear styrene–butadiene–styrene (SBS) triblock thermoelastic polymer; 25% polystyrene; molecular weight: 102,000. Shell Chemical Co. LeBlanc, (1977)
Neoprene W®	A grade of chloroprene rubber for general-purpose applications (E. I. DuPont & Co.)
Neoprene WRT®	A grade of chloroprene rubber used in general mechanical goods requiring optimum resistance to crystallization (E. I. DuPont & Co.)

TABLE IV (*continued*)

SBR-1500	Emulsion styrene–butadiene rubber, polymerized at 50°F and below. Target-bound styrene (23.5%)
Solprene 406, 411, and 4115®	Star-shaped styrene–butadiene block copolymers with 38, 31, and 40% polystyrene, respectively, and molecular weights of 208,000, 258,000, and 153,000, respectively. Phillips Petroleum Co. (Leblanc, 1977)
Sylgard-186®	A low-viscosity two-component methyl silicone. Weight average molecular weights are 31,000 (resin) and 55,000 (catalyst). Dow Corning Co.

Cure System Components

Altax®	Benzothiazyl disulfide (R. T. Vanderbilt Co.)
GMF	*p*-Quinonedioxime
MBT	2-Mercaptobenzothiazole
MBTS	2,2′-Benzothiazyl disulfide
MOR	(2-morpholinothio)benzothiazole
SP-1055®	Bromomethylated and methylolated alkyl phenol–formaldehyde resin (Schenectady Chemical Co.)
Sulfasan R®	4,4′-Dithiodimorpholine (Monsanto Chemical Co.)
Tellurac®	Tellurium diethyldithiocarbamate (R. T. Vanderbilt Co.)
Tuads®	Tetramethylthiuram disulfide (R. T. Vanderbilt Co.)

[a] Information sources: (1) "Rubber World Blue Book." Bell Communications, Inc., New York, 1978; (2) M. J. Haward, ed., "Elastomeric Materials." Cordura Publications, Inc., La Jolla, California, 1977. (3) Literature references cited in this chapter.

TMA unit is the equipment used to perform a particular type of thermomechanical analysis, and a TMA creep analysis may signify, for example, an isothermal creep analysis performed via a TMA unit. The latter terminology is intended to focus on the wide variety of analyses that can be conducted with basic thermal analysis equipment.

III. Experimental Considerations

The thermal analysis instrumentation and procedures that have been found applicable to elastomer systems are standard in many respects. Basic operating principles and procedures associated with the subject DTA, DSC, TG, DTG, DMA, and TMA have been described in detail in a variety of sources, including Chapters 1 and 2 of this book. Accordingly, this summary is limited to scoping information regarding key experimental variables. The literature references should be consulted for in-depth treatment of specific techniques or elastomer systems.

A. ATMOSPHERE

Nitrogen and argon are common for many studies. Air is acceptable for some analyses (e.g., transitions and viscoelasticity studies) and is required for others (e.g., oxidative stability, vulcanizate composition, or carbon black analysis). Helium has also been employed for its high conductivity, especially for TMA, but is less common. Stepwise decomposition in nitrogen and air is used for vulcanizate composition analysis.

B. SAMPLE SIZE

The size of the sample varies substantially depending on the type of analysis. For example, 0.2- to 15-mg samples have been used in the identification of elastomers, whereas 20- to 100-mg samples have been used in the analysis of carbon black in vulcanizates. Typical sample sizes are in the 10- to 30-mg range.

C. HEATING RATES

Typical heating rates are in the 10 to 30°C/min range. Lower values (2 to 5°C/min) have been used for determining transition temperatures via TMA. Evaluation of a range of heating rates is desirable in the analysis of new or unknown elastomer compositions. This provides information regarding optimum conditions for detection and resolution of minor transitions as well as procedures for more routine analysis of a given system.

D. SAMPLE HISTORY

It has been amply demonstrated that significant variations in T_g, T_m, and degree of crystallinity can be produced by manipulation of crystallization and annealing temperatures or by relative cooling and heating rates. Additional types of sample history effects may also be encountered. One of these is due to the viscoelastic characteristics of the polymer. Samples that have been pressed, molded, or extruded may exhibit appreciable elastic "memory," which may result in eventual severe distortion of the sample, which would be more rapid when it is heated. Such effects have been noted in TG, DSC and TMA experiments. Another type of time-dependent sample history variation can be encountered with thermoelastic block copolymers. Here changes in dimensional stability and/or polymer "transitions" may occur due to aggregation of the hard-block segments and resultant formation of various "supermolecular" structures.

IV. Analysis of Single-Elastomer Systems

The approach that will be used in this review is to consider first the simplest systems, namely,. those based on single elastomers, and then progress to elastomer compounds, and finally to vulcanized (cured) elastomer systems. The following key points will be developed throughout this analysis:

1. Straightforward applications of thermal analysis can provide, for some elastomers, rather extensive information regarding glass transition temperature, crystallinity and/or melting, thermal stability, mechanical properties, rheology, and viscoelasticity.

2. Many of these same features can be detected in elastomer compounds and vulcanizates, thus forming the basis for a variety of qualitative and quantitative analyses of practical elastomer systems, including those involving various blends of polymers and filler systems.

3. Basic thermal analytical procedures for glass transition and melting point analyses, coupled with careful thermal history manipulation of the system, can provide a wide variety of information, ranging from polymer or copolymer microstructure to thermoelastic block copolymer characteristics, polymer blend composition, quality, and compatibility (see also Chapter 4).

A. GLASS TRANSITION TEMPERATURE (T_g)

A clear, well-defined glass transition temperature (T_g) obtained by DTA, is shown for a random ethylene–propylene copolymer in Fig. 1. Also illustrated is a type of variation in T_g thermogram characteristics due

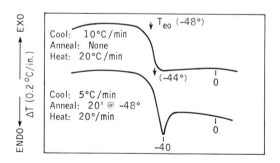

Fig. 1   Influence of low-temperature annealing on T_g region of ethylene–propylene (EP) copolymer (Maurer, 1968).

to thermal history variations, as described in detail by Wunderlich, in Chapter 2 of this book. DTA of phase transitions in gutta percha is described by Rootare and Powers (1977), and Castillo *et al.* (1979) have evaluated the specific heats versus temperature for Guayule and natural rubber. Many factors can influence the location and appearance of the T_g for polymers. Instrumental effects are discussed by Garn and Menis (1977), and Peyser and Bascom (1977) describe the effect of fillers and cooling rate. Wrasidlo (1974) treats the theoretical aspects of the problem, and Gupta (1980) discussed the relationship between DMA loss peak temperature and T_g. In addition, a variety of practical considerations apply. Thus it is necessary to consider thermal history effects, residual solvents or very-low-molecular-weight species from the polymerization, and polymer composition. The latter area includes level and uniformity of copolymer composition and microstructure as well as consideration of mixtures of homopolymers and/or copolymers formed during exploratory polymerizations, by nonuniform reactor conditions, or by catalyst systems involving more than one mechanism and/or active species. T_g is changed only slightly by vulcanization and loading for many elastomer systems, thus enabling analysis of practical rubber products.

The ability to determine and interpret T_g values for polymers is an important polymer characterization capability. The reason is that T_g values have been used as a primary basis or key component for a remarkable range of polymer studies, including product uniformity and quality control, polymer identification, structure and uniformity, plasticizer efficiency, degree of curing, blend compatibility, phase composition and supermolecular structure in block polymers and ionomers. Each of these areas will be touched on in this chapter.

B. Crystallization and Melting

An ideal elastomer would have the ability to crystallize on stretching in order to develop high tensile strength but would not exhibit appreciable crystallinity in the unstretched state in order to minimize the temperature dependence of properties, enable high degrees of extension and recovery, etc. In practice, a number of elastomers exhibit a significant degree of crystallinity in the unstretched state. In some cases, crystallization and melting can occur near, or above, room temperature [various ethylene–propylene–diene rubbers (EPDM) and Neoprenes, respectively], as will be discussed subsequently. In other cases, crystallization and melting occur well below room temperature [e.g., butadiene rubber (BR), natural rubber (NR), and the polydimethylsiloxane system in Fig. 2]. The latter system illustrates the importance of cooling rate (in this case supercooling) in developing the maximum information about polymer crystallinity. Figure 2 indicates the detection of T_g, crystallization, and variations in two forms of crystallinity at about -120, -100, -60, and $-40°$, respectively.

Two additional cases of interest are BR and EPDM, both of which are semicrystalline polymers; however, they differ substantially in melting characteristics. By means of low-temperature annealing, BR melting and degree of crystallinity can be evaluated free of complications due to crystallization, and well separated from T_g (Fig. 3). EPDMs, however, represent additional complexity in that the endotherm due to melting may overlap with the T_g region, giving rise to complex thermograms that have been

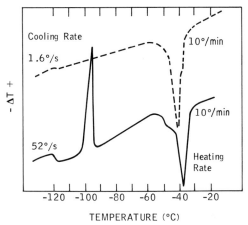

Fig. 2   Influence of cooling rate on transition behavior of polydimethylsiloxane (Helmer and Polmanteer, 1969).

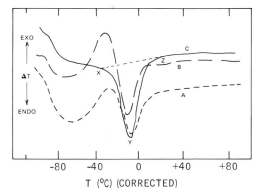

Fig. 3   Effect of pretreatment on BR crystallization: A, untreated; B, heated and quenched; C, heated and cooled slowly (Sircar and Lamond, 1973b).

designated Type 3 T_g intervals (Fig. 4). It will be shown that crystalline regions of these types can be used to advantage in certain qualitative and quantitative analyses of blends containing these elastomers.

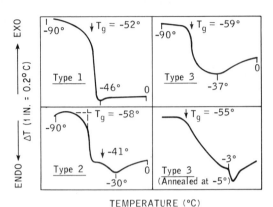

TEMPERATURE (°C)

Fig. 4 Main T_g intervals in ethylene–propylene copolymers indicate microstructure variations (Maurer, 1969a).

C. Thermal and Oxidative Degradation

Polymer degradation by thermal or thermooxidative reactions is accompanied by absorption or evolution of heat. Sircar and Lamond (1972) have used DSC to characterize these processes for many common elastomers. They classify elastomers into three main groups, based on degradation characteristics in nitrogen: (a) those exhibiting an endotherm; (b) those exhibiting an exotherm, and (c) those that show multiple exo- and/or endotherms. Figure 5 compares representative members of these categories. Physical transitions (T_g, T_m) and oxidative degradation characteristics were used to aid in elastomer identification, in some cases. The exo- or endothermic nature, shape, and temperature region that is exhibited by a particular elastomer in a given environment depends on the chemical composition, molecular structure, and configuration. As a result, degradation characteristics can be used to identify many polymers, including some that are difficult to identify by other techniques.

Both qualitative and quantitative analyses of some elastomers and blends can be accomplished by conventional and/or derivative thermogravimetric analysis (TG and DTG, respectively). As shown in Fig. 6, several important commercial elastomers degrade almost completely in nitrogen atmosphere, and differ from one another based on chemical composition. Treatments have also been developed to handle polymers that thermally degrade via a char-forming reaction (Fig. 7). DTG has been

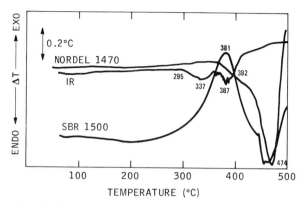

Fig. 5 Representative thermograms for elastomers exhibiting exothermic (SBR 1500), endothermic (EPDM), mixed endothermic–exothermic (IR) decomposition in nitrogen (Sircar and Lamond, 1972).

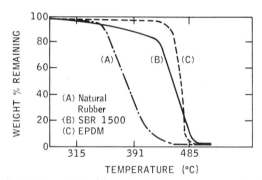

Fig. 6 Thermal stability variations for selected elastomers: natural rubber, styrene–butadiene rubber (SBR 1500) and ethylene–propylene–diene rubber (EPDM) (Maurer, 1969a).

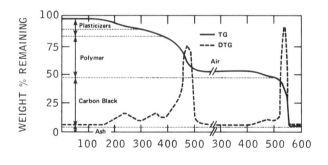

Fig. 7 Thermogravimetric curves for nitrile–butadiene rubber (NBR) formulation showing weight losses for plasticizers, rubber, and carbon black (Swarin and Wims, 1974).

demonstrated to aid in both quantitative and qualitative analysis of some
elastomer blends (e.g., Figs. 8 and 9).

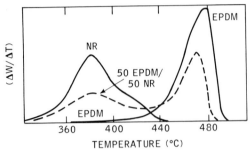

Fig. 8 Derivative TG of natural rubber (NR)/ethylene–propylene–diene rubber
(EPDM) systems (Maurer, 1974a).

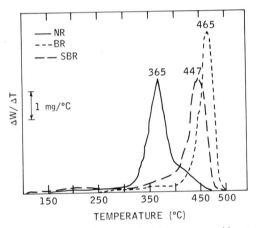

Fig. 9 DTG curves of different elastomers in nitrogen: peroxide-cured, sample weight
9.20–9.24 mg (Sircar and Lamond, 1975a).

The use of DSC and DTG as a means of distinguishing natural rubber
(NR) and isoprene rubber (IR) vulcanizates has been described by Sircar
and Lamond (1972, 1975a), Sircar (1977), and Brazier and Nickel (1975b).
In essence, the method is based on the observation that sulfur-cured,
black-reinforced IR and NR exhibit significantly different DTG and DSC
curves, in nitrogen. The raw elastomers and peroxide-cured gum vulcanizates, however, have identical thermograms of this type. Brazier and
Schwartz (1978) show that differentiation of IR from NR by DTG is a
function of sulfur level, carbon black type and level, and accelerator system. DSC analysis of cured carcass compounds (Sircar, 1977) indicated
that the heat of reaction and the corresponding exotherm area is greater

for IR than for NR, whereas peak temperatures vary as follows: NR > 50/50 blend > IR. These results are shown in Fig. 10. DTG of

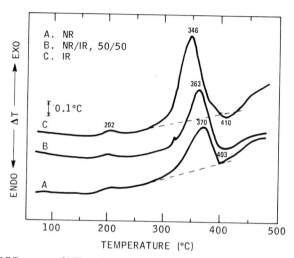

Fig. 10   DSC curves of NR and IR carcass compounds: A, 14.90 mg; B, 14.22 mg; C, 14.20 mg (Sircar, 1977).

these same compounds (Fig. 11) shows that the IR system has two sharp peaks (370 and 420), whereas the NR system has a sharp 370° peak and an inflection near 430°C. The blends of NR and IR were intermediate.

Schwartz and Brazier (1978b) describe studies directed toward differentiating and quantitatively analyzing NR and IR in *unknown* vulcanizates. They concluded that DTG, DSC, or DTG-GC could not give an

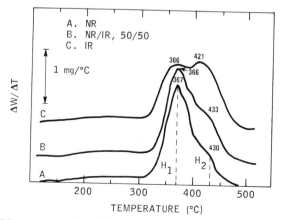

Fig. 11   DTG curves of NR and IR compounds: A, 9.28 mg; B, 9.87 mg; C, 9.85 mg (Sircar, 1977).

unequivocal answer alone unless information was available regarding the sulfur level, type of accelerator, and type and level of carbon black in the vulcanizate. However, it was noted that NR and IR could be differentiated (a) by DTG, if the sample is vulcanized and contains carbon black, and (b) by DSC on the basis of observed exothermic heat, if the sample is vulcanized.

Sircar (1978) describes the thermal analysis of NR and IR in blends of SBR and BR.

Estimation of butadiene content of styrene butadiene rubber (SBR) and butadiene rubber (BR) vulcanizates by DSC has been proposed by Sircar and Lamond (1973a). This method is based on evaluation of the DSC exothermic peak at about 380°C in nitrogen, which was shown to be unaffected by vulcanization or compounding ingredients (Fig. 12). The

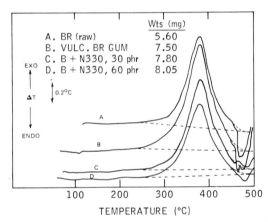

Fig. 12 Thermographs of vulcanized BR with increasing carbon black loading (Sircar and Lamond, 1973a).

linear relationship shown in Fig. 13 indicates that the method should be useful to estimate BR content of SBR and BR vulcanizates. The method is said to supplement and extend other methods for BR analysis that depend on estimation of styrene content. Available evidence indicated that the exothermic reaction on which this analysis is based is due to cyclization of BR.

Schwartz and Brazier (1978b) describe an analysis of the effect of heating rate on the BR degradation process. It is noted that BR degrades in two distinct weight loss steps. The first is said to be almost exclusively due to volatile depolymerization products; the second is attributed to degradation of a residue due to cyclized and cross-linked BR. Increasing heating rate or sample size increased the contribution of the depolymerization step to the overall process. Kinetics of the degradation processes

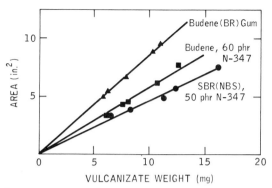

Fig. 13 Exotherm area versus weight of SBR and BR vulcanizates (Sircar and Lamond, 1973a).

are analyzed via the DTG data and suggestions are made regarding implied mechanisms for the overall BR degradation reaction scheme. Additional studies of this area are presented by McCreedy and Keskkula (1978, 1979), who showed that the degradation mechanism of BR differs substantially for isothermal versus dynamic heating. It was further shown that polymers that do not thermally cyclize or cross-link, such as polystyrene, degrade similarly during dynamic vs. isothermal heating.

D. Stabilization of Elastomer Systems

Stabilizers are commonly added to synthetic elastomers during the manufacturing process in order to protect the polymer from thermal and/or oxidative breakdown during finishing, packaging, and storage. (In some cases they may also serve a similar purpose during the polymer mastication stage of commercial mixing of elastomer compounds). A stabilizer or stabilizer package is customarily also added to elastomer *compounds*. The purpose in this case is usually to ensure that the finished (cured) rubber part meets the long-term stability requirements of the intended end use.

Analysis of the stability and stabilization of elastomer systems is inherently a complex and formidable task, as described by Shelton (1957), Ambelang *et al.* (1963), Kuzminskii (1966), Cunneen (1968), and Dunn (1974, 1976, 1978). Complications may arise from the following sources:

(a) contaminants in the gum rubber (e.g., catalyst residues, impurities in the monomers).

(b) conventional compounding ingredients and curatives (e.g., carbon black, sulfur, accelerators, plasticizers).

(c) positive or negative synergism due to the interaction of elastomer system components and the antioxidants.

(d)   instability of the polysulfidic linkages in the cured elastomer composition.

Definitive studies of the stability of practical systems are rarely undertaken, in part because of this complexity. Instead, the extent and effect of the aging process(es) are commonly assessed by monitoring some physical property as a function of time, temperature, load, flexing, etc.

1. Illustration of Applicable Thermal Analytical Techniques

Several thermal analytical techniques have been utilized to study or evaluate various aspects of the stability and stabilization of elastomer systems, ranging from gum rubber to cured commercial formulations. For example, TG has been employed to evaluate antioxidant content of polymer systems via the induction time to weight gain and/or rapid weight loss. Induction time methods based on DSC have also been used. Both methods are described by Bair in Chapter 9. Gedemer (1974) has used DTG to correlate weight loss in electrical insulation with the results from more conventional methods of assessing thermal stability of these compositions. Krizanousky and Mentlik (1978) used thermal analysis to predict the life of electrical insulating materials.

A different approach to monitoring the extent or effects of aging and degradation is offered by TMA. When cross-linking, embrittlement, or scission occur in the elastomer or elastomer network, one might expect corresponding changes in the hardness and apparent modulus of the system. As a result, techniques such as the penetrometer or solvent swelling procedures described in Sections IV and VII may prove useful for at least semiquantitative monitoring of the aging process, and hence of "stability." Restriction or enhancement of flow in a parallel plate plastometer measurement would in principle offer similar opportunities for monitoring stability of some systems and/or the relative effectiveness of various stabilizers. Interpretation of such TMA studies would have to be done with care to ensure that *all* potential contributors to observed "modulus" changes are accounted for. For example, loss of a volatile plasticizer or oxidation of a resin might lead to "hardening" of an elastomer system in which neither the polymer backbone nor the cross-linked network had as yet suffered any degradation.

A different approach to evaluate stabilizers for polymers is reported by Braddon and Falkehag (1973). These authors employed torsional braid analysis to study modified kraft lignin as a stabilizer for raw polymer(s). Here the "stabilized" polymer sample is coated on a braid. Changes in the polymer during the aging process are reflected in the torsional rigidity of the sample. The influence of antioxidant volatility was also evaluated.

It is also possible to make use of differences in the relative stabilities of various elastomers (Fig. 6) or of elastomers versus the other ingredients in the formulation (Fig. 14) to evaluate the basic composition of a

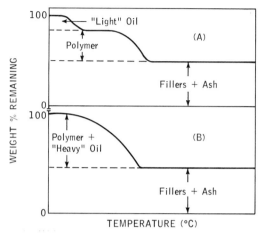

Fig. 14   Characteristics of oil loss regions (Maurer, 1969b).

vulcanizate. In some cases elastomer blend composition and carbon black blend composition can also be estimated.

2. Evaluation of Gum Elastomers

Thermal analytical techniques can, in principle, provide information regarding (a) the temperatures at which elastomers can be safely processed, stored, or utilized; (b) the type and quantity of stabilizer to be employed; and (c) concentration of a known antioxidant in a given polymer system.

Preliminary information about temperatures at which polymers can be processed without undergoing degradation can be obtained from TG analysis in air and nitrogen. As noted in Fig. 6, thermal stability of an elastomer is a function of its chemical composition. Calorimetric studies of elastomer degradation can also be used to advantage as shown by Sircar and Lamond (1972). Figure 5 compares the stability of several different elastomer types as measured by DSC. Goh and Lim (1979) have used thermal analysis to study the influence of inhibitors on metal ion catalyzed oxidation of natural rubber.

Both DSC and DTA have been used to describe the inherent thermal stability of elastomers and evaluate the influence of stabilizer type and concentration on polymer stability. Two examples of this approach will be described to illustrate the parameters of the experiment and the general type and complexity of the data that can be encountered. It will be apparent that various approaches can be taken to utilizing thermal analytical techniques for such investigations.

3. Evaluation of Butadiene-Containing Polymers

Smith and Stephens (1975) conducted a DSC evaluation of the energetics of oxidation and degradation of conventional styrene–butadiene rubber (SBR) and a stereospecific polybutadiene. The relative effectiveness of phenol, amine, or phosphite-type antioxidants on these processes was also compared.

In an oxidizing atmosphere, two exotherms are evident in the DSC thermograms of unsaturated elastomers as illustrated for polybutadiene in Fig. 15. The initial exotherm, which occurs in the 180–200°C region, was

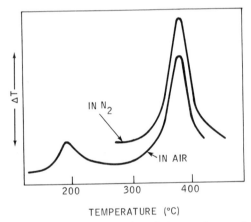

TEMPERATURE (°C)

Fig. 15 Influence of atmosphere on DSC degradation exotherms of polybutadiene rubber (Smith and Stephens, 1975. (Reprinted courtesy of The Journal of Elastomers and Plastics, Technomic Publishing Company, 265 Post Road West, Westport, CT 06880.)

related to an oxidative cross-linking reaction. The second exotherm, which occurs in the 350–400°C region, was attributed to polymer degradation or chain scission. It was concluded, in agreement with Bauman and Maron (1956), that both cross-linking and chain scission could be involved, to some extent, in both the low-temperature and the high-temperature processes. Similar ΔH values were obtained for the higher-temperature process in air vs. nitrogen atmosphere, thus suggesting that overlap is minimal.

The effect of antioxidant type and concentration on polybutadiene oxidation characteristics in air atmosphere is summarized in Table V. In-

TABLE V

DSC Analysis of Polybutadiene in Air[a]

Sample	Antioxidant concentration (phr)	Peak temp. (°C)	ΔH_1 (cal/g)	ΔH_2 (cal/g)
A	0.000	192	31.1	108.3
	DBPC[b]			
B	0.005	192	29.0	106.1
C	0.010	198	32.2	105.0
D	0.030	195	27.4	104.2
E	0.050	198	29.2	121.0
F	0.100	199	26.4	121.0
	PBNA[c]			
G	0.005	195	27.0	115.2
H	0.010	195	26.8	107.5
I	0.030	200	31.8	121.0
J	0.050	200	34.3	138.0
K	0.100	208	16.8	–
	TNPP[d]			
L	0.500	192	25.0	96.0
M	1.000	194	22.7	103.8
N	3.000	202	23.1	106.0
O	5.000	208	17.8	112.8
P	10.000	233	7.35	124.2

[a] From Smith and Stephens (1975). Reproduced by courtesy of The Journal of Elastomers and Plastics, Technomic Publishing Co., Inc., 265 Post Road West, Westport, CT 06880.
[b] 2,6-ditertiary butyl-*p*-cresol.
[c] Phenyl-β-naphthyl amine.
[d] Tri(nonylated phenyl)phosphite (Polygard®).

creasing antioxidant concentration (a) shifted the lower-temperature peak to higher temperatures, (b) reduced the enthalpy (ΔH_1) of the low-temperature exotherm, and (c) generally increased it (ΔH_2) for the higher-temperature exotherm. This latter effect suggested that antioxidant products from the ΔH_1 process are not necessarily inert, as commonly believed.

The differences among these three antioxidants are even further evident when an oxygen atmosphere is used. More complex thermograms are obtained as illustrated for DBPC stabilization of polybutadiene (Fig. 16). A shoulder at 315°C becomes larger as DBPC concentration increases,

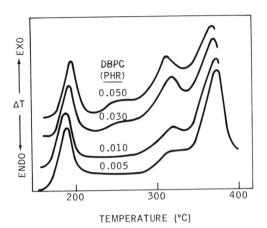

TEMPERATURE (°C)

Fig. 16 Influence of DBPC concentration (phr) on DSC thermograms of stabilized poly-butadiene rubber (Smith and Stephens, 1975. Reprinted courtesy of The Journal of Elastomers and Plastics, Technomic Publishing Company, 265 Post Road West, Westport, CT 06880.)

eventually becoming larger than the $\triangle H_2$ exotherm at concentrations near 3.0 phr (not shown in Fig. 16). A new shoulder developed (240°C) and increased with DBPC content. Apparently, DBPC stabilization of polybutadiene is not a simple process; rather, a complex degradation process is suggested in which antioxidants and their oxidation products play a major part. The influence of stabilizer type and level on $\triangle H$ for the low-temperature exotherm is summarized in Fig. 17.

Goh (1977) applied DSC to study the effect of 2,6-ditertiary butyl-*p*-

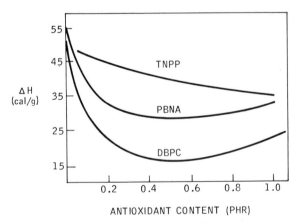

ANTIOXIDANT CONTENT (PHR)

Fig. 17 $\triangle H$ (low-temperature exotherm) versus antioxidant content (phr): polybutadiene rubber (Smith and Stephens, 1975. Reprinted courtesy of The Journal of Elastomers and Plastics, Technomic Publishing Company, 265 Post Road West, Westport, CT 06880.)

cresol (DBPC); N-isopropyl-N'-p-phenylenediamine (IPPD); N-phenyl-β-naphthylamine (PBNA); zinc diethyldithiocarbamate (ZDEC); and nickel dibutyldithiocarbamate (NDBC) on the activation energy (E_a) of oxidation of natural rubber (NR) and determine whether E_a was a useful criterion for measuring antioxidant effectiveness. The equation developed by Kissinger (1957) was considered adequate for E_a determination. Sample preparation involved (a) addition of antioxidant to a benzene solution of natural rubber (2%); (b) evaporation of a few drops of this solution in the sample pan to give a 0.5–1 mg film, about 10^{-3} cm thick. Acetone extraction of natural rubber was employed to remove naturally occurring stabilizers. The exotherm peak temperature of raw rubber is about 28° higher than that of extracted rubber. In nitrogen atmosphere, exotherms were not detected up to 600 K; thus the exotherm below 600 K in oxygen is attributed to oxidation of the sample.

The influence of antioxidant type on the peak temperature for natural rubber oxidation is shown in Fig. 18. Observed E_a decreases at higher an-

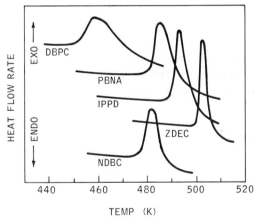

Fig. 18   DSC curves of acetone-extracted natural rubber containing 2 phr of various antioxidants in oxygen atmosphere (heating rate = 16 K/min) (Goh, 1977. Reprinted courtesy of The Journal of Elastomers and Plastics, Technomic Publishing Company, 265 Post Road West, Westport, CT 06880.)

tioxidant concentration (in other studies) were attributed to pro-oxidation effects due to the inhibitors. Comparison of the observed E_a data with the known relative antioxidant effectiveness of amines versus phenols indicated that E_a calculated from thermal analysis can provide a reasonable estimation of the effectiveness of an antioxidant. Goh and Phang (1978) provide additional information concerning the catalysis of rubber oxidation by metallic ions.

4. Studies of Natural Rubber Systems

Poncez-Velez and Campos-Lopez (1978) employed DSC to compare the thermal oxidation of natural rubber from two sources, Hevea brasiliensis (Hevea) and Parthenium argentatum (Guayule). This study illustrates the strong influence of experimental conditions on the nature and characteristics of the DSC thermograms. Using a DuPont 990 thermal analysis system, the oxidation characteristics of these two polymers were explored as follows: oxygen atmosphere (110 ml min^{-1}); 393–473 K temperature range; sample size (14–15 mg versus 3–4 mg) and heating rate (2–50°C min^{-1}).

The principal focus of this paper is the comparison of three different mathematical approaches for determining the energy of activation (E_a), apparent reaction order X, and the rate constant K for the oxidation process. The methods considered were the heat evolution method of Borchardt and Daniels (1957), the diffusion-controlled method of Jander and Bloom (1949), and the heating rate method of Kissinger (1957) as modified by Ozawa (1976). Best agreement was achieved by assuming first order in the Borchardt and Daniels (1957) method and three-dimensional diffusion with the Jander model (1949), with sample weights in the order of 3–4 mg. E_a values (in kilocalories per mole) were as follows: Borchardt–Daniels (16–17), Kissinger–Ozawa (18–19), and Jander–Bloom (23–24). They are in good agreement with the 20 kcal mole^{-1} obtained by Field et al. (1955). Additional data concerning the influence of stabilizers on the thermal oxidation of several elastomers, including Guayule and NR are presented by Gonzalez (1980, 1981).

5. High-Pressure DTA of Polymer–Antioxidant Systems

May et al. (1968) utilized DTA and high-pressure DTA to study the oxidation of styrene–butadiene rubber (SBR), and the influence of several types of antioxidants (phenolic, amine, and phosphite). Results obtained at 300 psi were considered more significant than those at atmospheric pressure. Addition of antioxidant was found to increase the energy of activation (E_a) and delay the onset of oxidation. Subsequently, May and Bsharah (1969) demonstrated a correlation between DTA peak height and tensile strength and ultimate elongation of oxygen bomb-aged vulcanizates.

The work of May et al. (1968) led to the following conclusions: (a) DTA enables rapid evaluation of the radical-trapping activity of rubber antioxidants, (b) DTA data support the conclusion that p-phenylenediamines are generally better than phenolic antioxidants, and (c) oxidation of rubber at higher temperatures may involve several simultaneous reactions.

The studies highlighted demonstrate that thermal analytical techniques offer a convenient and effective approach to develop detailed information concerning the stability and stabilization mechanisms of elastomer systems. Some of these analyses (e.g., screening of antioxidant type vs. E_a) appear to relatively straightforward tasks. Others (e.g., determining the stabilization mechanisms in a given system) are complex and require considerable care in both the experimental and interpretive phases. In the latter regard, it is of critical importance to establish the relationship between the aging mechanism(s) operative in the thermal analysis experiment and those operative in the environment that a given elastomer system will encounter in service.

E. THERMOMECHANICAL CHARACTERIZATION

Thermomechanical analysis (TMA) broadens significantly the types of polymer characterization that can be achieved by thermal techniques, as described by Barrall *et al.* (1974a), Prime (1978), Brazier and Nickel (1978), and Maurer (1978). Several opportunities will be illustrated here; more complete details are presented elsewhere in this review or the various references cited.

TMA of elastomers via a penetrometer probe produces thermograms that closely resemble the master curves obtained by conventional time–temperature superposition of modulus data. Figure 19 shows these data for isobutene–isoprene rubber (IIR). Several types of information can be obtained from such data, including the T_g value and, in certain cases, molecular weight and cross-linking variations (Maurer, 1978; Rosen, 1971). Further, it has been noted [e.g., Noshay and McGrath (1977)] the random versus block copolymers can be distinguished by the type of variation in

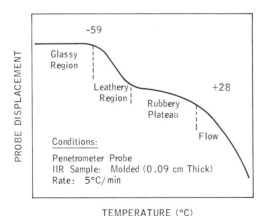

TEMPERATURE (°C)

Fig. 19 TMA (penetrometer) data resemble polymer master curves (Maurer, 1978).

the master curve that accompanies variations in polymer composition. For random polymers, T_g changes in a regular fashion with composition. For block polymers, the T_g values of the two constituents remain constant and the height of the modulus plateau varies with composition.

TMA equipment also provides a convenient method for analysis and comparison of polymer viscoelasticity via isothermal shear creep recovery curves (Maurer, 1978). The classical creep experiment is conducted by rapid application of a constant stress to the sample and subsequently monitoring the time-dependent strain over relatively long time periods. On application of the load, the resultant creep curve exhibits, in order, instantaneous elastic response, delayed elastic response, and, finally, viscous flow. On removal of the load, a recovery curve is generated that exhibits an instantaneous elastic recovery followed by delayed elastic recovery. A significant degree of nonrecoverable permanent set, due to viscous flow, is observed in many cases.

Viscoelastic response of an elastomer or elastomer compound may be influenced by a variety of parameters (molecular-weight distribution, branching, gel, transitions, interaction between polymer and filler, type and level of fillers, and plasticizers). In principle, then, TMA creep analysis should offer a sensitive and rapid method for detecting at least major variations in these parameters for a given elastomer system.

An illustration of this type of analysis is presented in Fig. 20, which compares the elastic recovery of a series of compounds based on isobutene–isoprene rubber (IIR), which differed in the die swell (independently measured) and elastic recovery of the compound as determined in the TMA creep recovery analysis. Remarkable reproducibility of the elastic recovery values was observed for independent measurements of a series of samples. It thus appears that TMA offers a rapid means for at least qualitative comparison of the viscoelastic characteristics of polymers, due in part to the small sample size requirements of the method.

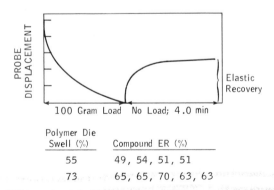

Polymer Die Swell (%)	Compound ER (%)
55	49, 54, 51, 51
73	65, 65, 70, 63, 63

Fig. 20   TMA creep analysis of elastomer (IIR) compound (Maurer, 1978).

Another type of analysis for which TMA appears to have high potential is polymer rheology by parallel plate viscometry. This approach is based on the techniques of Dienes and Klemm (1946), who established a basis for evaluating melt viscosity of various polymeric systems by analysis of the flow characteristics of polymer samples, contained between parallel plates, under the application of an applied load. Figure 21 shows

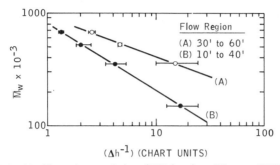

Fig. 21   Flow characteristics of IIR fractions (Maurer, 1978).

the relationship between $\bar{M}_w$ of a series of IIR fractions versus flow characteristics of small samples of these materials in a simple, "microscale" TMA parallel-plate rheometer (Maurer, 1978). Cessna and Jabloner (1974) previously reported successful, TMA-based, parallel plate rheology determinations of polymer systems. Their equipment appears to overcome certain apparent limitations of the relatively primitive parallel plate rheometer used to generate the data of Fig. 21.

Additional opportunities for the application of TMA to elastomer systems are discussed elsewhere in this chapter. Several of these relate to various approaches for evaluating the state of cure, or cross-linking in a vulcanizate, as described by Barrall *et al.* (1974a,b,); Barrall and Flandera, (1974), and Prime (1978). Linear expansion measurement for the prediction of postmolding shrinkage of mechanical goods is also reviewed based on the work of Fogiel *et al.* (1976). Detection and monitoring of cure state in finished rubber parts by TMA (penetrometry and expansion) has also been reported by Collins (1970). Finally, a survey of TMA-related studies of elastomer systems is presented by Brazier and Nickel (1978).

F. Dynamic Mechanical Analysis

1. Introduction

Mechanical properties (e.g., elongation, impact strength, yield point, hardness, softening temperature, and many allied properties) are important factors in determining the end-use performance characteristics of polymers. As a consequence, evaluation of mechanical properties of polymer systems is a key component in a variety of fundamental or applied investigations. These include (a) developing polymer synthesis–structure-property relationships; (b) determining the influence of compound ingredients, cure-system components, and cure state on mechanical properties via laboratory test procedures; and finally (c) relating these laboratory data to end-use performance characteristics of the system.

The theoretical and applied aspects of the mechanical properties of polymers have been developed in considerable detail. Excellent treatments of these subjects can be found in the following references. The list includes introductory-level presentations as well as more advanced texts for those with previous background in the area. In the former category are textbooks by Billmeyer (1971) and Aklonis *et al.* (1972). In the latter are works by Meares (1965), Nielsen (1962), Ward (1971), Bateman (1963), and Ferry (1970).

The following discussion, which will focus on elastomer systems, briefly treats the following topics:

(a) the viscoelastic nature of polymers and its influence on their mechanical properties,

(b) the types of mechanical property information that can be developed using conventional modulus versus temperature analysis,

(c) the utility of dynamic mechanical analyses of polymer systems, and finally,

(d) selected applications of several common DMA techniques to elastomer systems.

2. The Viscoelastic Nature of Polymers

Polymers are commonly described as viscoelastic to reflect their ability to display both viscous and elastic response. It has been shown that, depending on the temperature and time scale of measurement, such materials may exhibit properties of a viscous liquid, an elastic rubber, or a brittle glass. As a consequence, data are required over a wide range of frequency (time) and temperature to describe the viscoelastic behavior of a polymer system. One experimental approach involves isothermal evaluations using several techniques to cover the required frequency range. Another involves changing the temperature of the experiment. This brings the relaxation processes of interest within the time scale of convenient experimental techniques. A time–temperature relationship is used in the latter case, as described in the following discussion.

3. Conventional Mechanical Property Analysis

The classical demonstration of the viscoelastic nature of polymers is based on a series of measurements of the stress-relaxation modulus as a function of time, as described in numerous works (e.g., Aklonis *et al.*, 1972, and Ferry, 1970). A superposition principle has been evolved (see, for example, Aklonis *et al.*, 1972), which demonstrates the equivalence of time and temperature regarding polymer viscoelasticity. Application of this principle to the series of isothermal curves mentioned earlier enables construction of a so-called master curve that describes the complete viscoelasticity spectrum of a given polymer or polymer system.

In principle, any modulus-related measurement should enable generation of the same type of master curve for a polymer system, providing a suitable range of temperature/time is traversed. This is illustrated for an isobutene–isoprene rubber (IIR) in Fig. 19. The penetrometer height versus temperature curve of this system, which is a measurement of a modulus-related characteristic of the sample, displays all the primary features of the polymer viscoelastic spectrum. Moving in the direction of increasing temperature, the penetrometer curve reveals the glassy region, the glass transition temperature (T_g), the viscoelastic region, the rubbery plateau, and, finally, the viscous flow region of the polymer.

Master curves of the type shown in Fig. 19 are a classical type of data that are developed via mechanical property measurements of polymers. The method of measurement utilized as well as the time and temperature regions explored will determine the portion of the master curve (viscoelastic spectrum) that is generated.

Various factors that influence the nature of such master curves, in particular T_g, the height of the rubbery plateau, and the existence and location of the viscous flow region, have been described by Rosen (1971). In principle, evaluation of such master curves provides a means for comparing samples regarding such variables as cross-link density, molecular weight, and plasticizer content.

In addition to the main glass transition temperature T_g, which involves a large change in modulus, a polymer usually exhibits secondary transitions that involve relatively small modulus changes (Fig. 22). Such transitions have been attributed to features of the polymer microstructure such as side group motions. These transitions are of interest with regard to both the characterization and practical performance of elastomer systems. *Dynamic* mechanical analysis has been shown to be a sensitive means of detecting secondary transitions and relaxations in polymers.

In summary, the general types of information that can be developed

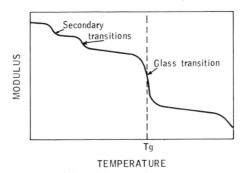

Fig. 22 Temperature dependence of modulus in a typical polymer (Ward, 1971).

from measurements of the modulus versus temperature relationship for polymer systems are illustrated in Fig. 22.

4. Dynamic Mechanical Analysis

Elastomer systems, particularly in the large-volume tire component applications, must perform under a wide range of conditions, which include both static and dynamic environments. Hence evaluation of the mechanical properties of elastomer systems should include dynamic mechanical analysis (DMA). An excellent introduction and background regarding DMA of polymer systems is given by Boyer (1977a) and Gillham (1977).

DMA is ideally suited for evaluating elastomer systems because of their viscoelastic nature. In essence, DMA is based on the different types of response exhibited by viscous and elastic elements when subjected to a sinusoidally varying strain. Polymers, because of their viscoelastic characteristics, exhibit a response intermediate between that of a purely viscous or a purely elastic system.

In the DMA experiment, the vector representing the dependent variable is resolved into components that are in phase or out of phase with the independent variable. Typical primary data from DMA are the following:

G' storage modulus; the in-phase component
G'' loss modulus; the out-of phase component
tan δ loss tangent; a measure of the phase angle between
           G'' and G' (tan $\delta = G''/G'$)

In general, DMA of polymer systems involves determining the variation of modulus and tan δ of the system as a function of frequency and/or temperature. Typical data are shown in Fig. 23. A single, usually low frequency is commonly employed and the sample is scanned over a wide temperature range.

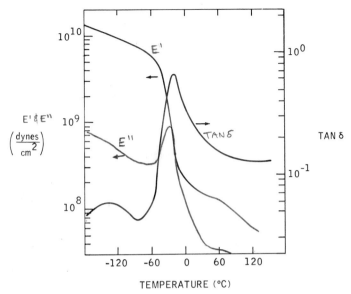

Fig. 23   Temperature dependence of storage modulus (E') and loss tangent (tan δ) of Estane 5714® at 110 Hz (Illinger *et al.*, 1972).

One of the strengths of DMA is that tan δ is a very sensitive method for detecting secondary transitions and relaxations in polymers. The use of DMA in evaluating polymer microstructure is illustrated in Fig. 24. Note the reduced temperature (T/T_g) that is used as a normalizing scale.

5. Representative Techniques for DMA of Polymer Systems

a. Torsion Pendulum and Torsional Braid.   Torsion pendulum experiments involve free-oscillation measurements. An initial torsional

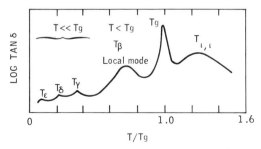

Fig. 24   Schematic relaxation spectrum for an amorphous polymer as commonly observed with a low-frequency torsion pendulum in the torsional braid mode (Boyer, 1977b. Copyright 1977 by John Wiley and Sons, Inc. Reprinted by permission of John Wiley and Sons, Inc.)

displacement is applied to the sample, and the frequency and amplitude decay of the oscillations are monitored on release.

Gillham (1974, 1977) describes the use of the torsion pendulum and torsional braid for the characterization of polymer systems. The latter technique, which employs small samples supported on a braid, enables rapid achievement of thermal equilibrium. This procedure enables measurements to be made through the load-limiting transition (T_g, T_m) regions in polymers. Manzione et al. (1981) have presented an extensive description of the use of torsional braid analysis in conjunction with electron microscopy and viscometry to study transitions, morphology, and mechanical properties of rubber-modified epoxies.

b. Dynamic Viscoelastometer (Rheovibron®). One of the most widely used instruments for DMA of polymer systems is the Rheovibron viscoelastometer. This forced-vibration, nonresonant technique was developed by Takayanaga (1965). In this procedure a sinusoidal displacement is applied to one end of a horizontally clamped sample and measured by a strain gauge. A second strain gauge measures the sinusoidal force generated at the other end of the sample. The phase angle (δ) between the stress and strain is read directly as tan δ. Storage modulus and loss modulus are calculated from the tan δ value and the amplitude of stress and strain.

Inherent limitations of the method are reviewed by Kenyon et al. (1977), who developed a fully automated Rheovibron-type system that reduced or eliminated these deficiencies. The greatly increased sensitivity of this automated system was reported to enable more accurate location and detection of minor relaxations and transitions. Voet and Morawski (1978) adapted the Rheovibron to enable measurement of dynamic properties of elastomers at large deformations, up to sample rupture. Their method involved superimposing small sinusoidal oscillations on statically extended samples. A Rheovibron study of the influence of process oil type on the dynamic mechanical properties of filled, cross-linked ethylene–propylene–diene elastomers is described by Byrne and Hourston (1979b). Hourston and Hughes (1981) report on a Rheovibron study of dynamic mechanical behavior to assess the degree of compatability in polyether ester–nitrile rubber blends.

c. Constant Oscillation Amplitude Procedure. An example of this type of procedure is the DuPont 981 DMA system. In this analysis, the sample is sinusoidally oscillated at its resonant frequency. The amplitude of oscillation is maintained constant via the addition, on each cycle, of an amount of energy equal to that dissipated by the sample. This make-up energy, which is a direct measure of sample damping, can be expressed as a relative damping value or as tan δ for quantitative analysis.

Resonant frequency data obtained with the 981 DMA system are converted to Young's modulus (*E*) of the sample. The energy dissipation (expressed as tan δ), has been related to properties such as impact resistance, brittleness, and noise abatement. Variations in frequency and energy dissipation have been used for both fundamental and quality control studies. Representative data have been described for thermoplastic, impact-modified thermoplastic, reinforced thermoplastic, elastomer, and thermoset systems (Blaine *et al.*, 1978).

d. Representative Additional Techniques. Haward (1973) discusses the use of dielectric relaxation, DMA, and nuclear magnetic resonance for studying secondary relaxations in polymers. He notes that the choice of the right method can be critical to the detection of the relaxation. Certain types of motions are not detectable by DMA. Information from at least three types of measurement was considered essential for the successful identification of the molecular motion responsible for the relaxation.

6. *Selected Applications of DMA to Elastomer Systems*

a. Damping Characteristics versus Elastomer Type. The inherent damping characteristics of an elastomer are an important consideration regarding its use in practical rubber compounds. For some uses, high damping is desired, as in vibration and noise control. In applications involving severe flexing, however, a high damping compound may be undesirable due to the accompanying high heat buildup. Basic elastomer properties are a major determinant of the damping characteristics of an elastomer compound as suggested by Fig. 25. DMA may be a valuable

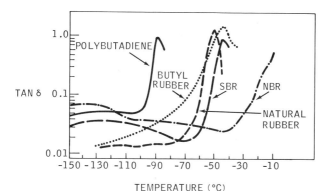

Fig. 25 Comparative damping of elastomers (E. I. Dupont & Co., 1977).

tool for monitoring the damping and heat buildup in elastomer compounds. This capability would prove valuable in both compound development and production/quality control programs. Byrne and Hourston (1979a) discuss various aspects of the thermal and dynamic mechanical properties of EPDM rubber.

   b. Practical Tire Component Formulations. DMA data have been reported for a series of tire formulations containing styrene-butadiene rubber (SBR) and/or polybutadiene rubber (BR) (E.I. DuPont & Co., 1977). The dynamic mechanical property data for these systems is complex as shown in Fig. 26. Both modulus and damping indicate the

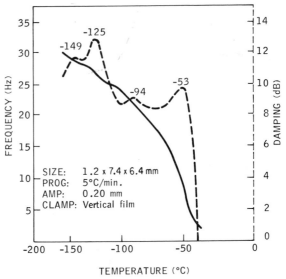

Fig. 26   Carbon black modified styrene–butadiene rubber (E. I. Dupont & Co., 1977).

presence of several components and/or secondary relaxation processes. The major damping peaks are attributed to the glass transitions of the polymers, allowance being made for possible shifts in T_g due to various factors (cross-linking, plasticization, filler interaction, and/or blending). The two minor peaks are attributed to secondary relaxations in the polymers, polymer–filler interactions, etc. It was suggested that such peaks may be relatable to heat buildup, vibration, and high-speed road noises.

   Elastomer formulations are inherently complex. In addition, there are potential complications due to variability in dispersion and interaction of the many ingredients. In view of these factors, and the multiple transitions exhibited by individual polymers, precise interpretation of

DMA data for such systems may prove to be a formidable task, particularly with regard to the secondary relaxation processes.

c. Thermoelastic Block Copolymers

(1) Ether-urethane copolymers. The use of DMA to elucidate the influence of polymer composition and temperature on the morphology of polyurethane–polyether block copolymers is described by Illinger *et al.* (1972). These systems were interpreted in terms of a polyether matrix containing segregated polyurethane domains, the latter serving as physical cross-links.

This study employed a model DDV-II Rheovibron system to explore block polymers of the general composition $MS[M(BM)_1S]_nM$. M represents 4,4'-methylene diphenyl diisocyanate (MDI), B is a chain extender (1,4-butanediol), and S is a low-molecular-weight polyether. The latter consisted of either poly(tetramethylene oxide) (PTMO) or a poly(propylene oxide) (PPO)–poly(ethylene oxide) (PEO) block copolymer.

The type of information accessible via such studies is illustrated by the following examples. The first involves a commercial polyether–polyurethane ("Estane®" series 5714). The soft segment of this polymer consists of PTMO of MW $\sim$ 2000 and the molar ratio of MDI:B:PTMO is 3.2:2:1. As shown in Fig. 23, this polymer exhibits a high glasslike modulus below about $-60°C$. The modulus decrease above this temperature is due to the glass–rubber transition of the polyether segments. A typical rubbery plateau is reached above 0°C followed by a gradual decrease in modulus above 120°C. Tan δ data indicate a major loss at $-30°C$ and a secondary relaxation process near $\sim -130°C$.

The second example concerns a polymer whose mechanical properties were remarkably different from the previous system, as shown in Fig. 27. The composition of this polymer, which was based on a PTMO block (MW $\sim$ 2,040), in terms of MDI:PTMO was 2.2:1.1.

Illinger *et al.* (1972) interpret this system as follows. The low urethane content of this system is insufficient to raise the polyether T_g by mobility restriction, hence the major modulus loss occurs at a low temperature $(-90°C)$, a value consistent with the expected T_g of PTMO homopolymer of MW 2000, "and subject to the imposed restraint." The modulus increase noted at $-30°C$ is attributed to partial crystallization of the polymer due to its composition and thermal history (a rapid quench followed by heating at 1°C min^{-1}). The polyether segments crystallize near $-30°C$ and subsequently melt, giving rise to the maximum in the modulus curve. Domain structure in this polymer, if present, is suggested to be poorly developed at best.

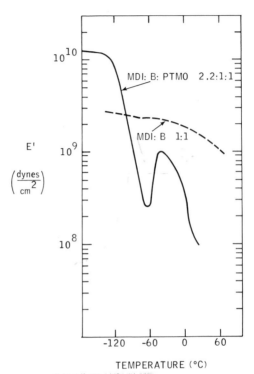

Fig. 27 Influence of polymer composition on storage modulus at 110 Hz (Illinger *et al.*, 1972).

The remaining prominent feature of these systems is the secondary relaxation observed near − 140°C, which was attributed to transitions in the polyether segments.

(2) Ester–ester copolymers. DMA data for "Hytrel®" have been described by Gill and Ikeda (1977). This polymer is a polyester thermoplastic elastomer consisting of poly(tetramethylene ether) glycol terephthalate (PTMEGT), soft segments, and tetramethylene tereph-thalate (4GT) hard segments. Their report presents a brief comparison of Rheovibron dynamic mechanical analysis of Hytrel 4056 by Nishi *et al.* (1975) versus a similar analysis via the DuPont 981 Dynamic Mechanical Analyzer. Both techniques yielded similar data for both the modulus (level and regions of change) and damping (Fig. 28) as a function of temperature. For this comparison, the 981 DMA unit offered the following advantages: shorter time (1.5 versus 4 h) and a *continuous* trace over the full temperature range spanned. The latter feature enabled detection of a relaxation by 981 that was absent in the Rheovibron data,

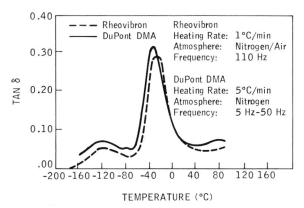

TEMPERATURE (°C)

Fig. 28 Comparative damping analysis for Hytrel 4056® (Gill and Ikeda, 1977).

possibly because of limited data *points* having been taken in the latter study. The major γ-relaxation and the α-relaxation were readily detected by both techniques. Kenyon *et al.* (1977) describe a modified Rheovibron system that reportedly enables improved detection of secondary transitions and relaxations in polymers.

d. Rubber-modified Polymer Systems

(1) Rubber–plastic blends. Dynamic mechanical analysis has been extensively used to seek correlations between the location and intensity of rubber phase loss peaks (usually secondary peaks) and mechanical properties of rubber–plastic blends. This activity followed pioneering studies by Buchdahl and Nielsen (1960), who reported an influence of low-temperature loss peaks on the impact strength of rubber-modified polystyrenes. A review of this subject by Boyer (1968), however, cites several examples that indicate that the presence of a low-temperature loss peak is neither a necessary nor a sufficient condition to ensure good room temperature impact strength in rubber-modified polystyrene. However, such peaks may serve to indicate the *presence* of the rubber. Additional treatments of transitions and relaxations in polymers are given by Boyer (1976, 1977b).

Haward (1973), in reviewing this general area, concludes that the presence of secondary relaxation regions does influence the mechanical properties of organic glasses. However, he observed that the extent to which secondary transitions (below T_g) influence yield and ultimate failure properties is difficult to determine. Boyer's (1968) review was cited as evidence that *some* influence of secondary relaxations on mechanical properties is possible in a given system.

Among other factors, Haward (1973) suggests the following reasons for the general lack of correlation between DMA data and failure properties of rubber-modified plastics:

(1) DMA is a small strain experiment whereas failure properties involve large strains.

(2) There is variability in structure of the dispersed phase.

(3) There are molecular-weight variations in the dispersed phase.

The complexity of studies of this type may be appreciated from Table VI from Haward (1973). Despite this complexity, he concludes that these

TABLE VI

VARIABLES OF RUBBER-MODIFIED GLASSY POLYMERS
WHICH AFFECT PROPERTIES[a]

Phenomenological variables	Structural variables involved
Matrix properties	(1) Chemical structure
	(2) Average molecular weight
	(3) Molecular weight distribution
Dispersed phase properties	(4) Chemical structure of rubber
	(5) Volume of rubber
	(6) Volume fraction of dispersed phase
	(7) Average particle size
	(8) Particle size distribution
	(9) Degree of cross-linking of rubber
	(10) Degree and structure of rubber matrix grafting
Interfacial adhesion	(11) Degree and structure of rubber matrix grafting

[a] From Haward (1973).

systems offer a unique opportunity for structure-property studies (versus crystalline polymers). Further, it was his opinion that the whole field is available for systematic studies, since previous work has not provided in-depth understanding.

Boyer (1968) has summarized instances in which changes in the physical properties of selected polymers may be related to secondary transitions and relaxations. Those properties that could be of interest to design engineers are shown in Table VII, which is taken from his 1968 paper.

Recent information indicates that the analysis of secondary, low-temperature relaxations may prove useful for specific rubber–plastic blends. For example, Gill and Hassel (1976) report DMA studies of an unspecified impact-modified polypropylene. Intensity of the tan δ value for a $-110°C$ transition was determined for a series of polypropylene samples of varying impact resistance. Correlation of tan δ peak values with ASTM Drop Weight Impact values is shown in Fig. 29. Advantages claimed for the DMA approach included shorter analytical time and easier sample prepa-

TABLE VII

Sᴏᴍᴇ Pʜʏsɪᴄᴀʟ Pʀᴏᴘᴇʀᴛɪᴇs Tʜᴀᴛ Cʜᴀɴɢᴇ ᴀᴛ Sᴇᴄᴏɴᴅᴀʀʏ Tʀᴀɴsɪᴛɪᴏɴs[a,b]

Modulus of elasticity	Thermal retraction
Coefficient of thermal expansion	Specific heat (C_p, DTA, etc.)
Impact strength	Rheological properties, apparent activation energy for melt flow
Tensile strength	Coefficient of friction
Toughness	Tackiness
Tearing energy	
Tensile breaking energy	
Creep rate	Solubility and diffusion of solvents

[a] From Boyer (1968).

[b] It is assumed that the secondary transition or relaxation process appears in a mechanical loss plot (tan δ) or a dielectric loss plot (ϵ'') or in a nuclear magnetic resonance plot, each as a function of temperature.

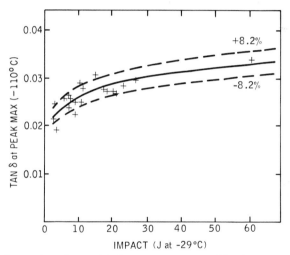

Fig. 29 Polypropylene drop weight impact versus tan δ. (Fitted curve ± 1 standard deviation) (Gill and Hassell, 1976).

ration. It was noted that careful preparation of the samples is essential to avoid complications with factors such as surface imperfections and variable dispersion of the impact modifier.

(2) Rubber–epoxy systems. Gillham *et al.* (1977) employed a torsional braid technique to monitor the changes and transitions that occur during and after cure of a rubber-modified epoxy resin. The system evaluated was a carboxy-terminated butadiene–acrylonitrile copolymer (CTBNX8®) prereacted and blended (5–40 wt %) with an epoxy resin

(EPON 1002®). Composite specimens were formed by polymerizing the blend on a polyimide resin (Kapton®) film.

The primary data from these experiments were relative rigidity $(1/p^2)$; where p is the period in seconds, and logarithmic decrement Δ (a measure of amplitude variation) versus temperature or time. These parameters are related to the modulus and loss characteristics of the sample as follows: Relative rigidity is directly proportional to the in-phase shear modulus (G'); logarithmic decrement is directly proportional to the ratio of the out-of-phase shear modulus G'' to G', and hence to tan δ. The observed influence of rubber content on the mechanical properties and morphology of these epoxy systems is illustrated by the following examples. Figure 30

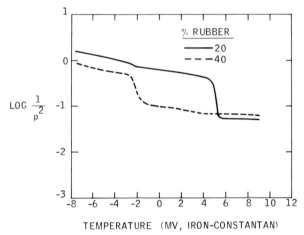

Fig. 30 Dependence of mechanical properties of rubber-modified epoxy resin on rubber content (torsional braid analysis) (Gillham *et al.*, 1977).

shows the in-phase modulus versus temperature for systems containing 20 versus 40% rubber. For the 20% rubber system a *weak* T_g due to the rubber is detected; the major T_g is that of the epoxy. When the rubber content was increased to 40%, however, the reverse situation was noted. These data suggest a phase inversion, i.e., epoxy is the continuous phase in the 20% rubber system and rubber is the continuous phase in the 40% rubber system.

A similar conclusion is suggested by the variation of the intensities of the logarithmic decrements for the epoxy and the rubber phases at their respective T_g values. As shown in Fig. 31, the dramatic increase in the intensity of the rubber Δ above about 20% rubber suggests the onset of a phase inversion. Confirming evidence was obtained by optical micros-

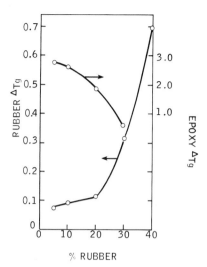

Fig. 31 Temperature and intensity of the glass transition of the epoxy phase and the rubber phase versus rubber content (Gillham *et al.*, 1977).

copy, which indicated that at 30 wt %, rubber is the continuous phase in these blends.

G. COPOLYMER COMPOSITION AND MICROSTRUCTURE ANALYSIS

Analysis of the relationship between copolymer composition and T_g is a rewarding task but can be an extremely complicated one as well. In the simpler case of random copolymers (i.e., a copolymer in which the sequence distribution of the monomer units in the polymer chain is random in nature), the T_g vs. composition relationship is often well described by the well-known GTW (Gordon–Taylor–Wood) equation:

$$T_g = K[(T_{g_2} - T_g)(W_2)/(1 - W_2)] + T_{g_1}$$

where T_g is the observed copolymer T_g, T_{g_1} the T_g of homopolymer 1, T_{g_2} the T_g of homopolymer 2, W_2 the weight fraction of polymer 2, and K is a constant that is characteristic of the particular copolymer.

The GTW equation is a common test of random copolymer characteristics, having been applied to thermoelastic compositions as well as more conventional polymers. Use of this equation as part of an examination of T_g versus composition relationships is presented next to aid in demonstrating the utility of thermal techniques (here, DTA) to aid in defining complex polymer microstructures.

An extensive study of the relationship between T_g and polymer composition has been reported for ethylene–propylene copolymers (EPM)

(Maurer, 1965). Analysis of samples prepared with a variety of catalyst systems, and of fractions separated from these polymers, disclosed several types of T_g intervals, or regions in EPM copolymers (Fig. 4). Examination of these systems in terms of the GTW equation led to the following conclusion:

(1)   Type 1 systems follow the GTW equation and thus are believed to be random copolymers.

(2)   Type 2 systems may be a type of block copolymer, since neither T_g fits the GTW relationship as a function of polymer composition.

(3)   Type 3 systems represent two types of complexity. First, the observed T_g is invariant with composition over a broad range of copolymer content. Second, low-temperature annealing studies suggest that the broad endotherm immediately following T_g is due to microcrystallinity in the polymer.

One final observation to be made about EPM copolymer systems is illustrated in Fig. 32. These data, which also demonstrate the value of

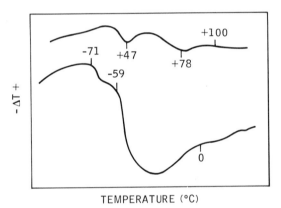

Fig. 32   Multiple transitions in ethylene–propylene copolymer fractions (Maurer, 1967b).

fractionation techniques in polymer characterization, reveal still additional complexity in EPM copolymer microstructure. Note, for example, the clearly resolved transition below the type 3 T_g, and the weak endotherms in the region between 0 and + 100°C. Clearly, very complex microstructure is possible in such copolymer systems and thermal analysis is an extremely valuable tool for detecting it.

The previous example was concerned with the detection of microstructure variations in samples prepared during exploratory synthesis studies. As a result, differences in polymer microstructure were expected. Such differences may also arise unexpectedly due to unknown features of new catalyst systems, variations of catalyst or conversion during polymerization, reactor mixing problems, etc. This kind of variability may sometimes go unnoticed unless detected during, e.g., fractionation studies, or suggested by physical property anomalies encountered during routine screening studies. An example of this kind of situation is reported by Collins *et al.* (1973) for nitrile–butadiene rubber (NBR). Their studies showed that all commercial, non-cross-linked NBR copolymers having less than 35% nitrile exhibit two transitions, a fact that had previously gone unnoticed (Fig. 33). The same type of relationship was shown by two

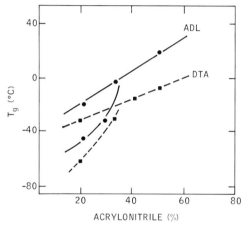

Fig. 33 Transition temperatures versus composition of cured butadiene–acrylonitrile copolymers (see Table IV for ADL and DTA identification) (Collins *et al.*, 1973).

independent methods, ADL rebound (shown in Fig. 33) and dynamic mechanical analysis. It was presumed that this work indicated the presence of two phases in samples containing less than 35% nitrile. DTA and Rheovibron analysis of polymer fractions verified that separation of the two transitions can be achieved by fractionation.

V. Analysis of Elastomer Blends

Blends involving two, three, or even four elastomers are used in practical rubber compounding to obtain improved combinations and/or balances of fabrication ease, cost, and end-use performance. Because of the complexity of these systems, it is a very difficult task to monitor blend uniformity or determine the cause of product variations encountered during production or end-use operations. Hence procedures that can identify potential factors involved in such product quality problems and can provide this information on a practical time scale will prove highly valuable. Several thermal analytical techniques have been found to provide significant advances in this area. The application of these techniques to the more routine type of product quality problems is covered in Sections VIII and IX. This section will deal with more general considerations regarding the use of thermal methods for (a) analysis of basic blend compatibility, (b) qualitative analysis of selected binary blends, and (c) use of T_g and/or T_m as novel indicators of blend quality (degree of mixing or dispersion) and of carbon black dispersion in specific types of binary elastomer blends.

A. EVALUATION OF BINARY BLEND COMPATIBILITY CONSIDERATIONS

A common method of testing for compatibility of a polymer blend is to determine whether two T_gs due to the constituent polymers are detected in the blend, in which case the blend is said to be incompatible. If a single T_g, intermediate between those of the component polymers, is detected, the blend is said to be compatible. "Mechanically compatible" blends represent a deviation from this general scheme, since they exhibit two T_gs but have finer morphology and are translucent.

Actually, various "compatible" systems have been found to exhibit either a T_g intermediate between the T_gs of the components, a broadened T_g interval, or a diffuse interval in which a strong T_g is not discernible. A typical study of this type is illustrated in Fig. 34, which shows the data obtained by Morris (1967) for gum vulcanizates of styrene–butadiene rubber (SBR)/butadiene rubber (BR) blends. Marsh *et al.* (1968) evaluated filled vulcanizates based on this blend system. A single T_g was observed, intermediate between those of SBR and BR, and varying with blend composition. This was taken as a measure of blend compatibility. The explanation given for the behavior of this blend is that due to covulcanization, a new "phase" is formed that behaves differently from SBR or BR. They also mention that for certain homogeneous blends, a single T_g, of one of the components, is observed, depending on the blend ratio. This observation, however, is probably not an unequivocal measure of blend compatibility, since it has been observed that detection of the T_g of a given blend component also depends on the *concentration* of the polymer in the blend (Maurer, 1969a; Sircar and Lamond, 1975d); and separation of the T_g values of the components. This is illustrated in the following discussion of the IIR/EPDM blend system.

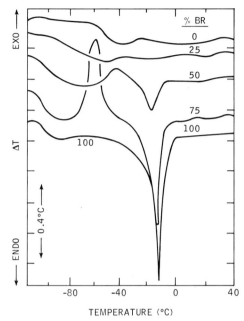

Fig. 34   Differential thermal analysis traces for cured BR/SBR blends of varying weight percent BR (shown on curves) (Morris, 1967).

B. Analysis of Isobutene–Isoprene Rubber (IIR)/Ethylene–Propylene–Diene Rubber (EPDM) Blend Composition

This blend system presents two potential routes to assessing compatibility of these elastomers. One of these is the T_g dependence on blend composition approach cited earlier. The other is based on potential use of the type 3 T_g interval of EPDM (Fig. 4) to detect the presence of this elastomer as described by Maurer (1969a). Thermograms for this blend system are shown in Fig. 35. These data reveal the appearance of T_g values due to one or both of the component polymers, depending on blend composition. Whether or not the single T_g systems unequivocally correspond to "compatible" polymer blends is open to question, based on the results of the following experiment. A mixture of 85 parts of isobutene–isoprene rubber (IIR) and 15 of type 3 ethylene–propylene–diene rubber (EPDM) was prepared by adding the solids to the DTA cell without attempts to mix them. The thermogram for this poorly mixed system exhibited only the T_g for the major component of the mixture (IIR). This experiment suggests

J. J. Maurer

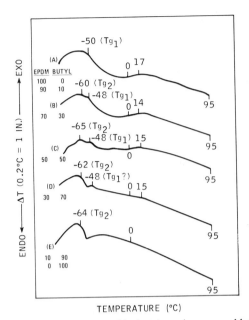

Fig. 35 Glass transition temperature of isobutylene–isoprene rubber/ethylene–propylene–diene rubber blends (Maurer, 1969a).

that the *concentration* of a polymer in a blend is a key consideration when interpreting T_g versus miscibility relationships for polymer blends.

The applicability of T_g measurement as a tool for polymer miscibility studies also depends on the difference between the T_g values of the specific blend components. Sircar and Lamond (1975b), for example, report T_g values of $-60°C$ for NR and $-55°C$ for EPDM and a single T_g for blends of these polymers, varying with the composition and nature of the EPDM. In contrast, microscopic studies (Gardiner, 1970) reportedly show distinct phases in accordance with predictions based on solubility parameters, solvent interaction parameters, and surface tension of the component polymers. Another factor to be considered is that the values of the components of a practical polymer blend may also be influenced to varying degrees by cross-linking, polymer–filler interaction, and plasticization as well as the method of analysis. Note, for example, the $-60°C$ T_g value reported for natural rubber (NR) (Sircar and Lamond, 1975b) and the $-65°C$ T_g reported for IIR (Maurer, 1969a). The T_g value for both these polymers is frequently reported as being in the -70 to $-72°C$ region (e.g., Aklonis *et al.*, 1972). A difference in rate of heating would give different T_g values for the same polymer. For this reason comparison should ideally be made at zero rate of heating by interpolation of T_g values from different rates of heating. This is, however, seldom done in practice.

Semiquantitative analysis of IIR/EPDM blend composition based on Fig. 35 have been successfully conducted. In one such case, the composition of an IIR–EPDM blend exhibiting off-specification properties was shown to be considerably different from that of a control compound based on DTA thermogram characteristics. Pyrolysis gas chromatography and TG analysis were used to confirm this finding. It should be noted that interpretation of this type of DTA data for commercial vulcanizates should be done with care in view of the general complexity of these systems. Particularly in doubtful cases, the analysis should be confirmed by another technique. As noted in Section IX, DSC thermograms of thermal degradation characteristics are useful for identification of various elastomers.

C. Identification of Elastomer Blends by DSC

The use of DSC for defining inherent differences in the thermal degradation characteristics of elastomers was described in Section IV. Sircar and Lamond (1973c) have used this technique to characterize a number of blends of commercial importance. Data are provided concerning the evaluation of both raw [polyvinylchloride (PVC)–nitrile–butadiene rubber (NBR) and acrylonitrile–butadiene–styrene (ABS)] and vulcanized blends [ethylene–propylene–diene rubber (EPDM)–butadiene rubber (BR), EPDM–styrene–butadiene rubber (SBR), natural rubber (NR)–BR, and NR–SBR]. It was found that the DSC patterns for a given elastomer are retained in the blend, thus providing a basis for identifying the polymers in the blend, in many cases.

Several commercial systems, each containing 60 phr of N-347 black, were studied in detail. The type of data obtained are illustrated by the chloro–isobutene–isoprene rubber (CIIR) blend systems shown in Fig. 36. The characteristic features of the component elastomers that are evident in the thermal degradation scans for the blends are the following: exotherms for NR, BR, and SBR around 202, 227, and 240°C, respectively; exotherms for BR and SBR near 375°C; and CIIR endothermic degradation (435°C). The BR blends also show an endotherm (482°C peak temperature) due to BR degradation and volatilization. This appears as a slight inflection in blends with SBR, and there is no effect in blends with NR. Analysis of the type and location of these DSC peaks in nitrogen and in air or oxygen provides information about the component elastomers in these systems. This approach may have considerable utility for the preliminary identification of many commercial elastomer blends.

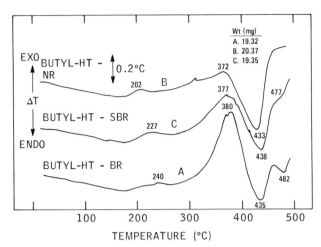

Fig. 36 Thermographs for vulcanized blends of chloro-isoprene–isobutene rubber (butyl-HT) with NR, SBR, and BR in nitrogen (Sircar and Lamond, 1973c).

D. CARBON BLACK TRANSFER IN BINARY ELASTOMER BLENDS

Evaluation of the influence of a noncrystalline component on the crystallization of a second component has been employed for several purposes. Systems where there was a high degree of interference were inferred to be highly compatible blends. Morris (1967) conducted such a study comparing the relative influence of styrene–butadiene rubber (SBR) versus isoprene rubber–synthetic (IR) on crystallization of high-cis butadiene rubber (BR).

Sircar and Lamond (1973b) take this approach one step further by using it to assess the degree of carbon black transfer in binary elastomer blends containing BR. This latter subject has been extensively studied, principally by electron microscopy (Callan *et al.*, 1971). Because of certain ambiguities in the microscopy results, a new approach to carbon black transfer was deemed appropriate. The approach used was to determine the presence of carbon black (CB) in BR by its effect on the crystallization of BR. This technique was supplemented by measurements of dynamic modulus and electrical conductivity at low shear stress. These approaches enable the use of high loadings and highly compatible blends. Also, since uncured stocks are used in DSC studies, potential sample changes due to treatments required in other techniques are avoided. In general, standard DSC procedures and commercially available samples were used in this study. Full details are given by Sircar and Lamond (1973b).

In the work of Ghijsels (1977), an annealing procedure was developed to ensure complete BR crystallization prior to DSC analysis. The need for this requirement is indicated in Fig. 3. As shown, a crystallization exotherm precedes the melting endotherm in nonannealed samples. This interferes with the crystallinity analysis, in addition to possibly indicating a system in which the maximum extent of crystallization has not been achieved.

As shown in Fig. 37, crystallinity is influenced by the presence of car-

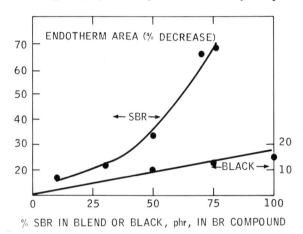

Fig. 37   Effect of SBR and carbon black on BR crystallization (Sircar and Lamond, 1973b).

bon black and to an even greater extent by SBR. The latter effect is considered to result from the high compatibility of these polymers. In effect, BR "captured" in the SBR phase cannot crystallize. This approach was used to determine the relative compatibility of elastomers in BR, as shown in Fig. 38, in terms of solubility parameters of the polymer. The effect of black incorporation on BR crystallinity is summarized in Table VIII.

Analysis of melting peak area vs. the means of carbon black addition provides a semiquantitative method for studying black transfer. Two examples from this table suffice to illustrate the analysis. First, based on analyses such as those shown in Fig. 37, and *assuming no carbon black transfer*, the loss in crystallinity of Sample 587 should be about 18% due to carbon black and 41% due to SBR, giving a total of 59% which agrees well with the observed value of 61. The expected values (18 and 41%) were determined via separate addition of carbon black and SBR, respectively, to BR. Thus no carbon black transfer takes place in this system. For Sample 588, however, the carbon black must be remaining in the SBR

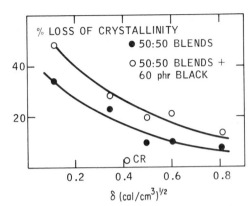

Fig. 38 Difference in solubility parameter as a function of loss of crystallinity of various BR blends (Sircar and Lamond, 1973b).

TABLE VIII

EFFECT OF BLACK INCORPORATION ON BR CRYSTALLINITY[a,b]

Sample number[c]	Polymer	Black added in	Melting peak area[d]/mg BR	Loss of BR crystallinity (%)
587	SBR/BR	BR	0.196	60.7
588	SBR/BR	SBR	0.480	3.8
589	SBR/BR	Preblend	0.272	45.5
590	SBR/BR	Both	0.361	28.7
180	NR/BR	BR	0.331	33.7
181	NR/BR	NR	0.426	14.6
182	NR/BR	Preblend	0.372	25.5
183	NR/BR	Both	0.386	22.6
529	BR/CIIR[e]	BR	0.426	14.6
516	BR/CIIR[e]	CI-IIR[e]	0.436	12.6
531	BR/CIIR[e]	Preblend	0.455	8.8
530	BR/CIIR[e]	Both	0.434	13.0

[a] From Sircar and Lamond (1973b).
[b] 50:50 blends, black loading 50 phr.
[c] Numbering system of Sircar and Lamond (1973b).
[d] Proportional to the heat absorbed.
[e] Chloro–isobutene–isoprene rubber.

phase, and the viscosity of this loaded SBR phase must be high enough to prevent the SBR from being encapsulated in the BR on a microscale; thus no effect of the SBR or BR crystallinity is observed. If carbon black became equally distributed in samples 587 and 588, a loss of crystallinity of about 36% would have been expected based on the data from separate carbon black or SBR addition studies mentioned earlier.

Similar analyses of blends of BR with chloro–isobutene–isoprene rubber (CIIR), however, suggested that a significant change occurred in the distribution of the carbon black. Note that crystallinity loss was independent of the manner in which carbon black is added.

This study provided strong evidence that carbon black does not transfer from SBR, BR, and natural rubber (NR) when conventional Banbury mixing procedures are used. However, black migration does take place from the low-unsaturation polymer chloro–isobutene–isoprene rubber (CIIR) toward the high-unsaturation BR in CIIR–BR blends. Electron micrographs showed that the carbon black did not actually transfer, but, rather, migrated to the interface, in agreement with the findings of Marsh *et al.* (1968) for NR–chloroprene rubber (CR) blends. Black migration in CIIR–BR was reduced by increasing CIIR–black interaction by heat treatment and by use of a promoter. This work highlights the importance of the sequence of addition of elastomers and carbon black, since carbon black generally does *not* migrate.

E. Degree of Dispersion of Elastomer Blends by Analysis of Crystallization Behavior

The use of DSC measurements of butadiene rubber (BR) crystallinity to assess the degree of mixing in BR–styrene–butadiene rubber (SBR) blends has been described by Ghijsels (1977) for unvulcanized, unloaded samples. The objective of this work was to define a quantitative, statistically based technique for determining the extent of mixing and hence relative effectiveness of various techniques and equipment. A critical factor in this analysis was the development of standard annealing procedures to ensure complete BR crystallization and avoid T_g and crystallization peak interference with crystallization measurements (Fig. 39). The approach is generally similar to that of Sircar and Lamond (1973b), but differs in the specific low-temperature treatment of the samples.

The data obtained with the BR-1220/SBR-1500 system is summarized in Table IX. Table IX shows that the spread in $\triangle H_f$ values would be a useful measure of blend composition among 20 mg randomly selected samples only for short mixing times. Only during the first stage of the blending process, therefore, could one use the statistical variation of $\triangle H_f$ as a reliable measure of blend quality.

In general, various functions of the standard deviation(s) are preferred as a measure of blend variability. Ghijsels (1977) employed an index, M, which ranges from 0 to 1 and is independent of composition. The assumption is made that the minimum variance S^2_{min} attainable for 20 mg spot samples is negligible compared to the maximum variance at the start of

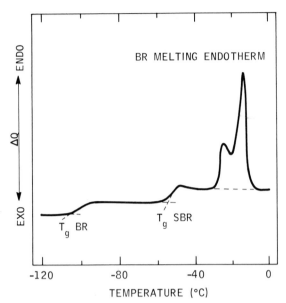

Fig. 39 Typical DSC thermogram of an unvulcanized BR/SBR blend when crystallized under conditions described in the reference (Ghijsels, 1977).

TABLE IX

FRACTIONAL HEAT OF FUSION OF BR-1220/SBR-1500 BLENDS[a,b]

Blend Ratio BR:SBR		Time of blending, s					
		5	15	30	45	120	300
50:50	Average	0.49	0.49	0.469	0.482	0.459	0.450
	Range	0.95	0.22	0.045	0.024	0.020	0.58
	Standard deviation	0.26	0.08	0.017	0.011	0.008	0.018
25:75	Average	—	0.23	0.238	0.208	0.093	0.054
	Range	—	0.48	0.070	0.043	0.031	0.031
	Standard deviation	—	0.14	0.026	0.014	0.011	0.008

[a] From Ghijsels (1977).
[b] Twelve observations for each blend.

mixing ($S_{\max}^2$). This leads to the following expressions for the mixing index, M, in terms of the experimental variance S^2:

$$M = 1 - 4S^2 \quad \text{for} \quad 50/50 \text{ blends}$$

$$M = 1 - \tfrac{16}{3} S^2 \quad \text{for} \quad 25/75 \text{ blends}$$

The semilogarithmic plots shown in Fig. 40 indicate that during the ini-

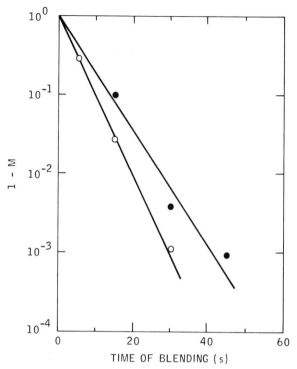

Fig. 40 Semilogarithmic plot of $1 - M$ as a function of time of blending for BR-1220/SBR-1500 blends: (open circle) 50/50 blends, (solid circle) 25/75 blends (Ghijsels, 1977).

tial stages, blending follows an exponential law. This suggests the possibility of describing the efficiency of polymer blending procedures in terms of a rate constant.

Additional information about the blending process can be obtained from analysis of the degree of interference of SBR with the crystallization of BR. As shown in Fig. 41, the average fractional heat of fusion decreases with increased blending time. Use of this decrease as a measure of degree of blending was proposed, especially for the latter stages of mixing where the index M cannot be used. Depending on the cause of this decrease, its magnitude is a measure of either interfacial area or particle size of BR.

An advantage cited for this general analytical approach is that it should be readily extended to filled systems, including filled vulcanizates, which are difficult to handle by other methods. It should also prove applicable to various blends of elastomers and/or plastics, providing one component can crystallize under conditions where the other one cannot.

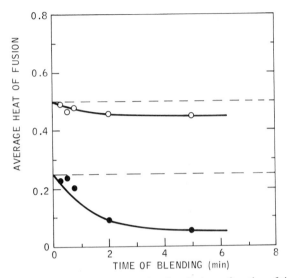

Fig. 41   Average heat of fusion (expressed as fraction) as a function of time of blending for BR-1220/SBR-1500 blends: (open circle) 50/50 blends, (solid circle) 25/75 blends (Ghijsels, 1977).

An interesting paper by Lee and Singleton (1979) describes the use of DSC and electron microscopy to relate processing characteristics and morphology in blends of styrene–butadiene rubber and *cis*-polybutadiene rubber with carbon black.

F. Elastomer–Plastic Blends

Blends of elastomers with either amorphous or crystalline thermoplastics are used commercially to improve toughness, impact strength, injection molding characteristics, etc. Numerous possibilities exist for the application of thermal analytical methods in these systems. These range from qualitative (T_g, T_m, TG, DTG) and quantitative (TG, DTG) analysis to mixing studies based on DSC evaluation of crystallinity variations, as described by Ghijsels (1977). Detailed consideration of this type of elastomer–plastic blend system is treated by Shalaby and Bair (see Chapter 4 of this book).

VI. Block Polymers

A. Introduction

Polymers that are composed of long sequences of chemically different monomers, joined at the ends, are termed block polymers. These systems are described in detail in Chapter 4 of this book and by Noshay and

McGrath (1977). The latter source provides a comprehensive overview of this class of polymers as well as numerous references. A block polymer composed of a pair of monomers A and B could have an AB, ABA or $(AB)_n$ microstructure, depending on synthesis conditions. The ABA and $(AB)_n$ structures are of special interest, since they form the basis for a new class of elastomer systems known as *thermoelastics*. These polymers exhibit rubbery behavior in the absence of chemical cross-linking due to the formation of physical cross-links.

Thermoelastics are composed of so-called hard blocks (T_g or T_m above room temperature) and soft blocks (T_g below room temperature). The soft block is the major component. Aggregation or association of the hard blocks leads to the formation of small domains that function both as physical cross-links and reinforcing filler. The key property of these systems is the thermally reversible physical cross-links. This enables the polymers to be melt processable (above the T_g or T_m of the hard block) in conventional plastics equipment. At lower temperatures, these materials exhibit excellent elastomeric properties, approaching those of chemically cross-linked elastomers in some, but not all cases.

A key feature of thermoelastics is the possibility for the formation of complex morphologies due to aggregation of the segmented polymers. The extent to which these "supermolecular structures" are formed depends on the chemical composition of the two monomers, molecular weight, crystallizability, volume fraction of the components, method of fabrication, and thermal history. The type of structure that is formed depends on volume fraction of the comonomers and mode of preparation of the sample. Spherical, rodlike, and lamellar structures have been reported.

Basic thermal analytical techniques should be useful for the characterization of block copolymers via the following types of investigations: solubility and/or swelling, mechanical and dynamic mechanical properties, rheology, elastic recovery, and selective degradation. As an example, DSC was shown to be an excellent tool for determining optimum mold temperature versus fatigue resistance of thermoplastic polyurethanes (Lawandy and Hepburn, 1980). Dynamic mechanical analysis and DSC have been used to study the physical aging of poly(acrylonitrile–butadiene–styrene) (Wyzgoski, 1980). These studies indicate that a simple time–temperature equivalence does not exist for physical aging in this system.

The objective of the following discussion is to illustrate the use of thermal analytical techniques in characterization studies of the three main structural types of commercial thermoelastics. Specific examples of styrene–diene ABA or radial polymers, ester–ether $(AB)_n$ polymers and urethane–ester $(AB)_n$ polymers will be presented.

B. Styrene–Butadiene–Styrene (SBS) Copolymers

Appropriate choice of synthesis conditions enables the preparation of linear or star-shaped SBS thermoelastic polymers. Commercial systems of both types, Shell's Kraton® and Phillips' Solprene®, respectively, are the subject of a study by LeBlanc (1977). The objectives of this work were to examine the tensile properties as a function of processing technique (molding vs. extrusion) and to compare the transition temperatures for these same systems, as determined by TMA expansion measurements.

The stress–strain (S/S) behavior of several different Solprenes was similar and exhibited a dependence on how the specimen was prepared and measurement direction. Higher S/S properties were observed for injection-molded samples and for a direction parallel to orientation during processing. Variations in S/S properties were observed during cycling experiments and as a function of annealing time following the initial extension cycle. These were interpreted in terms of an SBS "superstructure" that is broken during the first extension. The high value of Young's modulus suggested that the polystyrene domains cannot move independently of each other at low extension levels. This structure change appears to be reversible, since annealing of stretched samples at 60°C leads to a gradual recovery of the initial S/S properties. The breaking of a rigid structure during the first elongation is not compatible with these recovery characteristics. Since this recovery occurs far below T_g of polystyrene, it was assumed that the deformation mechanism does not concern the polystyrene domains as such. These observations suggested the existence of diffuse interfacial regions between the polybutadiene matrix and the polystyrene domains. Hence stress softening and yielding are interpreted as the result of a reversible deformation of a "superstructure" that depends on the processing technique used to form the sample.

It was anticipated that TMA differences would parallel the tensile behavior of these systems. This was confirmed in a series of expansion measurements involving a flat probe and 0.5×0.5 cm samples cut from tensile dumbbell specimens. Comparisons of raw pellets and extruded versus injection-molded specimens were made for a series of samples. Typical results for the star-shaped Solprene 406 polymer are shown in Fig. 42 and for the linear Kraton 1101 polymer in Fig. 43. The apparent melting region exhibited by the raw pellets is attributed to the collapse of their foamed structure. As noted in the figure, thermal expansion is greater for the injection-molded samples, presumably due to higher stress storage during the injection molding process. Extruded samples, however, may exhibit transitions not seen in molded samples. Transitions in both types of specimens occur between room temperature and 100°C; no transition is evident at the polystyrene T_g (+100°C). These characteristics were taken as

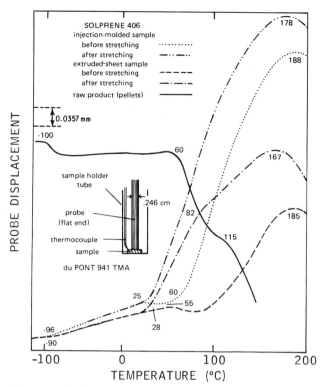

Fig. 42 Thermomechanical analysis of Solprene 406, at heating rate of 10°C/min; load, 10 g (Leblanc, 1977).

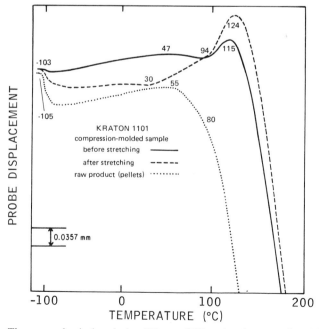

Fig. 43 Thermomechanical analysis of Kraton 1101, at heating rate of 10°C/min; load, 10 g (Leblanc, 1977).

confirmation of the hypothesis that the mechanical properties are due to a complex interfacial structure rather than glassy polystyrene domains.

Comparison of the effect of sample stretching on TMA curves is also of interest. Above 20°C, certain transitions observed in unstretched samples are absent in stretched specimens. Injection-molded Solprene 406 in the unstretched state, for example, exhibits expansion up to 25°C, a plateau up to 60°C, and further expansion to 188°C. After stretching, the 25°C transition is followed by direct expansion to 178°C. Less-pronounced differences are noted in compression-molded Kraton 1101 (Fig. 43), but the effect of stretching on transition characteristics is clearly evident. Thus TMA shows that stretch-induced structural changes relate to thermomechanical transitions *below* polystyrene T_g, thereby supporting the proposed reversibly deformable, diffuse interfacial structure model.

C. POLYETHER–POLYESTER THERMOPLASTIC ELASTOMERS

This study, by Lilaonitkul and Cooper (1977), is concerned with polymers composed of random sequences of polytetramethylene ether (PTMEG) soft segments and polytetramethylene terephthalate (4GT) hard segments. The morphology of this system has been interpreted on the basis of two models. One assumes continuous and interpenetrating crystalline and amorphous regions physically "cross-linked" by randomly oriented and interconnected lamellar hard segments. The other model suggests a spherulitic morphology composed of 4GT radial lamellae and interradial amorphous regions; the latter are assumed to be a mixture of PTMEG soft segments and uncrystallized 4GT hard segments.

Lilaonitkul and Cooper (1977) sought an improved definition of this system by determining the effects of copolymer composition, modification of 4GT hard segment, and sample processing methods on the morphology, crystallinity, microphase separation and deformation mechanism of these polymers. Analytical techniques included DSC in combination with small-angle light scattering, optical microscopy, dynamic mechanical analysis and simultaneous stress-strain dichroism. The role and utility of DSC in this kind of study is indicated by the following examples.

The effect of 4GT content on T_g and T_m is shown in Fig. 44. The observed variation in T_g suggested that solubilized hard segments are included in the amorphous phase. Analysis of this relationship in terms of the Gordon–Taylor–Wood equation is presented in Fig. 45. The best fit to the GTW relationship was obtained when correction was made for the hard phase in the amorphous material. However, the increasing breadth

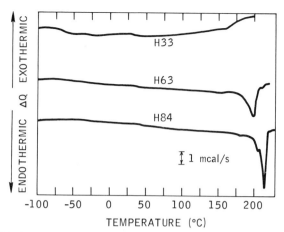

Fig. 44 DSC thermograms of compression-molded samples of varying weight percent hard segment content as shown on curves. Heating rate 20°C min⁻¹ (Lilaonitkul and Cooper, 1977).

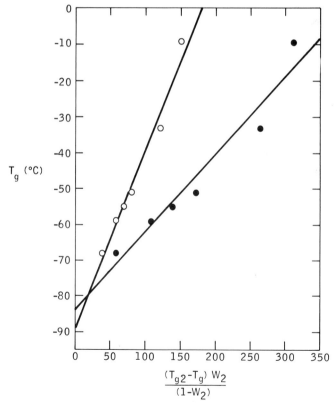

Fig. 45 T_g data plotted according to Gordon-Taylor equation. (open circle) W_2' defined as weight fraction amorphous 4GT; (solid circle) W_2 as weight fraction 4GT (Lilaonitkul and Cooper, 1977).

of the T_g region as 4GT content increases suggests a partially heterogeneous system.

DSC measurements provided additional information about morphology via analysis of melting behavior. T_m was observed to increase with hard segment concentration; and annealing produced peaks due to poorly ordered crystals as well as new peaks related to the annealing process. Higher degrees of crystallinity were observed in solvent cast films. These analyses suggested the occurrence of secondary crystallization as well as the melting and recrystallization of unstable small crystals. Increased annealing temperature shifted T_m to higher values; this suggested an extrapolation method for determining the true T_m as illustrated in Fig. 46.

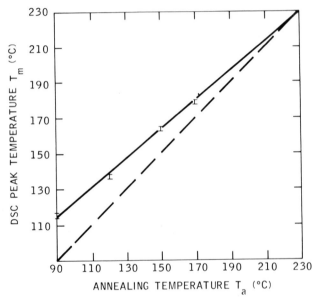

Fig. 46 Relation between the DSC annealing peak and the sample annealing temperature. Each point represents data for samples covering the 50–84 wt % hard segment range (Lilaonitkul and Cooper, 1977).

The complex nature of thermoplastic elastomers is evident from these data, which are only part of the complete study. It is evident that a wide variety of techniques is needed for complete characterization of the system; thermal analysis is clearly a valuable component. In this case, straightforward determination of T_g and T_m via DSC has enabled basic insight and conclusions regarding the morphology of these complex systems.

D. Transition and Phase Changes in Toluene Diisocyanate-Based Polyurethanes

The combined use of TMA and DSC for definition of complex transition behavior and phase segregation in polyurethanes is described by Schneider and Paik-Sung (1977). An extensive series of segmented polyurethanes based on 2,4-toluene diisocyanate (2,4-TDI) or 2,6-TDI butanediol, and polytetramethylene oxide of 1000 or 2000 molecular weight (PTMO-1000 and -2000, respectively) or polybutyleneadipate (PBA-1000 and PBA-2000) soft segments was evaluated. These studies, illustrated later, indicated that increasing the soft-segment molecular weight promotes phase segregation and suggest that phase mixing is more pronounced in polyester than in polyether samples.

Unlike conventional elastomers, thermoplastic elastomers may exhibit substantial differences in transition characteristics as measured by DSC and TMA. Further complexity may be introduced by variation of thermal or processing history of the sample, the extent varying with polymer composition. Typical DSC and TMA thermograms for 2,4-TDI-PBA 2000 samples are shown in Figs. 47 and 48, respectively. The first DSC scan reveals the soft segment T_g and melting of polyester crystallinity. A rerun of this sample after cooling reveals only the T_g, shifted to a lower temperature. The initial high temperature TMA scan (Fig. 48) shows transitions similar to DSC; however, a new transition appears above the soft-segment T_g in the second run. This is attributed to the hard-segment T_g, which had previously been concealed by melting of polyester crystallinity. The results for this system are summarized in Table X. Significant features are the increase in soft-segment T_g (attributed to mixing of the hard-

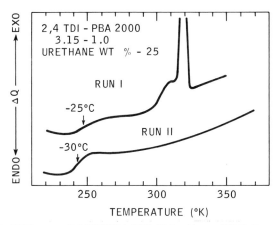

Fig. 47   DSC scans, 2,4-TDI-PBA 2000 (Schneider and Paik Sung, 1977).

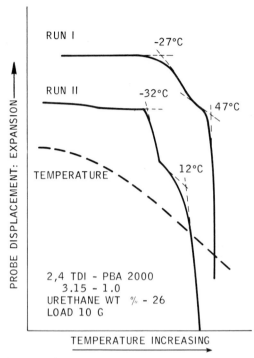

RUN I

-27°C

RUN II

-32°C

47°C

12°C

TEMPERATURE

2,4 TDI - PBA 2000
3.15 - 1.0
URETHANE WT % - 26
LOAD 10 G

PROBE DISPLACEMENT: EXPANSION

TEMPERATURE INCREASING

Fig. 48 TMA scans, 2,4-TDI-PBA 2000 (Schneider and Paik Sung, 1977).

TABLE X

<small>Transition Temperatures (°C) for 2,4-TDI-Polyester Polyurethanes[a]</small>

Samples	Urethane (Wt %)	T_g (DSC)	Upper transitions (TMA)	
			Run 1	Run 2
PBA 1000				
2	31	−13	39, 103	39
3	42	1.0	53	50(+)[b]
4	50	14	60, 162	63(+)
5	56	25	66, 182	75(+)
6	61	36	180	78(+)
PBA 2000				
2	19	−40		
3	27	−31		12
4	33	−23		20
5	39	−15		45
6	44	−6		64

[a] From Schneider and Paik-Sung (1977).
[b] (+) indicates gradual softening above the transition.
[c] All PBA 2000 samples, only soft segment melting observed on first run, $T = 40$–48°C.

block urethane in the soft-segment matrix); a regular increase in hard-segment T_g with increasing urethane content; and the 160–180°C transition attributed to allophonate and biuret bonding.

DSC studies of 2, 6-TDI–PBA 1000 (Fig. 49) indicate another morpho-

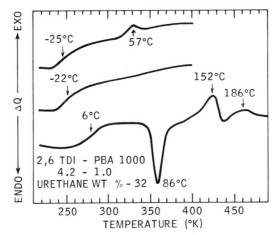

Fig. 49   DSC scans, 2,6-TDI-PBA 1000 (Schneider and Paik Sung, 1977).

logical feature. Specifically, the T_g value increases by more than 30°C during repeat runs after the sample is heated above the melting point. This suggested that enhanced phase segregation occurred during compression molding of the sample. Finally, comparison of the glass transition temperatures as a function of urethane content (Figs. 50 and 51) enables

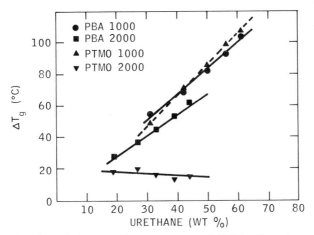

Fig. 50   Elevation of glass transition temperature, 2,4-TDI polyurethanes (Schneider and Paik Sung, 1977).

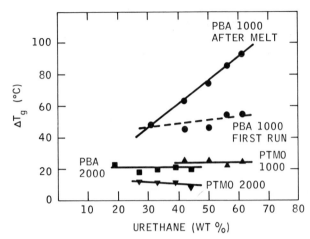

Fig. 51 Elevation of glass transition temperature, 2,6-TDI polyurethanes (Schneider and Paik Sung, 1977).

deductions concerning the effect of polymer parameters on phase mixing. The ordinate was chosen to compensate for the difference in T_g of PBA and PTMO. Noteworthy features of these relationships are the major change due to the increase in PTMO molecular weight (Fig. 50), and the extensive phase mixing in the 2,6-TDI–PBA 2000 samples despite the crystallization tendency of the hard segment.

The examples cited clearly indicate the considerable role of thermal analytical techniques for evaluating phase segregation and transition behavior in thermoplastic elastomers. Regarding the importance of information concerning transition characteristics, Schneider and Paik-Sung (1977) indicate that the glass transition temperature in combination with the copolymer equation provides a useful approximation for estimating the degree of phase separation.

Additional examples of the application of thermal analysis to polyurethane systems have recently been reported. Guise and Smith (1980) employed DSC and TMA to determine the influence of polymer composition on T_g and melting of cast films. Kaneko *et al.* (1980) employed DSC in a study of the influence of fatigue processes on the degree of order in polyurethanes.

VII. Analysis of Cure Characteristics

A. BACKGROUND INFORMATION

Curing, or vulcanization as it is commonly termed, is the process that converts a rubber compound from a weak, thermoplastic state to a tough, elastic state. This is accomplished by the formation of chemical cross-links between the elastomer molecules. The most frequent means for curing many elastomer compounds is the use of one or more accelerators plus sulfur at elevated temperature and pressure. Detailed analysis of the curing process is, in general, a formidable task for the following reasons:

(1)   The cure system consists of several components (e.g., ZnO, stearic acid, sulfur, one or more accelerators, and possibly cure retarders or promotors).

(2)   There are a variety of additives in the system, some of which may be capable of becoming involved in the overall reactions attending cross-linking of the polymer [stabilizer, process aids, plasticizer, and/or process oils, fillers (carbon blacks and/or mineral fillers), etc.].

(3)   Certain antioxidants may affect the cure rate of the system.

(4)   Commercial elastomer compounds frequently contain blends of polymers which may have substantially different cure rates.

(5)   There are a variety of chemical reactions that take place during the curing process, and these in turn lead to a variety of structures (disulfide, polysulfide, cyclic sulfide, etc., as shown in Fig. 52; Bateman, 1963).

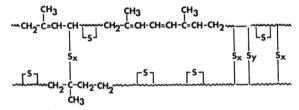

Fig. 52   Structural features of an unaccelerated sulfur–NR vulcanizate network (Bateman, 1963). Copyright 1963 by John Wiley and Sons, Inc. Reprinted by permission of John Wiley and Sons, Inc.

Further complexity is introduced by the fact that these structures may change (e.g., polysulfide → disulfides) as a function of time and temperature.

The potential applicability of thermal analysis to aid in elucidating the many complexities of vulcanization chemistry, particularly in practical rubber formulations, has attracted the attention of several workers. These activities, which may conveniently be classified into qualitative and quantitative approaches to vulcanizate analysis, have demonstrated that a remarkable array of information can be gained about such systems. As indicated later and in Section IX, thermal analysis clearly offers major potential for a variety of studies ranging from laboratory research to commercial quality control activities. Although much remains to be done, major steps have been taken toward the development of practical "microscale" techniques for guiding research, development, and commercial-scale elastomer curing studies.

B. QUALITATIVE ANALYSIS OF COMPOUNDS AND VULCANIZATES

1. Analysis of DTA Thermogram Characteristics

Several opportunities for the application of DTA to accelerated compounds and/or cured vulcanizates are summarized in Fig. 53 (E. I. DuPont

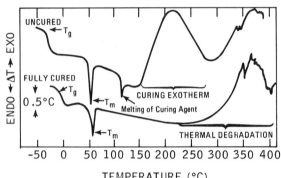

Fig. 53 DTA characteristics of an elastomer formulation (E. I. DuPont & Co., 1963). Reprinted with permission from Analytical Chemistry. Copyright by the American Chemical Society.)

& Co., 1963). The thermogram for the uncured formulation of unknown composition, reveals the following features: the glass transition temperature (T_g) of the "system," melting of cure system ingredient(s), melting of polymer crystallinity, exothermic reactions associated with the overall cure process, and, finally, decomposition of the polymer. Each feature may yield important information about the system. The amount of information that can be extracted solely from such a thermogram will depend on the amount of background information available about the system. Thus, if it is intended to assess the potential of the DTA technique for quality control of accelerated (uncured) and/or cured formulations, it would be highly useful, indeed mandatory in some cases, to develop the following type of background information about the systems:

(a) DTA scans for each ingredient in the formulation,

(b) the influence of cure state and plasticizer concentration on the T_g of the polymer,

(c) the concentration at which detection of each ingredient becomes possible, and

(d)   the extent of overlap of various features of the individual components.

Because of this wide variety of factors that contribute to the thermogram observed with a given formulation, plus the added complexities possible in commercial rubber formulations, the greatest potential for this type of analysis is for monitoring the uniformity of known systems. The degree of utility for "general unknown" vulcanizates will often be much more limited because of uncertainties regarding the variety of factors that could give rise to a particular thermogram characteristic, or the absence of same.

The DTA thermogram for the cured sample in Fig. 53 demonstrates the following:

(a)   an apparent effect of cross-linking on polymer T_g (when this effect is large enough, and well defined, it offers a potential for estimating the degree of cure),

(b)   the absence of curatives (presumably due to decomposition or interactions during the overall curing process),

(c)   polymer melting (note that this would be influenced by several factors, including the thermal history of the sample),

(d)   absence of the "curing exotherm," again indicating the decomposition or interaction of the curatives,

(e)   the decomposition pattern of the polymer [as indicated in Section VIII, this may sometimes provide useful information about the type of polymer(s) in the formulation].

2. Analysis of Sponge Formulation

An almost ideal opportunity for the application of thermal analysis to an elastomer system is presented by consideration of sponge formulations as shown by Maurer (1967a). In addition to the general cure system components described earlier, sponge formulations contain a blowing agent designed to produce the voids (cellular structure) in the cured vulcanizate. The key requirement here is to match the temperature regions where network formation and blowing agent decomposition–gas generation occur. Examples of a poor (curve C) and a good match (curve D) are shown in Fig. 54.

3. Consideration of Exotherm Areas

An additional point of considerable potential interest is also shown in Fig. 54. Curves A and B were obtained as part of a program to determine the potential utility of the DTA technique for assessing the cure chemistry and cure state in elastomer formulations. Note the similarity of the

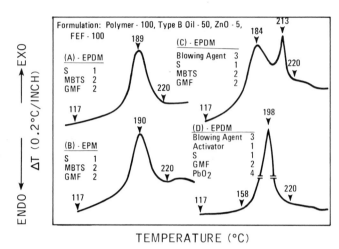

TEMPERATURE (°C)

Fig. 54 Analysis of curing and blowing regions of ethylene–propylene–diene (EPDM) and ethylene–propylene (EPM) rubber sponge compounds (Maurer, 1967a).

exotherms for system B, which contains a *saturated* ethylene–propylene rubber, and for system A, which contains an *unsaturated* (and hence vulcanizable) ethylene–propylene–diene elastomer. Comparison of these curves indicates (a) that there is a readily determinable exothermicity involved with this type of cure system and (b) that this activity may be largely related to decomposition and/or reactions among the cure system components rather than to cross-linking per se. It was observed that this situation suggested the potential for quality control (QC) type analyses of accelerated stocks.

Comparison of these exotherms also indicates the potential utility of various characteristic temperatures (e.g., initiation, peak, and terminal) for comparing the activity of different cure systems. Of course, evaluation of the exotherm area is also indicated as a potential route to quality control regarding the cure system. DTA studies of the cure process were limited by a variety of complicating factors (gas evolution during the cure reactions, reaction of curatives with thermocouple, etc.). Circumvention of these problems is reported by Brazier and Nickel (1975a), who describe the successful QC analysis of cure system composition in practical formulations (see Sections VII and IX) via differential scanning calorimetry techniques. Their procedures include analyses of exotherm area and characteristic temperatures similar to those considered earlier. Both isothermal and scanning calorimetric analysis of cure processes have been reported by Brazier *et al.* (1980).

Thermal analysis of the dicumyl peroxide curing of elastomers is described in recent papers by Brazier and Schwartz (1980) and Seidov *et al.* (1980).

C. Quantitative Analysis of Compounds and Vulcanizates

1. DTA Studies of the Hard Rubber Reaction

Bhaumik *et al.* (1962, 1965a,b,c) conducted an extensive DTA study of the hard rubber reaction. They determined the heat of reaction ($\triangle H_v$) for a series of rubber–sulfur mixtures and systematically evaluated the influence of "compounding" variables (metallic oxides, accelerators, fillers, etc.) on $\triangle H_v$. Among the major findings of this work were the following:

(a)   Heat evolution was nearly linear from 7 to 30% sulfur.

(b)   Metallic oxides did not significantly affect $\triangle H_v$, but metallic oxide–accelerator combinations did, the effect varying with the type of both ingredients.

(c)   Fillers influenced $\triangle H_v$, initiation temperature, and slope values to varying degrees.

These studies represent one of the early, detailed quantitative investigations of elastomer curing by thermal analytical methods. A very similar DSC study of conventional rubber compounds by Brazier and Nickel (1975a) is reviewed in more detail in the following section. A DTA study of interest is a report by Festisova and Shabanova (1980), who describe the properties of sulfur melts with vulcanization accelerators.

2. DSC Studies: Unaccelerated Sulfur Vulcanization (S-V)

The quantitative DSC analyses of Brazier and Nickel (1975a) represent a significant step forward in the application of thermal and analytical procedures to elastomer systems. The first part of their work involved a detailed examination of the exothermic heat of reaction of the *unaccelerated* sulfur vulcanization of natural rubber (NR) (the so-called hard rubber reaction). The key assumption involved in the $\triangle H_v$ determination is that all of the added sulfur is transformed to combined sulfur, per the studies of Bhaumik *et al.* (1962). The absence of a curing exotherm, when a sample is reheated in the DSC following the initial vulcanization run, is taken as confirmation of this. A further indication was the inability to detect sulfur in a benzene extract of a vulcanized sample.

Analysis of the dependence of $\triangle H_v$ on sulfur concentration (Fig. 55) revealed a change in slope that is taken as an indication of a change in mechanism (since it is not correlatable with a T_g increase due to curing).

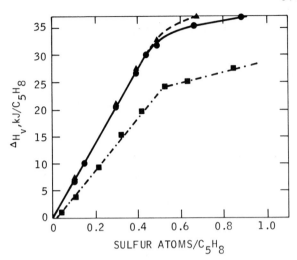

Fig. 55 Heat of vulcanization dependence on sulfur concentration in unaccelerated natural rubber vulcanization. (Open square) 20°C/min (Brazier, 1977); (open circle) isothermal calorimetry 25°C, and (open triangle) isothermal calorimetry 155°C (Bekkedahl and Weeks, 1969); Brazier, 1977).

The $\triangle H_v$ is attributed to the total set of series and consecutive reactions in the vulcanization process plus contributions from so-called maturation of some cross-link types during the high-temperature DSC environment (Fig. 52). Consequently, meaningful kinetic analysis of the hard rubber curing reaction is considered impossible.

3. DSC Studies: Accelerated Sulfur Vulcanization (S-V)

Accelerators in combination with stearic acid and ZnO improve the rate of overall sulfur utilization via formation of a complex. The mechanism is complex as reviewed in the text by Bateman *et al.* (1963). In essence, the effect is to produce a change in the overall distribution of sulfide structures (the actual distribution depending on the type of accelerator and the sulfur/accelerator ratio).

Based on the $\triangle H_v$ versus [S] relationship shown in Fig. 55, Brazier and Nickel (1975a), concluded that quantitative analysis of $\triangle H_v$ can be made for accelerated S-V in commercial elastomer formulations. The first step in such a program is to evaluate the effect of each ingredient on $\triangle H_v$. The effect of several thiazole accelerators on cure characteristics of a carbon black-filled NR compound is summarized in Table XI. The influence of sulfur and accelerator concentration is also shown as a calibration for subsequent analyses.

A major observation is that, since there is no effect on $\triangle H_v$ above a certain minimum accelerator concentration, $\triangle H_v$ can be used as a mea-

TABLE XI

EFFECT OF THIAZOLE ACCELERATORS ON DSC VULCANIZATION CHARACTERISTICS[a]

Accelerator	T_0 (°C)[b]	T_p (°C)[b]	T_c (°C)[b]	ΔH_v (J/g)	
Formulation: natural rubber 100, HAF-LS 50, zinc oxide 5, stearic acid 3, sulfur 2, accelerator 1.2					
None	192	222	245	9.4	
MBTS	168	201	238	6.6	
MBT	168	201	240	7.4	
MOR	164	193	218	7.2	
Formulation (phr): NR 80, BR 20, N326 black 50, stearic acid 3, zinc oxide 3, antioxidant 3, oil/plasticizer 17, cure system as shown					
Sulfur variation—MOR constant 1.2 phr					
Sulfur, phr	1.20	1.68	2.24	2.80	3.36
ΔH_v, (J g^{-1})	5.02	7.45	11.25	13.43	18.36
T_p (°C)	207	207	207.5	207.5	207
MOR variation—sulfur constant 2.24 phr					
MOR, phr	0.4	0.8	1.2	1.6	2.0
ΔH_v, (J g^{-1})	10.75	10.88	11.25	10.88	11.25
T_p (°C)	221	210	207	205	203

[a] From Brazier (1977).
[b] T_0, T_p, and T_c are the onset, peak, and return-to-base-line temperatures, respectively, for the DSC curing exotherm.

sure of sulfur concentration. Estimation of "accelerator" was accomplished via T_p value. A T_p reproducibility of ± 0.5°C enabled MOR to be estimated to 0.1 phr. In an earlier publication these authors report that T_p reduction depends on both the type and concentration of accelerator.

Systematic evaluations of stearic acid (little effect on ΔH_v) and zinc oxide were also conducted (Table XII). ZnO increased ΔH_v above about 3 phr. Therefore, evaluation of ZnO content is required in order to assess the effect of other factors on ΔH_v. The TG method proposed by Maurer (1969b) was used for estimation of ZnO content. Finally, the influence of carbon black content (Fig. 56) and natural rubber (NR)/butadiene rubber (BR) blend ratio was also determined. The effects were large enough to warrant determination of those quantities, via TG and DTG techniques similar to those developed by Maurer (1974a).

The analytical procedures just described are an excellent illustration of the strength of using DSC in combination with TG/DTG procedures. In this case, the latter techniques provide required monitoring and corrections of key system components, thus enabling the rapid, practical DSC studies. It is important to note that appropriate background information must be developed for each particular system of interest. The extent to which these procedures apply may be expected to vary depending on

TABLE XII

Effect of Stearic Acid and Zinc Oxide on Natural Rubber
Vulcanization Exotherm[a,b]

Zinc oxide	Stearic acid	ΔH_v (J/g vulcanizate)	T_o (°C)[c]	T_p (°C)[c]
5	0	8.62	194	221.5
5	3	6.44	194	221.5
5	5	6.44	194	221.5
0	3	7.78	198	226.0
1	3	5.73	196	222.5
3	3	5.85	196	222.0
8	3	9.71	195	220.0

[a] From Brazier (1977).

[b] *Formulation:* phr natural rubber 100, sulfur 2; HAF-LS carbon black 50.

[c] T_o and T_p are the onset and peak temperatures, respectively, of the DSC curing exotherm.

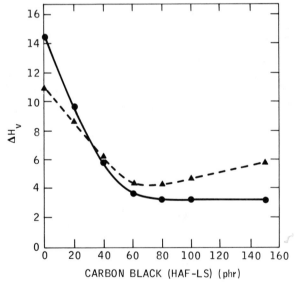

Fig. 56 Dependence of heat of vulcanization on carbon black level. Formulation: NR 100, ZnO 5, stearic acid 3, sulfur 2, MBTS 0.5, (open circle) in J/g of vulcanizate; (open triangle) in 10^{-2} J/C_5H_8 (Brazier, 1977).

applicability of the TG/DTG methods, the specific cure system and carbon blacks, etc.

A different application of DSC and TGA is described by Simpson and Percival (1979), who evaluated the exothermic decomposition of litharge–elastomer dispersions.

4. *TG Analysis of Cross-link Density via Solvent Swelling*

Evaluation of cross-link density by equilibrium solvent swelling is an established technique based on the well-known Flory–Rehner (1943) relationship. The conventional solvent-swelling procedure is lengthy and tedious. A swollen sample is removed from the solvent, placed between pieces of filter paper, and weighed in a weighing bottle. It is then removed from the bottle, which is weighed again. Swollen sample weight is obtained by difference. Swollen sample weight may change over long periods (days, weeks, or months), necessitating extrapolation procedures in order to determine "zero-time" equilibrium swell values. Prime (1978) has developed a microtechnique for rapid measurement of the solvent swell of cross-linked polymers. The key feature of this technique is the elimination of solvent evaporation from the sample, thus enabling the use of thin samples which equilibrate rapidly. A complete description of the procedure appears in Chapter 5.

The essential features of Prime's procedure are repeated here because of the apparent applicability to a variety of elastomers and thermosets. Thin samples (0.1—0.4 mm thick) of silica-filled poly(dimethylsiloxane) elastomers were swollen in hexane and toluene. The swollen sample, suspended from a Cahn balance, was weighed in an atmosphere saturated with the swelling solvent. These conditions enabled both rapid equilibrium (Fig. 57) and precise measurement of swell ratio. For the samples studied, precision was ± 1–2% for swell ratio and ± 10–12% for sol fraction. Accuracy of the technique was established by comparison with a conventional method, for samples ranging in cross-link density from 7 to 35×10^{-5} moles cm^{-3} and having from 1 to 4% sol fraction. A direct

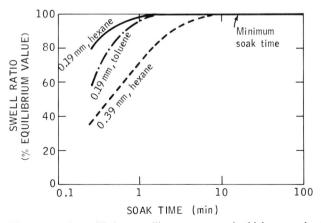

Fig. 57 Time to reach equilibrium swelling versus sample thickness and solvent type (Prime, 1978).

correlation between Young's modulus and swell ratio was established (Fig. 58), thus demonstrating an unambiguous analysis for monitoring cross-link density. Based on the speed and precision of the technique, Prime foresees wide applicability in research characterization, quality control, aging, and degradation of cross-linked polymers.

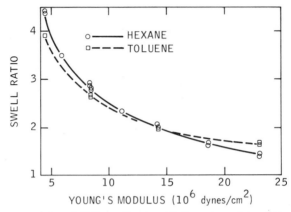

Fig. 58 Correlation of TMA swell ratio with Young's modulus (Prime, 1978).

Banerjee *et al.* (1979) also employed TGA to study vulcanization. In their work, weight losses were used to detect interactions of amines and other components in thiuram vulcanized natural rubber.

5. Analysis of the Cross-linking of Methyl Silicone Rubber

The cross-linking reaction in a two-component methyl silicone rubber (Sylgard-186®) was studied by DSC (Perkin Elmer DSC-1B) and TMA (DuPont 940). This paper by Barrall and Flandera (1974) is concerned with evaluating effective cross-links due to chemical bonds between polymer chains, as well as looped or entangled interactions. From the viewpoint of thermal analysis applications, key features of this paper are the use of DSC to measure heat of reaction and of a TMA unit to determine elastic modulus via penetration measurements.

Generally, TMA was run at room temperature on the same DSC samples used for heat of reaction. Elastic modulus (E_m) was calculated from the penetrometer measurements via an equation developed by Gent (1958).

$$E_m = (F/p^{1/2}) (9/16r^{1/2}),$$

where E_m is the elastic modulus, F the load, P the penetration, and r the probe radius. Additional treatments of the use of TMA for evaluation of modulus have also appeared, for example, Wood and Roth (1963), Jopling and Pitts (1965), Hwo and Johnson (1974), and Stiehler *et al.* (1979).

Cross-link density measurements were evaluated by conventional solvent swelling procedures and also from the TMA elastic modulus E_m via

$$e = E_m/3RT$$

where $T = 239$ K and $R = 8.314 \times 10^7$ ergs mole^{-1} K^{-1}. Heat of reaction was calculated from the total area under the DSC curve, and the energy of activation of the curing process was determined by the Arrhenius equation and reaction rates calculated by two methods. Method I (Barrett, 1967) involved a DSC curing area/exotherm height relationship; Method II (Rogers and Morris, 1966) involved only exotherm height as a function of temperature.

Heat of reaction, energy of activation, and cross-link density (CLD) were evaluated in terms of prepolymer concentration, dilution, and swelling. Several features of these relationships indicated complexity in both the curing reaction and the resultant network. An initial indication of this was the observation that catalyst concentration has a large effect on the shape, magnitude, and temperature of the DSC exotherm. Major reflections of this appear in the heat of reaction (Fig. 59) and the energy of

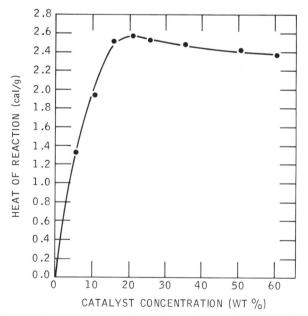

Fig. 59   Effect of catalyst concentration on the exothermal heat of reaction (Barrall and Flandera, 1974).

activation (Fig. 60). These relationships suggest that a second reaction can occur in the presence of excess catalyst.

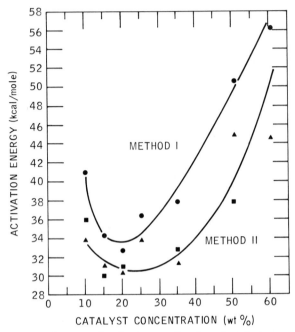

Fig. 60 Exothermal energy of activation as a function of catalyst concentration. (Solid squares) Calculated from ΔH and cross-link density (corrected) (Barrall and Flandera, 1974).

It has been assumed to this point that effective cross-links are due only to chemical bonds whose formation is detectable thermally. TMA evaluation of cross-link density of networks formed in the swollen state, however, suggests other possibilities. Assuming simple volume additivity, a linear relationship between cross-link density and dilution ratio would be expected. Instead a major departure from linearity is observed (Fig. 61). The tentative explanation proposed for this effect, following the work of Johnson and Mark (1972), is that some of the cross-links in the undiluted sample are due to entangled but not overlapping chains. Upon dilution with a good solvent, these chains disentangle and subsequently do not contribute the same number of effective cross-links. Based upon the studies summarized earlier, it was concluded that "the use of combined thermal and mechanical methods has been demonstrated to be a powerful tool for the exploration of cross-linked networks."

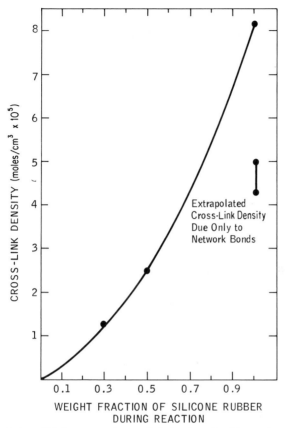

Fig. 61   Variation of TMA measured cross-link density with silicone oil dilution for 20% catalyst samples (Barrall and Flandera, 1974).

VIII. Evaluation of Vulcanizate Composition by TG and DTG

A. INTRODUCTION

As noted in Table I, elastomer compounds are complex mixtures of polymer, carbon black and/or mineral filler, curatives, plasticizers, and miscellaneous additional ingredients. This complexity is increased by the frequent use of blends of carbon blacks and/or elastomers. Because of this situation, it is a challenging task for the manufacturer to determine the cause of various problems that may arise in the compounding, processing, and curing operations leading to the finished product, and also for the manufacturer and customer to conduct quality control analyses on the finished rubber products.

Methods exist for the analysis of rubber formulations (American Society for Testing and Materials, Part 37 1979b), but these are often too expensive or time-consuming to be justified or useful for routine application. Consideration of the relative thermal stability and volatility of the basic components in elastomer compounds suggested that thermogravimetry (TG) had the potential to provide an effective and relatively rapid analysis of the "basic composition" (oil, polymer, carbon black, and mineral filler and/or ash) (Maurer, 1970b). Indeed, some of the "steps" in the proposed analysis had previously been demonstrated. For example, extensive literature exists for measuring and comparing various aspects of the thermal stability of polymers per se. Different techniques involving variations in atmosphere and methods of heating have been used to determine the influence of polymer structure, composition, and stabilizer systems on thermal stability. Chiu (1966) has mentioned or demonstrated other components of the proposed analysis in connection with plastics formulations. However, prior to the original work by Maurer (1969a), there was no published literature regarding TG as a route to achieve the proposed analysis for "basic composition" of elastomer compounds and vulcanizates.

Prior to a demonstration of the intended analysis, some comments are in order with regard to the general type of study being considered as well as the scope and general applicability of the methods. The basic TG experiment consists of recording the weight of a sample as it is heated in a controlled environment either isothermally or at a constant heating rate. The experimental record is a plot (thermogram) of some form of the weight change (e.g., actual weight or percent lost) versus time or temperature of the sample. It will be shown that extensive information about practical elastomer systems can be obtained by direct application and/or appropriate manipulation of these experimental parameters. In addition, the simple additional step of using the derivative of the primary weight change curve (DTG) extends the capability and scope of the analysis.

It should also be noted that the analysis will not always be completely accurate because of the wide range of formulations that may be encountered in commercial systems (i.e., a complete material balance will not be obtained). This is due to factors such as overlap of low-molecular-weight volatile products with polymer decomposition, decomposition of polymer blend components in a similar temperature region, etc. These complications limit the absolute accuracy of the analysis with regard to certain components; however, a wealth of information is obtainable. The maximum utility of these procedures will most likely arise in cases where extensive background information about the formulation is available, as in a laboratory or factory compounding environment. In addition, quality control evaluations of incoming articles based on a common formulation will enable assessment of product uniformity and hence degree of correlation with various practical characteristics of the finished parts. Thus there are practical limits to the kind and degree of information that can be extracted from such analyses of unknown vulcanizate compositions. On the other hand, the techniques clearly represent a major advance regarding routine and/or QC-type studies of even such complex systems as practical rubber formulations.

The approach to be used in describing the TG and DTG methods for vulcanizate analysis will be to proceed systematically from the simplest system [the polymer(s)] to more complex features of a practical single polymer formulation, i.e., oil, polymer, filler, and ash. Following this, more complex features of the system will be considered, i.e., polymer blends, characteristics of fillers, filler blends, factors influencing carbon black decomposition, etc.

B. Degradation Characteristics of Gum Elastomers

When elastomers are thermally degraded in an inert atomosphere, a variety of thermograms may result. In some cases, multistep weight loss is observed as with Hypalon 20®, chlorosulfonated polyethylene (CSM) (Smith, 1966a,b, 1968), where the three weight loss regions of the thermogram have been related to SO_2Cl, HCl, and main chain degradation. Careful analysis would be required to evaluate this system properly if encountered in an unknown vulcanizate formulation.

Despite the complexity of the Hypalon 20® decomposition pattern, it is interesting to note that only a small portion of the sample remains as a residue at 500°C. This characteristic is also shown by various commercial elastomers (Fig. 6) that exhibit simpler decomposition characteristics compared with CSM. In addition, the degradation patterns for these polymers differ significantly when evaluated by this common, controlled procedure. These two features of thermograms form the basis for analysis of the total polymer content in formulations based on many elastomer types. In addition, the inherent thermal stability differences shown in Fig. 6 suggest that quantitative analysis of polymer blend composition may be possible in some cases.

Another type of decomposition is encountered for elastomers that contain noncarbon functionality. For these polymers, decomposition leads to formation of a char that interferes with the analysis of polymer content. This problem can be overcome, however, as shown for nitrile–

butadiene rubber (NBR), by Swarin and Wims (1974) (Fig. 7), for chloroprene rubber (CR) and NBR by Schwartz and Brazier (1978a) and by Sircar and Lamond (1978). Thermal gravimetric analysis of the degradation
of polymers containing chlorine is reported by Jaroszynska *et al.* (1980).

C. VOLATILES, PLASTICIZERS, AND OILS

Several types of information about a gum, compounded, or vulcanized
elastomer can frequently be obtained from a TG thermogram in the temperature region before major weight loss due to polymer degradation
takes place. The conditions required for this analysis will vary depending
on the relative volatilities of these ingredients compared with the poly-
mer(s) (and possibly other nonpolymeric additives as well).

The simplest case is that of residual volatiles (water, polymerization
diluent, etc.). Interest in these ingredients relates to their potential influence on glass transition temperature, cure behavior, and vulcanizate quality (physical properties, porosity, etc.). Analysis for water or process solvents could best be accomplished by holding the sample at isothermal
conditions (inert atmosphere), and possibly under reduced pressure, as a
first step in the TG procedure. As shown by Brazier and Nickel (1975a,b),
this analysis of a compound or vulcanizate may also be accomplished and
facilitated by use of DSC (Figure 62) or DTG (Fig. 63).

Of greater interest with regard to the analysis of compounds and vulcanizates is the process oil or plasticizer(s) in the formulation. These materials are added to improve processability of the compound, reduce cost,
improve specific vulcanizate properties (e.g., low-temperature perfor-

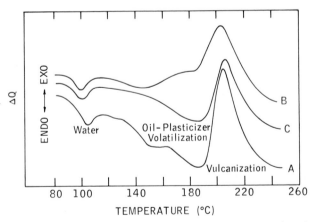

Fig. 62 DSC scans of typical tire stocks (formulations for A, B, C given in reference).
DSC scan rate 20°C min⁻¹, nitrogen atmosphere (Brazier and Nickel, 1975a).

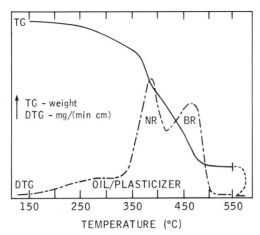

Fig. 63 TG and DTG thermograms of compound C (given in reference). ——TG, — · — · DTG. Heating rate 10°C min^{-1}; nitrogen atmosphere to 550°C, then oxygen atmosphere. DTG loop above 550°C omitted for clarity (Brazier and Nickel, 1975b).

mance), etc. Depending on the type of process oil and/or plasticizer used, as well as the type of elastomer(s) in the compound, several types of thermograms may be encountered during a standard run under dynamic heating rate conditions (Maurer, 1969a). If the oil and/or plasticizer is of "high" volatility relative to the polymer, then a thermogram, such as Fig. 14a, results; and oil content can be estimated directly from the thermogram. A more frequent case, however, is that in which volatilization of the oil overlaps the weight loss region due to polymer decomposition (Fig. 14b). The following approaches have been used to treat this type of sample.

(a)   Isothermal analysis at a temperature below that at which polymer decomposition occurs.

(b)   Use of a reduced pressure to aid in removing oil at a temperature below that where polymer weight loss becomes significant.

(c)   Extraction of the sample to remove oil, etc., prior to TG analysis. This provides a reasonable estimate of oil/plasticizer content if corrected for various low-molecular material removed with the oil (Maurer, 1969a).

(d)   Establishing a "correction curve" based on a reference temperature for a given polymer compound. In essence, this amounts to (1) selecting a temperature T_R for which the observed weight loss is due to all of the oil and a small amount of polymer decomposition; (2) developing a plot of the polymer weight loss at T_R versus polymer content remaining at T_R in a series of extracted vulcanizates containing varying levels of polymer; (3) using this plot to establish the true polymer content of an oil-containing vulcanizate; and (4) determining the oil content by subtracting the corrected polymer content from the total weight loss at T_R. This procedure gave results within about 2% of the known value for a series of isobutene–isoprene rubber (IIR) vulcanizates (Maurer, 1969a).

(e)   Swarin and Wims (1974) describe three methods for graphical resolution of oil and polymer weight loss. Their methods appear to work well with various ethylene–propylene–diene rubber (EPDM) formulations, but it should be noted that the overlap of oil and polymer is small in these systems because of the relatively high thermal stability of EPDM compared to other elastomers, and the degree of overlap will depend on the type of oil in the formulations.

Their three approaches were (1) overlay a TG curve of an unextended EPDM polymer obtained at the same experimental conditions as the oil-extended sample; estimate oil ''by difference''; (2) use a graphical extrapolation technique based on the intersection of linear regions of the TG thermogram due to polymer and polymer plus oil; (3) use the simultaneous DTG curve to determine when oil loss is complete; read oil content from the TG trace at this point. Monomeric plasticizers in nitrile–butadiene rubber (NBR) formulations were successfully analyzed by Swarin and Wims (1974) based on differences in DTG peak temperatures. A correlation curve was developed by gas chromatography (GC) analysis of plasticizers extracted from ten commercial samples and three standard samples (Table XIII). The GC versus DTG relationship for the ten samples that contained single monomeric plasticizers is shown in Fig. 64, which indicates the feasibility of using the DTG peak temperature for identification of single plasticizers in NBR formulations. When more than one monomeric plasticizer is present in the NBR compound (e.g., the last three samples of Table XIII), the DTG peak temperature is not as useful for analysis of plasticizer content.

Additional utility of DTG peak temperature was observed for systems containing different extender oils. In this case, good correlation was demonstrated for DTG peak temperature versus ASTM D1160 (American Society for Testing and Materials, 1979a) distillation range for several oils.

D. Polymer Content

1. Single-Polymer Systems—Minor or No Char

Determination of the polymer content in this type of system may be readily accomplished for a variety of polymers, as previously noted. The basic presumption is that all the weight loss in nitrogen is due to polymer, plasticizer, and/or oil. As previously mentioned, this does not account for

TABLE XIII

RESULTS OF GAS CHROMATOGRAPHIC AND DERIVATIVE THERMOGRAVIMETRIC
ANALYSIS OF PLASTICIZERS[a] IN NBR FORMULATIONS[b]

| Sample | Gas chromatography | | Derivative TG peak temperature (°C) |
	Relative retention time	Identity	
I	1.00	DBP	200
C	1.32	BF	233
D	1.32	BF	230
B	1.32	BF	230
F	1.90	DOA	253
Standard A	1.90	DOA	254
Standard B	2.18	DOP	264
Standard C	2.18	DOP	263
L	2.46	NODA	267
J	2.70	DOS	282
H	1.82	BBP	
		Mixture	260
	2.18	DOP	
G	1.00	DBP	
	1.32	BF Mixture	234
	1.82	BBP	
E	1.82	BBP	
	2.36	Unknown	
		Mixture	267
	2.70	DOS	
	3.02	Unknown	

[a] DBP = dibutyl phthalate, BF = bis(butoxyethoxyethyl) formate, DOA = di(2-ethyl-hexyl) adipate, DOP = di(2-ethylhexyl) phthalate, NODA = N-octyl, N-decyl adipate, DOS = di(2-ethylhexyl) sebacate, and BBP = butylbenzyl phthalate.
[b] From Swarin and Wims (1974).

various accelerators, stabilizers, waxes, etc., in the formulation. One factor that may complicate matters is the presence of various resins whose thermal decomposition extends over a broad temperature range.

Depending on the relative volatility and/or thermal stability characteristics of the resin and other ingredients in the formulation, the resin may significantly interfere with the analysis for basic vulcanizate composition. The problem will also, of course, be related to the resin content in the system. For known systems various correction procedures may enable compensation for these resin degradation characteristics. An interesting procedure to evaluate in this regard is the DTG technique used to overcome the char problem with nitrile–butadiene rubber (NBR) formulations (Swarin and Wims, 1974).

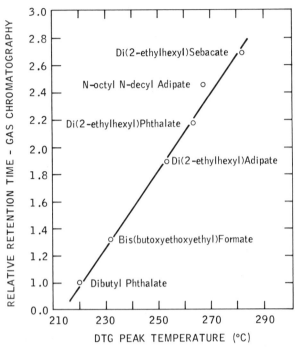

Fig. 64 Correlation of DTG data and gas chromatographic data for monomeric plasticizers. Commercial *N*-octyl, *N*-decyl adipate contains three peaks; the major peak is shown
here (Swarin and Wims, 1974).

2. Single-Polymer Systems—Char Forming

Swarin and Wims (1974) described the successful analysis of formulations based on nitrile–butadiene rubber (NBR), a polymer that leaves a
significant char during thermal degradation in nitrogen. This residue complicates analysis of both polymer and carbon black. This problem was
overcome by using two different atmospheres in the TG procedure and
simultaneous TG and DTG curves. The sample is first heated in a nitrogen
atmosphere to a temperature where most of the polymer has decomposed
and volatilized (about 550°C). Next the furnace is cooled to 300°C, air
flow is started, and the sample is heated again at 10°C min⁻¹. As shown in
Fig. 7, an initial gain in sample weight is followed by a small weight loss
beginning at 450°C. Similar evaluations of pure NBR demonstrated that
this weight loss corresponds to the decomposition and volatilization of
NBR "char." The DTG peak corresponding to this event was used to determine the amount of char and thus enable accurate determination of
both NBR and carbon black content in the formulation. It should be noted
that there is an overlap in the DTG peaks due to these two processes

which would lead to some error in estimation of the char oxidation "end point" via DTG. This does not appear to lead to serious errors as shown by their analysis of prepared NBR formulations.

The relative oxidation characteristics of the char and of the carbon black will influence the degree to which DTG will enable resolution of these events as shown by Swarin and Wims (1974), Pautrat *et al.* (1976), and Schwartz and Brazier (1978a). The last investigators added a new feature to this analysis, namely, use of a dilute oxygen atmosphere to control oxidation rate of the components. However, not all char-forming polymers can be differentiated from all carbon blacks in this manner. Thus, N110 and N330 type carbon blacks (see Table III) were found to be oxidized at the same temperature as the NBR residue (and thus could not be resolved from the char by DTG). However, N770 and N990 oxidized at different temperatures from the NBR char and thus could be differentiated from the char.

Pautrat *et al.* (1976) treated the char formation in NBR vulcanizates in a different manner. Their approach was to establish a correlation curve between the "graphitic residue" and nitrogen content of the polymer. This relationship, plus knowledge of the NBR composition and content, allowed acceptable estimates of polymer and carbon black content to be obtained. This approach enabled determination of the carbon black content of an NBR compound to be determined within 2%. Sircar and Lamond (1978) used a similar approach in which NBR composition was determined from a calibration curve relating T_g to percent acrylonitrile in NBR.

3. Binary Polymer Blends

The potential for quantitative analysis of binary elastomer blends is treated by Maurer (1974a), who describes two approaches for such an analysis. It should be noted here that the general applicability of the methods to be described will depend on the degree of difference in the thermal degradation characteristics of the components of the blend. Those with widely different thermal stabilities should be readily analyzed by inspection of the primary TG curve, since there would be little or no overlap of the degradation regions for the component elastomers. In the more common case, however, there is significant overlap of these degradation regions. It is this case that is considered here.

An example of a system with relatively minor overlap is the ethylene–propylene–diene ruber (EPDM)/natural rubber (NR) system shown in Fig. 65. Despite the overlap, the presence of more than one polymer component is clearly indicated. Two approaches to quantitative analysis of this system have been developed. The first is based on the weight percent

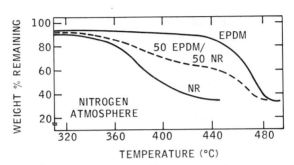

Fig. 65 TG analysis of EPDM/natural rubber systems; nitrogen atmosphere (Maurer, 1974a). Reprinted by courtesy of Marcel Dekker, Inc.)

polymer remaining at a particular reference temperature T_R. T_R, chosen by inspection of the TG curves for the component polymers, is a temperature at which one of the polymers (here, NR) has decomposed and volatilized. A plot of weight remaining at T_R thus gives a good estimate of EPDM and, by difference, NR. Successful application of the T_R approach to the NR/EPDM system is summarized in Table XIV.

TABLE XIV

TG Analysis of Vulcanized Natural Rubber/EPDM Blends[a]

Composition Wt % EPDM	Wt % remaining (410°C)	DTG peak (in) (NR)
0	8.5	2.95
20	26.3	2.20
50	48.8	1.30
80	63.3	0.78
100	84.7	–

[a] From Maurer (1974a). Reprinted by courtesy of Marcel Dekker, Inc.

An alternative approach for the analysis of this blend is shown in Fig. 8. Comparison of the DTG curves for a series of NR/EPDM formulations suggests quantitative analysis via DTG peak height versus blend composition. The data of Table XIV indicate a linear relationship between these variables; thus there are two measures of blend composition in this system.

Brazier and Nickel (1975b) present a systematic discussion of the factors that must be considered in a general analysis of polymer blend composition by TG/DTG. Review of this analysis is useful, since it illustrates a methodology for approaching new or unknown systems. In their case the treatment was applied to natural rubber (NR) blends with isoprene rubber–synthetic (IR), butadiene rubber (BR), or styrene–butadiene rubber (SBR).

The first step in this sequence was DTG of the raw polymers. Major observations here were that NR and IR have the same temperature of maximum degradation ($T_{\max}$), which allows their identification and estimation in many blends. An interesting note is that when carbon black (CB) filled, NR differs from IR, the latter having two degradation stages, the second of which is influenced by CB level and type. Silica also has a large effect (Brazier and Schwartz, 1978). Important aspects of BR degradation are that there are two stages that can be separated by increasing the heating rate. This feature is reported to be useful for analysis of BR–EPDM and SBR–EPDM blends whose peaks overlap at the normal 10°C/min heating rate. Detailed analyses are also presented for a series of commercial SBRs.

The next step was an evaluation of the DTG of blends, as exemplified by NR/BR (Fig. 63). These authors note, in agreement with earlier observations, that separate and distinct DTG peaks are encountered for only a limited number of blends. Treatment of the blend component characteristics was accomplished via consideration of a response factor ϕ, measured at $T_{\max}$ for a given elastomer in the blend (Table XV). In their quantitative

TABLE XV

DTG RESPONSE FACTORS FOR QUANTITATIVE ANALYSIS[a]

NR	BR	ϕNR (mg cm$^{-1\,b}$)	$T_{\max}$ (NR) (°C)	ϕBR (mg cm$^{-1\,b}$)	$T_{\max}$ (BR) (°C)
100	0	0.394	373	–	–
90	10	0.409	376	shoulder peak	
80	20	0.421	374	0.311	440
60	40	0.437	374	0.327	454
40	60	0.465	374	0.339	459
20	80	c	373	0.354	463
10	90	c	373	0.362	464
0	100	—	—	0.370	461
Polyisoprene		0.394	373		
Polyisoprene with 55 phr HAF-LS		0.547 / 0.822	372 / 420		
SBR 1500		0.413[d]	440		
EPDM (Royalene 301)		0.268[c]	461		

[a] From Brazier and Nickel (1975b).

[b] $\pm$ 0.008 mg/cm.

[c] Values not given because of interference from BR contribution.

[d] ϕSBR.

[e] ϕEPDM.

DTG studies, Brazier and Nickel first determined peak heights (in centimeters) for calibration blends, normalized to 0.079 mg/min-cm. The response factor (ϕ) for each component was then calculated in units of milligrams of the component per centimeter. Ideally, T_{max} would be constant at all blend ratios. Further, if no interaction between components occurred, ϕ, measured at T_{max}, would be independent of blend composition. Figure 66 and Table XV indicate considerable interaction in this blend.

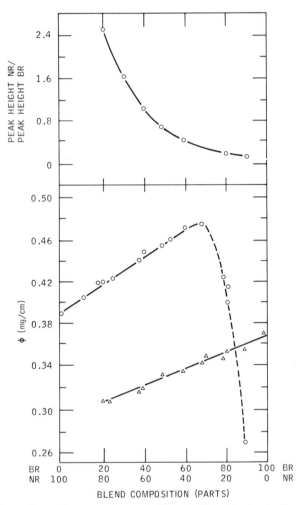

Fig. 66 Dependence of response factors for NR and BR on blend ratio. Top, variation of NR:BR peak height ratio with blend composition. Bottom, dependence of response factors on blend ratio. (Open circle) natural rubber, (open triangle) polybutadiene (Brazier and Nickel, 1975b).

For these blends, T_{max} (NR) was constant over the full composition range but T_{max} (BR) was not. It was found, for many blends, that stability of the more stable component is reduced (possibly because of attack by radicals generated during decomposition of the less stable component). No effects due to CB (up to 150 phr) or vulcanization were noted, except for the IR versus NR variation mentioned earlier. It was observed that the method is applicable in this system only over the 80–20 to 20–80 blend range; outside these limits, the peaks merge. Since both ϕ_{BR} and ϕ_{NR} were found to be functions of blend composition, it is necessary to determine the ratio for unknown blends before quantitative analysis via response factors is possible (Fig. 66). Similar results were reported for NR/SBR, in which system there was less interference of SBR compared to BR. It is worth noting that this work demonstrates successful quantitative analyses in a system in which interaction between the components occurs.

4. Ternary Blends

Maurer (1974a) extended the successful TG/DTG analysis described earlier for NR/EPDM blends to a ternary system in which one-component, ethylene–propylene–diene rubber (EPDM), was held constant while the others, styrene–butadiene rubber (SBR) and natural rubber (NR), were varied. Characterization of this system in terms of DTG peak heights and T_R is shown in Table XVI which suggests that quality control

TABLE XVI

TG Analysis of Vulcanized EPDM/NR/SBR Blends[a]

Composition (phr)			Weight % remaining (400°C)	DTG peak (in)	
EPDM	NR	SBR		NR	SBR
20	80	0	40.5	2.81	1.35
20	50	30	58.0	1.85	1.85
20	30	50	66.5	1.28	2.35
20	0	80	80.0	0.60	3.25

[a] From Maurer (1974a). Reprinted by courtesy of Marcel Dekker, Inc.

analysis of this system is feasible. Here again, maximum utility would be achieved when the formulation is known.

Brazier and Nickel (1975b) also discuss this triblend, as a representative tire sidewall system. In their case, however, NR was held constant. Figure 67 presents the calibration plots for this system. NR was found to be easily estimated since ϕ_{NR} and T_{max} were constant, but severe overlap of the SBR and EPDM resulted in a single DTG peak. Under these experimental conditions, ϕ_{SBR} is significantly different from ϕ_{EPDM}; however,

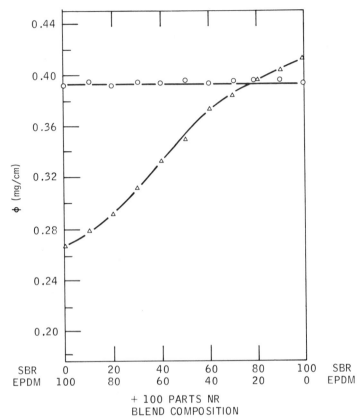

Fig. 67 Dependence of response factors for NR, SBR, and EPDM on blend ratio. NR constant at 100 parts: (open circle) ϕ NR; (open triangle) ϕ(SBR + EPDM) (Brazier and Nickel, 1975b).

there are too many variables to get the ratio of EPDM/SBR from Fig. 67. It was concluded that an independent measure of the EPDM/SBR ratio was required (e.g., pyrolysis IR).

This ternary system serves as an excellent illustration of the fact that the extent and type of information obtainable via TG/DTG can be heavily influenced by the specific composition being evaluated. Detailed background information is required for each specific system of interest.

E. Carbon Black Filler Analysis

Reinforcement of elastomers by carbon black is an extremely important process in the rubber industry (Kraus, 1965). Key properties of elastomer compounds and vulcanizates are influenced or controlled by the quantity and type(s) of carbon blacks in the system. A recent study by

Sircar *et al.* (1980) describes the electrothermal analysis of carbon black-loaded polymers. Techniques based on thermogravimetric analysis have made it possible readily to measure the carbon black content in many commercial formulations. Quality control techniques based on this advance are now common in the rubber industry. In view of these important aspects of carbon black analytical procedures, this topic will be presented in some detail. An important point to be made is that for some elastomer formulations, it appears possible to obtain at least semiquantitative analysis of carbon black blends. There are also hints that monitoring of carbon black and cure system uniformity may be possible in some cases.

1. Total Carbon Black Content

Detailed consideration of the experimental conditions for determining total carbon black content in vulcanizates as well as the factors that influence precision and accuracy of the analysis has been reported by Maurer (1969b, 1970a,b). In essence, the basic experimental procedure is quite simple, merely involving a change from a nitrogen to an air atmosphere while the sample is either heated at a constant rate or held isothermally at a temperature above the thermal degradation region of the polymer(s) and other organic material in the formulation, and high enough to oxidize the CB to CO_2. On introduction of air into the system, a weight loss takes place as indicated in Fig. 68. Assuming that all organic matter due to poly-

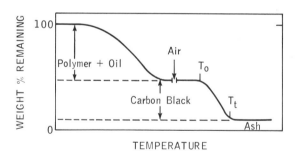

Fig. 68   Thermal gravimetric analysis of vulcanizate (Maurer, 1970a).

mer and other ingredients in the formulation has been completely volatilized prior to the introduction of air at 550°C, this weight loss is due to oxidation of the carbon black in the system. Assuming that the polymer does not leave a char, the residue is either ash or mineral filler plus ash. Jaroszynska *et al.* (1978) describe TG of vulcanizates containing mineral fillers.

Early data of Maurer (1969a), relating to the precision and accuracy of such analyses are presented in Tables XVII and XVIII. Table XVII intentionally includes a sample where significant discrepancy in the "oil" and

TABLE XVII

TG Results for Selected Unextracted Compositions[a,b]

Composition weight (%)	Oil	Polymer	Carbon black	Mineral filler + ash	Ash	Classification
Known	11.5	56.5	28.0	—	4	Successful
TG	12.5	56.5	28.0	—	3	
Known	12.5	50.0	35.0	—	2.5	Successful
TG	13.0	49.5	34.5	—	3.0	
Known	2.1	47.0	8.6	42.2	—	Partially
TG	9.5	39.5	9.0	42.0	—	Successful[c,d]
Known	26.1	29.0	14.5	30.5	—	Partially
TG	23.5	32.5	15.0	29.0	—	Successful[c,d]

[a] From Maurer (1969b).
[b] Two cases that present difficulty are intentionally shown.
[c] Note agreement of polymer + oil, carbon black, and mineral filler + ash.
[d] DTA can sometimes assist in detecting such problem compounds.

TABLE XVIII

Precision and Accuracy of Basic Composition Analysis[a]

System	Composition (wt. %)	Oil	Polymer	Carbon black	Ash
I[b]-Extracted[g]	Known	16.0	40.0	40	2^d–3.8^e
(Five extractions)	TG	16.9 ± 0.05[f]	40.6 ± 0.33	-------- 42.6 ± 0.33 --------	
I[b]-Unextracted	TG	13.7 ± 0.63	43.4 ± 0.77	41.1 ± 0.30	1.8 ± 0.24
(Ten analyses)					
II[c]-Unextracted	Known	12.3	49.1	34.4	2.5^d–4.2^e
(Six analyses)	TG	13.2 ± 0.8	50.5 ± 1.03	33.8 ± 0.24	2.5 ± 0

[a] From Maurer (1969b).
[b] Butyl—100 phr, Black—100, Oil—40, ZnO—5, Accelerators—4.5.
[c] Butyl—100 phr, Black—70, Oil—25, ZnO—5, Accelerators—3.5.
[d] ZnO only.
[e] ZnO + accelerators.
[f] Standard deviation, in all cases.
[g] Ten analyses.

"polymer" estimates occurs. Several factors that could contribute to this situation have been treated by various authors during development of these basic TG procedures. One factor is overlap of the process oil and polymer decomposition regions. Several routes to solve such a problem were discussed earlier.

Another factor that could contribute to the type of discrepancy noted in Table XVII is the presence of waxes, resins, etc., that decompose or volatilize in the same region as oil and polymer. DTA or DSC evaluation may prove useful for detecting the presence of these ingredients, depending, of course, on their concentration and inherent thermal characteristics.

As noted in Tables XVII and XVIII, good agreement between known and calculated carbon black content is obtained even for these butyl rubber systems in which there is an unusual level of discrepancy in the "polymer tail" analysis. Swarin and Wims (1974) report similar success with a series of EPDM and NBR formulations. Pautrat *et al.* (1976) describe the quantitative analysis of HAF, SRF and MT blacks in ethylene–propylene–diene rubber (EPDM), isobutene–isoprene rubber (IIR), and natural rubber (NR) as well as HAF in styrene–butadiene rubber (SBR). Total carbon black content by TG was generally between 3 and 6% higher than the known value, in agreement with similar findings of Cole and Walker (1970). Maximum difference from the mean value was 2.4% for a series of nine evaluations on the same size sample.

a. Additional Information from Dynamic TG of Carbon Blacks. During a systematic TG study of the influence of cure system composition on the analysis of basic vulcanizate composition, Maurer (1970a) noted that both physical and chemical factors influenced the detailed characteristics of carbon black decomposition during this type of analysis. In general, this area can be subdivided into consideration of carbon black characteristics versus the influence of other ingredients in the formulation, principally the cure system components. The carbon black area can be subdivided into surface area versus "other" (the latter referring to structure, functionality of the carbon black surface, and dissimilarities for similar types of carbon black produced at different locations and/or by different manufacturers). The following survey of this area will consider first the kinds of information that can be obtained by straighforward application of dynamic TG experiments. Following this, consideration will be given to the types of additional information that can be obtained via *isothermal* oxidation of residues from thermal degradation of vulcanizates.

2. Surface Area Considerations: Standard Formulation

A systematic TG analysis of the influence of compound ingredients on the basic compositional analysis of a standard isobutene–isoprene rubber (IIR) vulcanizate has been reported; see Maurer (1970a,b). Interesting variations were noted in the temperature region during which carbon black oxidation occurs. The general thermograms (Fig. 68) exhibited the following features: (a) an initial area of low weight loss, (b) a point (T_0) at which weight loss increases sharply (c) an intermediate region that may either be essentially linear or exhibit an inflection point, and (d) a terminal region ending at T_t. The characteristics of some of the thermograms suggest that at some point in the T_0 to T_t region, the heat of reaction may have increased the sample temperature at a rate higher than the programmed rate. In addition, a significant variation in T_0 and T_t was noted for different types of carbon blacks, all other compound ingredients being the same.

It was also noted that some carbon blacks began to degrade very soon after air was introduced, whereas a considerable lag was evident with other types. In the former case, it was not clear whether T_0 might lie below the temperature at which air was introduced in the standard TG procedure. The following modification was devised because of this problem. Following thermal degradation of the polymer, oil, etc., in nitrogen, the sample was allowed to cool in the flowing nitrogen atmosphere to 275°C. Air was then introduced and the analysis was completed in the standard manner. Data obtained by this approach is illustrated in Fig. 69.

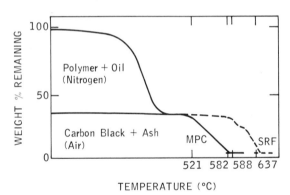

Fig. 69 Detection of carbon black differences in "standard" formulation (Maurer, 1970a).

Note that T_0 and T_t for MPC are, respectively, 67 and 55°C lower than for SRF. Also of interest is the inflection in the SRF curve, which suggests a two-step degradation process. The observed variation in T_0 and T_t suggested that TG might provide a novel approach for studying interactions of ingredients in compounds and vulcanizates as well as a new qualitative analysis of practical formulations.

Before undertaking detailed studies of the observed T_0 and T_t variations, the reproducibility of these parameters was examined for the set of carbon blacks shown in Table XIX. Experimental details are also presented in the table. As shown, these carbon blacks could be distinguished from one another at the 95% confidence level by means of T_0. T_t could assist in identification of the black when used in combination with T_0, but

TABLE XIX

TG Degradation Characteristics of Carbon Blacks[a,b]

Carbon black type[e]	T_0 (°C)[c] Mean	σ	T_t (°C)[c] Mean	σ
MPC	514.8	3.97	582.0	7.48
HAF	542.5	2.07	607.0	3.63
FEF	567.1	3.06	621.5	4.37
SRF	585.0	2.68	632.8	3.60
FT	615.8	13.36[d]	684.0	15.56[d]

[a] From Maurer (1970a).

[b] Formulation (phr): IIR—100, carbon black—50, ZnO—5.0, stearic acid—2.0, TMTDS—1.5, MBT—1.0, sulfur—1.25.

[c] Six determinations: three original compounds plus three extracted compounds.

[d] Large σ due to difference in extracted versus unextracted samples. For unextracted samples, a σ of 2.93 has been obtained for T_{15} %.

[e] See Table III for ASTM nomenclature.

is less useful alone. The larger values for T_t are consistent with the exothermic effects previously noted.

Further improvement of the procedure was undertaken to provide for the occasional run where the sharp rate of weight loss change is not evident at T_0, thus leading to less precise estimtes of this temperature. Instead of T_0 and T_t, temperatures corresponding to a fixed carbon black weight loss, here T 15%, and T 50%, were evaluated and found to yield improved precision for the MPC and HAF systems. The statistical reliability of this approach was established during the program summarized in Table XIX.

Both physical and chemical effects were considered as potential causes of the observed T_0 and T_t variations. In either case the effects may be enhanced by the fact that, following completion of the "polymer decompositon" step in nitrogen, the carbon black residue retains the shape of the original sample. This residue is probably a very "porous" structure that permits good contact with air as well as rapid diffusion of gaseous products out of the sample. Surface area and particle size were considered as potential factors in the observed CB oxidation characteristics. Figure 70 indicates, for these systems, a high correlation between BET surface area and T 15%. Similar behavior was observed for T_t and, to a lesser degree, for particle size. Thus a definite physical contribution to T_0 and T_t variations is indicated for this standard vulcanizate system.

3. Analysis of Carbon Black Blends by TG

The relationship shown in Fig. 70 suggested that carbon black blends whose components differed significantly in surface area might be at least

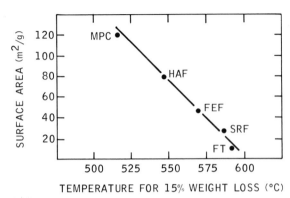

Fig. 70   Carbon black decomposition versus surface area (Maurer, 1970a).

semiquantitatively analyzed by programmed temperature TG. The feasibility of this approach was evaluated by comparison of the residue of a laboratory formulation (A) known to contain 25% small (HAF) and 75% large (MT) particle size blacks and those from two similar commercial development compounds that differed appreciably in key properties versus compound (A). The comparison shown in Fig. 71 suggests that these samples may differ significantly with respect to carbon black blend ratio and possibly accelerator systems as well (note T_0 values). Another means of analyzing such systems is presented in the section dealing with isothermal TG.

4. Confirmation of Surface Area Relationship for Additional Polymer Systems

Pautrat *et al.* (1976) confirmed and extended the applicability of the surface area relationship described by Maurer (1970a). These workers utilized four carbon blacks commonly employed in general-purpose and specialty rubbers and five elastomer systems [natural rubber (NR), isobutene–isoprene rubber (IIR), ethylene–propylene–diene rubber (EPDM), styrene–butadiene rubber (SBR), and nitrile butadiene rubber (NBR)]. The latter (NBR) extends the analysis to a polymer that graphitizes on thermal degradation in nitrogen. Further extension of the analysis involved systems that contained mineral filler plus carbon black.

The first step in this work was to confirm the T_{15} versus surface area relationship for "raw" EPC, HAF, SRF, and MT as shown in Fig. 72. Approximately 30 mg of sample was used, corresponding to the residue from TG of 100 mg of a tire tread compound. The degradation step in nitrogen was continued until constant weight was achieved, usually at about 600–650°C, a higher temperature than the 550°C generally employed by

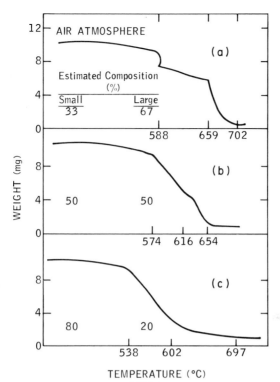

Fig. 71 Dynamic analysis of mixed blacks in TGA residue; air atmosphere (Maurer, 1970a).

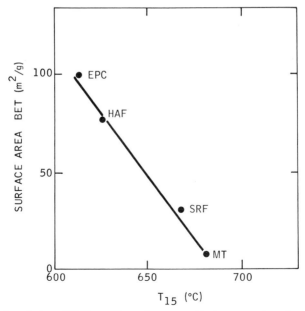

Fig. 72 Correlation of BET specific surface area with thermogravimetric T_{15} (Pautrat *et al.*, 1976).

Maurer (1970a). Similar to the method of Maurer, the sample was cooled to 275°C before introducing air and oxidizing the carbon black.

Subsequent evaluation of these blacks (30–150 phr) in NR, SBR, IIR, or EPDM compounds indicated that T_{15} for the carbon black residue was noticeably lower than for "raw" black. This was due to combined effects of chemical action of compounding ingredients and thermal treatments of the sample, in agreement with Maurer (1970a). Using the TG methods outlined earlier, it was found that these carbon blacks could be differentiated in these vulcanizates as shown in Table XX. The observed variabil-

TABLE XX

INFLUENCE OF CARBON BLACK[a] TYPE ON T_{15} VALUES[b]

Carbon black type	T_{15} value (°C)
MT	615–630
SRF	580–590
HAF	555–570
EPC	510–530

[a] 30–150 phr, in NR, SBR, IIR or EPDM.
[b] From Pautrat et al. (1976).

ity in the T_{15} values reflects, in part, the wide range of elastomer type and carbon black loading in the various compounds analyzed.

It was observed that the spread in T_{15} values was usually large for large-surface area blacks; thus it was difficult to distinguish EPC versus SAF in vulcanizates despite their different surface areas. The T_{15} of FEF, however, could be determined to be 570°C, which, as expected from its surface area, was intermediate between those of SRF and HAF types. The precision of the method was determined from 10 analyses of a vulcanizate containing 35 phr HAF in NR; deviations from the mean (550°C) were small—provided that conditions were rigorously standarized. Identification of carbon black in the presence of mineral fillers was also demonstrated. The T_{15} value for SRF (590°C) in an NR/30SRF/30 Whiting vulcanizate was well within the expected range. It was also possible to identify this black at higher loadings of Whiting, although, for unspecified reasons, quantitative determination became more uncertain.

The char that forms during thermal degradation of NBR vulcanizates complicates the identification of carbon black type, as noted by Pautrat et al. (1976), since it oxidizes in the same temperature range as the fine particle-sized blacks (EPC, HAF, SAF, etc.). For this reason it was not possible to characterize these blacks in NBR compounds; however, it was possible to identify the larger blacks (MT and SRF) since the char is oxidized at a much lower temperature than is required for such blacks.

a. Limitations on Carbon Black Identification by TG Methods. As previously noted, the variation in the oxidation characteristics of carbon black residues from TG of vulcanizates may be influenced by a variety of chemical and physical factors. The extensive studies described earlier indicate that the possibility of identifying carbon black and carbon black blend composition exists for some compounds and vulcanizate systems. It is clear, however, that this approach will not be able to serve as a means for identifying all carbon blacks in all possible vulcanizate systems. Among the reasons for this are (1) the problem with discriminating between blacks of similar surface area and (2) the char problem in NBR and Neoprene vulcanizates, which prevents detection of EPC, HAF, SAF type black. In addition, there are other complications relating to accelerator systems and carbon black characteristics other than surface area, as discussed in the following sections.

5. Influence of Accelerator System on Carbon Black Oxidation

The degree to which chemical effects can influence carbon black decomposition has been demonstrated by comparing several different types of cure systems in combination with each of several different carbon black types (Maurer, 1970a). The substantial influence of the cure system components on carbon black oxidation characteristics is apparent from the data in Table XXI. As shown, the cure system did not have a major effect on the basic vulcanizate composition or polymer weight loss characteristics.

The cure system effect on carbon black oxidation is important for two reasons. First, the general analysis of carbon black type via TG of *unknown* vulcanizates will not be possible, since the cure system effect on T_0, T_{15}, etc., can apparently dominate the surface area effects previously noted for a standard vulcanizate. However, this does not negate the potential use of the method for comparisons of samples of the same formulation, or of known systems. Thus possible utility exists for these methods in quality control analysis by both the suppliers and purchasers of rubber goods.

A second type of utility for this type of data is suggested by comparison of the shapes of, for example, the FEF black decomposition regions for the different accelerator systems. Specifically, the complex nature of some of these thermograms (Fig. 73) suggests the possibility of TG moni-

TABLE XXI

EFFECTS OF CURE SYSTEM AND BLACK TYPE ON DECOMPOSITION
OF BUTYL RUBBER VULCANIZATES[a]

| Cure system[b] | Average values | | | Polymer decomposition (°C) | | Carbon black decomposition (°C) | |
	% Polymer	% Black	% Ash	Temp. for 50% loss	Terminal temp.	Onset temp.	Terminal temp.
A	65.0	31.0	4.0	420	456	497	553
	65.2	30.8	4.0	418	455	529	582
	64.9	31.4	3.8	416	452	550	596
	64.8	31.8	3.3	412	454	551	605
B	65.0	32.0	3.3	418	453	513	558
	66.0	30.3	3.7	417	455	543	588
	65.1	31.3	3.6	415	451	567	608
	65.0	32.0	3.0	411	450	581	623
C	65.3	31.8	3.0	421	457	528	561
	65.8	31.3	2.8	421	464	601	612
	64.8	32.5	2.8	420	455	597	617
	64.0	33.0	3.3	416	452	595	628
D	63.3	30.0	6.7	417	464	427	461
	63.0	30.0	7.0	413	448	468	479
	62.8	30.4	6.9	414	453	464	497
	62.5	31.0	6.8	407	442	461	529

[a] From Maurer (1970b).
[b] Cure systems (phr):
   A: Altax (1.0), Tellurac (1.5), sulfur (1.0), ZnO (5.0), stearic acid (2.0).
   B: Sulfasan R (2.0), Tuads (2.0), ZnO (5.0), stearic acid (2.0).
   C: SP-1055 (12.0), ZnO (5.0), stearic acid (2.0).
   D: Altax (4.0), GMF (1.5), red lead (5.0), ZnO (5.0), stearic acid (2.0).

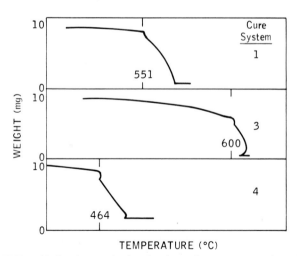

Fig. 73 FEF oxidation in standard vulcanizate (cure systems given in reference) (Maurer, 1970b).

toring of cure system uniformity. In addition, unique features of some systems (e.g., the oxidative effect of cure system 4) may enable their detection in unknown systems. The strong two-step regions in some systems are also interesting; they suggest studies to determine whether each region relates to one accelerator component.

6. Carbon Blacks That Deviate from the Surface Area Correlation

There have been several reports of additional factors that limit the use of TG for qualitative analysis of carbon black types. For example, A. Voet (personal communication, 1969) noted differences in oxidation characteristics for similar carbon blacks produced in different locations. Further, Spacsek *et al.* (1977) did not confirm the relationship between surface area and the temperature at which 15% of the carbon black weight loss occurred for various carbon blacks in vulcanizates and observed poor correlation with carbon blacks alone. Further data would be required to determine the significance of these results, particularly in view of the results of Pautrat *et al.* (1976).

More recently, Schwartz and Brazier (1978a) reported various blacks that do not follow a "t 20% versus surface area" correlation. The degree of this effect is noted in Fig. 74, which compares eight carbon blacks recovered from an SBR vulcanizate of fixed composition other than carbon black type. Two things are noteworthy about this system and set of blacks: (a) three of the blacks deviate significantly from this relationship

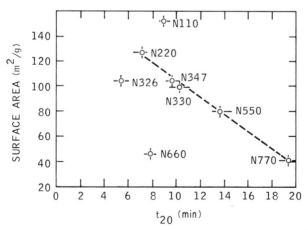

Fig. 74 Specific surface area of carbon black versus t_{20} for oxidation of virgin carbon blacks. Oxidation carried out isothermally at 600°C in 7% oxygen in nitrogen flowing at 100 ml min^{-1} (Schwartz and Brazier, 1978a).

and (b) five of the blacks show good correlation of t 20% decomposition versus surface area.

The general qualitative analysis of carbon blacks in *unknown* vulcanizates by oxidation characteristics alone is therefore not feasible. However, it will be shown that *in some cases* semiquantitative analysis of carbon black blends is possible (Maurer, 1974a).

Schwartz and Brazier (1978a) also provide data that support A. Voet's (private communication, 1969) contention that similar carbon blacks manufactured at different locations can have different oxidative characteristics. Among other factors that could account for this effect, differences in salt or metal oxide content in the blacks are suggested. It was noted that metal oxides are known to catalyze the oxidation of carbon black (Papierer *et al.*, (1967). A further point of interest is that any changes in carbon black aggregate size that occur during mixing, and variation of this factor among blacks from different sources, can further complicate qualitative analysis of carbon blacks in unknown vulcanizates. This suggests a study of the influence of mixing conditions on carbon black decomposition characteristics for a given filled system. If a significant effect due to carbon black aggregate size reduction is observed, this might prove a useful way to monitor and/or detect extreme variations in mixing efficiency in a given compound.

   a. Characterization of Rubber-Grade Carbon Black.   Spacsek *et al.* (1977) applied thermal analytical techniques (TG, DTG, and DTA) to carbon blacks from a different perspective. Their general objective was to assess the potential utility of these procedures for the *characterization* of carbon blacks. The specific approach used was to seek correlations among conventional carbon black parameters, thermal analytical characteristics of the blacks, and practical features of the vulcanizates containing these blacks (modulus, hardness, elasticity, etc.).

The initial phase of this study consisted of evaluation of the "raw" carbon blacks in two temperature intervals: up to 250°C (1.25°C/min) and up to 1000°C (5°C/min). both air and nitrogen environments were employed. The key result from this work was an observed correlation between the loss in weight at 250°C (air or nitrogen) and specific surface area of the black. These workers did not observe the correlation between carbon black surface area and oxidation characteristics demonstrated by Maurer (1970a), Pautrat *et al.* (1976), and, in part, by Schwartz and Brazier (1978a). Instead, a high degree of scatter was reported for the "raw" carbon blacks and *no* correlation was observed for carbon black residues from an SBR vulcanizate. The information given in the paper is insufficient to determine the reason for this discrepancy.

The second part of their study involved an evaluation of these blacks in a simplified styrene–butadiene rubber (SBR) compound to demonstrate reinforcing effects free of other complications. Correlations were established between the loss in carbon black weight measured up to 250°C in nitrogen and Young's modulus, hardness, and elasticity of the vulcanizates. These correlations were ascribed to (a) volatiles content being proportional to carbon black surface area and (b) previously established correlations between carbon black surface area and reinforcing effects in vulcanizates.

7. Isothermal Analysis of Carbon Black Residues from Vulcanizates

Isothermal degradation was evaluated in detail by Maurer (1970a) to determine whether oxidation rate differences might aid in distinguishing among different blacks. Figure 75 indicates the large difference in oxidation rate observed for two "raw" blacks that differ appreciably in particle size and surface area. Similar results were obtained using residues from standard vulcanizates (Fig. 76). The major observations regarding these data are the following: degradation characteristics of the blacks vary, with

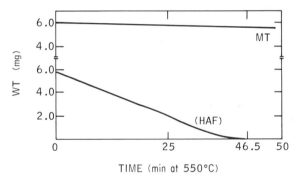

Fig. 75 Degradation characteristics of selected carbon black types (Maurer, 1970a).

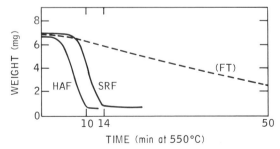

Fig. 76 Oxidation of carbon blacks in residues from standard formulation (Maurer, 1970a).

MT being the most extreme; there is an apparent "induction period" before rapid oxidation commences; oxidation is more rapid for carbon black residues than for the "raw" black (note HAF data). This latter observation suggests that (a) the "porous" nature of these residues causes them to degrade more rapidly than the "raw" black, or (b) there may be chemical effects influencing carbon black oxidation. Examples of chemical factors that might be involved in these oxidation rate differences include functionality of the carbon black surface, accelerators and other compounding ingredients, vulcanization by-products, and by-products from the thermal degradation step.

Evaluation of the ingredients in, and the procedures used to prepare, the standard vulcanizate was conducted to assess their influence on T_0 and T_t. Neither T_0 nor T_t was influenced by polymer alone, heat treatment with polymer at curing temperature, the presence of zinc oxide and stearic acid, or the presence of accelerators (and possibly sulfur also, although this point has not been fully studied).

All of the experiments were conducted over a short period of time and are self-consistent. However, the T_0 and T_t values were lower for this FEF compound than for a similar compound previously evaluated. Comparison of the thermograms suggested that a large part of the variation was due to a shift of the entire thermogram on the temperature axis. Shifting the data back to a common reference point would bring it into much closer agreement. This type of problem could occur because of repositioning of thermocouples during replacement, or changes in temperature scale calibration. In order to avoid such problems, temperature scale calibration and checks of standard vulcanizate compositions are periodically conducted. Inherent precision of the method is indicated by the analysis of random samples from two extracted commercial systems shown in Table XXII.

The relative oxidation characteristics shown in Fig. 76 suggested that at least semiquantitative analysis of binary carbon black mixtures may be possible if one component is a thermal black. This idea was evaluated as described in Section VIII and Fig. 71.

Isothermal TG (Fig. 77) may aid in confirming indications from dynamic TG. Other potential approaches for analyzing carbon black blends include DTG and reference temperature methods similar to those successfully employed in the determination of oil content and polymer blends. Maximum utility of the approach is anticipated in cases where a reference set of compounds differing only in black ratio has been evaluated by both dynamic and isothermal TG. Such data will provide the maximum opportunity for devising calibration curves for estimation of carbon black blend composition.

TABLE XXII

Precision of T_0 for Commercial Vulcanizates[a],
Carbon Black Onset Temperature (T_0), °C

Formulation A	Formulation B
569	563
574	559
568	551
565	554
567	557
569	557
567	555
568	554
568	553

[a] From Maurer (1970a).

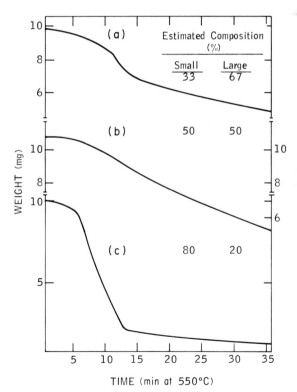

Fig. 77   Isothermal analysis of mixed blacks in TGA residue; air atmosphere (Maurer, 1970a).

a. Systems Containing Carbon Black and a Char-Forming Polymer.
Two-stage TG involving an isothermal condition followed by a 20°C min⁻¹
heating rate was used by Schwartz and Brazier (1978a) to aid in separating
oxidation of carbonaceous residues from that of carbon black. Application of this technique to a Neoprene W® vulcanizate is shown in Fig. 78.

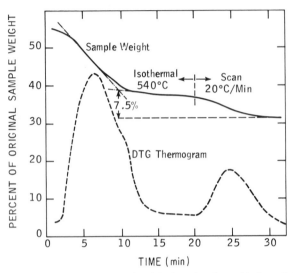

Fig. 78 Weight loss and derivative of weight loss for the oxidation of the residue remaining from the pyrolysis in nitrogen of a Neoprene W vulcanizate containing 9% of N550 black and 36.5% inorganics. Oxidation carried out isothermally at 540°C in 7% oxygen in nitrogen flowing at 100 ml min⁻¹ (Schwartz and Brazier, 1978a).

Oxidation of the "char" remaining from thermal degradation of the chloroprene rubber (Neoprene W®) is seen to be much faster at 540°C than is the carbon black. Carbon black content was estimted as 7.5% from this analysis (known value: 9.0%). Material remaining after carbon black oxidation is inorganic filler. The purpose of the 20°C min⁻¹ heating rate was merely to complete the analysis in a shorter time (i.e., versus isothermal oxidation).

b. Carbon Black Blends of Similar Surface Area. A striking illustration of the ability of TG/DTG procedures to provide detailed analysis of complex elastomer systems is shown by Schwartz and Brazier (1978a). They describe the successful analysis of a carbon black blend in a vulcanizate that contained a blend of polymers, one of which (Neoprene W®) leaves a char during isothermal degradation of the polymer components. Isothermal analysis at 540°C was used to separate chloroprene rubber (Neoprene WRT®) char from the carbon black blend residue in this com-

plex system (Fig. 79). DTG of the carbon black residues provided estimates of 12.5% N326 and 7% N990, compared to the added values of
13.5% and 6.7%, respectively. Similar results were reported for NBR vulcanizates that also are char forming.

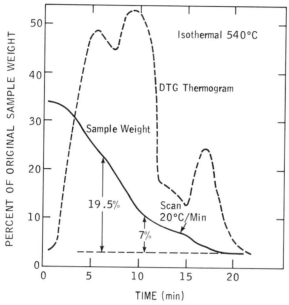

Fig. 79   Weight loss and derivative of weight loss curves for the oxidation of the residue
from the pyrolysis in nitrogen of a 50:50 NR/Neoprene WRT blend containing 13.5% N326
and 6.7% N990 carbon blacks. Oxidation conditions: same as Figure 78 (Schwartz and Brazier, 1978a).

Thus it has been shown that control of the oxidation technique used in
conjunction with DTG enabled the oxidation of medium and high surface
area blacks to be separated from that of the char. These authors emphasize, however, that "as in the case of noncarbonizing polymers the
method would not be infallible in identification of the type of black."

Application of isothermal TG to vulcanizates containing carbon black
blends of similar surface areas has been described by Maurer (1974a). A
series of FEF/SRF blends was analyzed in an isobutene–isoprene rubber
(IIR) vulcanizate, a polymer system that leaves little or no char following
thermal degradation in nitrogen. As shown in Fig. 80, which is based on
four analyses of each composition, major variations in carbon black blend
composition should be readily detected for this system. Smaller variations
may be more difficult to detect by this method, particularly for blends rich
in one compound. It is apparent from Fig. 80 that discrimination between

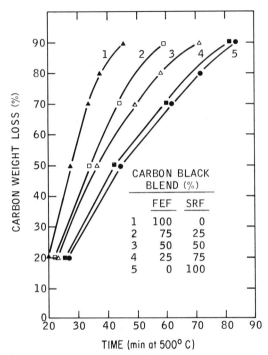

Fig. 80 Analysis of SRF/FEF in vulcanizate residues. Each curve is average of four analyses. (Maurer, 1974a. Reprinted by courtesy of Marcel Dekker, Inc.)

the blacks may be achievable by considering the times to reach a high degree of weight loss. Presumably, the residue in the latter stages of degradation is primarily SRF black. Since these data were obtained, high-quality DTG instrumentation has become commercially available as part of various thermal analysis system. This approach may prove useful for analysis of certain carbon black blends as it has for some polymer blends.

8. Graphical Analysis of Carbon Black Oxidation Data

Visual inspection of the oxidation characteristics exhibited by FEF in this elastomer system (Fig. 81) suggested an additional approach to carbon black analysis. Specifically, the high degree of symmetry of the curves suggested that the data could be transformed into a linear relationship via a normal probability plot (Fig. 82). This plot suggests the feasibility of graphical or computer analysis of carbon black blend content for carbon blacks that exhibit oxidation characteristics similar in form to FEF but different from each other. For such systems, one would expect the individual blacks to exhibit different straight lines on the normal probability plot. In the ideal case a blend of these blacks would appear between,

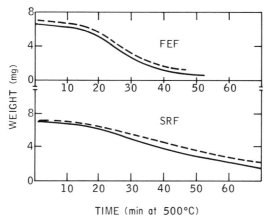

Fig. 81 Oxidation of carbon black residues from vulcanizates (reproducibility). (Maurer, 1974a. Reprinted by courtesy of Marcel Dekker, Inc.)

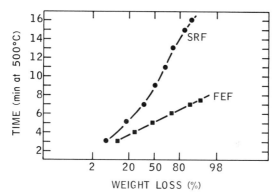

Fig. 82 Decomposition of carbon blacks from vulcanizates: probability plot. (Maurer, 1974a. Reprinted by courtesy of Marcel Dekker, Inc.)

and asymptote, the lines. Another interesting question is whether the different accelerator residues in a given vulcanizate system may influence oxidation of the carbon black to different degrees. In this case it would be interesting to determine whether (particularly in systems containing a single carbon black) (a) a bimodal probability plot is detected and (b) *accelerator* ratio can be deduced from this plot. Such analyses, if feasible, would be complicated by any factors that influence the basic uniformity of oxidation characteristics for a given carbon black [e.g., the variation with source of manufacture, as described by Schwartz and Brazier (1978a)].

The type of plot obtained for SRF (Fig. 82) is quite interesting. Assuming that treatment of SRF oxidation characteristics by normal probability analysis is appropriate, this plot suggests that SRF behaves like a bimodal blend. This possibility merits further consideration in view of previous work involving oxidation via the dynamic TG method (Maurer, 1970a), which also indicated that SRF oxidation characteristics are inherently more complex than those of several other carbon black types.

IX. Application of Thermal Analyses to Commercial Elastomer Systems

It has been amply demonstrated that the basic thermal analytical techniques (DTA, DSC, TG, DTG, TMA, and DMA) can provide a wide range of information concerning the cure system, polymer, filler, and ash content of commercial elastomer systems. Such analyses can be extended, in some cases, to include quantitative determination of polymer blend and carbon black blend content. The potential utility of these techniques for quality control analyses of elastomer systems by both the suppliers and purchasers of rubber goods has been discussed (Maurer, 1970b). This potential is based on the types of information that can be obtained, the simplicity of the techniques, and the practical time scale associated with these procedures. The general potential of these techniques has been recognized by the industry as evidenced by the wide variety of end-use formulations that have been evaluated (Table XXIII). In addition, TG of reclaimed rubber compounds is discussed by Chakraborty (1978). The objective of the following discussion is to provide a more detailed introduction to a spectrum of practical applications of thermal analysis to elastomer systems. These procedures have been found useful in many phases of rubber research, development and end-use applications.

A. QUALITY CONTROL ANALYSIS OF FACTORY MIXES

Brazier and Nickel (1975b) present an excellent example of the quality control capability of thermal analysis via an in-depth evaluation of 20 consecutive factory mixes of the compounds shown in Table XXIV. The objective of this analysis was to determine the variability of the compounds due to such factors as weighing errors which would lead to, presumably, relatively minor deviations from specification limits. Evaluation of the cure characteristics of these same formulations has also been reported by Brazier and Nickel (1975a). A typical DSC scan for this type of system is shown in Fig. 62.

Evaluation of natural rubber (NR) butadiene rubber (BR) blend composition by DTG is discussed in Section VIII. Application of these techniques to compound C is shown in Table XXV and Fig. 83. Sulfur and

TABLE XXIII

Practical Formulations Successfully Analyzed by Thermal Methods

Formulation	Elastomer	Method			Type of analysis		Reference
		DSC	TG	DTG	Quantitative	Qualitative	
Inner tube	IIR		×		×		Maurer (1970b)
Weatherstrip	EPDM		×		×		
Sponge	EPDM		×		×		
Engine mount	IIR		×		×		
Electrical insulation	EPDM		×		×		
Tire sidewall	NR/EPDM, NR/SBR/EPDM		×	×	×		Maurer (1974a)
Tire tread and black sidewall	NR, IR, BR, SBR, and blends	×	×	×		×	Sircar and Lamond (1975a)
White sidewall compounds	EPDM, EPDM/NR, EPDM/SBR, EPDM/NR/SBR, EPDM/NR/CIIR, EPDM/SBR/CIIR, EPDM/NR/SBR/CIIR	×		×		×	Sircar and Lamond (1975b)
Tire white sidewalls	CR/OE-BR, CR/OE-SBR, CR/NR, CR/NR/CSM	×	×	×		×	Sircar and Lamond (1975c)
Tire inner liners	NR/SBR/CIIR or BIIR, NR/CIIR, NR/BIIR, SBR/CIIR	×	×	×		×	Sircar and Lamond (1975d)
O-Rings	NBR		×	×	×		Swarin and Wims (1974)
Gaskets	NBR		×	×	×		
"Commercial formulations"	EPDM		×	×	×		

TABLE XXIV

Formulations of Two Typical Tire Components[a]

Compound	B phr	B wt %	C phr	C wt %
Natural rubber	80	45.71	60	33.05
Polybutadiene	20	11.43	40	22.03
Carbon black	50[b]	28.57	55[c]	30.30
Stearic acid	1.8	1.03	2.5	1.38
Zinc oxide	3	1.71	3	1.65
Antioxidant	2	1.14	2.25	1.24
Oil/plasticizer	11.5	6.57	16.0	8.8
Sulfur	2.8[d]	1.6	1.5[e]	0.82
Accelerator	1.2	0.69	1.0	0.55
Inhibitor	—	—	0.3	0.17
Bonding agent	2.7	1.54	—	—

[a] From Brazier and Nickel (1975b).
[b] N326 (HAF-LS) carbon black.
[c] N339 carbon black.
[d] Added as 2.8 phr oiled Crystex (80% sulfur in oil).
[e] Elemental sulfur (rubber grade).

accelerator content were determined by DSC; polymer blend composition was evaluated via DTG peak heights. Carbon black, zinc oxide, and oil content were read directly from the TG thermogram. In order to save time, the weight loss at 300°C was used as a measure of the oil content. This procedure gave results within ± 2 phr of the known value for a series of calibration samples. Various approaches for determination of oil content are considered in Section VIII. It should be noted that a complete mass balance of the compound is not achieved. Specifically, weight losses due to water, stearic acid, the cure system, and antioxidants are not determined by these methods but do make small contributions to the thermograms.

The data shown in Table XXV indicate that no major compounding error took place during this series of factory mixes. A consistently low NR content was noted in factory mixes but not for laboratory formulations; the cause was not described. Estimated carbon black loadings were 1–2% higher than expected, probably because of elastomer, oil, and plasticizer carbonization products. Talc was suggested as the cause of the higher than expected ash values.

A quality control-type analysis of the polymer blend composition is shown in Fig. 83. Here the results are presented as an envelope within which the results for all 20 mixes would fall when it was superimposed on

TABLE XXV

COMBINED DTG/DSC ANALYSIS FOR COMPOUND C, TABLE XXIV[a]

Batch no.	% NR[b]	% BR[b]	Oil[c]	% Carbon black[d]	% Ash[e]	% Sulfur[f]	% Accelerator[f]
1	31.3	21.4	8.6	31.7	2.1	0.82	0.55
2	32.8	23.1	8.0	31.1	2.3	0.80	0.55
3	31.7	21.0	8.9	32.2	2.2	0.71	0.50
4	31.6	21.4	8.8	31.6	2.4	0.79	0.55
5	32.9	21.9	8.9	31.8	2.2	0.95	0.55
6	31.1	21.9	8.5	32.5	2.4	0.89	0.55
7	30.7	21.4	8.1	31.9	2.5	0.81	0.60
8	30.8	22.4	8.9	32.3	2.1	0.86	0.60
9	31.4	22.5	8.5	32.2	2.2	0.82	0.60
10	31.6	22.4	8.7	32.2	2.3	0.88	0.44
11	32.4	22.7	9.4	32.1	2.1	0.81	0.44
12	31.6	22.0	8.9	32.2	2.2	0.84	0.50
13	33.2	22.4	9.3	31.9	2.2	0.81	0.50
14	32.2	22.5	8.5	31.8	2.4	0.90	0.55
15	33.2	23.3	8.6	32.5	2.5	0.76	0.55
16	31.0	22.3	8.6	31.7	2.2	0.82	0.50
17	32.0	23.4	8.3	32.5	2.1	0.74	0.55
18	31.9	22.6	8.4	32.5	2.1	0.71	0.55
19	33.2	22.2	8.4	32.0	2.3	0.80	0.55
20	32.0	23.3	8.5	32.1	2.4	0.85	0.50
Nominal	33.0	22.0	8.8	30.3	1.65	0.82	0.55

[a] From Brazier and Nickel (1975b).

[b] ±0.5%.

[c] Wt % loss at 300°C.

[d] Wt % loss at 550°C in nitrogen less ash.

[e] Residual weight at 575°C in oxygen.

[f] Values from Brazier and Nickel (1975a). Accelerator based on 1°C change per 0.1 phr accelerator for the observed cure exotherm peak temperature.

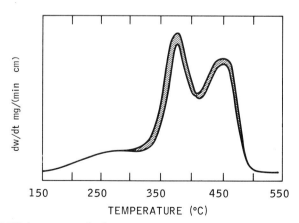

Fig. 83 DTG thermograms for factory compounds (formulation given in reference). Envelope is area into which the DTG of twenty consecutive mixes fell with sample size normalized to 16.00 mg at a sensitivity of 0.079 mg min⁻¹ cm⁻¹ (Brazier and Nickel, 1975b).

thermograms for the individual mixes. For convenience, sample weights between 15 and 20 mg were used and results were normalized to 16.00 mg. Another means of displaying the data for this type of analysis is shown in Fig. 84 for compound B. Results are presented in terms of a distribution

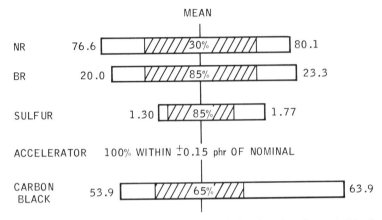

Fig. 84 Analysis of typical factory mixes (formulation given in reference). Distribution of each component about the mean for twenty mixes. Absolute range at end of each bar. Shaded areas represent the percentage of batches falling within the following limits of the mean: NR and BR ± 1.0 phr; carbon black at 2.0 phr; sulfur, accelerator 0.15 phr (data from Brazier and Nickel, 1975a) (Brazier and Nickel, 1975b).

about the mean value, and the variation within arbitrarily defined limits around these means. The variations for this compound were higher than for compound C; however, conventional quality control tests (e.g., a rapid tensile modulus determination) did not detect the variations in either formulation.

Brazier and Nickel (1975b) note that all the major components of a fully compounded stock were analyzed by these methods on a time scale consistent with routine quality control procedures. DSC gave information about cure characteristics, sulfur, and accelerator levels in about 5 min. The other components shown in Table XXV were determined in about 35 min via TG/DTG at a heating rate of 10°C min⁻¹. It should be recalled that this analysis relates to a *known* elastomer system. A more limited compositional analysis would be obtained for *unknown* formulations since calibration information regarding the polymer blends, accelerator system, etc., would not be available.

B. Expansion, Transition, Hardness, and Low-Temperature Properties via TMA

The various TMA modes have been successfully used for the determination of a variety of practical characteristics of polymer systems, several of which are discussed in this book. These applications include cure state, shrinkage, thermal history effects on formed or shaped samples, swelling and dissolution of plastics, elasticity of elastomer compounds, etc. References to many of these applications are given by Brazier and Nickel (1978). These authors describe the use of various TMA modes for evaluating a series of properties of elastomer system. Their data were obtained with the DuPont 942 system, as described by Hwo and Johnson (1974).

An illustration of the types of TMA information that can be developed is shown in Fig. 85 for chloroprene rubber (Neoprene) vulcanizates evaluated over a temperature range from -100 to $+20°C$ at a heating rate of

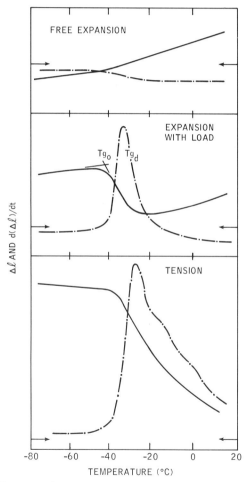

Fig. 85 TMA-free expansion, indentation and tension thermograms: Neoprene vulcanizate 5°C min^{-1}, 50 g load in indentation and tension. —— Δl versus temperature; — · — · $d(\Delta l)dt$ versus temperature (Brazier and Nickel, 1978).

5°C min⁻¹. These expansion and penetration data enable coefficient of expansion and glass transition temperatures to be evaluated. Note the use of the derivative signal for both the loaded expansion and the tension measurements. The derivative plot has been shown by various authors to be a sensitive means to aid in detecting minor transitions and relaxations and also to define a reproducible transition point in cases where a diffuse transition region is encountered. Below T_g, coefficients of expansion agree for both the loaded and unloaded probe. Above T_g, indentation occurs, the extent varying with hardness of the compound as discussed below. DTMA and tension measurements appear to be a sensitive technique for evaluating polymer blends since several "transitions" are reproducibly detected by this method but not by other modes. Illustrative TMA data for a variety of polymer systems are given in Table XXVI.

Analysis of T_g of a compound is sometimes a useful means of establishing plasticizer effectiveness and concentration. In these cases TMA offers a rapid means for this type of evaluation. It has been noted by various workers (e.g., Brazier and Nickel, 1978; Maurer, 1967a) that in some cases the addition of plasticizers/process oils may have no effect on the T_g of the compound. In those cases T_g may sometimes prove useful for evaluation or monitoring of polymer type. The influence of plasticizing resins on nitrile–butadiene rubber (NBR) vulcanizates was evaluated by TMA by Brazier and Nickel (1978). It was observed that both the extent of indentation and the temperature range over which it occurred varied with plasticizer content for different resins. Further, the shape of the thermogram appeared to vary with hardness of the compound (Fig. 86), thus suggesting that TMA penetration measurements may provide a technique for hardness measurements.

The possibility of using TMA for evaluating the low-temperature properties of vulcanizates was also explored by Brazier and Nickel (1978). A linear relationship between Gehmann rigidity modulus and T_{g_o} or T_{g_d} (via indentation or in tension) was observed for a wide range of vulcanizates as illustrated in Fig. 87. Exact temperature equivalence is not observed due to differences in the methods. Several T_{g_d} values are observed in blends. For systems that exhibit a single transition, TMA/DTMA has been able to provide this type of low-temperature property information in as little as 4 min, following cooling of the sample.

C. Evaluation of Shrinkage of Molded Vulcanizates by TMA

Mold shrinkage, the difference between mold cavity dimensions and those of the vulcanized molded product, is an important factor in the

TABLE XXVI

THERMOMECHANICAL ANALYSIS OF VULCANIZATES[a]

Vulcanizate	α (°C)$^{-1}$ × 10^4 ± 0.03 below T_g ± 0.1 above T_g	Expansion (°C)		Tension (°C)		Gehman[c]
		T_{g_d}[c]	T_{g_o}[b]	T_{g_d}	T_{g_o}	G_T
Polyacrylate	1. 0.54 (−120 to −40), 2.3 (+80)	−8	−16	0	−11	−11
	2. 0.53 (−120 to −60), 1.8 (+60)	−11, 22	−21, 16	−4	−15.5	−15.5
Medium nitrile NBR	1. 0.45 (−80), 1.8 (+40)	−18	−24	−15	−21	−21
	2. 0.51 (−80), 1.8 (+40)	−22	−30	−18	−26	−26
Neoprene WRT	0.49 (−60), 1.69 (+40)	−33	−40	−27.5	−36	−34.5
Low nitrile NBR	0.45 (−120 to −60), 1.47 (+40)	−32	−40	−28	−36	−39.5
BR/SBR/NR	0.46 (−110), 1.57 (+20)	−74	−89	−58, −38, −5	−70	−65
BR/SBR	0.60 (−110), 1.53 (+40)	−78	−70 ± 5	−57, −27, −6	−70	−75

[a] From Brazier and Nickel (1978).

[b] T_{g_o} = glass transition temperature calculated via conventional extrapolation procedures.

[c] T_{g_d} = temperature of DTMA peak at maximum rate of indentation or tensional strain.

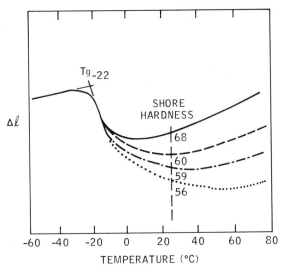

Fig. 86 Effect of plasticizer resin on indentation of nitrile vulcanizate, 5°C min^{-1}. ——— , No plasticizer; 10 phr — — coumarone indene resin; — — — 10 phr hydrocarbon resin; — — — 10 phr hydroxy resin. Formulation: Krynac 36:40 SP 100, MT black 85, FEF black 15, Betanox Special 2.0, antioxidant MB 2, DiCup 40C 3.5 (Brazier and Nickel, 1978).

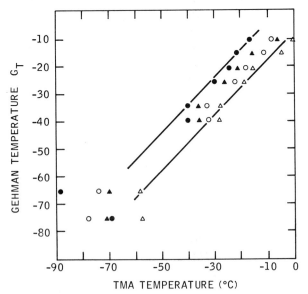

Fig. 87 TMA versus Gehman results. (Vulcanizates described in reference.) 5°C min^{-1}. Tension: (open triangle) T_{g_d}; (solid triangle) T_{g_o}. Indentation: (open circle) T_{g_d}; (solid circle) T_{g_o} (Brazier and Nickel, 1978).

production of precision-molded rubber goods. Molding conditions and the specific cure systems employed are key variables regarding this parameter. Fogiel *et al.* (1976) developed a fundamental definition of the factors that control the three-dimensional shrinkage of fluorelastomers (FKM). This comprehensive study is based on measurements made with the DuPont 941 TMA system operating at 5°C min^{-1} with a 1 g load; samples (4.8 mm wide and 6.4 mm thick) were cut from a cured pad.

A noteworthy feature of their analysis was the care required to calibrate the system so that reliable expansion and shrinkage measurements could be made. Vulcanized elastomer standards of known expansion coefficient were required because metal standards, because of their low expansion coefficient, give TMA displacements too small for reliable calibration over the desired temperature range 25–200°C). The volumetric expansion coefficients of these standards, obtained by dilatometry, were used to develop an equation $[\alpha = a + b(t - 25)]$ that relates the coefficient of linear expansion (α) to temperature (t); a and b are calibration constants. This equation was used to calibrate the TMA with the vulcanized standards previously described. The calibration constant was the average of the results for the three dimensions. Precision of the analysis was $\pm 2\%$ (95% confidence limits) over the range (25–200°C).

These expansion coefficients were considered to be equilibrium values based on the following criteria: (a) the same α was obtained for repeat runs on a given sample, including the use of probe load and heating-rate variations; (b) α was not affected by the repeated heating, over a 2-week period; (c) both stepwise and constant-rate heating gave the same results.

The calibrated TMA system was used to evaluate average shrinkage and anisotropy in three directions. These results were then compared with those predicted based on expansion data. Based on theoretical considerations, the shrinkage $\overline{S}$ of press-cured materials averaged over three dimensions is given by

$$\overline{S} = (\bar{\alpha}_c - \bar{\alpha}_m)(t - 25)/[1 + \alpha_c(t - 25)]$$

where $\bar{\alpha}_c$ and $\bar{\alpha}_m$ refer to similarly averaged α's for the cured compound and the mold, respectively, at the press cure temperature t. In general, good agreement with this relationship was observed, results being better for the 177°C cure than for the 150°C cure.

The influence of MT carbon black, which is commonly used in FKM compounds, was evaluated for comparison with that predicted based on assumed additivity of the components, i.e.,

$$\bar{\alpha}_c = \bar{\nu}_r\bar{\alpha}_r + (1 - \nu_r)\bar{\alpha}_f$$

where $\bar{\alpha}_r$; $\bar{\alpha}_f$ are linear isotropic α's for rubber and filler, respectively, and

ν_r is the volume fraction of rubber. Calculated values for α_c were found to be in good agreement with experimental values. Prediction of average shrinkage values for filled vulcanizates is therefore possible based on $\bar{\alpha}_r$ values for gum stocks as shown in Fig. 88. A variety of experimental and theoretical considerations that could explain the highly anisotropic nature of FKM compounds are treated in the paper.

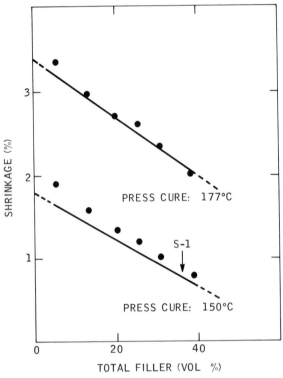

Fig. 88 Predictability of shrinkage from $\bar{\alpha}$ of unfilled FKM vulcanizate and volume fraction of filler. Lines: calculated from equations in reference. Points: experimental. Bottom line and points displaced downward by 1% (Fogiel *et al.*, 1976).

It was concluded that (a) TMA, properly calibrated, is "eminently suitable for the measurement of the thermal expansion coefficient of rubbers," and (b) "the relative simplicity and precision of the TMA method for the determination of the coefficient of isotropic expansion, and hence, of volumetric expansion of rubbers, makes this important thermodynamic parameter easily accessible experimentally."

D. QUALITATIVE ANALYSIS OF ELASTOMERS IN TIRE COMPONENTS

Analysis of formulations containing elastomer blends is a challenging task because of the general complexity of the system as previously discussed. This is further complicated by the difficulty of identifying the type(s) and relative proportions of the polymers in the system. Sircar and Lamond (1975a–d) undertook the task of identifying the polymers in tire components via the use of a series of basic thermal analytical techniques (DSC, TG, and DTG) to develop what they termed a "total thermal analysis" of the formulation. Their treatment of a system involves (a) DSC for T_g as well as thermal and thermooxidative degradation characteristics, (b) TG for basic polymer stability analysis, and (c) DTG to distinguish differences in the degradation rates of polymer blend components. This approach was applied to a wide variety of polymer blends and tire components as indicated in Table XXIII. A complete description of this type of analysis is contained in the various references by Sircar and Lamond (1975a–d). The following discussion illustrates the ability of this approach to provide highly useful qualitative analyses of elastomer composition in various binary, ternary, and even quaternary blends in use in the tire industry.

1. Tread and Black Sidewall Formulations

DTG curves (nitrogen atmosphere) are shown in Fig. 9 for natural rubber (NR), styrene–butadiene rubber (SBR), and butadiene rubber (BR), the major polymers most commonly used in these applications. The relative position of these curves suggests that NR should be readily detected in blends with SBR or BR; however, identification of the other component would be difficult. This analysis is illustrated for peroxide-cured NR/BR blends (Fig. 89a) and for black sidewalls from commercial tires (Fig. 89b). Although NR was readily identified, it was not possible to tell whether the higher temperature peak was due to SBR or BR.

Polymer blend composition for the systems in Fig. 89 was established by infrared analysis. DTG peak heights were observed to vary with composition in a regular manner for peroxide-cured blends, but not for SBR/NR black sidewall compounds. Variation in black loading was suggested as a possible factor. The use of DTG peak heights for quantitative analysis of polymer blend composition is also described by Maurer (1974a) and Brazier and Nickel (1975b). The latter describe a number of compositional factors that influence applicability of the approach for a given system.

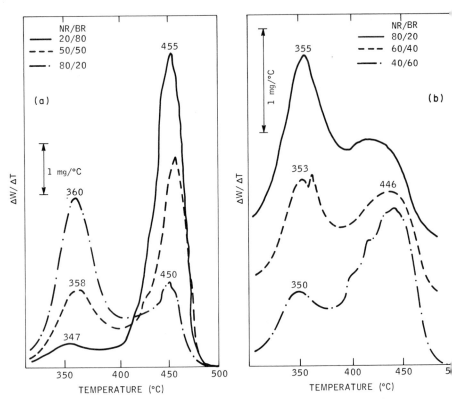

Fig. 89   (a) DTG curves of peroxide-cured blends in nitrogen: NR/BR (20:80, 9.36 mg; 50:50, 9.65 mg; 80:20, 9.66 mg) (Sircar and Lamond, 1975a). (b) DTG curves of blends from tire sidewalls in nitrogen: NR/BR (80:20, 9.77 mg; 60:40, 9.70 mg; 40:60, 9.59 mg) (Sircar and Lamond, 1975a).

Analysis of T_g via DSC can also be useful for determining the composition of certain elastomer blends. The T_g versus blend composition relationship is shown for a peroxide-cured and a commercial SBR/BR tread section in Fig. 90. The feasibility of analyzing blend composition in commercial treads by this approach is suggested, since such formulations contain less than 50% BR. In practice, blend analyses were sometimes inaccurate, possibly because of microstructure differences in the various commercial BR polymers.

It was concluded that a "reasonably accurate" identification of polymers in treads or black sidewalls could be achieved by total thermal analysis. DSC (nitrogen) distinguished SBR, BR, or blends from NR or isoprene rubber–synthetic (IR). T_g was used to distinguish between these polymers, except in the case of SBR/BR blends containing a high per-

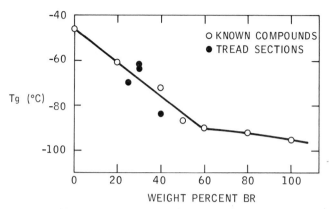

Fig. 90   Glass transition temperature of SBR/BR blends (Sircar and Lamond, 1975a).

centage of BR. Confirmation of DSC and T_g analyses was achieved by DTG.

2. White Sidewall Compounds (EPDM and Blends)

Extension of the total thermal analysis scheme to ternary systems is illustrated in Fig. 91 for blends of constant styrene–butadiene rubber (SBR) composition; ethylene–propylene–diene rubber (EPDM), and chlorinated isobutene–isoprene rubber (CIIR) varied from 20–50% of the total. Data for conventional white sidewall formulations are shown.

Based on DSC in nitrogen (Fig. 91a) SBR, CIIR, and EPDM are readily identified by the exotherm at 370°C and the endotherms at 438 and 484°C, respectively; the latter varying in size according to the ratio of the components in the blend.

DTG analyses in nitrogen (Fig. 91b) show that CIIR is readily identified by a peak at about 440°C.SBR and EPDM both appear near 490°C. However, the latter pair of polymers can be readily distinguished by DSC in nitrogen. Extension of these analyses to NR-containing ternary and quaternary blends is illustrated in Fig. 92. DTG in nitrogen reveals the NR, EPDM, and CIIR peaks around 370, 480, and 440°C. SBR and EPDM, which show peaks in the same area, can be distinguished from their DSC curves and thus do not present a problem in the overall scheme of analysis.

In practice, it was reported that total thermal analysis could identify all the polymers in various commercial white sidewall formulations. Good agreement with infrared analysis of blend composition was reported.

White sidewall blend compounds containing chloroprene rubber (CR) are a special case, as described by Sircar and Lamond (1975c). DSC

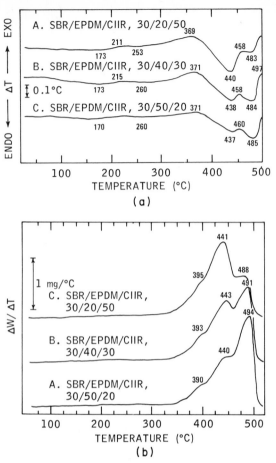

Fig. 91 (a) DSC curves of white sidewall compounds of SBR/EPDM/CIIR in nitrogen (formulation given in reference). A, 7.00 mg; B, 6.48 mg; C, 4.54 mg. (b) DTG curves of white sidewall compounds of SBR/EPDM/CIIR in nitrogen (formulation given in text). A, 9.42 mg; B, 9.60 mg; C, 9.37 mg (Sircar and Lamond, 1975b).

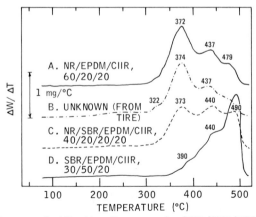

Fig. 92 DTG curves of white sidewall compounds of NR/SBR/EPDM/CIIR in nitrogen. A, 9.45 mg; B, 9.48 mg; C, 9.60 mg; D, 9.42 mg (formulations given in reference) (Sircar and Lamond, 1975b).

curves of sulfur-cured CR (Fig. 93) differed from those of peroxide-cured (Fig. 94) vulcanizates with respect to both the shape and peak temperature of the exotherm. The reaction that gives rise to this peak has been attributed to dehydrochlorination of CR followed by cross-linking. This proposal was supported by the results of TG and DTG analyses.

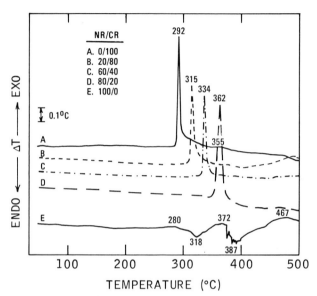

Fig. 93   DSC curves of NR/CR blends in nitrogen. (Formulation given in reference.) A, 0.46 mg; B, 0.48 mg; C, 0.40 mg; D, 1.18 mg; E, 7.45 mg (Sircar and Lamond, 1975c).

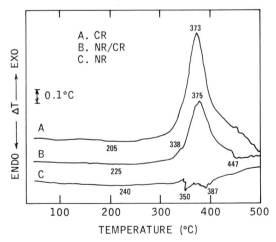

Fig. 94   DSC curves of NR and CR in nitrogen (peroxide-cured). A, 5.59 mg; B, 5.14 mg; C, 4.96 mg (Sircar and Lamond, 1975c).

3. Inner-Liner Compounds (Halobutyl Blends)

This study focused on the identification of isobutene–isoprene rubber (IIR), chlorinated isobutene–isoprene rubber (CIIR), and bromo–isobutene–isoprene rubber (BIIR) in inner-liner compounds by the total thermal analysis approach. Another objective was to determine whether these polymers could be differentiated from one another in this type of formulation. Typical data obtained by DSC (nitrogen) is illustrated in Fig. 36. Characteristic features of these thermograms were discussed in Section V. Of particular interest here is the CIIR endothermic degradation peak. This is the main endotherm for degradation of IIR, CIIR, and BIIR, which generally occurs in the 430–440°C range for carbon-black-reinforced vulcanizates. Similar TG curves are also observed for these three polymers.

DTG curves (nitrogen) for a natural rubber (NR)/CIIR blend in an inner-liner recipe are shown in Fig. 95. Characteristic peaks for NR

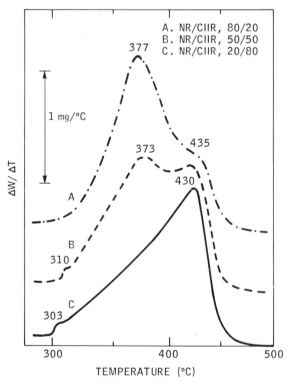

Fig. 95   DTG curves of blends of CIIR with other elastomers in nitrogen (inner-liner formulation). NR/CIIR: 80/20, 9.52 mg; 50/50, 9.75 mg; 20/80, 9.52 mg (Sircar and Lamond, 1975d)

(370°C) and CIIR (430°C) are noted except for compound C, which exhibited only the CIIR peak. All blends containing greater than 40% CIIR also showed a small peak near 300–310°C, characteristic of halogenated butyl polymers. Similar analysis of the ternary NR/CIIR/styrene–butadiene rubber (SBR) system is shown in Fig. 96. These curves generally resem-

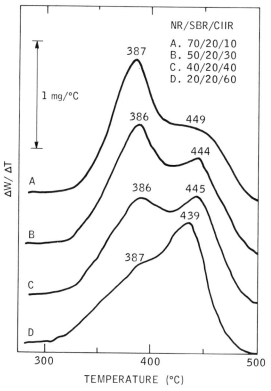

Fig. 96 DTG curves of inner-liner compounds of NR/SBR/CIIR triblends in nitrogen. A, 9.54 mg; B, 9.44 mg; C, 9.43 mg; D, 9.57 mg (Sircar and Lamond, 1975d).

ble those of the NR/CIIR binary blends except for slightly higher peak temperatures.

These studies indicated that IIR, BIIR, and CIIR are easily identified as isobutene–isoprene-type polymers, in these systems, by combined DSC and DTG procedures. However, these methods do not distinguish between BIIR and CIIR. [D. W. Brazier (personal communication, 1978) indicates that CIIR and BIIR can be differentiated in the pure state by the combined use of DTG and GC techniques.] Additional points of interest in Sircar and Lamond's (1975c) paper include consideration of the effect of blend components on the DTG peak temperatures for a given elastomer. A directional correlation with the exo- or endothermic nature of the degradation process for each polymer was suggested.

E. PROBLEM SOLVING VIA THERMAL ANALYSIS

The thermal analytical procedures described in this chapter have also been effectively employed to diagnose and/or solve a variety of problems encountered with a wide range of elastomer systems. These situations relate to the full range of elastomer research, development, production, and quality control activities. Among the types of questions or problems that have been addressed are the following:

(a) Anomalous properties of compounds raised questions about accuracy of the formulation.

(b) Apparent heterogeneity led to concerns about contamination of a mixed stock.

(c) Unexpected performance characteristics raised questions, in a development program, regarding unidentified compounding or processing approaches used in a factory evaluation of a development compound.

(d) Poor correlation of laboratory and tire test was completely unexpected. Questions were raised about whether a particular polymer was omitted from the compound.

(e) An error in coding a multisegment tire test was suspected as the cause of anomalous results.

The following examples illustrate the kinds of practical information that thermal analytical techniques can provide.

An interesting problem that has been encountered was concerned with identifying material that plugged a screen pack during processing of an accelerated master batch. The following possibilities were under consideration as potential causes of the problem: poor dispersion of compounding ingredients, scorched (prematurely cured) polymer, scorched compound, and/or contamination by a foreign ingredient. The data presented in Table XXVII indicate substantial differences in composition, T_g, and cure characteristics of the contaminant, thus suggesting introduction of a "foreign" compound into the master batch, possibly during the mixing operations.

A similar problem involved undispersed small particles in an unaccelerated master batch. The particles were similar in size and shape but differed in hardness. TG analyses of a set of these particles (Fig. 97) indicated that there were probably two factors involved in the problem. Poor mixing is suggested by curves A and B. In addition, contamination by a material having substantially different thermal stability characteristics is suggested by curve C. This latter material was most likely a resin.

TABLE XXVII

ANALYSIS OF CONTAMINANT BY TGA AND DTA[a]

Thermal gravimetric analysis	*Oil + polymer*	*Carbon black*	*Ash*
Base compound	63	33.5	3.5
Contaminant	50	41	9
Differential thermal analysis	T_g	*Curing exotherm (peak)*	
Base compound	−72	140°C (weak)	
Contaminant	Nonevident	207°C (very strong)	
	(−100−+25°C)		

[a] From Maurer (1969b).

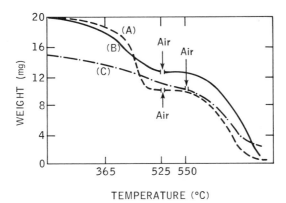

TEMPERATURE (°C)

Fig. 97 Analysis of complex dispersion problem. A, B, and C are undispersed particles taken from factory Banbury-mixed compound. A and B reveal poor mixing; C is a contaminant (Maurer, 1969a,b).

A problem that is not so straightforward is indicated in Fig. 98. In this case two samples that were supposed to be identical exhibited significantly different properties. Basic TG analysis indicated that these materials appeared to have the same "polymer plus oil," total carbon black, and ash contents. Isothermal oxidation of the carbon black residues indicated significantly different characteristics, which suggested the following possible differences in the two systems: different type of carbon black, different source of carbon black, a blend of carbon blacks in one sample, and/or a difference in the accelerator system in one of the samples. The bases for these suggestions are covered in the discussion of carbon black analysis by TG techniques (Section VIII).

The final example to be considered is concerned with a multisegment-tire test of four laboratory compounds. The test involved two tires

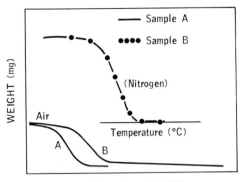

TIME (min at 532°C)

Fig. 98 Isothermal analysis of carbon black residues from two commercial vulcanizates whose properties unexpectedly varied substantially (Maurer, 1969c).

containing three segments per tire: a control and two experimental compounds that differed in polymer blend composition. In addition, the latter two systems were to be evaluated at each of two different levels of an additive. A question arose as to whether each tire contained a control plus *one* experimental compound at each of two additive levels, or a control plus each compound at one additive level. Comparison of DTG thermograms of samples from the tires versus known formulations readily identified which of the two evaluation procedures had been used.

X. Thermal Analysis of Elastomer Systems: Status and Outlook

The many examples cited in this chapter clearly establish that thermal analysis techniques (DTA, DSC, TG, DTG, TMA, and DMA) are broadly applicable in elastomer research, product development, manufacturing, and quality control or assurance programs. Additional evidence of this applicability can be found in recent reviews by Brazier (1980a,b) and Krishen (1981). The emergence of quality assurance procedures, e.g. the work of Harris (1978), based on thermal analytical methods indicates that *purchasers* of finished rubber goods have recognized the valuable role these techniques can play in the rubber industry. However, these techniques may prove to be of greatest value to the rubber goods *manufacturer* who can establish reference data to define the influence of individual components of the formulations on thermogram characteristics, and hence, use these procedures for "troubleshooting" and quality control (QC) purposes.

Instruments oriented toward QC usage have appeared on the market.

These are a step in the right direction in making available specific QC-related analyses at reduced equipment costs versus those of more versatile, but higher-priced units. Another important advance is the emergence of microprocessor-based systems. These facilitate thermal analyses requiring routine variations of thermal history and sample environment. This latter capability, of course, should also prove useful in various research and product development programs.

Additional applications of thermal analysis in elastomer research and development can be envisioned. Some of these could come from combining thermal methods with other techniques for analyzing volatile components generated by heating or pyrolyzing polymer compounds or vulcanizates under controlled conditions. Further advances may come from enhanced sensitivity and resolution capability of the equipment, with an attendant improvement in analyzing complex mixtures. An intriguing consideration is the expanded use of computer analysis of data, e.g., from the TG experiment. This approach might prove especially useful with regard to major end-use applications. In such cases it might be advantageous to establish the volatilization or decomposition kinetics of at least the major components in the formulation. This background information may enable enhanced computer resolution analysis of TG data resulting from overlapping processes and hence an improved analysis of elastomer system composition and uniformity. A particularly interesting example concerns analysis of carbon black(s) in a commercial formulation. It has been shown that the oxidation characteristics of blacks may be influenced by the type of cure system used in the formulation. One wonders whether for certain types of known systems TG/DTG data can one day routinely supply information about the total carbon black content, the ratio of carbon blacks, and the presence (and perhaps the ratio) of certain accelerator systems in a formulation. The groundwork for testing this approach has been presented in this review.

It seems likely, at this stage, that important additional advances will be made in the application of thermal analytical methods to elastomer systems. Irrespective of this, it seems justifiable to claim that proven applications in both research and development activities have already earned thermal analysis a place among the important procedures for characterizing elastomer systems.

Acknowledgments

As with any such endeavor, this review has benefited from and been strengthened by contributions from many sources. It is a pleasant task to give recognition to these contributors. Dr. Anil K. Sircar, J. M. Huber Corp., provided critical and exceptionally thorough reviews of both the original and the revised version of the text. Reviews, critique, and suggested revisions of the original text were also provided by Dr. David W. Brazier, Dunlop

Corp., Dr. Luis G. Roldan, J. P. Stevens & Co., Inc., and Dr. Edith A. Turi, Allied Corp. Reviews of the polymer stability and dynamic mechanical analysis sections were provided by Dr. Edward N. Kresge and Dr. Gary VerStrate, respectively, of Exxon Research and Engineering Co. Important reference material was supplied by Dr. R. B. Prime, Dr. D. W. Brazier, Dr. A. K. Sircar, and Dr. E. A. Turi, and also by E. I. duPont de Nemours & Co., Inc., Linseis Instrument Co., Mettler Instrument Corp., and Perkin-Elmer Corp.

Dr. Edith A. Turi merits special commendation for her many contributions, which included basic organization of the book, surveys and copies of background literature, several reviews of the chapter, boundless patience with the author, and constant support of the project from inception to completion. It has been a privilege and a pleasure to be associated in this task with someone so dedicated to the advancement of thermal analysis.

References

Aklonis, J. J., MacKnight, W. J., and Shen, M. (1972). "Introduction to Polymer Viscoelasticity." Wiley (Interscience), New York.

Alliger, G., and Sjothun, I. J. (1964). "Vulcanization of Elastomers." Van Nostrand-Reinhold, Princeton, New Jersey.

Ambelang, J. C., Kline, R. H., Lorenz, O. M., Parks, C. R., Wadelin, C., and Shelton, J. R. (1963). *Rubber Chem. Technol.* **36,** 1497–1541.

American Society for Testing and Materials (1979a). *Annu. Book ASTM Stand.* Part 23 (D-1160).

American Society for Testing and Materials (1979b). *Annu. Book ASTM Stand.* Part 37 (D297).

American Society for Testing and Materials (1979c). *Annu. Book ASTM Stand.* Part 37 (D1418).

Banarjee, B., Chakravarty, S. N., Kamath, B. V., and Biswas, A. B. (1979). *J. Appl. Polym. Sci.* **24,** 683–692.

Barrall, E. M., III, and Flandera, M. A. (1974). *J. Elastomers Plast.* **6,** 16–25.

Barrall, E. M., III, Flandera, M. A. F., and Logan, J. A. (1974a). *Thermochim. Acta.* **5,** 415–432.

Barrall, E. M., III, James, P. A., Dawson, B., and Logan, J. A. (1974b). *J. Macromol. Sci. Chem.* **A8,** 135–155.

Barrett, K. E. J. (1967). *J. Appl. Polym. Sci.* **11,** 1617–1626.

Bateman, L., ed. (1963). "The Chemistry and Physics of Rubber-Like Substances." Wiley, New York.

Bateman, L., Moore, C. G., Porter, M., and Saville, B. (1963). Chemistry of Vulcanization, *in* "The Chemistry and Physics of Rubber-Like Substances" (L. Bateman, ed.). Wiley, New York.

Bauman, R. G., and Maron, S. H. (1956). *J. Polym. Sci.* **22,** 203–212.

Bekkedahl, N., and Weeks, J. J. (1969). *J. Res. Natl. Bur. Stand., Sect. A* **73A,** 221–231.

Bhaumik, M. L., Banerjee, D., and Sircar, A. K. (1962). *J. Appl. Polym. Sci.* **6,** 674–682.

Bhaumik, M. L., Banerjee, D., and Sircar, A. K. (1965a). *J. Appl. Polym. Sci.* **9,** 1367–1384.

Bhaumik, M. L., Banerjee, D., and Sircar, A. K. (1965b). *J. Appl. Polym. Sci.* **9,** 1731–1742.

Bhaumik, M. L., Banerjee, D., and Sircar, A. K. (1965c). *J. Appl. Polym. Sci.* **9,** 2285–2296.

Billmeyer, F. W., Jr. (1971). "Textbook of Polymer Science." Wiley (Interscience), New York.

Blaine, R. L., Gill, P. S., Hassel, R. L., and Woo, L. (1978). *Appl. Polym. Symp.* **34,** 157–171.

Borchardt, H. J., and Daniels, F. (1957). *J. Am. Chem. Soc.* **79,** 41–46.

Boyer, R. F. (1968). *Polym. Eng. Sci.* **8,** 161–185.

Boyer, R. F. (1976). *Polymer* **17,** 996–1008.

Boyer, R. F. (1977a). "Symposium on Automated Dynamic Mechanical Techniques in Polymer Research and Development." Society of Plastics Engineers, Princeton, New Jersey.

Boyer, R. F. (1977b). *Encycl. Polym. Sci. Technol. Suppl.* **2,** 745–839.

Braddon, D. V., and Falkehag, S. I. (1973). *J. Polym. Sci., Polym. Symp.* **40,** 101–104.

Brazier, D. W. (1977). *Thermochim. Acta* **18,** 147–160.

Brazier, D. W. (1980a). *Rubber Chem. Technol.* **153,** 437–511.

Brazier, D. W. (1980b). *In* "Applications of TG in the Study and Analysis of Elastomers and Elastomer Compounds" (J. P. Redfern, ed.) (in preparation).

Brazier, D. W., and Nickel, G. H. (1975a). *Rubber Chem. Technol.* **48,** 26–40.

Brazier, D. W., and Nickel, G. H. (1975b). *Rubber Chem. Technol.* **48,** 661–677.

Brazier, D. W., and Nickel, G. H. (1978). *Thermochim. Acta* **26,** 399–413.

Brazier, D. W., and Schwartz, N. V. (1978). *Rubber Chem. Technol.* **51,** 1060–1074.

Brazier, D. W., and Schwartz, N. V. (1980). *Thermochim. Acta* **39,** 7–20.

Brazier, D. W., Nickel, G. H., and Szent-Györgi, Z. (1980). *Rubber Chem. Technol.* **53,** 160–175.

Buchdahl, R., and Nielsen, L. E. (1960). *J. Appl. Phys.* **21,** 482–487.

Byrne, L. F., and Hourston, D. J. (1979a). *J. Appl. Polym. Sci.* **23,** 1607–1617.

Byrne, L. F., and Hourston, D. J. (1979b). *J. Appl. Polym. Sci.* **23,** 2899–2908.

Callan, J. E., Hess, W. M., and Scott, C. E. (1971). *Rubber Chem. Technol.* **44,** 814–837.

Castillo, J. E., Hernandez, R. M., and Campos-Lopez, E. (1979). *Thermochim. Acta* **33,** 323–329.

Cessna, L. C., Jr., and Jabloner, H. (1974). *J. Elastomers Plast.* **6,** 103–113.

Chakraborty, K. K. (1978). *J. Indian Chem. Soc.* **55,** 405–408.

Chiu, J. (1966). *Appl. Polym. Symp.* **2,** 25–43.

Cole, H. M., and Walker, D. F. (1970). *Rev. Gen. Caoutch. Plast.* **47,** 751–752.

Collins, E. A., Jorgensen, A. H., and Chandler, L. A. (1973). *Rubber Chem. Technol.* **46,** 1087–1102.

Collins, W. E. (1970). *Pittsburgh Conf. Anal. Chem. Appl. Spectrosc. 21st Cleveland, Ohio, 1970.*

Cunneen, J. I. (1968). *Rubber Chem. Technol.* **41,** 182–208.

Dienes, G. J., and Klemm, H. F. (1946). *J. Appl. Phys.* **17,** 458–471.

Donnet, J. B., and Voet, A. (1976). "Carbon Black." Dekker, New York.

Double, J. S. (1966). *Plast. Inst., Trans. J.* **34,** 73–82.

Dunn, J. R. (1974). *Rubber Chem. Technol.* **47,** 960–975.

Dunn, J. R. (1976). *Rubber Chem. Technol.* **49,** 978–991.

Dunn, J.R. (1978). *Rubber Chem. Technol.* **51,** 686–703.

E. I. DuPont & Co. (1963). *Anal. Chem.* **35,** Issue No. 7, 101A.

E. I. DuPont & Co. (1977). "Thermal Analysis Review: Dynamic Mechanical Analysis." Prod. Bull. E. I. DuPont & Co., Wilmington, Delaware.

Eirich, F. R., ed. (1978). "Science and Technology of Rubber." Academic Press, New York.

Ferry, J. D. (1970). "Viscoelastic Properties of Polymers," 2nd/Ed. Wiley, New York.

Festisova, L. M., and Shabanova, L. D. (1980). *Kauch. Rezina.* **12,** 49–51.

Field, J. E., Woodford, D. E., and Gehman, S. J. (1955). *J. Polym. Sci.* **15,** 51–67.

Flory, P. J., and Rehner, J., Jr. (1943). *J. Chem. Phys.* **11,** 521–526.

Fogiel, A. W., Frensdorff, H. K., and MacLachlan, J. D. (1976). *Rubber Chem. Technol.* **49,** 34–42.

Gardiner, J. B. (1970). *Rubber Chem. Technol.* **43,** 370–399.

Garn, P. D., and Menis, O. (1977). *J. Macromol. Sci., Phys.* **B13**(4), 611–630.

Gedemer, T. J. (1974). *In* "Polymer Characterization by Thermal Methods of Analysis" (J. Chiu, ed.), pp. 95–104. Dekker, New York.

Gent, A. N. (1958). *Trans., Inst. Rubber Ind.* **34,** 46–57.

Ghijsels, A. (1977). *Rubber Chem. Technol.* **50,** 278–291.

Gill, P. S., and Hassel, R. L. (1976). "Polypropylene Impact Resistance by Dynamic Mechanical Analysis" Prod. Bull. TA-76. E. I. DuPont & Co., Wilmington, Delaware.

Gill, P. S., and Ikeda, R. M. (1977). "Advantages of the DuPont Dynamic Mechanical Analyzer for Polymer Characterization" Appli. Brief No. TA-77. E. I. DuPont & Co., Wilmington, Delaware.

Gillham, J. K. (1974). *AIChE J.* **20,** 1066–1079.

Gillham, J. K. (1977). "Symposium on Automated Dynamic Mechanical Techniques in Polymer Research and Development." Society of Plastics Engineers, Princeton, New Jersey.

Gillham, J. K., Glandt, C. A., and McPherson, C. A. (1977). *In* "Chemistry and Properties of Crosslinked Polymers" (S. S. Labana, ed.), pp. 491–520. Academic Press, New York.

Goh, S. H. (1977). *J. Elastomers Plast.* **9,** 186–192.

Goh, S. H., and Lim, Y. B. (1979). *Thermochim. Acta* **32,** 81–85.

Goh, S. H., and Phang, K. W. (1978). *Thermochim. Acta* **25,** 109–115.

Gonzalez, V. (1980). *Rubber Chem. Technol.* **53,** 378 (abstr.).

Gonzalez, V. (1981). *Rubber Chem. Technol.* **54,** 134–145.

Guise, G. B., and Smith, G. C. (1980). *J. Appl. Polym. Sci.* **25,** 149–161.

Gupta, A. K. (1980). *Makromol. Chem., Rapid Commun.* **1,** 201–204.

Harris, J. (1978). *Elastomerics* **110,** 48–49.

Haward, R. N. (1973). "The Physics of Glassy Polymers." Wiley, New York.

Helmer, J., and Polmanteer, K. (1969). *J. Appl. Polym. Sci.* **13,** 2113–2118.

Hourston, D. J., and Hughes, I. D. (1981). *Polymer* **22,** 127–129.

Hwo, C. H., and Johnson, J. F. (1974). *J. Appl. Polym. Sci.* **18,** 1433–1441.

Illinger, J. L., Schneider, N. S., and Karasz, F. E. (1972). *Polym. Eng. Sci.* **12,** 25–29.

Jander, G., and Bloom, C. (1949). *Z. Anorg. Chem.* **258,** 205–220.

Jaroszynska, D., Kleps, T., and Gdowska-Tutak, D. (1978). *"Polimery"* (*Warsaw*) **23,** No. 10, 354–357.

Jaroszynska, D., Kleps, T., and Gdowska-Tutak, D. (1980). *J. Thermal Anal.* **19,** 69–78.

Johnson, R. M., and Mark, J. E. (1972). *Macromolecules* **5,** 41–45.

Jopling, D. W., and Pitts, E. (1965). *Br. J. Appl. Phys.* **16,** 541–549.

Kaneko, Y., Watabe, Y., Okamato, T., Iseda, Y., and Matsunega, T. (1980). *J. Appl. Polym. Sci.* **25,** 2467–2478.

Kennedy, J. P., and Tornqvist, E. G. M., eds. (1969). "Polymer Chemistry of Synthetic Elastomers." Wiley (Interscience), New York.

Kenyon, A. S., Grote, W. A., Wallace, D. A., and Rayford, M. C. (1977). *J. Macromol. Sci., Phys.* **B13**(4), 553–570.

Kissinger, H. E. (1957). *Anal. Chem.* **29,** 1702–1706.

Kraus, G., ed. (1965). "Reinforcement of Elastomers." Wiley (Interscience), New York.

Krishen, A. (1972). *Anal. Chem.* **44,** 494–497.

Krishen, A. (1981). *Anal. Chem.* **53,** 159R–162R.

Krizanousky, L., and Mentlik, V. (1978). *J. Therm. Anal.* **13,** 571–580.

Kuzminskii, A. S. (1966). *Rubber Chem. Technol.* **39,** 88–111.

Lawandy, S. N., and Hepburn, C. (1980). *Elastomerics* **112,** 24–26.

Leblanc, J. L. (1977). *J. Appl. Polym. Sci.* **21,** 2419–2437.

Lee, B., and Singleton, C. (1979). *J. Appl. Polym. Sci.* **24,** 2169–2183.

Lilaonitkul, A., and Cooper, S. L. (1977). *Rubber Chem. Technol.* **50,** 1–23.

McCreedy, K., and Keskkula, H. (1978). *J. Appl. Polym. Sci.* **22,** 999–1005.

McCreedy, K., and Keskkula, H. (1979). *Polymer* **20,** 1155–1159.

MacKillop, D. A. (1974). *Anal. Chem.* **40,** 607–609.

Manzione, L. T., Gillham, J. K., and McPherson, C. A. (1981). *J. Appl. Polym. Sci.* **26,** 889–919.

Marsh, P. A., Voet, A., Price, L. D., and Mullen, T. J. (1968). *Rubber Chem. Technol.* **41,** 344–372.

Maurer, J. J. (1965). *Rubber Chem. Technol.* **38,** 979–990.

Maurer, J. J. (1967a). *Div. Rubber Chem., Am. Chem. Soc., 21st Meeting Montreal, 1967.* [Abstract; *Rubber Chem. Technol.* **40,** 1592–1594 (1967).]

Maurer, J. J. (1967b). *East. Anal. Symp., Am. Chem. Soc., New York, 1967.*

Maurer, J. J. (1968). *In* "Analytical Calorimetry" (R. S. Porter and J. F. Johnson, eds.), Vol. 1, pp. 107–118. Plenum, New York.

Maurer, J. J. (1969a). *Rubber Chem. Technol.* **42,** 110–158.

Maurer, J. J. (1969b). *In* "Thermal Analysis" R. F. Schwenker and P. D. Garn, eds.), Vol. 1, pp. 373–386. Academic Press, New York.

Maurer, J. J. (1969c). *Div. Rubber Chem. Am. Chem. Soc., 95th Meeting, Los Angeles, 1969.* [Abstract: *Rubber Chem. Technol.* **42,** 1490–1491 (1969).]

Maurer, J. J. (1970a). *Rubber Age* **102,** 47–51.

Maurer, J. J. (1970b). *NBS Spec. Publ. (U.S.)* **338,** 165–185.

Maurer, J. J. (1974). *J. Macromol. Sci., Chem.* **A8,** 73–82.

Maurer, J. J. (1978). *In* "Thermal Methods in Polymer Analysis" (S. W. Shalaby, ed.), pp. 129–161. Franklin Inst. Press, Philadelphia, Pennsylvania.

May, W. R., and Bsharah, L. (1969). *Ind. Eng. Chem., Prod. Res. Dev.* **8,** 185–188.

May, W. R., Bsharah, L., and Meerifield, D. B. (1968). *Ind. Eng. Chem., Prod. Res. Dev.* **7,** 57–61.

Meares, P. (1965). "Polymers: Structure and Bulk Properties." Van Nostrand-Reinhold, Princeton, New Jersey.

Morris, M. C. (1967). *Rubber Chem. Technol.* **40,** 341–349.

Morton, M., ed. (1973). "Introduction to Rubber Technology." Van Nostrand-Reinhold, Princeton, New Jersey.

Nielsen, L. E. (1962). "Mechanical Properties of Polymers." Van Nostrand-Reinhold, Princeton, New Jersey.

Nishi, T., Kwei, T. K., and Wang, T. T. (1975). *J. Appl. Phys.* **46,** 4157–4165.

Noshay, A., and McGrath, J. E. (1977). "Block Copolymers, Overview and Critical Survey." Academic Press, New York.

Ozawa, T. (1976). *J. Therm. Anal.* **9,** 369–373.

Papierer, E., Donnet, J. B., and Schultz, A. (1967). *Carbon* **5,** 113–125.

Pautrat, R., Metrivier, B., and Morteau, J. (1976). *Rubber Chem. Technol.* **49,** 1060–1067.

Peyser, P., and Bascom, W. D. (1977). *J. Macromol. Sci. Phys.* **B13,** 597–610.

Ponce-Vélez, M. A., and Campos-López, E. (1978). *J. Appl. Polym. Sci.* **22,** 2485–2497.

Prime, R. B. (1978). *Thermochim. Acta* **26,** 165–174.

Rogers, R. N., and Morris, E. D. (1966). *Anal. Chem.* **38,** 412–414.

Rootare, H. M., and Powers, J. M. (1977). *J. Dent. Res.* **56,** No. 12, 1453–1462.

Rosen, S. L. (1971). "Fundamental Principles of Polymeric Materials for Practicing Engineers." Barnes & Noble, New York.

Schneider, N. S., and Paik-Sung, C. S. (1977). *Polym. Eng. Sci.* **17**, 73–80.

Schwartz, N. V., and Brazier, D. W. (1978a). *Thermochim. Acta* **26**, 349–359.

Schwartz, N. V., and Brazier, D. W. (1978b). *J. Appl. Polym. Sci.* **22**, 113–124.

Seidov, N. M., Aliguliev, R. M., Guseinov, F. O., Ibragimov, Kh. D., Ovanesova, G. S., and Talaybova, T. N. (1980). *Vysokomol. Soedin., Ser B* **22**(4), 154–157.

Shelton, J. R. (1957). *Rubber Chem. Technol.* **30**, 1251–1290.

Simpson, M. B. H., and Percival, W. C. (1979). *Rubber Chem. Technol.* **52**, 899.

Sircar, A. K. (1977). *Rubber Chem. Technol.* **50**, 71–82.

Sircar, A. K. (1978). *Thermochim. Acta* **27**, 337–367.

Sircar, A. K., and Lamond, T. G. (1972). *Rubber Chem. Technol.* **45**, 329–345.

Sircar, A. K., and Lamond, T. G. (1973a). *J. Appl. Polym. Sci.* **17**, 2569–2577.

Sircar, A. K., and Lamond, T. G. (1973b). *Rubber Chem. Technol.* **46**, 178–191.

Sircar, A. K., and Lamond, T. G. (1973c). *Thermochim. Acta* **7**, 287–292.

Sircar, A. K., and Lamond, T. G. (1975a). *Rubber Chem. Technol.* **48**, 301–309.

Sircar, A. K., and Lamond, T. G. (1975b). *Rubber Chem. Technol.* **48**, 631–639.

Sircar, A. K., and Lamond, T. G. (1975c). *Rubber Chem. Technol.* **48**, 640–652.

Sircar, A. K., and Lamond, T. G. (1975d). *Rubber Chem. Technol.* **48**, 653–660.

Sircar, A. K., and Lamond, T. G. (1978). *Rubber Chem. Technol.* **51**, 647–654.

Sircar, A. K., Lamond, T. G., and Wells, J. L. (1980). *Thermochim. Acta* **37**, 315–324.

Smith, D. A. (1966a). *J. Polym. Sci., Part B* **4**, 215–221.

Smith, D. A. (1966b). *Kautsch. Gummi, Kunstst.* **19**, 477–481.

Smith, D. A. (1968). *Rubber J.* **150**, 33–35, 37–38.

Smith, R. C., and Stephens, H. L. (1975). *J. Elastomers Plast.* **7**, 156–172.

Spacsek, K., Somolo, A., and Soos, I. (1977). *J. Therm. Anal.* **11**, 211–219.

Stiehler, R. D., Decker, G. E., and Bullman, G. W. (1979). *Rubber Chem. Technol.* **52**, 255–262.

Swarin, S. J., and Wims, A. M. (1974). *Rubber Chem. Technol.* **47**, 1193–1205.

Takayanaga, M. (1965). *Proc. Int. Congr. Rheol., 4th, 1963* pp. 161–175.

Voet, A., and Morawski, J. C. (1978). *Rubber Chem. Technol.* **47**, 758–777.

Ward, I. M. (1971). "Mechanical Properties of Solid Polymers." Wiley (Interscience), New York.

Wood, L. A., and Roth, F. L. (1963). *Rubber Chem. Technol.* **36**, 611–620.

Wrasidlo, W. (1974). *Fortschr. Hochpolym. Forsch.* **13**, 1–99.

Wyzgoski, M. G. (1980). *J. Appl. Polym. Sci.* **25**, 1443–1467.

CHAPTER 7

Fibers

*M. JAFFE**

Fiber Industries Incorporated
Charlotte, North Carolina

I. Introduction

The thermal analysis of fibers is the thermal analysis of oriented, semi-crystalline polymers. The uniqueness of fibers lies only in their anisotropy: anisotropy of microstructure, of properties, and of geometry. Whether the fiber is natural or synthetic in origin, the key to the solution of fiber problems is to be found in the chemistry and physics of polymeric molecules.

It is shown throughout this book that thermal analysis is an accurate and convenient tool for monitoring the response of polymers to temperature, stress, and environment. As these are three of the primary variables of the fiber formation process, thermal analysis (TA) can offer information for the setting of fiber process parameters. Thermal analysis allows the chemical and structural "fingerprinting" of fibers, the monitoring of important end-use properties, the study of interactions of fibers with additives and coatings, and, ultimately, insight

---

* Present address: Celanese Research Company, Summit, New Jersey.

into the process–structure–property spectrum available to a given fiber-forming material. The words *process, structure,* and *properties* are used throughout this chapter in the most general sense. *Process* implies any and all events in which the fiber is subjected to the influences of heat, stress, and/or chemical agents prior to analysis. *Structure* implies the morphology of all phases comprising the fiber and the manner by which the phases are interconnected, as well as the interactions of all phases with foreign materials (i.e., additives, finishes, etc.) present in or on the fiber. *Properties* include any parameters of interest to the fiber user or analyst. Before proceeding to the discussion of these topics in depth, however, a brief description of fiber technology and terminology is in order.

A fiber may be defined as a structure whose length is much greater than its cross-sectional dimensions, an arbitrary ratio being about 100:1. Fiber size is conveniently characterized in units of linear density; the accepted SI unit is the tex, defined as the weight in grams of 1000 m. As until recently the common-use linear density unit was the denier, d, defined as the weight in grams of 9000 m, the SI unit most often used is the decitex, dtex, equal to the weight in grams of 10,000 m and close in value to denier. Typical fiber decitexes range from about 1 to 10; 1 g of a 1 dtex filament is over 5 miles in length. In most published literature, fiber mechanical properties are given in the specific unit of grams-force per denier (g/d); the accepted SI unit is centinewtons per decitex (cN/dtex). The breaking strength of a fiber is termed its tenacity. A listing of fiber units and conversion factors is given in Table I and more complete listings can be found in the reference list.

For most end uses, fibers are brought together to form a yarn (multiplicity of filaments); yarn size is specified by the total denier of the assemblage and the number of fibers comprising it, i.e., 70/35 (70 decitex yarn, 35 filaments—each filament being 2 dtexpf-dtex per filament). A filament yarn is a yarn composed of essentially infinite length fibers; staple yarns are composed of short (several inches) fiber lengths cohesively intermingled. Most natural fibers (fiber of plant or animal origin) can be obtained only as staple (wool of infinite length would require the invention of a novel sheep); synthetic or man-made yarns may be converted to staple through a chopping process. The processes of forming synthetic polymers into filaments and of converting staple fibers into yarns are both termed spinning; care must be taken to insure clarity as to which is meant when a spinning process is discussed.

The detailed microstructure and hence the specific property set of any given fiber is determined by the conditions under which the filaments

TABLE I

FIBER UNITS AND CONVERSION FACTORS[a]

A. Linear Density (Yarn Count) Systems

Yarn number system	Decitex[b]
Cotton count (c.c.)	$\dfrac{5\ 905.41}{\text{c.c.}}$
Denier (d)	$d/0.9$
Grains/120 yards (gr/120 yd)	$\text{gr}/120\ \text{yd} \times 5.905\ 41$
Jute count (j.c.)	$\text{j.c} \times 344.482$
Linen lea (l.l.)	$\dfrac{16\ 535.2}{\text{l.l}}$
Metric count (m.c.)	$\dfrac{10\ 000.0}{\text{m.c.}}$
Woolen cut (w/c)	$\dfrac{16\ 535.2}{\text{w/c}}$
Woolen run (w.r.)	$\dfrac{3\ 100.34}{\text{w.r.}}$
Worsted count (w.c.)	$\dfrac{8\ 858.12}{\text{w.c.}}$

B. Fiber, Yarn, and Fabric Properties

Property to be measured	Current unit	Conversion factor[c]	SI units and other units and terms to be used with SI	Symbol
Linear density	various yarn numbering systems	see part A	decitex	dtex
			kilotex	ktex
Twist+	turns per inch	×3.9370 07	turns per decimeter	turns/dm
	turns per inch	×0.393 700 7	turns per centimeter	turns/cm
Twist factor	turns per inch ÷ cotton count	×95.67 340	(turns per meter) . tex	(turns/m)
				tex

(*continued*)

TABLE I (continued)

Property to be measured	Current unit	Conversion factor[c]	SI units and other units and terms to be used with SI	Symbol
Cut length	inches	×25.4	millimeters	mm
Breaking load	pounds-force	×4.448 222	newtons	N
	pounds-force	×444.822 2	centinewtons	cN
	grams-force	×9.806 65	millinewtons	mN
Tear strength	pounds-force	×4.448 222	newtons	N
Tenacity or yarn tension	grams-force per denier	×88.259 85	millinewtons per tex	mN/tex
	grams-force per denier	×8.825 985	centinewtons per tex or millinewtons per decitex	cN/tex mN/dtex
	grams-force per denier	×0.882 598 5	centinewtons per decitex	cN/dtex
	pond (p/dTex)	×0.794339	centinewton per decitex	cN/dtex
Work of rupture	grams-force centimeter	×0.009 806 650	newton centimeters	n-cm
Initial modulus	grams-force per denier	×88.259 85	millinewtons per tex	mN/tex
	grams-force per denier	×8.825 985	centinewtons per tex or millinewtons per decitex	cN/tex mN/dtex
Bending/twisting rigidity	grams-force square millimeter	×9.806 65	nanonewton square meters	nN-m²
Bursting pressure	pounds-force per square inch	×6.894 757	kilopascals	kPa
Woven fabric thread counts				
Warp	ends per inch	×0.393 700 7	ends per centimeter	ends/cm
Weft (filling)	picks per inch	×0.393 700 7	picks per centimeter	picks/cm
Knitted fabric Thread counts				
Wales	wales per inch	×0.393 700 7	wales per centimeter	wales/cm
Courses	courses per inch	×0.393 700 7	courses per centimeter	courses/cm

Property	U.S. unit	Metric unit	Conversion factor	SI symbol
Sewing				
Stitch and course length	stitches per inch	stitches per centimeter	×0.393 700 7	stitches/cm
	inches	millimeters	×25.4	mm
Mass per unit area	ounces per square yard	grams per square meter	×33.905 75	g/m²
Cover factor (woven fabrics)	threads per inch ÷ cotton count	threads per centimeter . tex	×9.567 340	(threads/cm) . tex
Tearing strength, grab strength, or seam strength	pounds-force	newtons	×4.448 222	N
Bursting pressure	pounds per square inch	kilopascals	×6.894 757	kPa
Bond peel Strength	grams-force per inch width	newtons per centimeter width	×0.003 860 858	N
Slippage resistance at seam	pounds-force	newtons	×4.448 222	N
Stitch density (knitting)	needles per 100 millimeters	needles per 100 millimeters		needles/100 mm

[a] References: (1) Antoine, V., "Guidance for Using the Metric System—SI Version." Society for Technical Communication, Washington, D.C., 1975. (2) "Guide to Metric Conversion for Textiles." Canadian Textile Institute, Montreal, Quebec, 1975. (3) Hopkins, R. A., "The International (SI) Metric System and How It Works." AMAJ Publ. Co., Tarzana, California, 1973. (4) "Metric Laws and Practices in International Trade." U.S. Department of Commerce, Washington, D.C., 1976. (5) O'Neill, P. J., "The Wiley Metric Guide." John Wiley & Sons Australasia Pty. Ltd., Sydney, Australia, 1976. (6) "SI Units and Recommendations for the Use of Their Multiples and of Certain Other Units, ISO 1000-1973 (E)." American National Standards Institute, New York, 1973. (7) "Standard for Metric Practice, E 380-76." American Society for Testing and Materials, Washington, D.C., 1976. (8) "Standard Recommended Practice for Designation of Yarn Construction, D 1244-69." American Society for Testing and Materials, Washington, D.C., 1975. (9) "Standard Recommended Practice for Use of the Tex System to Designate Linear Density of Fibers, Yarn Intermediates, Yarns, and Other Textile Materials, D 861-76." American Society for Testing and Materials, Washington, D.C., 1976. (10) "Standard Tables of Equivalent Numbers for Yarns Measured in Various Numbering Systems, and of Recommended Conversion Factors, D 2260-76." American Society for Testing and Materials, Washington, D.C., 1976. (11) "TRI Reports, The International System of Units." Textile Research Institute, Princeton, New Jersey, 1976.

[b] The conversion factors are based on the following relationships given in ASTM Standard for Metric Practice E 380-76: 1 yd = 0.9144 m, exactly, 1 lb (avoirdupois) = 0.453 592 37 kg, exactly, and 1 gr = 6.479 891 × 10⁻⁵ kg, exactly. The conversion factors containing fewer than six significant digits are exact values.

[c] All conversion factors with less than seven significant figures are exact.

are produced. For fibers of plant (cotton, linen) or animal (wool, silk) origin these conditions are determined by species and environment (see, for example, the treatment of fibers in the Encyclopedia Britannica, 1966, or Billmeyer, 1962) and are beyond the scope of this discussion. None of the natural fibers are thermoplastic, all are partially crystalline, and all are hydrophilic. For synthetic fibers the producer has control of the formation conditions and a brief description of the principal fiber-spinning and modifying processes is necessary.

The simplest fiber-forming process is the melt-spinning of thermoplastic polymers. The polymer melt is extruded through a plurality of orifices (spinneret), and the resulting fibers are solidified in the spinning column and wound up on a bobbin. Although the engineering variables of the process are many and complex, the principal parameters controlling the as-spun fiber structure and properties of a given polymer are the applied stress and the rate of cooling (see, for example, Ziabicki, 1976). Nonthermoplastic polymers are spun from solution (termed a dope): if the solvent is volatile and removed in the spinning column, the process is called dry spinning; if the fibers must be coagulated in a nonsolvent, the process is called wet spinning. The solution spinning processes possess all the complexities of melt-spinning with the additional complication of the mass transfer of the solvent out of the fiber (Ziabicki, 1976). These processes are shown diagrammatically in Fig. 1.

In most cases as-spun synthetic fibers are unoriented and possess little crystallinity or crystallinity markedly different from that of the final product. They must be further processed to develop end-use properties, especially modulus and tenacity. This is achieved by stretching the fibers (drawing) under conditions where significant molecular orientation is imparted. The principal variables are draw ratio (ratio of initial to final cross-sectional area), temperature, and strain rate. The polymer may be plasticized by residual solvent or an imposed environment, and liquid media may be used to improve heat transfer. Drawing processes may be in-line with spinning or may be separate steps (Ziabicki, 1976). They may involve a single or multiple drawing steps and may be coupled with twisting and/or intermingling steps. Drawn yarns are usually stabilized by an in-line annealing (crystallization) step under either constant length or relaxation (draw ratio < 1) conditions. The drawing process is shown schematically in Fig. 2.

For textile applications it is desirable to produce yarns that are bulky (crimped) to improve aesthetics, comfort, and/or insulating qualities and to allow bending with intrinsically bulky natural fibers. This is accomplished by texturing processes, which modify the fiber shape and structure such that the fiber will have a random, meandering configuration in

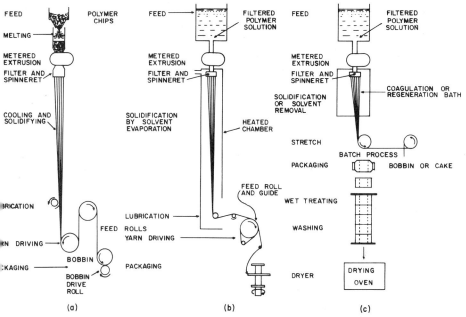

Fig. 1   Schematic diagram of the three principal methods of spinning fibers: (a) melt spinning, (b) dry spinning, (c) wet spinning. [Work of Riley (1956).]

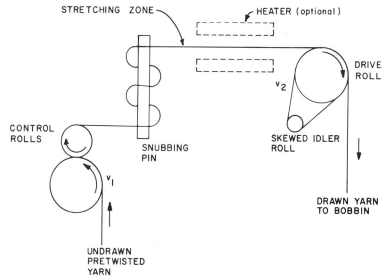

Fig. 2   Schematic diagram of fiber drawing: $V_1 < V_2$, drawing; $V_1 > V_2$, relaxation. [From Billmeyer (1962).]

the yarn, i.e., a shorter end-to-end distance than the "flat" or straight configuration imparted by spinning and drawing. This can be achieved by three basic methods: (a) an additional step to nonuniformly deform the yarn under conditions of temperature and stress such that a memory of nonlinear fiber configuration is retained (false-twist texturing, stuffer box, gear crimping processes, etc.—see Fig. 3), (b) spinning of multicomponent fibers, each component possessing different shrinkage characteristics causing bending of the fibers during relaxation, and (c) spinning or drawing yarns under controlled, nonuniform conditions, also to promote differential shrinkage. Method a is the most common and it should be noted that all of these processes tend to complicate and mask the intrinsic yarn structure.

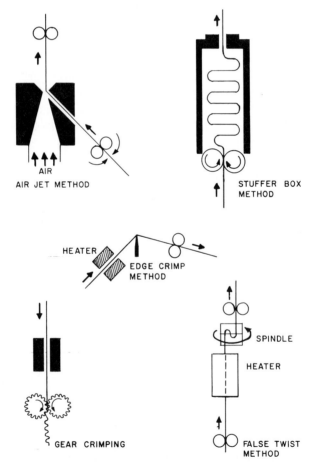

Fig. 3 Schematic diagrams of the principal methods of fiber texturing.

Common name[a]	Chemical name	Method of production	Nature of yarn	Major uses
Wool	Keratin	Animal Origin—sheep	Staple	Wearing apparel, carpets
Silk	Fibroin and sericin	Animal Origin—silkworm	Filament, staple	Wearing apparel
Cotton	Cellulose	Plant origin	Staple	Wearing apparel, home furnishings
Cellulose acetate (CA)	15–92% acetylated cellulose	Dry spun	Filament, tow, staple	Wearing apparel, cigarette filters
Cellulose triacetate (CTA)	>92% acetylated cellulose	Dry spun	Filament, tow, staple	Wearing apparel
Rayon	<15% acetylated cellulose	Wet spun	Filament, tow, staple	Wearing apparel, home furnishings, industrials
Acrylic (PAN)	Polyacrylonitrile ≥ 85% by weight[b]	Wet or dry spun	Filament, tow, staple	Wearing apparel, carpets
ModAcrylic (PAN)	Polyacrylonitrile 35–85% by weight[b]	Wet or dry spun	Filament, tow, staple	Deep pile fabrics, carpets, wigs
Nylon 6	Polycaprolactam	Melt spun	Filament, tow, staple	All fiber applications
Nylon 6,6	Poly(hexamethylene adipamide)	Melt spun	Filament, tow, staple	All fiber applications
Polyester (PET)	Poly(ethylene terephthalate)	Melt spun	Filament, tow, staple	All fiber applications
Polyolefin	Polypropylene	Melt spun	Filament, tow, staple	Carpet, home furnishings

| Common name[a] | Typical properties | | | Safe ironing |
	Tenacity (cN/dtex)	T_m (°C)	Moisture level	temperature (°C)
Wool	0.9–1.8	132 (degrades)	10–15	107
Silk	2.7–5.4	170 (degrades)	10–15	121
Cotton	2.7–5.4	240 (degrades)	6–10	182
Cellulose acetate (CA)	1.1–1.4	230	6	180
Cellulose triacetate (CTA)	1.1–1.3	288	3.2	250
Rayon	0.6–5.4	205–240 (degrades)	13	215
Acrylic (PAN)				
ModAcrylic (PAN)	1.8–3.2	250–320	1.3–2.5	148–176
Nylon 6	1.8–3.2	188–250	0.4–4	130–148
Nylon 6,6	2.2–8.6	212	4.5	171
Polyester (PET)	2.7–8.6	250	4–4.5	149
Polyolefin	2.2–8.6	252	0.4–0.8	180
	4.3–6.3	165	0	149

[a] For trade names, see, for example, American Association for Textile Technology (1977).

[b] Copolymerized with monomers such as vinyl acetate, vinyl chloride, styrene, etc.

[c] For additional information see Encyclopedia Britannica, 1966; Man-Made Fiber Producers Association, 1970, 1971; Textile Institute, 1965; Mark et al., 1967, 1968; Moncrieff, 1975; Morton and Hearle, 1975; Stout, 1970; American Association for Textile Technology, 1977.

At this stage, both natural and synthetic yarns undergo further processing, resulting ultimately in end-use articles such as fabrics, carpets, cords, threads, and ropes. Typically, the yarn is knitted, woven, or formed into a fabric, and the fabric is scoured, dyed, and then structurally and dimensionally stabilized by a heat-settting procedure. The order of these steps is a function of the particular yarn and the desired end product. A multiplicity of conditional variations exist for each type of process mentioned. For details the reader is referred to any of several excellent fiber science texts and monographs (Encyclopedia Britannica, 1966; Hearle and Peters, 1963; Moncrieff, 1975; Morton and Hearle, 1975; Stout, 1970; Ziabicki, 1976). What all of these processes have in common is the modification of the fiber microstructure through additional imposed mechanical and thermal history.

Table II lists the common names, method of spinning, typical properties, and end-use applications of the major fibers. It is significant to note the range of properties achievable in most fibers (especially the synthetics); this range is realized through control of the overall fiber formation process, which, in turn, specifies the fiber microstructure.

To describe fiber microstructure, a series of questions must be answered:

1. What is the degree of crystallinity?
2. What is the size and shape of the crystalline units?
3. What is the orientation of the crystalline units?
4. What is the orientation of the noncrystalline areas?
5. How are these microstructural units interconnected?
6. Is the structure homogeneous throughout the fiber?

The range of structures possible in polymers (as well as their response in thermal and analysis) has already been treated by Wunderlich in Chapter 2. for in-depth treatments of polymer structure the reader is urged to consult polymer science texts (Billmeyer, 1962; Geil, 1963; Van Krevlen, 1972; Wunderlich, 1973, 1976). Attempts at models of fiber structure may be found in the work of Becht and Fisher (1970), Hearle (1977), Hearle and Peters (1963), Lindenmeyer (1980), Peterlin (1972a,b, 1975), Prevorsek (1971), Prevorsek *et al.* (1973) Reimschuessel and Prevorsek (1976), Samuels (1974), Spruiell and White (1975a,b), Statton (1967, 1970), Statton *et al.* (1968), Valk *et al.* (1980), Wasiak and Ziabicki (1975), and Ziabicki (1976).

Figure 4 diagrammatically summarizes the major thrust of this brief introduction. In the ensuing sections of this chapter the utility of thermal analysis in providing the information necessary to fill in the various legs of the process–structure–property triangle of fiber-forming polymers will be discussed.

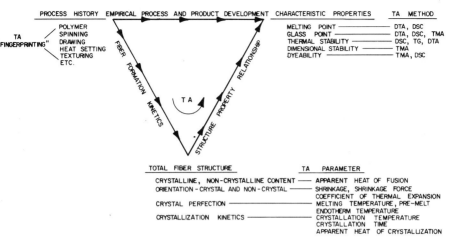

Fig. 4 The process–structure–property interrelation approach to fiber problem solving with examples of TA utility.

II. Characterization of Fibers by TA

A. General Considerations

The biggest problem presented by fibers to the thermal analyst is the fiber geometry. Sample weights of even 5 mg can represent extremely long lengths (e.g., 5 mg of a 100 dtex yarn is equal to 50 cm). Care must be taken in the compaction of these samples to avoid introducing structural changes through deformation and to insure thermal equilibrium during heating. Monofils, tows, and fabrics all present additional problems of sample fabrication, each of which requires individual solution. It is often prudent to use several different sample preparation techniques to allow the identification and elimination of artifacts before standardizing procedures.

When reporting TA data, as much sample information as possible should be included. Molecular weight, additive content, finishes, comonomer levels, and moisture level can all significantly alter a fiber's TA response. For meaningful date interpretation, some knowledge of the fiber processing history is vital (compare, for example, the DSC traces of as-spun and drawn and crystallized PET, Figs. 5, 14), and details of the sample origin can also be useful. The use of appropriate supporting techniques to verify interpretations cannot be overemphasized and a listing of some of the more useful may be found in Table III.

Over the past 10 to 15 years, a number of reviews of the applications of TA to fibers have been written. These vary from illustrations of typical fiber responses to more comprehensive manuals of technique and

TABLE III

SOME USEFUL FIBER STRUCTURE CHARACTERIZATION TECHNIQUES[a]

Technique	Information
*O*ptical *M*icroscopy	Size and shape of morphological units, filament uniformity, average chain orientation (birefringence), identification and location of dyes, additives, voids, etc., within fiber microstructure—size scale 10^{-4}–10^0 cm
*T*ransmission *E*lectron *M*icroscopy	Morphological details of fiber microstructure, location, and orientation of crystalline units—size scale 10^{-8}–10^{-4} cm
*S*canning *E*lectron *M*icroscopy	Morphological details of fiber microstructure, especially surfaces, location, and identification of additives and foreign inclusions—size scale 10^{-7}–10^0 cm
Wide-*A*ngle *X-R*ay *D*iffraction	Unit cell parameters, average crystallite size, average molecular chain orientation (crystalline)—size scale 10^{-8}–10^{-7} cm
*S*mall-*A*ngle *X-R*ay *D*iffraction	Average size, shape, and orientation of crystallites, average void size and orientation—size scale 10^{-7}–10^{-5} cm
Light scattering	Average size, shape, and orientation of morphological units—size scale 10^{-6}–10^{-4} cm
Tensile testing	Fiber mechanical properties (modulus, tenacity, elongation, etc.), indirect information about orientation, tie molecules, fiber uniformity
Dynamic mechanical testing	Dynamic mechanical properties (G',G''—real and imaginary modulus components, tan δ), phase transition temperatures and kinetics
Density, dilatometry	Transition temperatures, phase transition kinetics, % crystallinity—% disorder
*I*nfra-*R*ed Spectroscopy	Specific molecular configuration, molecular orientation, crystalline and noncrystalline amounts, tie molecule concentration and configuration, identification of dyes additives and finishes

[a] Italic letters indicate commonly used abbreviations.

interpretation (Becht and Fisher, 1970; Berndt and Heidemann, 1972; Buchanan and Hardegree, 1977; Cassel, 1976; Csete and Levi, 1976; Heidemann and Berndt, 1976; Textile. Institute, 1965; Levy, 1970; McAdie, 1969; Porter and Johnson, 1968, 1970; Schwenker, 1967; Schwenker and Chattergee, 1972; Seves and Vicini, 1974; Song, 1977; Valk, 1972; Valk *et al.*, 1971; Zimmermann and Budisch, 1969). Particularly valuable is the review "Thermal Behavior of Textiles" by Slater (1976), which summarizes the literature through 1974 (1436 references). All serve to increase the on-hand TA response file so important to the user of thermal analysis.

In the ensuing sections the application of various TA techniques to fibers is dealt with in detail. The examples used illustrate the type of behavior and responses commonly observed in fiber evaluation, independent of the particular chemical composition of the fiber employed. Although the importance of chemical composition should not be underestimated and will certainly cause shifts in specific parameters such as solvent sensitivity or the location of the glass transition or melting point on the temperature axis, the interpretation of the given TA responses shown in this section is generally applicable to all fibers. In short, the macromolecular nature of fibers more than their exact chemistry determines most fiber responses. Those effects and parameters peculiar to a given fiber will be discussed in Section III.

B. DTA/DSC—EXPERIMENTAL ASPECTS

The applications of DTA/DSC to fibers fall into three broad categories: (1) The monitoring of physical transitions such as the glass transition, melting, and crystallization; (2) the monitoring of chemical reactions such as decomposition, and (3) kinetic studies of either physical or chemical phenomena. In DTA only the temperatures associated with these events may be accurately evaluated; DSC allows evaluation of both the temperature and the energetics of the processes investigated (see Chapter 2). As with any other application of DTA or DSC, heating rate, sample size, and atmosphere have major effects on the resulting thermograms. Care must be taken to record these parameters and to eliminate artifacts due to experimental technique (see Chapter 1). In our own laboratories we have standardized on a 20°C/min heating rate for both DSC and DTA and 5–10 mg sample weights for DSC.

The preparation of fiber samples for reproducible results presents special problems, because of the long lengths involved and the tendency of most fibers to shrink on heating. Any deformation of the fiber during sample preparation will change its microstructure and hence the observed TA response. Miller (1971) has suggested a new DTA/DSC instrument design where the sample is wound on the outside of a grooved aluminum cylinder. In conventional instrumentation the choice is between cutting the fiber into short lengths or producing a small coil by carefully winding the necessary fiber length around a form (such as tweezer tips). In our laboratories we have found the latter to be the more satisfactory (faster, easier to control sample weight). In DSC instruments using aluminum pans the sample must be packed such that the entire pan area is filled and the pan bottom remains flat for good thermal contact. Pan lids should always be crimped to prevent spurious effects due to sample movement during heating. If a special atmosphere is employed, holes may be punched in the lid,

or lids may be punched from aluminum or gold screening. Several researchers have noted differences in the melting response of samples held
constrained (constant length) when compared to unconstrained (free-to-
shrink) samples (Mead and Porter, 1976; Miyagi and Wunderlich, 1972;
Samuels, 1975; Todoki and Kawaguchi, 1977b; Tashiro *et al.*, 1980). Con-
straining the sample retards the relaxation of previously imposed molecular orientation and tends to slow crystallization and crystal-perfecting
processes. In comparative studies of fibers produced under varying process conditions it is important that the degree of sample constraint be held
constant to observe accurate parameter shifts. Constraint is accomplished
either by using restraining devices or by fixing the fiber ends. Such effects
are most prominent in highly oriented samples and are consistent with the
mechanistic differences noted by Statton *et al.* (1970) in the annealing of
drawn fibers under constant length and free-to-shrink conditions.

Figure 5a shows a typical fiber thermogram, that of an as-spun PET
yarn heated at 20°C/min. Proceeding from low temperatures, we note the
glass transition, the exotherm associated with "cold crystallization"
(crystallization on heating from the glass), and the melting endotherm. On
cooling at 20°C/min (Fig. 5b) crystallization from the melt is observed and
on reheating (Fig. 5c) cold crystallization is absent. In the second heating,
the sample is allowed to decompose (high-temperature endothermic
peak). Useful parameters to be obtained from these traces are the glass
transition temperature, (T_g), the peak temperatures of melting and crystallization $(T_m, T_{x\text{cold}}, T_{x\text{melt}})$, the enthalpies of melting and crystallization
$(\Delta H_m, \Delta H_{x\text{cold}}, \Delta H_{x\text{melt}})$, and the peak temperature and enthalpy of decomposition $(T_D, \Delta H_D)$. For details of parameter calculation and needed
constants, see Chapter 2. The accuracy and reproducibility of such results
is, of course, a function of both machine technique and sample uniformity; with reasonable care temperature values to ± 0.5°C and enthalpy
values of $\pm 5\%$ are achievable. Utilizing such parameters, Philip (1972)
has developed a simple scheme for the identification of synthetic fibers,
based on small (0.3–3.0 mg) sample sizes. The method uses the first heating of the sample to erase processing history and depends on the second
heating parameters for sample identification.

The uniform processing of fibers and the achievement of significant
draw ratios usually requires the precursor yarn to be heated above its
glass transition temperature T_g. T_g also marks the temperature at which
sufficient mobility is imparted to the molecular chains for processes such
as thermal shrinkage and crystallization to occur. It has recently been observed, in bulk and film samples, that the age, orientation, and thermal
history of the glass significantly affect T_g, as well as the conditions necessary for stable processing (Mininni *et al.*, 1973). Siegmann and Turi

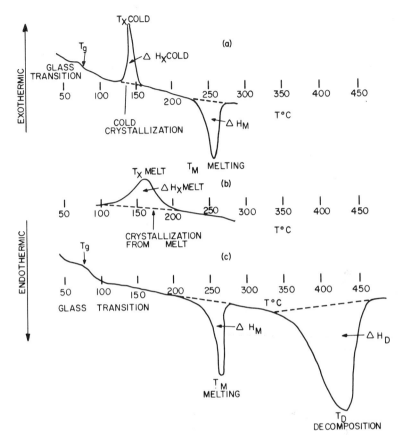

Fig. 5 Typical DSC thermograms of as-spun poly(ethylene terephthalate) fiber: (a) first heating, (b) first cooling, (c) second heating. 0.9IV PET, low-speed spinning conditions. Perkin-Elmer DSC-II, 8 mg sample weight, coiled sample, N_2 atmosphere, 2 min hold at 280°C between heating and cooling, 20°C/min heating and cooling rate. (Courtesy of B. Morris, Celanese Research Company.)

(1974), and Hagege (1977), have confirmed that similar phenomena occur in as-spun PET yarns as illustrated by the DSC traces of *P*artially *O*riented *Y*arn (spin-oriented PET yarn) shown in Fig. 6 (data of Hagege, 1977).

Perhaps no use of DTA/DSC techniques gives more information about fiber structure than melting experiments. The complexity of the thermal and stress history under which fibers are produced is directly reflected in the complex and varied melt phenomena observed. In addition, the melting point sets the ultimate use temperature of a fiber, helps to define annealing conditions for increasing fiber stability (which in turn will be

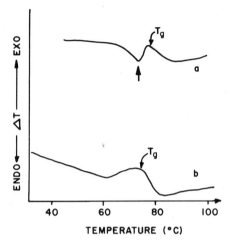

Fig. 6 DSC thermograms of the glass transition of partially oriented PET yarns (POY) as a function of yarn age and processing. Conditions: (a) sample wound up at 3300 m/min and stored 3 months at 20°C, 65% RH; (b) sample wound up at 1100 m/min and stored 3 days at 20°C, 65% RH. DuPont 900 DSC, 5 mg sample, 20°C/min heating rate. [From Hagege (1977).]

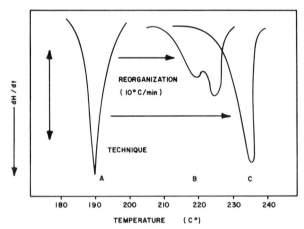

Fig. 7 DSC thermograms of the melting of Nylon 6 yarn as a function of sample preparation: (a) sample irradiated with x rays and cut into 3 mm lengths; (b) sample cut into 3 mm lengths; (c) thin wire attached to each end of sample, sample wound tightly around an aluminum plate (6.5 × 2.5 × 0.4 mm) and the wires tied to prevent sample shrinkage. Nylon 6 yarn drawn 5.0 using cold pin/hot plate process. Perkin-Elmer DSC-1B, 10°C/min heating rate. [From Todoki and Kawaguchi (1977b).]

reflected in melting-point shifts) and is certainly a necessary parameter for determining how a given polymer is to be melt-spun.

Wunderlich has shown (Chapter 2) that metastable polymer crystals will tend to perfect themselves on heating or annealing. The nature of this perfecting process will be a function of both the starting structure and the experimental conditions employed, and care must be taken to keep experimental conditions constant to prevent confusing and irreproducible results. Todoki and Kawaguchi (1977a,b) have shown that the origin of double-melting endotherms in drawn Nylon 6 is partial melting and recrystallization and that the phenomenon can be eliminated if the sample is restrained during heating or irradiated prior to heating. Figure 7 compares the melting of drawn Nylon 6 samples as a function of sample preparation technique. The behavior observed with any of these techniques is a valid reflection of the fiber microstructure and hence its processing history. Multiple melting peaks are commonly observed in fiber thermograms, and the application of several experimental techniques serves to clarify the particular multiple melting phenomena under investigation (Berndt and Heidemann, 1972; Bhatt and Bell, 1976; Heidemann and Berndt, 1976; Hybart and Platt, 1967; Jaffe, 1978; Lengyel *et al.*, 1975; Prati and Seves,

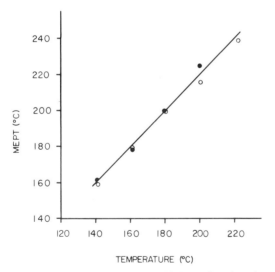

Fig. 8 Middle endotherm peak temperature (MEPT) as a function of annealing temperature (zero load, silicon oil) at various exposure times: (○) 20 min; (●) 40 min, poly(ethylene terephthalate) yarn drawn 4:1 over a 80°C heated pin and a 180°C heated block. Data generated with a DuPont 990DSC, 15 mg sample, inert atmosphere, 20°C/min heating rate. [From Oswald *et al.* (1977).]

1973; Prokopchuk *et al.*, 1976; Sweet and Bell, 1972; Todoki and Kawaguchi, 1975, 1977a,b; Valk, 1972; Wiesner, 1968).

Small endotherms at temperatures lower than the main melting event have been observed in fiber thermograms by many authors (Berndt and Bossman, 1976; Jaffe, 1978; Oswald *et al.*, 1977). In work performed on drawn and crystallized PET fibers, Valk *et al.* (1971), Valk (1972), Oswald *et al.* (1977), and Wiesner (1974, 1976a,b) ascribe these peaks to the melting of small crystallites formed in the amorphous regions of the fiber microstructure during annealing, an interpretation of the mechanism of premelt endotherm formation that is valid for all fibers. Figure 8 shows how the location of this premelting endotherm in PET varies with annealing temperature. In an extensive study of PET textile yarns, Berndt and coworkers (1973; Berndt and Bossman, 1976) relate the position of the premelt endotherm to the time, temperature, tension, and media seen by the yarns in aftertreatments such as dyeing and heat setting. Defining the premelting peak temperature as T_{eff}, they show the peak to reflect yarn structural features related to end-use properties such thermal stability and dyeing behavior. If more than one such treatment is performed on the yarn, two T_{eff}s can be observed, if the second treatment is at a lower (effective) temperature than the first. Figure 9a and b shows that treatment times of less than 1 sec (such as experienced, for example, in texturing processes) are sufficient for the structures melting in premelt endotherms to form. Treatment in aqueous media (Heidemann *et al.*, 1977) or solvents (Berndt *et al.*, 1973) shifts the position of T_{eff} to higher temperatures, all other conditions being equal. The utility of T_{eff} as a process-monitoring parameter and as a diagnostic tool for problems of yarn nonuniformity is self-evident.

The effect of heating rate and restraint on the melting of PET tire cord was investigated by Miyagi and Wunderlich (1972) who, as shown in Fig. 10, found the behavior to vary from reorganization (T_m drop with increasing heat rate) to superheating (T_m rise with increasing heating rate) as a function of sample preparation conditions. Berndt and Bossman (1976) describe the heating rate effects associated with the premelt endotherm in PET textile yarns. Rate-dependent melting of unrestrained as-spun polypropylene yarns was investigated by Jaffe (1978), who found the nature of the reorganizations observed to strongly reflect spinning conditions. The morphological information that can be obtained from rate-dependent melting studies has been described by Wunderlich in Chapter 2; the possibility of interpretative errors involved in neglecting heating-rate effects cannot be overemphasized.

It has been shown that fast-heating techniques can separate melting and decomposition phenomena in fibers composed of polymers such as

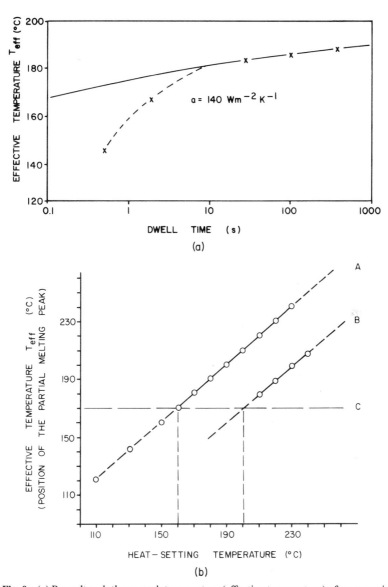

Fig. 9 (a) Premelt endotherm peak temperature (effective temperature) of commercially drawn poly(ethylene terephthalate) fiber, as a function of heat setting time at 180°C and zero tension data generated with a DuPont 900 DSC, 30°C/min heating rate. [From H. J. Berndt and A. Bossman (1976), *Polymer* **17,** 241. By permission of the publishers, IPC Business Press Ltd. ©]. (b) Influence of the heat-setting temperature, time of treatment, and high-temperature dyeing on the premelt endotherm peak temperature (effective temperature) of commercially drawn poly(ethylene terephthalate) fibers: A, heat set 20 sec in hot air, B, heat set 0.2 sec in hot air, C, high-temperature dyeing at 130°C, 90 min. Conditions as in (a). [From H. J. Berndt and A. Bossmann (1976), *Polymer* **17,** 241. By permission of the publishers, IPC Business Press Ltd. ©]

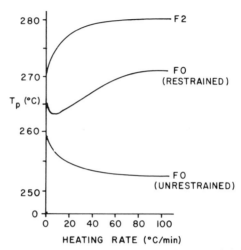

Fig. 10 DTA melting peak temperature of poly(ethylene terephthalate) tire yarn as a function of heating rate and sample preparation technique: FO unrestrained—sample place in capillary tube after cutting to short lengths; FO restrained—sample tightly wound around thermocouple and the ends tied to prevent shrinkage; F2—sample FO annealed at 250°C for 24 hour in vacuo experiment performed unrestrained. DuPont DTA 1–2 mg sample weight, heat rate varied. [From Miyagi and Wunderlich (1972).]

PAN (Dunn and Ennis, 1971; Hinrichsen, 1971; Layden, 1971) and poly(p-phenyleneterephthalamide) (Brown and Ennis, 1977), where the temperatures associated with these processes are similar. Traces of PAN melting as a function of heating rate are shown in Fig. 11. The recognition of this type of irreversible melting (melting occurring simultaneously with a chemical transformation) is invaluable for deducing microstructural models of semicrystalline fibers that yield little morphological information with conventional heating techniques.

Crystallinity determinations of fibrous samples by DSC can be carried out with standard techniques (see Chapter 2), although the structural changes usually observed in fibers during heating requires the careful separation of crystallinity induced during heating from crystallinity of the starting sample. This can be accomplished by studying peak size as a function of heating rate and using ancillary crystallinity measurement techniques such as density and x-ray diffraction.

Figure 12 shows the change in size and location of the cold crystallization exothermic peak in as-spun PET yarns as a function of spinning conditions. Increased spinning speeds lead to increased molecular orientation in the spun yarn prior to solidification, which, in turn, causes increased crystallization rates and increased crystallinity development during the spinning process (Smith and Steward, 1974; data of Heuvel and Huisman, 1978). The shift in $T_{x\text{cold}}$ to lower temperatures is essentially linear with

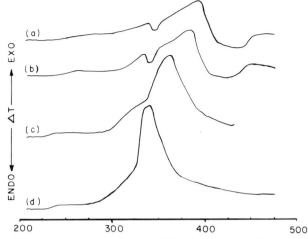

Fig. 11 Effect of heating rate on the DSC thermograms of polyacrylonitrile: (a) 100°C/min heating rate, (b) 80°C/min heating rate, (c) 40°C/min heating rate (2×), (d) 20°C/min heating rate (4×). Commercial filament yarn sample, finish removed, DuPont 900 DSC, sample cut into short lengths. [From Dunn and Ennis (1970).]

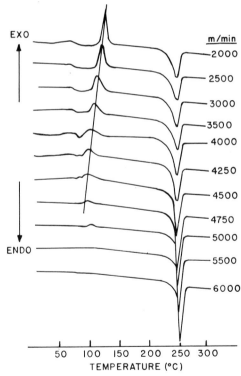

Fig. 12 DSC thermograms of as-spun poly(ethylene terephthalate) yarns as a function of yarn wind-up speed. Yarns spun to constant count of dTex 167/30. Spin temperature constant at 290°C, 250 μm hole size. DuPont 990 DSC, 20°C/min Heating Rate. [From Heuvel and Huisman (1978).]

increasing orientation and can be used as a measure of spun-yarn orientation by correlating the peak temperature with an orientation parameter such as birefringence or sonic modulus. To determine the crystallinity of such samples by DSC, the enthalpy of cold crystallization must be separated and subtracted from the enthalpy of melting.

To understand better the effect of stress on the crystallization of fibers during melt-spinning, Jaffe (1978) has investigated the recrystallization of as-spun polypropylene fibers produced under different levels of melt stress by varying the melt temperature or melt time in the DSC and keeping the crystallization temperature fixed. The results indicated that the higher bulk crystallization rates observed in high-stress-spun fibers was a consequence of a higher concentration of crystal nucleating species (produced in the original spinning). The broad range of melt treatment conditions employed was never sufficient to reduce the bulk crystallization rate of high-stress-spun fibers down to the levels of samples that never saw a stress crystallization history (see Fig. 13). This and other work emphasizes the importance of consistent melt history in DSC crystallization experiments, because of the persistent memory of process history in fiber samples. Turi and Sibilia (1978) have found that the spinning performance (based on spinning break rate) of PET polymer can be predicted from the bulk crystallization kinetics of the starting of PET chip. Poor performance is associated with fast crystallization rates as manifested by $T_{x\text{melt}}$ values obtained at a 10°C/min cooling rate from the melt ($T_{x\text{melt}}$ "poor" 15–20°C > $T_{x\text{melt}}$ "good" for 0.94 IV polymer).

The chemical stability and the effects of atmosphere and additives on

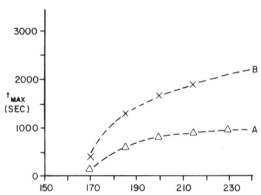

Fig. 13 Time to reach the maximum crystallization rate (t_{max}) as a function of melt holding temperature for variously spun polypropylene yarns: A, sample spun under high stress (2.5×10^{-2} g/d); B, sample spun under low stress ($\sim 2.5 \times 10^{-3}$ g/d). Perkin-Elmer DSC-IB, crystallization isothermal at 130°C, melt hold time constant at 5 min. See reference for details. [From Jaffe (1978).]

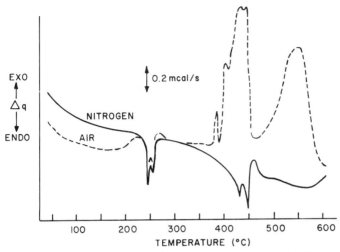

Fig. 14 DSC thermograms of flame-retarded poly(ethylene terephthalate) yarns as a function of atmosphere: (solid line) N2, (dashed line) air. DuPont 990 DSC, 15 mg sample weight, 10°C/min Heating Rate. [From Hassel (1977).]

the burning and decomposition of fibers is also amenable to study by DTA/DSC techniques. Figure 14 illustrates the markedly different response of flame-retarded PET to heating in air and nitrogen [data of Hassel (1977)]. Basch and Lewin (1973a,b, 1974), and Rodrig *et al.* (1975) have studied the effects of physical structure and various chemical agents (Basch and Lewin, 1975) on the decomposition of cellulosics. The range of changes in behavior that can be observed is shown in Fig. 15. Dimov *et al.* (1973) have shown that a correlation exists between the size of the thermooxidative exothermic peak of Nylon 6 yarns (which reflects the extent of molecular-weight decay) and the mechanical strength retention of these yarns after thermal treatment. These types of experiments serve not only to define conditions for the safe usage of fibers, but also to provide a convenient means of screening the effectiveness of various additives for improving ultimate performance.

This summary of DTA/DSC applications to fibers is not meant to be all-inclusive; its purpose is only to illustrate the breadth of applicability and the kinds of information that can be obtained. Additional examples of specific fiber behavior can be found in Section III of this chapter.

C. TMA—EXPERIMENTAL ASPECTS

An operational definition of the thermal mechanical analysis of fibers is the measurement of dimensional changes and/or the measurement of the forces generated during dimensional changes in fibrous systems as a

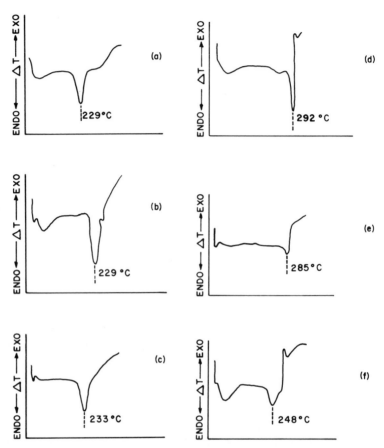

Fig. 15 The influence of flame-retardant systems and fine structure on DSC thermograms of cellulosic fabrics: (a) Cotton sulfated with sulfuric acid, (b) rayon sulfated with ammonium sulfamate, (c) ramie impregnated with ammonium bisulfate, (d) cotton phosphorylated with urea-phosphoric acid, (e) ramie impregnated with diammonium phosphate, (f) rayon phosphylated with urea-phosphoric acid. DuPont 900 DSC, 10°C/min heating rate. [From Basch and Lewin (1975).]

function of temperature. Four TMA instruments are currently commercially available: the Textechno Thermofil designed to measure dimensional change and shrinkage force of fibers (design based on the Textilforschung Krefeld instrument), the Kanebo Thermal Stress Tester Type KE-2 designed to measure the shrinkage force of fibers, the Perkin-Elmer Model TMS-2 Thermomechanical System designed to measure dimensional changes of materials in any geometry (specially designed fiber appliance), and the DuPont 943 Thermomechanical Analyzer designed to measure dimensional changes and shrinkage forces of materials in any geometry (specially designed fiber appliance). The preceding definition is termed operational because the older techniques of creep, stress relaxation, temperature-dependent tensile testing, and dynamic mechanical testing, although all feasible with existing TMA equipment (Berndt, 1975; Gill *et al.*, 1975) and all certainly thermal mechanical methods applicable to fibers, will not be included in this discussion. A method for calculating classical creep parameters from TMA data has recently been proposed by Forgács (1978). In addition, measurements performed continuously on bobbins of yarn rather than on relatively short discrete lengths will be mentioned only briefly, as will length change measurements where the environment causing the length change and the length change monitoring device are not integral (i.e., oven shrinkage of fibers as monitored with a cathetometer). For discussion of these ancillary thermal mechanical techniques the reader is referred to polymer physics (Bueche, 1962; Nielson, 1962; Rodriguez, 1970; Samuels, 1974; Schultz, 1974; Ward, 1971, 1975) and fiber science (Hearle, 1977; Mark *et al.*, 1967; Moncrieff, 1975; Morton and Hearle, 1975) texts.

The principal applications of TMA to fibers are

(1) the "fingerprinting" of samples (TMA spectra are more detailed than DTA/DSC spectra hence TMA is the preferred technique if only one method is used);

(2) the direct measurement of end-use properties such as the coefficient of linear thermal expansion (CTE), thermal shrinkage S_T, and the shrinkage force (SF_T—usually reported in units of stress, i.e., cN/dtex, etc.);

(3) the monitoring of physical transition temperatures such as T_g and T_m; and

(4) the investigation of the kinetics of shrinkage and shrinkage force phenomena.

The key variables in TMA experimentation are time/temperature, atmosphere, and applied load. As with any TA technique, the observed response is intimately related to the sample history, and, conversely, once an understanding of the response is achieved, sample history can be accurately deduced. As pointed out by Valk (1972), this is especially true for TMA since the parameters monitored are the same (temperature, length change, stress) as the major variables of fiber processing. A similar conclusion was reached by Addyman and Ogilvie (1979) in their recent review of the utility of TMA in the solution of complex problems in fiber processing.

Experimentally, the largest problem encountered in TMA is to obtain accurately and reproducibly the desired level of fiber loading, especially zero load. If load effects are suspected, experiments may be run as a function of increasing decitex (decreasing applied stress) and a convenient decitex yielding maximum shrinkage values chosen as standard (this decitex value will change with samples of different modulus). For most samples we have found $\simeq 100 - 250$ decitex to be a safe range. In instruments using crimping devices to mount samples, gauge length reproducibility is facilitated by constructing simple jigs to hold the devices (split balls, rods, etc.) at fixed length during sample preparation. In our own laboratories we have standardized on 0.5 in. gauge length $\simeq 200$ yarn dtex and 10°C/min heating rate. Care to eliminate moisture should be exercised for measurements in the neighborhood of 0°C.

For most materials the coefficient of linear thermal expansion is a positive (on heating), reversible constant for a given physical state. Oriented, polymeric materials, although always possessing a positive coefficient of volume expansion, CVE, are an exception to this generalization for two reasons: (1) the inherent anisotropy of bonding along and between chains causes the CTE in the chain direction of perfectly ordered polymeric molecules to be negative, and (2) the metastability of most oriented polymeric structures results in a strong driving force toward disorientation above T_g; hence the CTE above T_g is irreversible until structural equilibrium (or at least steady state) is achieved. The behavior described in (2) is shrinkage and will be discussed separately in detail; the effect described in (1) results in the CTE measurement yielding not only an important end-use property, but a simple monitor of fiber molecular orientation as well. The maximum value of CTE, indicative of complete disorientation (isotropic fiber), is equal to one-third the coefficient of volume expansion ($\frac{1}{3}$CVE); the minimum value, indicative of perfect orientation, is equal to the CTE of the molecular chain axis of the crystallographic unit cell. CVE data may be found in most handbooks (see, for example, Brandrup and Immergut, 1966) and CTE values for many polymer crystals have been collected (Wakelin *et al.*, 1960); hence the limits of the CTE–orientation correlation are readily available. Intermediate points can be filled in through use of orientation monitoring techniques such as birefringence, sonic modulus, or wide-angle x-ray diffraction. As CTE depends much more strongly on chain orientation than crystallinity (Jaffe, 1977), the former methods are preferred for most synthetic fibers. A plot of CTE versus birefringence for PET is shown in Fig. 16. Kimmel (1971) shows CTE decreasing as a function of draw ratio in Fig. 17 for drawn acrylic yarns. A correlation of CTE with mechanical properties for these same yarns is shown in Fig. 18. Porter *et al.* (1975) have measured negative values of

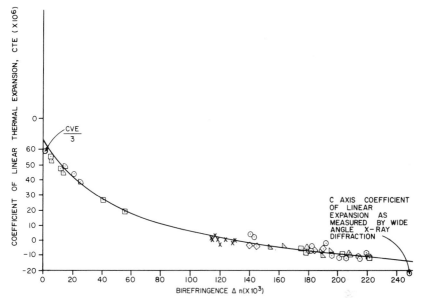

Fig. 16 The relationship of the coefficient of linear thermal expansion (fiber axis direction) to the birefringence of various processed poly(ethylene terephthalate) yarns. IV range, 0.67–0.95. Crystallinity range 0–62%. DuPont 900 TMA, −180 to −20°C measurement range 0.5 in gauge length, 5 g stabilizing load. 10°C/min heating rate. [From Jaffe, unpublished.]

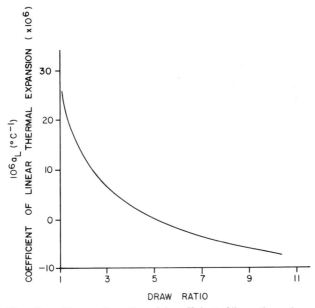

Fig. 17 The effect of draw ratio on the axial coefficient of linear thermal expansion for a typical acrylic yarn. Measurement conditions as in Fig. 16. [From Kimmel (1971).]

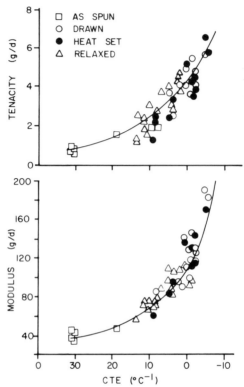

Fig. 18 The correlation of the axial coefficient of linear thermal expansion to the mechanical properties of typical acrylic yarns of various process histories. Measurement conditions as in Fig. 16. [From Kimmel (1971).]

CTE in ultraoriented polyethylene fibers. Porter also points out that the ability to control CTE in polymeric fibers allows the matching of expansion properties between fiber and matrix in composite applications, thus reducing the probability of failures due to thermal stress generation at fiber–matrix interfaces.

Few fiber properties are as important as the thermal shrinkage in determining the end-use potential of a given fiber (safe ironing temperature of textiles, performance of reinforcing cords in running tires, etc.) and few properties reflect the process history or the structural state of the fiber as completely. Figure 19 shows diagrammatically a typical shrinkage thermogram at zero load for a drawn partially crystalline fiber. Also included in Fig. 19 is a trace of the first derivative of the shrinkage dS/dT. Starting at room temperature, four regions of length change behavior are noted: (1) From room temperature up to T_g the sample undergoes reversible thermal expansion, although small amounts of irreversible shrinkage may occur if

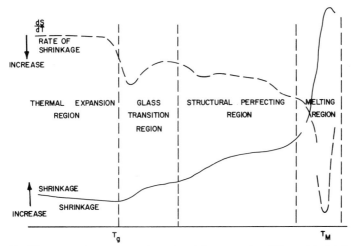

Fig. 19 Diagrammatic representation of a shrinkage and a differential shrinkage (DS) TMA thermogram of a drawn, partially crystalline fiber showing the four length change regions.

included solvents, or moisture, are expelled from the fiber structure. The value of dS/dT is constant in this region and if the apparatus is properly calibrated the value of dS/dT yields CTE directly. (2) At T_g the fiber undergoes a rapid irreversible shrinkage process due to the relaxation of oriented amorphous chains not bound in crystalline regions (providing a simple method of measuring T_g). The corresponding peak observed in the dS/dT trace corresponds to the temperature of the maximum rate of this relaxation. (3) In this intermediate temperature region between T_g and T_m shrinkage takes place due to reorganization, chain folding, recrystallization, and general perfecting of the fiber structure. In addition, molecular chain relaxation processes (i.e., entropy shrinkage) are continuing as in region 2. This shrinkage is most highly dependent on process history, and parameters equivalent to the T_{eff} described earlier can be obtained from the dS/dT trace. (4) The rapid shrinkage prior to sample failure is a consequence of melting; tie molecules are pulling out of crystalline units and disorienting, as are the crystalline units themselves. The start of this rapid shrinkage is about equivalent to the start of melting in DTA/DSC, the dS/dT peak reflects that temperature at which the sample is sufficiently molten so as not to be able to support its own weight. The high-temperature dS/dT peak is a useful measure of T_m, but may be confounded if chemical changes are also taking place. This mechanistic interpretation of the fiber shrinkage is shown diagrammatically in Fig. 20. Not all of the shrinkage regions described will be evident with every sample; each

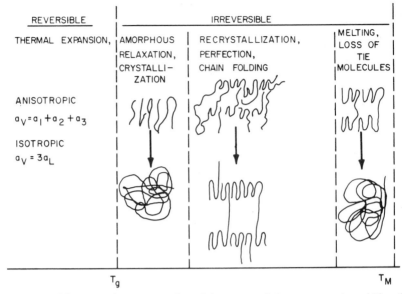

REVERSIBLE | IRREVERSIBLE

THERMAL EXPANSION, | AMORPHOUS RELAXATION, CRYSTALLI-ZATION | RECRYSTALLIZATION, PERFECTION, CHAIN FOLDING | MELTING, LOSS OF TIE MOLECULES

ANISOTROPIC
$a_V = a_1 + a_2 + a_3$

ISOTROPIC
$a_V = 3a_L$

T_g T_M

Fig. 20 Diagrammatic representation of the structural changes occurring within a fiber in the four length change regions.

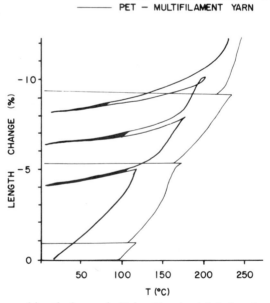

——— NYLON 66 MULTIFILAMENT YARN
——— PET — MULTIFILAMENT YARN

Fig. 21 Observed length changes in Nylon 66 and poly(ethylene terephthalate) textile filament yarns during cycled heating and cooling. German Textile Research Center designed TMA 0.2×10^{-3} N/tex preload, 30°C/min heating and cooling rate. [From Berndt and Heidmann (1977).]

region reflects an aspect of the fiber process history and structure, which can differ widely from sample to sample. Varying views on the nature of the shrinkage process and several attempts to describe shrinkge kinetics mathematically can be found in the published literature (Bhatt and Bell, 1976; Bosley, 1967; Pokrovskaya and Uterskii, 1972; Prevorsek et al., 1974; Ribnick, 1969a,b; Samuels, 1974; Shishoo and Bergh, 1977; Wilson, 1974).

Berndt and Heidemann (1977) have studied the reversibility of Nylon 66 and PET yarn shrinkage and find, as shown in Fig. 21, that after heating to a temperature $T > T_g$, shrinkage up to that temperature is eliminated;

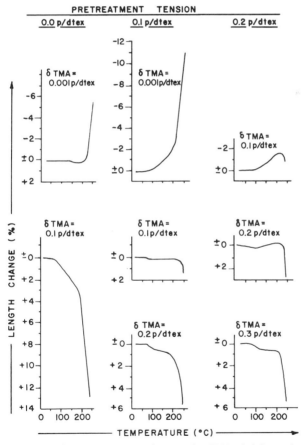

Fig. 22 The effect of pretreatment tension on the TMA shrinkage thermograms of poly(ethylene terephthalate) textile filament yarn as a function of the applied stress. Pretreatment tension, 0, 0.1, 0.2 P/dtex. Pretreatment temperature 200°C. DuPont 941 TMA, 2 cm gauge length, 15°C/min heating rate. [From Berndt and Heidemann (1972).]

on reheating the cooled sample, the CTE of the relaxed polymer is observed up to the temperature *T*, above which shrinkage recommences. Hence only length changes associated with inherent thermal expansion are reversible; all shrinkage processes are irreversible. It is possible, however, to perform the heating–cooling experiment at rates faster than the disorientation rate of the fiber structure, resulting in some shrinkage (or shrinkage forces) still being observed during a second heating (Berndt and Heidemann, 1974).

The interaction of stress applied during the TMA shrinkage experiment and the stress previously seen by yarns in processing is shown in Fig. 22 (Berndt and Heidemann, 1972). Figure 23a and b show the effect of pretreatment temperature and stress on the essentially zero load shrinkage and differential shrinkage response of PET yarns (Berndt and

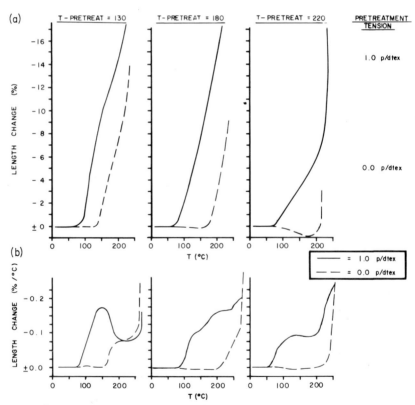

Fig. 23 (a)The effect of pretreatment tension and temperature on the zero load (0.001 P/dtex) TMA shrinkage thermograms of poly(ethylene terephthalate) textile filament yarns. Pretreatment tensions 0 (solid line), 1.0 (dotted line) P/dtex. Pretreatment temperatures 130, 180, 220°C. Conditions as in Fig. 22. [From Berndt and Heidemann (1972).] (b) The derivative TMA thermograms of the experiment described in (a).

Heidemann, 1972). These results illustrate the "fingerprinting" effectiveness of the TMA experiment. Heidemann and Berndt (1976) define the stress applied in the S or dS/dT experiment that prohibits shrinkage up to T_{eff} (see section on DTA/DSC in this chapter) as the effective stress (σ_{eff})

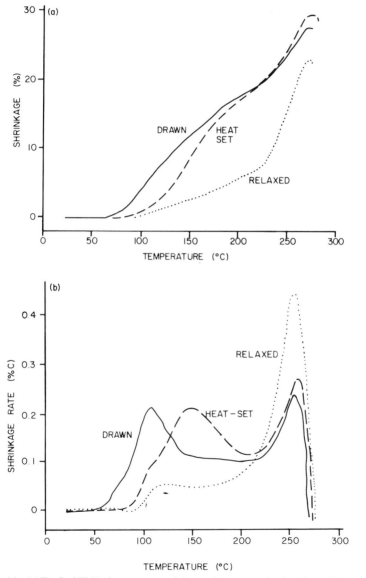

Fig. 24   (a) Typical TMA thermograms of drawn, heat set, and relaxed acrylic yarn. Conditions as in Fig. 16 without stabilizing load. (b) Typical derivative (dS/dT) thermograms of the acrylic yarns shown in (b). [From Kimmel (1971).]

seen by that fiber in processing. The effective stress σ_{eff}, which may be directly measured in SF experiments, will also be discussed in the next section.

Kimmel (1971) has extensively studied the shrinkage behavior of acrylic yarns, finding the shape and magnitude of the shrinkage thermograms strongly to reflect processing conditions, as shown in Fig. 24a and b. In Fig. 25 Kimmel correlates fast heating DSC endotherm data for

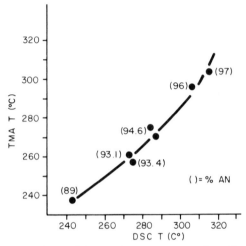

Fig. 25 The correspondence of the melting temperature of acrylic copolymer yarns as determined by TMA and DSC. TMA temperature ≡ temperature of maximum; shrinkage rate at 10°C/min heating rate; DSC temperature ≡ peak temperature of endothermic peak at 75°C/min heating rate. DuPont 990 TMA and DSC—TMA conditions as in Fig. 24, DSC Conditions—flowing N_2 atmosphere, coil sample. Number in parenthesis = % acrylonitrile in copolymer. [From Kimmel (1971).]

acrylics of varying comonomer content with the high-temperature dS/dT peak observed in TMA shrinkage experiments. On the basis of this correlation he suggests TMA to be a method complementary to DSC or DTA for monitoring the melting of fibers that melt irreversibly in the sense previously described in the DTA/DSC section. Fast shrinkage temperature parameters have been interpreted by Jaffe (1977) to imply melting in other semicrystalline fibers that undergo chemical transformations at elevated temperatures such as poly(p-phenylene terephthalamide) (Kevlar®), and poly(2,2'-m-phenylene-5,5'-bibenzimidazole) (PBI). It is possible that for the chemical transformation to proceed in these types of semicrystalline polymers, the molecular mobility introduced by melting is a prerequisite.

The shrinkage behavior of as-spun PET yarns as a function of spinning speed has been investigated by Heuvel and Huisman (1978) with the re-

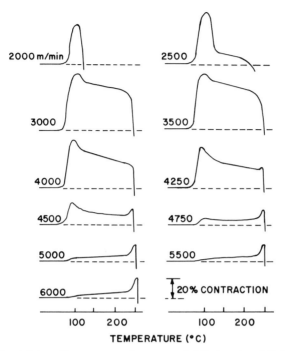

Fig. 26   TMA shrinkage thermograms of as-spun poly(ethylene terephthalate) yarns as a function of yarn wind-up speed. Yarn formation conditions as in Fig. 12. Perkin-Elmer TMS-1 TMA, 0.04 GN/tex Applied Load, 20°C/min heating rate. [From Heuvel and Huisman (1978).]

sults shown in Fig. 26. As spinning speed (melt stress) increases, there is first an increase, then a decrease in the shrinkage just above T_g. The increase is due to increasing noncrystalline orientation, the decrease due to the onset of crystallization during spinning. The appearance of shrinkage in the region of T_m at still higher speeds indicates that an oriented structure stabilized by crystalline units is set up in high-speed spinning, and at the highest speeds investigated the curves are similar to the curves shown for drawn yarn (see Fig. 19). As a generalization, increased orientation (see the work of Samuels, 1974) will tend to increase shrinkage and increased crystallinity will tend to decrease shrinkge, independent of how the changes or orientation and crystallinity are accomplished. A useful comparison of TMA and DSC data is the comparison of Figs. 26 and 12, showing the respective TMA and DSC thermograms of the same as-spun PET yarns.

The measurement of the shrinkage force (SF) of fibers, also referred to as thermal stress analysis (TSA), is complementary to the shrinkage

measurement and is similarly related to processing history and fiber microstructure. The magnitude of the SF reflects the level of contraint with
which stress transmitting, disorienting species are opposed by the fiber
microstructure at a given temperature; i.e., it is a measure of the stability
and tautness of the structural units (tie molecules, etc.) preserving the orientation "frozen" into the fiber structure during processing. Figure 27
shows a typical SF trace (zero pretension) for a drawn and crystallized
yarn, separated into the four regions of behavior described earlier in the
analysis of the S and dS/dT traces of this same material (Fig. 19 and 20).
The structural analysis offered for the shrinkage behavior holds equally
well for the shrinkage force. Note that the maximum in SF occurs prior to
melting; the temperature and magnitude of the SF maximum are strongly
dependent on the stresses and temperatures the fiber experienced in processing. Heidemann and Berndt (1976) define an effective stress σ_{eff}, analogous to the T_{eff} discussed in Section II. B, that is the stress under which
the fiber will be at its original length at T_{eff} in a TMA shrinkage experiment, i.e., the stress necessary to balance the shrinkage force generated
at T_{eff}. To measure σ_{eff} in a shrinkage force experiment, the preloading on
the fiber is increased until an SF thermogram with a zero slope at T_{eff} is
achieved. The value of SF in the zero slope region is equal to σ_{eff} and the
necessary preload to obtain this response is greater than σ_{eff}. Just above

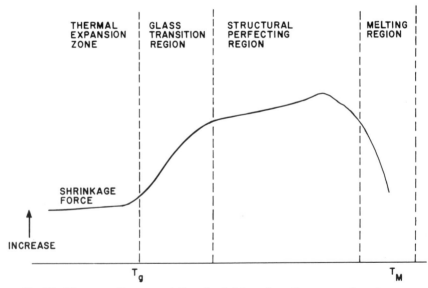

Fig. 27 Diagrammatic representation of a shrinkage force thermogram for a drawn, partially crystalline fiber showing the four length change regions described in Figs. 19 and 20.

T_{eff} a rapid decay of SF will be observed, indicating that the structure retaining the frozen-in stress has been destroyed. This experiment thus allows the simultaneous determination of σ_{eff} and T_{eff}. The stress σ_{eff} is a reflection of the total stress history of the fiber just as T_{eff} is a reflection of the fiber's total thermal history. A new technique for obtaining T_{eff} and σ_{eff} in a single experiment was described by Berndt and Heidemann (1980) at the sixth ICTA conference. The method involves making small strain adjustments to the sample while heating in the SF apparatus (the textechno Thermofil was used) such that stress relaxation and shrinkage force development (stress retardation) are balanced. In this manner an equilibrium of relaxation and retardation processes is obtained, the shrinkage force will maximize at T_{eff}, and this stress level is σ_{eff}. These parameters may be utilized to identify probable process causes of fiber nonuniformity or to predict when fibers produced differently will behave similarly in further process steps such as dyeing or texturing. The sensitivity of SF to processing history is illustrated with the data of Kimmel (1971) in Fig. 28, which shows the effect of heat setting and relaxation on

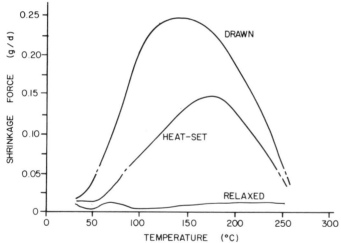

Fig. 28 Typical shrinkage force thermograms of drawn, heat-set, and relaxed acrylic yarns. "Home-made" shrinkage force analyzer utilizing an Instron tensile tester and a Du-Pont 900 temperature programmer controlling the heating of a clam-shell type oven with a 0.5 in bore, 6 in gauge length, 10°C/min heating rate [From Kimmel (1971).]

the SF trace of a drawn acrylic yarn. The sensitive effect of processing temperature on the SF response of a drawn acrylic is shown in Fig. 29. Figures 30 and 31 show the effect of processing strain and stress on the SF response of PET yarns as measured by Heidemann and Berndt (1974). In Fig. 32 the expected increase of SF with increasing draw ratio (increasing

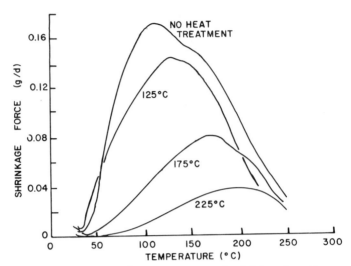

Fig. 29 The effect of heat setting temperature on the shrinkage force thermogram of a typical drawn acrylic yarn. Heat setting time = 5 min. Equipment and conditions as in Fig. 28. [From Kimmel (1971).]

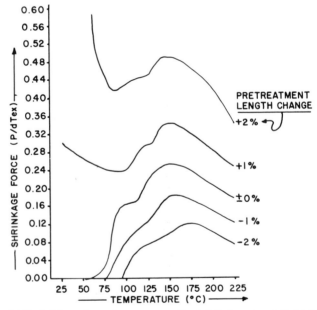

Fig. 30 TMA shrinkage force thermograms of poly(ethylene terephthalate) textile filament yarns as a function of the length change imposed during pretreatment. [From Heidemann and Berndt (1974).]

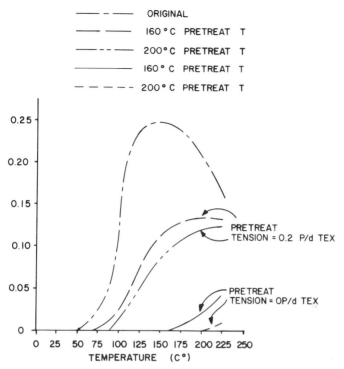

ORIGINAL ——— - ———
160 ° C PRETREAT T ——— · ———
200° C PRETREAT T ——— - - ———
160 ° C PRETREAT T ———————
200 ° C PRETREAT T — — — — —

PRETREAT
TENSION = 0.2 P/d TEX

PRETREAT
TENSION = OP/d TEX

Fig. 31 TMA shrinkage force thermograms of poly(ethylene terephthalate) textile filament yarns as a function of pretreatment tension and temperature. 12°C/min heating rate.
[From Heidemann and Berndt (1974).]

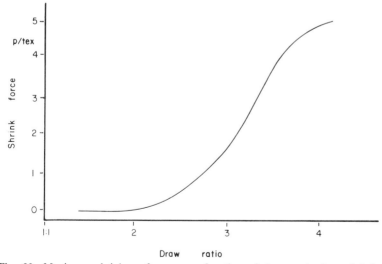

Fig. 32 Maximum shrinkage force as a function of draw ratio for poly(ethylene
terephthalate) yarns. See reference for spinning and drawing details. Shrinkage force measured with a "statigraph" strength tester equipped with a temperature control device, 0.5
P/tex starting stress, 200 mm gauge length device run isothermally at series of temperatures
for sufficient time to record maximum developed shrinkage force. [From Hoffrichter (1973).]

orientation) is shown for a series of polyester yarns investigated by Hof-frichter (1973). The kinetics of SF buildup and decay has been investigated by Berndt and Heidemann (1974). Like shrinkage, shrinkage force will tend to increase with increasing orientation and decrease with increasing crystallinity. The relationship between shrinkage and shrinkage force is not unique, although at any given temperature they are related by a microstructure and time-dependent modulus.

Results have been described that illustrate the utility of SF as a process diagnostic tool. Structurally, SF lends insight into the nature of the tie molecules in a given fiber structure, an area extremely difficult to study experimentally. Reviews of the applications of SF analysis to textile yarns have been written by Buchanan and Hardegree (1977) and the Kanebo Company. Kitakawa (1980) has summarized the effects of draw texturing conditions and precursor yarn types on the resulting textured yarn properties through detailed evaluation of SF spectra.

In concluding this discussion of the applications of TMA to fibers, no better final illustration of the utility of the technique could be offered than a summary of the research performed at the Textile Research Institute at

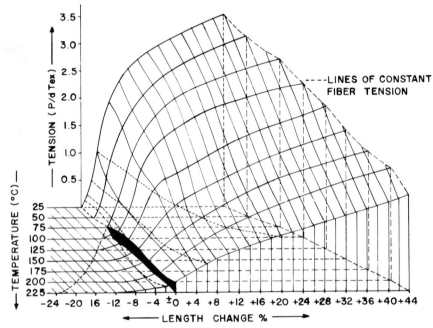

Fig. 33 Tension–length change diagram for textile filament poly(ethylene terephthalate) yarn (Diolen 100 d/tex 36 fil) as a function of temperature. Time of each measurement point constant at 20 sec. [From Valk *et al.* (1971).]

Krefeld, West Germany (Valk, Berndt, Heidemann), into the relationship of after processing (texturing, heat setting, etc.) parameters to the product performance of PET yarns. Figure 33 shows the effect of after-process temperature and stress (treatment time 20 sec) on the length change behavior of the resulting PET yarns. Figure 34 amply indicates the sensitivity of dye uptake (many other end-use responses could be substituted) to subtle process variations. Given the complex process history most fibers receive, a scheme for identifying yarns in equivalent structural states is

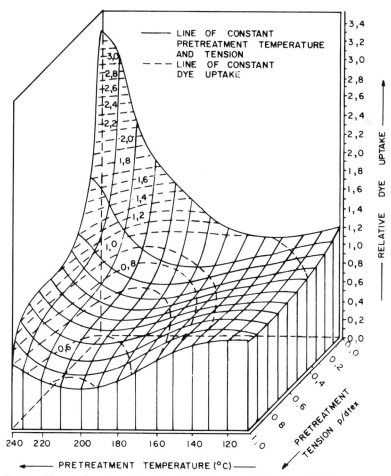

Fig. 34   The relative dye uptake of poly(ethylene terephthalate) textile filament yarn as a function of pretreatment temperature and tension; treatment time constant at 20 sec, dyeing conditions 90 min, 130°C, free to shrink, 3% resolinmarine blue, dye bath ratio 1 : 500. [From Valk *et al.* (1971).]

paramount, either for the overall definition of optimum process parameters or for the characterization of nonrepresentative yarn samples.

The parameters T_{eff} and σ_{eff}, which the Krefeld group has related both to property-controlling features of the fiber microstructure and to the sum of the fiber process history, are a major step in this direction. As T_{eff} and σ_{eff} are readily measured by TMA, the case for the utility of TMA in the investigation and solution of fiber problems is made. A schematic representation of the Krefeld approach, again emphasizing the interactions between process history, structure, and fiber TA response, is shown in Fig. 35.

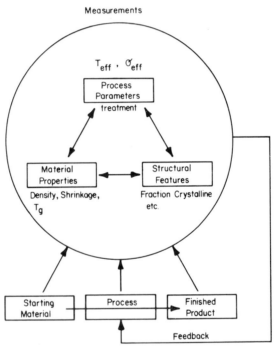

Fig. 35 Logic diagram illustrating the Krefeld group approach to the solution of fiber problems emphasizing the utilization of thermal analysis. [From Valk (1972).]

D. TG—EXPERIMENTAL ASPECTS

The response of fibers to TG is more a function of chemical constitution than physical microstructure; hence the emphasis placed on fiber processing in previous sections shifts here to the details of the molecular species present in or on the fiber-forming material. The exceptions to this statement are studies of gaseous reactions with, or chemical transforma-

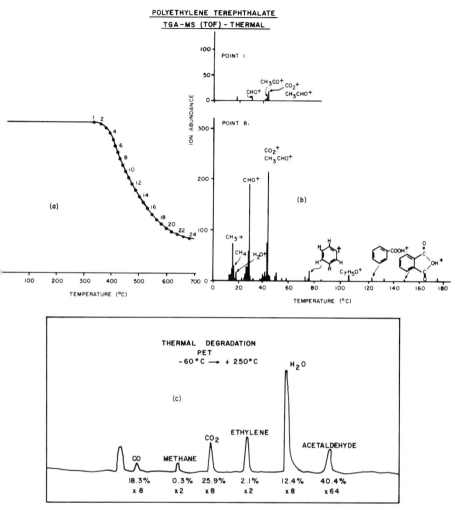

Fig. 36 The thermal degradation of poly(ethylene terephthalate) as clarified by interfacing TG, vapor phase chromatography, and mass spectrometry techniques: (a) TG, (b) vapor phase chromatography trace, (c) mass spectra at two temperatures. DuPont 900 TG, 2 mg sample weight, N_2 atmosphere, 15°C/min heating rate. (Courtesy of A. DiEdwardo, Celanese Research Company.)

tions in, fibers causing weight changes at temperatures below which melting or backbone degradation takes place. Major applications of TG to fibers include (1) measurement of moisture, solvent, or volatile additive or finish content; (2) monitoring of degradation behavior, in either inert or reactive atmospheres; (3) measurement of degradation kinetics; and (4) studies of the influence of additives or fiber blends on relative thermal stability. The amount of useful information obtained in TG can be greatly enhanced by interfacing the TG instrument with analytical instrumentation capable of identifying the chemical species released during the TG experiment. Figure 36 shows the information obtainable on a PET fiber when the TG unit is interfaced with a mass spectrometer or a vapor-phase chromatograph.

TG, more than any other thermal technique, suffers from irreproducibility of results; this is especially true in the comparison of oxidative degradation rate constant data obtained in different laboratories. The major cause of this problem is the extreme sensitivity of TG results to the surface-to-volume ratio of the sample investigated, a particularly difficult parameter to control in nonthermoplastic fiber samples. The magnitude of the effect is illustrated in Fig. 37; the solution to the problems lies in the careful grinding of samples so that efficient and reproducible packing may be achieved in the TG sample holder. In reporting TG data, details of sample weight and preparation, heating rate, and atmosphere should always be included. Sample weights of about 2 mg and a heating rate of 15°C/min are standard in our laboratory.

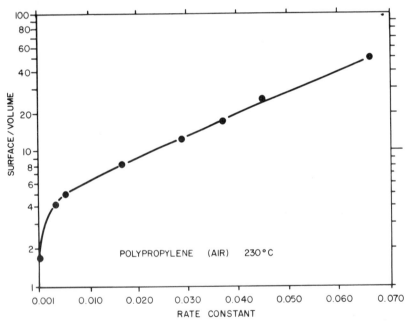

Fig. 37 The effect of sample surface-to-volume ratio on the measured degradation rate constant (K) of a polypropylene yarn heated in a TG unit in air. DuPont 900 TG. (Courtesy of A. DiEdwardo, Celanese Research Company.)

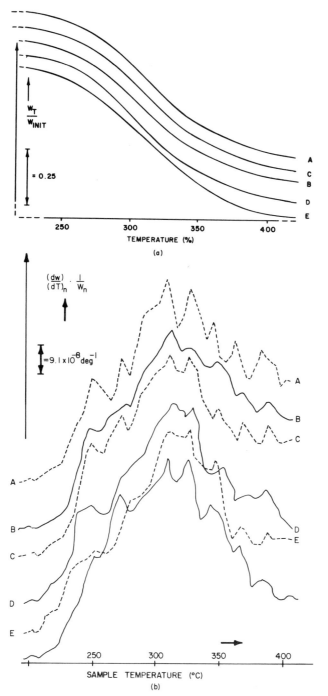

Fig. 38 (a) TG thermograms of predried keratin fibers: A, Merion 70's (TOP); B, Lincoln 44's (TOP); C, Blackface 36's (TOP); D, cashmere (NOIL); E, angora fur. See reference for instrumentation details—10 mg sample weight, fiber cut to pass through 60 mesh sieve, continuous evacuation ($p_{Initial} = 10^{-2}$ torr) 4°C/min heating rate. (b) Derivative TG (DTG) thermograms of predried keratin fibers. samples and conditions as in (a). [From Crighton and Hole (1976).]

One straightforward application of TG to fibers is the measurement of moisture content, an important parameter in fabric comfort. Mansrekar (1973) has compared the techniques of TG, Karl Fischer titration, and oven drying for moisture determination in a variety of fabrics and concluded the TG method is the best compromise between accuracy and convenience. Crighton and Hole (1976) have shown that differences in the thermal degradation behavior of various wool samples provides a convenient tool for their characterization. As illustrated in Fig. 38, the derivative rather than integral curves are most effective for this purpose (a TG "fingerprinting" technique).

The emergence over the past decade of PAN as the precursor yarn of choice for the production of graphite fiber has spurred many studies of the kinetics and mechanism of PAN pyrolysis (Fitzer and Mueller, 1971, 1973; Grassie and McGuchan, 1972). Major variables in these studies include atmosphere, comonomer content, and yarn pretreatment conditions, with the overall objective of defining the optimum precursor yarn in terms of carbon yield, process efficiency, and level of graphitic properties. Grassie and McGuchan (1972) have compared the pyrolysis of a variety of PAN copolymers and show that both the nature and amount of co-

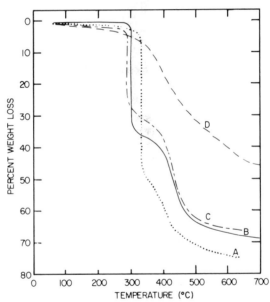

Fig. 39 Vacuum TG thermograms of polyacrylonitrile yarn as a function of thermal and irradiation history. Low comonomer American Cyanamid produced yarn: A, as Received; B, C_6^{60} irradiated; C, C_6^{60} and e^- irradiated; D, C_6^{60} irradiated and preoxidized. See reference for instrumentation details. 5°C/min heat rate. [From Gaulin and McDonald (1971).]

monomer significantly affect the rate and mechanism of the reaction. Gaulin and McDonald (1972) have studied the effect of radiation treatments on PAN pyrolysis; a comparison of the effects of various treatments on the resulting dynamic weight loss behavior of an American Cyanamid PAN yarn is shown in Fig. 39. A. DiEdwardo (private communication, 1974) in the Celanese Research Company Laboratories applied a combined TG—gas chromatography—mass spectrometer technique to a homopolymer PAN with a typical result shown in Fig. 40. TG has played an important dual role in the development of graphite fiber precursors, providing effective screening of candidates while simultaneously providing the kinetic and mechanistic data needed for the understanding of the complex reactions taking place.

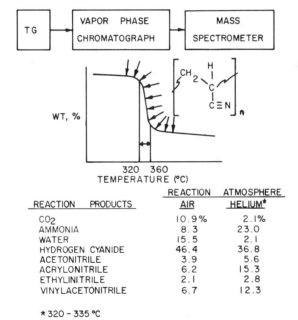

REACTION PRODUCTS	REACTION ATMOSPHERE	
	AIR	HELIUM*
CO_2	10.9%	2.1%
AMMONIA	8.3	23.0
WATER	15.5	2.1
HYDROGEN CYANIDE	46.4	36.8
ACETONITRILE	3.9	5.6
ACRYLONITRILE	6.2	15.3
ETHYLINITRILE	2.1	2.8
VINYLACETONITRILE	6.7	12.3

* 320 – 335 °C

Fig. 40 A diagrammatic representation of a TG thermogram of a homopolymer polyacrylonitrile yarn showing the reaction products as determined by in-line gas chromatography and mass spectrometry. (Courtesy of A. DiEdwardo, Celanese Research Company.)

An extensive investigation of the effects of crystallinity and molecular orientation on the vacuum and oxidative pyrolysis of cellulose has been performed by Basch and Lewin (1973a,b, 1974). Noting the importance of sample purity, an extensive physical characterization and sample preparation procedure was undertaken prior to kinetic studies. Crystallinity was shown to stabilize cellulose toward pyrolysis; the apparent activation

M. Jaffe

energy E_A of the reaction was found to increase from an extrapolated value of about 30 kcal/mole for noncrystalline regions, to about 60 kcal/mole for crystalline regions, as shown in Fig. 41. For the details of the calculation of activation energies from TG data see Chapters 1 and 2. The effect of orientation is more complex, slowing the reaction in air and accelerating it in vacuum. An example of the isothermal weight loss behavior in vacuum of 20 and 50% stretched cellulosic fibers is shown in Fig. 42. The rate decrease in air is attributed to a lowering of accessibility of oxygen to the structure, the rate of increase in thermal pyrolysis attrib-

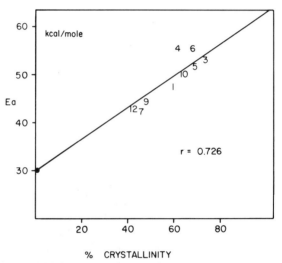

Fig. 41 The apparent activation energy of the pyrolysis of cellulosic yarns as a function of yarn crystallinity. Activation energies measured by TG using the method of Horowitz and Metzger (1963). DuPont 950 TG, 5 mg sample weight, 10°C/min heating rate. Crystallinity determined by wide-angle x-ray diffraction.

Sample	Crystallinity (%)	E_a (±2)
1 Cotton, deltapine	63.7	48.3
3 Cotton, pima	70.8	52.1
4 Cotton, pima mercerized	65.3	55.0
5 Ramie	70.3	51.7
6 Ramie, hydrolyzed	69.5	55.0
7 Tire yarn	57.1	43.3
8 High modulus yarn	67.5	45.2
9 Vincel 64	54.9	44.9
10 Vincel 28	66.0	48.9
11 Textile rayon	51.2	47.8
12 Eylan	49.8	44.2

[From Basch and Lewin (1973a).]

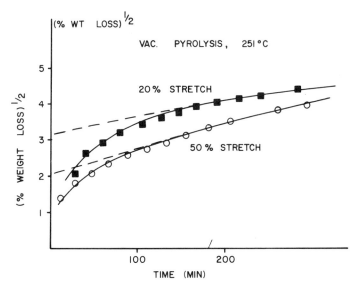

Fig. 42   The weight loss of rayon yarns versus the time of vacuum pyrolysis at 251°C as a function of degree of yarn orientation (degree of stretch). DuPont 950 TG, continuous vacuum. [From Basch and Lewin (1974).]

uted to the increased proximity of cross-link-forming groups. The general import of this study lies in its illustration of the interaction of chemistry and physics in the determination of fiber behavior.

The problem of fiber flammability has led to a number of investigations aimed at correlating thermal stability as monitored by TG with burning test results. Bingham and Hill (1975), have catalogued the thermal behavior of a variety of materials known to be flame resistant. Although they do not correlate their results with flame tests, typical results comparing flame-resistant cottons with untreated cotton (Fig. 43) show major differences in weight loss behavior under oxidative degradation conditions. Carroll-Porczynski (1973) has shown that TG can be used as a screening tool for predicting the flammability of fabrics composed of polymers blends. The method is to construct an additive TG curve based on the behavior of individual blend components and compare the result to the TG curve of the actual composite. If the additive curve shows greater weight loss than the actual composite, the flammability performance of the blend is synergistic and usually satisfactory. Results for a good (50/50 wool–Nomex®) and a bad (50/50 wool–polypropylene) blend are shown in Fig. 44. An alternate approach to predicting the performance of flame-retardant additives in fibers was attempted by Davidson and Roberts (1980), who used TG and DSC to characterize a series of aromatic bromine-

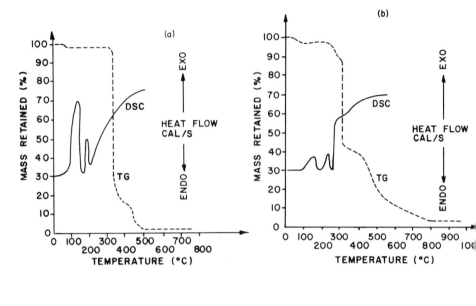

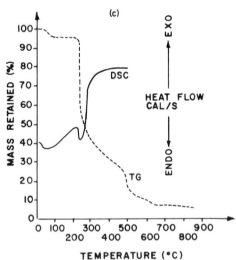

Fig. 43 The Oxidative Decomposition of Untreated and Flame-Retarded Cotton as Monitored by TG and DSC Techniques: (a) cotton, (b) cotton pyrovatex, (c) Durelle. Staton Model HT-SM thermoloalame, 100–200 mg sample weight, air atmosphere (static), 6°C/min heating rate. Perkin-Elmer DSCO1B, 5–6 mg sample weight, perforated pan lids, flowing air atmosphere, 16°C/min heating rate. [From Bingham and Hill (1975).]

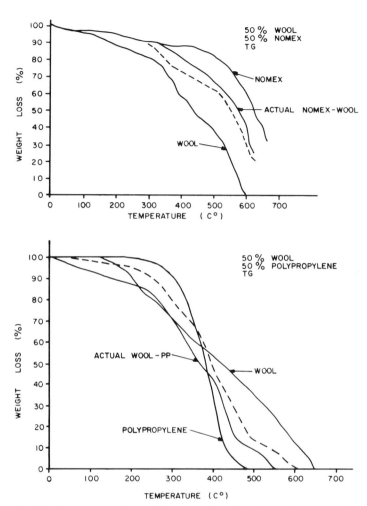

Fig. 44 TG thermograms of wool/Nomex and wool/polypropylene composite fabric: (solid line) actual curves of component fibers and composite, (dashed line) additive curve based on component fiber results. TR-02/STA TG apparatus, flowing air. Atmosphere (500 ml/min), 10°C/min heating rate. [From Carrol-Porczynski (1973).]

containing compounds. The sought-after common thermal property that predicted the end-use effectiveness of these additives as fiber flame retardants could not, however, be identified.

Given the similarity of the response to TG of fibers and other polymeric articles, the fiber-motivated reader is strongly urged to ignore chapter titles in the search for examples of TG applications to fiber problem solving. Most of the examples cited here emphasize combining or interfacing TG with other techniques, thus transforming TG observations into investigations of the mechanism and kinetics of specific chemical reactions. The importance of developing such detailed base lines before drawing conclusions on the basis of weight loss data cannot be overstated.

III. Specific Fibers

In the preceding sections, a base line of familiarity with the applications of thermal analysis to fiber science has been constructed. The purpose of this section will be to review the pertinent current literature in depth, so as to provide the thermal analyst a more complete picture of the fruitfulness of the TA–fiber marriage. The volume of work done on a particular fiber seems to be a function of its commercial import and the timing of its discovery, i.e., PET > nylon > PAN > cellulosics > wool. For the researcher, the areas of omission may prove more useful than the areas of past concentration; for the technologist a listing of new solutions to old problems will, it is hoped, be apparent.

A. Fibers of Animal Origin

1. Wool*

Wool is a proteinaceous fiber composed of keratin, as are all the other animal fibers with the exception of silk. Obtained from sheep, wool is complex in structure and fiber properties vary between breeds, within the same breed as a function of environment and diet, and even from position to position on the same animal. The surfaces of wool fibers are covered with overlapping scales, an important feature that allows the fibers to interact mechanically and form dense structures such as felts. Wool is naturally crimped, is highly resilient, has a moisture regain of up to 18%, and is an excellent insulator. Principal uses include carpets and textiles.

Crighton et al. (1968, 1971, 1972, 1976, 1977, 1978) have made extensive use of thermal analysis (DTG, TG) in the characterization of various keratin-based fibers (see Fig. 38). Crighton stresses the need for purified samples and the use of supporting techniques for the assignment of observed thermal or chemical transitions. In an extensive DTA study of treated wools (to improve shrink resistance and washability) Crighton and Hole (1978) have shown that the thermograms obtained change with treatment, while stressing that DTA alone is not sufficient for treatment characterization. In the course of this and earlier work they identify the seven peaks reproducibly observed in wool DTA as follows: (1) 160–168°C—re-

---

* See Encyclopedia Britannica (1966) and Stout (1970).

versible transition, correlated to the onset of side-chain motion; (2) 171–194°C—chemical origin; (3) 197–205°C—small shoulder, with origin unspecified; (4) 235°C; and (5) 243°C—doublet due to physical transitions that correlate with loss of α-helical content and shrinkage; (6) 292°C—decomposition processes; and (7) 321°C—"Apparent peak in the profile of the degradation endotherm." Konda *et al.* (1973) have studied the change in the 230°C endothermic peak observed in variously drawn wool fibers. Utilizing DSC in conjunction with wide-angle x-ray diffraction, they assigned this peak to the disappearance of the α form of wool keratin on heating. DSC traces as a function of applied draw and relaxation are shown in Fig. 45. Brazauskas *et al.* (1973) have used DTA to optimize the annealing and dyeing of wool and find temperature, time, and dye pH to have significant effects. Both DTA/DSC and TG have been utilized in assessing wool flammability (see Chapter 8).

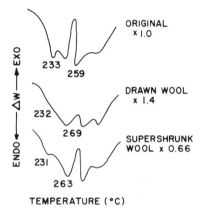

Fig. 45 DSC thermograms of wool keraton fibers as a function of drawing and relaxation treatments. Drawing performed in water at 20°C. Relaxation (supershrinking) in cuprammonium solution. Rigaku DSC, 3 mg sample weight, N_2 atmosphere. [From Konda *et al.* (1973).]

2. Silk*

Silk is the one continuous-filament natural fiber, spun from the spinneret of the cultivated silkworm, *Bombyx mori*, in the forming of its cocoon. As-spun silk consists of two strands of the protein fibroin, bonded together by a second protein, sericin (study of silkworms = sericulture). After suitable treatment a number of cocoons are unwound (reeling) and formed into a yarn with the average take-up length being 900 m (hence the definition of denier). Silk not amenable to continuous take-up is chopped into staple, and spun yarns are formed (spun silk). Historically, silk has been weighted, i.e., treated with tin or lead salts to improve body and reduce cost (not common in silk used in the United States). Silk possesses high tensile strength, high elasticity, a moisture regain of up to 30%, and

* See Encyclopedia Britannica (1966) and Stout (1970).

excellent aesthetics (drape, hand, feel, luster). Silk is used almost exclusively for textile applications.

Magoshi and Nakamura (1975, 1978) and Magoshi *et al.* (1977, 1979) have investigated the thermal transitions of silk fibroin by DSC, finding the glass transition at 175°C, cold crystallization of the μ random coil form to β crystalline form peaking at 212°C and degradation of the β form peaking at 280°C, as illustrated in Fig. 46. TG experiments confirm the latter

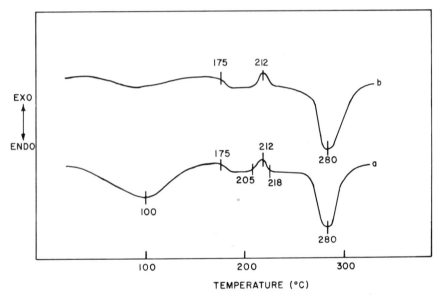

Fig. 46   DSC Thermograms of amorphous silk fibron with random coil conformation: (a) untreated, (b) stored in desiccator containing silica gel for two weeks. Perkin-Elmer 1B-DSC, N_2 atmosphere, 8°C/min heating rate. [From Magoshi and Nakamura (1975).]

assignment, showing decomposition to begin at T_g and $T_{x\text{cold}}$ to shift up in temperature with increasing heating rate. Thermal expansion measurements performed on a homemade apparatus at 3°C/min showed an abrupt increase in sample length beginning at about 165°C, confirming the T_g (extrapolation yielded $T_g = 175$°C). DTA investigation of the properties of sericin by Ichizo and Takeuchi (1971) showed endothermic decomposition behavior to vary as a function of sericin source (T_D range 200–314°C). Sakaguchi *et al.* (1972) utilized DTA in an investigation of the mechanism of the tin weighting of silk. DTA and TG were used by Huh and Soong (1976) in studies of silk flammability.

3. Collagen*

Collagen is the highly ordered, proteinaceous fiber found in the supportive and connective tissues of animals: tendons, the dermis layer of the skin, the tissues surrounding joints. Although the use of thermal analysis in the study of collagen is very much in its infancy, a few researchers (Finch and Ledward, 1973; Hellauer and Winkler, 1975) have studied the effects of media (salt solutions, pH) on its denaturation behavior by DSC. The length changes of collagen as a function of temperature and stress have been investigated by Mnuskina *et al.* (1974), who define stress parameters useful in the characterization of previous chemical history.

4. Fibers of Plant Origin—Cellulosic †

Cellulose $(C_6H_{10}O_5)x$, is the basic constituent of all plant fibers and serves as the raw material source of several man-made fibers. Cotton is the most prevalent of the plant fibers, grows in lengths from 1 to 6.5 cm, and has the appearance of a twisted tape (in actuality a collapsed tube). Other plant fibers of commercial import include jute, flax, ramie, and hemp. Wood is a fiber-reinforced composite with cellulosic fibers sitting in a lignin matrix.

The properties of plant fibers vary greatly from species to species and usages include textile (cotton, flax), cordage (jute, hemp), and industrial products. All are reasonably strong, have high moisture regains (6–12%), and perform well when wet. After spinning into a yarn, cotton may be treated with alkali, to improve luster and reduce residual shrinkage (mercerization) and cross-linked to improve wrinkle resistance. All the plant fibers are morphologically complex and differ in molecular-weight parameters, crystallinity, and chain orientation.

Cotton linters (fibers too short to be spun) and wood pulp are the raw material for two man-made fibers, rayon and acetate. Rayon is defined as a cellulosic fiber in which less than 15% of the hydroxyl groups have been replaced. Termed *regenerated cellulose,* all rayon is wet-spun. Rayon may be used as a filament yarn (textured or flat), may be chopped into staple for spun yarns, or may be drawn to improve tensile properties for industrial applications. Acetate is produced by acetylating cellulose, followed by dry spinning. If more than 92% of the hydroxyl groups are acetylated, the product is triacetate. The advantage of triacetate is its increased crystallinity, which stabilizes the microstructure and allows its use as a permanent press fiber. Available as filament or staple, acetate

---

* See Encyclopedia Britannica (1966).
† See Encyclopedia Britannica (1966), Moncrieff (1975), Morton and Hearle (1975), and Stout (1970).

fibers are amenable to texturing and/or drawing processes. The high moisture regain, wearing comfort, and excellent aesthetics of the man-made cellulosic fibers render them important textile fibers.

The great majority of reported thermal analysis studies of cellulosic fibers are concerned with either pyrolysis or flammability. The extensive DSC and TG investigations of Basch and Lewin (1973a,b, 1974), Lewin and Basch (1972), and Rodrig *et al.* (1975), on the effects of crystallinity, orientation, and cross-linking on the pyrolytic behavior of cellulosics has already been described (see Section II.D). Von Hornuff and Müller (1975) have investigated the effects of various finishes, additives, bound water, and carboxyl group concentration on the degradation temperature and energy of cellulose by DTA, finding all to have a significant effect on the parameters monitored. Broido *et al.* (1973) utilized TG in a study of the molecular-weight drop occurring early in cellulose pyrolysis (1% weight loss). Their results indicate that the initial molecular-weight loss is severe and occurs by the rupturing of chain at the crystal–amorphous interface. Using DSC and TG, Kokta and Valde (1972) show that the grafting of various polyacrylates onto cellulose causes major shifts in both the temperature and heat of degradation. Havens *et al.* (1971) used DSC and rate-dependent TG (20–160°C/min) to show that rate of heating was an unimportant parameter for modelling the pyrolysis behavior of wood (oak and pine).

Rayon, like polyacrylonitrile, has been used as a precursor yarn for the production of graphite fiber. TG studies have proved useful in the optimization of pregraphitization conditions (Baver *et al.*, 1973; Fialkov *et al.*, 1968) (stabilization treatment at 150–300°C) and the effects of chemical treatments (chlorosilanes) on the ultimate carbon yield (Duffy, 1969).

The effects of additives (Basch and Lewin, 1975; Godfrey, 1970; Smith *et al.*, 1970), fine structure (Basch and Lewin, 1975), dyeing (Timpa *et al.*, 1974), and poly(vinylidene chloride) coating (Nagano, 1976) on the flame retardance of cellulosics have been characterized by DSC and TG techniques. Langley *et al.* (1980) utilized TG and DSC to elucidate the mechanism by which aromatic phosphates and phosphoramides flame-retard cotton cellulose, concluding the phosphoramide compounds are more effective flame retardants because of their increased reactivity (more exothermic reactions) with cellulosic moieties. Weiss *et al.* (1974) have monitored the length changes occurring in treated and untreated cotton by TMA. As shown in Fig. 47, flame retardants tend to increase the mechanical integrity of degraded cellulose; the shrinkage observed is useful both to assess the effectiveness of a given additive and to predict how the burning garmet will interact with the body of the wearer.

Hall and Godwin (1970) have studied flat and textured cellulose triacetate fibers by TMA, identifying 175°C as T_g. By coupling TMA with

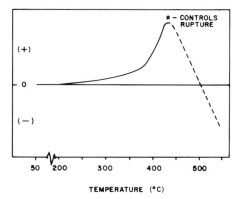

TEMPERATURE (°C)

Fig. 47 TMA thermogram of cotton fabrics. Solid line represents control and flame-retardant cottons. Broken line represents only flame-retardant cotton. See reference for sample details. DuPont TMA, 10 gm load, fabric five wrap threads wide, 13.7–15.2 mm gauge length, flowing 0_2 atmosphere, 10°C/min heating rate. [From Weiss *et al.* (1974).]

direct sample observation with a stereomicroscope, they developed a straightforward technique for monitoring the crimp retention of textured yarns as a function of temperature. The low-temperature (-100 to $-5°C$) thermal expansion of cellulose was investigated by Ioelovics *et al.* (1975), who find the CTE increases with increasing amorphous content and decreases with increasing orientation. The shrinkage of cellulosic fibers has been studied by Ganchuk *et al.* (1973), who conclude variation in results on multiple heating cycle experiments to be due to changing yarn moisture content and/or the relaxation of residual strain.

Chatterjee and Schwenker (1972) have shown DTA to be effective in characterizing the nature of cellulose derivatives as well as their degree of substitution. The cross-linking of cellulose by chemical reactions similar to those occurring in cotton permanent press treatments has been studied by Gilbert and Rhodes (1970). DSC results indicate that the *N*-methylol compounds investigated may cross-link the cellulose directly; evolve formaldehyde, which in turn will form cross-links, or self-condense prior to cross-linking, depending on reaction conditions, catalysis, and the particular *N*-methylol employed.

B. SYNTHETIC FIBERS

1. Nylons*

Nylon is a generic term applied to linear polyamides. Aliphatic nylons are described chemically by the number of carbon atoms between nitrogen

* See Black and Preston (1973), Encyclopedia Britannica (1966), Moncrieff (1975), Morton and Hearle (1975), and Stout (1970).

in the chain, i.e., the nylon formed from hexamethylene diamine and adipic acid is Nylon 66, the polymer formed from caprolactam is Nylon 6, etc. Nylon 66 and Nylon 6 are by far the most important commercially. Differing primarily in melting point (66 ~ 265°C, 6 ~ 225°C), the common nylons are melt-spun. Depending on processing details (draw ratio, crystallization, texturing), nylons can be tailored in properties to satisfy markets as diverse as women's hosiery, tire cord, carpets, and ropes. The moisture regain of nylon is on the order of 4%, but the hydrogen bonding of water with the amide linkage can have a profound influence on measured properties (moisture acts as a plasticizer, lowering T_g). The more important properties of nylon include high flexibility and resilience coupled with high strength (3.6–7.2 cN/dtex) and heat settability. Nylon is available as a monofilament, continuous-filament yarn, and staple.

In the past few years, several specialty nylons have been introduced by the DuPont Company under the tradenames of Qiana®, Nomex®, and Kevlar®. Qiana is the polyamide of bis-p-aminocyclohexylmethane and dodecanedioic acid. It is melt-spun to form a luxury apparel yarn possessing excellent, silklike aesthetics. Nomex and Kevlar are aramids (aromatic polyamides); Nomex is poly(m-phenyleneisophthalamide), Kevlar poly(p-phenyleneterphthalamide). Nomex is dry-spun from dimethyl–formamide–LiCl$_4$ solution, is capable of being steam-drawn, and is used primarily as a nonflammable fiber. Kevlar is wet-spun from 100% sulfuric acid into water (points of special interest in the Kevlar system are that the spinning dope is a nematic liquid crystal and the wet-spinning process is modified by removing the spinneret from the coagulation bath), forming fiber of extremely high strength and modulus (tenacity up to 27 cN/dtex, modulus up to ~900 cN/dtex) utilized in high-performance applications such as tire belts and fiber-reinforced composites.

The glass transition of 31 nylon homopolymers and copolymers as a function of molecular structure and moisture content has been investigated by Buchanan and Walters (1977). Using DSC and dynamic mechanical techniques, they show the T_g lowering due to moisture is dependent only on the average spacing between amide groups along the chain and derive a procedure for accurately predicting dry and wet T_g behavior. Turska and Gogolewski (1975) have employed DSC as a sample characterization method in a study of the "crystalline memory" effect on the crystallization kinetics on Nylon 6. Precursor crystallinity was found to influence both the rate and nature of further crystallization profoundly.

The melting behavior of Nylon 6 (Todoki and Kawaguchi, 1975, 1977a,b) and 66 (Hybart and Platt, 1967; Sweet and Bell, 1972) yarns as a function of drawing and annealing history has received considerable attention in the thermal analysis literature. The already reviewed work of

Todoki *et al.* (1975, 1977a,b) summarizes both the observations and mechanism involved. Artunc (1976) and Artunc and Egbers (1976) have used DTA to study in detail the effect of draw ratio, annealing time and temperature, measurement atmosphere, and heating rate on Nylon 6 thermograms. They conclude increasing heating rate tends to increase the observed T_g (47–58.5°C, 2–50°C/min), and broaden the dehydration peak. Stretching increases the melting point (draw ratio of 3.12 increases T_m from 221 to 225°C), free to shrink annealing of drawn yarns tends to lower T_m, and the atmosphere employed (air, nitrogen, vacuum) can cause significant T_m shifts. Figure 48a,b,c, and d show examples of the thermograms on which these conclusions are based. Arakawa *et al.* (1968) have shown that 170°C methoxy-methylation of drawn (3.2X) Nylon 6 yarns retards crystalline reorganization during DSC heating with a resultant T_m lowering from 220°C to 195°C. The 195°C melting is identified as representative of the crystals present in the original yarn structure. Wishiewska (1972) has utilized DSC-determined heats of fusion and crystallinities of variously drawn Nylon 6 yarns to define the overall crystalline perfection of the fiber structure. The specific heat of drawn Nylon 6 has been measured by Neduzhii *et al.* (1971), and Gyori *et al.* (1976) have used DTA to monitor the effects on fiber structure of false-twist texturing and dyeing, correlating an exothermic peak at 142–150°C with process and structural parameters. Several authors (Fujimoto and Yamashita, 1971; Hirschler *et al.*, 1971; Kunugi *et al.*, 1975) have monitored Nylon 6 shrinkage as a function of after-processing history. Anton (1973) has combined DSC, shrinkage, and shrinkage force measurements with infrared, NMR, and x-ray techniques in an investigation of the relaxation effects occurring in Nylon 66 and 6–66 copolymers. Emphasizing the complexity of relaxation in nylons (moisture effects, crystallization, etc.), Anton concludes that no one technique is sufficient to characterize the observed behavior completely.

The method of polymerization, melt or interfacial, of 66 and 610 nylon has been investigated by Jasse and Moutte (1972), who show by DSC and other techniques that different molecular-weight distributions are obtained that will manifest in fibers as property differences. DTA and DSC were utilized by Hampson and Manley (1974) to determine the nature and amount of elastomer present on coated nylon fabrics and Arons and Macnair (1970) to study the grafting of polyacrylic acid onto Nylon 66. A number of authors (Dimov *et al.*, 1973; Okuhashi, 1970; Petakhova *et al.*, 1975) have employed DTA in studies of nylon thermal stability.

In a DSC study of Qiana fiber, Hatayama and Kanetsuna (1970) find T_g to occur over a broad temperature range (90–150°C), attributing the broadness to motion of CH_2 groups. T_g and cold crystallization behavior

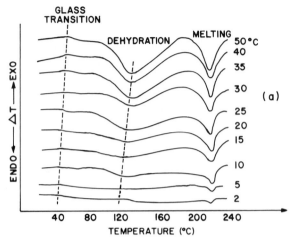

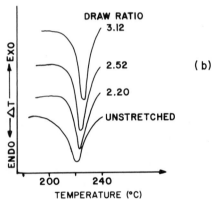

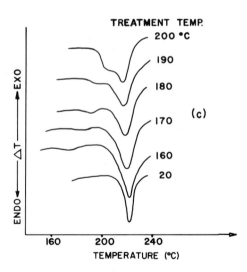

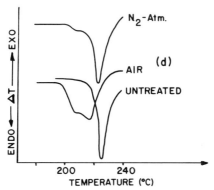

Fig. 48 DTA thermograms of Nylon 6 yarns. (a) Effect of heating rate on the observed glass transition temperature, dehydration temperature, and melting temperature of unstretched yarns. (b) Influence of draw ratio on the melting of Nylon 6 yarns. Flowing N_2 atmosphere (0.5 l/min), 30°C/min heating rate. (c) The effect of annealing temperature on the melting of Nylon 6 yarns. Samples annealed in air, free-to-shrink for 20 min. Samples drawn 3.12:1. Flowing N_2 atmosphere 0.5 l/min, 30°C/min heating rate. (d) Effect of atmosphere during annealing of the melting of Nylon 6 yarns. Samples annealed at 150°C. Flowing N_2 atmosphere (0.5 l/min), 30°C/min heating rate. DuPont 900 DTA, reference material —glass powder. [From Artunc and Egbers (1976).]

of a series of Nylon 66 copolymers were investigated by Kiyotsukuri and Nagasawa (1976). DTA, TG, and TMA have been applied to the characterization of Aramid fibers by Brown and Ennis (1977), Mikolajczyk and Ratujczyk (1975), Jaffe (1977), and Penn and Larsen (1979). Brown and Ennis (1977) identify T_g at about 300°C for both Kevlar and Nomex, endothermic decomposition peaks at 590°C and 440°C for Kevlar and Nomex, respectively, and an ~560°C melting point for Kevlar. Figure 49 from the data of Jaffe (1977) shows typical TG, DSC, and TMA thermograms for a poly(p-phenyleneterephthalamide yarn); note that the three traces viewed together yield much more information than each viewed separately. A variety of sample atmospheres in the TG studies were used by Chatfield *et al.* (1979a,b) and Zhmaeva *et al.* (1977) to define the thermal stability of chlorinated aromatic polyamides, aromatic polyamide fabrics and aromatic polyazoamide fibers, respectively. Silver and Dobinson (1977) utilize TG and DTA to conclude that poly(p-xylyleneterephthalamide) fiber could not successfully compete commercially with other aramids because of poor thermooxidative stability.

2. Poly(Ethylene Terephthalate) (Polyester, PET)*

Poly(ethylene terephthalate) is the condensation polymer of ethylene glycol and terephthalic acid with a melting point of about 260°C and a

---

* See Encyclopedia Britannica (1966), Kolshak and Vinogradova (1965), Moncrieff (1975), Morton and Hearle (1975), and Stout (1970).

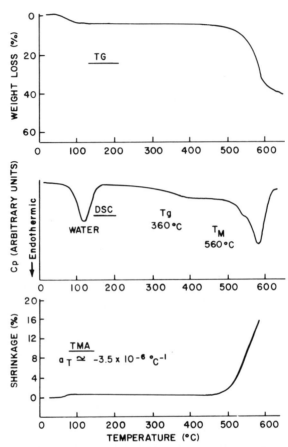

Fig. 49 TG, DSC, and TMA thermograms of a typical high-strength aramid [poly(*p*-phenylene terephthalamide)] yarn. DuPont 990 TG, DSC, TMA. TG—N_2 atmosphere, 2 mg sample, 15°C/min heating rate. DSC—N_2 atmosphere, 10 mg coiled sample, 20°C/min heating rate. TMA—zero load, 0.5 in gauge length 10°C/min heating rate. [From Jaffe (1977).]

glass transition at about 75°C. Like nylon, it is melt-spun and can be drawn and textured to a broad property range suitable for diverse end-use applications. The moisture regain of polyester is quite low (<1%), rendering yarn properties essentially unaffected by moisture except at high temperatures, where hydrolytic degradation can be detrimental. The low-moisture regain leads to comfort problems in apparel applications, a deficiency overcome by blending textured staple with cotton. Other key properties of polyester include high tensiles (especially modulus) and the ease of achieving shape permanence in textiles through heat setting. It has been recently found that spin-oriented polyester spun yarns (POY-partially oriented yarns) are particularly suitable precursors for texturing processes. Over the past decade polyester has become the largest-volume synthetic fiber produced.

The thermal analysis literature concerned with PET fiber is considerably richer than that found for other fibers. There are probably two reasons for this: (1) The development of PET as an important commercial fiber parallels in time the emergence of thermal analysis as a polymer characterization technique, and (2) the slow crystallization rate of PET results in particularly detailed and informative thermograms. Many of the applications examples cited earlier in this chapter (Section II) have been drawn from polyester investigations. The sections dealing with DTA/DSC and TMA rely especially heavily on polyester data. Several reviews of thermal analysis dedicated to polyester fiber have been published. (Berndt and Heidemann, 1972; Dimov *et al.*, 1970b; Heidemann and Berndt, 1976; Valk, 1972; Yamazaki, 1971).

The specific heat of PET, a property of fundamental interest to thermal analysts, has been studied by a number of authors (Gotze and Winkler, 1967; Smith and Dole, 1956) the most recent being an investigation by Haly and Snaith (1969) of the PET-water system. Glass transition phenomena have been investigated by Siegmann and Turi (1974) (structural effects, effect of annealing below T_g), Hagege (1977) (POY yarns), Halip (1971) (drawing effects) and Vanicek (1975) (commonomer content). All agree T_g shifts somewhat with fiber structure and that further yarn processing conditions should reflect these shifts; all place T_g at 75°C ± 5 with the exception of Vanicek, who extrapolates to 86°C. By similar extrapolation procedures Vanicek identifies 286°C as the equilibrium melting point of PET. Jeziorny (1970) has compared crystallinity values of PET fibers as determined by DSC and x-ray diffraction and finds reasonable agreement between the techniques. The kinetics of cold crystallization of undrawn, unoriented yarns has been investigated by Miller (1967), who found the process to be first order between 100 and 115°C with an activation energy of 44 kcal/mole. Melt crystallization kinetics were studied by Mitsuishi and Ikeda (1966), who found double crystallization exotherms under certain conditions. They ascribe this observation to the formation of separate morphologies (folded chain and fringe micelles). Miyagi and Wunderlich (1972) have analyzed the rate-dependent melting of PET tire cords, with results as described in Section II.B.

The effects of spinning conditions on the properties of POY yarns have been investigated by both DSC and TMA. The detailed analysis by Heuvel and Huisman (1978) (spinning-structure model) and the observations on T_g by Hagege (1977) have already been discussed (Section II.B). Shimizu and co-workers (1977, 1978a,b) used a variety of techniques, including DSC, to evaluate the effect of spinning speed, melt drawdown and throughput, and molecular weight on the structure, crystallinity, and chain orientation of PET POY yarns. A similar study of high-speed PET spinning utilizing DSC and TMA data was presented by Perez and Le-Cluse (1979). The shrinkage behavior of POY yarns was investigated by Warwicker and Vevers (1980) and the crystallization kinetics of POY yarns was studied by Ikeda (1980). This latter study concludes that shrinkage measurements are more sensitive than DSC for detecting the early stages of such rapid crystallization events. Quynn (1972) has shown that increasing spun yarn orientation (birefringence) leads to a decrease in the observed peak temperature of cold crystallization (increased crystallization rate), correlating this effect with the reduced tendency of the more oriented spun yarns to stick to heated metal surfaces. In a followup note, MacLean (1974) points out that the cold crystallization peak temperature is heating-rate dependent, but at any given heating rate his data confirm the observations of Quynn. Figure 50 shows the influence of spin-

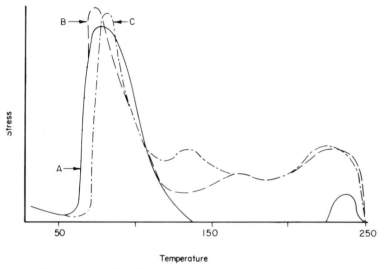

Fig. 50   TMA (shrinkage force) thermograms of as-spun polyester yarns as a function of take-up speed. (A) 45 m/sec (2700 m/min), (B) 58.3 m/sec (3500 m/min), (C) 61.7 m/sec (3700 m/min). Kanebo thermal stress tester, 3°C/min heating rate [From Buchanan and Hardegree (1977).]

ning speed (orientation) on the shrinkage force of as-spun PET. The investigators, Buchanan and Hardegree (1977), ascribe the position of the lower temperature peak to molecular orientation and yarn age and associate the upper temperature peak with increasing yarn crystallinity.

A drop in the CTE measured in the stretching direction as a function of increasing draw ratio for PET samples drawn at 7.5%/min at 75°C was observed by Sidorovich and Kuvshinskii (1961). This observation, reported in 1961, is the earliest data showing the CTE-orientation correlation in polymers. Andrianova and co-workers (1978) in a particularly interesting study used classic calorimetry to determine the heat effects and structural changes occurring in amorphous PET during uniaxial drawing. Seyfurth *et al.* (1972) used a multiplicity of structure-monitoring techniques such as density, birefringence and infrared spectroscopy to follow the structural changes taking place during polyester drawing and conclude DTA to be the most sensitive to small changes in chain orientation. With increasing draw ratios and draw temperatures up to 80°C, the peak temperature and amount of observed cold crystallization decreases and the total enthalpy of melting increases. The total analysis shows that increasing draw ratio reduces chain molecular motions, with chain mobility showing drastic decreases at stretches greater than 150%. Cieniewska and Rutajska (1979) show that as the draw ratio and heater temperatures are increased in the draw texturing of POY yarns both the melting temperature and heat of fusion of the textured yarns increase, as monitored by DSC. The reduction of the temperature and magnitude of cold crystallization as a function of increasing draw ratio (similar to the observations made on oriented spun yarns) has been reported by Coppola and Frediani (1973) and Schauler and Liska (1975). Huh and Choi (1977) find multiple melting peaks in drawn PET fibers by DTA, and use heating rate techniques to ascribe the peaks to the rearrangement of unstable crystals. Kitakawa (1980) utilized shrinkage force spectra to evaluate the effect of draw texturing variables on the performance of the resulting crimped PET yarns. Backer and Yang (1976) employed TMA data to aid in their elucidation of the factors controlling PET threadline stability during draw texturing. The strain-induced reversible crystal–crystal transition of poly(butylene terephthalate) yarns has been studied by Tashiro *et al.* (1980) using DSC under tension, x-ray diffraction, and infrared spectra techniques. They find the zero heating rate melting point for the α (unstrained) crystal structure to be 225°C and 219°C for the β (strained) crystal structure. The effects of increasing draw ratio and draw temperature on the shrinkage force for PET yarns has been investigated by Hoffrichter (1973) and Buchanan and Hardegree (1977), who show increasing amounts of stretch and decreasing draw temperature serve to increase the SF levels observed. Higher draw temperatures shift the SF peak maxima to higher temperature values, implying that a more thermally stable, oriented structure is being produced. Hassel (1977) has used shrinkage force to differentiate textured and untextured PET yarns and to show the general utility of the TMA technique.

More than any other facet of PET fiber technology, the after-processing steps, which convert the drawn crystalline yarns produced by the primary manufacturers to finished end-use products, have been evaluated by thermal analysis. Using yarns obtained from commercial sources, these studies aim at correlating the effects of annealing, deformation, and relaxation to yarn structure and end-use performance through the identification of thermal analysis parameters that reflect both the process and property parameters of interest. Most comprehensive of these studies are the DTA and TMA investigations of Berndt, Heidemann, and Valk, already reviewed in Section II. Annealing has been shown by many authors (Berndt et al., 1973; Berndt and Bossman, 1976; Heidemann and Berndt, 1976; Heidemann et al., 1977; Maeda and Kanetsuna, 1966; Oswald et al., 1977; Prati and Seves, 1973; Valk, 1972; Wiesner, 1968) to result in premelt endotherms in DSC thermograms. The exact position of the peak depends on annealing time, temperature, stress, and media, as well as spinning history. All these variables are combined in the Krefeld group (Berndt and Bossman, 1976; Heidemann and Berndt, 1976; Valk, 1972) concept of T_{eff}. It has similarly been shown that the temperature, rate, and extent of deformation experienced by a yarn will define the dimensional and structural stability of the resultant yarn as monitored directly by TMA (Aleksandriiskii et al., 1968; Berndt and Heidemann, 1972, 1974, 1977; Heidemann and Berndt, 1973, 1974, 1976; McGregor et al., 1977; Mikhailov et al., 1970; Simov et al., 1973a,b; Tanaka and Nakajima, 1972a,b; Valk, 1970). Examples have already been shown (Section II.C) of how shrinkage and shrinkage force parameters (such as $\sigma_{effective}$) may be used to define the structure, properties and process history of a polyester yarn. An additional example, the effect of the texturing parameters of heater temperatures and amount of overfeed (relaxation) on the shrinkage force of the textured yarn is shown in Fig. 51. Results such as these have been utilized as a process diagnostic, to define aspects of fiber structure such as orientation and crystallinity and to predict end-use performance characteristics as diverse as textile dyeing behavior (McGregor et al., 1977; Valk et al., 1971), heat setting effectiveness (Valk, 1972), and the fatigue behavior of tire cords (Mikhailov et al., 1970).

Quinson et al. (1973) have utilized DSC and TG to study water vapor diffusion and sorption in textured polyester and show by comparison of the observed behavior with that of the precursors yarn that a temporary surface modification occurs in the texturing process. Simov and co-workers (1973a) showed that at temperatures less than 170°C an azo dye (Terasil Marine Blue SGL) aids fiber crystallization but tends to retard crystallization at higher temperatures. The thermooxidative stability of polyester under conditions seen in spinning has been investigated by Dimov et al. (1970a, 1971) using TG and DTA techniques. The effects of

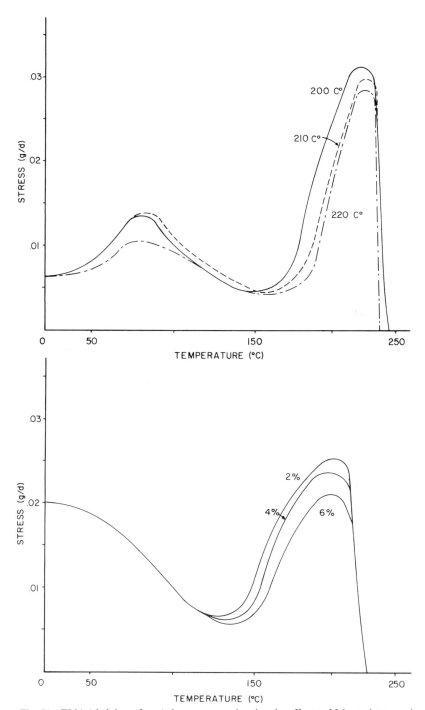

Fig. 51 TMA (shrinkage force) thermograms showing the effects of false twist texturing variables on the resultant traces for polyester feed yarn: (a) effect of the first heater temperature; (b) effect of overfeed (relaxation). Kanebo thermal stress tester, 3°C/min heating rate. [From Buchanan and Hardegree (1977).]

copolymerization on the thermal stability of chemically modified PET fibers has been investigated by Budin and Vanicek (1977) (Na salt of 5-sulfoisophthalic acid as comonomer, i.e., cationically dyeable PET) and Varma *et al.* (1979) (hexanediol comonomer). Using TG and DSC techniques, both groups find the fiber melting and decomposition temperatures decreasing as the comonomer content increases. The crimpability of polyethylene terephthalateisophthalate copolymers was investigated by Repina *et al.* (1973) and Krapotkin *et al.* (1973) using both shrinkage and shrinkage force measurements. Although several DSC and TG studies relating to polyester thermal stability and flammability are present in the literature, they tend to be general in nature and will be treated later in a brief review of fiber flammability investigations. For a detailed treatment of fiber flammability see Chapter 8.

A new class of all aromatic polyesters exhibiting liquid crystalline behavior is now receiving a great deal of scientific attention as both a fiber former and a thermoplastic. In contrast to the lyotropic Kevlar® system, these thermotropic polymers offer the potential simplicity of melt processing while retaining the extraordinary mechanical and thermal properties associated with the aramids. Although most of this activity is appearing through the patent literature (search patents assigned to DuPont, Celanese, Eastman Kodak, etc.), several papers have appeared in the open literature. A general review of the concepts and structures associated with liquid crystalline polymers has been published by Wendorff *et al.* (1978). Menczel and Wunderlich (1980) have investigated the glass transition behavior of poly[(ethylene terephthalate-*co-p*-oxybenzoate) (copolymer of PET and poly(*p*-hydroxy benzoic acid)] and find an absence of hysteresis on heating and cooling through T_g. In a study of thermotropic anthraquinone-containing polyesters, Warner (1980) concludes this polymer exhibits enthalpic relaxation on annealing below T_g, similar to conventional polymeric and small-molecule glasses. An investigation of the crystallization kinetics of thermotropic polyesters by Warner and Jaffe (1980) concludes that techniques and concepts similar to those used with flexible macromolecules can provide insight into the behavior of the thermotropics. Few polymer developments have offered the combination of commercial and scientific excitement or raised as many complex materials physics questions as the rapidly expanding area of thermotropic polymer technology.

3. Acrylics (PAN)*

The acrylics are co or ter polymers of acrylonitrile with other vinyl monomers such as vinyl acetate, vinyl chloride, styrene, etc. ($\geq 90\%$

* See Encyclopedia Britannica (1966), Moncrieff (1975), Morton and Hearle (1975), and Stout (1970).

acrylonitrile), although homopolymer polyacrylonitrile is commercially manufactured. Acrylics are always solution spun, with both wet- and dry-spinning processes being practiced, followed by drawing and usually texturing. Most commonly available as staple, acrylics find their largest applications in knitted textiles because of the fiber's woollike aesthetics. The tensile properties of acrylic yarns are low (2.7–3.6 cN/dtex tenacity) and, similar to nylon, are quite sensitive to moisture content. The physical structure of acrylic fibers has long been an area of interest to polymer physicists because the polymer, although atactic, does show regions of high order difficult to describe. Strong interchain hydrogen bonding slows structural perfecting processes once formed, further complicating structural analysis. Acrylics have found an additional application as the precursor fiber in the production of graphite fiber in the past decade.

Thermal analytical investigations of acrylic fibers fall into two groups, the first aimed at structure-property clarification, the second at pyrolysis behavior. In the first group, the most comprehensive study is contained in the Fiber Society lecture presented by Kimmel in 1971, which unfortunately has never appeared in written form. Illustrations from this lecture, showing the utility of TMA and DSC in the evaluation of acrylic behavior, have been used extensively in Section II. The second area of activity, pyrolysis behavior, has been spurred by interest in the mechanisms of transforming acrylics to graphite fibers.

In a DTA study of the acrylic glass transition, Wegener and Merkle (1971) show that T_g is a strong function of residual solvent and annealing conditions. Dimethyl formamide (spinning solvent) containing yarns showed T_g's at 55, 71, 115, and 140°C. Solvent-free yarns show two T_g's, at 84 and 140°C. Annealing at temperatures equal to or greater than 140°C brought all the yarns to a similar T_g of 71–73°C and suppressed the 140°C transition. Effects of heating rate, atmosphere, and annealing tension are also included in this work. The important discovery that fast heating techniques can allow the melting of homo and copolymer acrylics to be observed by DTA has already been discussed (Dunn and Ennis, 1970, 1971; Hinrichsen, 1971; Layden, 1971). Dunn and Ennis (1971) have shown that fast heating DTA experiments are an effective characterization tool for identifying commercial acrylic variants, as illustrated in Fig. 52. In an investigation of drawn acrylics, Schauler and Kashani (1975) show draw ratio tends to sharpen the exothermic decomposition peak and change the observed shrinkage behavior. They identify two glass transitions in the homopolymer, at 60° and 100°C, respectively. A number of authors (Bechev and Kiril, 1973; Dobretsov *et al.*, 1969; Petrosyan *et al.*, 1977; Yunusov, 1974) have reported thermomechanical data for various acrylics (shrinkage, shrinkage force), including studies of the effects of cross-linking (Petrosyan *et al.*, 1977) and ultrahigh molecular weight (Dobretsov *et*

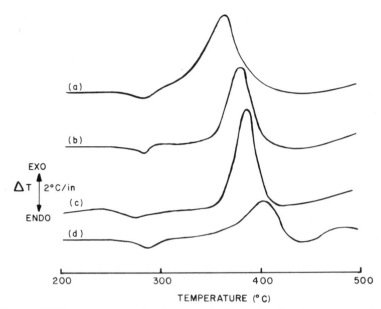

Fig. 52 DSC thermograms of various acrylic fibers showing the melting region: (a) Orlon, (b) Cashmillion, (c) Crelan 61B, (d) Acrilan 16. DuPont 900 DSC, chopped fiber encapsulated sample (2–3 mg). Air atmosphere, 100°C/min heating rate. [From Dunn and Ennis (1971).]

al., 1969). A problem in the acrylic thermal analysis literature is the lack of specification of comonomer species and amount, water concentration, and residual solvent content. Without these qualities being known, temperatures and magnitudes of transition and relaxation phenomena become meaningless, other than as simple observations on a given sample.

DTA and TG have been employed by several authors (Gorbacheva *et al.*, 1973; Grassie and McGuchan, 1972; Minagawa *et al.*, 1974) in studying the mechanism and kinetics of acrylic degradation. Factors such as monomer content, spinning method, and additives have all been investigated, as well as experimental parameters such as heating rate, atmosphere, and sample size. In general, comonomers tend to lower the temperature of exothermic degradation, and the results observed are extremely sensitive to oxygen concentration, surface-to-mass ratio of samples, and additives or other chemical species present in the fiber.

The oxidative stabilization process (preoxidation), which transforms polyacrylonitrile fiber into a structure capable of maintaining its integrity at the high temperatures (1000–3000°C) necessary for carbon fiber production, is exceedingly complex. Concurrent phenomena taking place include cyclization of the PAN structure, a decrease in the number of nitrile groups, and increased conjugation due to CN bond formation (manifest in the fiber changing color from white to black), formation of oxidative cross-links, dehydrogenation of the cyclized structure, melting of the PAN crystalline domains, and a shrinkage of up to 30%. The rate and order at which these chemical and physical changes take place is a function of atmosphere, comonomer content, temperature, applied stress, and yarn structure. In an extensive study, Fitzer and co-workers (Fieldler *et al.*, 1971; Fitzer and Mueller, 1971, 1973; Mueller *et al.*, 1971) have applied DTA, TG, and TMA to the mechanistic clarification of the stabilization process. They conclude that the high-temperature shrinkage process is purely chemical in origin, resulting from the compaction of the PAN during cyclization to a ladder polymer type of structure. As all oriented semicrystalline polymers are known to undergo shrinkage simultaneously with melting, and given the evidence for PAN melting (Dunn and Ennis, 1970; Hinrichsen, 1971; Layden, 1971) and the correlation of the high temperature shrinkage rate maximum with melting temperature shown by Kimmel (1971), it is likely that, although chemical rearrangements are very important, physical effects play a role. The data necessary to settle fully the stabilization mechanism question in PAN may already exist in the abundant graphite process patent literature and a detailed review of the information therein contained would be a substantial contribution to this important, developing technology. TG and DTA have been used by several authors (Gaulin and McDonald, 1971, 1972; Picklesimer, 1972) to define the effectiveness of a variety of treatments aimed at improving PAN as a graphite precursor. These treatments include irradiation (Gulin and McDonald, 1971) and the substitution of thioamide groups for pendant cyano groups (Picklesimer, 1972). Other general studies of PAN preoxidation may be found in the work of Gupta and Trehan, (1976), Gaulin and McDonald (1972), and A DiEdwardo (private communication, 1974).

4. Polyolefins*

Polyolefins are the addition polymers of olefinic monomers, the most important being polyethylene and isotactic polypropylene. Although low melting points and poor aesthetics restrict the use of these polymers in many fiber applications, low cost, good tensile properties, chemical resistance, and durability have led to the usage of polypropylene fiber in indoor–outdoor carpeting, upholstery fabric, ropes and packaging materials. Both polyethylene and polypropylene have been extensively studied

* See Encyclopedia Britannica (1966), Moncrieff (1975), Morton and Hearle (1975), Spruiell and White (1975a,b), and Stout (1970).

as model semicrystalline polymers and have been predominant in analyses of melt extrusion, drawing, and annealing processes. Currently, they are receiving renewed attention in investigations of oriented polymer crystallization (e.g., flowing melt and solution studies), and the morphologies so produced. Although the suitability of the polyolefins as models for the commercially important fiber-forming polymers can be argued, the potential impact of these studies on the fiber industry, in terms of new products and processes, should not be ignored. Although this review will not deal directly with much of this work, the serious student of fiber structure and properties is urged to become familiar with the polyolefin literature. [See, for example, the work of Keller (1975a,b), Peterlin (1972a,b, 1975), Ward (1971, 1975), Spruiell and White (1975a,b), Geil (1963), Pennings*et al.* (1970, 1973a,b), Porter *et al.* (1975), Mead and Porter (1976), and Mead *et al.* (1979).]

Given the magnitude of the polyolefin research reported in the literature, the paucity of thermal analysis investigations of polyolefin fibers is surprising, although much relevant work may have been done on films. A study of polypropylene spinnability by Halip and Chirica (1975) indicates that an isotacticity of at least 95% and a DTA crystallinity of at least 50% is necessary for fiber production. Samuels (1975) has investigated the melting of restrained, drawn polypropylene fibers and isothermally crystallized bulk material and finds multiple melting peaks in both cases. Detailed data analysis and sample characterization failed to identify the cause of the phenomena other than to indicate that it was the result of several morphologies being present. A correlation of the higher-temperature melting peak and the amorphous orientation level of the drawn fibers was found, however, and extrapolation of the higher peak temperatures led to the identification of 220°C as the equilibrium melting point of polypropylene. Multiple melting peaks were also observed by Lengyel *et al.* (1975) and Jaffe (1978). Based in part on the rate-dependent melting of polypropylene fibers spun at various stress levels and their subsequent recrystallization kinetics, Jaffe (1978) suggests high-stress spun fibers contain two crystalline morphologies, fibrillar and lamellar, with no truly extended chain material present in either crystalline phase. The effect of heat treatment on the degradation kinetics of polypropylene was investigated by Baltenas *et al.* (1971), who showed all treatments to cause the observed activation energy of degradation to increase (precursor, 52.9 kcal/mole; steam treat 125°C, 30 min, 73.5 kcal/mole, dry heat, 160°C, 40 sec, 54.6 kcal/mole). The shrinkage force behavior of as-spun polypropylene as a function of spinning speed has been reviewed by Buchanan and Hardegree (1977) as shown in Fig. 53: Note the similarity to the

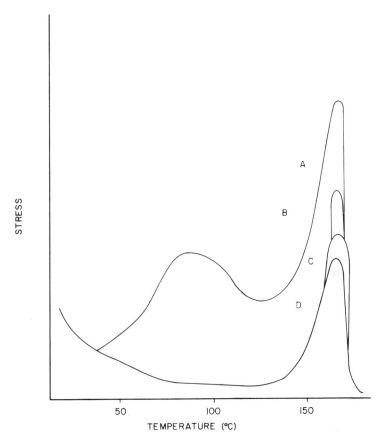

Fig. 53 TMA (shrinkage force) thermograms for as-spun polypropylene yarns as a function of spinning take-up speeds. A, 4000 m/min; B, 3000 m/min; C, 2000 m/min; D, 1000 m/min. [From Buchanan and Hardegree (1977) after Gerasimova *et al.* (1971).]

poly(ethylene terephthalate) data shown earlier (Fig. 50) and the shift toward crystalline region effects to be expected with the much-faster-crystallizing polypropylene system. For polypropylene shrinkage data the papers and books of Samuels (1973, 1974) are the best source, although the data were obtained by oven shrinkage techniques rather than, by TMA.

Andrienko *et al.* (1976) have shown that wear causes structural degradation of polypropylene fabrics as evidenced by loss of crystallinity and reduction of melting point by DTA. Ikeda and co-workers (1973a,b), have investigated the effect of grafting acrylic acid onto polypropylene fabrics by DSC and TG and find both the heat of fusion and the degradation temperature of the fabric decreases with increasing acrylic acid content.

5. Miscellaneous Fibers

Some of the less commercially important fiber-forming polymers to receive attention in the thermal analysis literature include polyvinyl alcohol (method of polymer preparation on fiber structure) (Kalinina *et al.*, 1970), (effect of phosphorous acid modification (Laszkiewicz, 1971), polyvinyl chloride (decomposition behavior as a function of drawing and annealing steps) (Stanciu, 1974), (pretreatments to improve graphite fiber precursor utility) (Lashina *et al.*, 1975), and phenolic fibers (pyrolysis, graphite fiber production) (Lin and Economy, 1973). In the past few years a large number of potential high use temperature, low-flammability polymers such as imides (Prokhorov, 1974; Prokopchuk *et al.*, 1976; Goel *et al.*, 1979), imidequinoxalines (Duffy and Augl, 1971a,b), hydrazides (Frazer *et al.*, 1966), benzimidazoles (Davis *et al.*, 1972; Jaffe, 1977), etc., have been screened as fiber formers, with DTA, TG, and occasionally TMA techniques being utilized to assess probable fiber performance. The list presented here is by no means all-inclusive and the interested reader

TABLE IV

SOME TYPICAL APPLICATIONS OF TA TO FIBER FLAMMABILITY

Reference	Problem treated	TA methods employed
Hassel (1977)	TA methods, polyester examples	DTA, DSC, TG, TMA
Perkins *et al.* (1966)	Chemically modified cotton, cotton–modified acrylic blends	DTA, TG
Hendrix *et al.* (1970)	Flame retardants, thermal decomposition; cotton, wool, Nomex	DTA, TG
Dipietro *et al.* (1971)	Flame retardants, mechanisms; cotton, polyester nylon	DSC, TG
Amigo and Chanh (1975)	Heat of combustion, textile fibers	DTA
Wiesner (1974)	Flame-retarded polyesters	DTA
Wiesner (1976a)	Flame-retarded rayon	DTA, TG
Wiesner (1976b)	Degradation mechanisms, cellulosics, polyester and polypropylene	DTA
Mack and Hobart (1975)	Flame-retarded polyester– cotton blends	DSC
Carroll-Potczynski (1973)	Fiber blends	TG
Bingham and Hill (1975)	Flame-retardant cotton	DSC, TG
Saglio *et al.* (1972)	Saturated chlorinated polyethylene	TG, TMA
Ordoyno and Rowan (1978)	Binary mixtures	DTA, TG
Martin and Miller (1978)	Polyester–wool blends	TG

should search government reports and the patent literature for further information. Fiber blends and copolymer fibers have received some attention, i.e., fiber blend characterization by DTA (Crighton and Holmes, 1972; Ordoyno and Rowan, 1978) amide–ester copolymers fibers (interchange kinetics) (Dimov and Georgiev, 1973), (shrinkage force) (Dietrich and Kaufmann, 1976) and the evaluation of rayon–polyester yarns in the fabric skins of honeycomb structures (Taylor, 1973).

DTA, DSC, and TG have been heavily utilized in studies of yarn and fabric flammability as a function of yarn molecular structure, yarn blend composition, and additive or coating effects. A listing of typical fiber-related references can be found in Table IV; procedural details and mechanistic interpretations are discussed in Chapter 8.

IV. Conclusions

The major strength of thermal analysis in the characterization of fibers lies in the amount of process, structure, and property information contained in easily obtained thermograms. Preceding sections have shown how DTA, DSC, TMA, and TG may be used to obtain parameters relating to crystalline and amorphous perfection and orientation, the nature of tie molecules, and the thermal and chemical stability of fiber structures of interest. The major weakness of thermal analysis in fiber studies is the same ease of obtaining data that is a principal strength; the interpretation of thermograms without a base line constructed from highly characterized, judiciously chosen and variously prepared samples, is, at best, speculative and difficult, and often impossible or misleading. Only after a depth of experience and a catalogue of thermal analysis spectra have been accumulated can the process, structure, and property performance of a fiber be "read" directly from thermal analysis data.

The published thermal analysis literature relating to fibers is rich in showing the breadth of DTA/DSC and TG thermograms possible for various fiber types, especially the melting behavior of synthetics and the thermal decomposition of common textiles. There is a dearth of papers, however, that attempt the valuable and difficult task of generalizing specific observations to models of fiber formation and performance by keying in on the similarities of response of oriented macromolecules rather than emphasizing the obvious differences caused by differing molecular structures. Such an approach would likely add increased understanding and utility to the thermal analysis of the more complex natural fibers, to date largely ignored by thermal analysts. TMA could be especially useful in this area because it is the thermal analysis technique most sensitive to molecular chain orientation, and high molecular chain orientation is the unifying feature of fiber structures. The general lack of TMA data in the literature can probably be ascribed to the newness of the technique and it is expected that as the technique becomes more widespread, its utility and effectiveness will be recognized and publications will follow.

The ability of thermal analysis to "fingerprint" fiber structure or fiber processing history implies a major role could be played by thermal analysis as a quality control tool in all aspects of fiber production. Although this is certainly true and there is no doubt that thermal analysis is underutilized in quality control applications, given the amount of information contained in thermal analysis spectra and the necessity of working with extremely small sample sizes, thermal analysis is better suited as a process diagnostic and research tool in the fiber industry. Most fiber problems are concerned with uniformity, within or between bobbins, and the optimum quality control techniques are those that can quickly scan bobbins of yarn on a continuous basis. Once a problem has been observed, however, thermal analysis techniques are unmatched for quickly providing the information necessary to eliminate the problem (identification of process temperatures, stress, etc.). Again, a data base showing the effect of process variables on thermograms of the resulting yarns is imperative. A simultaneous study of the range of end-use properties caused by these same process variables completes the definition of the process–structure –property relationship spectrum (see Fig. 4) and extends the utility of the study from the diagnosis of an existing process to the efficient identification of new processes and products.

Most current, commercial thermal analysis equipment designed for small sample weights (1–10 mg) is suitable for fiber research. Welcome additions to these product lines would include sample holders and sample preparation aids designed to eliminate the difficulties presently associated with mounting the necessarily long lengths of fibrous samples under conditions of reproducible weight and stress in existing machine geometries. Such equipment when coupled to an access to fiber samples of known history and a background in polymer science will provide a valuable tool for the study and clarification of the physics, chemistry, and engineering of fiber-forming systems.

ACKNOWLEDGMENTS

The author is indebted to Dr. E. Turi, Dr. L. Roldan, and Dr. G. Valk for the hours spent in reviewing, discussing, and offering valuable suggestions for improving the organization and technical content of this chapter. The many useful comments of my Celanese and Fiber Industries colleagues, especially Mr. B. Morris, Mr. A. DiEdwardo, and Dr. A. Buckley, are also highly appreciated. My thanks to the Celanese and the Fiber Industries organizations for providing clerical and drafting help and, most of all, to my wife for putting up with it all.

References

Addyman, L., and Ogilvie, G. D. (1979). *Br. Polym. J.* **11**(3), 15.
Aleksandriiskii, S. C., Aizenshtein, E. M., and Petukhov, B. V. (1968). *Mekh. Polim.* **4**(2), 369.
American Association for Textile Technology (1977). "Textile Fibers and Their Properties," Managr. No. 109. A.A.T.T., Greensboro, North Carolina.
Amigo, J. M., and Chanh, N. B. (1975). *J. Therm. Anal.* **7**, 183.
Andrianova, G. P., Arutyunov, G. A., and Popov, YU. V. (1978). *J. Polym. Sci., Polym. Phys. Ed.* **16**, 1139.
Andrienko, P. P., Shastuk, T. S., and Pavlov, A. I. (1976). *Izv. Vyssh. Uchebn. Zaved., Tekhnol. Legk. Prom-sti.* **6**, 15.
Anton, A. (1973). *Text. Res. J.* **43**, 524.
Arakawa, T., Nagatoshi, F., and Arai, N. (1968). *Polym. Lett.* **6**, 513.
Arons, G. N., and Macnair, R. W. (1970). *U.S. Army Natick Lab., Rep.* No. 71-17-CE (TS-174).
Artunc, H. (1976). *Lenzinger Ber.* **40**, 10.
Artunc, H., and Egbers, G. (1976). *Chemiefasern/Text.-Ind.* **26**(6), 510.
Backer, S., and Yang, W. (1976). *Text. Res. J.* **46**, 599.
Baltenas, R., Monkeviciute, G., and Paskevicius, V. (1971). *Nauchno-Issled. Tr. Litov. Nauchno-Issled. Inst. Tekst. Prom-sti.* **1**, 237.
Basch, A., and Lewin, M. (1973a). *J. Polym. Sci., Polym. Chem. Ed.* **11**, 3071.
Basch, A., and Lewin, M. (1973b). *J. Polym. Sci., Polym. Chem. Ed.* **11**, 3095.
Basch, A., and Lewin, M. (1974). *J. Polym. Sci., Polym. Chem. Ed.* **12**, 2053.
Basch, A., and Lewin, M. (1975). *Text. Res. J.* **45**, 246.
Baver, A. I., Korovkina, L. A., Matveeva, N. V., and Sukhodrovskaya, K. A. (1973). *Tr. Inst. Goryuch Iskop., Moscow* **29**(1), 99.
Bechev, K., and Kiril, D. (1973). *Tekst. Prom-st. (Sofia)* **22**(5), 242.
Becht, J. Jr., and Fischer, H. (1970). *Kolloid Z. Z. Polym.* **240**, 766.
Berndt, H. J. (1975). *Melliand Textilber.* **56**, 928.
Berndt, H. J., and Bossman, A. (1976). *Polymer* **17**, 241.
Berndt, H. J., and Heidemann, G. (1972). *Dtsch. Faerber-Kal.* **76**, 408.
Berndt, H. J., and Heidemann, G. (1974). *Melliand Textilber. Int.* **55**(6), 548.
Berndt, H. J., and Heidemann, G. (1977). *Melliand Textilber.* **1**, 83.
Berndt, H. J., and Heidemann, G. (1980). *Proc. Int. Conf. Therm. Anal., 6th, 1979*, p. 345.
Berndt, H. J., Schulz, H., and Heidemann, G. (1973). *Melliand Textilber.* **54**, 773.
Bhatt, G. M., and Bell, J. P. (1976). *J. Polym. Sci., Phys.* **14**, 575.
Billmeyer, F. W. (1962). "Textbook of Polymer Science." Wiley (Interscience), New York.
Bingham, M. A., and Hill, B. J. (1975). *J. Therm. Anal.* **1**, 347.
Black, B. W., and Preston, J., ed. (1973). "High Modulus Wholly Aromatic Fibers." Dekker, New York.
Bosley, D. E. (1967). *J. Polym. Sci., Part C* **20**, 77.
Brandrup, J., and Immergut, E. H., ed. (1966). "Polymer Handbook." Wiley (Interscience), New York.
Brazauskas, V., Paskevicius, V., and Monkeviciute, G. (1973). *Tekst. Prom-st. (Moscow)* **6**, 74.
Broido, A., Javier-son, A. C., Ouano, A. C., and Barrall, E. M., II (1973). *J. Appl. Polym. Sci.* **17**, 3627.
Brown, J. R., and Ennis, B. C. (1977). *Text. Res. J.* **47**, 62.
Buchanan, D. R., and Hardegree, G. I. (1977). *Text Res. J.* **47**, 732.
Buchanan, D. R., and Walters, J. P. (1977). *Text Res. J.* **47**, 398.

Budin, J., and Vanicek, J. (1977). *Chem. Vlakna* **27**(1), 14.
Bueche, F. (1962). "Physical Properties of Polymers." Wiley (Interscience), New York.
Carroll-Porczynski, C. Z. (1972). *Therm. Anal., Proc. Int. Conf., 3rd, 1971* Vol. 3, p. 273.
Cassel, B. (1976). *Perkin-Elmer Bull.* No. L-475.
Celanese Corporation (1975). "Man-Made Fiber and Textile Dictionary." Celanese Corp.
Chatfield, D. A., Einhorn, I. N., Mickleson, R. W., and Futrell, J. H. (1979a). *J. Polym. Sci., Polym. Chem. Ed.* **17**, 1353.
Chatfield, D. A., Einhorn, I. N., Mickleson, R. W., and Futrell, J. H. (1979b). *J. Polym. Sci., Polym. Chem. Ed.* **17**, 1367.
Chatterjee, P. K., and Schwenker, R. F. (1972). *Tappi* **55**(1), 111.
Cieniewska, M., and Ratajska, M. (1979). *Tekst. Prom-st. (Sofia)* **39**, 30.
Coppola, G., and Frediani, P. (1973). *Nuova Chim.* **49**(10), 37.
Crighton, J. S., and Findon, W. M. (1977). *J. Therm. Anal.* **11**, 305.
Crighton, J. S., and Happey, F. (1968). *Symp. Fibrous Proteins* [*Pap.*], 1967 p. 409.
Crighton, J. S., and Hole, P. N. (1976). *Proc. Int. Wolltextil-Forschungskonf., 5th, 1975* p. 499.
Crighton, J. S., and Hole, P. N. (1978). *Thermochim. Acta* **24**, 327.
Crighton, J. S., and Holmes, D. A. (1972). *Proc. Therm. Anal. Int. Conf., 3rd, 3*, 411.
Crighton, J. S., Findon, W. M., and Happey, F. (1971). *Appl. Polym. Symp.* **18**, 847.
Csete, A., and Levi, D. W. (1976). PLASTEC *Note* **29**.
Davidson, T. E., and Roberts, C. W. (1980). *J. Appl. Polym. Sci.* **25**, 1491.
Davis, H. J., Model, F. S., Jaffe, M., Sieminski, M. A., Boom, A. A., and Lee, L. A. (1972). *U.S. Dep. Inter., Off. Saline Water, Final Rep.* Contract No. 14-30-2810.
Dietrich, K., and Kaufmann, S. (1976). *Faserforsch. Textiltech.* **27**(9), 491.
Dimov, K. and Georgiev, J. (1973). *Faserforsch. Textiltech.* **24**(3), 120.
Dimov, K., Aleksandrova, I., Lazarova, R., and Bechev, K. (1970a). *Tekst. Prom-st. (Sofia)* **19**(3), 24.
Dimov, K., Betschev, C., Lazarova, R., and Aleksandrova, I. (1970b). *Faserforsh. Textiltech.* **21**(11), 492.
Dimov, K., Aleksandrova, I., Lazarova, R. and Bechev, K. (1971). *God. Nauchnoizsled. Inst. Khim. Prom-st.* **8**(1), 297.
Dimov, K., Georgiev, J., and Bechev, C. (1973). *Faserforsh. Textiltech.* **24**(8), 337.
Dipietro, J., Barda, H., and Stepniczka, H. (1971). *Text. Chem. Color.* **3**(2), 45.
Dobretsov, S. L., Lomonosova, N. V., and Stel'mukh, V. P. (1969). *Vysokomol. Soedin., Ser. B* **11**(4), 782.
Duffy, J. V. (1969). *U.S. Nav. Ordinance Lab. Rep.* No. NOLTR 69-113.
Duffy, J. V., and Augl, J. M. (1971a). *U.S. Nav. Ordinance Syst. Rep.* No. NOLTR 70-245.
Duffy, J. V., and Augl, J. M. (1971b). *U.S. Nav. Ordinance Syst. Rep.* No. NOLTR 70-247.
Dunn, P., and Ennis, B. C. (1970). *J. Appl. Polym. Sci.* **14**, 1975.
Dunn, P., and Ennis, B. C. (1971). *Thermochim. Acta* **3**, 81.
Encyclopedia Britannica (1966). William Benton, Chicago, Illinois.
Fialkov, A. S., Kuchinskaya, O. F., Kabardina, V. A., and Zaychikov, S. G. (1968). *Khim. Tverd, Topl. (Leningrad)* **3**, 116 (Translation: Foreign Translation Division, Report No. FTD-MT-24-117-69).
Fieldler, A. K., Fitzer, E., and Mueller, D. J. (1971). *Org. Coat. Plast. Chem.* **31**(1), 380.
Finch, A., and Ledward, D. A. (1973). *Biochim. Biophys. Acta* **295**(1), 301.
Fitzer, E., and Mueller, D. J. (1971). *Makromol. Chem.* **A4**, 117.
Fitzer, E., and Mueller, D. J. (1973). *Polym. Prep., Am. Chem. Soc., Div. Polym. Chem.* **14**(1), 386.
Forgács, P. (1978). *J. Polym. Sci., Polym. Lett. Ed.* **16**, 1.

Frazer, A. H., Memeger, W., Jr., and Wilson, D. R. (1966). *Tech. Rep. AFML-TR—Air Force Mater. Lab. (U.S.)* **AFML-TR-65-221,** Part II.

Fujimoto, F., and Yamashita, M. (1971). *J. Text. Mach. Soc. Jpn.* **17**(4), 124.

Ganchuk, L. M., Fainberg, E. Z., and Mikhailov, N.V. (1973). *Khim. Volokna* **15**(1), 31.

Gaulin, G. A., and McDonald, W. R. (1971). *Air Force Rep.* **SAMSO-TR-71-61,** Vol. 1.

Gaulin, C. A., and McDonald, W. R. (1972). *Space Missile Sys. Organ., Rep.* **TR-0059(6250-40)-5,** Vol. I.

Geil, P. H. (1963). "Polymer Single Crystals." Wiley (Interscience), New York.

Gerasimova, L. S., Sukharev, N. I., Mezhirova, S. Ya., Lavrov, B. B. Pakshver, A. B., and Filbert, V. D. (1971). *Fibre Chem. (Engl. Transl.)* **1,** 98, 101.

Gilbert, R. D., and Rhodes, J. H. (1970). *J. Polym. Sci., Part C* **30,** 509.

Gill, P. S., Hassel, R. L., and Woo, L. (1975). *Coat. Plast., Prepr. Pap. Meet. (Am. Chem. Soc., Div. Org. Coat. Plast. Chem.)* **35**(2), 400.

Godfrey, L. E. A. (1970). *Text. Res. J.* **40,** 116.

Goel, R. N., Varma, I. K., and Varma, D. S. (1979). *J. Appl. Polym. Sci.* **24,** 1061.

Gorbacheva, V. O., Mikhailova, T. K., Fedorkina, S. G., Konnova, N.F., Azarova, M. T., and Konkin, A. F. (1973). *Khim. Volokna* **15**(5), 16.

Gotze, W., and Winkler, F. (1967). *Faserforsch. Textiltech.* **18,** 385.

Grassie, N., and McGuchan, R. (1972). *Eur. Polym. J.* **8,** 865.

Gupta, P. K., and Trehan, J. C. (1976). *Indian J. Technol.* **14**(3), 133.

Gyori, K., Bialo, G., and Rusznak, I. (1976). *Magy. Textiltech.* **29,** 457.

Hagege, R. (1977). *Text. Res. J.* **47,** 229.

Halip, V. (1971). *Mater. Plast. (Bucharest)* **8**(2), 80.

Halip, V., and Chirica, L. (1975). *Mater. Plast. (Bucharest)* **12**(3), 157.

Hall, J. H., and Godwin, R. W. (1970). *Instrum. News* **21**(2), 1.

Haly, A. R., and Snaith, J. W. (1969). *Text. Res. J.* **39,** 906.

Hamson, F. W., and Manley, T. R. (1974). *Chem. Ind. (London)* **16,** 660.

Hassel, R. L. (1977). *Am. Lab. (Fairfield, Conn.)* **9,** 35.

Hatayama, T., and Kanetsuna, H. (1970). *Sen'i Gakkaishi* **26**(2), 89.

Havens, J. A., Welker, J. R., and Sliepcevich, C. M. (1971). *J. Fire Flammability* **2,** 321.

Hearle, J. W. S. (1977). *J. Appl. Polym. Sci.* **31,** 137.

Hearle, J. W. S., and Peters, R. H., ed. (1963). "Fibre Structure." Butterworth, London.

Heidemann, G., and Berndt, H.-J. (1973). *Melliand Textilber.* **54,** 546.

Heidemann, G., and Berndt, H.-J. (1974). *Chemiefasern Text.-Anwendungstech./Text.-Ind.* **24**(1), 46.

Heidemann, G., and Berndt, H.-J. (1976). *Melliand Textilber. (Engl. Ed.)* **6,** 485.

Heidemann, G., Berndt, H. J., and Stein, G. (1977). *Melliand Textilber.* **3,** 250.

Hellauer, H., and Winkler, R. (1975). *Connect. Tissue Res.* **3**(4), 227.

Hendrix, J. E., Anderson, T. K., Clayton, T. J., Olson, E. S., and Barker, R. H. (1970). *J. Fire Flammability* **1,** 107.

Heuvel, M., and Huisman, R. (1978). *J. Appl. Polym. Sci.* **22,** 2219.

Hinrichsen, G. (1971). *Angew. Makromol. Chem.* **20,** 121.

Hirschler, R., Mihalik, B., and Mikesi, G. (1971). *Kolor. Ert.* **13**(9-10), 223.

Hoffrichter, S. (1973). *Faserforsch. Textiltech.* **24**(7), 289.

Horowitz, H. H., and Metzger, G. (1963). *Anal. Chem.* **35,** 1464.

Huh, Y. W., and Choi, Y. Y. (1977). *Sumyu Konghakhoe Chi* **14**(2), 49.

Huh, Y. W., and Soong, S. K. (1976). *Sumyu Konghakhoe Chi* **13**(1), 33.

Hybart, F. J., and Platt, J. D. (1967). *J. Appl. Polym. Sci.* **11,** 1449.

Ichizo, A., and Takeuchi, T. (1971). *Sen'i Gakkaishi* **27**(11), 486.

Ikeda, R. M. (1980). *J. Polym. Lett. Ed.* **18,** 325.

Ikeda, T., Tsuji, W., and Ikeda, Y. (1973a). *Sen'i Gakkaishi* **29**(6), T243.
Ikeda, T., Tsuji, W., and Ikeda, Y. (1973b). *Sen'i Gakkaishi* **29**(7), T267.
Ioelovics, M., Slysh, L. I., Tkachenko, F. F., and Kozlov, P. V. (1975). *Tezisy Dokl.—Vses. Konf. Khim. Fiz. Tsellyul., 1st, 1975* Vol. 2, p. 135.
Jaffe, M. (1977). *Therm. Anal. Symp., Am. Phys. Soc., APS Prepr., 1977* p. 1.
Jaffe, M. (1978). *In* "Thermal Methods in Polymer Analysis" (S. W. Shalaby, ed.) p. 93. The Franklin Institute Press, Philadelphia, Pennsylvania.
Jasse, B., and Moutte, A. (1972). *Bull. Soc. Chim. Fr.* **6**, 2251.
Jeziorny, A. (1970). *Polimery (Warsaw)* **15**(2), 71.
Kalinina, T. N. Afanaseva, G. N., Vol'f, L. A., Meos, A. I., Kremer, E. B., Frenkel, S. Ya., and Mnatsakanov, S. S. (1970). *Vysokomol. Soedin., Ser. B* **12**(9), 661.
Kanebo Engineering Ltd. (1979). "Practical Application of Kanebo Thermal Stress Tester." Kanebo Eng. Ltd., Osaka, Japan.
Keller, A. (1975a). *J. Polym. Sci., Polym. Symp.* **51**, 7.
Keller, A. (1975b). *Plast. Polym.* **43**(163), 15.
Kimmel, R. M. (1971). Fiber Society Lecture.
Kitakawa, T. (1980). *Fiber Prod.* **8**, 50.
Kiyotsukuri, T., and Nagasawa, Y. (1976). *Sen'i Gakkaishi* **32**(5), T-187.
Kokta, B. V., and Valde, J. L. (1972). *Tappi* **55**(3), 375.
Kolshak, V. V., and Vinogradova, S. V. (1965). "Polyesters" (Engl. ed.) Pergamon, Oxford.
Konda, A., Tsukada, M., and Kuroda, S. (1973). *J. Polym. Sci., Polym. Lett. Ed.* **11**, 247.
Krapotkin, V. P., Fantina, G. I., Geller, V. E., and Aizenshtein, E. M. (1973). *Khim. Volokna* **15**(5), 12.
Kunugi, T., Moriya, M., Yamamoto, Y., and Hashimoto, M. (1975). *Nippon Kagaku Kaishi* **9**, 1593.
Langley, J. T., Drews, M. J., and Barker, R. H. (1980). *J. Appl. Polym. Sci.* **25**, 243.
Lashina, L. V., Bogdanova, V. A., Levit, R. M., and Temnikov, K. L. (1975). *Khim. Volokna* **3**, 31.
Laszkiewicz, B. (1971). *J. Appl. Polym. Sci.* **15**, 437.
Layden, G. K. (1971). *J. Appl. Polym. Sci.* **15**, 1283.
Lengyel, M., Mathe, K., and Bodor, G. (1975). *Kem. Kozl.* **43**(1), 73.
Levy, P. F. (1970). *Am. Lab.* **2**, 46.
Lewin, M., and Basch, A. (1972). *Natl. Bur. Stand. [Rep.] (U.S.)* **NBS-GCR4.**
Lin, R. Y., and Economy, J. (1973). *Appl. Polym. Symp.* **21**, 143.
Lindenmeyer, P. H. (1980). *Text. Res. J.* **50**, 395.
McAdie, H. G., ed. (1969). "Proceedings of the Third Toronto Symposium on Thermal Analysis." Toronto Section, Chem. Institute of Canada.
McGregor, R., Grady, P. L., Montgomery, T., and Adeimy, J. (1977). *Text. Res. J.* **47**, 598.
Mack, C. H., and Hobart, S. R. (1975). *U.S., Agric. Res., Serv. South. Reg. [Rep.]*, **ARS-5-60**, 89.
MacLean, D. L. (1974). *J. Appl. Polym. Sci.* **18**, 625.
Maeda, K., and Kanetsuna, H. (1966). *Kogyo Kagaku Zasshi* **69**(9), 1784.
Magoshi, J., and Nakamura, S. (1975). *J. Appl. Polym. Sci.* **19**, 1013.
Magoshi, J., and Nakamura, S. (1978). *Therm. Anal. [Proc. Int. Conf.], 5th, 1977* p. 10.
Magoshi, J., Magoshi, Y., Nakamura, S., Kasai, N., and Kakudo, M. (1977). *J. Polym. Sci., Polym. Phys. Ed.* **15**, 1675.
Magoshi, J., Mizuide, M., Magoshi, Y., Takahashi, K., Kubo, W., and Nakamura, S. (1979). *J. Polym. Sci., Polym. Phys. Ed.* **17**(3), 515.
Man-Made Fiber Producers Association (1970). "Guide to Man-Made Fibers." Man-Made Fiber Prod. Assoc., New York.

Man-Made Fiber Producers Association (1971). ''Man-Made Fiber Fact Book.'' Man-Made Fiber Prod. Assoc., New York.

Mansrekar, T. G. (1973). *Chem. Color.* **5**, 186.

Mark, H. F., Atlas, S. M., and Cernia, E. (1967). ''Man-Made Fibers,'' Vol. 1. Wiley (Interscience), New York.

Mark, H. F., Atlas, S. M., and Cernia, E. (1968). ''Man-Made Fibers,'' Vol. 2. Wiley (Interscience), New York.

Martin, J. R., and Miller, B. (1978). *Text. Res. J.* **48**, 97.

Mead, W. T., and Porter, R. S. (1976). *J. Appl. Phys.* **47**(C10), 4278.

Mead, W. T., Desper, C. R., and Porter, R. S. (1979). *J. Polym. Sci., Polym. Phys. Ed.* **17**, 859.

Menczel, J., and Wunderlich, B. (1980). *J. Polym. Sci., Polym. Phys. Ed.* **18**, 1433.

Mikhailov, N. V., Tokareva, L. G., Gorbacheva, V. O., Fainberg, E. Z., Terekhova, G. M., Khokhlova, N. S., and Karkova, N. G. (1970). *Khim. Volokna* **5**, 9.

Mikolajczyk, W., and Ratujczyk, J. (1975). *Przegl. Wlok.* **29**(12), 602.

Miller, B. (1967). *J. Appl. Polym. Sci.* **11**, 2343.

Miller, B. (1971). *Thermochim. Acta* **2**, 225.

Minagawa, M., Okamoto, M., and Ishizuka, O. (1974). *Nippon Kagaku Kaishi* **2**, 387.

Mininni, R. M., Moore, R. S., Flick, J. R., and Petrie, S. E. B. (1973). *J. Macromol. Sci., Phys.* **B8**(1-2), 343.

Mitsuishi, Y., and Ikeda, M. (1966). *J. Polym. Sci., Polym. Phys. Ed.* **4**, 283.

Miyagi, A., and Wunderlich, B. (1972). *J. Polym. Sci., Polym. Phys. Ed.* **10**, 1401.

Mnuskina, N. K., Babloyan, O. O., and Kutin, V. A. (1974). *Kozh. Obuvna Prom-st.* **16**(11), 36.

Moncrieff, R. W. (1975). ''Man-Made Fibres,'' 6th ed. Newnes-Butterworth, London.

Morton, W. E., and Hearle, J. W. S. (1975). ''Physical Properties of Textile Fibres,'' 2nd ed. Textile Institute-Heinemann, London.

Mueller, D. J., Fitzer, E., and Fiedler, A. K. (1971). *Int. Conf. Carbon Fibres, Their Compos. Appl., Pap., 1971* Vol. 2, p. 1.

Nagano, M. (1976). *Sen'i Kako* **28**(1), 9.

Neduzhii, I. A., Labinov, S. D., Shimchuk, T. Ya., and Vishenskii, S. A. (1971). *Simp. Fiz.-Khim. Polim.* **9**, 73.

Nielson, L. E. (1962). ''Mechanical Properties of Polymers.'' Van Nostrand-Reinhold, Princeton, New Jersey.

Okuhashi, T. (1970). *Kobunshi Kagaku* **27**(304), 562.

Ordoyno, N. F., and Rowan, S. M. (1978). *Thermochim. Acta* **23**, 371.

Oswald, H. J., Turi, E. A., Harget, P. J., and Khanna, Y. P. (1977). *J. Macromol. Sci., Phys.* **B13**(2), 231.

Penn, L., and Larsen, F, (1979). *J. Appl. Polym. Sci.* **23**, 59.

Pennings, A., Van der Mark, J. M. A., and Kiel, A. M. (1970). *Kolloid Z. Z. Polym.* **237**(2), 336.

Pennings, A., Zwijnenburg, A., and Lageveen, R. (1973a). *Kolloid Z. Z. Polym.* **237**(2), 336.

Pennings, A., Zwijnenburg, A., and Lageveen, R. (1973b). *Kolloid Z. Z. Polym.* **251**(7), 500.

Perez, G., and LeCluse, C. (1979). *Proc. Int. Chem. Fiber Conf., 18, 1979* p. 1.

Perkins, R. M., Drake, G. L., Jr., and Reeves, W. A. (1966). *J. Appl. Polym. Sci.* **10**, 1041.

Petakhova, N. N., Goncharova, V. A., Smirnov, L. N. and Litvinov, V. P. (1975). *Khim. Geterotsikl. Soedin.* **6**, 770.

Peterlin, A. (1972a). *Text. Res. J.* **42**(1), 20.

Peterlin, A. (1972b). *Appl. Polym. Symp.* **20**, 269.

Peterlin, A. (1975). *Adv. Chem. Ser.* **142**, 1.

Petrosyan, V. A., Gabrielyan, G. H., and Rogovin, Z. A. (1977). *Prom-st. Arm.* No. 2, p. 36.

Philip, W. M. S. (1972). *J. Forensic Sci.* **17**(1), 132.

Picklesimer, L. G. (1972). *Tech. Rep. AFML-TR—Air Force Mater. Lab.* (*U.S.*) **AFML-TR-72-14**, Part I.

Pokrovskaya, L. V., and Uterskii, L. E. (1972). *Khim. Volokna* **14**(2), 10.

Porter, R. S., and Johnson, J. F., eds. (1968). "Analytical Calorimetry," Vol. 1. Plenum, New York.

Porter, R. S., and Johnson, J. F., eds. (1970). "Analytical Calorimetry," Vol. 2. Plenum, New York.

Porter, R. S., Weeks, N. E., Capiati, N. J., and Krzewki, R. J. (1975). *J. Therm. Anal.* **8**, 547.

Prati, G., and Seves, A. (1973). *Tinctoria* **70**(8), 267.

Prevorsek, D. C. (1971). *J. Polym. Sci., Part C* **32**, 343.

Prevorsek, D. C., Harget, P. J., Sharma, R. F., and Reimschuessel, A. C. (1973). *J. Macromol. Sci., Phys.* **8**(1), 127.

Prevorsek, D. C., Tirpak, G. A., Harget, P. J., and Reimschuessel, A. C. (1974). *J. Macromol. Sci., Phys.* **9**(4), 733.

Prokhorov, O. E. (1974). *Mater. Resp. Konf. Tekst. Khim., 3rd, 1974* p. 86.

Prokopchuk, N. R., Bessonov, M. I., Korzhavin, L. N., Baklagina, Yu. G., Kuznetsov, N. P., and Frenkel, S. Ya. (1976). *Khim. Volokna* **6**, 44.

Quinson, J. F., Chabert, B., Chauchard, J., Soulier, J. P., and Edel, G. (1973). *Bull. Sci. Inst. Text. Fr.* **2**(5), 1.

Quynn, R. G. (1972). *J. Appl. Polym. Sci.* **16**, 3393.

Reimschuessel, A. C., and Prevorsek, D. C. (1976). *J. Polym. Sci., Phys.* **14**, 485.

Repina, L. P., Vorob'eva, R. V., and Aizenshtein, E. M. (1973). *Khim Volokna* **15**(6), 53.

Ribnick, A. (1969a). *Text. Res. J.* **39**, 428.

Ribnick, A. (1969b). *Text. Res. J.* **39**, 742.

Rodrig, H., Basch, A., and Lewin, M. (1975). *J. Polym. Sci., Polym. Chem. Ed.* **13**, 1921.

Rodriquez, F. (1970). "Principles of Polymer Systems." McGraw-Hill, New York.

Saglio, N., Berticat, P., and Vallet, G. (1972). *J. Appl. Polym. Sci.* **16**, 2291.

Sakaguchi, T., Hirabayashi, K., Sawaji, y., Shinichi, T., Kakegawa, E., and Tsukada, M. (1972). *Nippon Sanshigaku Zasshi* **41**(4), 263.

Samuels, R. J. (1973). *J. Macromol. Sci., Phys.* **138**(1-2), 41.

Samuels, R. J. (1974). "Structured Polymer Properties." Wiley (Interscience), New York.

Samuels, R. J. (1975). *J. Polym. Sci., Polym. Phys. Ed.* **13**, 1417.

Schauler, W., and Kashani, U. (1975). *Faserforsch, Textiltech.* **26**(6), 270.

Schauler, W., and Liska, E. (1975). *Faserforsch. Textiltech.* **26**(5), 225.

Schultz, J. (1974). "Polymer Material Science." Prentice-Hall, Englewood Cliffs, New Jersey.

Schwenker, R. F., and Chattergee, P. K. (1972). *Differ. Therm. Anal.* **2**, 419.

Schwenker, R. F., Jr. (1967). *Proc. Toronto Symp. Therm. Anal. 2nd, 1967* p. 59.

Seves, A., and Vicini, L. (1974). *Textilia* **50**(11), 15.

Seyfurth, H. E., Henkel, H., Langner, K., Schonherr, F., and Wiesener, E. (1972). *Faserforsch. Textiltech.* **23**(6), 235.

Shimizu, J., Toriumi, K., and Tumai, K. (1977). *Sen'i Gakkaishi* **33**(5), 208.

Shimizu, J., Okui, N., Kaneko, A., and Toriumi, K. (1978a). *Sen'i Gakkaishi* **34**(2), 64.

Shimizu, J., Okui, N., Kikutani, T., and Toriumi, K. (1978b). *Sen'i Gakkaishi* **34**(3), 35.

Shishoo, R., and Bergh, K. M. (1977). *Text. Res. J.* **47**, 56.

Siegmann, A., and Turi, E. (1974). *J. Macromol. Sci. Phys.*, **B10**(4), 689.

Sidorovich, A. V., and Kuvshinskii, E. V. (1961). *Vysokomolekularnye Soedinehiya* **3**, 161.

Silver, F. M., and Dobinson, F. (1977). *J. Polym. Sci. Polym. Chem. Ed.* **15**, 2535.

Simov, D., Fakirov, S., Seganov, I., and Mikhailov, M. (1973a) *God. Softii Unik Khin. Fak.* **65**, 347.

Simov, D., Fakirov, S., Mikhailov, M., and Petrenko, P. (1973b). *Vysolkamol. Soyed.* **A15**(8), 1775.

Slater, K. (1976). *Text. Prog.* **8**(3), 1.

Smith, C. W., and Dole, M. (1956). *J. Polym. Sci.* **20**, 37.

Smith, F. S., and Steward, R. D. (1974). *Polymeri* **15**, 283.

Smith, J. K., Rawls, H. R., Felder, M. S., and Klein, E. (1970). *Text. Res. J.* **40**, 211.

Song, S. K. (1977). *Sumyu Konghakhoe Chi* **14**(2), 84.

Spruiell, J. E., and White, J. L. (1975a). *Appl. Polym. Symp.* **27**, 121.

Spruiell, J. E., and White, J. L. (1975b). *Polym. Eng. Sci.* **15**(9), 660.

Stanciu, I. (1974). *Ind. Usoara* **25**(10), 536.

Statton, W. O. (1967). *J. Polym. Sci., Part C* **20**, 117.

Statton, W. O. (1968). *Z. Kristallogr., Kristallgeom., Kristallphys., Kristallchem.* **127**, 229.

Statton, W. O., Koenig, J. L., and Hannon, M. (1970). *J. Appl. Phys.* **41**(11), 4290.

Stout, E. E. (1970). "Introduction to Textiles," 3rd ed. Wiley, New York.

Sweet, G. E., and Bell, J. P. (1972). *J. Polym. Sci., Polym. Phys. Ed.* **10**, 1273.

Tanaka, N., and Nakajima, A. (1972a). *Bull. Inst. Chem. Res., Kyoto Univ.* **50**(2), 65.

Tanaka, N., and Nakajima, A. (1972b). *Bull. Inst. Chem. Res. Kyoto Univ.* **50**(2), 70.

Tashiro, K., Naki, Y., Kobayashi, M., and Tadokoro, H. (1980). *Macromolecules* **13**, 137.

Taylor, R. F. (1973). *U.S.A.E.C.* **BDX-613-783**.

Textile Institute (1965). "Identification of Textile Materials," 5th ed. Text. Insti., Manchester, U. K.

Timpa, J. D., Segal, L., and Drake, G. L., Jr. (1974). *Text. Res. J.* **44**, 858.

Todoki, M., and Kawaguchi, T. (1975). *Kobunshi Ronbunshu* **32**(2), 112.

Todoki, M., and Kawaguchi, T. (1977a). *J. Polym. Sci., Polym. Phys. Ed.* **15**, 1067.

Todoki, M., and Kawaguchi, T. (1977b). *J. Polym. Sci., Polym. Phys. Ed.* **15**, 1507.

Turi, E., and Sibilia, F. (1978). *In* "Thermal Methods in Polymer Analysis" (S. W. Shalaby, ed.), p. 77. Franklin Inst. Press, Philadelphia, Pennsylvania.

Turska, E., and Gogolewski, S. (1975). *J. Appl. Polym. Sci.* **19**, 637.

Valk, G. (1972). *Lenzinger Ber.* **33**, 1.

Valk, G., Berndt, H.-J., and Heidemann, G. (1971). *Chemiefasern* **5**, 1.

Valk, G., Jellinck, G., and Schroder, U. (1980). *Text. Res. J* **50**, 44.

Vanicek, J. (1975). *Chem. Prum.* **25**(8), 423.

Van Krevlen, D. W. (1972). "Properties of Polymers." Am. Elsevier, New York.

Varma, D. S., Kumar, K. V. A. R., and Veena, A. (1979). *Man-Made Text. India*, 73.

von Hornuff, G., and Müller, G. (1972). *Faserforsch. Textiltech.* **23**(11), 466.

Wakelin, J. H., Sutherland, A., and Beck, L. R. (1960). *J. Polym. Sci.* **42**, 278.

Ward, I. M. (1971). "Mechanical Properties of Solid Polymers." Wiley (Interscience), New York.

Ward, I. M. (1975). "Structure and Properties of Oriented Polymers." Wiley, New York.

Warner, S. (1980). *Macromolecules* **13**(2), 450.

Warner, S., and Jaffe, M. (1980). *J. Cryst. Growth* **48**, 184.

Warwicker, J. O., and Vevers, B. (1980). *J. Appl. Polym. Sci.* **25**, 977.

Wasiak, A., and Ziabicki, A. (1975). *Appl. Polym. Symp.* **27**, 11.

Wegener, W., and Merkle, R. (1971). *Chemiedasern Text. Anwendungstech.* **21**(6), 575.

Weiss, L. C., Wade, R. H., and Andrews, F. R. (1974). *Text. Res. J.* **44**, 892.

Wendorff, J., Finkelman, H., and Ringsdorf, H. (1978). *J. Polym. Sci. Polym. Symp.* **63**, 245.

Wiesner, E. (1968). *Faserforsch. Textiltech.* **19**(7), 301.

Wiesner, E. (1974). *Chem. Vlakna* **24**(1), 1.
Wiesner, E. (1976a). *Termanal 76, Celostatha Konf. Term. Anal.* [*Pr.*], 7th, 1976 0-69-0-75.
Wiesner, E. (1976b). *Chem. Vlakna* **26**(3-4), 146.
Wilson, M. P. W. (1974). *Polymer* **15**, 277.
Wishiewska, W. (1972). *Polimery* **17**(10), 517.
Wunderlich, B. W. (1973). "Macromolecular Physics," Vol. 1. Academic Press, New York.
Wunderlich, B. W. (1976). "Macromolecular Physics," Vol. 2. Academic Press, New York.
Yamazaki, Y. (1971). *Sen'i Kako* **23**(9), 628.
Yunusov, B. (1974). *Kh. Zavod. Lab.* **40**(4), 470.
Zhmaeva, I. V., Tokurev, A. V., Vdovina, Z. A., Rudinskuya, G. Ya, Serova, L. D., Lisit-syn, A. P., and Efremov, V. Ya. (1977). *Fibre Chem.* (*Engl. Transl.*) **9**(3), 250.
Ziabicki, A. (1976). "Fundamentals of Fibre Formation," Chapter 3. Wiley, New York.
Zimmerman, H., and Budisch, J. (1969). *J. Therm. Anal.* **1**(1), 107.

CHAPTER 8

Thermal Analysis in Polymer Flammability

E. M. PEARCE
Department of Chemistry
Polytechnic Institute of New York
Brooklyn, New York

Y. P. KHANNA
Corporate Research and Development
Allied Corporation
Morristown, New Jersey

D. RAUCHER*
Department of Chemistry
Polytechnic Institute of New York
Brooklyn, New York

I. Introduction

Polymers are being increasingly utilized as construction materials in a number of areas, such as home furnishings, domestic and industrial buildings, appliances, fabrics, and transportation vehicles. An expanded growth of polymers concurrent with the proliferation of safety

* Present address: Monsanto Company, St. Louis, Missouri.

793

standards being set by government and private agencies has indicated that reducing the flammability of polymeric materials is of primary importance. As a result, the polymer industry has continued to be interested in laboratory scale techniques that can provide some measure of flammability. However, the complicating factor in polymer flammability is that it is still poorly understood and that consequently, anomalies exist in relation to mechanisms of flame retardation, test methods, and meaningful code regulations.

The ideal flame-retardant polymer system has been described as having a high resistance to ignition and flame propagation, low rate of combustion, low rate and amount of smoke generation, low combustibility and toxicity of product gases, retention of reduced flammability during use, acceptability in appearance and properties for specific end uses, no environmental or health safety impact, and little or no economic penalty. A flame-retardant treatment, in addition to being formulated from efficient, economic chemicals, must be applied without the use of unusual processing conditions, in commercial equipment, reproducibly, with no effect on other processing steps, and must be durable under all use conditions. All of these requirements have dictated the type of evaluation necessary for flame-retardant systems.

Thermoanalytical techniques are useful adjuncts for the preliminary evaluation of a number of flammability parameters. These methods are fast, reproducible, and sensitive, and they provide fundamental information. Although thermal methods of analysis have had the deficiencies associated with small-scale tests, they still provide valuable insights and useful laboratory scale preliminary evaluation.

In this chapter we have reviewed the information that could be obtained from thermal analysis and described how to apply it to the polymer flammability area. In particular, we have concentrated on the results obtained from thermal analysis in relation to the flame retardation of important polymeric systems. No attempt has been made for an exhaustive review of this subject; instead, examples illustrating the general applicability of thermal methods in flammability evaluation have been presented.

II. Polymer Flammability

This section deals briefly with the basics of polymer flammability and approaches to flame retardation. A familiarity with both of them is essential for the successful application of thermal analysis in flammability evaluation.

A. COMBUSTION PROCESS

The factors associated with polymer flammability have been recently reviewed by Pearce and Liepins (1975). Fundamentally, four processes are considered to be involved in polymer flammability: (1) preheating, (2) decomposition, (3) ignition, and (4) combustion and propagation.

Preheating involves heating of the material by means of an external source that raises its temperature at a rate dependent on the thermal intensity of the ignition source, the thermal conductivity and specific heat of the material, and the latent heat of fusion of the substance if it happens to be crystalline.

When sufficiently heated, the material reaches a characteristic temperature where it begins to degrade (i.e., it loses its original properties as the weakest bonds begin to break). Gaseous combustible products may then be formed, with a rate dependent on factors such as the intensity of external heat, temperature required for decomposition, and rate of decomposition.

Flammable gases increase until a concentration is reached that follows their sustained oxidation in the presence of the ignition source. The ignition characteristics of the gas and the availability of oxygen are two important variables in any ignition process.

After ignition and removal of the ignition source, combustion becomes self-propagating if sufficient heat is generated and radiated back to the material to continue the decomposition process. The combustion process is governed by such variables as the rate of heat generation, rate of heat transfer to the surface, surface area, and the rate of decomposition.

B. RETARDATION APPROACHES

Polymer combustion, a highly complex process, is believed to be composed of a vapor phase in which the reactions responsible for the formation and propagation of the flame are taking place and a condensed phase in which fuel for the gas reactions is being produced. Flame retardancy, therefore, can be improved by appropriately modifying either one or both of these phases (Frisch, 1979).

The approaches toward reducing the flammability of polymer systems could be grouped into the following three categories:

(a)   Vapor Phase: In the vapor-phase approach, a flame retardant or the modified polymer unit, on heat exposure, releases a chemical agent that inhibits free radical reactions involved in the flame formation and propagation. Flame retardation could be implemented by incorporating fire-retardant additives, impregnating the material with a flame-retardant substance, or using flame-retardant comonomers in the polymerization or grafting.

(b) Condensed Phase: In condensed-phase modification the flame retardant alters the decomposition chemistry, which favors the transformation of the polymer to a char residue. This could be achieved by the addition of additives that catalyze char rather than flammable product formation or by designing polymer structures that favor char formation.

(c) Miscellaneous: These approaches include dilution of the polymer with nonflammable materials (e.g., inorganic fillers), incorporation of materials that decompose to give nonflammable gases such as CO_2, formulation of products that decompose endothermically, etc.

Although there are several test methods for evaluating the burning behavior of different polymers, Fenimore and Martin's (1966) limiting oxygen index (LOI) will be used to illustrate the relative flammability of materials. This test measures the minimum concentration of oxygen in an oxygen–nitrogen atmosphere that is necessary to initiate and support a flame for >3 min:

$$\text{LOI} = \frac{\text{volume of } O_2}{\text{volume of } O_2 + \text{volume of } N_2} \times 100$$

III. Thermal Analysis and Flammability Evaluation

Thermal analysis has been extensively employed in the polymer flammability area. In addition to the characterization of the degradation patterns of the non-flame-retarded and flame-retarded polymers, much other important information can be derived through thermoanalytical methods.

Probably the most extensively used thermal technique for the evaluation of flammability characteristics of polymeric materials is thermogravimetric analysis (TG). The technique is used through its display of the degradation profile of the material to obtain information on the fundamental process of a burning system, i.e., thermal decomposition. TG can be used to decipher whether flame-retarding mechanisms appear to occur in the vapor or condensed phase. Differential thermal analysis (DTA) and differential scanning calorimetry (DSC) comprise other major thermal methods and are mainly used to observe the changes in the polymers prior to and during the decomposition. Less commonly used techniques include thermal evolution analysis (TEA), thermomechanical analysis (TMA), and dynamic mechanical analysis (DMA). Whereas TEA is employed to identify the gaseous decomposition products, TMA and DMA measure the mechanical properties of the polymeric materials before and after the flame retardation modification.

In order to exemplify the versatility of thermal analysis in flammability evaluation, each of the commonly used techniques (TG, DTA, and DSC) will be described in detail with emphasis on the type of information obtainable.

A. MECHANISM OF FLAME RETARDATION

Flame retardancy is believed to result from the presence of one or more key elements in the retardant—such as phosphorus, nitrogen, chlorine, bromine—or of a volatile compound (e.g., water of hydration). Although the question of how various flame retardants work is still, in some cases, highly controversial, major advances have been made in this field during the past decade. Despite wide disagreement about how some of these substances suppress burning, the following are some of the widely accepted theories and their supporting thermoanalytical data:

1. Vapor-Phase Mechanism

The flame retardants operating by the vapor-phase mechanism are known to dissociate into species that can remove free-radical intermediates from the flame reaction or that can replace the major propagating species with those that do not propagate the flame as readily.

Halogenated flame retardants (e.g., chlorinated paraffins, chlorocycloaliphatics, and chloro- and bromo-aromatic additives) have been commonly used in flame-retarding plastics. These additives are postulated to function primarily by a vapor-phase flame inhibition mechanism, which according to Hastie (1973) is characterized by the following criteria:

(a) Flame-retardant element is lost from the substrate.
(b) Flame inhibition is insensitive to substrate structure.
(c) Flame retardance is sensitive to oxidant, e.g., O_2 or N_2O.
(d) Flame retardant does not change the composition or amount of volatiles.

Considering some of the characteristics of vapor-phase flame retardation, it appears that thermal analysis should be helpful in detecting vapor-phase activity of the flame retardant. This is illustrated, for example, by the work of Reardon and Barker (1974), who studied the pyrolysis and combustion of Nylon 6 with and without organobromine compounds.

These authors employed TG and DTA to study the thermal decomposition of both treated and untreated Nylon 6. A comparison of the TG thermograms of the untreated sample and the sample treated with hexabromobiphenyl indicated that the latter exhibited about 50°C lower initial decomposition temperature (Fig. 1). Both the materials underwent major degradation in the 300–400°C region and left almost no residue at 500°C.

The DTA curves of the two samples have been shown in Fig. 2. Untreated Nylon 6 exhibited a melting peak at about 240°C [it is rather high for Nylon 6 and may be attributed to experimental conditions (e.g., higher heating rate, higher sample size, temperature calibration, etc.)] and a decomposition endotherm at ~450°C. Although the sample of Nylon 6 treated with hexabromobiphenyl showed no change in the melting behavior, it exhibited a much larger endotherm at ~425°C. The latter, we believe, resulted from a simultaneous process involving the catalytic reaction of the additive with polymer and the sample degradation.

Gas chromatography and infrared studies on the pyrolysis of both treated and untreated Nylon 6 showed that ε-caprolactam was the main product. By coupling this information with the TG and DTA data, Reardon and Barker (1974) concluded that the brominated flame retardants acted as decomposition catalysts for Nylon 6. In view of these results one would expect the flame-retarding efficiency of these additives to be considerably lower for Nylon 6. The authors, however, did not report any flammability measurements.

Although the fire-retardant effectiveness of organobromine compounds in Nylon 6 was not quoted, the work of Reardon and Barker (1974), in our opinion, clearly demonstrates that the flame-retardant element was lost from the substrate (from TG) and that it did not alter the composition or amount (from GC, IR, and TG) of volatiles. These observations strongly suggest that the fire-retardant effect of organobromine compounds on the flammability of Nylon 6, if any, must be in the vapor phase.

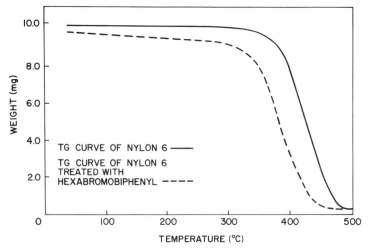

Fig. 1 TG curves of Nylon 6 and Nylon 6 treated with hexabromobiphenyl at a heating rate of 6°C/min., [From Reardon and Barker (1974).]

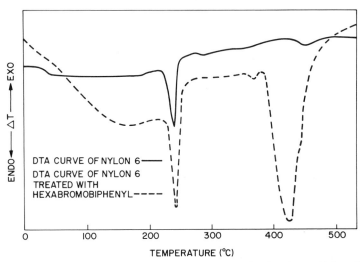

Fig. 2   DTA curves of Nylon 6 and Nylon 6 treated with hexabromobiphenyl. [From Reardon and Barker, (1974).]

Halogenated additives are the flame retardants known to function predominantly in the vapor phase. Since the preceding section deals with the use of thermal analysis in detecting vapor-phase activity, it may be appropriate to mention briefly how the flame retardants function in the vapor phase.

Burning of most polymers can be regarded as a hydrogen-containing combustion system that is generally described and quantified in terms of the H_2–O_2 reaction scheme (Petrella, 1978). The hydrogen–oxygen combustion scheme contains the following reactions:

$$\cdot H + O_2 \rightleftharpoons \cdot OH + O\cdot$$

$$\cdot O + H_2 \rightleftharpoons \cdot OH + H\cdot$$

and these dominate the combustion process because of their chain-branching nature (the species like $\cdot$ H and H_2 originate from polymer pyrolysis). Halogenated flame retardants (MX) are postulated to function primarily by the following mechanisms, which produce HX, the actual flame-inhibiting species,

$$M - X \rightleftharpoons M\cdot + X\cdot$$

$$RH + \cdot X \rightleftharpoons R\cdot + HX$$

or

$$H_2 + \cdot X \rightleftharpoons H\cdot + HX$$

The flame-inhibiting effects of HX are manifested through reactions that inhibit the chain-branching step of the hydrogen–oxygen combustion systems, such as

$$\cdot H + HX \rightleftharpoons H_2 + X\cdot$$

and

$$\cdot OH + HX \rightleftharpoons H_2O + X\cdot$$

Larsen (1974) has proposed, however, that the halogenated materials suppress the flame by a physical mechanism involving the undecomposed agents. This theory has generally not been favored.

2. Condensed-Phase Mechanism

The flame retardants that function by this mechanism have the ability to increase the conversion of polymeric materials to a char residue (a carbonaceous product comprised mainly of carbon) during pyrolysis and thus decrease the formation of flammable, carbon-containing gases. The char helps to protect the substrate by interfering with the access of oxygen. Formation of char on the polymer surface can also be postulated to act as a heat shield for the base material (Neuse, 1973). This is illustrated by Fig. 3.

Phosphorus-based additives are typical examples of flame- retardants that could act by a condensed-phase mechanism. According to Hastie (1973), the following are the indicators of a condensed-phase operation for a flame retardant:

(a) Flame-retardant results in enhanced char formation.
(b) Flame-retardant element is retained in the substrate.
(c) Flame-retardant element is often ineffective in the vapor phase.
(d) Flame retardance is sensitive to the substrate structure.
(e) Flame retardance is insensitive to oxidant (e.g., N_2O or O_2).
(f) Flame retardant changes the composition of volatiles.

Thermogravimetric analysis is a powerful tool for detecting condensed-phase activity. By this technique one measures the residue remaining at an arbitrary temperature above the major degradation region of the treated and untreated samples.

To show the importance of thermogravimetric analysis in this area of flame retardation we have considered the studies on cellulose flammability made by Tang and Niell (1964). They studied the thermal stability of untreated α-cellulose and compared it with those of 2% $NH_4H_2PO_4$, $AlCl_3 \cdot 6H_2O$, $Na_2B_4O_7 \cdot 10H_2O$, and $KHCO_3$ treated α-cellulose samples (Fig. 4). Generally a decreased threshold temperature of 270°C for un-

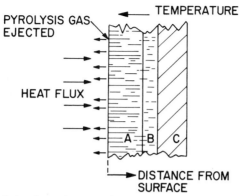

Fig. 3   Effect of char formation on polymer combustion: A, char layer; B, transition zone of incomplete charring; C, virgin polymer material.

treated α-cellulose and an increased char yield was noted for all the samples. Although a decrease in the threshold temperature of the treated polymer did not necessarily imply any fire-retardant effectiveness, it resulted from the catalytic effect of the additive that altered the polymer degradation route, leading to higher char residues (condensed-phase activity). The authors made these studies in vacuum, although similar information could be derived through inert atmospheric TG studies. The

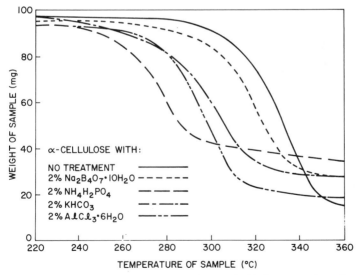

Fig. 4   TG curves of α-cellulose untreated and treated with flame retardants, in vacuum. [From Tang and Niell (1964).]

experiments carried out in air may not prove useful, because of the oxidation of the char. Actual comparisons as to flame-retarding efficiencies cannot be made, since it would not be meaningful to compare the flame retardants on the weight percent basis used in these studies. However, it seems possible to evaluate the relative efficiencies of flame-retardant additives with respect to charring ability and hence flame retardancy using thermogravimetric analysis, provided the additives are added on an equal molar basis.

Finally, one must realize that the problem of determining whether a retardant operates mainly by a condensed-phase or a vapor-phase mechanism is complicated for many systems. For example, although halogenated additives and phosphorus-based flame retardants are known to function predominantly by the vapor-phase and condensed-phase mechanisms, respectively, several instances are known where the opposite may be true, and where some flame retardants may involve both mechanisms.

Consider, for example, the thermal degradation and flame retardancy of tetrabrominated epoxy resin studied by Nara and Matsuyama (1971). One would normally expect that bromination of the epoxy resin results in flame retardation through a vapor-phase mechanism. Surprisingly, thermogravimetric analysis in this case (Fig. 5) suggested that a condensed-phase mechanism was also operating, since the nonbrominated resin exhibited no char at 600°C, compared to about 10% residue for the bromin-

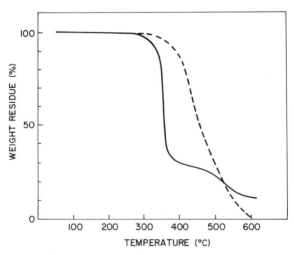

Fig. 5   TG curves of the nonbrominated (– – – –)) and brominated (———) epoxy resins, in air, at 15°C/min heating rate. [From Nara and Matsuyama (1971).]

ated resin. Nara and Matsuyama (1971) extended their degradation studies to identifying the decomposition products of brominated epoxy resin. They were able to identify the presence of HBr in the degradation products using IR and gas chromatographic methods. This information when coupled with thermal analysis suggested that in this case, vapor-phase as well as condensed-phase mechanisms could have been operating in the flame-retardation process.

In similar situations where more than one mechanism could operate, thermal analysis in conjunction with elemental analysis, gas chromatography, mass spectrometry, and infrared spectroscopic techniques might prove to be a powerful tool for flame retardation studies.

3. Miscellaneous Mechanisms

Apart from vapor-phase and condensed-phase mechanisms described previously, flame retardants could also function in a number of other ways. The following are some other alternative modes of flame inhibition:

(i) Large volumes of noncombustible gases may be produced that dilute the oxygen supply to the flame and/or dilute the fuel concentration needed to sustain the flame.

(ii) The endothermic decomposition of the fire retardant could lower the polymer surface temperature and retard pyrolysis of the polymer.

(iii) The flame retardant may act as a thermal sink to increase the heat capacity of the combustion system or to reduce the fuel content to a level below the lower limit of flammability.

Hydrated alumina ($AL_2O_3 \cdot 3H_2O$) is a unique filler for plastics in the sense that it also acts as a flame retardant additive. This material suppresses combustion in a number of ways. The efficiency of hydrated alumina as a flame-retardant filler in reinforced polyester resins has been described by Bonsignore and Manhart (1974). The TG and DTA curves for the decomposition of alumina hydrate as reported by these authors are shown in Fig. 6. The material contained 34.6% chemically bound water that was lost in the 230–600°C temperature range (TG). Simultaneous DTA showed that the water loss was endothermic with the measured enthalpy of 470 cal/g of the hydrate. Bonsignore and Manhart (1974) compared the thermal behavior of a general-purpose polyester resin with and without hydrated alumina using DSC (Fig. 7). The polyester showed exothermic oxidative degradation at 220–230°C corresponding to the onset of weight loss by TG. However, in the presence of 60% alumina hydrate the overall decomposition appeared to be endothermic, implying flame retardation. Oxygen index studies confirmed that flame resistance increased with increasing alumina hydrate content. Reduced flammability in this

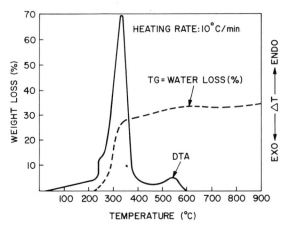

Fig. 6 TG and DTA curves of alumina trihydrate in air. [From Bonsignore and Manhart (1974).]

case could be attributed to factors such as reduction in the enthalpy of polymer combustion and reduced fuel content and oxygen supply in the flaming zone.

Other additives that could act by a similar mechanism include $NaHCO_3$, $KHCO_3$, $NH_4H_2PO_4$, etc. These materials decompose to give noncombustible gases that dilute and cool the combustible gases and the substrate because of endothermic decomposition.

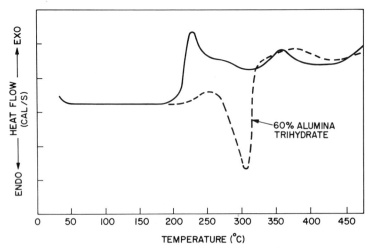

Fig. 7 DSC of "general purpose" polyester alone and filled with 60% alumina trihydrate, in air. [From Bonsignore and Manhart (1974).]

B. SYNERGISM IN FLAME RETARDATION

The effect of a mixture of two or more flame retardants may be additive, synergistic, or antagonistic. Synergism is the case in which the effect of two or more components taken together is greater than the sum of their individual effects. The concept of synergism is very important, since the development of synergistically efficient flame retardants can lead to less expensive polymer systems with reduced effects on other desirable properties. One of the classical illustrations of synergism observed in flame retardation is the addition of antimony (III and V) oxide to halogen-containing polymers. Here we have described the antimony (III)–halogen synergism and the role played by thermal analysis in substantiating its mechanism.

In polymer systems containing Sb_2O_3 and halogen, it has often been suggested that fire retardance is mainly due to the gas phase flame inhibition by the volatile antimony trihalide, since maximum flame retardancy is observed when the mole ratio Sb/X is $1:3$ (Lyons, 1970; Tesoro, 1971; Rhys, 1969; Schmidt, 1965; Einhorn, 1970). It is believed that HCl formed by the degradation of the chlorinated additive reacts with Sb_2O_3 to yield SbOCl as an intermediate, and the subsequent reactions are responsible for the increased flame retardancy. The formation of an intermediate by the combination of Sb_2O_3 and a chlorine source was also confirmed by thermoanalytical data (Touval, 1971); for example, the indivdual TG's of Sb_2O_3 and a chlorinated paraffin showed a 0 and 67% weight loss, respectively, in the RT–375°C temperature range. Assuming that there was no interaction between the components of this system, a 34% weight loss would be expected from a 50:50 mixture. Instead, a 72% weight loss was observed in the temperature range under consideration.

In a typical substrate of R·HCl (chlorine source) + Sb_2O_3 + polymer, the formation of SbOCl has been illustrated by the following equations:

$$R\text{—}H + Cl\cdot \xrightarrow{\sim 250°C} R\cdot + HCl$$

$$2HCl + Sb_2O_3 \xrightarrow{\sim 250°C} 2SbOCl + H_2O$$

Working under the assumption that SbOCl was the primary product of the interaction between Sb_2O_3 and a chlorine source, Pitts *et al.* (1970) studied the thermal decomposition of SbOCl by TG and DTA. They proposed the following reaction sequence in order to account for the three weight loss steps on the TG curve.

	Theoretical % wt loss	% Wt loss by TG

Step I: $5SbOCl(s) \xrightarrow{245-280°C}$
$Sb_4O_5Cl_2(s) + SbCl_3(g)$ 26.4   24.1

Step II: $4Sb_4O_5Cl_2(s) \xrightarrow{410-475°C}$
$5Sb_3O_4Cl(s) + SbCl_3(g)$ 6.9   8.4

Step III: $3Sb_3O_4Cl(s) \xrightarrow{475-565°C}$
$4Sb_2O_3(s) + SbCl_3(g)$ 10.6   9.5

A simultaneous DTA thermogram further indicated that the three weight loss steps on the TG curve were endothermic.

The endothermic release of $SbCl_3$ during the formation of flammable degradation products was considered to be responsible for the flame-retardance effect observed in flexible urethane foams containing SbOCl (Pitts, 1972). The thermal decomposition of SbOCl occurred within an appropriate temperature range that happened to match the degradation temperature range of the foam itself. Pitts (1972) also evaluated the effect of various metal oxides on the flame-retardant effect of SbOCl. Metal oxides

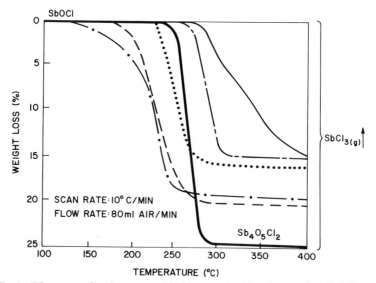

Fig. 8 TG curves of antimony oxychloride–metal oxide mixture. (━) SbOCl, (— · —) 8SbOCl + 2CuO, (– – –) 8SbOCl + 2Fe$_2$O$_3$, (· · · · · · · · ·) 10 SbOCl + 5TiO, (– —— –) 10 SbOCL + 5CaO, (———) 10 SbOCl + 5ZnO. [From Pitts (1972). Reproduced by courtesy of the Journal of Fire and Flammability, Technomic Publishing Co., Inc. 265 Post Road West, Westport, CT 06880, USA.]

such as CaO and ZnO were found to have a negative effect when used as partial replacements for SbOCl. The thermogravimetric analysis data revealed that both of these oxides raised the decomposition temperature of SbOCl by 25–50°C. On the other hand, Fe_2O_3 and CuO proved to be quite useful even in catalytic amounts. Interestingly, the TG study indicated that both Fe_2O_3 and CuO lowered the decomposition temperature of SbOCl by 50–100°C (Fig. 8). However, the TG curves of the foams containing these oxides and Sb_2O_3 showed no significant differences. These observations suggested that both Fe_2O_3 and CuO enhanced the flame retardancy by lowering the energy barrier for the transformation of SbOCl to $SbCl_3$. Therefore, this was consistent with a model that in Sb_2O_3–halogen synergistic reactions, SbOCl was formed as an intermediate that on thermal decomposition evolved $SbCl_3$, the actual flame retardant.

Papa and Proops (1972) have also suggested a phosphorus–bromine synergism based on their charring studies on flexible polyurethane foams. Although these workers used a different charring procedure, one could also employ thermogravimetric analysis to generate char residues at various temperatures for better quantitative investigations of this and similar reaction systems.

C. Condensed-Phase Processes

The amount of char residue formed on the thermal degradation of a polymer is a measure of its flame resistance. Van Krevelen (1975) had shown that there was a correlation between char residue and the oxygen index of polymers; the higher the char, the lower the flammability (Fig. 9). He also pointed out that char formation and hence flame resistance of a polymer could be predicted from its structure. In general, an increase in aromaticity yields high char residues and has been shown to correlate with oxygen index (Parker *et al.*, 1973). Also substituents in polymers have a considerable influence on char residue; the latter depends on the type of functional group, polymer, and the position of the substituent.

Khanna and Pearce (1978) studied the flammability and char formation of copolymers of styrene and vinylbenzyl chloride. This study was aimed at reducing the flammability of polystyrene by introducing chloromethyl ($—CH_2Cl$) groups on to the benzene rings in order to promote cross-linking and therefore char formation on degradation. The results are shown in Fig. 10. The TG of polystyrene indicated that the polymer degraded completely to volatiles by 500°C, leaving no char residue. The degradation products included substantial amounts of styrene and other combustible volatiles, thus making polystyrene quite flammable, with an oxygen index of 19. However, when the benzene rings in polystyrene contained

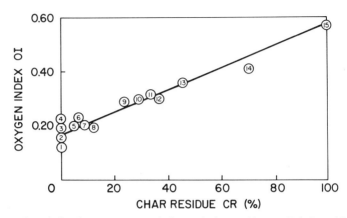

Fig. 9 Correlation between oxygen index and char residue: 1, Polyformaldehyde; 2, polyethylene, polypropylene; 3, polystyrene, polyisoprene; 4, nylon; 5, cellulose; 6, polyvinyl alcohol; 7, PETP; 8, polyacrylonitrile; 9, PPO®; 10, polycarbonate; 11, Nomex®; 12, polysulfone; 13, Kynol®; 14, polyimide; 15, carbon [Reproduced from Van Krevelen, D. W. (1975), *Polymer* **16**, 615, by permission of the publisher, IPC Business Press Ltd. ©].

—CH$_2$Cl groups, the char yield at 500°C was 55%. This would mean a significant enhancement in the flame retardation. The predicted flame retardation of polystyrene due to the presence of —CH$_2$Cl substitutents was, in fact, reflected in an increase of oxygen index from 19 to 50. In addition, Fig. 10 revealed that the copolymers yielded higher char residues than the theoretically predicted values based on an additive effect. This was indicative of a synergistic effect of the —CH$_2$Cl groups with the benzene rings in promoting the char yield of polystyrene.

Although it is well known that the polymers based on aromatic and other cyclic structures produce high char yields on thermal degradation, there is no generalization about the substituent effect on the char yield of polymers. For instance, a —Cl group on the benzene ring may not affect the char yield (e.g., in polystyrene), or may increase the char (e.g., in epoxies), and in some cases (e.g., aromatic polyamides) the char yield may even be reduced. The influence of a substituent on char residue depends on its type and position, the polymer, and its mode of degradation.

Recently, Chaudhuri *et al.* (1980) studied the effect of various substituents (e.g., —CH$_3$, —OCH$_3$, —OH, —COOH, —SO$_3$H, —NO$_2$, —Cl) on the char-forming ability of aromatic polyamides like poly(1,3-phenylene isophthalamide) and poly(1,4-phenylene terephthalamide). Their results indicated that the introduction of these substituents on to the polymer backbone led to a decrease in both the thermal stability and the char yield of the parent materials. On a comparative basis, only the chlorosubstitution exhibited a small reduction in the thermal stability and either a

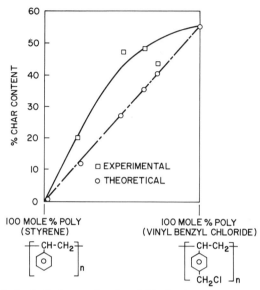

Fig. 10   Effect of substitution on the char yield of polystyrene at 500°C, in argon atmosphere.

small decrease or almost the same char yield as that of the unsubstituted polymers. Y. P. Khanna and E. M. Pearce (unpublished results, 1980) then carried out a systematic study of the degradation of poly(1,3-phenylene isophthalamide) and poly(1,4-phenylene terephthalamide) and their chloroderivatives with regard to the char yield, flammability, and mode of flame retardation. The preliminary results shown in Table I (Khanna and Pearce, unpublished results, 1980) have special significance, as described later.

The thermal stability of unsubstituted polyamides was impaired on chlorosubstitution probably as a result of decreased chain stiffness. The char yield at 700°C, as measured by TG, was either slightly decreased or remained essentially the same, after chlorosubstitution. Consequently, recalling the linear correlation of char residue with flame retardancy, they did not necessarily expect any improvement in flame retardation of these unsubstituted aramids as a result of chlorosubstitution.

However, based on the TG data and elemental analysis of the polymer and its char at 700°C, Y. P. Khanna and E. M. Pearce (unpublished results, 1980) predicted an enhanced flame retardancy of these aramids when they contained chlorogroups. For instance, the emission of C, H, N, and O elements from poly(1,3-phenylene isophthalamide) and poly-(1,4-phenylene terephthalamide) by 700°C was greatly reduced when these

TABLE I

Variations in the Elemental Composition of Aromatic Polyamides upon Degradation

Polyamide		η_{inh}^{b}	% of Original Element Present ^{c} at 700 C				
Code	Structure		C	H	O	N	Cl
PMI	$[HN-\bigcirc-NH-CO-\bigcirc-CO]_n$	0.98	73.3	33.3	38.4	60.7	–
PCI	$[HN-\bigcirc_{Cl}-NH-CO-\bigcirc-CO]_n$	0.79	81.2	44.1	41.7	64.2	1.5
PPT	$[HN-\bigcirc-NH-CO-\bigcirc-CO]_n$	1.74	68.6	28.6	41.6	45.1	–
PCT	$[HN-\bigcirc_{Cl}-NH-CO-\bigcirc-CO]_n$	0.83	78.6	36.1	42.0	63.6	3.3

^{a} From Y. P. Khanna and E. M. Pearce, unpublished results (1980).
^{b} Determined on a 0.5 g/dL solution in concentrated H_2SO_4 at 25°C.
^{c} Calculated from the elemental analysis of the polymer and its char at 700°C, and the total weight loss up to 700°C by TG.

polymers had a chlorine substitutent. Moreover, the chlorine-bearing aramids lost almost all their chlorine by 700°C. Based on these results, they predicted a flame-retarding effect of chlorine on poly(1,3-phenylene isophthalamide) and poly(1,4-phenylene terephthalamide) by a twofold method:

(a) *Vapor-phase mechanism*: Since most of the chlorine was lost by 700°C, the degradation products may have had considerable flame resistance due to the flame-quenching action of chlorine.

(b) *Condensed-phase mechanism*: Since the emission of flammable products forming elements (C, H, N, O) was reduced in the case of chloropolymers, a condensed-phase activity of the chlorine must have existed.

These predictions of flame retardation were in fact supported by a 10–15 higher oxygen index for the chloroaramids.

Thus it has appeared that thermogravimetric analysis coupled with elemental analysis could provide useful information about flame retardation and the mode of flame retardancy in some polymeric systems where flame retardation was not obvious.

Recent studies of Kambour (1981) suggest that char formation in all polymers raises LOI for reasons beyond the fact that char is the fuel withheld from the flame. For example, working with bisphenol-A carbonate–dimethylsiloxane block polymers (Kambour *et al.*, 1981), the authors have proposed that the synergism in LOI is correlated with a rise in pyrolytic char and an improvement in char oxidation resistance. Thus, in addition to the increase in char residue, the transport barrier (oxidative resistance) property of the pyrolytic char was improved. Although Kambour (1981) has not described the use of thermoanalytical techniques in these studies, TG could be a powerful method not only to evaluate the pyrolytic residue but also to measure the thermo-oxidative resistance of the char. The latter can easily accomplished by TG studies in inert and oxidative atmospheres.

D. SELECTION OF FIRE-RETARDANT ADDITIVES

The choice of flame retardants is dependent on the nature of the polymer, the method of processing, the proposed service conditions, and economic considerations. Although the processing, service, and economic factors are important, flame retardancy potential of an additive is of primary importance, and this can be readily evaluated by thermal analysis.

Einhorn (1971) has described the use of TG in selecting the appropriate fire retardants. The technique involves matching the degradation of candidate additives with that of the polymer under consideration. Figure 11 represents the TG thermogram for a hypothetical polymer that exhibits a simple unimolecular degradation process. Point *A* represents the region of initial decomposition and point *B* represents the temperature of maximum rate of degradation.

Fire retardants are screened so as to select a material having thermal

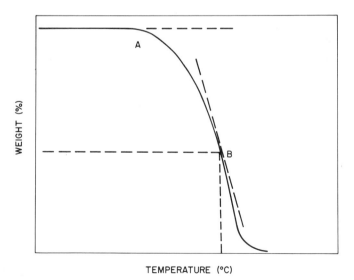

Fig. 11   TG thermogram for a hypothetical polymer. [From Einhorn (1971).]

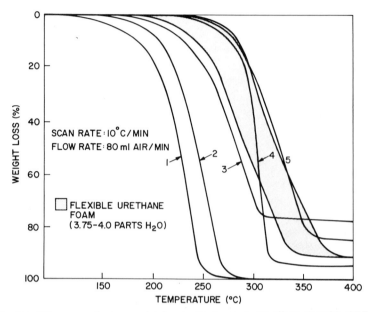

Fig. 12 TG curves of commercial flame retardants: 1, Stauffer's FYROL CEF-tris(2-chloroethyl) phosphate; 2, Stauffer's FYROL FR-2-tris(2,3-dichloropropyl) phosphte; 3, Monsanto's PHOSGARD C-22-R; 4, Michigan Chemical's FIREMASTER T-23-P-tris(2,3-dibromopropyl) phosphate); 5, Monsanto's PHOSGARD 2XC-20. [From Pitts (1972). Reproduced courtesy of the Journal of Fire and Flammability, Technomic Publishing Co., Inc. 265 Post Road West, Westport, Ct 06880, USA.]

characteristics similar to those in Fig. 11. The efficiency of matching the degradation curve of the polymer with the volatilization or degradation characteristics of the fire retardant has been cited as the key to effective flame retardancy (Schmidt, 1965). If the flame-retardant additive possesses low thermal stability compared to that of the polymer, it will be lost before its function is needed, and if the additive has greater stability, it may be intact at the time its function is needed. The matching of degradation curves of flammable substrate and the retardant additive is quite evident when the current commercial flame-retardant additives were screened for polyurethane foam (Fig. 12). As expected, it was found that the fire retardants 3–5 were ideally suited for this case. Depending on the thermal degradation characteristics of the flammable substrate, more than one fire-retardant additive could also be employed.

Several manufacturers or suppliers of flame retardants have listed TG weight loss data in order to facilitate the selection of appropriate fire-retardant additives. These data are reported in Table II only for those additives with identifiable structures.

TABLE II[a]

TG Weight Loss Data for Various Flame Retardants

Flame retardant	TG weight loss at (°C)		
	1%	5%	10%
Alumina, hydrated	—	—	290
Analine,2,4,6-tribromo-	121	156	174
Barium metaborate	200	350	1000
Benzene, hexabromo-	232	265	280
Benzene, pentabromoethyl-	180	217	232
Biphenyl, hexabromo-	—	—	299
Biphenyl, octabromo-	—	—	336
Bisphenol-A, tetrabromo-	245	284	298
Bisphenol-A, tetrabromo-, bis(methylether)	244	280	296
Bisphenol-A, tetrabromo-, bis(2,3-dibromopropylether)-	284	315	322
Bisphenol-A, tetrabromo-, bis(2-hydroxyethyl ether)-	284	322	337
Bisphenol-A, tetrabromo-, bis(allyl ether)-	220	245	261
Bisphenol-A, tetrabromo-, bis(2,3-dibromopropyl carbonate)-	—	—	328
Bisphenol-A, tetrabromo-, diacetate	247	278	283
Bisphenol-S, tetrabromo-, butenediol, dibromo-	240	—	—
Carbonate, bis(2,4,6-tribromophenyl)-	227	287	308
Cyclododecane, hexabromo-	—	230	255
Cyclohexane, pentabromo-, chloro-	175	200	235
Diphenylamine, decabromo-	260	277	351
Diphenyloxide, pentabromo-	—	247	—
Diphenyloxide, octabromo-	274	325	340
Diphenyloxide, decabromo-	317	357	373
Methylenedianiline, tetrabromo-	—	—	268
Neopentyl alcohol, tribromo-	—	—	160
Neopentyl glycol, dibromo-	115	135	150
Phosphate, tris(2-chlorothyl)-	—	170	190
Phosphate, tris(β-chloropropyl)-	—	155	175
Phosphate, tris(2,3-dibromopropyl)-	215	270	285
Phosphate, tris(dichloropropyl)-	—	210	225
Phosphonate, bis(2-chloroethyl)-vinyl-	—	135	155
Phosphonate, diethyl N,N-bis-(2-hydroxyethyl) amino methyl-	—	165	180
Phosphonate, dimethyl methyl-	—	60	75
Phosphonium bromide, ethylene bis,-tris(2-cyanoethyl)-	250	—	285
Phosphonium bromide, tetrakis(2-cyanoethyl)-	253	—	307

[a] Compiled from various product data literature.

E. Heat of Combustion and Flammability

As far as polymer flammability is concerned, more meaningful results are obtained from the TG measurements than from the DSC measurements. The DSC, however, is useful in observing changes in a polymer

prior to the onset of decomposition. As the polymer degradation progresses, the interpretation of DSC thermograms becomes very complex,
but an important parameter, namely, heat of combustion, could be obtained from the DSC measurements in air. Hassel (1977) has described the
use of a pressure DSC cell for determining the heat of combustion ($\triangle H_c$)
of a polymer. Although the relative $\triangle H_c$ values for the retarded and
nonretarded materials should provide the necessary information, the absolute heats of combustion can also be obtained using a suitable standard
and sufficient determinations (Hassel, 1976).

There are conflicting views on the significance of $\triangle H_c$ data. It was
mentioned in Section II.A that after ignition and removal of the ignition
source, combustion became self-propagating if sufficient heat was generated and radiated back to the material to continue the decomposition
process. Therefore, one might expect an enhanced flame retardancy as a
result of a lower heat of combustion. Ohe and Matsuura (1975) have developed a relationship between the heat of combustion and the oxygen
index of polymeric materials. A fairly good, direct relationship existed between these two parameters for several polymers, although some failures
were also observed. As another example, the fire-retardant polyethylene
has been reported to have about 11% lower $\triangle H_c$ than the polyethylene
composition that was not fire retardant (Hindersinn and Wagner, 1967).
However, Krekeler and Klimke (1965) point out that no convincing correlation existed between $\triangle H_c$ and fire retardancy. For example, highly combustible cellulose nitrate has a low heat of combustion (4134 cal/g),
whereas a self-extinguishing grade of polyethylene has a relatively high
heat of combustion (9970 cal/g).

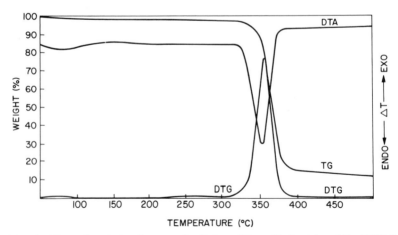

Fig. 13   Thermal analysis of untreated cellulose. [From Shafizadeh and Fu (1973).]

Therefore, the preceding discussion might lead one to conclude that there is no definite correlation between the fire retardancy and the heat of combustion. However, we believe that the heat of combustion could be a significant parameter in the flammability studies, provided it is used carefully. For instance, while screening fire-retardant cross-linking agents for a polyester resin, Alsheh and Marom (1978) found *o*-bromostyrene to be most effective in flame retardation (ASTM 635 test), and this additive also resulted in the lowest $\triangle H_c$. In similar efforts of additive screening for a particular polymer, it may generally be expected that a fire retardant that results in a lower $\triangle H_c$ is more effective. However, it must be realized that a total reliance on this parameter may be misleading in some cases and that the conclusions on the relative flammability of materials should be supplemented by other measurements.

F. APPLICATION TO SPECIFIC POLYMER SYSTEMS

The previous sections were aimed at providing a variety of information obtainable by thermal analysis regarding the flammability characteristics of materials. This section, on the other hand, will deal with the use of thermal techniques in the flame retardation of some specific polymers. Since the purpose of this chapter was to emphasize the role of thermal analysis in flammability evaluation, we shall restrict ourselves to some of those polymeric materials that are either commonly used or have specialized applications. However, the basic approaches used in these illustrative examples could always be extended to other systems.

1. Fibers

Here we have described the thermal characterization of fiber-forming polymers like cellulose and wool (natural polymers); Nylon 6, Nylon 66, and polyethylene terephthalate (synthetic polymers); Nomex®, Durette®, and Kynol® (synthetic polymers having high-temperature and flame-retardant properties) before and after flame-retardant treatment.

a. Cellulose. Cellulosic fibers amount to almost half of the fibers used in the world today, although they are considered flammable because of their rapid ignition and consumption after ignition. Before discussing the flame retardation of cellulose, it is pertinent to understand its thermal degradation. The thermal analysis of cellulose shown in Fig. 13 (Shafizadeh and Fu, 1973) indicated that the thermal decomposition of pure cellulose began at ~300°C as an endothermic process (TG and DTA). As the temperature increased, the degradation proceeded very rapidly, reaching a maximum rate of weight loss at 360°C (DTG) and leaving a 12% residue. Isothermal experiments at 300°C accompanied by simultaneous identification of the products showed that the pyrolytic products consisted of char, a tar fraction, and volatile degradation products.

Theories about the pyrolysis mechanisms for cellulose have been based on the origin of two main pyrolysis products: tar and char. Madorsky *et al.* (1956, 1958) investigated the degree and rate of tar formation during cellulose pyrolysis and showed by infrared and mass spectroscopy that levoglucosan (1,6-anhydro-β-D-glucopyranose) was the major tar component. A large number of other materials have been identified as cellulose pyrolysis products. However, Glassner and Pierce (1965) demonstrated that the chromatograms obtained from pyrolyzed cellulose and pyrolyzed levoglucosan were identical, suggesting that the observed peaks from cellulose were actually secondary decomposition products of levoglucosan. Madorsky *et al.* (1956, 1958) also suggested that the preceding pyrolytic path that resulted in levoglucosan formation competed with a dehydrative process that gave rise to char. This idea was reinforced by the results of Kilzer and Broido (1964). Using TG and DTA techniques, they found that cellulose, when heated at temperatures below 250°C for extended periods of time, lost up to 10% of its weight as water and on subsequent high-temperature decomposition yielded an increased proportion of char. The postulated pyrolysis paths can be depicted as shown in the accompanying diagram.

$$
\text{cellulose}
\begin{cases}
\xrightarrow[\substack{\text{slightly}\\\text{endothermic}}]{200-280°C} \text{dehydrocellulose} + H_2O \to H_2O + CO + CO_2 + \text{char, etc.} \\[2em]
\xrightarrow[\text{endothermic}]{280-340°C} \text{tar (primarily levoglucosan)}
\end{cases}
$$

Of greatest importance to one interested in the fire retardation of cellulose is the dehydration of cellulose leading to the formation of char. An efficient flame retardant reduces the proportion of the volatile products to the amount of char formed. The reduction of tars and the corresponding increase in char correlates well with decreased flammability as pointed out by OI values.

The pyrolysis of pure cellulose in vacuum yields from 6 to 23% char, depending on the nature of the cellulose (Basch and Lewin, 1974). Apart from the fine structural considerations, the char yields of cellulosics can be influenced by many factors. For example, prolonged heating at low temperatures increases the final char yield (Broido and Nelson, 1975), whereas rapid heating and removal of volatile products favor tar formation (Eventova *et al.*, 1975). The effects of pyrolysis temperature and atmosphere on char properties have been studied spectroscopically (Miller and Gorrie, 1971).

Several acid-type flame retardants are known for cellulose, and they are believed to act via dehydration to produce water and char at the expense of flammable tars (Lyons, 1970). Consider, for example, the thermal stability of cellulose treated with 5% antimony trichloride (Lewis acid catalyst). This treatment caused gradual dehydration and decomposition of cellulose at relatively low temperatures, leaving about 40% char at 500°C (Fig. 14). Untreated material under identical conditions gave a substantial amount of flammable levoglucosan and its condensation products (Fig. 13). Barker (1974) has also shown by DTA that increasing amounts of phosphoric acid caused a decrease in the temperature of endothermic pyrolysis for cotton fabrics, which then led to a subsequent increase in the oxygen index.

Several workers (Perkins *et al.*, 1966; Gilliland and Smith, 1972) have reported the thermal characterization of cellulosic fabrics treated with flame retardants based on tetrakis (hydroxymethyl) phosphonium (THP) structure. The TG and DTA curves of cotton fabric treated with three flame retardants, namely, tris(l-aziridinyl)phosphine oxide (APO), urea-NH_3, and methylolmelamine (MM) mixed with tetrakis(hydroxylmethyl) phosphonium chloride (THPC), are shown in Fig. 15 (Perkins *et al.*, 1966).

The three DTA curves in nitrogen had a similar pattern, each having two sharp exothermic peaks. The APO–THPC-treated fabric showed peaks at the lowest temperatures, 257 and 295°C; the THPC–MM showed

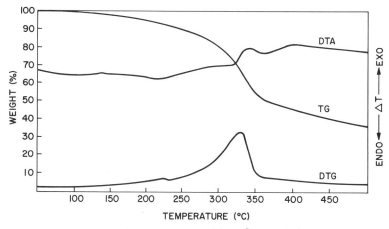

Fig. 14 Thermal analysis of SbCl$_3$-treated cellulose. [From Shafizadeh and Fu (1973).]

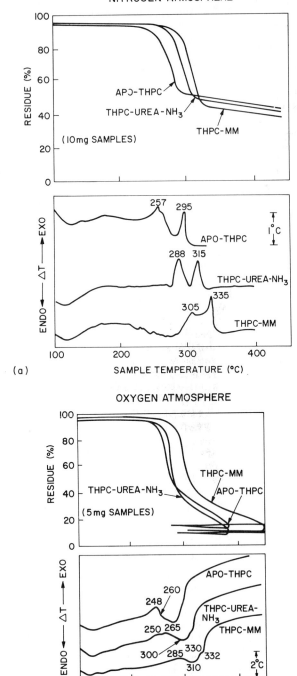

Fig. 15 TG and DTA thermograms of THPC resin-treated cotton fabrics in nitrogen and oxygen atmospheres at a heating rate of 15°C/min. [From Perkins *et al.* (1966).]

peaks at the highest, 305 and 335°C. Analysis of the TG curves also revealed a lower temperature degradation of the APO–THPC sample; however, it yielded the maximum char consistent with the previously made conclusions that a lower decomposition temperature favored char formation. The double peak DTA pattern in nitrogen was typical for these flame-retardant resins; all three lowered the decomposition temperature of the sample and increased the residue. The first peak was believed to be due to the exothermic decomposition of the treated fabric as evidenced by the TG curve; the second peak probably represented further cross-linking reactions in the char. For a better interpretation one would require the TG–DTA analysis of these flame retardants separately.

In an oxygen atmosphere, the TG of these samples showed uncontrolled degradation beyond 300°C. This was caused by the excessive heat liberation most likely due to the ignition of the pyrolytic products in oxygen. Comparison of the DTA thermograms of these three materials in nitrogen and oxygen revealed that the first peak due to the exothermic degradation occurred at lower temperatures in oxygen. The second peak in oxygen was not as well defined as in nitrogen, since in oxygen the char residue was volatilized on oxidation and therefore caused a shift in the base line.

Recently Brady and Langer (1980) have reported that the flame-retardant effect of phosphate additives on wood can be readily evaluated by TG. They found that a higher char residue and a higher temperature of the maximum rate of char combustion were associated with lower flammability.

b. Wool. Wool has been regarded as a relatively safe fiber from the flammability point of view (Tesoro and Meiser, 1970; Perrot, 1971; Crawshaw *et al.*, 1971). The natural flame resistance properties originate from the presence of disulfide cross-links, its relatively high nitrogen and moisture content, high ignition temperature, low heat of combustion, and relatively high oxygen index of about 25. Another important property of wool fiber is that it does not undergo melt and drip—a common problem with synthetic fibers and responsible in many cases for skin injuries or the propagation of actual fires. Although wool is regarded as a relatively safe fiber when compared with common natural and synthetic fibers, it could be flame retarded to a higher degree if required.

The flame propagation of wool has been the subject of many investigations (Benisek, 1975). Although very little is known about the actual mechanism of thermal degradation and combustion of wool, several workers have applied thermal analysis to characterize the untreated wool and the flame-retarded material.

Based on DTA data, Manefee and Yee (1965) postulated that in wool, an amide cross-linking at about 160°C, melting at 235°C, endothermic disulfide cleavage at 230–250°C, and general decomposition above 250°C took place. Carrol-Porczynski (1971), using DTA, TG, and mass spectrometry, had shown that wool produced an endothermic peak at low temperatures associated with the loss of water. An endothermic region around 247°C was associated with the release of sulfur compounds owing to the breaking of the sulfur bonds in the wool protein, CS_2 and COS being the main products at this temperature. SO_2 was formed at 315°C and CO_2 release also started in this endothermic region.

Hendrix et al. (1970) showed that a large improvement in fire resistance of wool could be achieved by treatment with 15% H_3PO_4. TG results indicated that phosphoric acid treatment of wool did not lower its decomposition temperature, but the char residue at 500°C was raised from 29.8 to 39.3%. The expected increase in flame retardancy due to H_3PO_4 treatment, based on TG data was supported by an increased oxygen index from a value of 25.5 for untreated wool to 47.1 for the treated wool.

Beck et al. (1976) have evaluated the effect of many flame retardants on wool using TG. They showed that weak acidic materials, such as boric acid/borax, ammonium sulfamate, dihydrogen phosphate, etc., were effective additives, since they lowered the weight loss and increased the char yield of wool.

c. Nylon 6. On exposure to ignition, nylons (6 and 66) undergo melting and dripping, which results in fire retardation as the molten polymer drip carries the flame away (Crawshaw et al., 1973).

Ammonium bromide and ammonia dihydrogen phosphate had been shown to be effective flame retardants (Douglas, 1957). Ammonium bromide may be fixed onto nylon with a urea–formaldehyde resin (Mosher, 1962) or with an aminotriazine–aldehyde condensate (Burnell, 1960). It should be noted that very few flame-retardant treatments are known for Nylon 6 and that none of them are used in practice on a reasonable scale (Crawshaw et al., 1973).

Although commercial flame retardants for Nylon 6 are not known, thermal analysis has been employed in screening several fire-retardant additives. Reardon and Barker (1974) investigated organobromine compounds as fire retardants for Nylon 6 (Figs. 1 and 2). As discussed before, based on TG, DTA, and gas chromatographic studies, it could be concluded that the organobromine compounds have a deleterious effect on the combustability of Nylon 6. This was explained by the catalytic activity of these additives toward the decomposition of Nylon 6. The TG thermogram indicated a vapor-phase activity of organobromine additives, whereas the degradative catalytic activity was evident from the DTA and gas chromatographic data. One must realize that this catalytic activity was entirely different from the condensed phase activity that we have described in the preceding sections.

Barker *et al.* (1972) have evaluated triphenylphosphine oxide, triphenylphosphine, triphenylphosphate, and triphenylphosphite as flame retardants for Nylon 6. The fire-retardant efficiency of these compounds exhibited dependence on the nature of the oxidant (N_2O or O_2), which was suggestive of vapor-phase activity for these additives. Thermal analysis of the untreated and treated Nylon 6 samples did not reveal any significant alteration in the decomposition patterns. Based on this piece of information, one could conclude that these types of thermally stable phosphorous compounds acted by a vapor-phase mechanism in flame-retarding Nylon 6.

d. Nylon 66. DiPietro *et al.* (1971) have described thermoanalytical studies on a Nylon 66 fabric treated with an unidentified experimental backcoating flame-retardant latex. The DSC thermogram of an untreated

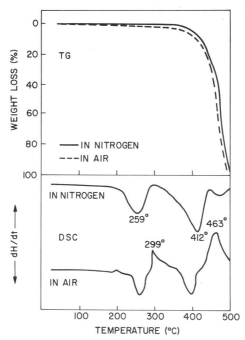

Fig. 16 TG and DSC curves of untreated Nylon 66 fabric in nitrogen and air. [From DiPietro *et al.* (1971).]

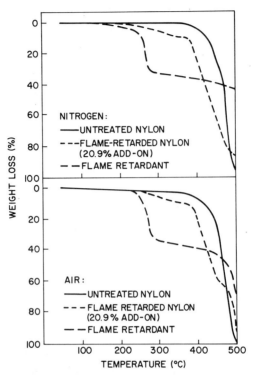

Fig. 17 TG curves of flame retarded and untreated nylon fabric and of the flame retardant in nitrogen and air. [From DiPietro *et al.* (1971).]

Nylon 66 fabric, under nitrogen atmosphere, exhibited an endotherm at 259°C indicating the polymer melt temperature (Fig. 16). The second endotherm at 412°C was ascribed to polymer decomposition, since it was in the temperature range for the major weight loss as shown on the TG thermogram. However, in air, two additional exotherms were noticed (299 and 463°C), and these most likely were related to the oxidation of the Nylon 66 melt and oxidation of the char, respectively. TG thermograms of the untreated Nylon 66 fabric in nitrogen and air showed similar patterns, although the decomposition temperature and char yield were lowered by 10°C and 4–5%, respectively, in air.

Flame-retarded Nylon 66 showed lower thermal- and thermooxidative stabilities as compared to untreated material (Fig. 17). For example, in a nitrogen atmosphere the temperature at 50% weight loss of flame-retarded Nylon 66 was 463°C compared with 485°C for the untreated one. In the presence of flame retardant, the char yield at 500°C was increased from

7.1% to 14% in nitrogen medium. The DSC thermograms supported the TG results.

In conclusion, the presence of flame retardant lowered the initial decomposition temperature and increased the char residue. This information derived from thermal analysis suggested an improved fire resistance, which was in fact supported by the increased oxygen index values.

e. Poly(Ethylene Terephthalate) (PET). Polyester fibers, because of their outstanding physical properties, ease of care, and low cost, have captured 40% of the total U.S. wearing apparel industry. Much of this fiber is used in outerwear, sleepwear, and carpets, and as a consequence the need for development of fire-resistant PET has become very important.

Flame retardation of PET has been recently reviewed by Lawton and Setzer (1975). This review indicated that halogen and/or phosphorus compounds added to the melt prior to spinning PET fibers were useful flame retardants. Thermal analysis has been extensively used in evaluating various types of flame retardants for PET, with regard to their effectiveness, mode of action, and the existence of any synergism.

Antimony–halogen synergism in flame retarded PET fibers has been well known. Figure 18 shows the TG thermograms of PET fibers containing 6 wt % bromine with and without the addition of 1.5 wt % antimony oxide. Incorporation of antimony oxide resulted in higher weight loss in the initial decomposition range of 350–400°C. As discussed in Section III.B, this lowering of the decomposition range was attributed to an inter-

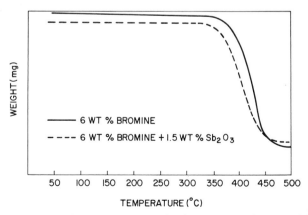

Fig. 18 Comparison of TG thermograms (in nitrogen atmosphere, 10°C/min heating rate) of PET fibers with and without antimony synergist. [From Lawton and Setzer (1975).]

action of antimony and the bromine source and was the cause of the well-recognized antimony–halogen synergism.

Phosphorus compounds are known to flame-retard polymeric materials predominantly by action in the condensed phase (Hilado, 1968, 1969). Recent results have indicated that phosphorus compounds may also be active in the gas phase (Hastie and Blue, 1973). In some polymeric materials the flame-retarding efficiency may depend on the oxidation state of the phosphorus in additives (Lyons, 1979; Papa and Proops, 1972), whereas in other polymers nearly any form of phosphorus is equally effective (Van Wazer, 1961). For PET, the effectiveness and mechanistic action of a phosphorus flame retardant depends on its type, and thermal analysis has been used to substantiate it.

The addition of red phosphorus to PET has been reported to decrease its flammability (Granzow *et al.*, 1977). TG experiments indicated increasing char with increasing phosphorus content. The oxygen index values were also increased; for 0, 4, and 12% phosphorus the OI numbers were 20.4, 27.5, and 32.6, respectively. It was concluded that condensed-phase reactions were responsible for the reduced flammability.

Barker (1974) has evaluated triphenylphosphine oxide as a flame retardant for PET. DTA and TG showed no apparent changes in the decomposition of the polymer with and without the additive. Also the compound was effective in an oxygen atmosphere but not in a nitrous oxide atmosphere. These results were typical of flame retardants acting in the vapor phase. Deshpande *et al.* (1977) confirmed this conclusion but also found that when the triphenylphosphine oxide moiety was incorporated into the backbone of the polymer, a condensed-phase activity also occurred as shown by TG. When used as an additive, triphenylphosphine oxide may simply volatilize and function in the vapor phase. However, when it was part of the polymer backbone, volatilization may not have been completely possible, and then, in some manner, the retardant could enhance the residue formation.

Koch *et al.* (1975) studied the reduced flammability of PET in the presence of a flame-retardant mixture of hexabromobenzene and triphenylphosphate. TG experiments indicated that the thermogram of the flame-retarded system was a simple composite, on a weight basis, of the TG thermograms for each of the individual components. This was used for interpreting the flame-retarding mechanism as being primarily in the vapor phase.

However, the vapor-phase activity of phosphorus-based flame retardants cannot be generalized. Consider, for example, the TG thermograms of PET fiber with and without an organophosphate additive (Fig. 19). The major decomposition of the fiber without additive occurred in the 400–

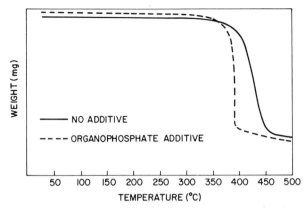

Fig. 19   Comparison of TG thermograms (in air, 10°C/min heating rate) of PET fibers with and without organophosphate additive. [From Lawton and Setzer (1975).]

450°C range as compared to the 350–400°C interval when 1.3 wt % of phosphorus was present. Volatilization of the phosphate additive in the 350–400°C region was too small to account for the difference; the additive may act as a catalyst for decomposition. Setzer and Lawton (1975) claimed that the mechanism of flame inhibition in this case may be many faceted.

It appears that various flame retardants in PET act by different modes and that thermal analysis could be used to suggest a particular mechanism of flame retardation.

f. Nomex®.   In the past decade several high-temperature man-made fibers have been made commercially available. These materials based on aromatic and heterocyclic structures are also flame resistant; the inherent flame retardancy is due to a heavy carbonaceous char that rapidly forms on the surface exposed to a flame. However, the high-temperature and flame-resistant fibers are rather expensive and the textile properties are more or less inferior when compared with common textile fibers; as a result, they are mainly used as protective clothing. A few examples of these materials are Nomex—aromatic polyamide; PBI—polybenzimidazole; Kynol®—cross-linked phenolic; Durette®—a chlorinated aromatic polyamide; Kermal®—polyamidimide; etc.

Nomex is known to be a poly(*m*-phenylene isophthalamide). Its thermal decomposition in relation to flammability was studied by a variety of techniques, including thermal analysis (Chatfield *et al.,* 1979). Initial TG experiments indicated weight loss at 100°C due to absorbed moisture. Dried samples showed initial weight loss at about 380°C in helium, air, or

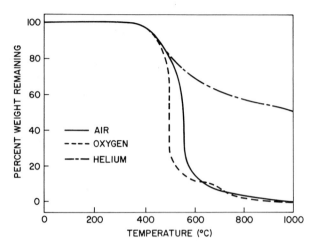

Fig. 20   Effect of medium on the TG thermogram of Nomex®. [From Chatfield *et al.* (1979).]

oxygen (Fig. 20). In the inert atmosphere, a 52% char yield was obtained at 1000°C. In air or oxygen medium, no residue was left as a consequence of the char oxidation. A similarity of the initial portions (10–15%) of the degradation curves of Nomex in all three media, in our opinion, suggests the formation of char on thermal exposure regardless of the atmosphere. The build-up of a fire-resistant char residue on thermal exposure in a fire situation is the primary reason for the inherent flame retardancy of this fabric.

Based on the TG thermogram, Chatfield *et al.* (1979) chose 550°C as the temperature for pyrolytic studies. The major pyrolytic products were identified as water, carbon monoxide, carbon dioxide, 1,3-dicyanobenzene, and 3-cyanobenzoic acid. The mechanism of decomposition was discussed in regard to these products.

g. Durette®.   Durette is made by chlorination or oxychlorination of poly(*m*-phenylene isophthalamide) at high temperatures.

Chatfield *et al.* (1979) have studied the thermal degradation mechanism of Durette. Their TG results showed that this material had slightly lower thermal stability than Nomex. The char residue at 1000°C in inert atmosphere was about 50%. It appeared that Durette may derive its superior flame resistance from the liberation of hydrogen chloride during its thermal decomposition.

Bingham and Hill (1975) have investigated the thermooxidative stabilities of two Durette samples, one without any treatment and the other heat treated. The heat treatment was given to the material (details of the process are not available) by the manufacturer to confer greater flame resistance. Thermal treatment raised the stability of the Durette fabric (e.g., the onset of decomposition was increased by 10°C and the rate of decomposition lowered). Comparison of the TG thermograms of the two samples (Figs. 21 and 22) showed that at 600°C, the treated material still retained about 25% of its original weight.

It is possible that during the thermal treatment, Durette underwent a cross-linking reaction that eventually resulted in a higher-heat-resistant char. An exothermic peak at about 400°C on the DSC thermogram of the untreated sample (Fig. 21) was probably associated with the cross-linking phenomenon, the latter being further supported by the absence of the exotherm in the DSC curve of the thermally treated Durette. The higher flame resistance of the heat-treated material could thus be attributed to the cross-linking reaction.

h. Kynol®. Kynol fabrics are composed of cross-linked phenolics, which have been well recognized as outstanding flame-resistant materials.

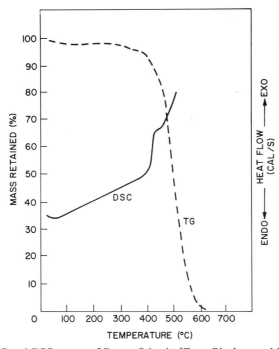

Fig. 21   TG and DSC curves of Durette® in air. [From Bingham and Hill (1975).]

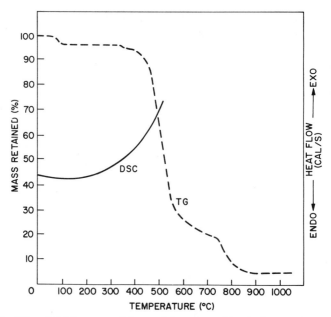

Fig. 22 TG and DSC curves of Durette® heat treated in air. [From Bingham and Hill (1975).].

The exceptional flame-retardant properties of phenolic fibers have been discussed in detail by Economy (1978). When exposed to a flame, phenolic resins produce heavy char and noncombustible products like H_2O and CO_2, which are desirable conditions for excellent flame retardancy.

Kynol fibers have been evaluated as possible precursors for carbon fibers (Lin and Economy, 1973; Economy and Lin, 1971). On exposure to elevated temperatures, phenolic fibers are rapidly carbonized in yields as high as 60%, without showing any loss in mechanical properties. TG studies of the carbonized fibers showed that their thermal stability depended on the temperature to which they were exposed; the higher the treatment temperature of phenolic fiber, the greater the thermal stability.

2. Plastics

Many new flame-retardant chemicals have been developed for use in thermoplastics. The majority of these flame retardants are of the additive type, usually halogen- and/or phosphorus-based compounds, and sometimes are used with a synergist. Many of the more recently introduced plastics such as polycarbonates and polysulfones fortunately have improved flammability behavior due to their aromatic character. In this section we shall discuss briefly the fire retardation of polyolefins, polystyrene, and polyvinyl chloride with reference to the role of thermal analysis.

a. Polyolefins. The polyolefins here are taken to include polyethylene and polypropylene only. TG studies have indicated that polyethylene and polypropylene degraded completely by 440° and 400°C, respectively, leaving no residue. Schmidt (1966) and Reichherzer (1966) identified the degradation products of polyolefins as being olefins, paraffins, and cyclic hydrocarbons. The formation of these highly combustible products explained the poor flammability behavior of these polymers, which have oxygen indices of about 18.

Several halogenated flame retardants have been utilized for polyethylene and polypropylene based on the TG data (Schwarz, 1973). The criterion for screening useful fire-retardant additives for a particular polymer have been described earlier. Touval (1975) has also reported on the use of TG for evaluation of antimony–halogen synergistic mixtures as flame retardants for polyethylene and polypropylene.

Peters (1979) has described the use of red phosphorus in flame-retarding high-density polyethylene (HDPE) and suggested its mode of action using thermal analysis as a major tool. He concluded that the fire retardancy by phosphorus could be accounted for by vapor- as well as condensed-phase modes of flame inhibition. The dependence of the retardant activity on the oxidant (O_2 or N_2O) implied the vapor-phase mechanism, and the existence of condensed-phase activity was shown by TG. The incorporation of red phosphorus into polyethylene increased the thermal and thermooxidative stabilities of the base polymer. For example, the TG

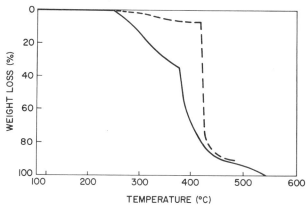

Fig. 23 Thermogravimetric analysis in air. (——) HDPE; (----) HDPE + 8% Red P. [From Peters (1979).]

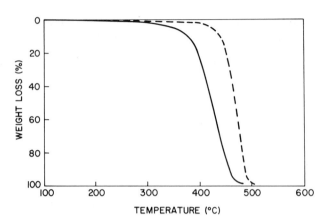

Fig. 24   Thermogravimetric analysis in nitrogen. (——) HDPE; (----) HDPE + 8% Red P. [From Peters (1979).]

of HDPE with 8% P in air exhibited a 6% weight loss at 400°C compared to a weight loss of about 70% for the base polymer. (Fig. 23). The increase in thermooxidative stability was also exemplified by the heats of oxidation as determined by DSC. Addition of 6%P lowered the heat of oxidation by a factor of more than 4. The thermal stability of HDPE was also significantly improved in the presence of red phosphorus. In nitrogen at 440°C, HDPE exhibited 70% weight loss compared to only 8% for a sample containing 8% red P (Fig. 24). Based on thermal and other studies, Peters (1979) suggested that red phosphorus increased the thermal-oxidative stability of HDPE by scavenging oxygen at the polymer surface, thus retarding the oxidative degradation processes. Also the increased thermal stability in the presence of red phosphorus has been attributed to the quenching of polymer radicals by phosphorus, which then prevented the free-radical chain degradative processes. These interactions between HDPE and P in the condensed phase led to the reduction in polymer flammability, as expected.

   b. Polystyrene.   Lindemann (1973) has reviewed the flame retardation of polystyrene. TG studies were reviewed in regard to demonstrating the degradation reaction for polystyrene, which produced large amounts of styrene monomer and rendered the material flammable.

   The most common technique of flame-retarding polystyrene is the incorporation of halogens either as additives or as part of the polymer. It is normally preferred that the halogen be a part of the polymeric chain, as the halogenated additives tend to be noncompatible and, in addition, can cause several other side effects.

Prins *et al.* (1976) have illustrated the effect of polymeric bromine on the flammability of polystyrene using thermal analysis. They studied the degradation characteristics of polystyrene (PS) and polybromostyrene (PBS) by differential TG, TG/MS, and DTA techniques. PBS decomposed about 10°C lower than PS, and in both cases the main products were their respective monomers regardless of the medium (argon or oxygen). However, in an oxygen atmosphere, a larger fraction remained as a residue in the PBS sample, but this was eventually oxidized above 500°C (Fig. 25). The DTA thermograms in argon (Fig. 26) showed endothermic decomposition peaks for both polymers in the same temperature region as in the TG. In an oxygen atmosphere, however, the situation was completely changed. The PS exhibited a large exothermic peak in about the same temperature range where the thermal degradation of the polymer took place. This strongly suggested that the exothermic peak was the result of combustion. On the other hand, PBS showed only a minute exotherm in its thermal degradation region and a major exothermic transition at about 500°C, with the latter being due to the oxidation of carbonaceous residue. Under the conditions of these experiments, one can surmise that the bromine present inhibits most of the oxidative chain reactions and thus the polymer continues to degrade to the monomer, but combustion is not supported. Thus DTA analysis in oxygen suggested an effect of the

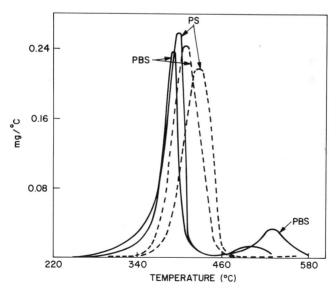

Fig. 25   Differential TG of polystyrene and polybromostyrene in argon and air at a heating rate of 6°C/min; ----argon; ———oxygen. [From Prins *et al.* (1976).]

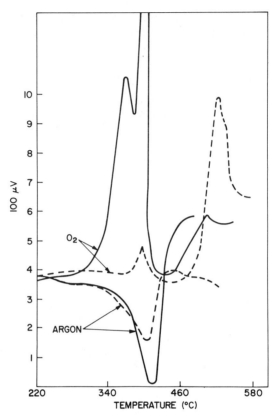

Fig. 26 DTA of polystyrene and polybromostyrene in argon and in oxygen at a heating rate of 6°C/min; ————PS;−−−−−PBS [From Prins *et al.* (1976).]

bromine inhibition on the chain reactions taking place during the combustion of PBS.

Khanna and Pearce (1978) have also described the flame retardation of polystyrene by introduction of the —CH$_2$Cl groups onto the benzene rings. The role of thermal analysis in this case has been explained in Section III.C.

Brauman (1979) has demonstrated that multifunctional benzyl chloride types of Friedel-Crafts alkylation reagents can be used with a Sb$_2$O$_3$–chlorine Lewis acid precursor in impact polystyrene (IPS) to promote the cross-linking and charring of the system. Increased char formation correlated well with improved combustion performance. Although Brauman (1979) used radiant pyrolysis apparatus for obtaining the char residues, one could also apply thermogravimetric analysis to do so. The author, however, used TG technique to study the thermal stability of the various chars. As shown in Fig. 27, the thermal stabilities of the chars made with 4,4′-bis(chloromethyl) diphenyl oxide (CMDPO), 2,4,6-tris(chloromethyl) mesitylene (Mes), and mellitic trianhydride (MAnh) were compared to that of a control char dervied from IPS, a chlorinated wax (CW) and Sb_2O_3. Although these pyrolysis chars underwent major weight losses between 415 and 475°C, they did exhibit differences in total amount of weight loss and the onset temperature of the weight loss. The MAnh char began to lose weight at a temperature lower than that of the control (225 vs. 260°C). The MAnh char also exhibited the greatest major weight loss (about 50%); others ranged from 25 to 35%. The Mes and CMDPO chars began to degrade at 330°C and 325°C, respectively. Determination of the flammability ratings using ASTM D635 test showed that only the MAnh sample burned; the others were classified as nonburning.

   c. Polyvinyl Chloride.   PVC has a high level of chlorine and as a result is considered inherently flame retardant. However, in many circumstances further improvement in flammability reduction of PVC is desired. An efficient way to enhance flame retardation of PVC is by the addition of Sb_2O_3 as a synergist. Additive retardants based on halogen and/or phosphorus are also employed.

   Sobolev and Woycheshin (1974) have pointed out that the burning rate of PVC could be lowered using $Al_2O_3 \cdot 3H_2O$ as a flame-retardant filler.

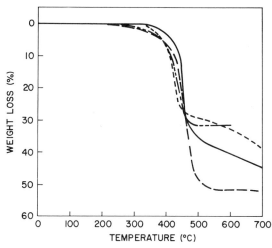

Fig. 27   TG thermograms of radiant pyrolysis chars from impact polystyrene Friedel-Crafts systems; 10°C/min, nitrogen, (—) IPS-CW-Sb_2O_3-CMDPO, (----) IPS-CW-Sb_2O_3-Mes, (· · · · ·) IPS-CW-Sb_2O_3, (— — —) IPS-CW-Sb_2O_3-MAnH. [From Brauman (1979).]

The TG and DTA curves of the hydrate were used to explain the flame-retardant action. The mechanistic action of $Al_2O_3 \cdot 3H_2O$ has been described in Section III.A.3.

Imhof and Steuben (1973) conducted DTA and TG studies in oxygen on polyvinyl chloride and polyvinylidene chloride (PVDC). Smoke and oxygen index studies were reported. Substitution of a chlorine for a hydrogen usually increases smoke generation (e.g., polyvinyl chloride and polyethylene), but the partial chlorination of PVC decreased smoke. This was related to the higher char residue between 500 and 600°C for PVDC as compared to PVC.

Boettner et al. (1969) have studied the volatile degradation products of vinyl plastics with the aid of DTA and TG coupled with IR and MS techniques. The major products appeared to be HCl, CO_2, and CO. The breakdown mechanisms were proposed.

3. Thermosets

Fire retardancy in thermosetting polymers is achieved largely by the use of reactive fire retardants, since the common fire-retardant additives lack permanence. The flammability of thermosetting materials can be reduced by the addition of inorganic fillers and/or reactive flame-retardant components; flame-retardant vinyl monomers or cross-linking agents could be used as well. Use of thermal techniques has been cited in characterizing the flame-retarded thermosetting resins. Some examples that demonstrate the importance of thermal analysis in the fire retardation of thermosets are described below:

a. Unsaturated Polyester Resins. A large number of halogen-containing, reactive diols, polyols, anhydrides, and other functional group-containing intermediates have been employed to produce flame-retardant polyester resins (Voygt et al., 1970); Learmonth et al., 1969; Learmonth and Nesbitt, 1972).

Prins et al. (1976) used bromostyrene as partial replacement of styrene for cross-linking polyester resins. Their results on thermooxidative stability correlated well with the standard flammability tests. Figure 28 shows the DTA thermograms of two polyester resins: one containing only styrene and another containing bromostyrene and styrene in a 80–20 mole ratio. In an argon atmosphere, the DTA thermograms exhibited complex endothermic degradation. On the other hand, in an oxygen atmosphere, the resin containing only styrene oxidized to give a large exotherm at about 300°C and a medium exotherm at about 500°C. The first exotherm was representative of the combustion of styrene primarily and the second exotherm was interpreted as related to the oxidation of carbonaceous res-

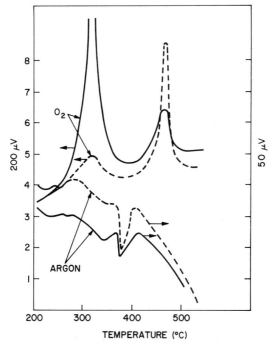

Fig. 28 DTA of polyesters crosslinked with styrene (—) and with 0.8 mole fraction bromostyrene and 0.2 mole fraction styrene (– – – –) at 4°C/min heating rate. [From Prins *et al.* (1976).]

idue that resulted from the decomposition of the resin. However, the resin containing 80 mole % bromostyrene and 20 mole % styrene showed only a small exotherm at 300°C and a larger exotherm at about 500°C. It appeared that bromostyrene, unlike styrene, did not burn, and thus resulted in a smaller exotherm at 300°C, but promoted char formation that eventually increased the size of the exotherm at about 500°C. Flame retardance predicted from these experiments was also supported by the heat of combustion data; styrene-containing resin evolved 1460 cal/g, whereas 80 mole % bromostyrene-based resin liberated only 280 cal/g on combustion. The oxygen index values for the styrene and bromostyrene crosslinked resins were 22.8 and 33.5, respectively.

b. Epoxy Resins. As in polyester resins, reactive halogen-containing fire-retardant compounds are most often used. Tetrabromobisphenol-A is perhaps the most widely used component for fire-retardant epoxy resins.

Nara and Matsuyama (1971) and Nara *et al.* (1972) investigated the thermal degradation and flame retardance of tetrabrominated bisphenol-A diglycidyl ether compared to the nonbrominated structure. As discussed earlier, (Fig. 5), their results indicated that bromine acted by vapor-phase as well as condensed-phase mechanisms of flame inhibition.

Kourtides *et al.* (1975) have reported that the base resin, the diglycidyl ether of bisphenol-A (DGEBA), gave a pyrolysis char yield of 23% at 700°C and an oxygen index of 21. Recently Pearce *et al.* (1978) studied the char formation of copolymers of DGEBA with the diglycidyl ether resins of phenolphthalein (DGEPP) and bisphenol-fluoreneone (DGEBF) cured with trimethoxybroxine. The char residues at 700°C from the TG experiments in an inert atmosphere showed a maximum at about 20% DGEBF for the DGEBA/DGEBF system. On the other hand, a linear increase in char residue and oxygen index was obtained as the concentration of DGEPP in the DGEBA/DGEPP copolymer system increased.

Stahly *et al.* (1974) have studied epoxy resins that were cured by phosphonitrilic chloride polymers $[(PNCl_2)_n]$ and cyclooligomers $[(PNCl_2)_3]$. DTA curves in the temperature range of 100–600°C suggested suppression of combustion because the usual large exotherm was minimized.

c. Polyurethanes. Although polyurethanes are used in a number of forms, fire retardancy in these polymers is primarily of commercial importance in foams. Urethane foams, in general, are fire retarded by the chem-

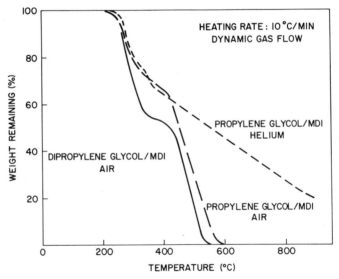

Fig. 29   TG curves for rigid foams prepared from propylene glycol and dipropylene glycol with MDI. [From Backus *et al.* (1968).]

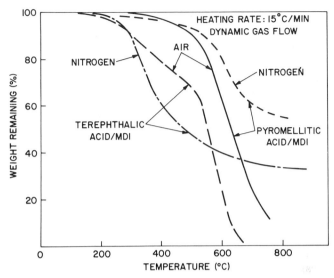

Fig. 30   TG curves for polymers of terephthalic acid and pyromellitic acid with MDI. [From Backus *et al.* (1968).]

ical incorporation of halogens, phosphorus, and/or nitrogenous compounds into the foam.

Backus *et al.* (1965) and Saunders and Backus (1966) have dealt with the thermal stability of common types of rigid urethane foams and an analysis of the factors affecting their flammability. TG experiments on both polyester and polyether rigid urethane foams showed that their decomposition started at about 225°C. The volatile components were held responsible for the foam flammability. Polyisocyanates prepared by phosgenation of aniline–formaldehyde condensates produced char residues rather than flammable gases.

Backus *et al.* (1968) extended their work to synthesize flame-retardant polyurethanes from aromatic polyisocyanates. Their basic approach was to use structural elements that were thermally stable and that formed heavy char residues. Figure 29 shows the TG curves of rigid urethane foams prepared from a polymeric isocyanate, 4,4'-diphenylmethane diisocyanate (MDI) with propylene glycol and dipropylene glycol. As indicated by the thermogram, propylene-glycol-based polymer on degradation in inert atmosphere produced over 30% char at 700°C. However, in air, a two-step weight loss pattern was indicated. The authors suggested that the first step was the loss of lower-molecular-weight polyols and aliphatic fragments that formed on polymer degradation. The char remaining

       E. M. Pearce, Y. P. Khanna, and D. Raucher

after the first step was derived primarily from more thermally stable polyisocyanate. The second step involved thermal and thermooxidative degradations of the char. The influence of increased aromaticity on char yields was further illustrated by the polymers of MDI with pyromellitic and terephthalic acids (Fig. 30). Backus *et al.* (1968) have also pointed out a relationship between flammability and thermal stability (Table III). The data clearly showed a good correlation of flammability with the temperature and extent of initial decomposition. The self-extinguishing (SE) poly-

TABLE III[a]

COMPARISON OF THE FLAMMABILITY AND THERMAL DECOMPOSITION
TEMPERATURES OF MDI-BASED POLYMERS

	COREACTANT	STRUCTURE OF REPEATING UNIT	TEMP. °C, OF TG WEIGHT LOSS IN AIR OF 20%	40%	FLAMMABILITY
1	DIPROPYLENE GLYCOL		280	320	SE(MELTS)
2	HYDROQUINONE		280	300	BURNED
3	p-XYLENEDIOL		310	340	BURNED
4	p-PHENYLENEDIAMINE		320	350	BURNED
5	m-XYLENEDIAMINE		360	390	SE
6	TEREPHTHALIC ACID		380	580	SE
7	PYROMELLITIC ACID		560	605	SE
8	Cl₄p-XYLENEDIOL		320	335	NONBURNING
9	Cl₄m-XYLENEDIAMINE		330	430	NONBURNING

[a] From Backus *et al.* (1968).

mers (except "1," which melted and dripped) exhibited 40% weight loss in the 390–350°C range; chlorinated polymers should be treated as a separate class due to their different flame inhibition mechanism.

Several other workers (Mickelson and Einhorn, 1970; Mickelson, 1972; Einhorn, 1972; Einhorn and Seader, 1972) have also reported on the use of thermal analysis in evaluating the flammability of urethane foams.

4. Elastomers

As in the case of other polymeric materials, burning of an elastomer is mainly governed by the amount and type of pyrolytic gases (i.e., its thermal decomposition behavior). Consequently, thermal analysis has been extensively used in this area also.

Systematic thermal degradation studies on polybutadiene and butadiene copolymers with styrene (SBR rubber) and acrylonitrile (nitrile rubber), and on polyisoprene have been carried out by Madorsky *et al.* (1949) Madorsky and Straus (1958), and Straus and Madorsky (1950). TG results in air showed that polyisoprene began to decompose between 250 and 320°C, whereas the corresponding temperatures for polybutadiene and SBR were 380 and 370°C, respectively. However, all three polymers were consumed by 500°C (Pap, 1970). DTA revealed both exothermic and endothermic processes. Exothermic processes were attributed to cyclization, oxidation, and/or recombination of radicals, whereas thermal bond scission and removal of volatile products accounted for the endothermic transitions.

Skinner and McNeal (1948) and Torii *et al.* (1958) have studied the thermal decomposition of polychloroprene. This rubber is known to lose HCl and undergo an exotherm at 377°C that results from cross-linking (Schuuenker and Beck, 1960). The loss of flame-inhibiting species, HCl, and the tendency to produce substantial amounts of cross-linked char on thermal degradation have made this elastomer most suitable where flame retardancy is desired.

IV. Conclusions

In various parts of this chapter it has been demonstrated that thermal analysis is a useful technique for assessing the flammability of polymeric materials. Thermoanalytical techniques can provide insights as to how a material might react on exposure to a fire [e.g., the nature of decomposition (endothermic or exothermic) and its temperature, amount of volatiles versus char residue, effect of medium on the degradation characteristics, etc.]. A knowledge of the decomposition pattern of the base polymer aids one in selecting appropriate flame retardants. In order to design the flame-retardant formulations on a more logical basis it is always helpful to understand the mode of action of a particular additive. A clue to the flame-retardant mechanistic action could be obtained using thermal analysis.

Large amounts of fire-retardant additives lead to detrimental effects on polymers. Thermal techniques have been employed to develop synergistically efficient flame retardants that not only reduce the cost but also minimize other undesirable side effects.

Toxicity of the gases evolved on degradation of either flame-retarded or non-flame-retarded polymers has been a serious hazard. In the past few years, a considerable amount of research has been directed in developing polymer compositions that yield high char residues and decreased volatile products. This helps, in many cases, not only to reduce toxicity but may also to reduce flammability. Condensed-phase processes leading to carbonaceous products appear to be the future direction of flame retardation. Thermal analysis is a powerful technique for studying the condensed-phase reactions occurring during polymer degradations.

It is to be noted that the thermal techniques for the evaluation of flammability have some limitations also. For example, the thermal methods involve small-scale testing under controlled conditions, which may differ widely from those in an actual fire situation. Even as an analytical tool it often requires other techniques, such as infrared and mass spectroscopy, gas chromatography and elemental analysis, to have a fuller understanding of the degradative processes of the polymer. At the same time, however, one should also realize that each flammability test usually has had some kind of drawback and that there is no single test capable of predicting the performance of a material in a real fire situation.

In spite of some drawbacks, thermoanalytical methods have been a significant laboratory tool for the study of the flammability behavior of polymers, and their applications to this area will continue to grow.

ACKNOWLEDGMENTS

The authors wish to express their sincere appreciation to the editor, Dr. E. A. Turi, without whom a timely and much needed book like this would not have been written. Her constant encouragement, concern, and helpful comments have in fact made the writing of this manuscript a real pleasure.

Our special thanks go to Professor R. H. Barker of Clemson University, whose critical evaluation and comments have resulted in a significant improvement of this manuscript. Helpful suggestions and comments of Dr. S. W. Shalaby of Ethicon, Inc., are also appreciated.

References

Alsheh, D., and Marom, G. (1978). *J. Appl. Polym. Sci.* **22,** 3177.

Backus, J. K., Darr, W. C., Gemeinhardt, P. G., and Saunders, J. H. (1965). *J. Cell. Plast* **1,** 178.

Backus, J. K., Bernard, D. L., Darr, W. C., and Saunders, J. H. (1968). *J. Appl. Polym. Sci.* **12,** 1053.

Barker, R. H. (1974). *NBS Spec. Publ. (U.S.)* **411.**

Barker, R. H., Bostic, J. E., Reardon, T. J., and Strong, R. A., (1972). *Abstr. Pap., 164th Natl. Am. Chem. Soc. Meet.* Cell-67.

Basch, A., and Lewin, M. (1974). *J. Polym. Sci., Polym. Chem. Ed.* **12,** 2053.

Beck, P. J., Gordon, P. G., and Ingham, P. E. (1976). *Text. Res. J.* **46,** 478.

Benisek, L. (1975). *Flame Retard. Polym. Mater.* **1,** 137.

Bingham, M. A., and Hill, B. J. (1975). *J. Therm. Anal.* **7,** 347.

Boettner, E. A., Ball, G., and Weiss, B. (1969). *J. Appl. Polym. Sci.* **13,** 377.

Bonsignore, P. V., and Manhart, J. H. (1974). *Proc. Annu. Conf.—Reinf. Plast./Compos. Inst., Soc. Plast. Ind.* **29,** Paper 23C.

Brady, T. P., and Langer, H. G. (1980). *Therm. Anal., Proc. Int. Conf., 6th, 1980* Vol. 2, p. 443.

Brauman, S. K. (1979). *J. Polym. Sci., Polym. Chem. Ed.* **17,** 1129.

Broido, A., and Nelson, M. A. (1975). *Combust. Flame* **24,** 263.

Burnell, M. R. (1960). U.S. Patent 2,953,480.

Carrol-Porczynski, C. Z. (1971). *Text. Inst. Ind.,* p. 188.

Chatfield, D. A., Einhorn, I. N., and Mickelson, R. W., and Futrell, J. H. (1979). *J. Polym. Sci., Polym. Chem. Ed.* **17,** 1353–1381.

Chaudhuri, A. K., Min, B. Y., and Pearce, E. M. (1980). *J. Polym. Sci., Polym. Chem. Ed.* **18**(10), 2949.

Crawshaw, G. H., Duffield, P. A., and Mehta, P. N. (1971). *J. Appl. Polym. Sci., Appl. Polym. Symp.* **18,** Part II, 1183.

Crawshaw, G. H., Delman, A. D., and Mehta, P. N. (1973). *Flame Retard. Polym. Mater.* **1,** 277.

Deshpande, A. B., Pearce, E. M., Yoon, H. S., and Liepins, R. (1977). *J. Appl. Polym. Sci., Appl. Polym. Symp.* **31,** 257.

DiPietro, J., Barda, H., and Stepniczka, H. (1971). *Text. Chem. Color.* **3**(2), 40.

Douglas, D. O. (1957). *J. Soc. Dyers Colour.* **73,** 258.

Economy, J. (1978). *Flame Retard. Polym. Mater.* **2,** 203.

Economy, J., and Lin, R. Y. (1971). *J. Mater. Sci.* **6,** 1151.

Einhorn, I. N. (1970). "Fire Retardance of Polymeric Materials." Presented at Polymer Conference Series on Flammability Characteristics of Polymeric Materials. University of Utah, Salt Lake City.

Einhorn, I. N. (1971). Reprint from Fire Research Abstracts and Reviews, Vol. 13, p. 3. National Academy of Sciences, Washington, D.C.

Einhorn, I. N. (1972). "Degradation of Polymer Properties," Proceedings of the Polymer Conference Series. University of Utah, Salt Lake City.

Einhorn, I. N., and Seader, J. D. (1972). "Physical Chemical Factors Governing the Combustion of Polymers," Polymer Conference Series. University of Utah, Salt Lake City.

Eventova, J. L., Rudenko, A. P., Kulakova, I. I., Kanovich, M. M., Gorbachera, V. O., Konkin, A. A. Volkova, N. S., and Erofeeva, N. F. (1975). *Khim. Volokna* **4,** 29(1974); *Chem. Abstr.* **82,** 258z, 32 (1974).

Fenimore, C. P., and Martin, F. J. (1966). *Mod. Plast.* **44**(3), 141.

Frisch, K. C. (1979). *Int. J. Polym. Mater.* **7**, 113.
Gilliland, B. F., and Smith, B. F. (1972). *J. Appl. Polym. Sci* **16**, 1801.
Glassner, S., and Pierce, A. R. (1965). *Anal. Chem.* **37**, 525.
Granzow, A., Ferrillo, R. G., and Wilson, A. (1977). *J. Polym. Sci.* **21**, 1687.
Hassel, R. L. (1976). *In* "Evaluation of Polymer Flammability by Thermal Analysis," Pittsburgh Conf. Cleveland, Ohio.
Hassel, R. L. (1977). *Am. Lab.* **9**(1), 35.
Hastie, J. W. (1973). *J. Res. Natl. Bur. Stand., Sect. A* **77a**, No. 0, 733.
Hastie, J. W., and Blue, G. D. (1973). *Org. Coat. Plast. Chem.* **33**, 484.
Hendrix, J. E., Anderson, T. K., Clayton, T. J., Olson, E. S., and Barker, R. H. (1970). *J. Fire Flammability* **1**, 107.
Hilado, C. J. (1968). *J. Cell. Plast.* **4**, 339.
Hilado, C. J. (1969). "Flammability Handbook for Plastics," p. 82. Technomic Publ. Co., Stamford, Connecticut.
Hindersinn, R. R., and Wagner, G. M. (1967). *Encycl. Polym. Sci. Technol.* **7**, 1.
Imhof, L. G., and Steuben, K. C. (1973). *Polym. Eng. Sci.* **13**, 146.
Kambour, R. P. (1981). *J. Appl. Polym. Sci.* **26**, 861.
Kambour, R. P., Klopfer, H. J., and Smith, S. A. (1981). *J. Appl. Polym. Sci.* **26**, 847.
Khanna, Y. P., and Pearce, E. M. (1978). *Flame Retard. Polym. Mater.* **2**, 43.
Kilzer, F. J., and Broido, A. (1964). "Speculations on the Nature of Cellulose Pyrolysis," WSS/CI Paper 64–4. U.S. Department of Agriculture, Washington, D.C.
Koch, P. J., Pearce, E. M., Lapham, J. A., and Shalaby, S. W. (1975). *J. Appl. Polym. Sci.* **19**, 227.
Kourtides, D. A., Parker, J. A., and Gilwee, W. J. (1975). *J. Fire Flammability* **6**, 373.
Krekeler, K., and Klimke, P. M. (1965). *Kunststoffe.* **55**(10), 758.
Larsen, E. R. (1974). *J. Fire Flammability/Fire Retard. Chem. Suppl.* **1**, 4.
Lawton, E. L., and Setzer, C. J. (1975). *Flame Retard. Polym. Mater.* **1**, 193.
Learmonth, G. S., and Nesbitt, A. (1972). *Br. Polym. J.* **4**, 317.
Learmonth, G. S., and Thwaite, D. G. (1969). *Br. Polym. J.* **1**, 154.
Learmonth, G. S., Nesbitt, A., and Thwaite, D. G. (1969). *Br. Polym. J.* **1**, 149.
Lin, R. Y., and Economy, J. (1973). *Appl. Polym. Symp.* **21**, 143.
Lindemann, R. F. (1973). *Flame Retard. Polym. Mater.* **2**, 1.
Lyons, J. W. (1970). In "The Chemistry and Uses of Fire Retardants," Wiley (Interscience), New York.
Madorsky, S. L., Straus, S., Thompson, D., and Williamson, P. (1949). *J. Res. Natl. Bur. Stand. (U.S.)* **42**, 199.
Madorsky, S. L., and Straus, S. J. (1958). *J. Res. Natl. Bur. Stand. (U.S.)* **61**, 77.
Madorsky, S. L., Hart, V. E., and Straus, S. J. (1956). *J. Res. Natl. Bur. Stand. (U.S.)* **56**, 343.
Madorsky, S. L., Hart, V. E., and Straus, S. J. (1958). *J. Res. Natl. Bur. Stand. (U.S.)* **60**, 393.
Manefee, E., and Yee, G. (1965). *Text. Res. J.* **35**, 801.
Mickelson, R. W. (1972). *Thermochim. Acta* **5**, 329.
Mickelson, R. W., and Einhorn, I. N. (1970). *Thermochim. Acta* **1**, 147.
Miller, B., and Gorrie, T. M. (1971). *J. Polym. Sci., Part C* **36**, 3.
Mosher, H. H. (1962). U.S. Patent 3,017,292.
Nara, S., and Matsuyama, K. (1971). *J. Macromol. Sci., Chem.* **5**(7), 1205.
Nara, S., Kimura, T., and Matsuyama, K. (1972). *Rev. Electr. Commun. Lab.* **20**, 159; *Chem. Abst., 77*, 75562h (1972).
Neuse, E. W. (1973). *Mater. Sci. Eng.* **11**, 121.

Ohe, H., and Matsuura, K. (1975). *Text. Res. J.* **45**(11), 778.

Pap, Z. (1970). *Muanyag Gumi* **7**, 243.

Papa, A. J., and Proops, W. R. (1972). *J. Appl. Polym. Sci.* **16**, 2361.

Parker, J. A., Fohlen, G. M., and Sawko, P. M. (1973). "Development of Transparent Composites and Their Thermal Responses," Paper presented at Conference on Transparent Aircraft Enclosures, Feb. 5–8, Las Vegas, Nevada.

Pearce, E. M., and Liepins, R. (1975). *Environ. Health Perspect.* **11**, 59.

Pearce, E. M., Lin, S. C., Lin, M. S., and Lee, S. N. (1978). *East. Anal. Symp. 1977* p. 187.

Perkins, R. M., Drake, G. L., and Reeves, W. A. (1966). *J. Appl. Polym. Sci.* **10**, 1041.

Perrot, P. (1971). *Text. Chim.* **27** (7/8), 20.

Peters, E. N. (1979). *J. Appl. Polym. Sci.* **24**, 1457.

Petrella, R. V. (1978). *Flame Retard. Polym. Mater.* **2**, 159.

Pitts, J. J. (1972). *J. Fire Flammability* **3**, 51.

Pitts, J. J., Scott, P. H., and Powell, D. G. (1970). *J. Cell. Plast.* **6**, 35.

Prins, M., Marom, G., and Levy, M. (1976). *J. Appl. Polym. Sci.* **20**, 2971

Reardon, T. J., and Barker, R. H. (1974). *J. Appl. Polym. Sci.* **18**, 1903.

Reichherzer, R. (1966). *Kunstst.-Rundsch.* **13**, 482.

Rhys, J. A. (1969). *Chem. Ind.* (London) p. 187.

Saunders, J. H., and Backus, J. K. (1966). *Rubber Chem. Technol.* **39**, 461.

Schmidt, J. (1966). *Plast. Kautsch.* **13**, 83.

Schmidt, W. G. (1965). *Trans. J. Plast. Inst.* p. 247.

Schuuenker, R. F., and Beck, L. R. (1960). *Text. Res. J.* **30**, 624.

Schwarz, R. J. (1973). *Flame Retard. Polym.* **2**, 87.

Setzer, C. J., and Lawton, E. L. (1975). Monsanto Triangle Park Development Center Inc. (unpublished results).

Shafizadeh, F., and Fu, Y. L. (1973). *Carbohyd. Res.* **29**, 113.

Skinner, G. S., and McNeal, J. H. (1948). *Ind. Eng. Chem.* **40**, 2303.

Sobolev, I., and Woycheshin, E. A. (1974). *In* "Flammability of Solid Plastics" (C. J. Hilado, ed.), Fire and Flammability Ser., Vol. 7, p. 263. Technomic, Westport, Connecticut.

Stahly, E. E., Johnson, R. D., and Rice, R. G. (1974). *In* "Flammability of Solid Plastics" (C. J. Hilado, ed.), Fire and Flammability Ser. Vol. 7, p. 295. Technomic, Westport, Connecticut.

Straus, S., and Madorsky, S. L. (1950). *J. Res. Natl. Bur. Stand. (U.S.)* **50**, 165.

Tang, W. K., and Niell, W. K. (1964). *J. Polym. Sci., Part C* **6**, 65.

Tesoro, G. L. (1971). "Synergism During Fire Retardation," Presented at Polymer Conference Series on Flammability Characteristics of Polymeric Materials. University of Utah, Salt Lake City.

Tesoro, G. L., and Meiser, C. H. (1970). *Text. Res. J.* **40**, 430.

Torri, C. H., Hoshii, K., and Isshiki, S. (1958). *J. Soc. Rubber Ind. Jpn.* **29**, 3.

Touval, I. (1971). "Halogen Synergists for Flame Retardation," Presented at Stevens Institute of Technology, Hoboken, New Jersey.

Touval, I. (1975). "Flame Retardants for Plastics," Lecture presented at Flame Retardant Polymeric Materials Course, Plastics Institute of America. Stevens Institute of Technology, Hoboken, New Jersey.

Van Krevelen, D. W. (1975). *Polymer* **16**, 615.

Van Wazer, J. R. (1961). "Phosphorus and Its Compounds," Vol. II, p. 1955. Wiley (Interscience), New York.

Vogt, H. C., Davis, P., Fujiwara, E. J., and Frisch, K. C. (1970). *Ind. Eng. Chem. Prod. Res. Dev.* **9**(1), 105.

CHAPTER 9

Thermal Analysis of Additives in Polymers

HARVEY E. BAIR

Bell Laboratories
Murray Hill, New Jersey

I. Introduction

The chemical and physical properties of most commercial plastics can be improved by blending one or more additives into a polymer. The choice of a particular additive or additives for a resin formulation is based on knowledge of a polymer's inherent weaknesses and the conditions under which it will be processed and used. Thermal analysis is useful not only to help the researcher and manufacturer in developing effective additive systems for polymers but also to aid the user in evaluating whether a polymeric material has the expected processing properties and performance characteristics. The additives discussed in

845

this chapter include antioxidants, stabilizers, lubricants, plasticizers, nucleating agents, fire retardants, and other materials.

Most of the analytical problems dealing with the quantitative determination of known additives in a polymer arise from at least two factors: the difficulty of separating an additive from its polymer matrix and the usually low concentration of additives present in most polymers. Fortunately, it is often possible to assay thermoanalytically an additive in a commercial resin without extracting it. If the additive is incompatible with the resin, it can be detected in a separate crystalline or glassy phase by either its T_m or T_g and measured quantitatively from ΔH_f determinations at T_m or ΔC_p measurements at T_g. Conversely, when an additive is soluble in a polymer its concentration can be estimated from shifts in T_m or T_g of the resin. Modern thermal instrumentation permits one to determine routinely the concentration of many additives to 1 wt %, and in special cases to less than 0.01 wt %.

In this chapter the author will show how various thermoanalytical measurements of the chemical and physical properties of additives in polymers can be made by DSC and TG. Wherever possible an attempt will be made to illustrate a particular thermal analysis (TA) method by solving a specific problem dealing with the incorporation or use of additives in polymer systems. In particular, the aspect of direct quantitative analysis of additives in polymers by TA will be emphasized, for the author believes that this approach to additive analysis has not been fully appreciated by many TA practitioners.

II. Protective Agents

Polymers vary widely in their ability to withstand deterioration when exposed to the same environment. These differences in stability are due primarily to chemical structure, but also may be caused by impurities which are often present in the polymer at trace levels. In a typical exposure environment polymeric degradation can be caused by both reactive chemicals like oxygen or water, and by various sources of energy such as heat, UV radiation, or mechanical stress. It is well known that heat accelerates polymeric degradation in the presence and absence of chemical reactants. In fact, most thermal analysis methods used to study polymer stability take advantage of this general phenomenon to acquire a lot of experimental data in a short time.

In order to protect a polymer from the deleterious effects of its environment, various additives, as, for example, antioxidants, antihydrolysis agents, UV absorbers, etc., are blended into the polymer. It is the objective of this section to show how thermal analytical techniques have been developed to study the behavior of antioxidants and stabilizers in protecting polymers.

The largest reported application of thermal analysis to investigate stabilization systems in polymers has been the protection of polyethylene. Thus we shall examine these results in detail as an example of what might be done with different polymer systems.

A. ANTIOXIDANTS

1. General Mechanism

Although polyethylene is a comparatively stable polymer, it is susceptible to oxidation during processing, storage, and use. The degradation of the polyolefin involves the breaking of primary chemical bonds. After reaction with less than 10 cm^3 of oxygen per gram of polymer the material's useful properties are destroyed (Hawkins *et al.*, 1959a,b; Hansen *et al.*, 1964).

The mechanisms of polyethylene oxidation have been reviewed and described as a free radical chain reaction (Shelton, 1972). Free radicals (R°) are formed that combine with oxygen to create peroxy radicals

$$R° + O_2 \rightarrow ROO° \tag{1}$$

These radicals can remove a hydrogen atom from another portion of the polymer to form hydroperoxide and a polymer free radical:

$$ROO° + RH \rightarrow ROOH + R° \tag{2}$$

Thus a repetitive cyclic chain reaction results with the addition of oxygen to the newly formed free radical. The hydroperoxides that have been generated along the polyethylene's backbone are unstable and decompose to form free radicals that initiate more chain reactions.

The oxidative degradation of polyethylene can be delayed by adding small amounts of labile-hydrogen donors such as hindered phenols or secondary aromatic amines. These antioxidants (AH) combine with peroxy radicals and yield a relatively stable radical

$$ROO° + AH \rightarrow ROOH + A° \tag{3}$$

Thus in this manner the peroxy radicals are rendered incapable of propagating the chain reaction. However, in time the stabilizer is consumed, and polyethylene is left unprotected.

The structural formulas of two typical hindered phenols whose thermoanalytical behavior will be examined in this section are as follows:

where / ≡ CH_3 and × ≡ $C(CH_3)_3$

The chemical identification of the upper formula is 4,4′-thiobis(3-methyl-6-t-butyl phenol) and of the lower structure is tetrakis [methylene 3-(3′,5′-di-t-butyl-4′-hydroxyphenyl) propionate] methane. These two compounds will be referred to throughout the remainder of this chapter as Phenol A and B, respectively.

The need for protecting polyethylene at elevated temperatures from oxidative and nonoxidative deterioration is shown in Fig. 1 where the rate of volatilization in oxygen and nitrogen of an unprotected branched polyethylene is plotted against temperature. The degradation rate of polyethylene was found to be approximately 100,000 times more rapid in oxygen than nitrogen at 260°C. Activation energies of 27 and 61 kcal/mol were calculated for degradation in oxygen and nitrogen, respectively (H. E. Bair, unpublished, 1971). In these isothermal studies the evolution of low-molecular-weight volatile fragments formed by the rupture of the polymer chain are detected thermogravimetrically as a loss of sample weight as a function of time. Samples weighing a few milligrams each were heated at 320°C/min to the test temperature in a Perkin-Elmer TGS-1 thermobalance (Chapter 1).

Normally, concentrations of 0.1% by weight of chain-terminating antioxidants such as phenols or amines are employed to retard the thermal oxidation of polyolefins. For more sensitive polymers and in certain applications such as cable insulation auxiliary stabilizers like peroxide decomposers and metal deactivators are used to protect the polymer.

Antioxidant effectiveness must be evaluated for two general periods in the lifetime of the polymer. The first phase is during the processing or fabrication when the polymer is formed into its intended shape at elevated temperatures. The second important period occurs when the polymer is exposed to the conditions of actual use. This latter phase also includes storage.

Thermal analysis methods are ideally suited to measure antioxidant ef-

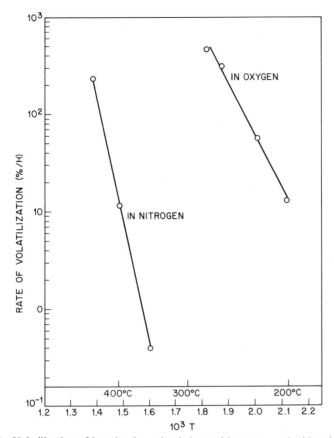

Fig. 1 Volatilization of low-density polyethylene without any antioxidant in nitrogen and in oxygen atmospheres (H. E. Bair, unpublished, 1971).

fectiveness at or near processing conditions, whereas the prediction of service life from accelerated aging can prove to be difficult, if not impossible. The strengths and weaknesses of these thermal techniques for measuring stabilizer effectiveness during processing and in the field will be presented in the next two sections.

2. High-Temperature Stability

From the early 1960s to the present, most thermal analysis techniques for studying polymeric stabilizer systems have been based on the antioxidant's ability to inhibit an oxidation reaction (see, for example, Rubin *et al.*, 1961; Howard, 1972, 1973b; Bair, 1973; Marshall *et al.*, 1973). Typically, a sample is heated to a specified temperature and the period of time before the onset of rapid thermal oxidation is monitored. This time interval is called the induction period. The end of the induction period is signaled by an abrupt increase in a sample's temperature, evolved heat, or weight gained and may be observed by DTA, DSC, or TG devices, respectively. All three of these thermal analytical techniques are similar in principal to the oxygen absorption method described by Hawkins *et al.* (1959a) and used extensively in research on the oxidative behavior of materials.

a. Detection of an Oxidative Induction Time (OIT)

i. OIT by DTA. Stabilized polyethylene samples varying in form from powder to an actual piece of insulated wire can be run in the typical DTA device. Perhaps the best samples to work with are disks cut from a thin (~10 mil) film. In the latter case if the film has been freshly pressed, the antioxidant should be uniformly distributed throughout the sample.

Obviously an oxidative induction time (OIT) can be measured by DTA in a number of ways. Let us examine an isothermal method that was developed within the Bell System for the evaluation of polyethylene wire and coating insulation (Howard, 1973b). This particular test method is currently being examined by ASTM for possible adoption as an industry standard in determining a polymer's induction period. Under this procedure the sample is placed in the DTA sample holder at room temperature on either aluminum or preoxidized copper dishes as required. The holder assembly is closed an an inert gas (nitrogen) is flushed across the sample at 200 ± 25 ml/min. A programmed controller heats the furnace chamber at a rate of 10°C/min up to the preselected temperature. After the unit has reached equilibrium at this temperature, the inert gas is replaced by oxygen flowing at the same rate. This automatically introduces a small shift in the base line of the recorder chart, marking zero time (Fig. 2). The unit is maintained isothermally at the preselected temperature until an abrupt departure from the base line signals the beginning of the oxidation exotherm. The end of the induction period is defined by the intersection with the base line of a line tangential to the rising portion of the exotherm.

It should be noted that some DTA instruments may require flow rates considerably below 200 cm³/min in order to function properly. This is true of the Perkin-Elmer DSC instruments, as will be pointed out in the next section.

Most commercial DTA apparatus should be capable of detecting the induction period from about 170 to 220°C. Using the DTA method variation in the OIT for an unprotected polyethylene sample ranges from about 1 h at 170° to less than a minute at 220°C where induction times become too short to be meaningful.

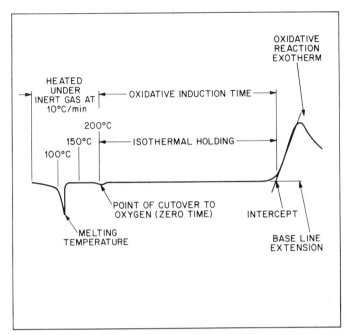

Fig. 2 Typical DTA scan for the determination of polyethylene's oxidative induction period (Howard, 1973b).

Lowering the test temperature not only lengthens the induction period, but also decreases the oxidation rate, which makes detection of the induction period more difficult to observe. Normally this sets a lower operating temperature of about 170°C.

An alternative to the isothermal stability tests is the continuous scanning method. In this technique samples are heated at a programmed rate in oxygen until oxidative breakdown begins. The sample with the highest degradation temperature is presumed to be the most effective stabilizer system, at least at elevated temperatures. Since the degradation rate increases exponentially with temperature, the method's ability to discriminate between different levels of antioxidant protection at high temperatures is reduced if not entirely eliminated. Because of this limitation the isothermal method is much more widely used than the continuous scanning technique for studying the oxidative behavior of protected polyolefins.

ii. OIT by DSC. Gray (1975) has shown that when proper conditions are selected there is no significant difference whatsoever in either the general nature of the results or in the actual induction times when sim-

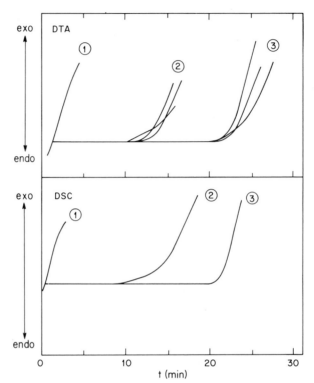

Fig. 3 Comparative DTA and DSC runs at 200°C of polyethylene with three levels of antioxidant concentration: 1, No antioxidant; 2, 0.05 wt % Santowhite (Monsanto Chemical Co.), a phenolic AO; 3, 0.10 wt % Santowhite (Monsanto Chemical Co.), a phenolic AO (Gray, 1975).

ilar samples are run on either a DTA or DSC unit, such as the Perkin-Elmer DSC-2 or 1 (see Chapter 1). However, in stability tests using the DSC-1 or 2 the sample holder covers *must be in place!* Without sample holders covers most of the heat liberated at the gas–sample interface will be lost to the relatively cold purge gas and thus not exchanged with the sample holder. With covers in place the oxidation is carried out at the desired reaction temperature. Even with the sample holders covers in place and the sample resting in an open pan, the atmosphere over the sample is nearly that of the oxygen purge gas even though it is not sweeping directly over the sample surface. In fact, comparable results were obtained for samples with three different levels of protection when the conventional DTA apparatus such as the Stone or DuPont instruments were run at 100 cc/min and the Perkin-Elmer DSC-2 was run at 40–16 cm³/min (Fig. 3).

Although the standard test calls for samples to be heated at 10°C/min from room temperature to the test temperature (usually 200°C for polyethylene) in an inert atmosphere to prevent premature oxidation of the sample, it is possible to heat samples at 100°C/min or faster to 200°C in oxygen without any appreciable decrease in the sample's induction period. The only requirements are that the thermal device come to equilibrium in less than 2 min without overshooting the intended test temperature. In fact, Westover has designed and built a DTA device that makes use of this effect (Howard, 1973b; Howard *et al.*, 1974).

The Westover instrument operates in a constant stream of oxygen at a fixed temperature of 200°C. Six samples can be loaded, run, and then removed from the device at one time. This DTA instrument, which was designed for rapid, routine quality control testing, measures induction periods with values close to those obtained for similar samples in a commercial single-specimen DTA.

iii. OIT by TG. In the TG method small samples weighing less than 10 mg are heated at 160°/min to fixed temperature (typically 200°C) in oxygen and the change in weight of the sample is monitored as a function of time. Figure 4 shows weight changes displayed by two samples, one with and the other with no stabilizer. The unprotected low-density polyethylene begins to absorb oxygen rapidly about 1 min after reaching 200°C, and its weight increases. Within 10 min chain scission is extensive enough to create volatile components and cause a maximum in the TG weight curve.

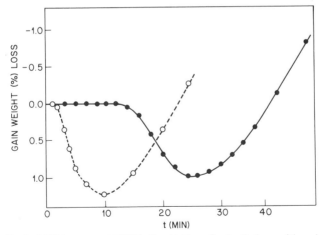

Fig. 4 Typical TGA curves at 200°C of oxidation of polyethylene with and without antioxidant in pure oxygen atmosphere: (broken line) without antioxidant; (solid line) polyethylene with 0.01 wt % Phenol A (Bair, 1973).

With protected polyethylene the onset of the initial weight gain is preceded by an induction period in which the weight remains constant. Under isothermal conditions the induction period is found to be proportional to the antioxidant concentration in the polymer. Although polyethylene with 0.03% Phenol A has an increase in the induction period of 33 min over the unprotected polymer, the two TG curves are similar in shape.

b. Induction Period as a Function of Antioxidant Concentration. By measuring the induction period of samples containing known amounts of antioxidant, a calibration curve such as that shown in Fig. 5 can be constructed. At 200°C, the induction time ranges from as low as 9 min at a concentration of 0.005% to 88 min at 0.100% Phenol A. The TG method is capable of detecting antioxidants at concentrations above 0.001 wt %. In addition, increased sensitivity in the analysis can be gained by simply lowering the test temperature, which will spread the time scale out and thus increase the difference in induction times between two samples with different antioxidant concentrations.

Unlike the case for DTA and DSC there is no lower temperature limit for the TG technique, only the practical consideration of how long one is willing to wait for induction data. This is true because the former methods must detect the exothermic reaction or temperature rise per unit time, whereas the thermogravimetric approach measures the cumulative weight gained during oxidation before the liberation of volatile degradation products has begun. As the temperature is lowered, less volatiles are evolved

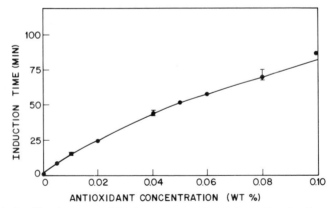

Fig. 5   A calibration curve of the time to oxygen absorption by low-density polyethylene versus concentration of Phenol A at 200°C in oxygen (Bair, 1973).

TABLE I

CONCENTRATION OF ANTIOXIDANT
IN FIELD SAMPLES[a]

Sample no.	Specified concentration (wt %)	Present concentration (wt %)		Age of sample (years)
		TG	UV	
1	0.060	0.072	0.075	2
2	0.060	0.059	0.061	2
3	0.060	0.047	0.051	2
4	0.060	0.065	0.054	2
5	0.060	0.057	0.072	2
6	0.10	0.035	0.028	7
7	0.06	0.038	–	7
8	0.06	0.055	0.055	7
9	0.10	0.01–0.03	–	9
10	0.10	0.35	–	9
11	0.10	0.026	–	7
12	0.06	0.060	0.065	3
13	0.10	0.007	<0.01	15

[a] From H. E. Bair (unpublished, 1972).

and the total weight gain is much greater than 1%. Weight changes of this order are readily detectable on most commercial TG equipment.

An example of the TG's ability to determine quantitatively the concentration of an antioxidant in aged field samples of polyethylene is shown in Table I (H. E. Bair, unpublished, 1971). Samples were taken from a point about 30 mils beneath the outer surface of the polyethylene. The amount of antioxidant in each sample was measured by the TG method and compared to the value found by ultraviolet spectrophotometric techniques. However, the latter method cannot analyze samples 9 and 10, which contain a butyl rubber compound whose UV absorption spectra overlays that of the antioxidant. Note that the agreement between the antioxidant concentration by the two methods was generally better than 0.01 wt %.

c. Additional Factors That Influence Induction Time.   Copper is known to catalyze the thermal-oxidative degradation of polyolefins (Hansen *et al.*, 1964; Reich and Stivala, 1969). Chan and Allara (1974a,b) have used infrared reflection studies to characterize degradation at copper–polymer interfaces. One striking example of this is the premature breakdown of polyethylene-insulated copper wire. Howard (1972, 1973a,b) has indicated that DTA induction times can be reduced by as much as one-third of the normal value when the sample is run in contact with copper. In order to assess the effect of copper on stability, fine copper screen (100 mesh) can be cut out and pressed into a thin film's upper surface and placed in a copper sample container or the actual insulated conductor can be run (Yamaguchi *et al.*, 1974).

It has been shown elsewhere that pigmentation can affect oxidative service life (Hansen *et al.*, 1964; Pusey *et al.*, 1971; Hawkins *et al.*, 1971). The adverse effect of pigments on high-temperature stability of polyethylene can be tested by DSC, DTA, and TG.

d. Estimated Lifetime at Processing Temperatures. As pointed out earlier, the thermal decomposition of polyethylene is a random chain scission process that generates free radicals. If an antioxidant is present, it will interact with the decomposing polymer until it is entirely consumed. Thus if one monitors the concentration of stabilizer in polyethylene as a function of time and processing temperature, an upper temperature limit can be found for the safe use of a particular polyethylene system in the melt.

As an example of how this thermal scheme can be employed to solve a practical problem, we will show how one can find the upper temperature limit for the thermal bonding of a polyethylene liner to the inside of a metal tube (H. E. Bair, unpublished, 1972).

In this case a branched polyethylene was used to line the inside of a tube. The liner in the form of a cylindrical plastic bag is inserted into the metal tube and forced against the metal walls by compressed nitrogen. It is then bonded in place by induction heating. Although the entire melt bonding operation is performed rapidly, the polyethylene is typically at temperatures between 200° and 280°C for 5–10 min.

The polyethylene liner contained about 0.025 wt % of a hindered phenol antioxidant whose chemical name is 1,3,5-trimethyl-2,4,6-tris(3,5-di-*t*-butyl-4-hydroxybenazyl) benzene. Hereafter, we shall refer to this particular stabilizer as Phenol C. It should be noted that the bonding operation is performed in a nitrogen atmosphere, for in the absence of oxygen the degradation temperature of polyethylene is increased by an enormous margin. This fact is illustrated in Fig. 6, where samples of polyethylene without an antioxidant were heated at 10°C per minute from room temperature until they degraded. In oxygen the onset of oxidation is observed at 183°C as a gain in weight and proceeds to a maximum of more than 1 wt % at 235°C. Beyond this temperature oxidative decomposition yields a spectrum of hydrocarbon fragments. Many of these fragments are volatile, and thus rapid weight loss occurs above 250°C (482°F). In contrast to this behavior, a sample of the same material in a nitrogen atmosphere had to be heated above 300°C (572°F) before thermal decomposition was ini-

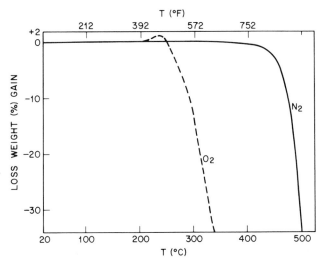

Fig. 6 Volatilization behavior of unprotected low-density polyethylene in nitrogen and oxygen as a function of temperature. Heating rate 10°C/min (H. E. Bair, unpublished, 1972).

tiated. At extremely high temperatures and in an oxidizing atmosphere, the addition of an antioxidant cannot significantly prolong the life of polyethylene. For instance, the addition of 0.04 wt % Phenol C, an antioxidant, in polyethylene will extend its lifetime in oxygen at 200°C from only 1 min to 25 min; however, at 260°C the same sample with Phenol C can only last 15 s before total stabilizer depletion. Since 0.04 wt % is the typical concentration of Phenol C found in the liner before processing, we obviously cannot depend on the antioxidant for protection against thermal oxidation in the bonding process. Rather one must take steps to insure the exclusion of oxygen from the process and then rely on the inherent strength of polyethylene to resist thermal decomposition.

Simulated bonding experiments were conducted by heating samples of the polyethylene in nitrogen from room temperature to a predetermined temperature T_f at approximately 800°C per minute; then the sample was held at T_f for 2.5 min and subsequently cooled at 20°C per minute to room temperature. Following the thermal cycle the samples were analyzed thermogravimetricaly for the amount of active antioxidant remaining in the polyethylene dielectric..

In Fig. 7 the amount of unspent Phenol C is plotted against T_f. Samples that were contained in Al pans had a slight decrease in the level of Phenol C from 0.026 wt % at room temperature to about 0.022 wt % at 300°C (572°F). At 300°C and above the concentration of active antioxidant decreased rapidly until only a trace (~0.001 wt %) is present after cycling

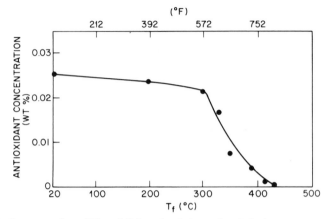

Fig. 7 Concentration of Phenol C in polyethylene after 2.5 min storage at a specified temperature T_f. Original polyethylene film contained 0.026 wt % Phenol C (H. E. Bair, unpublished, 1972).

to 418°C. Note that the inception of rapid antioxidant loss begins above 300°C.

A maximum temperature of 390°C was attained in the simulated bonding cycle before total antioxidant depletion. Note that this particular antioxidant has an appreciable vapor pressure above 350°C, and a portion of the loss can be attributed to antioxidant volatilization. In addition, dielectric measurements that are sensitive to any polar groups created by the degradation indicate that the polyethylene liner can be heated to 321°C without any deleterious effects on the electrical loss of the material. Heating to 343°C caused a 40% increase in the liner's 30 MHz dissipation factor. These results support our contention that the polyethylene liner can be processed at temperatures in the vicinity of 300°C for short periods of time without significant degradation if stringent measures are taken to exclude oxygen.

e. Prediction of Long-Term Performance of Polyethylene Stabilizer Systems at Service Temperatures. The most direct way to evaluate the long-term performance of a stabilized polyethylene is to examine it periodically under actual service conditions (Gilroy, 1974; Howard and Gilroy, 1975). Since most polyethylene formulations are expected to have useful lives that are measured in terms of years, the prediction of the service life of a protected polymer is often based on data collected at elevated temperatures, usually above the polymer's melting point, in order to accelerate testing. In predicting the service life of a polyethylene–stabilizer system, induction periods are determined at several temperatures

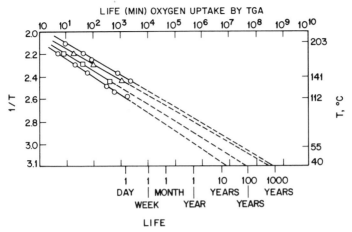

Fig. 8 A plot of TGA induction time against reciprocal temperature for mixtures of polyethylene and additives; ●, polyethylene and 0.01 wt % Phenol A; △, polyethylene and 0.01 wt % Phenol A and 0.75 wt % blue colorant; □, polyethylene and 0.01 wt % Phenol A embedded in copper screen; ○, polyethylene (Bair, 1973).

by any of the previously mentioned techniques. An Arrehenius-type plot is then constructed that relates the induction period to the reciprocal of the absolute temperature (Fig. 8).

By extrapolation of the high temperature (135–203°C) data in Fig. 8 to 40°C one would estimate that it would take about 700 years before a polyethylene sample that contained about 0.01 wt % of Phenol A would undergo thermal oxidation. Actual low-temperature tests and data from the field have shown that embrittlement as a consequence of thermal oxidation of compounded polyethylene can occur in 4 or 5 years at temperatures near 40°C. These field failures are for applications where polyethylene has been used as insulation over copper wire at thicknesses of about 6 mils. The antioxidant is the same as was used in the high-temperature TG induction period tests, but the antioxidant concentration is about 6 times greater in the thin-wire insulation than in the TG samples! The reasons for this nonlinear behavior can only begin to be understood by studying the complex chemical and physical interactions that take place in the solid state and will be reviewed in the next section. Nevertheless, in an attempt to circumvent these difficulties supplementary tests are carried out at temperatures below 100°C for extended periods of time. These tests include the periodic monitoring of antioxidant effectiveness by DSC or DTA of samples aged in static or circulating air oven or from pedestal enclosures (Chan *et al.*, 1978; Gesner *et al.*, 1974). When results such as

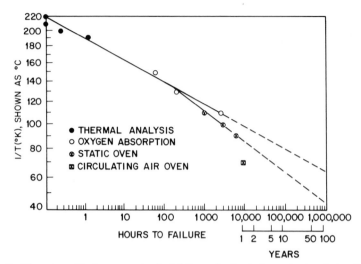

Fig. 9 Times to oxidation or mechanical failure of polyethylene wire insulation (Chan *et al.*, 1978).

these are combined with high-temperature data, the nonlinear behavior of the induction period can readily be seen (Fig. 9).

In practice the low-temperature DSC or DTA tests are used as a conservative guide as to when the stabilizer system has become ineffective. When the DTA–DSC tests indicate that the antioxidant has lost its effectiveness, mechanical tests are begun at room temperature. One such test is to wrap samples of insulated wire around a steel mandrel and examine it visually for signs of cracking (Gilroy, 1974). These tests are carried out periodically until the first sign of cracking occurs. Thus by using thermal analysis in conjunction with supplementary aging tests, one can select stabilization systems that should provide excellent long-term performance.

3. Factors Affecting Long-Term Aging

a. Blooming of Antioxidant Additives. Previous work on polymer stabilization has been largely devoted to understanding the chemical mechanisms of polymer oxidation and the manner in which antioxidants can be used to retard these degradative processes. However, in addition to antioxidant depletion by chemical consumption, polymer stabilizers may be lost by physical processes. Bair (1973) has shown by surface microscopy studies that Phenol A, as well as other antioxidants such as Phenol B, is exuded readily from polyethylene to its surface at temperatures of 70°C and lower.

In the field this exudation process can contribute to premature stabilizer loss through the loss of the antioxidant by evaporation, water extraction, or physical handling. The basic parameters that govern the rate and extent of exudation of antioxidants are the equilibrium solubility and the diffusion coefficient of antioxidants in polyethylene. Roe *et al.* (1974) have developed a thermogravimetric technique that permits one simultaneously to measure the solubility and diffusion coefficient of antioxidants in polyethylene. This TG method will be illustrated in the next section.

b. Physical Properties Related to Antioxidant Blooming

i. Melting and glass transitions. In its pure crystalline form Phenol A has a melting temperature of 114°C with an enthalpy of fusion ΔH_f, equal to 6880 cal/mole (H. E. Bair, unpublished, 1972). However, when Phenol A is cooled rapidly from the melt, it vitrifies with a glass temperature T_g of 22°C (Fig. 10). the Phenol antioxidants B and C have a T_m of 247 and 124°C and a T_g of 92° and 43°C, respectively (H. E. Bair, unpublished, 1972).

Evidence for the partial immiscibility of Phenol A in polyethylene at a concentration of 0.10 wt % has been detected by DSC measurements (Bair, 1973). Two melting endotherms were observed at 110 and 164°C, which were attributed to fusion of separate phases of polyethylene and Phenol A, respectively. It was estimated from ΔH_f associated with the melting of the antioxidant that about 90 wt % was insoluble in the polyethylene at room temperature. Since such small amounts ($<0.10\%$) of antioxidant are usually present in the polymer, the melting point

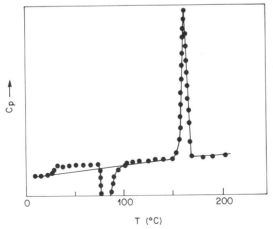

Fig. 10 Comparative C_p curves for Phenol A in crystalline form (unbroken line) and glassy state (filled circles) (H. E. Bair, unpublished, 1972).

technique cannot be employed quantitatively to examine stabilizer solubility because of the lack of adequate sensitivity. In addition, the T_ms of a few of the antioxidants and T_gs of most stabilizers are masked by the broad low-density polyethylene fusion curve, which can range as low as 35°C for the inception of melting to above 100°C for its termination.

Based on the results of a series of thermal and microscopy studies the following picture of antioxidant exudation has emerged (Roe *et al.*, 1974). The antioxidant is completely soluble in the polyethylene melt. On cooling from the melt, polyethylene crystallizes and the antioxidant is forced into the amorphous regions of the polymer. Probably in the mixed state the insoluble portion of the antioxidant remains in the liquid or glassy state until it exudes to the surface of the polyethylene and crystallizes.

ii. Equilibrium solubility and diffusion coefficient. As an aid in evaluating the ability of an antioxidant to stay in a polymer and be effective in retarding oxidation, a technique was developed by Roe *et al.* (1974) to measure simultaneously the solubility and diffusion coefficient of an antioxidant in polyethylene. The method is based on analyzing the concentration profile across a stack of polyethylene sheets that the antioxidant has been forced to diffuse through. An example of the use of this method to measure the solubility and rate of movement of Phenol A through low-density polyethylene is given. The experimental setup is depicted schematically in Fig. 11. Molded disks of unprotected branched polyethylene, DYNK, of 5 mil thickness and 2 in. diameter are stacked in the center of the diffusion device. Between 13 and 27 disks are employed during a run. A disk of 2 in. diameter and 60 mil thickness molded from the same DYNK polyethylene but containing 2 wt % of the antioxidant is placed above and below the stack and serves as a reservoir of antioxidant. The assembly is clamped between two heavy brass plates, and a pressure of about 30 psi is applied to the polyethylene through six bolts fitted with compression spring sleeves. The whole device is now placed in a vacuum oven for a period of time at a constant temperature. As time elapses the excess antioxidant originally contained in the thicker, outer disks exudes out and diffuses into the thinner, inner disks. At the end of the run, the polyethylene disks are peeled apart, and the concentration of the antioxidant in each disk is analyzed by the thermogravimetric method. The primary data thus obtained are the values of concentration as a function of position after a fixed length of time. By comparing the concentration profile with master curves obtained by solution of the diffusion equation with appropriate boundary conditions, one can determine the solubility and the diffusion coefficient, as described in detail later in the next section.

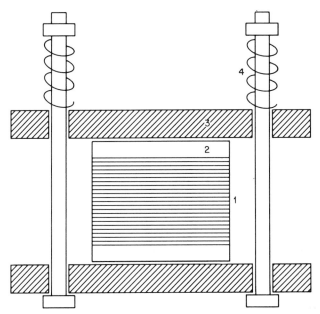

Fig. 11   Schematic illustration of diffusion cell: 1, stack of polyethylene sheets, each 5 mil thick and 2 in. in diameter; 2, polyethylene disks 1/16 in. thick, containing large excess of antioxidant; 3, brass plates; 4, bolts and nuts with compression springs (Roe *et al.,* 1974). Reprinted by permission of John Wiley & Sons, Inc.

In the diffusion cell depicted in Fig. 11, we have a diffusion process within a solid bounded by two parallel surfaces that are very large in area compared to the distance between the surfaces. The concentration of antioxidant within the solid interior is zero initially, while the concentration at the surface is maintained essentially at the equilibrium solubility because of the large excess of antioxidant available in the outer, thicker disks. The solution of Fick's diffusion equation under the preceding boundary conditions (with the diffusion coefficient D independent of concentration) is (Carslaw and Jaeger, 1959)

$$C(x,t) = C_0 - \frac{4}{\pi} C_0 \sum_{n=0}^{\infty} \frac{(-1)^n}{2n + 1}$$

$$\chi c^{-(2n+1)^2 \pi^2 K/4} \cos \frac{(2n + 1)\pi x}{2l} \qquad (4)$$

where 2l is the total thickness of the stack of thin disks, x is the position coordinate with the origin at the center of the stack, $C(x, t)$ is the

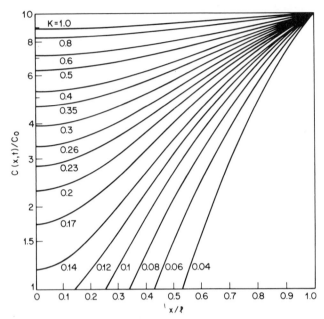

Fig. 12 Solutions of diffusion equation with boundary conditions corresponding to those prevailing in the cell shown in Fig. 11: x is the position coordinate from the center of the polyethylene stack; 1 is half the total thickness of the stack; $C(x, t)$ is the concentration at position x at time t; C_0 is the equilibrium solubility. The curves are calculated for different values of K indicated, where $K = Dt/l^2$ (Roe *et al.*, 1974). Reprinted by permission of John Wiley & Sons, Inc.

concentration of antioxidant at position x at time t, C_0 is the equilibrium solubility, and K is given by

$$K = Dt/l^2 \tag{5}$$

A semilog plot of $C(x, t)/C_0$ against x/l caluated for a number of values of K is shown in Fig. 12 (Roe *et al.*, 1974).

The concentration of antioxidant determined experimentally is plotted on similar semilog paper against x/l where x is taken to be the coordinate of the center of the 5 mil disks. An example of such a plot is shown in Fig. 13. The first and last thin disks were not analyzed for concentration because of the possibility of physical contamination of their outer surfaces by antioxidant from the reservoir disks. The experimental plot, such as Fig. 13, is then overlaid on the computed set of curves in Fig. 12. The best match of all the experimental points to a single calculated curve was found by sliding the plot vertically. From the amount of vertical shift

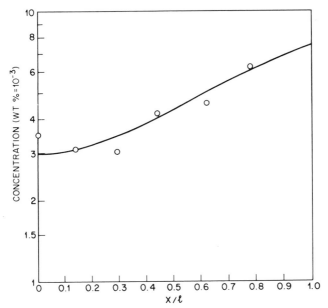

Fig. 13 Example of concentrate profile across the stack of polyethylene sheets, which was attained after Phenol A had been allowed to diffuse for 11 days at 45°C. Solid curve is the solution of diffusion equation with $K = 0.3$ as shown in Fig. 3. The solubility is given by the value of the theoretical curve at $x/l = 1.0$ (Roe *et al.*, 1974). Reprinted by permission of John Wiley & Sons, Inc.

one can then calculate the solubility C_0, and from the value of K of the best matching curve, one can obtain D according to Eq. (5).

The sensitivity for the determination of D is best when K is in the vicinity of 0.15, while C_0 can be determined more accurately at larger values of K. The duration time of diffusion and the number of 5 mil disks (hence l) were adjusted for each run so as to give a K value of 0.2–0.3 whenever an approximate estimate of D was possible.

Figure 14 shows the values of the diffusion coefficient of Phenol A obtained at three temperatures. The straight lines drawn in the figure correspond to an activation energy of 12.4 kcal/mole, which was obtained previously by Jackson *et al.* (1963) from measurement of the diffusion of a few antioxidants in branched polyethylene by a radioactive traces technique. The agreement between the two sets of data is excellent.

The solubility of Phenol A was determined by the diffusion method. The precision of the solubility values determined by the method is somewhat less than that for the diffusion coefficient. This is because one needs to know only the relative variation of concentration in the

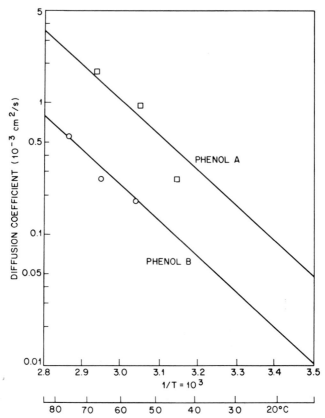

Fig. 14 Observed diffusion coefficients plotted against reciprocal absolute temperature. The two straight lines are drawn with a slope corresponding to the activation energy of 12.4 kcal/mole (Roe *et al.*, 1974). Reprinted by permission of John Wiley & Sons, Inc.

determination of a diffusion coefficient, whereas an absolute value of concentration is required for the solubility. The accurate determination of antioxidant concentration in polyethylene at levels below 1 part in 10,000 is difficult. It will be shown that the measured concentrations of antioxidant are always within a factor of 2 of the solubility values predicted on the basis of the data in lower-molecular-weight hydrocarbon solvents.

In Fig. 15 the solubility of Phenol A as determined by a visual technique in the hydrocarbon solvents is plotted against temperature. From this behavior the solubility of the antioxidant in polyethylene at 23°C was estimated. The broken lines in Fig. 15 are drawn to pass through these predicted solubilities and to parallel the hydrocarbon solubility

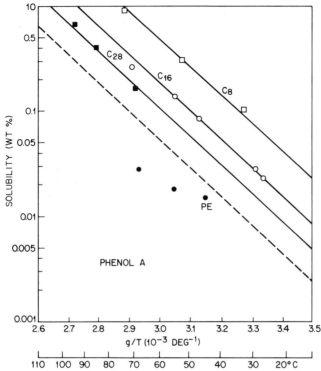

Fig. 15   Solubility of Phenol A in hydrocarbon media presented here as a function of temperature and size of hydrocarbon solvent molecules. Filled circles shown as the solubility in polyethylene were taken from the solubility data obtained by the diffusion experiment (see Table II), but were multiplied by 2 to correct for the degree of crystallinity (around 50%). Broken line is the solubility of Phenol A in polyethylene predicted from the regular solution theory on the basis of its solubility in lower hydrocarbon solvents (Roe *et al.*, 1974). Reprinted by permission of John Wiley & Sons, Inc.

data. Also the TG determinations of the antioxidant solubility in polyethylene are multiplied by 2, as depicted in this figure.

The factor of 2 is necessary because the polyethylene samples we used were about 50% crystalline. Although the observed values deviate from the expected values shown by the broken lines sometimes by as much as a factor of 2, the agreement is still considered very satisfactory, especially in view of the extreme difficulty of determining small concentrations of antioxidants in polyethylene.

iii.   Loss of accumulated antioxidant from the polymer's surface.
For physical depletion of an antioxidant from a polymer to occur, two

events must take place: first, exudation of the antioxidant to the polymer's surface by diffusion and, second, the subsequent loss of the antioxidant from the polymer's surface. For loss of the accumulated surface antioxidant, three possible modes are examined: physical handling, water extraction, and evaporation.

If one assumes that Phenol A stays in polyethylene as a supersaturated solution until it diffuses out to the surface, the average concentration of the antioxidant that remains inside polyethylene can be predicted from solubility and diffusion data that was collected in the past section. In Fig. 16 the concentration of Phenol A that remains in a 5 mil thickness of polyethylene is plotted against time (R. J. Roe, unpublished, 1972). The original sheet had 0.10 wt % of Phenol A. The final concentration represents the equilibrium solubility at the indicated temperature. The prediction of such a rapid rate of exudation is consistent with microscopic observations made by Bair (1973).

The physical depletion of antioxidant is complete only when the material that has exuded to the surface and accumulated there becomes lost. We will discuss some of the possible ways by which the loss of surface antioxidant can occur. The discussion is, however, only

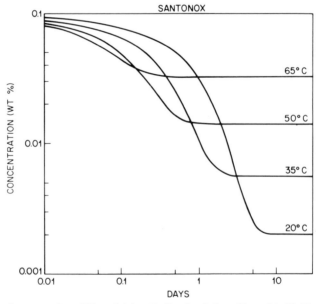

Fig. 16 Concentration of Phenol A in a 5 mil polyethylene film at 20, 35, 50, and 65°C as a function of time. Film originally contained 0.10 wt % Phenol A (R. J. Roe, unpublished, 1972).

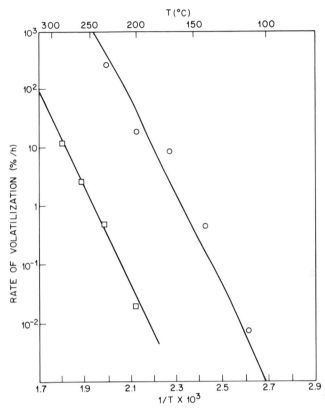

Fig. 17   Volatilization of Phenol A (□) and Phenol B (○) plotted against reciprocal absolute temperature (R. J. Roe, unpublished, 1972)

speculative because what occurs in practice depends very much on unknown and mostly uncontrollable factors present in the field usage.

As is seen from Fig. 16, almost all the antioxidant originally added to polyethylene would have exuded out to the surface in less than 2 weeks while in storage unless the antioxidant was kept in reserve inside the polyethylene by some specific means. At the time of use, the antioxidant accumulated on the surface could easily be rubbed of by handling.

There are possibly two ways by which water can carry away surface antioxidant. The first is by slow dissolution. This will depend on the solubility of antioxidant in water. We have determined that the solubility of Phenol A in water at room temperature is about 10 ppm or about the same as that of water in polyethylene (H. E. Bair, unpublished, 1972). Thus removal of Phenol A from polyethylene by a water extraction process is unlikely.

Another possible effect of water is a simple act of washing away. No experimental data are, however, available so far to substantiate such a conjecture.

The rate of evaporation of a solid depends on the sublimation vapor pressure of the material and the pattern and rate of air flow around it. H. E. Bair (unpublished, 1972) has measured the rate of evaporation of a few antioxidants at high temperature by thermogravimetric analysis (TG). Some of his results are reproduced in Fig. 17 for easy reference. These data were obtained in the same apparatus with identical furnace geometry and a constant rate of flow of nitrogen gas. As a result the rate of evaporation given in Fig. 17 is expected to be proportional to the vapor pressure.

In order to correlate the laboratory data of sublimation rate with the rate of loss of antioxidant from a sample, one needs the knowledge of mass transport patterns both at the TG furnace in the laboratory and at a given field installation. Both of these are difficult to determine, and in the following discussion we will attempt to obtain a crude approximation. First we convert the data in Fig. 17, given in the units of percent weight loss per hour, into a flux in units of weight lost per unit area per unit time. Nitrogen gas was blown steadily in and around the furnace all the time, and we may take either the opening of the furnace (cross-sectional area about 0.5 cm²) or the surface of the sample pan (exposed area 0.1 cm²) as the boundary plane across which the transport of vapor molecules occurs at the slowest rate.

The amount of sample used in the TG measurement was typically 1.5 mg, and therefore the rate of weight loss of 1% per hour shown in Fig. 17 would correspond to 0.3 to 1.5×10^{-4} g/cm²-h. Extrapolation of Phenol A data in Fig. 17 to 60°C leads to the sublimation weight loss of $2 \times 10^{-6}\%$/hr or a flux of 0.6 to 3×10^{-10} g/cm²-h. If all the antioxidant, originally milled into the concentration of 0.1 wt % exudes to the surface of 5 mil polyethylene insulation, the total accumulation will be about 1.2×10^{-5} g/cm². If there is a free air flow around the wire so that Phenol A vapor is carried away readily from the surface, and if the flux is equal to that indicated by the TG experiment, then all the accumulated Phenol A will be lost in 4.7 to 23 years at 60°C.

The preceding calculations are very approximate, and the numerical results may not be given too much reliance. It is still reasonable, however, to conclude that the possiblity of losing all the Phenol A from the wire surface by sublimation is marginal but cannot be ruled out. In the case of Phenol B, in view of its much lower vapor pressure, one can safely conclude that the loss by sublimation need not be considered at all.

c. Complex Chemical and Physical Interactions. The rapid loss of antioxidant effectiveness at 70°C and lower cannot be explained simply by antioxidant exudation and subsequent loss from the surface. Recent studies have shown that the exuded antioxidant molecules are chemically degraded by a mechanism different from that which has been observed at temperatures in the melt (Bair *et al.*, 1974). The former degradation mechanism has an activation energy of about 10 kcal/mole, whereas the latter process is characterized by an activation of 30 kcal/mole. The low-temperature antioxidant depletion was observed by measuring the residual antioxidant content by TG, in comparison with results obtained by UV and liquid chromatography. An important finding in this work was that one can differentiate between "active and dead" antioxidant molecules through TG and UV analysis, respectively. The UV method essentially determines the total concentration of phenol groups present and cannot discriminate clearly between the active antioxidant and its reaction products (Luongo, 1965). The latter molecules cannot react with peroxy radicals to yield relatively stabile radicals (Eq. 3); hence these AO molecules are called dead. In contrast, the TG technique measures only the active antioxidant molecules that can react to retard the polymer's oxidation.

4. Selection of Antioxidants

The evaluation of a stabilization system for all polyolefins utilizes all the thermal analysis techniques we have reviewed in this chapter plus other tests. The thermal techniques such as the dynamic and static tests are well suited to aid in the initial screening of an antioxidant system because they require small sample sizes and can be carried out rapidly. However, the long-term performance of a particular antioxidant system must be performed at temperatures below the polymer's melting point in order to avoid the nonlinear effects due to different oxidative reactions. Finally, quantitative thermal evaluations must be supplemented by other analytical measurements and mechanical testing.

Fundamental studies of the chemical and physical aging of stabilized polyethylene can lead to new directions in the developement of future antioxidant systems.. Knowledge of the low solubility and severe exudation of hindered phenols from polyethylene has led to attempts to increase the solubility of antioxidant additives in polyethylene. Albarino and Schonhorn (1974) have increased the compatibility of Phenols A and B by the use of silane coupling agents and, thereby, reduced antioxidant exudation. Kaplan *et al.* (1973) have chemically bonded a phenol-type antioxidant to several polyolefins and measured the resulting material's antioxidant behavior by oxygen uptake and TG. Chan and Johnson (1980) have reported the use of carbon black–antioxidant formulations to suppress thermal oxidation and antioxidant migration in polyethylene cable jackets in the solid state.

When acrylonitrile–butadiene–styrene (ABS) is weathered outdoors or aged thermally indoors, it deteriorates most readily through the polybutadiene (BD) component (Kelleher *et al.*, 1965; Shimada and Kabuki, 1968). Oxidation of the BD in ABS leads to an increasing and broadening T_g as well as a decrease in ΔC_p at T_g (Bair *et al.*, 1980a). The latter quantity can be used to assay quantitatively the amount of unoxidized rubber in a photo- or thermal-oxidized sample, and thus in this way a program may be designed to select the most effective antioxidant for a rubber-toughened plastic. These DSC studies of aged ABS samples have revealed also the effectiveness of carbon black as a screen against photoinduced degradation and as an additive to retard oxidation at or near room temperature (Bair *et al.*, 1980a).

B. STABILIZERS

1. Function

Poly(vinyl chloride) is one of the leading polymers both in diversity of application and total weight of finished products. PVC acceptance by industry for such widespread usage is due in part to the successful development of stabilizers for PVC, which is the least naturally stable polymer in commercial use. As Grassie (1966) pointed out, "Had this polymer been discovered at the present stage of development of the plastics industry, it would almost certainly have been eliminated as useless because of its general instability to all the common degradative agents."

Since most polymers are used in air, thermal oxidation is usually the route that leads to the chemical breakdown of the material. This was certainly the case for polyolefins which we have just reviewed. However, many plastics are processed in ways that partially exclude oxygen and at temperatures where thermal degradation may occur for particularly heat-sensitive polymers such as PVC. Unfortunately, classic thermal analysis techniques may not be well suited to the study of PVC degradation for several reasons. These limitations will be described in detail in the next section.

PVC degrades by the elimination of HCl without main chain scission. TG studies coupled with mass spectrometer measurements show that the main reaction taking place during pyrolysis below about 350°C is dehydrochlorination (Stromberg *et al.*, 1959; Guyot *et al.*, 1962; Madorsky, 1964). In Fig. 18 the first plateau in the volatization curve is at 60 wt %,

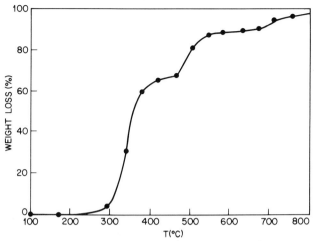

Fig. 18   Volatilization of poly(vinyl chloride) at 20°C/min in nitrogen (Salovey and Bair, 1970). Reprinted by permission of John Wiley & Sons, Inc.

which corresponds to complete elimination of HCl from the polymer (Salovey and Bair, 1970). Practically all the HCl comes off at temperatures as low as 220°C. Sequential HCl removal leaves a colored polymer containing long conjugated sequences. Thus one major function of a stabilizer for a PVC resin is acid absorption. However, we shall not dwell on any of the details concerning dehydrochlorination and stabilization of PVC but rather refer the reader to a review of this subject by Loan and Winslow (1972).

2. Heat Stability Tests

a. Color Changes.   The color changes that accompany dehydrochlorination proceed as follows: clear or colorless → yellow → yellow-orange → red-orange → red → brown. PVC is sensitive to mild heating with the first visible sign of discoloration after several hours at temperatures in the range of 100–140°C. Although there is some difficulty in measuring color in a meaningful way, many static and dynamic tests have been developed to monitor color changes as a function of oven aging or a variety of mixing conditions. Typically, oven-aged or processed samples' color is compared with standard color chips. Details pertaining to these test procedures may be found in ASTM Recommended Practice D-2115 and in various stabilizer manufacturers' technical data (Nass, 1963; Struber, 1968). Probably the most universal test is the congo red test in which resin sample is heated and the time required for a congo red paper held in the effluent gas to change color is measured.

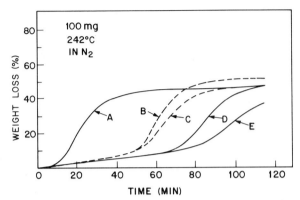

Fig. 19 Thermogravimetric study of stabilization of poly(vinyl chloride): A, PVC control; B, 100 g PVC, 1 g Epoxol 9-5, and 1 g Thermolite 17; C, 100 g PVC, 1 g Epoxol 9-5, and 1 g Thermolite 13; D, 100 g PVC, 1 g Epoxol 9-5, and 1 g Thermolite 35; E, 100 g PVC, 1 g Epoxol 9-5, 1 g Thermolite 31 (Chiu, 1966). Reprinted by permission of John Wiley & Sons, Inc.

b. Conventional Thermal Methods. The effect of various stabilizers on the decomposition of PVC can sometimes be observed thermogravimetrically as is shown in Fig. 19 (Chiu, 1966). In this instance the relative effectiveness of the various stabilizers is clearly shown. However, in actual practice it is almost always necessary that any considerations involving techniques to measure thermal stability be applied to the total PVC composition rather than the resin alone.

Commercial formulations usually have a large number of compounding ingredients, such as plasticizers, impact modifiers, and mold lubricants, in addition to stabilizers. These additional additives can often volatilize at relatively low temperatures and mask the loss of HCl in a thermogravimetric study. Also, compounded PVC resins usually have reduced melt viscosities that facilitate processing by enabling lower melt temperatures to be used and by reducing the amount of shear heating. Both of these effects will tend to minimize degradation during processing. However, the superior processing stability of these PVC formulations cannot be determined directly by DTA or TG techniques; whereas they can be detected readily by instruments sensitive to melt viscosity. Conversely, TG can rank stabilizers based on their relative isothermal volatilization rates without being influenced by whether the particular stabilizer is soluble or not in the resin (H.E. Bair, unpublished, 1980).

PVC degradation is accompanied by cross-linking, which causes the resin's melt viscosity to increase markedly. Thus, PVC compounds that are more stable at processing temperatures will remain longer at elevated temperatures during mixing without an increase in viscosity. There are numerous publications on the use of various torque rheometers to evaluate the processing and heat stability of PVC compounds (DeCoste, 1965; Heiberger *et al.*, 1969).

Scalco and Bair (unpublished, 1979) have made a comparative study of the stability of blends of various additives with PVC that has been reclaimed from scrap wire and other field equipment. The process stability of these mixtures was determined on a Brabender torque rheometer at 205°C with blades operating at 100 rpm. Each sample's stability was taken as the amount of time it could be worked at 205°C without cross-linking occurring. The same samples were heated in a TG system at 10°C/min from room temperature to the temperature where 1 wt % of material volatilized. The results of these two tests are listed in Table II.

Note that with each addition of an additive the amount of time before cross-linking began increased; however, there is no consistent pattern to the temperatures associated with 1% weight loss. other than a slight increase in temperature with the addition of 3 phr of tribasic lead sulfate (stabilizer #1 in Table II). Obviously the TG method cannot detect the advantage of 1 phr petroleum wax in the mixture although the processing stability time is improved by 28%.

TABLE II

Comparison of Stability of PVC Compounds
by Rheometer and TG Tests[a]

Sample components	SAMPLES[b]					
	1	2	3	4	5	6
Reclaimed PVC	100	100	100	100	100	100
Plasticizer	–	–	8	8	8	8
Stabilizer (#1)	–	3	–	3	3	3
Fire retardant	–	–	–	–	3	3
Stabilizer (#2)	–	–	–	–	15	15
Mold lubricant (#1)	–	–	–	–	1	1
Mold lubricant (#2)	–	–	–	–	–	1
Time, min[c]	20	58	42	102	107	150
Temperature, °C[d]	211	224	219	219	220	223

[a] From Scalco and Bair (unpublished, 1979).

[b] Sample weights in parts of additives to 100 parts of PVC.

[c] Time to viscosity increase due to cross-linking in Brabender torque rheometer.

[d] Temperature where sample has lost 1 wt pt in TG run.

The resolution of the degradation of PVC by DTA or DSC is not an easy task, for commercial PVC resins are slightly crystalline and melt in the same temperature range as the dehydrochlorination process (Gouinlock, 1975). In addition, many of the additives are volatile in this temperature range. Because of these complications there are few if any successful attempts to study PVC stability by DSC or DTA.

3. Conclusion

It appears that TG analysis is better suited to PVC stability studies than either DTA or DSC.

Although the thermal stability of many PVC formulations under processing conditions is not readily amenable to standard thermal analysis practices much useful information about the compatibility and permanence of some of these additives can be determined, as will be demonstrated in the following sections. The paucity of papers in this area indicates the need for development of some new thermoanalytical methods to characterize PVC stability.

C. UV ABSORBERS

1. Function

Plastics are degraded by exposure to sunlight. It is the UV portion of the solar spectrum that has sufficient energy ($\sim 80-90$ kcal/mole) to rupture many of the molecular bonds present in polymer structures. These broken bonds cause the rapid oxidation of plastics and produce discoloration, loss of ductility, and finally embrittlement. Plastics can be protected against photodegradation by the use of UV absorbers.

Ultraviolet absorbers function by preferential absorption of the UV energy striking the polymer's surface and are able to dissipate it to the surrounding environment as harmless thermal energy. The selection of a UV light stabilizer for incorporation into a polymer is normally a complex task. Not only must the portion of the UV spectrum that is primarily responsible for degradation of a particular polymer be considered, but also the additive's compatibility, volatility, and heat stability must be evaluated. These latter properties can be determined by thermal analysis. In the following section a practical example of how several thermal techniques can be used to characterize the physical behavior of a UV absorber for polycarbonate will be given (H. E. Bair, unpublished, 1976).

2. Thermophysical Property Measurements

Tinuvin P, a substituted hydroxyphenol benzotriazole, is a typical UV absorber used to protect polycarbonate (PC) from photodegradation. Several tenths of a percent by weight of Tinuvin P is added to PC in the "dry crumb" state. The resulting blend is pelletized and formed into 3/8 in sheets. In some optical applications these PC plies are laminated into 3/4 in or thicker pieces. Usually the forming operations are carried out above T_g near 170°C. After processing, it is not unusual to observe optical defects that have formed at the interfaces of the PC plies. Samples taken from the blemished areas were analyzed by mass spectrometry and found to contain excess Tinuvin P.

The key parameters governing the accumulation of excess additive on a surface are the additive's solubility and volatility. In what follows we shall demonstrate what thermal analysis techniques can be used to measure these properties.

a. Volatility. The volatilization behavior of Tinuvin P was measured thermogravimetrically using a Perkin-Elmer TGS-1 (H. E. Bair, unpublished, 1976). Samples weighing about 4 mg, each were heated at 20°C/min in air from 23 to 300°C. In Fig. 20 the vaporization of Tinuvin P is plotted against temperature. Under these conditions the UV absorber began to vaporize significantly above 160°C. A 0.25 mg sample of the additive vaporized completely at 180°C in 5 min.

PC films, 10 mils thick, with 10 and 40 wt % of Tinuvin P lost at least

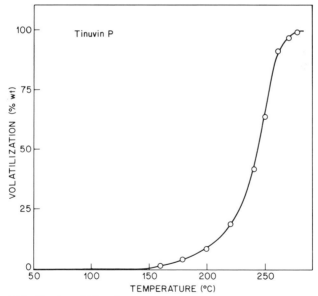

Fig. 20 Volatilization of Tinuvin P in air. Heating rate 20°C/min (H. E. Bair, unpublished, 1976).

90% by weight of Tinuvin P in 25 min at 180°C. A film with 1 wt % of the UV absorber appeared to lose most of the additive. However, at the 1% level we are not certain if the major portion of the loss is due to the UV stabilizer or other trace components.

b. Solubility. The melting temperature T_m of a sample of Tinuvin P from Ciba-Geigy was determined by DSC to be 131°C. However, after recrystallization of the additive T_m occurred repeatedly at 109°C. Apparently Tinuvin P has more than one crystalline modification.

The compatibility and solubility of Tinuvin P in PC was assessed by the presence or absence of the UV absorber's melting transition in blends of the two materials. Three blends containing 1, 10, and 40 wt % Tinuvin P in PC were prepared by dissolving both the UV stabilizer and PC in chloroform. After dissolution the chloroform was removed by evaporization. There was no sign of melting near 130°C at the 1% level of Tinuvin P. However, the films that contained 10 and 40% Tinuvin P have endothermic transitions near 130°C. These results indicate that the UV absorber is soluble in PC to a level of 1 wt % but is immiscible at some point between 1 and 10 wt %.

3. Conclusion

These results indicate that the excess Tinuvin P that has been found at the interfaces of the PC is due to the relatively high vapor pressure of the additive at processing temperatures of 170 to 180°C. Unless the temperature and time of the extrusion and forming operations can be carried out at shorter and lower values, a less volatile UV stabilizer should be sought.

III. Plasticizers

A. INTRODUCTION

1. Definition

A plasticizer is a material that is blended into a polymer to improve its processibility and flexibility. Usually the plasticizer behaves like a solvent when mixed into a polymer and results in the lowering of the melt viscosity, the glass temperature (T_g), and the tensile modulus. These alterations permit a wide variety of uses for plastics ranging from toys to telecommunication products.

Although plasticizers are used in many types of materials, the main usage of organic plasticizers is in polymeric materials. And in the plastics industry it is PVC by far that consumes the bulk with about 80% of all plasticizer production going into various PVC formulations. Typical commercial PVC compounds contain from 20 to 50 wt % plasticizer. With such heavy emphasis on the use of plasticizers in PVC we shall limit our discussion to the thermal characterization of plasticized PVC systems.

A plasticizer must meet a number of requirements before it is selected for use in a specific PVC formulation. However, one particularly desirable feature that it should have is permanence. In other words, the plasticizer should not only remain in the PVC compound throughout its expected life but should also resist any chemical change. Thus, for a plasticizer to be retained in a PVC compound it should exhibit the following characteristics: compatibility, low volatility, and resistance to migration and extraction.

Another important property is plasticizer efficiency, which is determined by finding how much plasticizer must be added to a polymer to give a required level of mechanical or physical response. One way to evaluate plasticizer efficiency is to measure T_g as a function of plasticizer concentration.

2. Compatibility

If the plasticizer is miscible with a polymer, then a mixture of the two substances should yield a homogeneous blend with a single glass temperature. Ideally the resulting T_g can be approximated by a simple relationship such as (Wood, 1958; Fox, 1956; Gordon and Taylor, 1952):

$$T_g = w_1 T_{g_1} + w_2 T_{g_2} \qquad (6)$$

where $T_{g1,2}$ = glass temperature of the mixture and the components (K), respectively, and $w_{1,2}$ = weight fraction of the plasticizer. Recently, Couchman (1978) has derived a relation to predict T_g based on calorimetric data. This relation yields T_gs for miscible polymer blends for values of T_g and ΔC_p at T_g for the unblended polymer components as given by the equation

$$\ln (T_g/T_{g_1}) = [w_2 \, \Delta C_{p2} \, \ln[T_{g2}/T_{g_1}]]/[w_1 \, \Delta C_{p1} + w_2 \, \Delta C_{p2}] \qquad (7)$$

Leisz et al. (1980) have applied this equation successfully to the case of four miscible polymer blends, including two where one component was PVC.

The T_g for PVC (Geon 103 EP from B. F. Goodrich Co.) based on specific heat measurements made on a Perkin-Elmer DSC-2 is illustrated in Fig. 21 (H. E. Bair, unpublished, 1976). This PVC is unplasticized and contains only 5 wt % of tribasic lead sulfate, a thermal stabilizer. The glassy to rubbery transition of the PVC sample is accompanied by a discontinuous change in C_p between about 70 and 115°C. Conventionally, T_g is taken at the halfway point in the transition between C_p of the glass (C_{p_g})

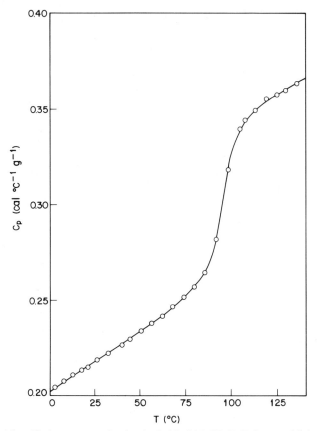

Fig. 21 Specific heat curve of poly(vinyl chloride) (H. E. Bair, unpublished, 1976)

and fluid (C_{p_f}) states, which would correspond to the temperature where there would be a change of slope in a volume versus temperature plot of PVC. By this definition, T_g equals 89°C.

Sometimes for simplicity and to minimize heat lag effects one may assign T_g as the temperature where the linear portion of the C_p curve in the transition interval intersects the C_p curve of the glass, or in this case 83°C. This latter value may be used to define the temperature T_{g_i}, where the glass transition begins. Conversely, the point T_{g_f} where the linear portion of C_p in the transition region extrapolates up to the C_p curve of PVC in the fluid state marks the termination of the glass transition. Thus the width of T_g is

$$\Delta T_g = T_{g_f} - T_{g_i} \tag{8}$$

In this example T_{g_f} equals 97°C and ΔT_g about 14°C at a heating rate of 20°C/min.

A sharp transition ($\Delta T_g < 15°$) is typical of a homogeneous material, at least on the scale of thermal measurement that implies heterogeneities < 50–100 Å. A broad transition, on the other hand, suggests a somewhat less than homogeneous blend. MacKnight *et al.* (1978) have discussed this concept in terms of blend compatibility. A compatible blend exhibits a sharp transition but a less compatible material shows a wider one. A totally incompatible material would, of course, show the two separate transitions, signifying little attraction of one component for another.

Another important parameter that can be obtained from these C_p measurements is the magnitude of the C_p increase, ΔC_p, at T_g where

$$\Delta C_p = C_{p_f} - C_{p_g} \qquad (9)$$

ΔC_p for this sample equals 0.078 cal g^{-1} °C. These measurements can aid in estimating the fraction of material participating in a particular phase of a multicomponent system (Bair, 1970b, 1974b; Bair *et al.*, 1980a).

3. Compatibility and Plasticizer Efficiency

Brennan (1977a) has determined the T_gs of mixtures of dioctyl phthalate (DOP) and PVC by DSC. DOP is one of the most common plasticizers used in PVC formulations. Although a single T_g is observed in each blend indicating miscibility, ΔT_g (Eq. 8) is much greater for the blends than for

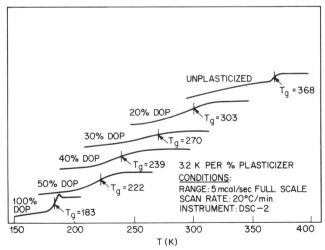

Fig. 22   DSC scans of glass transitions of DOP and PVC and four blends containing 50, 40, 30, and 20 wt % DOP in PVC (Brennan, 1977a).

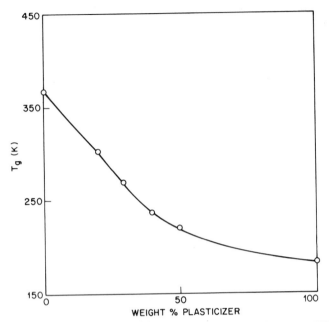

Fig. 23 T_g plotted against concentration of DOP in PVC (Brennan, 1977a).

the individual components (Fig. 22). Note the severe dependence of T_g on the concentration of DOP. In Fig. 23 T_g is plotted versus plasticizer content. The results show a linear relationship from zero to 45 wt % plasticizer. As discussed previously, one measure of plasticizer efficiency is the amount T_g is depressed as the level of plasticizer is increased. From the slope of the line between 0 and 40 wt % in Fig. 23 an efficiency value of about 3°C/wt % DOP was calculated.

Although Brennan does not mention it, the experimentally determined T_gs are as much as 22°C lower than values calculated from theories applicable to polymer–plasticizers systems. This may mean that DOP does not form a simple miscible blend with PVC but instead forms a heterogeneous structure. C_p measurements of blends of a phthlate di-iso-decylphthlate (DIDP), a phosphate tricresylphosphate (TCP), and an azelate, di-2-ethylhexylazelate (DOZ), plasticizers and commercial PVC resins have revealed the presence of two distinct glassy phases, in addition to a small crystalline one (Bair and Warren, 1979, 1980, 1981). The T_g of the lower temperature phase decreased markedly with increasing plasticizer content in the usual manner, whereas the previously unreported T_g of the high-temperature phase was less sensitive to plasticizer concentration.

This phenomenon was attributed to the preferential solvation of PVC by the plasticizer. In addition, the breadth of T_g (Eq. 8) was used to characterize blend homogeneity. Small additions of plasticizer efficiently reduced T_g but had little effect on ΔT_g of the lower transition. The plasticizer was apparently extremely tightly bound and dispersed at this point, portraying all the qualities of a well-solvated, homogeneous material. At some point, however, further addition of plasticizer exhibited the properties of a less compatible material, the broadened DSC T_g showing a plasticizer-rich region at the upper region and a gradient in relative component concentration between these extremes.

4. Plasticizer Permanence

a. Master Curve. Once a master curve of T_g versus plasticizer concentration has been prepared for a particular PVC composition, it can be used to determine the amount of plasticizer in an unknown formulation of the particular plasticizer. P. C. Warren and H. E. Bair (unpublished, 1977) have prepared a master curve for blends of DIDP and PVC (Fig. 24). Note that ΔT_g ranges from 17 to 61°C as the phr of DIDP changes from 0 to 35 phr (Table III). Plasticizer efficiency for this system is about 3°C/phr DIDP.

b Effects of Aging. Samples containing 30 parts DIDP per hundred parts resin (PHR) (21.9 wt % DIDP) were oven-aged at 121°C for varying periods of time. The loss of weight as a function of time in the oven was monitored and assumed to be equal to the loss of plasticizer through

TABLE III

T_g for Blends of DIDP in PVC[a]

DIDP concentration		T_{g_i}[b]	ΔT
(Phr)[c]	(wt %)	°C (K)	(°C)
0	0	84 (357)	17
5	4.5	68 (341)	19
10	8.5	53 (326)	25
15	12.3	35 (308)	33
20	15.7	27 (300)	31
25	18.9	9 (282)	41
30	21.9	−4 (269)	56
35	24.6	−12 (261)	61

[a] From Warren and Bair (unpublished, 1977).
[b] T_{g_i} is taken as the onset of the glass transition discontinuity at a heating rate of 20°C/min.
[c] Parts of DIDP per 100 parts of PVC resin.

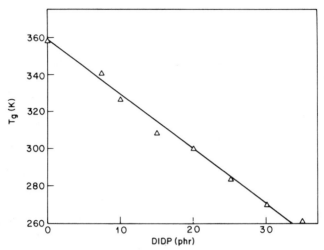

Fig. 24 T_g plotted versus concentration di-iso-decyl phthalate in PVC (P. C. Warren and H. E. Bair, unpublished, 1977).

volatilization. In addition the T_gs of the oven-aged samples were determined by DSC and used to estimate the amount of plasticizer from the master curve (Fig. 24). The agreement between the concentration of DIDP in PVC as determined from weight loss measurements and T_gs is good to within a few percent (Table IV). Thus the effects of aging on the composition of PVC formulations can be determined from thermogravimetric and calorimetric measurements.

TABLE IV

ESTIMATION OF DIDP LOST FROM PVC DURING OVEN AGING AT 121°C[a,b]

Time at 121°C d_a	Plasticizer lost (wt %)	Concentration DIDP calculated (phr)[c]	T_g (K)	Conc. DIDP estimated (phr)[d]
0	0	30.0	275	28
3	12.8	26.2	287	24
6	18.1	24.6	290	23
10	23.1	23.1	299	20
14	35.6	19.3	305	18
24	54.0	13.8	315	14

[a] From Warren and Bair (unpublished, 1977).
[b] Sample originally contained 30 phr DIDP.
[c] Concentration based on weight loss measurements.
[d] Concentration from T_g data and Fig. 24.

B. PLASTICIZED PVC SYSTEMS

1. Introduction

One common deficiency of low-molecular-weight plasticizers is their susceptibility to migration and volatilization. The advent of polymeric plasticizers has attracted a lot of interest in their potential use because they will provide superior permanence in terms of physical properties in contrast to their low-molecular-weight counterparts. In this section we shall study the thermal characterization of a monomeric and a polymeric plasticizer in blends of PVC.

The monomer tetraethylene glycol dimethacrylate (TEGDMA) is a reactive material that can be polymerized by heat or irradiation (Pinner 1959, 1960; Miller, 1959; Salmon and Loan, 1972). TEGMDA acts as a plasticizer when mixed with PVC and thus lowers the melt processing temperatures. However, post extrusion irradiation will cause the monomer to polymerize and form a three-dimensional network with PVC grafted to it. In this way a product can be formed at safe processing temperatures and irradiated subsequently to yield a high modulus material at elevated temperatures.

Recently, Hammer (1977) and Tordella *et al.* (1976) have reported that a terpolymer (TP) of ethylene, vinyl acetate, and carbon monoxide (Elvaloy 741) is miscible with PVC. X-ray morphological, dielectric, dynamic mechanical, and calorimetric studies have been carried out to gain insight into the compatibility and physical properties of blends of the terpolymer with PVC (Bair *et al.*, 1977; Anderson *et al.*, 1979). We shall review here how these thermal measurements were made on both the high- and low-molecular-weight systems.

2. Blends of TEGDMA, a Monomer in PVC

a. Absorption–Desorption Behavior. Figure 25 exhibits absorption–desorption experiments based on weight changes (Bair *et al.*, 1972). PVC absorbed 7 wt % TEGDMA at room temperature. The equilibrium amount increased rapidly as the temperature increased and reached 60% by weight at 90°C. The TEGDMA absorbed at elevated temperatures does not redistribute itself, at room temperature, into two phases: a pure TEGDMA phase and a phase containing 7% TEGDMA. However, as indicated in Fig. 25, the amount of TEGDMA desorbed when the temperature was lowered was insignificant. The data imply that once equilibrium is reached at a higher temperature, the swollen PVC retains nearly all the absorbed TEGDMA when cooled to room temperature.

b. Calorimetry

i. Identification of multiple phases. TEGDMA monomer has a

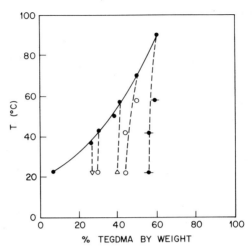

Fig. 25 Absorption–desorption of TEGDMA by PVC: (solid line) absorption curve, (dashed lines) desorption curves. [Reprinted with permission from Bair *et al.* (1972). Copyright by the American Chemical Society.]

glass transition temperature of −82°C. The polymer of TEGDMA undergoes a small second-order transition at about 92°C; the temperature range of this glass transition is twice as broad as is usually observed for other polymers. The values of T_g for the TEGDMA polymer is to be compared with the glass temperature of two similar polymers; 130°C for poly(ethylene glycol dimethacrylate) and 82°C for poly(decamethylene glycol dimethacrylate). The T_g of the unplasticized PVC used in these studies was 87°C.

A mixture of 33 wt % monomer and 67 wt % PVC was mixed in a high-speed blender at 80°C. Samples were irradiated using 1 MeV electrons from a van de Graaff generator. In Fig. 26, plots of changes in specific heat as a function of temperature are constructed for seven TEGDMA-PVC samples with the following radiation doses: 0, 0.04, 0.05, 0.06, 0.08, 0.15 and approximtely 1 Mrad. For the purpose of comparison, all the ΔC_p measurements for each sample were superimposed on the C_p curve of the 0.15 Mrad sample. The specific heats of the latter sample in the temperature range −38 to 122°C are found elsewhere (Bair *et al.,* 1972).

The unirradiated sample has three transitions at −64, −33, and 40°C, respectively. They are designated as transitions I_B, II_B, and III_B, in order of increasing temperature. The transition at −33°C seems to be the major one because it is accompanied by a large increase in C_p. Irradiation up to

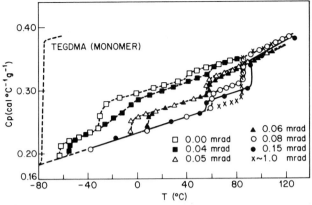

Fig. 26 Specific heat curves of a blend of TEGDMA (33 wt %) and PVC (67 wt %) with different doses of irradiation. [Reprinted with permission from Bair *et al.* (1972). Copyright by the American Chemical Society.]

0.04 Mrad shifted each of the three transition to higher temperatures, namely, -56, -10, $60°C$. However, only two transitions appeared at -7 and $58°C$ when the dose was 0.05 Mrad; at this dosage, gel started to form rapidly. The 0.06 Mrad specimen also yielded two transitions, but at slightly higher temperatures. The 0.08 and 0.15 Mrad, respectively. Further increase in dosage to ~ 1 Mrad resulted in the occurrence of only one transition at $84°C$ which appeared to be unreasonably low when compared with the higher T_g of $92°C$ for the 0.15 Mrad sample. The low value is probably attributable to radiation-induced degradation of PVC, as suggested by the extensive discoloration of the specimen. The transition temperatures for each specimen are listed in Table V together with ΔC_p values for each transition.

It was also found that when the 0.06 Mrad sample, with two transitions at 8 and $55°C$, was extracted in ether to remove soluble materials subsequent calorimetric experiment revealed a single transition at $52°C$. The change in C_p was about the same magnitude as the sum of the two prior ΔC_p values at 8 and $55°C$ (Table V). It appears that low-molecular-weight polymer and/or residual monomer are still present in the 0.06 Mrad sample.

Dynamic mechanical studies of the same samples show similar results except that the preceding room temperature phase transition cannot be detected (Bair *et al.*, 1972). However, Davis and Slichter (1973) have observed all three phase transitions in these same materials in NMR studies although each transition is shifted about $20°C$ above the DSC-determined

TABLE V

Transition Temperatures[a]

Sample	Dose	T_g (°C)	ΔC_p	Remarks
TEGDMA		−82	0.198	
Poly(TEGDMA)[b]		92	0.050	Broad
PVC		84	0.055	
TEGDMA-PVC	0	−64	0.024	
		−33	0.045	
		40	0.010	
	0.04	−56	0.222	
		−10		
		60	0.023	
	0.05	−7	0.022	
		58	0.017	
	0.06	8	0.022	
		55	0.031	
	0.06	52	0.062	After ether extraction
	0.07			
	0.08	60	0.025	
		87	0.036	
	0.15	55	0.015	
		92	0.045	
	1.0	84	0.065	
	5.0			

[a] From Bair *et al* (1972).
[b] Polymerized at 80°C overnight with benzoyl peroxide as catalyst.

T_gs. These shifts in T_g are due to the high frequency of the NMR measurements. In addition, the amount of material in the unirradiated mixture's high-temperature phase was estimated to be 13 wt % from ΔC_p data (see next section) and 10 wt% from the NMR relaxation studies.

ii. Quantitative analysis of each phase. From the ratio of ΔC_p for each transition to the total change in ΔC_p for all the transitions in a given mixture, one can estimate the weight fraction of the phase (or phases) responsible for each transition (Table V). The results are plotted in Fig. 27. It is apparent that the weight fraction for phases II_B plus III_B increases, while I_B decreases, and that phase I_B decreases as the level of irradiation increases until finally the former reaches 100%. These experimental observations show that the monomer-rich phase (I_B) is converted continuously to phase having much higher transition temperatures as the dose increases. Finally, all phases converge to one detectable transition temperature, because the completely polymerized phases are all expected to

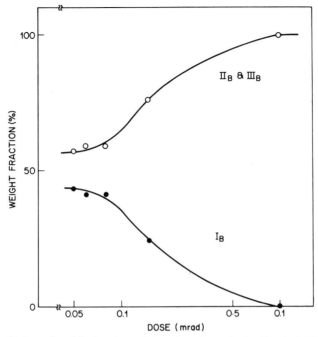

Fig. 27 Estimated weight fractions of the microphases in the blend of TEGDMA and PVC. [Reprinted with permission from Bair *et al.* (1972). Copyright by the American Chemical Society.]

have transition temperatures within narrow range of 85 (PVC) and 92°C [poly(TEGDMA)] that are undistinguishable in the DSC measurements. The weight fractions for samples with 0 and 0.04 Mrad dose were not estimated because there is a large difference in ΔC_p values for TEGDMA and its polymer and a large error will be introduced by the same calculation.

Unsaturation and gelation studies have shown that nearly all the monomer as polymerized at a dose of 0.1 Mrad produced an insoluble gel. The DSC results are in good agreement with these results for they show at 0.15 Mrad that the T_g of the major phase is approaching a limiting value and network formation is near completion. Loshaek (1955) has shown that additional irradiation and cross-link density will not raise T_g appreciably.

One final note of interest, electron microscopy studies of the state of mixing of TEGDMA and PVC before irradiation indicate that the DSC measurements are capable of detecting minute phases as small as 100 Å in diameter. These microscopy studies make use of Kato's (1965, 1967) method of staining the double bonds in the monomer with osmium tetraoxide (O_sO_4).

3. Mixtures of Elvaloy 741, a Polymeric Plasticizer, in PVC

a. DSC Heating Experiments. T_gs, T_ms, and apparent heats of fusion ΔQ_f were measured with a Perkin-Elmer DSC-2 that employed a Scanning Auto Zero accessory to produce an essentially flat base line from -100 to $150°C$. Specific heat calculations were carried out on a Tektronix Model 31 calculator that was interfaced to the DSC unit. DSC determinations of C_p on a standard Al_2O_3 sample in the temperature range of -30 to $120°C$ were found to be within 1% of the values obtained on an adiabatic calorimeter at the National Bureau of Standards.

The lower curve in Fig. 28 indicates that C_p increases linearly with temperature from -60 to $80°C$. Between 87 and $93°C$, the glass transition is evident from the discontinuous change in C_p. This T_g corresponds to the α-transition that has been observed on these same materials by dielectric and dynamic mechanical measurements (Bair *et al.*, 1978a; Anderson *et al.*, 1979). Above T_g, C_p increases smoothly to $120°C$.

The C_p curve of the terpolymer is more complex (upper curve, Fig. 28). It increases linearly from -20 to $40°C$, but above $-34°C$ increases abruptly and a broad endothermic peak, typical of polymeric melting, occurs between 50 and $77°C$. The initial discontinuing in C_p at $-34°C$ is attributed to the onset of T_g and the subsequent increases in C_p to fusion. The inception of melting of the terpolymer crystals can be estimated by extrapolating the C_p curve of the liquid between 80 to $150°C$ to lower

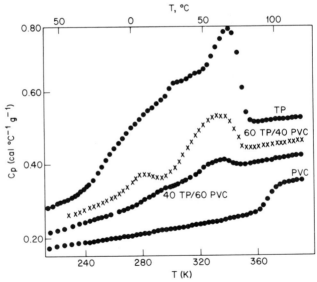

Fig. 28 Specific heat as a function of temperature for TP, 40 TP/60 PVC, 60 TP/40 PVC and PVC (Anderson *et al.*, 1979).

temperatures until it intersects the C_p curve near $-15°C$. The broad temperature range of fusion, from -15 to $77°C$, reflects the presence of lamellae of varying thickness. In this way we estimate ΔQ_f from the area under the C_p curve between the beginning of fusion at $-15°C$ and its completion at $77°C$. The value of ΔQ_f is 8.3 cal/g. Since linear polyethylene heat of fusion is about 68 cal/g and the terpolymer crystallizes in a slightly expanded polyethylene-type unit cell, it would appear that the terpolymer's level of crystallinity is of the order of 10% (Bair *et al.*, 1977).

Blending PVC with the terpolymer not only raises T_g but also depresses T_m and ΔQ_f. C_p of two blends, 60% TP/40% PVC (open squares) and 40% TP/60% PVC (filled circles) is depicted in Fig. 28. The onset of T_g begins at -19 and $-6°C$ and T_m occurs at 73 and 71°C for the blends containing 60 and 40 wt % of the terpolymer, respectively. In addition, ΔQ_f is lowered to 2.8 and 0.6 cal/g for these same samples from the terpolymer's original value of 8.33 cal/g. Please note that only a single T_g is detected for each blend, which indicates miscibility of the amorphous portion of the terpolymer with PVC. In addition, the T_gs as determined by DSC at 20°/min agreed within 5° of T_gs as measured by dielectric and mechanical techniques at 100 Hz (Anderson *et al.*, 1979).

b. DSC Cooling Experiments. Samples containing 60 wt % of the terpolymer were cooled from the melt at 40°C/min. However, at 40 wt % TP there was no detectable crystallization exotherm, although a small amount of melting occurred upon reheating. Dynamic crystallization studies where 40% TP/60% PVC samples were cooled from the melt at 20, 10, and 5°C/min indicated that the onset of crystallization T_c occurred at about 15°C, which is just prior to vitrification. Blends with 20 and 10 wt % TP did not show any sign of crystallization.

A plot of T_c versus T_g indicates that the former decreased with increasing PVC content whereas the latter increased (Fig. 29). Extrapolation of the T_c data in Fig. 29 to the point where $(T_c - T_g)$ reached 0°C occurs at 25% TP/75% PVC. In other words, blends containing 25% or less TP will not normally have any crystallinity because vitrification greatly restricts chain mobility and hence retards crystal growth. Crystallization can be induced by lowering the blends T_g by using a solvent (Bair *et al.*, 1977; Anderson *et al.*, 1979).

C. SUMMARY

We have reviewed how thermal analysis measurements particularly utilizing a DSC unit can be used to elucidate the complex behavior that exists between mixtures of both high- and low-molecular-weight plasticizers and PVC. DSC studies are well suited to investigate not only com-

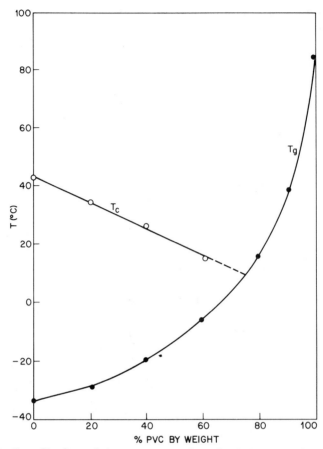

Fig. 29 Crystallization and glass temperatures in TP/PVC blends as a function of PVC concentration (Bair, 1974b).

patibility but also plasticizer permanence and the effects of aging on plasticizer content. This writer notes the lack of data reported in the literature in this area and hopes the reader will be encouraged to attempt more work in this field.

IV. Other Additives

A. Fire Retardants (FR)

1. Introduction

Polymers can be stabilized against burning in two ways. In the first of these, new materials with unusually high thermal and oxidative stability can be synthesized. Unfortunately, the prohibitive cost and absence of other desirable properties often limits the use of these special polymers to low-volume applications. In the second, additives are combined with a normally flammable polymer. It is this latter approach that is usually the most practical and economical way to obtain a fire-retardant polymer.

Commercial ABS resins, which are composed of a rubbery polybutadiene (BD) phase and a glassy matrix of poly(styrene-*co*-acrylonitrile) (SAN), can be made fire retardant by mixing antimony oxide (Sb_2O_3) and an organic halide into the base resin. Warren (1972) has described the mechanism by which this synergistic combination of inhibitors is believed to function. Chapter 8 reviews the numerous procedures for evaluating the behavior of FR additives. In this section we show only how DSC and TG can be used to assay quantitatively this type of FR-ABS system.

A large number of the organic halides used in FR-ABS formulations are soluble in either the BD or SAN phase or in both (Bair, 1974b). Most of these FR additives form glasses when cooled from the melt and will lower or raise the BD or SAN T_g, depending on the respective position and relative difference in T_g between the additive and the polymeric component. Thus in this case the FR additive is separated from the ABS resin and identified. Subsequently, the effect of the additive on the BD and SAN T_gs as a function of concentration must be determined. Sometimes, the FR additive is immiscible in the ABS, in which case it can be assayed from measurement of either its ΔC_p at T_g or ΔH_f at T_m depending on whether it is in a glassy or crystalline state. In what follows one specific example of a FR ABS resin will be given; however, the thermal techniques employed can be used on all of these types of additives.

2. Quantitative Thermal Analysis of FR-ABS

a. DSC. In Fig. 30 C_p of ABS without any FR additives is compared with that of a FR-ABS over a temperature range of -110 to $150°C$. The two T_gs for the BD and SAN phases of the ABS resin were detected at -88 and $109°C$, respectively. In contrast to this behavior the FR-ABS resin has not only a lower C_p over the entire temperature span because of the FR additives but also the BD T_g was shifted $12°$ higher to $-76°C$ and the SAN transition was lowered $11°$ to $93°C$. Bair (1970a,b,1974b) has shown how the quantity of BD and SAN can be determined from ΔC_p measurements. In addition, the FR-ABS has a small melting transition near $40°C$ with ΔH_f equal to 0.13 cal/g. This endothermic transition is possibly due to several percent of a chlorinated hydrocarbon. The shifts in T_g indicate that this particular FR is soluble in both phases of the ABS resin. Next, the organic halide was separated and identified by GC-mass spectroscopy to be hexabromobiphenyl (FR-I). The T_g of FR-1 was found to be $35°C$. The effect of FR-I on the T_gs of BD and SAN as a function of

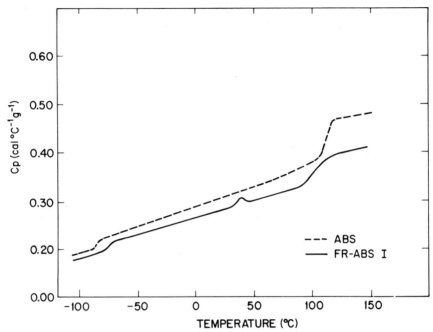

Fig. 30 Comparative plot of specific heat of ABS and FR-ABS (Bair, 1974b).

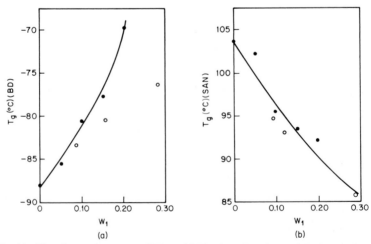

Fig. 31 The glass temperature of BD and SAN plotted against weight fraction w_1 of FRI
(a) ● FRI/ABS blends, ○ FRI/BD blends. (b) ● FRI/ABS blends, ○ FRI/SAN blends.

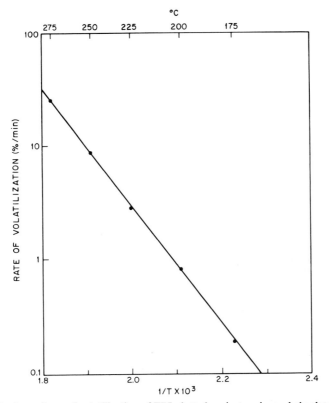

Fig. 32   Log of rate of volatilization of FRI plotted against reciprocal absolute temperature (Bair, 1974b).

concentration was studied after milling known amounts of FR-I into an ABS resin. The data are plotted in Fig. 31 and when examined indicated that about 14 wt % FR-I had been added to this particular FR-ABS resin to produce the observed shifts in T_g.

b. TG.   The FR concentrates used in producing self-extinguishing ABS resins depend on an organic halide compound that vaporizes before the degradation of the ABS structure. The vaporization behavior of the FR additive used in this FR-ABS is no exception to this rule. In Fig. 32 the rate of volatilization of FR-I has been plotted against reciprocal temperature. Note that at processing temperatures (~230°C) and atmospheric pressure, FR-I volatizes at 10 wt % per minute.

This appreciable rate of volatilization can be used to gain an independent measure of the amount of FR-I in the FR-ABS. In Fig. 33 the

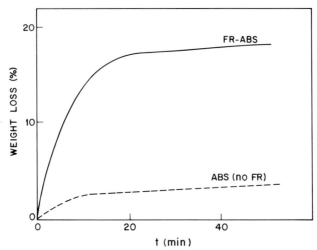

Fig. 33 Weight loss of FRI and ABS at 275°C in nitrogen plotted against temperature (Bair, 1974b).

difference in amount of weight loss that occurs at 275°C after 30 min between FR-ABS and ABS is 14 wt % and is attributed to the loss of FR-I. This value agrees well with our earlier estimates based on shifts in T_g.

Since T_m of the inorganic oxide exceeds the degradation temperature of all the organic components in the resin, the amount of Sb_2O_3 in FR-ABS was determined thermogravimetrically by pyrolyzing the resin at 450°C. The remaining residue equaled 6.8 wt % of original sample. Nondispersive x-ray (NDX) spectrometry of the residue indicated mainly the presence of Sb atoms. Since O atoms are not detectable by this instrument, it is presumed that the Sb is present as an Sb_2O_3.

3. TMA of FR-ABS

The lowering of the T_g of the SAN matrix of FR-ABS should produce a softer material with lower flexural strength. These inferences are borne out in a flexure test. Using a thermomechanical analyzer (Perkin-Elmer TMS-1), small (20 mil thick and 100 mil wide) specimens of FR-ABS and ABS were supported on knife edge supports of a flexure platform at a maximum fiber stress of 264 psi. The deflection of the specimen was monitored continuously at a heating rate of 5°C/min. At 60°C FR-ABS began to deflect more than ABS. This difference increased with increasing temperature (Fig. 34).

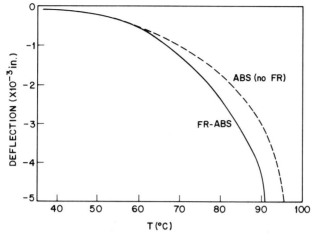

Fig. 34 The deflection of 20 mil films of ABS and FR-ABS under a maximum fiber stress of 264 psi plotted versus temperature. Heating rate 5°/min (Bair, 1974b).

4. Conclusion

It has been shown that direct quantitative analysis of a FR-ABS resin can be performed by DSC and TG. However, the principles demonstrated here should be applicable, in general, to nearly all fire-retardant additives used in a variety of polymer systems. An important aspect of FR additives, not covered in this section, is their effect on the polymer's physical properties (see Chapter 8).

B. NUCLEATING AGENTS

The mechanical and optical properties of semicrystalline polymers are intimately connected with the morphology that develops during cooling from the melt. In the case of polypropylene, it is well known that rapid cooling from the melt produces numerous small spherulites and creates a finished product with improved clarity and toughness (Lane, 1966). However, quenching only works with maximum effect when the fabricated plastic has thin sections. Thus numerous studies have been initiated by Beck and Ledbetter (1965) and others to find additives that will act as nucleating agents and accelerate a polymer's crystallization rate.

The development of crystallinity in bulk polymers is inherently tied to the nucleation and growth of spherulitic structures. Price *et al.* (1961) has shown, in general, that spherulites are nucleated on heterogeneities in the melt. These heterogeneities can be such things as catalysts residues which are active at low degrees of supercooling. Beck and Ledbetter (1965)

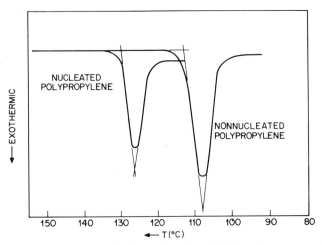

Fig. 35 Thermograms of crystallization of nucleated (0.25 wt % aluminum dibenzoate) polypropylene and polypropylene with no nucleating additives (Beck and Ledbetter, 1965). Reprinted by permission of John Wiley & Sons, Inc.

reported on how to speed up the primary rate of nucleation by adding foreign materials that function as heterogeneous nuclei for polymer crystallization.

In earlier work Beck (1966, 1967) has described how DTA can be used to evaluate the relative ability of various nucleating agents to affect the crystallization behavior of polypropylene (PP) compositions. A typical thermogram of PP nucleated with 0.25 wt % aluminum dibenzoate and nonnucleated PP is shown in Fig. 35. A 20°C increase in peak temperature (i.e., supercooling) was observed upon nucleation of PP by the aluminum dibenzoate as compared to the polymer without any nucleating additives. Beck reported a steeper slope on the high-temperature side of the crystallization exotherm for the nucleated sample, which reflects the occurrence of more rapid crystallization. Finally, the difference in peak heights in Fig. 35 is due simply to variations in sample mass and not to nucleation effects. Beck (1966) has shown by DTA that the nucleating effect of aluminum dibenzoate on PP is nearly independent of molecular weight.

By measuring a nucleating agent's ability to decrease a polymer's supercooling, Beck (1975a) has evaluated a series of compounds mixed with PP and poly(vinylidene chloride–vinyl chloride) copolymers. In this work, the higher the freezing point of the mixture, the better the compound is considered to be as a nucleating agent. Beck (1975b) has also suggested that the efficiency of nucleating agents for isotactic polystyrene can be tested by measuring the change in the polymer's ΔH_f. Positive nucleation was considered to have occurred whenever ΔH_f for the nucleated polymer was higher than that of the neat polymer.

Gallez *et al.* (1976) have studied the action of plasticizers and nucleating additives on the crystallization for bisphenol-A polycarbonate (PC). One surprising result of this work was the development of PC crystals with unusually high T_m (300°C) with the use of nucleating salts. The melting point of solution-crystallized polycarbonate (PC) has been reported to be as high as 250–260°C when crystallized slowly from methylene chloride (O'Reilly *et al.*, 1964) and as low as 181°C when crystallized rapidly in an acetone solution (Kambour *et al.*, 1966). It appears that the high T_m of the nucleated PC crystals is associated with the growth of relatively thick lamellae at small supercoolings.

Additional and supplementary data concerning the crystallization behavior of nucleated polymers can be gained by dilatometric and microscopic techniques. These types of studies have been carried out on Nylon 6 by Inoue (1963) and on polyethylene terephthalate by Przygocki and Wlochowicz (1975). It would appear that DSC could permit crystallization kinetics to be determined at larger supercoolings than by dilatometry because of the more rapid temperature equilibration of the smaller sample masses used in DSC measurements.

C. Blowing Agents

Foamed plastics are of great commercial interest because of savings that result when a gas is used to replace a relatively high-priced polymer. The foamed polymer is produced by the gaseous decomposition of a chemical compound such as azodicarbonamide at extrusion temperatures. The ideal blowing agent should decompose over a finite and narrow temperature range.

Several methods have been used to investigate the decomposition of blowing agents. Hansen (1962) has shown that DTA can determine the temperature and whether the process is endo or exothermic. Also the heat of transition is readily detectable by DSC. May (1977) has found that hermetically sealed sample containers minimize DSC base-line drift that occurs during analysis of plastisols containing such volatiles as plasticizers and diluents.

The thermal decomposition of blowing agents may be studied by thermogravimetric analysis. TGA weight loss curves are characteristic of individual blowing agents and in principle could be used to identify an unknown material. The time–temperature dependency of the volatilization of azodicarbonamide, a blowing agent, can be observed at various heating rates (Fig. 36). The breakdown of the azodicarbonamide is complex and apparently involves several reactions.

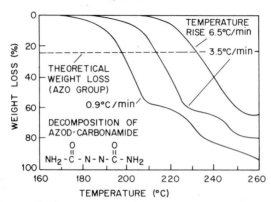

Fig. 36 TG curves for the decomposition of a blowing agent azodicarbonamide at three different heating rates (Hansen, 1962).

Renshaw (1976) has reported that in order to achieve the optimum results in vinyl plastisol foam processing, blowing agent decomposition must be controlled so that expansion takes place after melting but before significant increases in melt viscosity develop. These are two approaches for evaluating these parameters in plastisol-based foams. One way is to evaluate various formulations by spreading the plastisols, fusing and expanding them in an oven, and comparing the final results. The other method is based on determining the gelation–fusion–melt viscosity behavior of a particular resin–plasticizer combination with a Brabender Plasti-Corder and matching these characteristics with TG of the appropriate stabilizer–blowing agent combinations.

D. Processing Aids: Mold Lubricants

A typical mold lubricant (ML) should act in two ways to aid processing. First, it should lower a resin's viscosity and, second, the ML should act as a mold release agent. If an ML additive is insoluble in a resin at room temperature, the DSC can be used to do qualitative and quantitative analysis of the additive.

Reed *et al.* (1974) have shown that a typical ML used in ABS resins separates by crystallizing during cooling of an ABS part. In Fig. 37 the preceding room temperature C_p versus T plot of a commercial resin shows the melting of this ML between 125 and 150°C with ΔQ_f equal to 3.60 cal/g. After this particular ML was separated and identified by other techniques, its heat of fusion ΔH_f was found to be 23.6 cal/g. One may then calculate the amount in weight percent of ML in this ABS from the following

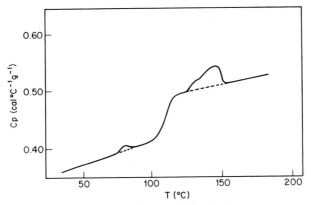

Fig. 37 Specific heat curve for an ABS resin (Reed *et al.*, 1974).

relationship:

$$C_{ML} = (\Delta Q_f / \Delta H_f) \times 100 \tag{10}$$

Thus in this case the ABS resin has 3.60 wt % of a ML. Blyler and Bair (1975) have shown that this amount of ML reduces the resin's viscosity by approximately 30%.

If the ML is to act as a release agent, it should come to the surface of a part during solidification in the mold. Since we can estimate the amount of ML in ABS from ΔQ_f, one can compare the amount of ML near the surface to the level in the interior of the part. Studies based on this thermal analysis technique have shown that there is about three times as much ML in the bulk as near or on the surface of an ABS part (Bair, 1974a).

The presence of about 1 wt % of a low-density polyethylene (LDPE) in Noryl resins has been detected by DSC measurements. The LDPE acts as mold lubricant in these materials (Bair, 1970a).

E. CARBON BLACK

Carbon black is often added to a plastic to improve its resistance to thermal and photoinduced degradation. The concentration of carbon black in a resin can be determined by TG. A routine procedure for this measurement is shown in Fig. 38 for the analysis of a mixture of carbon black and polyethylene (Brennan, 1977b). A sample weighing a few milligrams is placed in the sample pan with an inert gas like nitrogen. After purging, the sample is heated rapidly (320°/min) to 600°C to decompose the polyethylene pyrolytically, leaving a residue of carbon black and possibly any inorganic materials. The presence of the latter material can be detected at this point by switching from nitrogen to either air or oxygen to

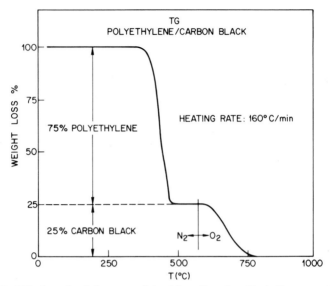

Fig. 38   TG of a polyethylene containing 25 wt % carbon black (Brennan, 1977b).

combust the carbon black. Inert fillers can be isolated in this way and then identified by spectroscopic analysis.

F. WATER

1. Introduction

Water is not an additive that is compounded purposely into a commercial plastic. However, since all polymers can absorb water from the atmosphere it is an important component of each and every piece of plastic. The presence of water in a polymer can lead to marked changes in a resin's chemical and physical state. For example, absorbed water in an epoxy can lower T_g by as much as 20°C or in polycarbonate lead to hydrolytic degradation (Morgan and O'Neal, 1977; Pryde and Hellman, 1979; Bair *et al.*, 1979). Thus the quantitative analysis of water in a polymer is a useful measurement in characterizing the behavior of a resin.

Bair and Johnson (1977) have reported a way to measure the amount of water that has associated to form microscopic water-filled cavities (clusters) in polyethylene at a level of 10 ppm and greater. By combining this calorimetric technique with a coulometric method, it is possible to differentiate between clustered water and the total water sorbed by any resin. It was found, in general, that clusters are formed when a polymer is saturated with water at an elevated temperature and is rapidly cooled to room temperature. During cooling the solubility of water in the material is lowered and some water condenses in the form of microscopic water-filled cavities, provided that the internal pressure generated by the excess water exceeds the strength of the polymer.

The difference between the total amount of water sorbed by a polymer and the concentration of clustered water equals the amount of bound water. The presence of these bound water dipoles moving in concert within their polymer environment manifests itself as an increase in the polymer's low-temperature dielectric loss behavior (Bair *et al.*, 1978b,c; Johnson *et al.*, 1980). In addition, bound water was found to plasticize

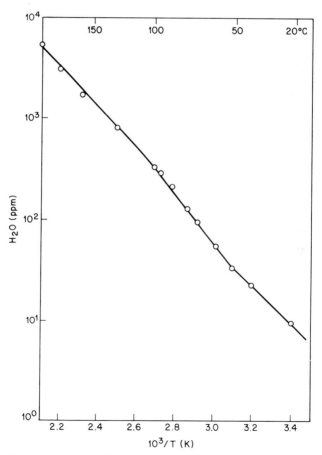

Fig. 39   Saturation concentration of water in 0.93 density polyethylene plotted against temperature (Bair and Johnson, 1977).

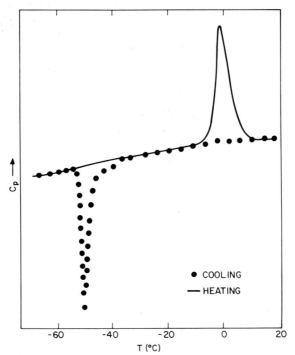

Fig. 40 DSC trace of polycarbonate after 800 h in 97°C water. Sample cooled to −70°C before reheating (Bair *et al.*, 1978c).

poly(vinyl acetate), causing T_g to decrease; whereas clustered water did not (Bair *et al.*, 1980b, 1981).

2. Total Water Measurement

The total water content of a polymer can be determined by a coulometric-type instrument, DuPont Moisture Analyzer Model 26-321A (Chapter 1). In this device the water within a polymer is driven off rapidly by heating the specimen above T_g or T_m. The liberated water is collected on the hydroscopic surfaces of P_2O_5 crystals and then electrolyzed to H_2 and O_2. The electrolytic current is proportional to the quantity of water in the sample. By simply housing the moisture analyzer in a dry box, water levels down to 1 or 2 ppm can be detected.

Utilizing the coulometric method, the equilibrium solubility of water in 0.93 density polyethylene was measured from 4 to 208°C (Fig. 39). Water concentrations ranged from as little as 4 ppm at 4°C to about 6000 ppm at 208°C. The results indicate that the water content of polyethylene is strongly temperature dependent.

3. Clustered Water Concentration

The calorimetric method for the qualitative and quantitative analysis of clustered water is depicted in Fig. 40. The polycarbonate sample had been in the 97°C water bath for 800 h and the total water content of the sample was determined coulometrically to be 0.92 wt %. When about 20 mg of this sample was placed in the DSC-2 and cooled at 20°C/min from room temperature to -140°C, the onset of crystallization of the clustered water was detected near -40°C and proceeded at a maximum crystallization rate at -50°C (filled circles, Fig. 40). We believe that the large undercooling is due to the microscopic size of the clusters and the absence of heterogeneities in the water. Under these conditions, water on the surface of the polycarbonate would freeze between -10 and -20°C. When the sample was reheated from $-$C°C to room temperature, a first-order transition was detected at -5°C with a heat of transition ΔH_{tr} of 0.215 cal/g or 0.27 wt % water in the clustered form.

If the concentration of clustered water in parts per million (ppm) is represented by C, then

$$C = \frac{10^6 \, \Delta H_{tr}}{\Delta H_f} \tag{11}$$

where ΔH_f is the heat of fusion of water or 79.7 cal/g. In this case, C equals 2690 ppm or 29% of the total water. ΔH_{tr} is independent of time at -55°C and lower. At undercoolings of -40°C or less ΔH_{tr} is time dependent. The large supercooling necessary for complete freezing is due to the submicron-sized cavities and water free of heterogeneities.

In the future, it may be possible to use the clustered water technique as a probe to estimate the void content of composite polymer systems.

V. Conclusion

Throughout this chapter we have reviewed numerous thermoanalytical techniques for the characterization of additives that play a key role in the behavior of most commercial polymer formulations. Although these additives perform such diverse functions in a resin as stabilizing, plasticizing, lubricating, and nucleating, we have shown that they may all be quantitatively assayed by similar DSC and TG methods. In addition, the measurement of many other important additive properties such as compatibility, solubility, and volatility has been demonstrated using the same thermal analysis instrumentation.

Prope. utilization of these thermal techniques and other complementary tests should make it easier for the manufacturer to evaluate the behavior of polymer formulations and develop the most effective additive systems. On the other hand, the plastics user may employ these same methods to set processing conditions and judge the resin's ability to retain its desirable properties over its intended lifetime. Unfortunately, in only a few areas such as the stabilization of polyolefins have extensive long-term studies been initiated. These investigations are aimed at understanding the complex sequence of physical and chemical changes that lead to the premature deterioration of a polymer under service conditions.

In the future, it seems that similar studies of the effects of physical and chemical aging on the properties of many other polymer–additive systems will be required, particularly as increased performance requirements for modern plastics are being constantly demanded from the plastics industry. It is hoped that these future needs will be met, in part, by the use of thermal methods discussed in this book and where necessary with the development of new thermal techniques.

ACKNOWLEDGMENTS

The author thanks Dr. E. A. Turi and Dr. W. L. Hawkins for helpful discussions and comments concerning the preparation of this manuscript. In particular, I should like to acknowledge Dr. Hawkin's critical review of the section on antioxidants. In addition, this writer is grateful to Professor R. D. Deanin for his critical review of the final manuscript.

References

Albarino, R. V., and Schonhorn, H. (1974). *J. Appl. Polym. Sci.* **18**, 635–649.
Anderson, E. W., Bair, H. E., Johnson, G. E., Kwei, T. K., Padden, R. J., Jr., and Williams, D. (1979). *Adv. Chem. Ser.* **176**, 413–433.
Bair, H. E. (1970a). *Anal. Calorim.* **2**, 51–60.
Bair, H. E. (1970b). *Polym. Sci. Eng.* **10**, 247–250.
Bair, H. E. (1971). Unpublished.
Bair, H. E. (1972). Unpublished.
Bair, H. E. (1973). *Polym. Eng. Sci.* **13**, 435–439.
Bair, H. E. (1974a). *Polym. Eng. Sci.* **14**, 202–205.
Bair, H. E. (1974b). *Anal. Calorim.* **3**, 797–806.
Bair, H. E. (1976). Unpublished.
Bair, H. E. (1980). Unpublished.
Bair, H. E., and Johnson, G. E. (1977). *Anal. Calorim.* **4**, 219–226.
Bair, H. E., and Warren, P. C. (1979) *Bull. Am. Phys. Soc.* [2] **24**(3), 288.
Bair, H. E., and Warren, P. C. (1980). *Abst. Int. Symp. PVC, 3rd, 1980* pp. 101–105.
Bair, H. E., and Warren, P. C. (1981). *J. Macromol. Sci. Phys.* (in press).

Bair, H. E., Matsuo, M., Salmon, W. A., and Kwei, T. K. (1972). *Macromolecules* **5**, 114–119.

Bair, H. E., Roe, R. J., and Gieniewski, C. (1974). *Soc. Plas. Eng. [Tech. Pap.]* **20**, 412–414.

Bair, H. E., Kwei, T. K., Padden, F. J., Jr., and Williams, D. (1977). *Coat. Plast. Prepr. Pap. Meet. (Am. Chem. Soc., Div. Org. Coat. Plast. Chem.)* **37**, No. 1, 240–245.

Bair. H.E., Anderson, E. W. Johnson, G. E., and Kwei, T. K. (1978a). *Polym. Prep. Am. Chem. Soc., Div. Polym. Chem.* **19**, No. 1, 143–148.

Bair, H. E., Johnson, G. E., and Merriweather, B. (1978b). *Soc. Plast. Eng. [Tech. Pap.]* **24**, 387–391.

Bair, H. E., Johnson, G. E., and Merriweather, B., (1978c). *J. Appl. Phys.* **49**, 4976–4984.

Bair, H. E., Falcone, D. R., Hellman, M. Y., Johnson, G. E., and Kelleher, P. G. (1979). *Polym. Prep., Am. Chem. Soc., Div. Polym. Chem.* **20**(2), 614–619.

Bair, H. E., Boyle, D. J., and Kelleher, P. G. (1980a). *Polym. Eng. Sci.* **20**, 995–1001.

Bair, H. E., Johnson, G. E., and Anderson, E. W. (1980b). *Polym. Prep., Am. Chem. Soc., Div. Polym. Chem.* **21**(2), 21–23.

Bair, H. E., Johnson, G. E., and Anderson, E. W. (1981). *Polym. Eng. Sci.* **21**(3).

Beck, H. N. (1966). *J. Polym. Sci., Polym. Phys. Ed.* **4**, 631–638.

Beck, H. N. (1967). *J. Appl. Polym. Sci.* **11**, 673–685.

Beck, H. N. (1975a). *J. Appl. Polym. Sci.* **19**, 371–373.

Beck, H. N. (1975b). *J. Appl. Polym. Sci.* **19**, 2601–2608.

Beck, H. N., and Ledbetter, H. D. (1965). *J. Appl. Polym. Sci.* **9**, 2131–2142.

Blyler, L. L., Jr., and Bair, H. E. (1975). *SPE Tech. Conf. Prep., Eng. Properties Struct. Div.* pp. 46–55.

Brennan, W. P. (1977a). *Perkin Elmer Therm. Anal. Appl. Study* No. 22, p. 5.

Brennan, W. P. (1977b). *Thermochim. Acta* **18**, 101–111.

Carslaw, H. S., and Jaeger, J.C. (1959). "Conduction of Heat in Solids," 2nd ed., p. 100. Oxford Univ. Press, London and New York.

Chan, M. G., and Allara, D. L. (1974a). *Polym. Eng. Sci.* **14**, 12–15.

Chan, M. G., and Allara, D. L. (1974b). *J. Colloid Interface Sci.* **47**, 697–704.

Chan, M. G., and Johnson, L. (1980). *Soc. Plast. Eng. [Tech. Pap.]* **26**, 494–496.

Chan, M. G., Gilroy, H. M., Heyward, I. P., Johnson L., and Martin, W. M. (1978). *Soc. Plast. Eng. [Tech. Pap.]* **24**, 381–383.

Chiu, J. (1966). *Appl. Polym. Symp.* **2**, 25–32.

Couchman, P. R. (1978). *Macromolecules* **11**, 1156–1160.

Davis, D. D., and Slichter, W. P. (1973). *Macromolecules* **6**, 728–733.

DeCoste, J. (1965). *J. Soc. Plast. Eng.* **21**, 764–768.

Fox, T. G. (1956). *Bull. Am. Phys. Soc.* [2] **1**, 123.

Gallo, F., Legras, R., and Mercier, J. P. (1976). *Polym. Eng. Sci.* **16**, 276–283.

Gesner, B. D., Shea, J. W., and Wight, F. R. (1974). *Proc. Wire Cable Symp. 23rd, 1974* pp. 7–10.

Gilroy, H. M. (1974). *Proc. Wire Cable Symp., 23rd, 1974* pp. 42–45.

Gordon, M., and Taylor. J. S. (1952). *J. Appl. Chem.* **2**, 493–498.

Gouinlock, E. V. (1975). *J. Polym. Sci., Polym. Phys. Ed.* **13**, 1533–1542.

Grassie, N. (1966). *Encycl. Polym. Sci. Technol.* **4**, 647.

Gray, A. P. (1975). *Perkin-Elmer Newsl.*

Guyot, A., Benevise, J. P., and Trambonze, Y. (1962). *J. Appl. Polym. Sci.* **6**, 103–108.

Hammer, C. F. (1977). *Coat. Plast. Prepr. Pap. Meet. (Am. Chem. Soc., Div. Org. Coat. Plast. Chem.)* **37**, 234–237.

Hansen, R. H. (1962). *SPE J.* **18**, 77–83.

Hansen, R. H., Russell, C. A., DeBenedictis, T., Martin, W. M., and Pascale, J. V. (1964). *J. Polym. Sci., Part A* **2**, 587–609.

Hawkins, W. L., Matreyek, W., and Winslow, F. H. (1959a). *J. Polym. Sci.* **41**, 1–11.

Hawkins, W. L., Hansen, R. H., Matreyek, W., and Winslow, F. H. (1959b). *J. Appl. Polym. Sci.* **1**, 37–43.

Hawkins, W. L., Chan, M. G., and Link, G. L. (1971). *Polym. Eng. Sci.* **11**, 377–380.

Heiberger, C. A., Phillips, R., and Canton, J. J. R. (1969). *Polym. Eng. Sci.* **9**, 445–451.

Howard, J. B. (1972). *Proc. Int. Wire and Cable Symp., 21st, 1972* pp. 329–341.

Howard, J. B. (1973a). *Soc. Plast. Eng. [Tech. Pap.]* pp. 408–412.

Howard, J. B. (1973b). *Polym. Eng. Sci.* **13**, 429–434.

Howard, J. B., and Gilroy, H. M. (1975). *Polym. Eng. Sci.* **15**, 268–271.

Howard, J. B., Gilroy, H. M., and Kokta, E. (1974). *Proc. Int. Wire Cable Symp., 23rd, 1974* pp. 46–52.

Inoue, M. (1963). *J. Polym. Sci., Part A* **1**, 2013–2020.

Jackson, R. A., Woodland, S. R., and Pajaczkowski, A. (1963). *J. Appl. Polym. Sci.* **12**, 1297–1309.

Johnson, G. E., Bair, H. E., Matsouoka S., Anderson, E. W., and Scott, J. E. (1980). *ACS Symp. Ser.* **127**, 451–468.

Kambour, R. P., Karasz, F. E., and Daane, J. H. (1966). *J. Polym. Sci., Part A2* **4**, 327–347.

Kaplan, M. L., Kelleher, P. G., Bebbington, G. H., and Hartless, R. L. (1973). *J. Polym. Sci. Polym. Lett. Ed.* **11**, 357–361.

Kato, K. (1965). *J. Electron. Microsc.* **14**, 220–221.

Kato, K. (1967). *Polymer* **8**, 33–39.

Kelleher, P. G., Boyle, D. J., and Gesner, B. D. (1965). *J. Appl. Polym. Sci.* **11**, 1731–1735.

Lane, J. E. (1966). *Br. Plast.* **39**, 9–14.

Leisz, D. M., Kleiner, L. W., and Gestenback, P. G. (1980). *Thermochim. Acta* **35**, 51–58.

Loan, L. D., and Winslow, F. H. (1972). *In* "Polymer Stabilization" (W. Hawkins, ed.) pp. 117–159. Wiley (Interscience), New York.

Loshaek, S. (1955). *J. Polym. Sci.* **15**, 391–404.

Luongo, J. P. (1965). *Appl. Spectrosc.* **19**, 117–124.

MacKnight, W. J., Karasz, R. E., and Fried, J. R. (1978). *In* "Polymer Blends" (D. R. Paul and S. Newman, eds.), Vol. 1, pp. 224–228. Academic Press, New York.

Madorsky, S. L. (1964). "Thermal Degradation of Organic Polymers," pp. 160–167. Wiley, New York.

Marshall, D. J., George, E. J., Turnispeed, J. M., and Glenn, J. L. (1973). *Polym. Eng. Sci.* **13**, 415–421.

May, W. (1977). *Plast. Technol.* p. 97.

Miller, A. A. (1959). *Ind. Eng. Chem.* **51**, 1271–1274.

Morgan, R. J., and O'Neal, J. E. (1977). "The Durability of Epoxies," McDonnell Douglas Res. Lab. Rep.

Nass, L. I. (1963). *In* "Modern Vinyl Compounding and Stabilization," pp. 11.1–11.24. Advance Div., Carlisle Chemical Works, Inc. Reading, Ohio.

O'Reilly, J. M., Karasz, F. E., and Bair, H. E. (1964). *J. Polym. Sci., Part C* **6**, 109–115.

Pinner, S. H. (1959). *Nature (London)* **183**, 1108–1109.

Pinner, S. H. (1960). *Br. Plast.* **25**, 35–39.

Price, F. P., Barnes, W. J., and Luetzel, W. G. (1961). *J. Phys. Chem.* **65**, 1742–1748.

Pryde, C. A., and Hellman, M. Y. (1979). *Polym. Prepr., Am. Chem. Soc., Div. Polym. Chem.* **20**(2), 620–623.

Przygocki, W., and Wlochowicz, A. (1975). *J. Polym. Sci.* **19**, 2683–2688.

Pusey, B. B., Chen, M. T., and Roberts, W. L. (1971). *Proc. Int. Wire Cable Symp., 20th, 1971* pp. 209–217.

Reed, T. F., Bair, H. E., and Vadimsky, R. G. (1974). *Polym. Sci. Technol.* **4**, 359–375.

Reich, L., and Stivala, S. (1969). "Autoxidation of Hydrocarbons and Polyolefins." Dekker, New York.

Renshaw, J. T. (1976). *Polym.—Plast. Technol. Eng.* **6**(2), 137–156.

Roe, R. J. (1972). Unpublished.

Roe, R. J., Bair, H. E., and Gieniewski, C. (1974). *J. Appl. Polym. Sci.* **18**, 843–856.

Rubin, A., Schreiber, H. P., and Waldman, M. A. (1961). *Ind. Eng. Chem.* **53**, 137–143.

Salmon, W. A., and Loan, L. D. (1972). *J. Appl. Polym. Sci.* **16**, 671–682.

Salovey, R., and Bair, H. E. (1970). *J. Appl. Polym. Sci.* **14**, 713–721.

Scalco, E., and Bair, H. E. (1979). Unpublished.

Shelton, J. R. (1972). *In* "Polymer Stabilization" (W. L. Hawkins, ed.), pp. 30–116. Wiley (Interscience), New York.

Shimada, J., and Kabuki, K. (1968). *J. Appl. Polym. Sci.* **12**, 655–669, 671–682.

Stromberg, R. R., Straus, S., and Achammer, B. G. (1959). *J. Polym. Sci.* **35**, 355–368.

Struber, U. R. (1968). *In* "Theory and Practice of Vinyl Compounding," p. 56. Argus Chemical Corp., Brooklyn, New York.

Tordella, J. P., Hyde, T. J., Gordon, B. S., and Hammer, C. F. (1976). *Mod. Plast.* **54**, No. 1, 64–68.

Warren, P. C. and Bair, H. E. (1977). Unpublished.

Warren, P. C. (1972). *In* "Polymer Stabilization" (W. L. Hawkins, ed.), pp. 313–352. Wiley (Interscience), New York.

Wood, L. A. (1958). *J. Polym. Sci.* **28**, 319–330.

Yamaguchi, K., Yoshikawa, T., Kishi, H., Masaki, M., and Sakamoto, N. (1974). *Soc. Plast. Eng.* [*Tech. Pap.*] **20**, 97–103.

List of Symbols

Page numbers refer to first occurrence or detailed definition

Chapter 1

T	temperature, 30
T_n	heating rate, 30
T_r	reference temperature, 46
T_s	sample temperature, 46
R	effective thermal resistance, 30
C_s	effective heat capacity of sample and holder, 30

Chapter 2

A	preexponential factor, 114; area 124
$[A]$	concentration of A, 113
$[B]$	concentration of B, 113
C	heat capacity, 125
$[C]$	concentration of C, 113
C_P	heat capacity at constant pressure, 126
C_V	heat capacity at constant volume, 126
c	linear crystal growth rate, 119; number of compounds, 195; concentration, 201.
$\bar{c}$	average speed, 115
D	Debye function, 133; absorbancy, 156
$[D]$	concentration of D, 113
d	total differential, 108
d_{AB}	collision diameter, 115
E	Young's modulus, 124
E_a	activation energy, 114
F	Helmholtz free energy, 109
f	tensile force, 124; degrees of freedom, 195
G	Gibbs free energy, 127
H	enthalpy, 126
h	Planck's constant, 116
I	intensity, 156
K	equilibrium constant, 116; Avrami constant, 118
k	Boltzmann constant, 110; packing fraction, 136; specific rate constant, 113
L	phenomenological coefficient, 112; heat of fusion per kg, 201

l	length, 121
$\mathbf{l}$	bond length vector, 98
M	mass, molecular mass, 131
M_o	repeating unit mass, 97
$\overline{M}_n$	number average molecular mass, 97
$\overline{M}_w$	weight average molecular mass, 97
$_eM$	mass flux, 112
$_iM$	mass production, 112
N	number of molecules, 97; normality of solutions, 103; number of moles, 108
$_eN$	matter flux, 111
$_iN$	matter production, 111
P	pressure, 108; probability, 212
p	steric factor, 115; number of phases, 195
Q	heat, 108
R	gas constant, 99
r	end to end distance, 98; fraction molten, 209
$\mathbf{r}$	end to end vector, 98
S	entropy, 109
$_eS$	entropy flux, 111
$_iS$	entropy production
T	temperature, 121
T_m	melting temperature, 143
T_m°	equilibrium melting temperature, 158
t	time, 111; path length, 156
U	total internal energy, 108
$_eU$	energy flux, 112
$_iU$	energy production, 112
V	volume, 108
v	specific volume, 156; volume fraction, 198
v^c	volume fraction crystallinity, 150
W	thermodynamic probability, 110
w	weight (mass) fraction, 97; work, 108
w^c	weight (mass) fraction crystallinity, 156
X	expected average, 119; extensive variable, 182
x	degree of polymerization, 95; mole fraction, 197; partial molar volume ratio, 198
Z_{AB}	collision number, 115
z	partition function, 116, coordination number, 199
α	thermal expansivity, 123; extension ratio, 139; incremental heat of fusion, 210
β	isothermal compressibility, 123
γ	thermal pressure coefficient, 124; surface free energy, 158
Δ	changes on fusion, chemical reaction etc. usually per mole
δ	small changes of quantities not functions of state, 108; solubility parameter, 191
∂	partial differential, 108
ϵ	absorbancy index, 156
ζ	internal variable, 112; sequence length, 211
Θ	theta temperature of a polymer molecule, 99; characteristic temperature for heat capacity, 133
θ	empirical temperature, 121
κ	transmission coefficient, 116
μ	reduced mass, 115; chemical potential, 185

ν	frequency, 116; number of placements, 182
π	3.14159
ρ	density, 119
Σ	sum
σ	Poisson's ratio, 125
τ	relaxation time, 112
χ	interaction parameter, 199
$\ddagger$	superscript for activated state, 115–117

Chapter 3

T	temperature, 239
T_g	glass transition temperature, 241
T_m	melting temperature, 241
$\overline{M}_n$	number-average molecular weight, 246
X_c	number of chain segments, 248
$\overline{M}_w$	weight-average molecular weight, 251
$\overline{M}_v$	viscosity-average molecular weight, 318
Å	Angstrom, 251
E_a	activation energy, 256
q	heating rate, 258
q/a	ratio of cation charge (q) to distance between cations and anions (a), 260
K	degree Kelvin, 259
Δt	the difference in Gibbs-DiMarzio flex energy, 260
k	Boltzmann's constant, 260
R	gas constant, 256
V_t	free volume, 262
P	pressure, 262
$\Delta C_p T_g$	specific heat increment at the glass gransition, 267
$T_{g(L)}$	the lower of two amorphous glass transitions, 264
$T_{g(U)}$	the upper of two amorphous glass transitions, 264
D	compliance, 315
T_c	premelting crystalline transition temperature, 264; or crystallization temperature, 274
$T_{c(c)}$	cold crystallization temperature, 318
γ	surface tension, 267
H_f	heat of fusion, 270
λ	draw ratio, 272
I_0 and I_c	light transmission, initial and final, 273
θ	fraction of transformed (crystallized) material, 273
T_{ce}	ceiling temperature, 283
S	entropy, 284
ΔH_1	heat of polymerization from liquid monomer to solid polymer, 289
ΔH_2	heat of volatilization of liquid monomer, 289
ΔH_3	heat of polymerization of gaseous monomer to solid polymer, 289
t	time, 295
$t_{1/2}$	half-time for crystallization, 295
$\tan \zeta$	loss tangent, 299

E' and E''	dynamic and loss moduli, 300
ΔH_{vol}	heat of volatilization, 301
C_p	heat capacity, 312
T_{ll}	liquid–liquid transition temperature in amorphous polymer, 337
N_c	critical number of carbon atoms, 343
K	thermal conductivity, 345
ΔH_c	heat of crystallization, 348
ΔS_A	entropy of activation, 302

Chapter 4

a	solvent activity, 409
B	constant in volume–pressure relationship, 380; constant related to interaction parameter, 415
C	constant in volume–pressure equation, 380
C_p	heat capacity or specific heat at constant pressure, 126, 413
ΔC_p	increase in C_p at glass transition temperature, 413
DP	degree of polymerization, 397
H_f	heat of fusion, 383
$\Delta H, \Delta H_f^\circ$	heat of fusion, 414
ΔH_f^*	apparent heat of fusion, 414
M	sample weight, 419
$\%M_1$	percent monomer 1, 395
$\bar{M}_w$	weight average molecular mass, 390
R	gas constant, 415; heating rate, 419
P	pressure, 380
$\%P$	percent phosphorus, 370
S	full scale power output (DSC), 419
T	temperature, 415
T_c	crystallization temperature, 370
T_g	glass transition temperature, 368
T_m	melting temperature, 369
T_m°	equilibrium melting temperature, 415
tan	loss tangent, 383
TH	transition height at mid-point of glass temperature, 419
ΔQ_f	observed heat of fusion, 419
V	molar volume, 415
V	specific volume, 380
v/v_0	ratio of specific volumes (homopolymer and copolymer), 380
x	mass fraction, 419
β	compressibility, 380
ϕ	volume fraction, 409
χ	interaction parameter, 409
Δ	fractional increase in primary DSC output, 419

Chapter 5

A	cure stage, 440; preexponential factor, 442; amine group, 443; amplitude, 468; dynamic method, 533

a	amine concentration, 444; method, 474
B	stoichiometric parameter, 439; cure stage, 440; dynamic method, 540
b	method, 474
C	cure stage, 440; dynamic method, 545; heat capacity, 502
c	hydroxyl concentration, 444
d	total differential, 441
E	activation energy, 442; epoxide group, 443
e	epoxide concentration, 444
F	force, 474
f	function, 441; functionality, 507
G	shear modulus, 468
H	heat, enthalpy, 446
h	thickness, 474
k	rate constant, 441
l	thickness, 507
m	reaction order, 442
n	reaction order, 442
N	number of moles, 534
P	period, 468
R	gas constant, 442; radius, 474
T	temperature, 439
t	time, 439
V	volume, 474
v	volume fraction, 507
W	weight, 508
x	epoxide concentration (consumed), 444
SR	swell rating, 507
α	extent of conversion, degree of cure, 439; linear thermal expansion coefficient, 497; transition, 490
β	transition, 492; linear hygroscopic expansion coefficient, 502
γ	transition, 491; cubical thermal expansion coefficient, 528
Δ	change (eg change on curing), 446
δ	delta (as in tan δ), 468
η	viscosity, 474
κ	bulk modulus
ν	number of moles, 507; Poisson's ratio, 529
ϕ	heating rate, 541
π	3.1416, 474
ρ	density, 507

see also Tables I (p. 439) and X (p. 527)

Chapter 6

T_g	glass transition temperature,
T_m	melting point
ΔH	enthalpy
E_a	activation energy
X	apparent reaction order
K	rate constant

$\overline{M}_w$	weight average molecular weight
G'	storage modulus; the in-phase component
G''	loss modulus; the out-of-phase component
$\tan \delta$	loss tangent; a measure of the phase angle between G'' and G' ($\tan \delta = G''/G'$)
T/T_g	reduced temperature scale for displaying secondary transitions or relaxations
δ	phase angle between G'' and G'
E	Young's modulus
MW	molecular weight
ΔH_f	enthalpy of fusion
ΔH_v	enthalpy of curing (vulcanization)
T_p	peak temperature for the DSC curing exotherm
T_R	reference temperature for use in TG studies to determine the polymer, polymer blend or oil content of elastomer blends, compounds or vulcanizates
T_{max}	temperature of maximum degradation of a polymer in a TG analysis
ϕ	response factor (mg of component per cm of peak height; DTG analysis of elastomer blend)
T_0	temperature at which the weight loss increases sharply in a TG analysis of carbon black oxidation
T_t	temperature at which the weight loss terminates in a TG analysis of carbon black oxidation
T_{15}	temperature at which 15% of the weight loss has occurred in a TG analysis of carbon black oxidation
T_{20}	temperature at which 20% of the weight loss has occurred in a TG analysis of carbon black oxidation

Chapter 7

T_g	glass transition temperature
T_m	melting temperature
$T_{x\,cold}$	"cold" crystallization temperature
$T_{x\,melt}$	melt crystallization temperature
ΔH_m	enthalpy of fusion
$\Delta H_{x\,cold}$	enthalpy of "cold" crystallization
T_D	decomposition temperature
ΔH_D	enthalpy of decomposition
T_{eff}	effective temperature
CTE	coefficient of linear thermal expansion
S_T	shrinkage at specified temperature
SF_T	shrinkage force at specified temperature
CVE	coefficient of volume expansion
ds/dt	rate of shrinkage with respect to temperature
σ_{eff}	effective stress

Chapter 8

no symbols used

Chapter 9

C	concentration, 863
C_0	equilibrium solubility, 864

D	diffusion coefficient, 864
K	diffusion function, 863
T_g	glass transition temperature, 879
ΔC_p	specific heat change at T_g, 879
T_{g_i}	temperature marking the onset of glass transition, 880
T_{g_f}	temperature associated with end of glass transition interval, 880
C_{p_f}	specific heat at T_{g_f}, 881
C_{p_g}	specific heat at T_{g_i}, 881
C_{ML}	concentration of mold lubricant, 901
ΔQ_f	observed heat of fusion, 901
ΔH_f	heat of fusion, 901
ΔH_{tr}	heat of transition, 905
x	position coordinate, 863
t	time, 863
$2l$	total thickness of disks
w	weight fraction

Conversion Table

To Convert From	To (SI Unit)	Multiply By[b]	
FORCE			
dyne	newton (N)	1.000 000	E − 05
kgf	newton (N)	9.806 650	E + 00
lbf	newton (N)	4.448 222	E + 00
FORCE PER UNIT LENGTH (SURFACE TENSION)			
dyne/cm	newton per meter (N/m)	1.000 000	E − 03
LENGTH			
angstrom	nanometer (nm)	1.000 000	E − 01
foot	meter (m)	3.048 000	E − 01
inch	meter (m)	2.540 000	E − 02
micron	micrometer (μm)	1.000 000	E + 00
mil	millimeter (mm)	2.540 000	E − 02
millimicron	nanometer (nm)	1.000 000	E + 00
MASS			
ounce (avdp)	kilogram (kg)	2.834 952	E − 02
pound (avdp)	kilogram (kg)	4.534 924	E − 01
ton (short)	kilogram (kg)	9.071 847	E + 02
MASS PER UNIT VOLUME			
g/cm^3	kilogram per cubic meter (kg/m^3)	1.000 000	E + 03
lb/ft^3	kilogram per cubic meter (kg/m^3)	1.601 846	E + 01
lb/US gal	kilogram per cubic meter (kg/m^3)	1.198 264	E + 02
MASS PER UNIT TIME (FLOW RATE)			
lb/h	kilogram per second (kg/s)	1.259 979	E − 04
ton (short)/h	kilogram per second (kg/s)	2.519 958	E − 01
MOLAR MASS			
lb/mol	kilogram per mol (kg/mol)	4.535 924	E − 01

To Convert From	To (SI Unit)	Multiply By[b]	
PRESSURE AND STRESS			
atmosphere (stand)	kilopascal (kPa)	1.013 250	E + 02
bar	pascal (Pa)	1.000 000	E + 05
dyne/cm^2	pascal (Pa)	1.000 000	E − 01
inch of mercury (32°F)	kilopascal (kPa)	3.386 38	E + 00
inch of water (60°F)	kilopascal (kPa)	2.488 4	E − 01
millimeter of Hg (0°C)	pascal (Pa)	1.333 22	E + 02
psi (lbf/in^2)	kilopascal (kPa)	6.894 757	E + 00
torr (mm Hg, 0°C)	pascal (Pa)	1.333 22	E + 02
TEMPERATURE			
degree Celsius	kelvin (K)	$t_K = t_{°C} + 273.15$	
degree Fahrenheit	degree Celsius (°C)	$t_{°C} = (t_{°F} - 32)/1.8$	
TORQUE			
dyne cm	newton meter (N m)	1.000 000	E − 07
lbf ft	newton meter (N m)	1.355 818	E + 00
VELOCITY			
ft/min	meter per second (m/s)	5.080 000	E − 03
ft/s	meter per second (m/s)	3.048 000	E − 01
VISCOSITY			
centipoise	pascal second (Pa s)	1.000 000	E − 03
centistokes	square meter per second (m^2/s)	1.000 000	E − 06
VOLUME			
ft^3	cubic meter (m^3)	2.831 685	E − 02
gallon (US liquid)	cubic meter (m^3)	3.785 412	E − 03
pint (US liquid)	cubic meter (m^3)	4.731 765	E − 04
quart (US liquid)	cubic meter (m^3)	9.463 529	E − 04
VOLUME PER UNIT TIME (FLOW RATE)			
gallon (US liquid)/min	milliliter per second (mL/s)	6.309 020	E + 01

PHYSICAL CONSTANTS

Avogadro's number	6.022 17	E + 23 molecules/mol
Boltzmann's constant	1.380 62	E − 23 J/K
Faraday's constant	9.648 67	E + 04 C/mol
Gas constant	8.314 34	E + 00 J/K mol
Gas constant	8.314 34	E + 00 Pa m³/K mol
Planck's constant	6.626 20	E − 34 J s

[a] From SI Metric Reference Card No. 1, by courtesy of Allied Corporation, Morristown, New Jersey.

[b] The conversion factor is followed by the letter E (for exponent), a plus or minus symbol, and two digits which indicate the power of 10 by which the number must be multiplied to obtain the correct value. For example, 9.290 304 E − 02 is $9.290\ 304 \times 10^{-2}$.

[c] The relationship between flux density B and field strength H in SI units is defined to be $B = \mu_0 H + J$ where $\mu_0 = 4\pi \times 10^{-7}$ H/m (permeability constant) and J is the magnetic polarization (expressed in teslas).

Author Index

Numbers in italics refer to the pages on which the complete references are listed.

A

Abbas, K. B., 308, 309, *354*
Abbot, G. L., 85, *89*
Abolafia, O. R., 539, *563*
Achammer, B. G., 872, *909*
Acitelli, M. A., 436, 447, 460, 461, 469, 479, 481, *563*
Adam, G. A., 171, *228*, 291, *354*
Adams, G. C., 255, 256, *354*
Adamson, M. J., 501, 530, *563*
Addyman, L., 733, *785*
Adeimy, J., 774, *788*
Afanaseva, G. N., 782, *788*
Aggarwal, S. L., 367, *428*
Ahlstrom, D. H., 309, 310, 350, *354*, *359*
Aizenshtein, E. M., 774, 776, *785*, *788*, *790*
Akiyama, J., 51, 52, 55, *90*
Aklonis, J. J., 600, 601, 618, *704*
Albarino, R. V., 871, *906*
Aldridge, H., 249, *363*
Aleksandriiskii, S. C., 774, *785*
Aleksandrova, I., 771, 774, *786*
Alexander, L. E., 105, *228*
Alfrey, T., 180, *228*, 274, 343, *357*
Aliguliev, R. M., 641, *708*
Alishoev, V. R., 349, *354*
Allan, A. L., 341, *359*
Allara, D. L., 855, *907*
Allegra, G., 189, 190, *228*, *232*, 304, 305, *354*, *360*
Allegrezza, A. E., Jr., 385, *428*
Alleman, T. C., 59, *87*, 447, 448, 537, *565*
Alliger, G., 573, *704*
Allport, D. C., 367, *428*

Aloisio, C. J., 312, *360*, 424, *431*
Alsheh, D., 815, *841*
Altamirano, J. O., 413, 415, *431*
Amano, T., 255, 350, *360*
Ambelang, J. C., 589, *704*
Ambrose, R. J., 396, *430*
Amigo, J. M., 782, *785*
Anagnostou, T., 459, 525, *563*
Anderson, E. W., 414, 415, 417, *428*, *429*, 885, 890, 891, 903, 904, *906*, *907*, *908*
Anderson, H. C., *563*
Anderson, T. K., 782, 787, 820, *842*
Andrejs, B., *563*
Andrews, F. R., 764, 765, *791*
Andrianova, G. P., 773, *785*
Andrienko, P. P., 781, *785*
Angelo, R. J., 371, *428*
Angeloni, F. M., 493, 498, 530, *563*
Anna, P., *567*
Antal, I., 517, 526, *563*
Anton, A., 767, *785*
Apinis, A. P., 318, *358*
Arada, B., 517, *563*
Arai, N., 767, *785*
Arakawa, T., 149, *229*, 767, *785*
Archar, B. N. N., 114, *229*
Arfi, C., 221, *232*
Ariyama, T., 336, 337, *354*
Armstrong, C., 455, 456, *564*
Arnold, K. R., 374, *428*
Arnold, M., 194, *229*
Arond, L. H., 274, *360*
Arons, G. N., 767, *785*
Arpin, M., 401, *428*
Artunc, H., 767, 769, *785*

922

Subject Index